Handbook of Sensory Physiology

Volume VII/6A

Comparative Physiology and Evolution of Vision in Invertebrates

A: Invertebrate Photoreceptors

By

H. Autrum M. F. Bennett B. Diehn K. Hamdorf
M. Heisenberg M. Järvilehto P. Kunze R. Menzel
W. H. Miller A. W. Snyder D. G. Stavenga M. Yoshida

Edited by

H. Autrum

With 314 Figures

Springer-Verlag Berlin Heidelberg New York 1979

H. Autrum

Zoologisches Institut der Universität, Luisenstr. 14, 8000 München, Germany

ISBN-13: 978-3-642-67001-5 e-ISBN-13: 978-3-642-66999-6
DOI: 10.1007/978-3-642-66999-6

Library of Congress Cataloging in Publication Data Main entry under title: Comparative physiology and evolution of vision in invertebrates. (Handbook of sensory physiology; v. VII/6) Bibliography: p. Includes indexes. CONTENTS: A. Invertebrate photoreceptors. 1. Vision. 2. Invertebrates—Physiology. 3. Physiology, Comparative. I. Autrum, Hansjochem. II. Series. QP351.H34 vol. 7/6A, etc. [QP475] 591.1'82'08s 78-21470 (v. A) [592'.01'832]

Softcover reprint of the hardcover 1st edition 1979

2122/3120-543210

Preface

In the comparative physiology of photoreception by the Protista and the invertebrates two aspects are emphasized: (1) the diversity of visual processes in these groups and (2) their bearing upon general mechanisms of photoreception. Invertebrates have evolved a far greater variety of adaptations than vertebrates—modifications aiding survival in the remarkably different biotopes they occupy. The number of species in itself suggests this multiformity; each of them has peculiarities of its own, in morphology as well as in physiology and behavior. But these special adaptations are variations on a few great themes.

Although the catalogue of invertebrate species is immense, the literature concerning them nearly rivals it in extent—even if one considers only that fraction dealing with visual physiology. Taxonomy proceeds by grouping the species, categorizing them in genera, families, orders, and progressively larger units. Similarly, comparative physiology aims at an analogous, more or less comprehensive, classification. This Part A of Volume VII/6, like Part B that follows it, emphasizes the broad questions that concern groups larger than the individual species; in some cases these questions have general applicability. The middle course between approaches that are too specialized and those that are too general is often elusive, but here we attempt to follow it. The vast number of special adaptations—probably, as we have said, as large as the number of species—is beyond the range even of a handbook.

Comparative physiologists continually ask to what extent the findings in one species can be generalized? Of the roughly 85,000 known species of Diptera, the visual physiology of less than half a dozen have been studied intensively. And even these have been chosen, as a rule, not because they displayed particularly interesting adaptations, but because they could be raised easily in the laboratory or were readily obtainable in large numbers. Physiologists have usually tried to establish the generality of their results by showing that a certain morphologic structure is associated with a certain function. With this aim geneticists, initially working with *Drosophila*, have recently developed new methods (cf. Ch. 12) with results having implications for animals other than just the insects.

Behavior, physiology, and morphology—the latter ranging from anatomy down to ultrastructure and even to the configuration of molecules—are inseparable. Data regarding the anatomic and light-microscopic structure of invertebrate visual organs are so extensive that in these volumes they can be mentioned only where they particularly assist in the understanding of physiologic processes and where the relationship between form and function is especially close. In the foreground, then, we shall stress biophysics, biochemistry, and behavior. Nevertheless, the number of species cited as examples is quite large. To guard against premature generalization, the particular species studied is named in all of the following discussions; it is clearly a bad habit to speak simply of "the fly". On the other hand, the available variety of animals should not be used as an excuse for multiplying the already far too numerous publications by repetitious experiments revealing only analogous results. Such studies of related species, if they are to be of interest, must pose new questions of their own—whether the approach employed derives from ecology, behavior, biochemistry, or biophysics.

Comparative physiology can of course not be limited to the invertebrates. In many chapters vertebrates are drawn into the discussion, either because they have evolved comparable structures or functions or because they differ. Correspondingly, invertebrates are also treated elsewhere in this handbook, in addition to these two parts of Vol. VII; these articles are listed at the end of the Preface. In all cases, repetition has been avoided wherever possible.

Part A is concerned primarily with the physiology of photoreceptor cells. It is impossible to draw a strict boundary between this aspect and the central nervous processes, for various reasons. First there is the obvious fact that photoreception does not stop with the photoreceptors. Second, in invertebrates, far more than in vertebrates, the study of vision has begun with and depends on behavioral experiments; as a rule, the biochemical and biophysical work has stemmed from prior analyses of behavioral physiology. Color vision, sensitivity to polarized and UV light, and extraocular photoreception are examples. Circadian periodicity, which is treated briefly, does not yet enjoy a genuine causal analysis; as such, it is a quite distinct behavioral phenomenon.

The attentive reader of the contributions to these volumes will find many unsolved problems; the further research progresses, the further the horizon recedes. Untilled fields and even impenetrable jungles come into view. Often the relationships between particular variants of molecular structure or biophysical events and the behaviour that must somehow be associated with them are impossible even to guess at, let alone document experimentally. In the area of invertebrate photoreception innumerable questions remain unanswered; in particular, though, challenging problems arise as soon as methods and results from other disciplines are applied to biologic phenomena. Here "interdisciplinary" implies not only the application of mathematical, physical, and chemical knowledge and techniques to a special topic, but also the manifold approaches within the field of biology itself; one may consider a particular phenomenon from several viewpoints, among them evolutionary origin and adaptation to a given biotope. Our understanding in this area is so incomplete, and speculative interpretations so outnumber facts and theories, that a consistent, well-founded presentation remains beyond our reach. No attempt to produce one will be made here.

Almost daily new results are published; some of them open entirely novel aspects of general significance. In view of this rapid advance, we may well ask whether a survey of the present state of knowledge in the form of a handbook is really appropriate. Will such writings not become outdated in a few years, perhaps even before they have left the author's desk? On the contrary, I think it is precisely in such periods of tempestuous development that one must from time to time seek the eye of the storm, taking the opportunity to look the situation over and to see which problems have been solved and which remain open. A handbook gives us this chance. To the beginner it opens the way to the earlier literature; those more advanced may learn the current status of areas related to their own. At the least it stands as a milestone and a prominence from which we survey both past and future. A handbook cannot replace the substance of the classical works—those of Johannes Müller, Exner, Hesse and von Frisch, to name but a few—but it can point the way to them. In studying such pioneering papers, the reader will find remarkably clear statements of central problems, hypotheses, theories, and suggestions. Many a dead-end road, and many a fruitless discussion, would be circumvented if researchers were more familiar with this early literature. The value of such familiarity is evident in a number of the points made here. These handbook volumes, then, serve both as a milestone and as an informative overview; we owe our special thanks to the publisher for bringing them into being.

München, den 12. Oktober 1978 H. AUTRUM

Contents

List of Contributors

AUTRUM, H., Zoologisches Institut der Universität, Luisenstr. 14, 8000 München, Germany

BENNETT, M. F., Department of Biology, Colby College, Waterville, Maine 04901, USA

DIEHN, B., Department of Chemistry, University of Toledo, 2801 Bancroft Street, Toledo, Ohio 43606, USA

HAMDORF, K., Tierphysiologisches Institut der Ruhr-Universität, Postfach 2148, 463 Bochum-Querenburg, Germany

HEISENBERG, M., Institut für Genetik und Mikrobiologie der Universität, Röntgenring 11, 8700, Würzburg, Germany

JÄRVILEHTO, M., Department of Physiology, University of Oulu, Kajaanintie 52 A, 90220 Oulu 22, Finland

KUNZE, P., Biologisches Institut der Universität, Ulmer Str. 227, 7000 Stuttgart 60, Germany

MENZEL, R., Freie Universität FB 23, Institut für Tierphysiologie, Grunewaldstraße 34, 1000 Berlin 41, Germany

MILLER, W. H., Department of Ophthalmology and Visual Science, Yale University School of Medicine, New Haven, Conn. 06510, USA

SNYDER, A. W., The Institute of Advanced Studies, Research Schools of Biological and Physical Sciences, Australian National University, Canberra, Australia

STAVENGA, D. G., Department of Physics, Laboratorium voor Algememe Natuurkunde, Rijksuniversiteit Groningen, Westersingel 34, Groningen, The Netherlands

YOSHIDA, M.*, Tamano Marine Laboratory, 2-3-3 Shibukawa, Tamano, Okyama, Japan 706

---

* The name and the address of the Laboratory from August 1979: Ushimado Marine Laboratory, Kashino, Ushimado, Okayama, Japan 701–43

Chapter 1

Introduction

H. AUTRUM, München, Germany

With 4 Figures

Contents

"One might say that nature has taken delight in accumulating contradictions."
H. v. Helmholtz

Comparative physiology is the biologic science which begins with man, in order to understand man. To understand man (biologically) means to know how evolution has brought him into existence, to reveal the functions of organs, their mechanisms of operation and their roles as parts of the organism, and to satisfy the human urge to discover the laws of nature and to create order where diversity prevails, whether for the sake of the laws themselves or for the sake of the power that knowledge brings.

A. Comparative Physiology and Evolution

I. The Darwinian Way

The history of man goes back more than 3×10^9 years. Neither the characteristics man has in common with other organisms nor those that set him apart can be understood except through comparisons. This was the method DARWIN and HAECKEL used, proceeding from the early, simple species to late, more complex forms; although DARWIN in 1859 (in *The Origin of Species*) did not yet dare to include man in this evolution, he gave a hint of things to come: "Much light will be thrown on the origin of man and his history."

II. The Way of Comparative Physiology

The way of comparative sensory physiology is not that of comparative morphology. The former discipline begins with man. It is initially from our own experience that we know what purpose sense organs serve, what their functions and roles are. Our knowledge of the human eye begins with what we perceive directly — that it is used for seeing, that it discriminates colors, detects forms and movements, assists spatial orientation, and mediates in our relationship with the environment and our recognition of friends and enemies. The method of comparative physiology moves outward from man, following a retrograde path in both anatomy and physiology *via* the complicated sense organs of the vertebrates to single sense cells like those in the skin of annelids (*Lumbricus:* HESSE, 1896; *Pontobdella:* HESSE, 1897). In these cases and many others, comparative physiologists discovered responses to light long before comparative morphologists had found the visual cells themselves; DARWIN (1881, and others before him) observed that earthworms respond to light. On the other hand, when HESSE (1897) discovered the striated border in the visual cells of several annelids and APÁTHY (1902) made a similar discovery in the Hirudinea, anatomical analysis was far in advance of physiologic study. In fact, many morphologic properties of visual organs were discovered and described long before their physiologic significance became known; in some cases, it is still not known [see, e.g., FAHRENBACH (1975); the ciliary structures in sense organs are discussed, for example, by ATEMA (1973)].

But the physiologic properties found in advance of their morphologic substrates were many — and, again, in some cases the morphology is still unclear. In 1882 LUBBOCK discovered that ants *(Formica, Lasius)* respond to UV light, and FOREL (1886; rev. see FOREL, 1910) showed that they detect it with their eyes. KÜHN (1927) and KÜHN and POHL (1921) showed that for bees UV has the quality of a color; AUTRUM and VON ZWEHL (1964) found bee visual cells that were sensitive only to UV; and finally HAMDORF et al. (1971) demonstrated that there is a particular rhodopsin in insects *(Ascalaphus)* that is specifically sensitive to UV. Sensitivity of the eyes to UV occurs in both vertebrates and invertebrates, but only in the latter are there UV-specific receptors (see this Volume, Chs. 4, 9, and 10).

Once this physiologic capability had been discovered, researchers became aware of the further adaptations it had made possible. For example, bees and other insects can (1) orient to UV as a color (see, e.g., DAUMER, 1958) and (2) use the polarization of light, which is especially pronounced in the short-wavelength skylight, for long-distance orientation (VON FRISCH, 1949, 1965). They have developed visual cells highly specialized for the latter function (MENZEL, 1975; see this volume, Chs. 4 and 10). These adaptations may be regarded, from a phylogenetic viewpoint, as variations on a basic theme.

A further aspect of a comparative approach to physiology is that, given the inexhaustible diversity of forms of organization in nature, it is generally possible to find one or more biologic preparations in which fundamental problems may be studied much more easily than in organs of greater anatomic and physiologic complexity. Comparative physiology thus provides *models for general problems*. One of many examples of such a model preparation is *Limulus* (HARTLINE, 1928; HARTLINE and GRAHAM, 1932a, b; HARTLINE and RATLIFF, 1972). *Limulus*, an animal systematically very distant from all vertebrates and one which has existed unchanged for almost 200 million years, might easily be relegated to the "flea market" of zoology. But certain fundamental properties of visual cells and of optic nerves (especially their individual axons), lateral inhibition and the cellular mechanisms underlying it, and the dynamic processes of excitation and inhibition have been most thoroughly studied in the *Limulus* eye (for a review see HARTLINE and RATLIFF, 1972; FAHRENBACH, 1975).

B. Invertebrates and Vertebrates: Comparative Physiology and Evolution

I. Variability of Ecological Adaptations

The comparative method itself reveals how carefully generalizations must be made, if indeed they are justified at all. It is not always or even usually true that phylogenetically closely related species, genera, or families are as similar in physiology as they are in morphology. The squirrel monkey *(Saimiri sciureus)* and its close relative, the capuchin monkey *(Cebus)* both belong to the diurnal Anthropoidea. But their color vision is like that of a protanomalous human (JACOBS, 1963; DE VALOIS and JACOBS, 1968; MALMO and GRETHER, 1947; for a review see AUTRUM and THOMAS, 1973). By contrast, the tree shrew *(Tupaia)*, with a systematic position between the primates and the insectivores, has color vision comparable to that of man (see this handbook, Vol. VII/3, p. 682). The ability to discriminate colors is more highly developed in bees than in some primates. Almost all insects and many invertebrates can see UV, but only a few vertebrates (toads, newts, lizards, hummingbirds, and homing pigeons; DIETZ, 1972; KREITHEN and EISNER, 1978).

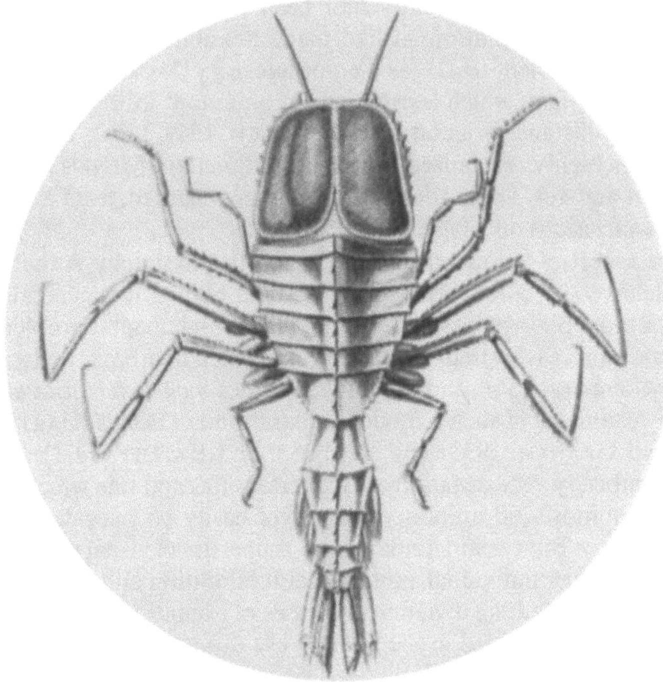

Fig. 1. *Cystisoma neptunus (Thaumops pellucida)* (Amphipoda), a crustacean from the Atlantic Ocean. The translucent animal is about 10 cm long. Note the large compound eyes. (WILLEMOUS-SUHM, 1875)

Photoreception is phylogenetically very old. It has thus been exposed to the action of very many selective factors, in the different phyla, orders, species and subspecies (e.g., those of *Gammarus oceanicus:* ALI and STEELE, 1961). The result has of course been a great variety of highly specialized adaptations, each of which takes many forms in both invertebrates and vertebrates (for a review see PLATE, 1924; DUKE-ELDER, 1958).

1. A sedentary habit can lead to a complete absence of eyes (this is probably the case, for example, in *Anodonta*, although this clam responds to a decrease in light intensity of about 10%; whereas other swimming lamellibranchs such as *Pecten* have highly developed eyes).

2. As animals evolved new activity patterns, changing from the diurnal habit to a preference for twilight and finally to life in the dark, a large number of adaptations developed, some of them quite striking. In *Thaumops (Cystisoma) neptunus* (G.-M.), an amphipod of the Indian and Atlantic Oceans, 10 cm long and as transparent as glass (Fig. 1), the eyes cover almost a third of the body surface (WILLEMOUS-SUHM, 1875). In other deep-sea animals, by contrast, the eyes have greatly or completely regressed.

3. The eyes of cavernicolous animals are often much reduced.

4. Frequently young, larval animals have well-developed eyes that later degenerate to a considerable degree.

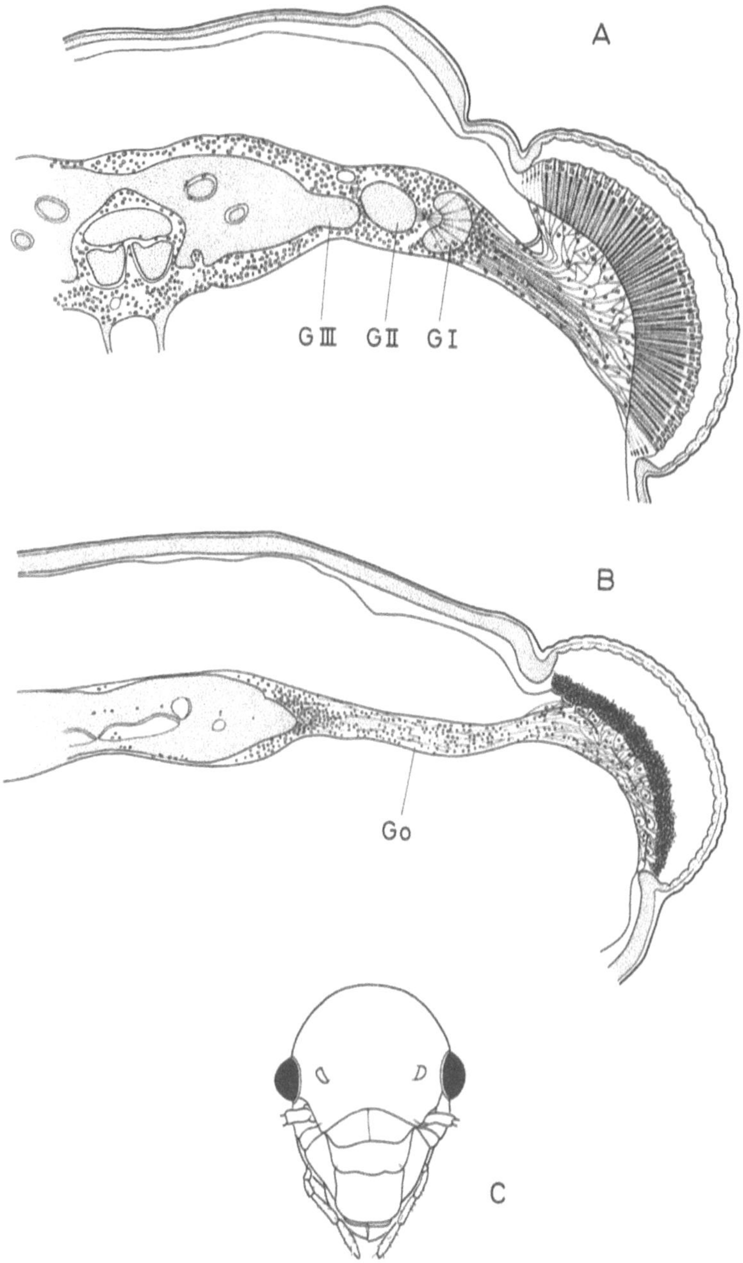

Fig. 2. *A* Section through the eye of a winged imago of *Termes* sp. Compound eye and optic ganglia (GI–III) are well developed. *B* Section through the eye of a queen. Crystal cones and retinula cells are absent; the optic ganglia (Go) degenerated. *C* External view of the head of a queen. The eyes look normal, the facets of the cornea are still preserved. (Redrawn from JÖRSCHKE, 1914)

Some insects, for example, the Polyctenidae (Heteroptera; parasites in fur) and many Anoplura (e.g., *Macropsylla*) — have no compound eyes at all. Among the termites, well-developed eyes are present in the winged males and females; after the wings are discarded the eyes and the optic ganglia degenerate, until finally only a simple layer of hypodermis cells containing black pigment remains (Fig. 2). The freely moving larval stages of many animals (e.g., the cypris larva of *Sacculina*) have eyes, that regress at the transition to parasitic life.

5. Many endoparasites completely lack eyes; this is true, for example, of all Cestodea. Among the nematodes, free-living species occasionally have eyes (see, e.g., BURR and WEBSTER, 1971; CROLL et al., 1972) while parasitic species do not.

Morphologic adaptations to the habitat, then, are extremely varied. Little is known about the physiologic significance of adaptations to certain biotopes and even less is known of their evolution.

Invertebrates have evolved a number of adaptations which are never found, or only rarely, among vertebrates. Here we shall cite only the ability to detect polarization of light and to orient by the pattern of polarization (see this volume, Ch. 9 and Vol. VII/6 B), the ability to see UV light (see Sec. III 1), and a high temporal resolving power (AUTRUM, 1950, 1958). The last of these abilities is found particularly in rapidly flying insects, some of which have critical fusion frequencies as high as 300/s. Fusion frequencies vary from one species to another, and evidently even among different visual cells of a single species; the visual cells R1-6 of flies have a flicker fusion frequency double that of the cells R7 and R8 (COSENS and WRIGHT, 1975). These specializations are of crucial importance in the vision and orientation (see, e.g., COLLETT and KING, 1975) of many invertebrates.

Another consequence of the phylogenetic antiquity of photoreception is its near ubiquitousness in the animal kingdom. The diversity of adaptations is less surprising than the extreme constancy of the morphology of a given type of visual organ, down to its molecular structure, and of the physiologic mechanisms. The accuracy of the pattern of connections of the first- and second-order neurons of the visual system of *Calliphora* has been described, for example, by HORRIDGE and MEINERTZHAGEN [1970; MEYER-ROCHOW (1972) describes structural irregularities, e.g., 1-5 cone cells, in the eyes of beetles (Staphylinidae)]. Another example is the constancy of the rhabdomere arrangement within a species, which contrasts with the remarkable variation among species or at least among families (see, e.g., EAKIN, 1972; AUTRUM, 1975). The slight variability of visual systems within a species is a sign of high selection pressure.

II. Structural Types of Visual Cells

Two basic forms underlie photoreceptor structure (EAKIN, 1963, 1966, 1968, 1972): either one cilium (or several) is present and the visual cell is modified in various ways, or the membrane of the visual cell is greatly enlarged by microvilli or lamellae.

In contrast to those of invertebrates, the visual cells of vertebrates are quite uniform; all of them contain ciliumlike structures with 9 filaments arranged cylindrically (producing a circle in cross section) in the connecting structure between the inner and outer segments. The outer segments consist of double discs stacked in a closely packed column. There are many different structural forms among the invertebrates (for a review see EAKIN, 1972). These fall into two categories: (1) the ciliary photoreceptors and (2) the rhabdomeric photoreceptors.

1. Visual Cells with Cilia

Since all vertebrate visual cells, but not all those of invertebrates, contain cilia or structures derivable from cilia, it is useful to categorize the ciliary photoreceptors, as EAKIN (1972) does in his excellent review. The membrane of the cilium is enlarged in a variety of ways:

1. Disclike processes of visual cells have been shown to exist in tunicate larvae (EAKIN, 1972; EAKIN and KUDA, 1971; BARNES, 1971; DILLY, 1961, 1964, 1969a). The discs, or lamellae, project into the ocellar cavity; by contrast to the discs of vertebrates they are oriented not at right angles but parallel to the long axis of the cilium. In the prosobranch mollusc *Pterotrachea*, several hundred such membranes are arranged roughly as in the vertebrate photoreceptors, perpendicular to the long axes of the cilia (DILLY, 1969; see also EAKIN, 1972, p. 629f.).

2. In the photoreceptors of *Branchiomma (Megalomma) vesiculosum* (KRASNE and LAWRENCE, 1966) and *Dasychone (Branchiomma) bombyx* each of 70–450 cilia at the distal end of the visual cell is surrounded by its own sacklike evagination of the membrane. This so-called striated border (Stiftchensaum) of the visual cells had already been described by HESSE (1899) and wrongly, as was later established, equated with the structures of similar appearance in arthropods (SCHULTZE, 1868). The ultrastructure of the border in Polychaetes, with one cilium per "sack", clearly distinguishes it from the microvilli of the arthropods. Another example of a ciliary photoreceptor arises in the distal visual cells of *Pecten* (BARBER et al., 1967).

3. In the ocelli of hydromedusae and sea stars there are visual cells with cilia (containing 9+2 microtubules) from which arise irregularly twisted microvilli (see EAKIN, 1972). Similar structures are found in the brains of nereids (DHAINAUT-COURTOIS, 1965, 1968); in this case it has not been shown that the cells are photoreceptors.

4. Lamellar processes of visual cells are found in the neural tube of *Branchiostoma* (EAKIN, 1963; NAKAO, 1964), in the dorsal ocelli of the pulmonate *Onchidium verriculatum* (YANASE and SAKAMOTO, 1965), and in the polyplacophoran mollusc *Onithochiton neglectus* (BOYLE, 1969).

5. Finally, the paraflagellar bodies of *Euglena* (LEEDALE, 1967; see also this volume, Ch. 2) should be mentioned.

2. Visual Cells with Rhabdomeres

The photoreceptors in this second group have rhabdomeres, which contain microvilli arranged in a remarkable variety of ways (see EAKIN, 1972; for insects: AUTRUM, 1975). These are also evaginations of the cell membrane, even where they appear to lie in the interior of the cell, as in the Hirudinea. In the photoreceptors of *Hirudo*, a narrow gap joins the phaosome to the cell surface. The microvilli are directed toward the phaosome (APÁTHY, 1888; LASANSKY and FUORTES, 1969).

3. Photoreceptors with Neither Cilia nor Rhabdomeres

In the *Limulus* ocellus there are cells with no rhabdomeres; they are situated at the bases of the retinula cells, which do have rhabdomeres. FAHRENBACH (1975) homologizes these with the eccentric cells of the compound eye, in which case they would be secondary neurons; this view is compatible with their fine structure and with the fact that they send dendrites, 100–150 μm long, between the rhabdomeres to the retinula cells (JONES et al., 1971; NOLTE and BROWN, 1972). But it seems not to have been proved beyond doubt that these rhabdomeric cells are directly light sensitive, particularly in view of their connections to other receptor cells.

In the 6th abdominal ganglion of decapod crustaceans there exists a paired photosensitive neuron (PROSSER, 1934; KENNEDY, 1963; WILKENS and LARIMER, 1976), which evidently lacks both optical structures to gather or filter light and specialized phototransductive segments. It has been found in a great number of freshwater and marine decapod crustaceans (WILKENS and LARIMER, 1976), but is lacking in *Homarus*, for example, and in other epigeal species (WILKENS and LARIMER, 1974). However, caudal photoreceptors do occur in the highly evolved troglobiotic species *Cambarus setosus* (Faxon) and *Orconectes pellucidus australis* (LARIMER, 1960; LARIMER et al., 1966), even though in *Cambarus setosus* the lamina ganglionaris and all retinal elements of the eyestalks are essentially nonexistent.

Neural photoreception is known to occur in numerous invertebrates: Polychaetes (*Nephthys*: CLARK, 1956), giant ganglion cells of *Aplysia* (ARVANITAKI and CHALAZONITIS, 1961; BROWN and BROWN, 1972), siphonal photoreceptors of the bivalves *Mya*, *Venus*, and *Spisula* (KENNEDY, 1958, 1960), and in the metasoma of scorpions (*Urodacus*: ZWICKY, 1968, 1970; *Heterometrus*: RAO and RAO, 1973). To the extent that these photoreceptors have been studied in detail, they contain neither microvilli nor ciliary structures.

Light sensitivity is not necessarily associated with cilia in protozoans (e.g., *Amoeba*).

4. Comments on the Evolution of Photoreceptors

The evolution of these extremely varied forms of photoreceptive cells—like that of all specialized functional adaptations—is very much a controversial subject.

EAKIN (1968, 1972) and EAKIN et al. (1977) infer, though with great caution, a dichotomy in the evolution of visual cells: one branch is formed by the ciliary photoreceptors, the other by the rhabdomeric receptors. But there are exceptions and contraindications to such a simple hypothesis (see WATERMAN, 1975). Ciliary homologies, or at least vestiges thereof (centrioles and striated rootlets), can be found in the compound eyes of beetles (HOME, 1972, 1975; TOMINAGA and KABU-

TA, 1973) and in the bee *Megachile* (WACHMANN and HENNIG, 1974). Similar structures have not been found in the many other arthropod rhabdomeres. However, it should be kept in mind that the development of these structures is studied in very few cases. Ciliary and rhabdomeric photoreceptors occur side by side in *Pecten* and *Lima*. The tunicates are of interest in this regard, because of their close relationship to the vertebrates; *Salpa* has microvillous photoreceptors, whereas those of the larvae of *Amaroucium* are ciliary (GORMAN et al., 1971). But the eyes of *Salpa* are not homologous, ontogenetically or phylogenetically, with those of the ascidian larva or of vertebrates (METCALF, 1893); the structure of the *Salpa* eye is therefore no argument against EAKIN's view.

VINNIKOV (1966, 1969, 1974) maintains that all photoreceptive structures are basically derivable from microvilli; he views the microvilli as the original structures, differentiation of which gave rise to the stereocilia. Microvilli are found in nearly all epithelial cells bordering free surfaces. VINNIKOV (1974) calls the vertebrate visual (and olfactory) cells with stereocilia "relict modifications". A fact supporting the hypothesis that the microvilli are the phylogenetically older elements is that in primitive forms—for example, turbellarians, rotifers, some polychaetes, and the Hirudinea—the photoreceptors have only microvilli, no cilia, although kinocilia are routinely present in other epithelial cells of these animals. On the other hand, among arthropods true cilia with 9+2 microtubules are found only in the male gametes (spermatids), whereas microvilli are of common occurrence, even apart from the photoreceptors; for example, they are found in the resorbing cells of the midgut (for a review see SMITH, 1968). Hence microvilli are more widely distributed than cilia.

In contrast to VINNIKOV (1974), VANFLETEREN and COOMANS (1976) concluded that the cilium is the universal and primary photoreceptoral organelle because it is believed to be the organizer for not only true ciliary photoreceptors but also rhabdomeric ones. EAKIN et al. (1977), however, explain the exceptions to EAKIN's theory as secondary or coenogenetic structures. In annelid eyes (HERMANS and EAKIN, 1969, 1970, 1974; ERMAK and EAKIN, 1976) most of the ciliary photoreceptors are integumentary structures, whereas in cerebral ocelli photoreceptors are rhabdomeric (EAKIN et al., 1977). Since photoreceptors are ectodermal in origin, "it is not surprising, that ciliary structures are commonly seen in eyes. It can be argued, that cilia when present in association with arrays of photosensitive microvilli are adventitious" (EAKIN et al., 1977).

Evolution is a fact, but the details of phylogenesis continue to be theories (or hypotheses) (MAYR, 1975) that can be correct or incorrect. It is a fact that the two types of photoreceptor in EAKIN's classification do in principle exist; it is a fact that there is much greater variability among the invertebrates than among the vertebrates; it is a fact that although we know certain things about the particular relationships between form, function and environment, we do not know very much. In sum, photoreceptors may have evolved in the first instance as microvilli, or the two groups of photoreceptors, ciliary and rhabdomeric, may have evolved independently. We are left with the unsolved problem of why so many different photoreceptors have ciliumlike structures and what their function is. The inference that they are only a "relict", as VINNIKOV (1974) maintains, is unsatisfactory indeed.

III. Primary Processes

1. Visual Pigments

The multifariousness in structure of eyes and photoreceptor cells contrasts sharply with the uniformity of the initial primary process; in all the vertebrates and invertebrates studied so far, the photopigments are chromoproteins, molecules with a protein and a chromophore component. The pigment is 11-*cis*-retinaldehyde, a fat-soluble carotenoid with 20 carbon atoms. This molecule (also called retinene) has been shown to be the prosthetic group of rhodopsin only in vertebrates (rev. see CRESCITELLI, 1972), in molluscs (HUBBARD and ST. GEORGE, 1958; HARA and HARA, 1972), and in an insect (*Ascalaphus:* PAULSEN and SCHWEMER, 1972) (see this volume, ch.4; GOLDSMITH, 1972). It may seem premature to conclude from these data on invertebrates that retinene is the sole, ubiquitous prosthetic group of rhodopsin. This is an unsolved problem for several reasons. (1) The vertebrates have been studied thoroughly, whereas relatively few invertebrates have been investigated. (2) In many lower vertebrates that have been examined in detail, at least two different retinenes (the aldehydes of vitamins A_1 and A_2) have been found (for a review see CRESCITELLI, 1972; and Vol. VII/5 of this handbook). (3) It is extraordinarily difficult to demonstrate the presence of other prosthetic groups in invertebrate photopigments because of the small quantities of these pigments in the eyes. Thus, such arguments as have been put forward are often indirect. (4) In *Euglena* and many plant photoreceptors flavines or flavoproteins are thought to be the photosensitive pigments (see DIEHN, this Volume, Ch.2; CHECCUCCI, 1976). Of course, the Protista diverged from the eukaryont line at a very early stage. The extent of this divergence is emphasized by the amino-acid sequences in their cytochrome c, which differ quite markedly from those of all other eukaryonts (PETTIGREW, 1973; LIN et al., 1973). Similarly, analysis of the amino acid sequences in the protein moieties of photopigments could well provide a good foundation on which to establish the phylogenetic relationships among the photoreceptors. (5) The surface membrane of the light-sensitive bacterium, *Halobacterium halobium*, contains bacteriorhodopsine (13-*cis*-retinaldehyde as chromophoric group), all-*trans* retinaldehyde (OESTERHELD and STOECKENIUS, 1971; BLAUROCK and STOECKENIUS, 1971), and a second photosystem (possibly a flavine with a λ_{max} at 370 nm), both controlling behavioral responses (HILDEBRAND and DENCHER, 1975).

Retinene is widespread if not ubiquitous, but the consequences of its isomerization, i.e., conversion from the *cis* into the *trans* form, are varied. In vertebrates rhodopsin bleaching involves several intermediate steps; finally, retinene is hydrolyzed and reduced enzymatically to vitamin A (for a review see YOSHIZAWA, 1972).

In invertebrates, on the other hand, rhodopsin is converted after illumination to only a colored intermediate product, metarhodopsin. Metarhodopsins (acid) of invertebrates are stable at room temperature. Their absorption maxima lie between 480 and 580 nm, depending on the species (*Calliphora:* HAMDORF and ROSNER, 1973; STAVENGA et al., 1973). This fact was first discovered by KROHN

(1840) and KRUKENBERG (1882); KRUKENBERG wrote: "The visual purple (of cephalopods) is not only resistant to light ... but also withstands a rather high temperature." HESSE (1897) described the striated border in *Planaria torva* as red in color and noted that this coloration became significantly paler upon illumination, but was not entirely bleached. A neutral or acid extract of cephalopod rhodopsin, however, at an ordinary temperature does not become bleached, but rather turns a deeper red. In the living cephalopod photoreceptors metarhodopsin is most probably in its acid form (HARA and HARA, 1972). When it is illuminated rhodopsin 493 is converted to the (acid) metarhodopsin 500. This causes negligible change in the color of the retina, by contrast to the result of illuminating the vertebrate retina. The rhodopsin and metarhodopsin of cephalopods have been studied in detail by HUBBARD and ST. GEORGE (1958). As far as is known, the invertebrates differ from the vertebrates in that in vivo the end product of the light-induced transformation of rhodopsin is metarhodopsin.

In invertebrates, illumination also brings about a reisomerization of metarhodopsin to rhodopsin in vivo (see this Volume, Ch. 4). This causes an extraordinary acceleration of light adaptation under normal light conditions (white light).

The concentration of rhodopsin in the discs of the outer segments in vertebrates is of about the same order of magnitude as that in the rhabdomeres of invertebrates (HAYS and GOLDSMITH, 1969; GRIBAKIN and GOVARDOVSKII, 1975). But the arrangement and mobility of the invertebrate photopigments, at least in the microvilli of some crustaceans and molluscs, differ from those in vertebrates. In the discs of the outer segments of vertebrates the axes of the pigments lie parallel to the surface of the discs (LIEBMAN, 1962) but not parallel to one another (CONE, 1972); within the discs the rhodopsin molecules can actually diffuse freely (POO and CONE, 1974; LIEBMAN and ENTINE, 1974; LIEBMAN, 1975; SNYDER and LAUGHLIN, 1975). In crustaceans (*Orconectes*, *Procambarus*), however, the metarhodopsin can neither freely diffuse nor (as in the vertebrates) freely rotate (WEHNER and GOLDSMITH, 1975). These differences may be due either to the curvature of the microvillar membranes, which allows free movement of the photopigments only along their long axes (GRIBAKIN and GOVARDOVSKII, 1975), or more probably to different lipid composition (see Sec. B III 2 b).

In invertebrates the problems of orientation of the rhodopsin molecules are more complex, not least because there is so much greater variety among invertebrates than in the relatively small, uniform group of vertebrates. The dichroic absorption of the microvilli in the eyes of flies (KIRSCHFELD, 1969) and of the crayfish *Orconectes* (WATERMAN et al., 1969) was shown to give a dichroic ratio of 2.0; this is compatible with the hypothesis that the rhodopsins are oriented parallel to the long axis of the microvilli (GOLDSMITH, 1975). Difficulties arise from the high sensitivity to direction of polarization, which can vary absorption by as much as a factor of 12 (SHAW, 1969; MOTE, 1974). This is a sure indication that the photopigments in crustaceans and cephalopods (HAGINS and LIEBMAN, 1963) tend to be oriented parallel to the long axis of the microvilli (see SNYDER and LAUGHLIN, 1975; GRIBAKIN and GOVARDOVSKII, 1975). Moreover, it is apparent that in other microvilli as well the orientation of the photopigments is, in part at least, considerably more orderly than is the case among vertebrates (SNYDER and LAUGHLIN, 1975; see this volume, Ch. 5).

2. Receptor Membranes

One aspect in which the photoreceptors of vertebrates and invertebrates differ markedly is the nature of the receptor membrane, both in its behavior with respect to inorganic ions and in its lipid structure (Fig. 3).

a) *Permeability (Resting Potential and Receptor Potential)*

In all the vertebrates so far studied, both rods and cones respond to light with hyperpolarization (TOMITA et al., 1967; BAYLOR and FUORTES, 1970). But the way this hyperpolarization comes about is different in the two elements. In cones, the membrane resistance of the outer segments increases as a result of a decrease in membrane permeability to sodium (for a review see TOMITA, 1972; RÜPPEL, 1975). In rods, however, at the peak of the photoresponse, the membrane resistance is decreased. The decreased resistance in rods results from an increase in the permeability of the membrane to ions, the equilibrium potential of which is more negative than the membrane potential in the dark (LASANSKY and MARCHIAFAVA, 1974; rods of *Ambystoma tigrinum*).

The resting potential of invertebrate photoreceptors in the dark is high (60–80 mV; but see the exceptions below), whereas the sodium conductance of the membrane is low. The resting potential is close to the K^+ equilibrium potential. Stimulation by light leads to an increase in conductance of the visual-cell membrane (by contrast to the situation in vertebrates). Visual cells of invertebrates, then, are, as a rule, *de*polarized (HARTLINE et al., 1952; BURKHARDT and AUTRUM, 1960). However, certain exceptions have been found; for example, the photoreceptors in the distal layer of the *Pecten* eye are hyperpolarized, whereas those of the proximal layer are depolarized (TOYODA and SHAPLEY, 1967; GORMAN and MCREYNOLDS, 1969; MCREYNOLDS and GORMAN, 1970a, b). Hyperpolarization also occurs in the visual cells of tunicates (GORMAN et al., 1971), in both the microvillous *(Salpa)* and ciliary *(Amaroucium)* form of photoreceptor. But it is universally the case among invertebrates—by contrast to vertebrates—that the receptor potential is accompanied by an *increase* in conductance of the membrane (MCREYNOLDS and GORMAN, 1970a, b; MCREYNOLDS and GORMAN, 1975). The resting potential (= prestimulus membrane potential) of visual cells of *Salpa democratica* in the dark is very low (~14 mV), as it is in vertebrates. Light causes a graded hyperpolarizing response of up to 70 mV, which is associated with an increase of membrane conductance. These eyes, however, are not homologous with the larval eyes of ascidian tunicates or the pineal or lateral eye of vertebrates (METCALF, 1893; GORMAN et al., 1971). In the neural photoreceptors, too, there occur both depolarization and hyperpolarization, the latter associated with an increase in K^+ permeability (BROWN and BROWN, 1972: giant neurons in *Aplysia*).

Since there is a great deal of active ion transport in the visual cells of vertebrates, excitability (and the dark current) of these cells disappears very rapidly, within ~60 s (HAGINS, 1965) when the ion pumps are blocked (e.g., by ouabain). Blocking occurs more quickly in the dark than under illumination (YOSHIKAMI and HAGINS, 1973). On the other hand, when the ion pumps in the visual cells of *Limulus* or crustaceans are inhibited by ouabain, the resting potential in the dark does not begin to decrease until 30 min or even more have passed (for a review see STIEVE, 1974), and the decrease is more rapid in light than in the dark.

b) Lipid Content and Fluidity

The photoreceptor membranes of vertebrates differ distinctly from those of invertebrates with respect to their lipids (content of cholesterol, distribution of phospholipids, types of fatty acids: MASON et al., 1973; ZINKLER, 1975; BENOLKEN et al., 1975). The membranes of *Limulus*, insects, and cephalopods are less fluid than those of vertebrates; differences also exist within each of these groups, although they are less pronounced.

c) Membrane Protein Turnover

In vertebrates (frog outer segments, rods of rhesus monkeys), the discs of the outer segments are continually renewed. Protein, synthesized in the inner segment, is assembled into discs at the base of the outer segment (YOUNG, 1967–1969, 1971b); discs at the tips of the photoreceptors are phagocytized by cells of the pigmented epithelium (YOUNG, 1971a; ISHIKAWA and YAMADA, 1970). The process is accelerated only slightly by light (YOUNG, 1967). In the microvilli of insects (the larval ocellus of *Aedes aegypti*) the protein turnover which proceeds in the dark after a period of illumination is 100 times as rapid as that of vertebrates; the membranes of the microvilli are broken down by the visual cells themselves, by endocytosis (WHITE, 1967, 1968; WHITE and LORD, 1975). This process is considerably accelerated by light. Similar results were reported by ITAYA for the rhabdomes of *Palaemonetes* (shrimps, Macrura natantia, Carididae); here, the production, utilization, and degradation of the microvillous photoreceptor membranes are discontinuous processes occurring only during changes from light to dark and dark to light (ITAYA, 1976; see also this handbook, Vol. VII/6, Part B, Ch. 4).

d) The Ca^{2+} Ions

The role of calcium ions in the visual cells of vertebrates appears to differ from that in invertebrates (see Fig. 3). This seems to be an established fact, but as far as the mechanism is concerned, there are at present more unproved hypotheses than convincing demonstrations (see, e.g., YOSHIKAMI and HAGINS, 1971; LIEBMAN, 1974; MASON et al., 1974; HAGINS and YOSHIKAMI, 1975; GOLDSMITH, 1975; FEIN and LISMAN, 1975; STIEVE, 1975; BAUMANN, 1975; WELLER et al., 1975; BEHRENS and KREBS, 1976; BROWN et al., 1977).

3. Rhodopsin Content and Sensitivity

There are also differences between vertebrates and invertebrates in the dependence of sensitivity of the receptors upon relative rhodopsin content (Fig. 4). The receptors of *vertebrates* follow, over a wide range, the Dowling-Rushton relationship:

$$\log I/I_0 = \alpha B,$$

where I_0 is the intensity of the threshold stimulus in the completely dark-adapted state, I is the intensity of a threshold stimulus during the transition from light

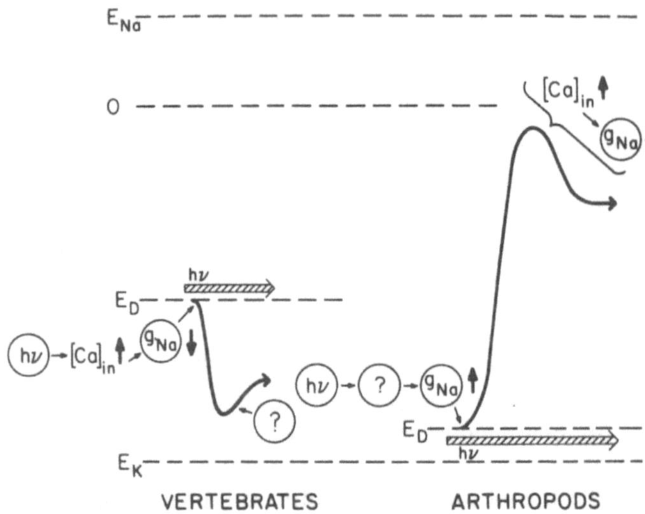

Fig. 3. Hyperpolarizing receptor potentials of vertebrate cones compared with the most common depolarizing type of response found in invertebrate photoreceptors. Horizontal lines indicate the relative positions of the sodium and potassium equilibrium potentials (E_{Na},E_K), as well as the approximate positions of the resting (dark) potentials (E_D). In vertebrate receptors E_D lies well below E_K because of high conductance of sodium (g_{Na}) in the dark. Light causes (in cones) g_{Na} to decrease; consequently the membrane potential increases, hyperpolarizes. The effect on g_{Na} is thought to be mediated by an increase in intracellular Ca^{2+}. Rods behave somewhat differently (see LASANSKY and MARCHIAFAVA, 1974). In most invertebrates E_D lies close to E_K in the dark. Light causes an increase in g_{Na} and consequently a depolarization. A secondary increase in internal Ca^{2+} causes a partial decrease in g_{Na}. (GOLDSMITH, 1975)

adaptation to dark adaptation, α is a constant which varies greatly (from species to species and even for rods and cones within a single species), and B is the amount in percent of visual pigment bleached (RUSHTON, 1961; ALPERN, 1971; RUSHTON and POWELL, 1972; DOWLING, 1960; DOWLING and RIPPS, 1970). The threshold intensity is thus logarithmically related to the concentration of visual pigment. But alongside this process, neural adaptation plays a considerable role in the photoreceptors of vertebrates. If the sensory cells are isolated from the pigmented epithelium, regeneration of rhodopsin becomes impossible in the dark; nevertheless, after partial bleaching of the visual pigments sensitivity can increase by more than 2 log units in the dark (see NORMAN and WERBLIN, 1974).

In *invertebrates*, sensitivity depends considerably less upon rhodopsin content. Adaptation (at least late adaptation) in the ventral eyes of *Limulus* does not depend on rhodopsin content at all, as was shown by FEIN and DEVOE (1973). They used the early receptor potential (ERP) as a measure of the amount of rhodopsin converted following a brief light stimulus. Simultaneous measurement of membrane potential and membrane resistance indicated that adaptation proceeds independently of all these parameters. These authors inferred the existence of (unknown) intermediate processes that determine adaptation. Electric adaptation of the visual cells of *Limulus* and insects (*Calliphora:* HAMDORF and ROSNER, 1973; *Drosophila:* STARK and ZITZMANN, 1976) involves processes other than change in photopigment concentration; transduction and adaptational mecha-

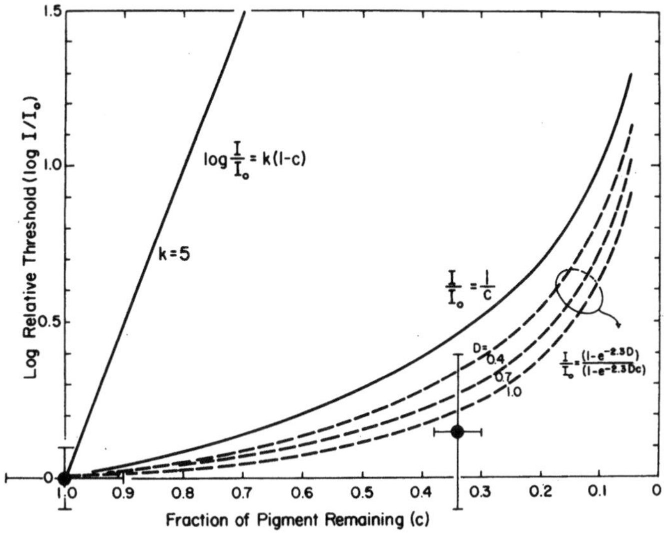

Fig. 4. Dependence of log threshold on rhodopsin concentration. The upper curve is the Dowling-Rushton relationship with a slope constant ($k=5$) appropriate for rat or skate rods. The linear relation between log threshold and fraction of pigment bleached (1–c) is observed with vertebrate receptors, but with wide variations in slope constant. The lower curves are theoretical functions that relate log threshold to c, assuming that the only effect of pigment loss is to decrease the probability of absorption. The *solid curve* represents no self-screening; the *broken curves*, varying amounts of self-screening corresponding to axial densities (D) of 0.4, 0.7, and 1.0 when c = 1.0. The two data points (± S.E.) show dark-adapted thresholds of normal lobsters and rhodopsin-deficient lobsters studied in late winter. The insignificant rise in log threshold associated with loss of $^2/_3$ of the rhodopsin is consistent with the hypothesis that fall of sensitivity is caused by decreased probability of absorption. (GOLDSMITH, 1975)

nisms are different. In other cases, where there is a correlation between the sensitivity of invertebrate photoreceptors and their rhodopsin content, the curve is much less steep than the comparable curve for vertebrates (see GOLDSMITH, 1975; HAMDORF, 1970; GLANTZ, 1971; BARNES and GOLDSMITH, 1977). GOLDSMITH (1975) and HAMDORF and SCHWEMER (1975) have used as a first approximation a relationship of the form $I/I_0 = C_0/C$, where C and C_0 are the concentrations of rhodopsin in the light- and dark-adapted states. Accordingly, the increase in threshold would be about inversely proportional to the concentration of rhodopsin (for details see this volume, Ch. 4).

C. Conclusions and Summary

It follows, from all that has been said, that the gaps in our knowledge of invertebrate photoreceptors are still much too large to permit generalization. The vertebrates are a relatively small, uniform group, and as EAKIN (1972) says, "from the standpoint of variation in architecture" (and as we can add, from the standpoint of comparative physiology) "they are dull and uninteresting". We are still suffering the aftereffects of the unfortunate events of 1793, when the Musée d'Histoire Naturelle in Paris was founded. Scholars were given responsibility for all departments but one; for the "animaux sans vertèbres" no qualified scientist could be found. As a result, the task of organizing these neglected animals was given to an amateur botanist, Jean Baptiste LAMARCK. Because of his efforts, relying more on administrative than on scientific considerations, the distinction based on presence or absence of a backbone became established for following generations (LAMARCK, 1815–1822). From the standpoint of comparative physiology the relationships between animals can appear quite different. This has been amply illustrated in this survey of animal photoreception; the most that can be said, in view of our very limited knowledge of visual physiology in the vast majority of animal species, is that the dichotomy between the vertebrate and invertebrate groups rests more on ignorance than on understanding.

As far as we know, both groups are similar in some respects and different in others. They have the following in common:

1. Numerous adaptations to their environment and habitat.
2. The well-known photopigments are retinaldehydes.

The photoreceptors of vertebrates differ from those of invertebrates in the following respects:

1. In the structure and function of their cellular membranes; in vertebrates conductance for Na-ions decreases during illumination; in invertebrates it increases, even in hyperpolarized cells.
2. In vertebrates the photopigments bleach down to vitamin A + opsin; in invertebrates the turnover goes only as far as metarhodopsin. In vertebrates a reisomerization from metarhodopsin to rhodopsin by light is of no (or almost no) importance. In invertebrates reisomerization from metarhodopsin to rhodopsin is in vivo of great importance.
3. Invertebrates have photoreceptors of the ciliary and microvillous type; in vertebrates there is always a cilium.
4. The dependence of the sensitivity of the photoreceptor on rhodopsin content is high in vertebrates and lower in invertebrates.

The author wishes to thank Miss I. Thomas for her engagement and help in the preparation of the text and Dr. M.A. Biederman-Thorson for the translation.

References

Ali, M.A., Steele, V.J.: Retinomotor responses of the amphipod *Gammarus oceanicus* from two latitudes to various light intensities and temperatures. Rev. Can. Biol. **20**, 665—674 (1961).

Alpern, M.: Rhodopsin kinetics in the human eye. J. Physiol. (Lond.) **217**, 447—471 (1971).

Apáthy, St.: Analyse der äußeren Körperform der Hirudineen. Mitt. Zool. Stat. Neapel **8**, 153 (1888).

Apáthy, St. von: Die drei verschiedenen Formen von Lichtzellen bei Hirudineen. Mit Demonstration von Neurofibrillenpräparaten nach der Hämatein- und der Nachvergoldungsmethode. Verh. 5. Congr. Zool. Berlin 1902, 707—728 (1902).

Arvanitaki, A., Chalazonitis, N.: Excitatory and inhibitory processes initiated by light and infrared radiations in single identifiable nerve cells (giant ganglion cells of *Aplysia*). In: Nervous inhibition (ed. W. Florey), pp. 194—231. Oxford-London-New York-Paris: Pergamon Press 1961.

Atema, J.: Microtubule theory of sensory transduction. J. Theor. Biol. **38**, 181—190 (1973).
Autrum, H.: Die Belichtungspotentiale und das Sehen der Insekten (Untersuchungen an *Calliphora* und *Dixippus*). Z. Vergl. Physiol. **32**, 176—227 (1950).
Autrum, H.: Electrophysiological analysis of the visual system in insects. Exp. Cell Res., Suppl. **5**, 426—439 (1958).
Autrum, H.: Les yeux et la vision des Insectes. In: Traité de zoologie (ed. P. P. Grassé), Vol. VIII, pp. 742—853. Paris: Masson & Cie. 1975.
Autrum, H., Thomas, I.: Comparative physiology of colour vision in animals. In: Handbook of sensory physiology, Vol. VII/3A (ed. R. Jung), pp. 661—692. Berlin-Heidelberg-New York: Springer 1973.
Autrum, H., Zwehl, V. von: Die spektrale Empfindlichkeit einzelner Sehzellen des Bienenauges. Z. Vergl. Physiol. **48**, 357—384 (1964).
Barber, V. C., Evans, E. M., Land, M. F.: The fine structure of the eye of the mollusc *Pecten maximus*. Z. Zellforsch. **76**, 295—312 (1967).
Barnes, S. N.: Fine structure of the photoreceptor and cerebral ganglion of the tadpole larva of *Amaroucium constellatum* (Verrill) (Subphylum: Urochordata; Class: Ascidiacea). Z. Zellforsch. **117**, 1—16 (1971).
Barnes, S. N., Goldsmith, T. H.: Dark adaptation, sensitivity, and rhodopsin level in the eye of the lobster, *Homarus*. J. Comp. Physiol. **120**, 143—159 (1977).
Baumann, F.: Electrophysiological properties of the honey bee retina. In: The compound eye and vision of insects (ed. G. A. Horridge), pp. 53—74. Oxford: Clarendon Press 1975.
Baylor, D. A., Fuortes, M. G. F.: Electrical responses of single cones in the retina of the turtle. J. Physiol. (Lond.) **207**, 77—92 (1970).
Behrens, M., Krebs, W.: The effect of light-dark adaptation on the ultrastructure of *Limulus* lateral eye retinular cells. J. Comp. Physiol. **107**, 77—96 (1976).
Benolken, R. M., Anderson, R. E., Maude, M. B.: Lipid composition of *Limulus* photoreceptor membranes. Biochim. Biophys. Acta **413**, 234—242 (1975).
Blaurock, A. E., Stoeckenius, W.: Structure of the purple membrane. Nature (New Biol.) **233**, 152—154 (1971).
Boyle, P. R.: Fine structure of the eyes of *Onithochiton neglectus* (Mollusca: Polyplacophora). Z. Zellforsch. **102**, 313—332 (1969).
Brown, H. M., Brown, A. M.: Ionic basis of the photoresponse of *Aplysia* giant neuron: K^+ permeability increase. Science **178**, 755—756 (1972).
Brown, J. E., Coles, J. A., Pinto, L. H.: Effects of injections of calcium and EGTA into the outer segments of retinal rods of *Bufo marinus*. J. Physiol. (Lond.) **269**, 707—722 (1977).
Burkhardt, D., Autrum, H.: Die Belichtungspotentiale einzelner Sehzellen von *Calliphora erythrocephala* Meig. Z. Naturforsch. **15b**, 612—616 (1960).
Burr, A. H., Webster, J. M.: Morphology of the eyespot and description of two pigment granules in the esophageal muscle of a marine nematode, *Oncolaimus vesicarius*. J. Ultrastruct. Res. **36**, 621—632 (1971).
Checcucci, A.: Molecular sensory physiology of *Euglena*. Naturwissenschaften **63**, 412—417 (1976).
Clark, R. B.: The eyes and the photonegative behaviour of *Nephthys* (Annelida, Polychaeta). J. Exp. Biol. **33**, 461—477 (1956).
Collett, T., King, A. J.: Vision during flight. In: The compound eye and vision of insects (ed. G. A. Horridge), pp. 437—466. Oxford: Clarendon Press 1975.
Cone, R. A.: Rotational diffusion of rhodopsin in the visual receptor membrane. Nature (New Biol.) **236**, 39—43 (1972).
Cosens, D., Wright, R.: Light elicited isolation of the complementary visual input systems in white-eye *Drosophila*. J. Insect Physiol. **21**, 1111—1120 (1975).
Crescitelli, F.: The visual cells and visual pigments of the vertebrate eye. In: Handbook of sensory physiology, Vol. VII/1 (ed. H. J. A. Dartnall), pp. 245—363. Berlin-Heidelberg-New York: Springer 1972.
Croll, N. A., Riding, I. L., Smith, J. M.: A nematode photoreceptor. Comp. Biochem. Physiol. **42A**, 999—1009 (1972).
Daemen, F. J. M.: Vertebrate rod outer segment membranes. Biochim. Biophys. Acta **300**, 255—288 (1973).
Darwin, Ch.: The formation of vegetable mould, through action of worms, with observations on their habits. London 1881.
Daumer, K.: Blumenfarben, wie sie die Bienen sehen. Z. Vergl. Physiol. **41**, 49—110 (1958).

De Valois, R. L., Jacobs, G. H.: Primate color vision. Science **162**, 533—540 (1968).
Dhainaut-Courtois, N.: Sur la présence d'un organe photorécepteur dans le cerveau de *Nereis pelagica* L. (Annélide polychète). C. R. Acad. Sci. [D] (Paris) **261**, 1085—1088 (1965).
Dhainaut-Courtois, N.: Étude histologique et ultrastructurale des cellules nerveuses du ganglion cérébral de *Nereis pelagica* L. (Annélide polychète). Comparaison entre les types cellulaires I—VI et ceux décrits antérieurement chez les Nereidae. Gen. Comp. Endocrinol. **11**, 414—443 (1968).
Dietz, M.: Erdkröten können UV-Licht sehen. Naturwissenschaften **59**, 316 (1972).
Dilly, N.: Electron microscope observations of the receptors in the sensory vesicle of the ascidian tadpole. Nature (Lond.) **191**, 786—787 (1961).
Dilly, P. N.: Studies on the receptors in the cerebral vesicle of the ascidian tadpole. 2. The ocellus. Q. J. Micr. Sci. **105**, 13—20 (1964).
Dilly, P. N.: Studies on the receptors in *Ciona intestinalis*. III. A second type of photoreceptor in the tadpole larva of *Ciona intestinalis*. Z. Zellforsch. **96**, 63—65 (1969a).
Dilly, P. N.: The structure of a photoreceptor organelle in the eye of *Pterotrachea mutica*. Z. Zellforsch. **99**, 420—429 (1969b).
Dowling, J. E.: Night blindness, dark adaptation and the electroretinogram. Am. J. Ophthalmol. **50**, 875—887 (1960).
Dowling, J. E., Ripps, H.: Visual adaptation in the retina of the skate. J. Gen. Physiol. **56**, 491—520 (1960).
Duke-Elder, St.: System of ophthalmology. Vol. I. The eye in evolution. London: H. Kimpton 1958.
Eakin, R. M.: Lines of evolution of photoreceptors. In: General physiology of cell specialization (eds. D. Mazia, A. Tyler). New York: McGraw Hill 1963.
Eakin, R. M.: Evolution of photoreceptors. Cold Spring Harbor Symp. Quant. Biol. **30**, 363—370 (1966).
Eakin, R. M.: Evolution of photoreceptors. In: Evolutionary biology, Vol. II (eds. T. Dobzhansky, M. K. Hecht, W. C. Steere). New York: Appleton-Century-Crofts 1968.
Eakin, R. M.: Structure of invertebrate photoreceptors. In: Handbook of sensory physiology, Vol. VII/1 (ed. H. J. A. Dartnall), pp. 625—684. Berlin-Heidelberg-New York: Springer 1972.
Eakin, R. M., Kuda, A.: Ultrastructure of sensory receptors in ascidian tadpoles. Z. Zellforsch. **112**, 287—312 (1971).
Eakin, R. M., Martin, G. G., Reed, C. T.: Evolutionary significance of fine structure of Archiannelid eyes. Zoomorphologie **88**, 1—18 (1977).
Ermak, T. H., Eakin, R. M.: Fine structure of the cerebral and pygidial ocelli in *Chone ecaudata* (Polychaeta: Sabellidae). J. Ultrastruct. Res. **54**, 243—260 (1976).
Fahrenbach, W. H.: The visual system of the horseshoe crab *Limulus polyphemus*. Int. Rev. Cytol. **41**, 285—349 (1975).
Fein, A., De Voe, R. D.: Adaptation in the ventral eye of *Limulus* is functionally independent of the photochemical cycle, membrane potential, and membrane resistance. J. Gen. Physiol. **61**, 273—289 (1973).
Fein, A., Lisman, J.: Localized desensitization of *Limulus* photoreceptors produced by light or intracellular calcium ion injection. Science **187**, 1044—1096 (1975).
Forel, A.: Das Sinnesleben der Insekten. München: E. Reinhardt 1910 (here cit. publication in Recueil Zoologique suisse, 1886).
Frisch, K. von: Die Polarisation des Himmelslichtes als orientierender Faktor bei den Tänzen der Bienen. Experientia **5**, 142—148 (1949).
Frisch, K. von: Tanzsprache und Orientierung der Bienen. Berlin-Heidelberg-New York: Springer 1965.
Glantz, R. M.: Peripheral versus central adaptation in the crustacean visual system. J. Neurophysiol. **34**, 485—492 (1971).
Goldsmith, T. H.: The natural history of invertebrate visual pigments. In: Handbook of sensory physiology, Vol. VII/1 (ed. H. J. A. Dartnall), pp. 685—719. Berlin-Heidelberg-New York: Springer 1972.
Goldsmith, T. H.: Photoreceptor processes: some problems and perspectives. J. Exp. Zool. **194**, 89—101 (1975).
Gorman, A. L. F., McReynolds, J. S.: Hyperpolarizing and depolarizing receptor potentials in the scalop eye. Science **165**, 309—310 (1969).
Gorman, A. L. F., McReynolds, J. S., Barnes, S. N.: Photoreceptors in primitive chordates: fine structure, hyperpolarizing receptor potentials, and evolution. Science **172**, 1052—1054 (1971).
Gribakin, F. G., Govardovskii, V. I.: The role of the photoreceptor membrane in photoreceptor optics. In: Photoreceptor optics (eds. A. W. Snyder, R. Menzel), pp. 215—236. Berlin-Heidelberg-New York: Springer 1975.

Hagins, W. A.: Electrical signs of information flow in photoreceptors. Cold Spring Harbor Symp. Quant. Biol. **30**, 403—418 (1965).

Hagins, W. A., Liebman, P. A.: The relationship between photochemical and electrical processes in living squid photoreceptors. Abstr. Biophys. Soc. Ann. Meeting ME 6 (1963).

Hagins, W. A., Yoshikami, S.: Ionic mechanisms in excitation of photoreceptors. Ann. N. Y. Acad. Sci. **264**, 314—325 (1975).

Hamdorf, K.: Correlation between the concentration of visual pigment and sensitivity in photoreceptors. Verh. Dtsch. Zool. Ges. **64**, 148—156 (1970).

Hamdorf, K., Rosner, G.: Adaptation und Photoregeneration im Fliegenauge. J. Comp. Physiol. **86**, 281—292 (1973).

Hamdorf, K., Schwemer, J.: Photoregeneration and the adaptation process in insect photoreceptors. In: Photoreceptor optics (eds. A. W. Snyder, R. Menzel), pp. 263—289. Berlin-Heidelberg-New York: Springer 1975.

Hamdorf, K., Schwemer, J., Gogala, M.: Insect visual pigment sensitive to ultraviolet light. Nature (Lond.) **231**, 458—459 (1971).

Hara, T., Hara, R.: Cephalopod retinochrome. In: Handbook of sensory physiology, Vol. VII/1 (ed. H. J. A. Dartnall), pp. 721—746. Berlin-Heidelberg-New York: Springer 1972.

Hartline, H. K.: A quantitative and descriptive study of the electrical response to illumination of the arthropod eye. Am. J. Physiol. **83**, 466—483 (1928).

Hartline, H. K., Graham, C. H.: Nerve impulses from single receptors in the eye of *Limulus*. Proc. Soc. Exp. Biol. Med. **29**, 613—615 (1932a).

Hartline, H. K., Graham, C. H.: Nerve impulses from single receptors in the eye. J. Cell. Comp. Physiol. **1**, 277—295 (1932b).

Hartline, H. K., Ratliff, F.: Inhibitory interaction in the retina of *Limulus*. In: Handbook of sensory physiology, Vol. VII/2 (ed. M. G. F. Fuortes), pp. 381—447. Berlin-Heidelberg-New York: Springer 1972.

Hartline, H. K., Wagner, H. G., MacNichol, E. F.: The peripheral origin of nervous activity in the visual system. Cold Spring Harbor Symp. Quant. Biol. **17**, 125—141 (1952).

Hays, D., Goldsmith, T. H.: Microphotometry of the visual pigment of the spider crab, *Libinia emarginata*. Z. Vergl. Physiol. **65**, 218—232 (1969).

Hermans, C. O., Eakin, R. M.: Fine structure of the cerebral ocelli of a sipunculid, *Phascolosoma agassizii*. Z. Zellforsch. **100**, 325—339 (1969).

Hermans, C. O., Eakin, R. M.: Sipunculan ocelli: fine structure in *Phascolosoma agassizii*. Proceed. Internat. Symp. Biol. Sipuncula and Echiura I (eds. M. E. Rice, M. Todorović), pp. 229—237. Kotor (1970).

Hermans, C. O., Eakin, R. M.: Fine structure of the eyes of an alciopid polychaete, *Vanadis tagensis* (Annelida). Z. Morph. Tiere **79**, 245—267 (1974).

Hesse, R.: Untersuchungen über die Organe der Lichtempfindung bei niederen Tieren. I. Die Organe der Lichtempfindung bei den Lumbriciden. Z. Wiss. Zool. **61**, 393—419 (1896).

Hesse, R.: Untersuchungen über die Organe der Lichtempfindung bei niederen Tieren. II. Die Augen der Plathelminthen. Z. Wiss. Zool. **62**, 527—582 (1897a).

Hesse, R.: Untersuchungen über die Organe der Lichtempfindung bei niederen Tieren. III. Die Sehorgane der Hirudineen. Z. Wiss. Zool. **62**, 671—707 (1897b).

Hesse, R.: Untersuchungen über die Organe der Lichtempfindung bei niederen Tieren. V. Die Augen der polychäten Anneliden. Z. Wiss. Zool. **65**, 446—516 (1899).

Hildebrand, E., Dencher, N.: Two photosystems controlling behavioural responses of *Halobacterium halobium*. Nature (Lond.) **257**, 46—48 (1975).

Home, E. M.: Centrioles and associated structures in the retinula cells of insect eyes. Tissue Cell **4**, 227—234 (1972).

Home, E. M.: Ultrastructural studies of development and light-dark adaptation of the eye of *Coccinella septempunctata* L., with particular reference to ciliary structures. Tissue Cell **7**, 703—722 (1975).

Horridge, G. A., Meinertzhagen, I. A.: The accuracy of the pattern of the first and second order neurons of the visual system of *Calliphora*. Proc. R. Soc. Lond. (Biol.) **175**, 83—94 (1970).

Hubbard, R., St. George, R. C. C.: The rhodopsin system of the squid. J. Gen. Physiol. **41**, 501—528 (1958).

Ishikawa, T., Yamada, E.: The degradation of the photoreceptor outer segment within the pigment epithelial cell of the rat retina. J. Electron Microsc. (Tokyo) **19**, 85—91 (1970).

Itaya, S. J.: Rhabdom changes in the shrimp, *Palaemonetes*. Cell Tissue Res. **166**, 265—273 (1976).

Jacobs, G. H.: Spectral sensitivity and color vision of the squirrel monkey. J. Comp. Physiol. Psychol. **56**, 616—621 (1963).
Jones, C., Nolte, J., Brown, J. E.: The anatomy of the median ocellus of *Limulus*. Z. Zellforsch. **118**, 297—309 (1971).
Jörschke, H.: Die Fazettenaugen der Orthopteren und Termiten. Z. Wiss. Zool. **111**, 153—280 (1914).
Kennedy, D.: Neural photosensitivity in *Mactra*. Biol. Bull. **115**, 338 (1958).
Kennedy, D.: Neural photoreception in a lamellibranch mollusc. J. Gen. Physiol. **44**, 277—299 (1960).
Kennedy, D.: Physiology of photoreceptor interneurons in the abdominal nerve cord of the crayfish. J. Gen. Physiol. **46**, 551—572 (1963).
Kirschfeld, K.: Absorption properties of photopigment in single rods, cones and rhabdomeres. In: Processing of optical data by organisms and machines (ed. W. Reichardt), pp. 116—136. New York-London: Academic Press 1969.
Krasne, F. B., Lawrence, P. A.: Structure of the photoreceptors in the compound eyespots of *Branchiomma vesiculosum*. J. Cell Sci. **1**, 239—248 (1966).
Kreithen, M. L., Eisner, Th.: Ultraviolet light detection by the homing pigeon. Nature **272**, 347—348 (1978).
Krohn, W.: Nachträgliche Beobachtungen über den Bau des Auges der Cephalopoden. Nova Acta Leopoldina (Halle) **19**, 41 (1839—42).
Krukenberg, C. F. W.: Über die Stäbchenfarbe der Cephalopoden. Unters. Physiol. Inst. Univ. Heidelberg **2**, 58 (1882).
Kühn, A.: Über den Farbensinn der Bienen. Z. Vergl. Physiol. **5**, 762—800 (1927).
Kühn, A., Pohl, R.: Dressurfähigkeit der Bienen auf Spektrallinien. Naturwissenschaften **9**, 738—740 (1921).
Lamarck, J.-B.: Histoire naturelle des animaux sans vertèbres, Vols. 1—7 (1815—1822).
Larimer, J. L.: The presence of a functional caudal photoreceptor in blind cavernicolous crayfish. Nature (Lond.) **210**, 204—205 (1966).
Larimer, J. L., Trevino, D. L., Asby, E. A.: A comparison of spectral sensitivities of caudal photoreceptors of epigeal and cavernicolous crayfish. Comp. Biochem. Physiol. **19**, 409—415 (1966).
Lasansky, A., Fuortes, M. G. F.: The site of origin of electrical responses in visual cells of the leech, *Hirudo medicinalis*. J. Cell Biol. **42**, 241—252 (1969).
Lasansky, A., Marchiafava, P. L.: Light induced resistance changes in retinal rods and cones of the tiger salamander. J. Physiol. (Lond.) **236**, 171—191 (1974).
Leedale, G. F.: Euglenoid flagellates. Englewood Cliffs, N. J.: Prentice Hall 1967.
Liebman, P. A.: In situ spectrophotometric studies on the pigments of single retinal rods. Biophys. J. **2**, 161—178 (1962).
Liebman, P. A.: Light-dependent Ca^{++} content of rod outer segment disc membranes. Invest. Ophthalmol. **13**, 700—701 (1974).
Liebman, P. A.: Birefringence, dichroism and rod outer segment. In: Photoreceptor optics (eds. A. W. Snyder, R. Menzel), pp. 199—214. Berlin-Heidelberg-New York: Springer 1975.
Liebman, P. A., Entine, G.: Lateral diffusion of visual pigment in photoreceptor disk membranes. Science **185**, 457—458 (1974).
Lin, D. K., Niece, R. L., Fitch, W. M.: The properties and amino-acid sequence of cytochrome c from *Euglena gracilis*. Nature (London) **241**, 533—535 (1973).
Lubbock, J.: Ants, bees, and wasps. (New Edition by Myers, J. G., New York: Dutton and Co. 1929) 1882.
Malmo, R. B., Grether, W. F.: Further evidence of red blindness (protanopia) in Cebus monkey. J. Comp. Psychol. Physiol. **40**, 143—147 (1947).
Mason, W. T., Fager, R. S., Abrahamson, E. W.: Characterization of the lipid composition of squid rhabdome outer segments. Biochim. Biophys. Acta **306**, 67—73 (1973).
Mason, W. T., Fager, R. S., Abrahamson, E. W.: Ion fluxes in disk membranes of retinal rod outer segments. Nature (London) **247**, 562—563 (1974).
Mayr, E.: Wie weit sind die Grundprobleme der Evolution gelöst? In: Evolution. Nova Acta Leopoldina (Halle) **42**, 171—177 (1975).
McReynolds, J. S., Gorman, A. L. F.: Membrane conductances and spectral sensitivities of *Pecten* photoreceptors. J. Gen. Physiol. **56**, 376—391 (1970a).
McReynolds, J. S., Gorman, A. L. F.: Photoreceptor potentials of opposite polarity in the eye of the scallop, *Pecten irradians*. J. Gen. Physiol. **56**, 392—406 (1970b).

McReynolds, J. S., Gorman, A. L. F.: Hyperpolarizing photoreceptors in the eye of a primitive chordate, *Salpa democratica*. Vision Res. **15**, 1181—1186 (1975).

Menzel, R.: Polarisation sensitivity in insect eyes with fused rhabdoms. In: Photoreceptor optics (eds. A. W. Snyder, R. Menzel), pp. 372—387. Berlin-Heidelberg-New York: Springer 1975.

Metcalf, M. M.: The eye of *Salpa*. Mem. Biol. Lab. Johns Hopkins Univ. **2**, 305—396 (1893).

Meyer-Rochow, V. B.: The eyes of *Creophilus erythrocephalus* F. and *Sartallus signatus* Sharp (Staphylinidae: Coleoptera). Light-, interference-, scanning electron-, and transmission electron microscope examinations. Z. Zellforsch. **133**, 59—86 (1972).

Mote, M. I.: Polarization sensitivity: phenomenon independent of stimulus intensity or state of adaptation in retinular cells of the crabs *Carcinus* and *Callinectes*. J. Comp. Physiol. **90**, 389—403 (1974).

Nakao, T.: On the fine structure of the *Amphioxus* photoreceptor. Tohoku J. Exp. Med. **82**, 349—369 (1964).

Nolte, J., Brown, J. E.: Electrophysiological properties of cells in the median ocellus of *Limulus*. J. Gen. Physiol. **59**, 167—185 (1972).

Norman, R. A., Werblin, F. S.: Control of retinal sensitivity. I. Light and dark adaptation of vertebrate rods and cones. J. Gen. Physiol. **63**, 37—61 (1974).

Oesterheld, D., Stoeckenius, W.: Rhodopsin-like protein from the purple membrane of *Halobacterium halobium*. Nature (New Biol.) **233**, 149—152 (1971).

Paulsen, R., Schwemer, J.: Studies on the insect visual pigment sensitive to ultraviolet light: retinal as the chromophoric group. Biochem. Biophys. Acta **283**, 520—529 (1972).

Pettigrew, G. W.: The amino acid sequence of cytochrome c from *Euglena gracilis*. Nature (Lond.) **241**, 531—533 (1973).

Plate, L.: Allgemeine Zoologie und Abstammungslehre, Vol. II. Jena: G. Fischer 1924.

Poo, M., Cone, R. A.: Lateral diffusion of rhodopsin in the photoreceptor membrane. Nature (Lond.) **247**, 438—441 (1974).

Prosser, C. L.: Action potentials in the nervous system of the crayfish. II. Responses to illumination of the eye and caudal ganglion. J. Cell. Comp. Physiol. **4**, 363—377 (1934).

Rao, G., Rao, P. K.: A metasomatic neural photoreceptor in the scorpion. J. Exp. Biol. **58**, 189—196 (1973).

Rüppel, H.: Membrane structure and transduction mechanism of visual receptors. In: Photoreceptor optics (eds. A. W. Snyder, R. Menzel), pp. 499—523. Berlin-Heidelberg-New York: Springer 1975.

Rushton, W. A. H.: Rhodopsin measurement and dark adaptation in a subject deficient in cone vision. J. Physiol. (Lond.) **156**, 193—205 (1961).

Rushton, W. A. H., Powell, D. S.: The rhodopsin content and the visual threshold of human rods. Vision Res. **12**, 1073—1081 (1972).

Schultze, M.: Untersuchungen über die zusammengesetzten Augen der Krebse und Insekten. Bonn: M. Cohen Co. 1868.

Shaw, S. R.: Sense-cell structure and interspecies comparisons of polarized light absorption in arthropod compound eyes. Vision Res. **9**, 1031—1041 (1969).

Smith, D. S.: Insect Cells. Their Structure and Function. Edinburgh: Oliver Boyd 1968.

Snyder, A. W., Laughlin, S. B.: Dichroism and absorption by photoreceptors. J. Comp. Physiol. **100**, 101—116 (1975).

Stark, W. S., Zitzmann, W. G.: Isolation of adaptation mechanisms and photopigment spectra by vitamin A deprivation in *Drosophila*. J. Comp. Physiol. **105**, 15—27 (1976).

Stavenga, D. G., Zantema, A., Kuiper, J. W.: Rhodopsin processes and the function of the pupil mechanisms in flies. In: Biochemistry and physiology of visual pigments (ed. H. Langer), pp. 175—179. Berlin-Heidelberg-New York: Springer 1973.

Stieve, H.: On the ionic mechanisms responsible for the generation of the electrical response of light sensitive cells. Mosbacher Coll. Ges. Biol. Chemie **25**, 79—105 (1974).

Stieve, H.: Steuert Calcium die Leitfähigkeit der Arthropoden-Sehzellmembran bei der Erregung? Verh. Dtsch. Zool. Ges. **67**, 19—23 (1975).

Tominaga, Y., Kabuta, H.: Fine structure of the compound eye of a ladybug, *Harmonia axigridis*. Fukuoka Univ. Sci. Rep. **2**, 87—99 (1973).

Tomita, T.: Light induced potential and resistance changes in vertebrate photoreceptors. In: Handbook of sensory physiology Vol. VII/2, (ed. M. G. F. Fuortes), pp. 483—511. Berlin-Heidelberg-New York: Springer 1972.

Tomita, T., Kaneko, A., Murakami, M., Pautler, E. L.: Spectral response curves of single cones in the carp. Vision Res. **7**, 519—531 (1967).
Toyoda, I. I., Shapley, R. M.: The intracellularly recorded response in the scallop eye. Biol. Bull. **133**, 490 (1967).
Vanfleteren, J. R., Coomans, A.: Photoreceptor evolution and phylogeny. Z. Zool. Syst. Evolutionsforsch. **14**, 157—169 (1976).
Vinnikov, Y. A.: Principles of structural, chemical, and functional organization of sensory receptors. Cold Spring Harbor Symp. Quant. Biol. **30**, 293—299 (1966).
Vinnikov, Y. A.: Special senses. In: The structure and function of nervous tissue, Vol. II, (ed. G. H. Bourne), pp. 265—392. New York: Academic Press 1969.
Vinnikov, Y. A.: Sensory reception. Cytology, molecular mechanisms, and evolution. Berlin-Heidelberg-New York: Springer 1974.
Wachmann, E., Hennig, A.: Centrioles and development of the compound eye in *Megachile rotundata* (F.) (Hymenoptera, Apidae). Z. Morphol. Tiere **77**, 337—344 (1974).
Waterman, T. H.: Expectation and achievement in comparative physiology. J. Exp. Zool. **194**, 309—343 (1975).
Waterman, T. H., Fernández, H. R., Goldsmith, T. H.: Dichroism of photosensitive pigment in the rhabdoms of the crayfish *Orconectes*. J. Gen. Physiol. **54**, 415—432 (1969).
Wehner, R., Goldsmith, T. H.: Restrictions on translational diffusion of metarhodopsin in the membranes of a rhabdomeric photoreceptor. Biol. Bull. **149**, 450 (1975).
Weller, M., Virmaux, N., Mandel, P.: Role of light and rhodopsin phosphorylation in control permeability of retinal rod outer segment disks to Ca^{2+}. Nature (Lond.) **256**, 68—70 (1975).
White, R. H.: The effect of light and light deprivation upon the ultrastructure of the larval mosquito eye. II. The rhabdom. J. Exp. Zool. **166**, 405—426 (1967).
White, R. H.: The effect of light and light deprivation upon the ultrastructure of the larval mosquito eye. III. Multivesicular bodies and protein uptake. J. Exp. Zool. **169**, 261—278 (1968).
White, R. H., Lord, E.: Diminution and enlargement of the mosquito rhabdom in light and darkness. J. Gen. Physiol. **65**, 583—598 (1975).
Wilkens, L. A., Larimer, J. L.: A survey of abdominal CNS photoreceptive interneurons in the decapod Crustacea. Am. Zool. **14**, 1280 (1974).
Wilkens, L. A., Larimer, J. L.: Photosensitivity in the sixth abdominal ganglion of decapod Crustacea: a comparative study. J. Comp. Physiol. **106**, 69—75 (1976).
Willemous-Suhm, R. von: On some Atlantic Crustacea from the "Challenger" expedition. Trans. Linn. Soc. (2) **1**, 23—59 (1875).
Yanase, T., Sakamoto, S.: Fine structure of the visual cells of the dorsal eye in molluscan, *Onchidium verrulatum*. Zool. Mag. **74**, 238—242 (1965).
Yoshikami, S., Hagins, W. A.: Ionic basis of dark current and photocurrent of retinal rods. Biophys. J. **10**, 60a (1971).
Yoshikami, S., Hagins, W. A.: Control of the dark current in vertebrate rods and cones. In: Biochemistry and physiology of visual pigments (ed. H. Langer), pp. 245—255. Berlin-Heidelberg-New York: Springer 1973.
Yoshizawa, T.: The behaviour of visual pigments at low temperatures. In: Handbook of sensory physiology, Vol. VII/1, (ed. H. J. A. Dartnall), pp. 146—179. Berlin-Heidelberg-New York: Springer 1972.
Young, R. W.: The renewal of photoreceptor outer segments. J. Cell Biol. **33**, 61—72 (1967).
Young, R. W.: Passage of newly formed protein through the connecting cilium of retinal rods in the frog. J. Ultrastruct. Res. **23**, 462—473 (1968).
Young, R. W.: A difference between rods and cones in the renewal of outer segment protein. Invest. Ophthalmol. **8**, 222—231 (1969).
Young, R. W.: Shedding of discs from rod outer segments in the rhesus monkey. J. Ultrastruct. Res. **34**, 110—203 (1971 a).
Young, R. W.: The renewal of rod and cone outer segments in the rhesus monkey. J. Cell Biol. **49**, 303—318 (1971 b).
Zinkler, D.: Zum Lipidmuster der Photorezeptoren von Insekten. Verh. Dtsch. Zool. Ges. **67**, 28—32 (1975).
Zwicky, K. T.: A light response in the tail of *Urodacus*, a scorpion. Life Sci. **7**, 257—262 (1968).
Zwicky, K. T.: The special sensitivity of the tail of *Urodacus*, a scorpion. Experientia **26**, 317 (1970).

Chapter 2

Photic Responses and Sensory Transduction in Motile Protists

B. DIEHN, Toledo, Ohio (USA)

With 18 Figures and 1 Table

Contents

A. Introduction

I. Rationale For Investigating Photomovement

The fact that some unicellular organisms respond to illumination by accumulating either at the light or dark side of the vessel in which they are contained, has been the subject of scientific study for more than 150 years. The reason for this interest is, of course, the realization that such organisms contain a sensory system that enables them not only to perceive light, but apparently also to determine its direction. As will be seen, recent work has shown that in most cases, the observed accumulation pattern is not in fact a consequence of truly directional responses. Nevertheless, the obvious ability of certain motile protists to exhibit a distinctive behavioral response to stimulation with light, has continued to fascinate researchers to this day.

There are several levels of sophistication at which the photomovement of microorganisms can be studied. One may simply wish to know how the cell moves in the presence of light as contrasted to its movement in the dark, or one may investigate the means by which the cell derives the energy which powers its photosensory system, to list but two avenues of inquiry.

This study will be concerned with efforts to describe the photosensory stimulus/response systems of unicells on the molecular level, i.e., it will deal with the chemical and physical basis of light-induced protistan behavior. For this kind of research a large variety of experimental approaches must be used to study the organism and consequently, a molecular description of photomovement behavior is beginning to emerge for only a few microorganisms.

The intent of this chapter, then, is not to compile a comprehensive description of all past and present studies dealing with protozoan photomovement. For such a compilation, the reader is referred to the reviews by BENDIX (1960) and NULTSCH (1975). The present chapter will, however, address problems emerging in other review articles, and incidentally also offer some answers to the questions posed by FEINLEIB and CURRY in the first volume of this series (1971). In keeping with this intent, the discussion will mention work done on individual aspects of the photobehavior of a variety of unicells, but will cover in detail only the photomovement of a few well-studied organisms.

II. Terminology

Historically, a wide variety of often contradictory terms have been used in the description of the photomovement responses of microorganisms. This has led to a situation where even in more recent literature, different terms have been used by different research groups to represent the same phenomenon, and consequently the field has been rather confusing to the uninitiated. To remedy this situation, an Ad Hoc Committee on Behavioral Terminology has recently recommended a set of guidelines (DIEHN et al., 1977). These guidelines, which provide a useful summary of the possible types of stimulus-induced movement behavior, are reproduced in the following:

1. Stimuli

A *stimulus* is defined as any quantity of energy or matter that, when interacting with the organism, can (but need not always) elicit a response.

a) The nature of the stimulus is indicated by prefixing the term for the response, e.g., kinesis with one of the following:

Prefix	*Stimulus*
photo-	radiant energy affecting specific receptor molecules
thermo-	thermal energy
galvano-	ionic-electric current
electro-	electric field
magneto-	magnetic field
geo-	gravitational force
mechano-	mechanical force
chemo-	molecular species acting upon specific receptor molecules

It is essential that the actual stimulus be identified properly since, for instance, both radiation and electric energy are capable of generating chemical substances that can serve as direct stimuli.

b) Where the nature or occurrence of a response depends upon an increase or decrease in stimulus intensity, an increase may be indicated by the use of the term *step-up* and a decrease by the term *step-down*. Thus, a *step-up photophobic response* is a phobic response that occurs upon an increase in light intensity.

2. Responses

A *response* is defined as any stimulus-induced alteration in the activity of the organism's motor apparatus that can (but need not always) result in an alteration of the movement or orientation of the organism. The responses are classified as follows:

a) *Kineses* are responses in which the steady-state rate of activity of the organism is controlled by the absolute magnitude of the stimulus intensity. The sign of a kinetic response is determined relative to a null value, i.e., the activity rate in the absence of a stimulus. In a *positive kinesis*, the activity rate is greater in the presence of the stimulus than in its absence; in a *negative kinesis*, the activity rate is lower in the presence of the stimulus than in its absence. Ideally, a given kinetic response level is maintained as long as the stimulus level remains constant, i.e., adaptation does not occur.

In most kinetic responses studied to date, the linear velocity of the organism is the activity parameter that is affected by the stimulus. Since this is not the only possible type of kinesis, the following (incomplete) subdivisions may be used:

α) The term *orthokinesis* indicates that the linear velocity of the organism is altered by the stimulus. A typical example of orthokinesis without adaptation is the photokinesis of blue-green algae, e.g., *Phormidium* (NULTSCH, 1962). In the literature on responses of microorganisms, the term kinesis, without prefix, is often used when describing orthokinesis.

β) In *klinokinesis*, the affected activity parameter is the frequency of directional change, i.e., the organism spontaneously changes its direction even in the absence of stimulation, and the stimulus modulates the frequency of such changes. *Direction* means the direction of net progress, e.g., in helically moving organisms, the direction of the helix's axis. An example of klinokinesis is found in the step-up chemoresponses of bacteria, e.g., *Salmonella* (MACNAB and KOSHLAND, 1972).

b) *Phobic Responses.* In this type of response, a change in stimulus intensity elicits a transient alteration in the activity of the organism (i.e., the response is controlled by a temporal stimulus gradient rather than by the absolute magnitude of the stimulus). The pattern of activity change depends on the morphology of a given organism, is determined genetically, and hence is independent of the nature of the stimulus. The only alternative to be considered is whether in a given case the step-up or the step-down signal elicits this unique response.

A typical phobic response consists of a change in linear velocity, normally a stop response, followed by a change of direction. Unlike the situation in a kinesis, the organism resumes its original behavior after some time by means of adaptation. A typical example is the photophobic response of *Euglena* (DIEHN et al., 1975). Phobic responses that have long adaptation times are nearly indistinguishable from kineses. In such cases, we suggest that the decision of whether one is dealing with one or the other be left up to the individual researcher.

3. Behavioral Consequences

a) *Taxis* is a movement oriented with respect to the stimulus direction; its essential characteristic, therefore, is active orientation of the organism with respect to the stimulus source. A taxis is not a response in the strict sense of the introductory definition, but rather the result of a response (or of a series of responses) of the motor apparatus.

If the orientation of the organism is toward the stimulus source, the taxis is *positive*; if away from it, *negative*. The term *transverse taxis* denotes perpendicular orientation of the organism with respect to the stimulus source.

b) *Accumulation/Dispersal.* Accumulation in or dispersal from a region with higher intensity of a stimulus, though often confused with a true behavioral response, is the most often observed macroscopic result of many of the phenomena discussed above.

Dispersal from a region of higher stimulus intensity, for instance, can be mediated by *negative taxis* (which directs the organism away from the stimulus source) (FEINLEIB and CURRY, 1971); by *step-up positive kinesis* (which results in the organism's spending less time in the region of higher stimulus intensity, because of increased linear velocity) (THROM, 1968); by *step-up phobic responses* (preventing the organisms from moving up a stimulus gradient) (DIEHN, 1969a); or by *step-up positive klinokinesis* (which shortens the organism's path as it moves in the direction of increasing stimulus intensity, and thus introduces a bias away from that direction) (TSANG et al., 1973).

The preceeding definitions come directly from the committee's terminology recommendations. These recommendations are deliberately incomplete with regard to the possible response types in order to facilitate, by introducing a minimum of new terms, the transition to a standardized terminology. In a rigorous version of this terminology all kinesis and phobic responses are subdivided according to the affected movement parameter by using the following prefixes:

α) *Ortho-:* linear velocity.

A *positive orthophobic response*, for instance, is expressed as a transient increase in linear velocity. Such a response is exhibited upon mechanical posterior stimulation of *Paramecium* (NAITOH and ECKERT, 1969) and of juvenile *Homo sapiens* (Any Parent).

β) *Klino-:* frequency of random directional changes.

This prefix always indicates that the stimulus introduces a directional bias into a random walk. An example of a *negative klinophobic response* is found in the step-up chemoresponse of *Escherichia coli* (BERG and BROWN, 1972).

γ) *Stropho-:* rate of turning or rotation.

The phobic response of *Euglena gracilis* (DIEHN, 1969a) upon a change in light intensity is an example of a *positive strophophobic response.*

δ) *Morpho-:* rate of change of body shape or outline.

A good example of a *morphokinesis* is exhibited by mechanically stimulated *Euglena* upon an increase in illumination (MIKOLAJCZYK and DIEHN, 1975).

Responses with the same prefix, such as *step-up negative klinokineses* and *klinophobic responses*, usually have similar behavioral consequences; in this example, accumulation toward the stimulus.

It is obvious that there is little sense in listing the full term every time the same response is mentioned in a publication. Nevertheless, using a proper term in the title and introduction serves not only to describe the response economically, but also to identify its relationship to other pertinent reactions of the organism.

This unequivocal terminology was not used consistently prior to 1975. Although this does not, of course, distract from the inherent value of prior studies, the best approach to reading this literature will be to ignore the terminology in general and to determine from the description what phenomenon is being investigated in each case. The following list of past and present usage will prove useful in interpreting the literature:

Present term	Previously used other terms
Step-up	Negative, direct
Step-down	Positive, inverse
Phototaxis	Phototopotaxis
Photoaccumulation	Positive phototaxis
Photodispersal	Negative phototaxis
Photophobic response	Photophobotaxis, stop or avoidance response

It is evident from the terminology guidelines that a variety of behavioral strategies can be employed by unicellular organisms to achieve net movement toward the light or away from it. This statement is not as anthropomorphic as it appears, since the great majority of photomotive microorganisms are photosynthetic and thus can in fact benefit from proper positioning with respect to light. Several examples will be given in this chapter of different mechanisms leading to photoaccumulation in different organisms.

At this point, a concise discussion of experimental methods for studying photomovement appears appropriate. There is no need for a detailed description of experimental apparatus, however, since this subject has been treated in a special review by CHECCUCCI (1973) and is also covered by NULTSCH (1975). Any of these experimental methods can yield useful data if they are employed as part of a well-considered investigative strategy.

III. Experimental Methods for Studying Photomovement

There are, in principle, three levels at which stimulus-induced movement can be investigated (all methods must of course include provisions for recording the onset and cessation of illumination changes):

1. Direct Observation

Observation under the microscope of the cell's motor apparatus constitutes the most unequivocal method for studying photomovement. When employed in real time, this technique is usually suitable only for qualitative studies since responses often occur in the millisecond time range. Yet, the first observation of photomovement of individual *Euglena* cells by STAHL (1880) was most important in pointing out that the direction of this organism's responses is primarily controlled by the intensity of the incident light. For quantitative studies, the responses under investigation must be recorded by cinemicrographic methods. High filming speeds are required to study flagellar responses in detail, and the evaluation of the recorded sequences becomes very tedious. For instance, in a quantitative study of the flagellar step-up photophobic response of *Euglena* by DIEHN et al. (1975), the analysis of 21 stimulation sequences required the evaluation of 20000 frames of 16 mm film.

2. Tracking Methods

Studies of individual cells can be simplified once one has established the consequences of a response of the cell's motor apparatus in terms of the movement of the cell as a whole. A photophobic response in *Euglena*, for instance, appears as a very distinctive turning of the cell around a lateral axis (DIEHN,

1969a). Thus, data on the responses can be gathered by tracking methods in which the path and position of the cell is recorded, either on film, e.g., by streak photography (FEINLEIB and CURRY, 1967) or on videotape. The latter method lacks sufficient resolution for direct visualization of flagella or cilia, but lends itself to computer processing and analysis (DAVENPORT et al., 1970). The above techniques allow simultaneous determination of the paths of several cells and thus greatly reduce the effort involved in collecting data, but they are accurate only as long as the cells remain in the focal plane. Thus, the measurements are restricted to cells moving two-dimensionally within the horizontal plane. This is objectionable only if the organisms are deliberately restricted to such movement, e.g., by confinement in a narrow cuvette (BRINKMANN, 1968). An automated tracking system, which follows a single cell in all three dimensions, has been described by BERG (1971). It consists essentially of three fiber-optics controlled servo loops that keep a microscope focused and centered on the cell under study. The movement of the microscope stage corresponds to the movement of the cell in three dimensions and is either recorded, or directly processed and analyzed by computer. This system was devised for studies of chemosensory behavior of bacteria and has not yet been utilized for photomovement investigations.

3. Mass-Movement Methods

In this type of experimental technique, the movement of individual organisms cannot be determined. Instead, the accumulation in or dispersal from an illuminated zone of the cell suspension is recorded. Historically, mass-movement methods were the first to be employed for studies of photomovement. Curiously enough, almost all contemporary individual researchers have entered the field by devising a mass-movement method, or redesigning an existing one, for measuring photoaccumulation. (This fascinating phenomenon will not be pursued further here, as this contribution is concerned with the behavior of unicells rather than with that of photobiologists.) Since photoaccumulation, though easily observed, is not a primary response, experimental methods that measure this phenomenon are of limited utility in studying photosensory processes. Nevertheless, such methods have the advantage of high statistical accuracy since very large numbers of organisms contribute to each measurement. Moreover, the major criticism of the method would be negated if, in a given apparatus under all experimental conditions, a single primary response leading to photoaccumulation could be established for an experimental organism.

Three variants of the mass-movement method exist that permit continuous electronic recording of the accumulation/dispersal phenomenon:

a) The *phototaxigraph* of LINDES et al. (1965) (Fig. 1) is essentially a light trap method in which those organisms that migrate toward the illuminated zone by true phototaxis are attracted by the light scattered by cells already in that region. A second probable mechanism leading to accumulation in the actinic region is trapping via step-down photophobic responses, at the light-dark interface, of cells that have entered this region through nonphotic movement. In the case of the unicellular alga *Euglena* (vide infra, see E,I2) experimental evidence suggests that

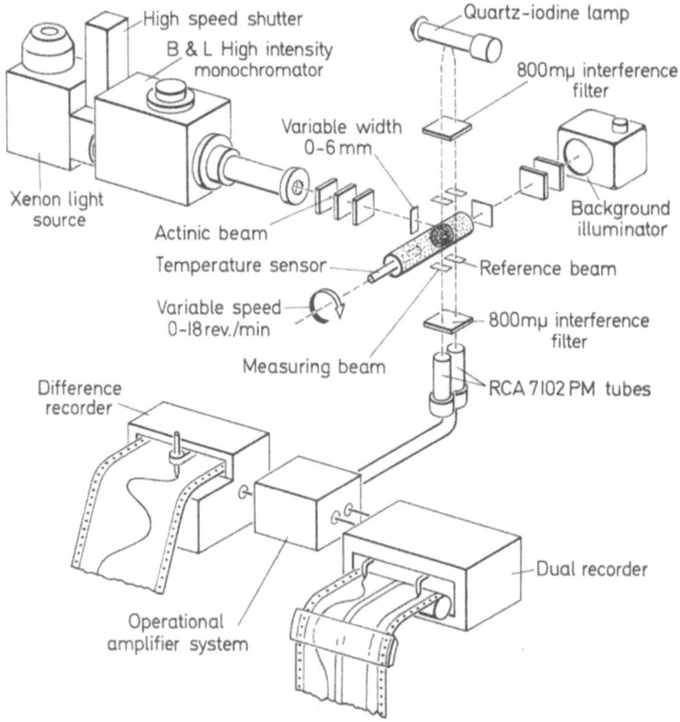

Fig. 1. Phototaxigraph. [After DIEHN (1969 b)]

phototaxis proper is a consequence of step-down phobic responses (DIEHN, 1969a) and thus phototaxigraph measurements probably reflect only a single response of *Euglena*, i.e., the step-down photophobic response. The mechanism of the photoaccumulation of *Euglena* in the phototaxigraph has been analyzed in detail, including computer simulation of the accumulation kinetics (DIEHN, 1970).

b) The *population method* of FEINLEIB and CURRY (1967)—a system that appears similar to the phototaxigraph but in which the stimulating light enters the cuvette parallel rather than perpendicular to the cuvette axis (Fig. 2). Users of this system argue that it measures only phototaxis. At first glance, this appears reasonable since there seems to be no possibility of photophobic trapping in the cuvette. However, the actinic light intensity is not really uniform in the cuvette; because of absorption and scattering of the stimulating light by the cells, the cells nearest the light source are more strongly illuminated than the cells in the rear. This is a potential problem, especially with dense suspensions. Later in this chapter (see E,I 2) some experimental results suggesting a phobic accumulation mechanism in this apparatus will be discussed.

c) *45° Apparatus*. In an ingenious modification of method (b), independently developed by THROM (1968), the stimulating light beam enters at all points along one side of the cuvette at a 45° angle (Fig. 3). Although internal light intensity

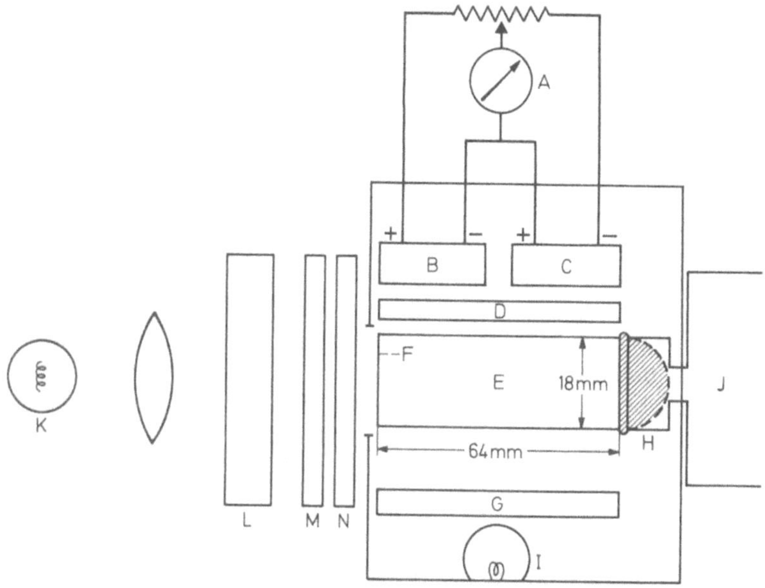

Fig. 2 A–N. *Schematic diagram of apparatus used in the population method, viewed from above.* (A) Amplifier—galvanometer. (B) Photocell 1. (C) Photocell 2. (D) Red filter. (E) Algal suspension. (F) No. 1 round covership. (G) Red filter. (H) Black paraffin. (I) Tungsten filament lamp. (J) 60 rpm motor. (K) 300 W tungsten filament lamp. (L) Water filter. (M) Neutral density filter. (N) Blue-green filter

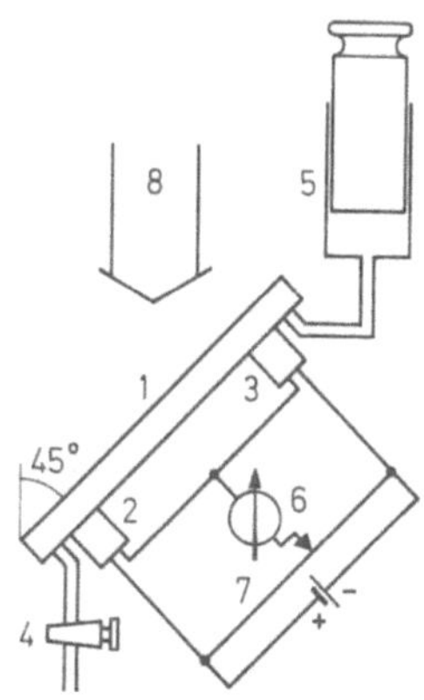

Fig. 3. "45° Apparatus" of Throm (1968), viewed from above. *1* Continuous flow cuvette for cell suspension. *2* and *3* Photoresistor. *4* Stopcock, to be closed for photomovement-measurements. *5* Syringe. *6* 6 μA recorder. *7* Balancing potentiometer. *8* Stimulating light simultaneously acting as measuring and reference beam

gradients in this arrangement are small at the beginning of an experiment when the cuvette is sufficiently narrow, positively responding cells soon encounter the cuvette wall and thus experience an additional mechanosensory stimulus. With *Chlamydomonas*, a high percentage of cells reverse their direction upon contact with the cuvette wall, particularly under conditions where their initial phototaxis is positive (RIEDL, 1976).

B. Cellular Organelles Involved in Sensory Transduction

I. The Photosensory Chain

By *sensory transduction* this author means the series of physiologic processes that begin with the absorption of light by the organism and ultimately end in a motor response. There is no organism among those studied to date for which all of the components of the sensory chain have been identified. For this discussion of the transduction processes I will use a reasonable subdivision of the sensory chain into functional units, rather than into postulated anatomic components which may or may not be found in subsequent work. The following functional elements must be present in any photic stimulus/response system:

1. Photoreceptor

In this organelle, the light is absorbed and converted into a physiologic signal that is chemical in nature. If the chemical species involved are ionic, an electric signal results. The photoreceptor can be an identifiable structure which is specialized for stimulus reception, or it may be an organelle with a primary function in the energy transduction system of the cell.

2. Processor

This functional element controls whether a stimulus, once it has been received, will indeed trigger a motor response. Thus, the processor is responsible for response thresholds, and in cells capable of exhibiting phobic responses (which require an ability of the cell to "remember" previous stimulus intensities), for temporal comparison of stimuli.

3. Effector

Upon command from the processor, this element effects a reorientation of the cell's propulsive apparatus. The effector is not to be confused with the cell's motor apparatus, which in most organisms propels the cell whether it is being stimulated or not. It is the alignment of the motor apparatus with respect to the cell body which is altered by the effector.

One can argue that a means is needed for transmitting the physiologic signal between the above three components of the sensory chain. This component is not included in the above list because its function (though not necessarily its molecular mechanism of action) is passive in nature.

Although in actuality any of the above functions may be combined in one anatomic unit of the cell, or one function may be split between several organelles, the above division facilitates rational analysis of the cell's sensory system.

A limited amount of information exists regarding the location of the photosensory apparatus within the cell of the components. At present no organelles have been identified with certainty that carry out the effector or processor functions in any unicellular organism. The effector, as a motile element, will probably be identified by its structure in those cells with responses involving active reorientation of cilia or flagella. To date, neither of these two sensory functions have received much experimental attention in photoresponsive organisms. The main reason for this is that the concept of information processing in primitive stimulus/response systems is relatively new, and consequently most of the experimental effort directed toward studying photosensory anatomy in unicells has been devoted in the past to localizing the photoreceptors. The latter work has met with some success, and for some organisms, experimental evidence has been developed showing that the photoreceptor function resides in specific organelles.

II. Photoreceptors in Protists

The most straightforward situation exists in purple bacteria. In these procaryotic organisms, both photoaccumulation and photodispersal are generally mediated by phobic responses, and the rate of photoaccumulation or photodispersal is a function of the rate photosynthesis (HAUPT, 1966). Thus, photoreception for photomovement occurs at the same site as for photosynthesis, i.e., in the photosynthetic apparatus. In contradiction to this, HARAYAMA and IINO (1976) reported a light-induced accumulation of photosynthetically incompetent *Rhodospirillum rubrum*. However, the threshold light intensity for the response is inordinately high (30 kW/m^2), and the effect they observe may very well reflect a perturbation of the cell's metabolism through light absorption in a component of its respiratory chain. This would be similar to the situation in *Salmonella*, where TAYLOR and KOSHLAND (1975) demonstrated control of klinokinesis by light of similar intensities.

In other procaryotes photoreception for the phobic responses may occur entirely in the photosynthetic apparatus, as in *Phormidium* (HAEDER, 1974); or in the photosynthetic organelle for one response (e.g., step-down) and elsewhere for the other response (e.g., step-up), as in *Halobacterium* (HILDEBRAND and DENCHER, 1975). In the few eucaryotic unicells where meaningful data are available specific photoreceptor organelles mediate stimulus reception for the phobic responses.

The photosynthetic alga *Euglena* is unquestionably the most extensively studied of these organisms. Figure 4 is a schematic drawing of *E. gracilis* showing the cellular organelles important for photosensory transduction. The *stigma* is a very prominent orange-red organelle situated on what has been defined as the dorsal side of the cell. Not surprisingly it was called the eyespot by early researchers; however, experimental evidence strongly suggests that it is not the photoreceptor organelle. The other candidate for the photoreceptor function is the paraflagellar body. ENGELMANN (1882) stated that *Euglena* exhibited a phobic response when a boundary between a light and a dark area was moved across the anterior level of

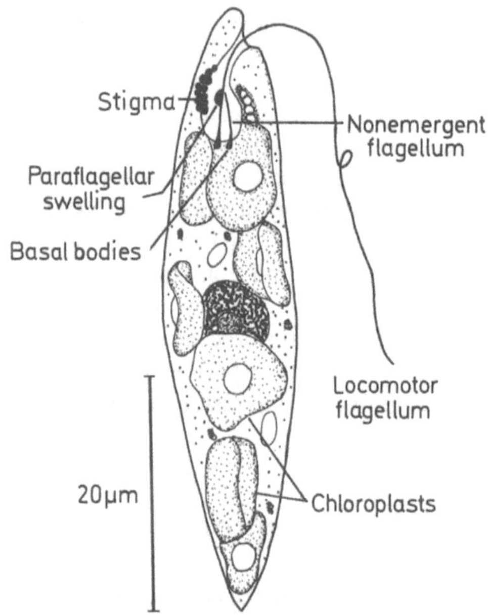

Fig. 4. *E. gracilis.* [After MIKOLAJCZYK and DIEHN (1976)]

the cell, but before the boundary reached the stigma. Proponents of the stigma-photoreceptor hypothesis generally claim that the reliability of these observations is suspect because of the close spatial proximity of the stigma and the paraflagellar body. This makes it difficult to convince critics that the boundary has crossed one but not the other. GOESSEL (1957) observed that a mutant of *Euglena* that contained no stigma still possessed the paraflagellar swelling. This mutant exhibited photophobic reactions, whereas *Astasia*, a relative of *Euglena* possessing neither stigma nor swelling, showed none. More recently, DIEHN (1973) and CHECCUCCI et al. (1976) also observed photoresponses in stigma-bleached mutants. Together with the discovery of BENEDETTI and CHECCUCCI (1975) that the paraflagellar body, but not the stigma, exhibits fluorescence, which is characteristic of the postulated receptor pigment (vide infra), these data very strongly support the hypothesis that the photoreceptor function for the phobic responses resides in the paraflagellar swelling.

The only other eucaryotic organism in wich the location of the photoreceptor is known with some confidence is the dinoflagellate *Gyrodinium dorsum.* The two flagella of this cell project from two prominent grooves, the *girdle* and the *sulcus* (Fig. 5). Studies by HAND and SCHMIDT (1975) indicate that when a photophobic stop response occurs, the sulcus/girdle junction always faces the light. These authors assert that the junction is shaded when the cell moves parallel to the light, and that a phobic response occurs when the anterior projection of the sulcus, acting as a light channel, admits a photic stimulus to the putative photoreceptor located at the sulcus/girdle junction.

Nothing is known about photoreception for the kinesis in eucaryotes. In fact, the only organism in which photokinesis has been studied in any detail is *Eu-*

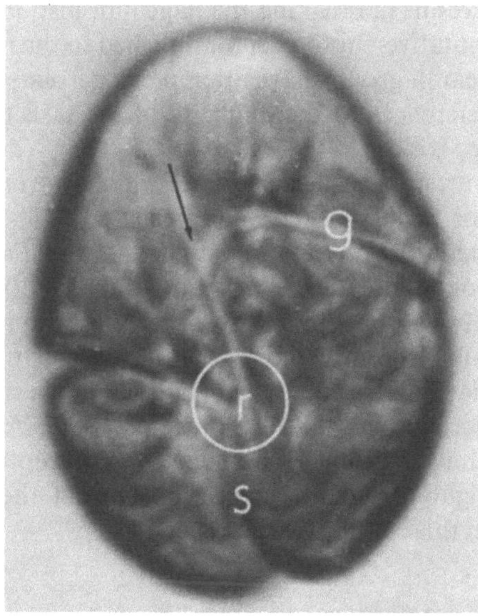

Fig. 5. Lateral view of *G. dorsum*. The position of the proposed receptor region (*r* in the *white circle*) is shown in relation to the sulcus (*s*) and girdle (*g*). The anterior light channel, which may function in the proper stimulus alignment during phototaxis, is marked by the *black arrow*. [After HAND and SCHMIDT (1975)]

glena (WOLKEN and SHIN, 1958). This may well reflect a lack of interest in the phenomenon on the part of the experimenters rather than on the part of eucaryotic cells. Since photokinesis persists over the duration of illumination, there is reason to expect that photoreception for the kinesis will be found in the photosynthetic apparatus.

C. Receptor Pigments and Molecular Mechanisms

I. Pigment Identification

Considering how little is known about the photoreceptor organelles of most unicells, the identification of the photoreceptor pigments clearly is a formidable task; yet it is an essential step in elucidating the molecular mechanism of stimulus transduction. It can be done unambiguously only by isolation of the photoreceptors followed by extraction of the pigments, and then proving that these pigments are involved in photoreception in vivo.

1. Pigment Isolation

The only photoreceptor which is a viable candidate for isolation, by virtue of having been characterized as a distinct organelle, is the paraflagellar body of *Euglena*. Yet while several techniques are known for removing the flagellum of

Euglena, all methods result in breaking the flagellum just *above* the paraflagellar swelling, leaving the putative photoreceptor attached to the flagellar stump within the canal. Several research groups have attempted to isolate the organelle, yet no more than oblique references to the difficulty of the task have filtered into the literature. The message is clear.

Pigment isolation attempts have been successful, however, in an organelle that is not a photoreceptor proper: BATRA and TOLLIN (1964) and BARTLETT et al. (1972) worked out methods for isolating stigmata from *Euglena* and identified their pigments as carotenoids. According to the former authors, the stigma contains 51% lutein, 32% cryptoxanthin, 8% β-carotene, and approximately 9% of an unknown carotenoid. Recently, the latter research group presented fluorometric data indicating that isolated stigma pigments contain a small amount of flavin in addition to the carotenoids (SPERLING et al., 1976). Although the authors suggest that this finding may give the stigma-photoreceptor hypothesis a new lease on life, the weight of other evidence in favor of the paraflagellar body, discussed elsewhere in this chapter, remains persuasive.

2. In vivo Microspectrophotometry

The situation just described calls for indirect methods of photoreceptor identification. If a microscope image is scanned by the analyzing beam of a spectrophotometer, the pigment content of the structure being observed can be determined via its light absorption. As in the previous method, the identity of the photoreceptor organelle must be known if this method is to be used for receptor pigment identification. In this respect, absorption microspectrophotometry offers no advantage and although organelle isolation is not required, screening by other structures in the cell can create serious problems. Absorption spectroscopy of the stigma of *Euglena*, for instance, was first performed by STROTHER and WOLKEN (1960). Even though this organelle is large and easily identified, their results are only in partial agreement with those of BATRA and TOLLIN (1964) who isolated stigmata and then determined their absorption spectrum. The in vivo spectra exhibited absorption maxima of 510, 530, and 630 nm which were not found in isolated stigmata, and were probably due to nonstigma pigments.

A better correlation between the in vivo and in vitro absorption spectra of stigmata was demonstrated in a recent study by BENEDETTI et al. (1976a), using a newly designed microspectrophotometer (BENEDETTI et al., 1976b). The authors found no evidence for differential absorption of polarized light by the stigma, indicating that the carotenoid molecules in this organelle ar not oriented as had been hypothesized by BOUND and TOLLIN (1967) and DIEHN (1969b). This makes sense also in view of BENEDETTI et al.'s estimate of 0.2 pg/stigma carotenoid content, which would give each pigment molecule a free space 50 times its own volume.

The problems of absorption microspectrophotometry intensify as smaller organelles are studied: targeting becomes more difficult, and screening by the rest of the cell more serious. For instance no one has yet reported an in vivo absorption spectrum of the paraflagellar body of *Euglena*.

A different approach is offered by the technique of in vivo fluorescence-emission spectroscopy. Although screening may still cause problems, selective excitation of fluorescence can allow the localization of structures bearing specific pigments. This method can also be used successfully with organelles that are too small, or too lightly pigmented, for absorption methods.

Since flavins had been implicated as the photoreceptor pigment by action spectrum measurements (DIEHN, 1969b, and others, vide infra), BENEDETTI and CHECCUCCI (1975) illuminated live *Euglenae* with excitation light of 360 and 450 nm. They observed strong green fluorescence, with a maximum around 525 nm, localized where one would expect to find the paraflagellar body. This fluorescence, which is characteristic of flavins, was found in green as well as chlorophyll-free cells. No such fluorescence was seen in the stigma.

The paraflagellar swelling of *Euglena* evidently contains flavins. If the flavin receptor pigment hypothesis is accepted, then the argument that the paraflagellar body is the receptor organelle would appear convincing. However, this argument is flawed in that a functioning receptor pigment should fluoresce only weakly, if at all. Since its primary function is the transduction of light energy into a chemical signal, a radiationless transition of the excited pigment molecule in live *Euglena* would be expected rather than efficient conversion of the excitation energy into fluorescence. Conceivably, the cells studied by BENEDETTI and CHECCUCCI were severely distressed from being compressed between microscope slide and cover slip.

3. Action Spectra

The state of affairs just described requires that other indirect methods be used to identify the photoreceptor pigments. The most generally applicable of these methods is the determination of action spectra. If one measures photomovement of an organism as a function of the wavelength and intensity of the stimulating light, one obtains an indication of how efficiently the photoreceptor converts light of a given wavelength to the primary physiologic stimulus. If there is no screening of the light before absorption in the photoreceptor, this action spectrum should accurately reflect the absorption spectrum of the receptor pigment. In practice, this also requires that one makes sure to measure a primary response of the cell and not a possible mixture of responses. The dual requirement for the absence of screening and a clearly characterized response limits the utility of mass-movement methods. On the other hand, single-cell observations are not only extremely tedious if statistical accuracy is desired, but also show poor resolution because eye-and-stopwatch quantification of behavioral responses is not very accurate. An automated cell-tracking technique would appear to be the method of choice, but in the only instance to date in which a tracking method has been used for determining an action spectrum (in *Gymnodinium*, FORWARD, 1974), the tracks were evaluated manually and the percentage of cells responding near threshold was determined rather than a dose-response curve. The latter approach, i.e., measuring the light intensity that elicits a given response in the middle of the dose-response curve, avoids aberrations due to saturation or threshold effects, and is the most appropriate one, both with mass-movement and with single-cell techniques.

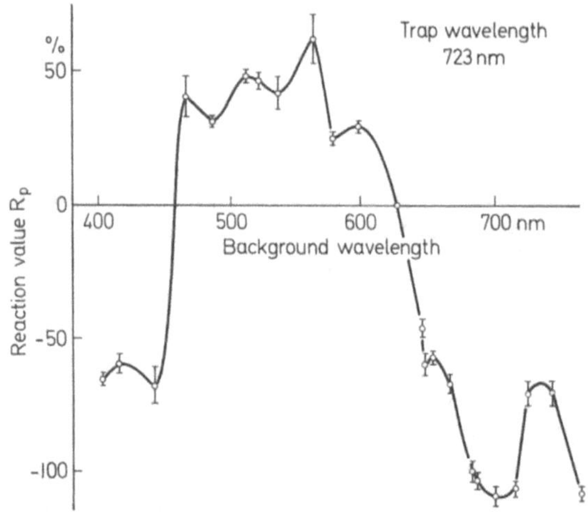

Fig. 6. Reaction value (R_p) of photoaccumulation (positive values) and photodispersal (negative values) of *Phormidium* relative to white light control, as a function of background wavelength using a constant trap wavelength of 723 nm. Intensity of both background and trap light 1 W/m^2. [After HAEDER (1974b)]

The fact has been mentioned already that in most procaryotic cells photoreception for photomovement occurs in the photosynthetic apparatus. The major evidence for this state of affairs comes from action spectrum measurements. The relevant material has been carefully reviewed by NULTSCH (1975). Most of it is not of interest in the context of this review because the intent of the studies did not extend beyond pigment identification.

A notable exception is HAEDER's work with *Phormidium uncinatum* (1973). He used an elegant variation of the action spectrum technique to elucidate the roles of photosystems I and II in controlling the photophobic responses of this organism.

In essence, a double irradiation technique was used in which all preparations were illuminated with homogeneous background light from above, and square light "traps" were projected onto them from below. Upon crossing the border of the trap, the cells in this situation experienced a transition from irradiation by two wavelengths simultaneously, to illumination by one wavelength. The following was observed (HAEDER and NULTSCH, 1973): With a background wavelength of 723 nm, which is mainly absorbed by photosystem I, an action spectrum for photomovement was found that resembles that of photosystem II (Fig. 6). Conversely, with background irradiation of 563 nm, which is predominantly absorbed by C-phycoerythrin (the main accessory pigment of photosystem II), the action spectrum for photomovement resembles that of photosystem I. The term photomovement was chosen judiciously in the preceding sentence: accumulation in the trap occurs between 440 and 700 nm, whereas dispersal from the trap is observed outside this wavelength range.

The observation that the background wavelength controls whether certain trap wavelengths lead to accumulation or dispersal in the trap prompted an investigation of photomovement as a function of both trap and background wavelength. A three-dimensional plot of the results (Fig. 7) show that at equal intensities (1 W/m^2) of both background and trap illumination, dispersal always occurs if the background wavelength is below 750 nm and the trap wavelength above 750 nm, i.e., when photosystem I is predominantly activated in the trap. Since this system accepts electrons from an intermediary electron transport chain (IETc), which has in turn received them from photosystem II (Fig. 8), HAEDER (1974) suggested that the phobic reaction which mediates dispersal is triggered by a decrease in the size of an "electron pool" located between the two systems. Since in biologic systems electrons cannot be stored per se, a less abstract formulation would be to say that the phobic response is triggered by a decrease in the ratio of the reduced (electron-rich) to the oxidized form of a member of the intermediate electron transport chain. We shall return to a discussion of this situation later.

The phobic reaction, incidentally, presents an interesting situation with respect to terminology: it is a step-up response with respect to one photoreceptor (photosystem I), but a step-down response with respect to the other. Clearly no terminology problem exists when the photoreceptor is specified—in hindsight an obvious necessity in cases where an organism has multiple receptors with different characteristics.

A procaryote with an unusual photosynthetic apparatus is *Halobacterium halobium*. This bacterium utilizes bacteriorhodopsin, instead of chlorophyll, as the light-harvesting pigment. The primary photochemical process, upon excitation of the bacterial retinal-protein complex, is a release of protons leading to the establishment of a pH gradient. This, in turn, is utilized for the synthesis of adenosine triphosphate (ATP) (RACKER and STOECKENIUS, 1974). The photosynthetic pigment exhibits an absorption spectrum typical of carotenoids and is responsible for the characteristic color of the cell's purple membrane (and hence of the salt evaporation ponds that are a familiar sight to anyone flying into San Francisco's International Airport).

HILDEBRAND and DENCHER (1975) studied the photophobic responses of this organism as a function of the excitation wavelength. They found that the action spectrum for the step-down response resembled the absorption spectrum of the purple membrane, in agreement with the observation that in procaryotes, photoreception for photomovement usually occurs in the photosynthetic apparatus.

The action spectrum for the step-up phobic response, however, was quite different. It had a peak in the near UV that was not found with the purple membrane, but coincided with the UV peak in the absorption spectrum of an aqueous riboflavin solution, as well as with the major absorption peak of a retinylidene protein that can be obtained by detergent treatment of the purple membrane. Although the identity of this pigment is still uncertain, it is clearly not involved in photosynthesis. Moreover, photoreception in the step-up system is probably significant behaviorally, since, in contrast to tumbling responses of other bacteria (See B,II), the response threshold is quite low (35 mW/m^2 vs 30 kW/m^2). *Halobacterium* thus promises to be one procaryotic cell in which photosensory transduction can be studied independently of energy transduction.

Among the eucaryotes there are only two organisms, *Stentor* and *Euglena*, in which action spectrum measurements have been followed up by systematic investigations of the physiologic events following photoreception.

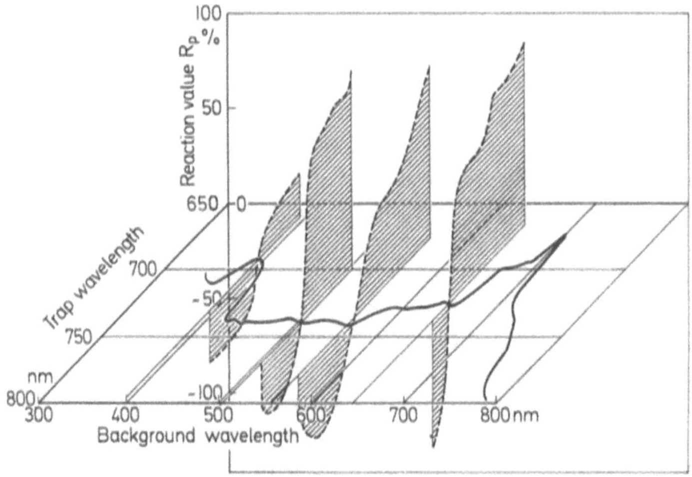

Fig. 7. Three-dimensional diagram showing accumulation or dispersal of *Phormidium* as a function of background and trap wavelength at equal intensities of 1 W/m². For the sake of clarity, only four background wavelengths are shown. The *solid line* connects all inversion points from photoaccumulation to photodispersal. [After HAEDER and NULTSCH (1973)]

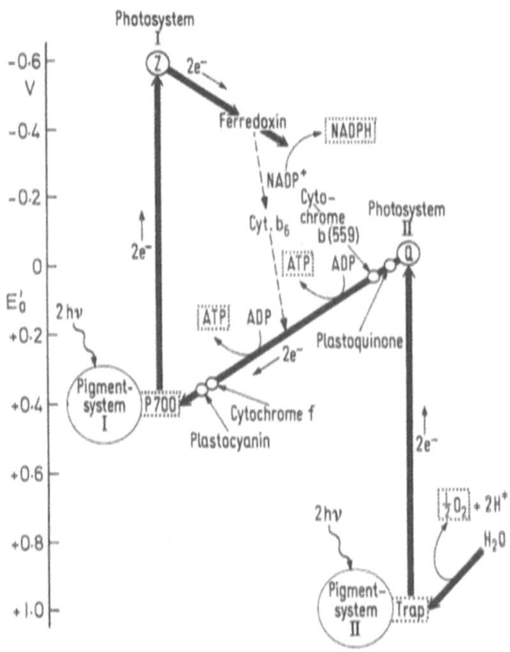

Fig. 8. Series arrangement of photosystems I and II. The two systems are connected by an electron-transport chain between acceptor *Q* of photosystem II and P700 of photosystem I. Non-cyclic electron flow employs both systems, starting from water and ending in NADPH. Cyclic electron flow requires only photosystem I; electrons boosted to Z in photosystem I can return to ground state P700 via a shunt provided by cytochrome b_6. The entire diagram is drawn to conform to the scale of reduction potentials at the left. Movement of electrons upward requires input of energy from light; movement downward proceeds spontaneously. End-products are in colored boxes. (From A. L. LEHNINGER, Short Course in Biochemistry, p. 288. By permission of Worth Publishers, Inc. New York, 1973)

The ciliate *Stentor coeruleus* exhibits a step-up photophobic response that is expressed as ciliary reversal and results in a change in the direction of swimming. The threshold action spectrum of this response agrees quite well (WOOD, 1976) with the reported absorption spectrum of the cell's major pigment, stentorin. It has been proposed that the latter is a member of the *meso*-naphthodianthrone group of compounds (MØLLER, 1962).

In cells that had been bleached by caffein treatment, the response threshold was elevated by a factor for 20. Correlated with this was a decrease in the in vivo absorption of *Stentor* at 620 nm, corresponding to the major absorption peak of stentorin, to 10%–30% of the control value.

Although no hypothesis exists as to how this type of pigment might function in the phototransduction process (and indeed *meso*-naphthodianthrones would represent an entirely new class of photoreceptor pigments), WOOD found that the light is eventually transduced to a graded receptor potential. These electrophysiologic phenomena will be discussed in the next section.

The flagellate *Euglena* is the only eucaryotic cell in which the molecular processes occurring upon photoreceptor pigment excitation have been clarified to some extent.

The early action spectrum measurements with this cell (BÜNNING and SCHNEIDERHÖHN, 1959; GOESSEL, 1957; WOLKEN and SHIN, 1958; DIEHN and TOLLIN, 1966) suffered from two major problems: (1) Photoaccumulation phenomena rather than primary motor responses of the cells, were being observed and, (2) the contributions of stigma and photoreceptor pigments were superimposed, at least in measurements with nonmutant cells. Consequenctly, the resulting spectra were poorly resolved and did not lend themselves to interpretations in terms of the identity of the receptor pigment.

A simultaneous solution for both problems was devised by DIEHN (1969b) upon learning of the observation by BOUND and TOLLIN (1967) that dispersal of *Euglena* occurs from the actinic region of the phototaxigraph upon illumination with polarized light. DIEHN's interpretation of his experimental results lead to the proper conclusions but was based on reasoning that is no longer entirely satisfactory in view of more recent work (CREUTZ and DIEHN, 1976). For this reason, I will at this point present an updated discussion of the premises and of those results obtained in 1969 that pertain to photoreceptor pigment identification:

In the phototaxigraph, accumulation in the illuminated zone can be the result both of photophobic trapping, via step-down responses, and of positive phototaxis of cells responding to light scattered by organisms already in the actinic zone. In our experience, native *E. gracilis* is rarely phototactic, though always capable of step-down photic responses. As a rough estimate, true phototaxis contributes no more than 20% to the accumulation in the illuminated region of the phototaxigraph—enough, however, to significantly impair action spectrum resolution. Dispersal from the actinic zone, on the other hand, is certainly not the result of a negative taxis, since here movement of the cells occurs at right angles to the stimulating beams rather than directly away from the actual light source, and the scattered light is not sufficiently intense to elicit a negative taxis.

The proper interpretation of the expulsion of *Euglena* by polarized actinic light is based on the realization that *any step-down phobic response, regardless of the organism involved or the nature of the stimulus, is a feedback-controlled mechanism for insuring maximum stimulation of the receptor. If the organism experiences a decrease in stimulus intensity it will undergo a phobic response with attendant changes in orientation, which ceases if the original stimulus intensity, or an intensity that most nearly approximates the original one, is experienced (again) within the adaptation time.*

With polarized light, maximum absorption in a receptor molecule occurs if the *e*-vector of the plane of polarization is oriented parallel to the magnetic dipole transition moment of the molecule. If an organism contained only one photoreceptor molecule, step-down phobic responses would maintain an alignment of the cell keeping the molecule parallel to the *e*-vector. Since in *Euglena*, there is clearly a net response of individual organisms to the direction of polarization of the light, there must be a preferential alignment of the receptor molecules within the cell. This alignment is such that their transition moments for absorption in the visible wavelength region are oriented at right angles to the long axis of the cell (for more details, see CREUTZ and DIEHN, 1976, and the discussion in Sec. E of this chapter). An alternative explanation is that stigma pigments are oriented parallel to the long axis of the cell, and that phobic responses both maximize photoreception and minimize stigma screening by keeping stigma molecules aligned perpendicularly to the *e*-vector of the stimulating light. Even though WALNE and ARNOTT (1967) have reported dichroism of the stigma, this latter hypothesis is becoming less attractive since additional evidence is still lacking that would indicate oriented stigma pigments (BENEDETTI et al., 1976a). By contrast, fluorescence from the photoreceptor organelle appears to be polarized (BENEDETTI and CHECCUCCI, 1976).

The dispersal response to polarized light in the phototaxigraph, reflecting pure step-down responses of individual cells was measured by DIEHN (1969b) as a function of wavelength. A very well resolved action spectrum resulted, but no expulsion by the polarized light was observed below 400 nm. Photoaccumulation however, did occur below this wavelength with nonpolarized light, thus indicating that the photoreceptor is sensitive in the near UV. Moreover strong expulsion with polarized light below 400 nm was observed upon rotating the plane of polarization by 90°. In this orientation, visible polarized light failed to cause expulsion of the cells.

One possible interpretation of these observations is that *Euglena* has two perpendicularly oriented photoreceptor pigments; another one is that there are two perpendicular screening pigments. The latter interpretation was initially favored by DIEHN (1969b). When the visible and near UV action spectra were combined (Fig. 9), the resulting composite spectrum strongly resembled a flavin absorption spectrum. Moreover, the UV and visible transition moments of flavins are known to be at a 60° angle to each other, thus explaining the differential response to perpendicularly polarized visible and UV light (in hindsight, it is evident that experiments with a 60° difference in the planes of polarization of visible and UV stimulating light would have been more useful than the actually studied angles of 45°, 90°, and 135°). The action spectrum obtained with polarized light is much better resolved than absorption spectra of aqueous solutions of flavins. The latter lack fine structure because of interaction of the solute molecules

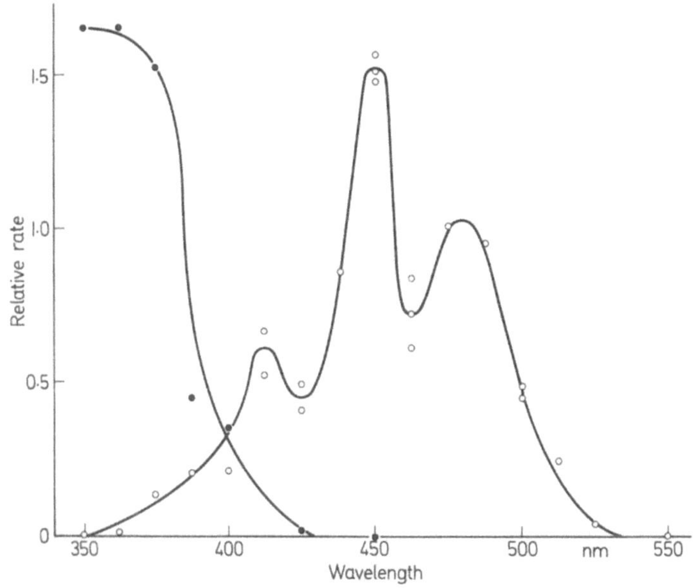

Fig. 9. Action spectrum of step-down response mediated photodispersal of *Euglena gracilis* by polarized light. *Ordinate:* Relative rate of dispersal. *Abscissa:* Wavelength (nm). *Solic circles: e*-vector of polarized light perpendicular to long axis of organism. *Open circles: e*-vector of polarized light parallel to long axis of organisms. [After DIEHN (1969b)]

with the highly polar solvent. A better resolved spectrum was obtained by DIEHN and KINT (1970), who measured the absorption spectrum of riboflavin in ethanol at 170° K. The agreement between this spectrum and the action spectrum of the step-down phobic response of *Euglena* is striking (Fig. 10), strongly suggesting that the photoreceptor pigment is indeed a flavin and, moreover, that an oriented array of pigment molecules is embedded in a hydrophobic matrix. Recent studies of the photophobic responses of detergent-treated cells (MIKOLAJCZYK and DIEHN, 1976) support this hypothesis further (Sec. C,II 1).

CHECCUCCI et al. (1976), perhaps unaware of the above-mentioned rationale for using polarized light, determined the action spectra of positive phototaxis (photoaccumulation) of green and dark-bleached *Euglena* cells and of negative phototaxis (photodispersal) of streptomycin-bleached cells in the phototaxigraph. The three action spectra, although poorly resolved, were rather similar and resembled those found by DIEHN (1969b) for photoaccumulation of green cells in nonpolarized light. The similarity of the action spectra is not surprising. As discussed above, accumulation is mostly, and dispersal entirely, mediated by phobic responses rather than taxis, and hence is only to a limited extent influenced by the presence or absence of potential screening structures. What is surprising is the fact that photoaccumulation was observed in dark-bleached and thus photosynthetically incompetent cells, which according to DIEHN and TOLLIN (1967) do not exhibit such behavior (see Sec. D).

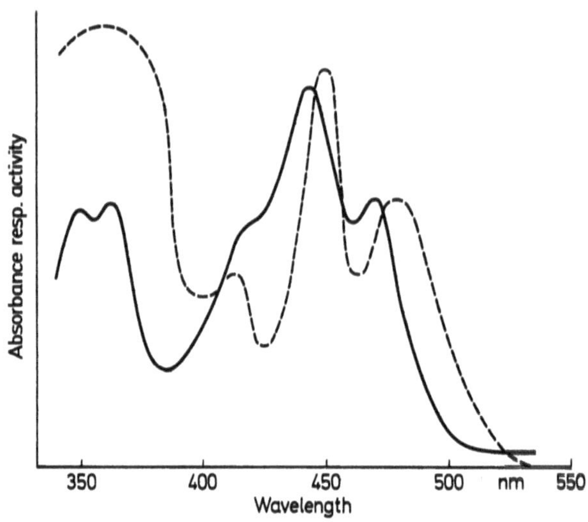

Fig. 10. Composite action spectrum of step-down photophobic response, assayed a photodispersal by polarized light in *E. gracilis (broken curve)*, and absorption spectrum of riboflavin in 95% ethanol at 170° K *(solid curve)*. [After DIEHN and KINT (1970)]

II. Primary Photoprocesses and Signal Conversion

1. Pigment Excitation

Once the photoreceptor pigment has been identified, the next questions requiring an answer are what the primary consequence of light absorption in the pigment is and what the nature of the resulting physiologic signal triggering the motor response is.

The first question has a straightforward answer in those organisms in which the photosynthetic apparatus serves as a photoreceptor for photomovement: light absorption leads to electronic excitation of the photosynthetic reaction centers, with subsequent oxidation of the chlorophyll, followed ultimately by the oxidation of water and utilization of the electrons thus generated. An exception is the photosynthetic apparatus of *Halobacterium*, where light absorption in bacteriorhodopsins leads to the generation of a hydrogen ion gradient.

Although the primary light reactions of photosynthetic pigments and of rhodopsin have been studied extensively and will not be discussed here (see CLAYTON, 1973; RADMER and KOK, 1975, for reviews), much less is known about the primary photoprocesses occurring in other types of photoreceptor pigments.

Once a good guess is available concerning the identity of the receptor pigment, the tentative identification can be supported experimentally by demonstrating a specific effect on photomovement of substances that are known to interfere with the probable mechanism of signal transduction by this pigment.

For the flavin photoreceptor of *Euglena*, DIEHN and KINT (1970) proposed that light absorption may lead to electronic excitation of the flavin to the triplet state. They proceeded to show that iodide, an efficient triplet quencher, selectively inhibited photoaccumulation of *Euglena*. These results agree with the hypothesis

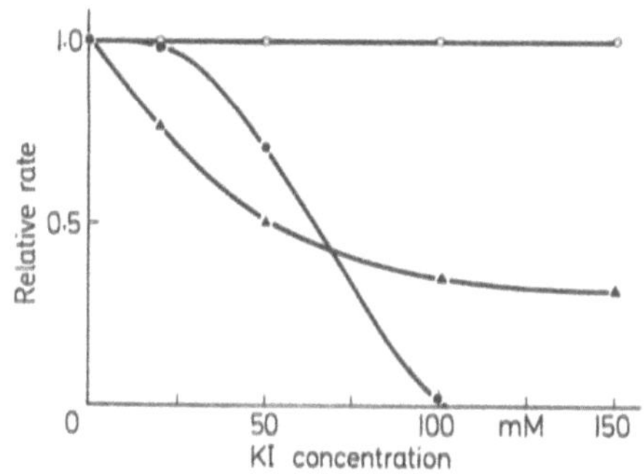

Fig. 11. Photomovement of *Euglena* as a function of KI concentration, relative to the control experiments. *Open circles:* Fraction of motile cells exhibiting the step-up photophobic response. *Solid circles:* Fraction of motile cells exhibiting the step-down photophobic response. *Solid triangles:* Rate of forward swimming. [After MIKOLAJCZYK and DIEHN (1975)]

that the primary photoact consists of flavin excitation to the triplet state, followed by transduction of the excitation energy in a nonradiative process. The authors suggested at that time that transduction into an electric signal could be accomplished via photoconduction by the excited pigment across a poised potential difference.

In an extension of this work, MIKOLAJCZYK and DIEHN (1975) studied the effect of iodide on the step-up and step-down photophobic responses separately. They found that in 150 mM iodide, the step-down response was completely inhibited, whereas the step-up response remained unaffected. The rate of forward swimming was also decreased by iodide, as was to be expected in view of its known effect as uncoupler of oxidative phosphorylation. However, the shapes of the inhibition vs concentration plots (Fig. 11), as well as the degree of inhibition, differed for both parameters, suggesting that both motility and the sensory system for the step-down photic response are affected by iodide at different sites.

MIKOLAJCZYK and DIEHN concluded that there are two photoreceptor systems in *Euglena*, one for the step-down response and one for the step-up response. They also suggested that the receptor pigment for the step-up response was most likely not a flavin—a somewhat rash conclusion that ought to have been challenged by reviewers of the manuscript. A more appropriate statement would have been that if the receptor pigment for the step-up response is a flavin, it is not accessible to externally added iodide.

Further support of the two-photoreceptor hypothesis comes from studies by MIKOLAJCZYK and DIEHN (1976b) of the effect of detergent treatment on the photophobic responses of *Euglena*. They found that the cationic detergent CTAB (cetyltrimethylammonium bromide) selectively interfered with the step-down phobic response at concentrations as low as 10^{-5}M. Even at 10^{-4}M, no effect was seen on the step-up response. Although one might conclude from these data that there are two photoreceptors, it is possible that the detergent affects a separate transduction path for the step-up response rather than a separate receptor.

The primary photochemical reaction of a receptor pigment can also be characterized through its kinetic parameters. CREUTZ et al. (1978) studied the effect on

Euglena's step-down photophobic response, as assayed by a light-trap method, of varying the magnitude of the light intensity gradient experienced by cells attempting to leave the trap. Their data indicated that the step-down response is a function of both the light intensity, I, and the light intensity change, ΔI, as separate parameters. A mathematical analysis of the results led the authors to a first-order kinetic model of photoreceptor pigment interconversion. The observed kinetics are consonant with the earlier hypothesis (DIEHN and KINT, 1970) postulating that pigment excitation triggers vectorial charge transport across the photoreceptor membrane, thus converting the stimulus into an electric signal.

2. The Physiologic Signal

I will continue with the discussion of the situation in *Euglena* before devoting the remainder of this section to *Stentor*. As mentioned above, an electric nature of the physiologic signal was suggested by DIEHN and KINT in 1970. No progress has been made in testing this hypothesis in vivo, but some potentially relevant results were obtained by FRÖHLICH and DIEHN (1974) in a bilayer lipid membrane (BLM) model system for *Euglena*'s photoreceptor. These authors incorporated riboflavin tetrapalmitate into an oxidized cholesterol bilayer. Upon illumination of this membrane, no effects were observed until an external potential (from 16 to 70 mV) was applied. At the physiologic pH values (for *Euglena*), of 6 and 7 illumination of an electrically biased membrane resulted in a transient current flow against the applied potential, followed by a return to the dark value of the current. When the light was turned off, a transient current pulse in the opposite direction, i.e., in the direction of the applied potential, resulted (Fig. 12). The qualitative resemblance of these effects to the step-up and step-down motor responses in *Euglena* is striking. The authors have proposed a mechanism of this phenomenon that accounts for the observed dependence of the effect upon experimental parameters such as magnitude of the applied potential, light intensity, etc. It will suffice to say that the results suggest a change in dipole moment of the electronically excited flavin, followed by electrostatic interactions with another moiety in the lipid or aqueous phase. Since similar effects were observed with an unbuffered ascorbate gradient across the membrane instead of an externally applied potential, and since those effects disappeared upon buffering, light-induced proton translocation might well be involved in this phenomenon. SONG (1968), in a pertinent contribution, has shown that the acidities of the electronically excited states of flavins considerably exceed that of the ground state.

WOOD (1976) has shown that the photophobic response for *S. coeruleus* is correlated with light-induced electric phenomena in the cell. *Stentor* assumes a trumpet shape when resting on a substrate (in which case the photophobic response consists of body contraction), and is pear shaped when in the swimming configuration. In either form, a microelectrode inserted in the vacuole detects a positive-going receptor potential upon illumination of the organisms, followed by an action potential and either contraction (in the sessile form) or ciliary reversal (in the swimming form). This action potential differs from classical action potentials in that its amplitude is variable rather than all-or-none, and hence it is more properly termed a *regenerative response* (Fig. 13). The average onset latency at the

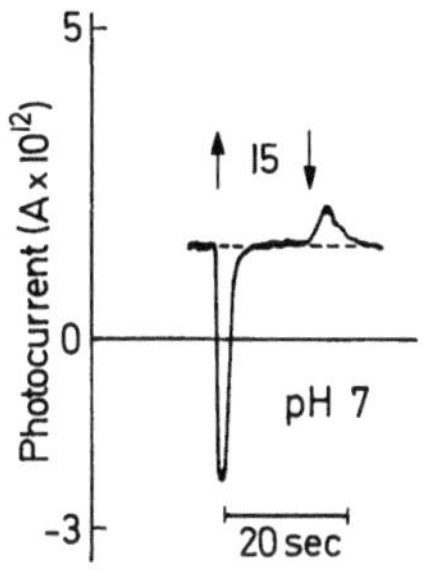

Fig. 12. Photocurrent observed upon illumination and darkening (arrows) of an oxidized cholesterol bilayer liquid membrane containing riboflavin tetrapalmitate. White light, 4 kW/m^2; applied potential 50 mV. [After FROEHLICH and DIEHN (1974)]

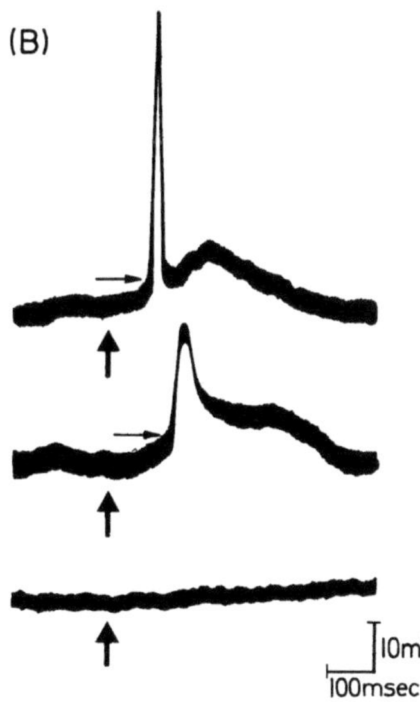

Fig. 13. The regenerative photoresponse of *Stentor coeruleus*. [After WOOD (1976)]. Successive records taken from an intravacuolar electrode when the stentor was exposed to photic stimuli of 2 (bottom record), 20 (middle record), and 200 (top record) mW/m^2. The stimulus onset occurred at the vertically oriented arrows and continued throughout the remainder of the records. During each of the top 2 records a regenerative response and ciliary reversal was elicited. This regenerative response is superimposed upon the photic receptor potential. Next to each of the top records a horizontally oriented arrow marks the approximate onset of the regenerative response. The steady state potential inside the vacuole was approximately −5 mv, but this should not be taken as the resting potential of the cell since the electrode is not in the cell cytoplasm and a large steady state potential may be present across the vacuole membrane

maximum stimulus intensity was 172 ms—of the same order of magnitude as the transduction time for the flagellar photophobic response in *Euglena* (DIEHN et al., 1975). The temporal correlation between regenerative responses and ciliary reversal in *Stentor* makes the assumption reasonable that the former trigger the latter, as they have been proven to do in mechanical stimulation of *Paramecium* by ECKERT and NAITOH (1972). The polarity and amplitude of the photic receptor potential in sessile *Stentor* was not altered by polarization of the plasma membrane. The photic receptor potential has the properties expected of a stimulated electrogenic pump mechanism.

It is not known by what molecular mechanism the photoreceptor pigment of *Stentor* initiates the electric regenerative response upon illumination, nor by what mechanism, in any organism, an electric signal might trigger an alteration in ciliary or flagellar motor behavior.

Although *Stentor* is the only unicellular organism for which an electric mechanism of photic stimulus transduction has been shown, the indirect evidence accumulating in other experimental cells, notably *Euglena*, makes it probable that similar responses are involved in their photic response systems and will be discovered and studied in due time.

D. Interactions of Photosensory and Energy-Transducing Systems

Since the great majority of photomotile unicells are photosynthetic, the question naturally arises whether there is a link between the systems mediating photosynthesis and photomovement. Even in those cells where the answer is obviously affirmative because the photosynthetic apparatus also functions as a photoreceptor for the movement responses, determining the exact point of interaction and the molecular mechanism of response modulation are far from trivial problems. Moreover, the question remains whether photomovement responses can be modulated by reactions in other energy-transducing pathways in the case of organisms that are not photosynthetically competent.

I. Photosynthesis

In the procaryotes, some evidence regarding the location of the linkage point between photosynthetic electron transport and photophobic responses has been obtained by HAEDER (1974a, 1976) for *Phormidium*. As mentioned previously (Sec. C,I 3) photophobic responses are triggered by removal of electrons from the IETC. For the discussion that follows, reference should be made to Figure 8.

HAEDER approached the problem of determining the location of the proposed electron pool controlling the phobic responses by studying the effects of metabolic inhibitors that are known to block electron flow at specific points in the IETC. Dichlorophenyl dimethylurea (DCMU), which reportedly inhibits electron flow between photosystem II and plastoquinone (CRAMER and HORTON, 1975), blocks photoaccumulation, as does dibromomethyl isopropyl benzoquinone (DBMIB), which prevents the reoxidation of plastoquinone by the next component in the IETC (BOEHME et al., 1971). KCN, inhibiting plastocyanin (BERG and KROGMANN, 1975), blocks photodispersal in concentrations at which photoaccumulation remains unaffected. Disalicyclidene propanediamine, an inhibitor of ferredoxin reduction (TREBST and BURBA, 1967), has an effect which is very similar to that of KCN. An inspection of Fig. 8 will show that the phobic responses of *Phormidium* must be linked to photosynthetic electron transport between plastoquinone and plastocyanin in the IETC.

In *Euglena*, photoreception does not occur in the photosynthetic system. Nevertheless, DIEHN and TOLLIN (1967) demonstrated control of photaccumulation, and thus of the step-down phobic response, by photosynthesis. Specifically, they tested the effect of a number of inhibitors and uncouplers of oxidative and photophosphorylation upon photoaccumulation. They observed that DCMU inhibited photoaccumulation within a few hours, after a transient stimulatory effect. This inhibition did not occur when external electron donors, such as reduced dichlorophenol indophenol (DCPIP), were added to the cell suspension simultaneously with the inhibitor. A similar effect was observed with other inhibitors of noncyclic electron transport, although substances that are believed to interfere with cyclic electron transport did not specifically inhibit photaccumulation. Hence, the point at which photosynthesis exerts control over the step-down phobic response of *Euglena* must be located between plastoquinone and the point in the IETC where electrons arrive from the cyclic electron transport pathway. Moreover, the gradual onset of inhibition indicates that it is not the concentration of a reduced or oxidized species in the IETC that controls the phobic response, since these concentrations respond instantly to changes in the rate of electron transport. More probably, control over the step-down phobic response is exerted by a substance generated as a result of noncyclic electron transport.

Indeed, the authors found that carbonyl cyanide trifluoromethoxyphenylhydrazone (CCCP), known to be a potent uncoupler of photophosphorylation (BAMBERGER et al., 1963), had nearly the same effect as DMCU. One can conclude that the phobic response is controlled by the level in a reservoir of a substance produced at the first site of photophosphorylation. TOLLIN and ROBINSON (1970), in further support of this hypothesis, found that ATP levels in live *Euglena* cells were correlated with their rate of photoaccumulation.

Euglena recovers from inhibition of photoaccumulation by any of the above substances within approximately 48 hrs, even if the cells are kept in contact with the inhibitors. Apparently, the metabolic block is bypassed at the specific site of action of the inhibitor or uncoupler. This phenomenon is interesting not only in the context of photomovement but also in the sense of revealing a repair mechanism of photosynthetic energy transduction which had not been previously discovered. In fact, photomovement studies, in organisms where sensory and energy-transducing systems interact, can be used to study nonbehavioral responses such as photosynthesis in vivo on the molecular level.

II. Other Metabolic Pathways

DIEHN and TOLLIN (1967) also studied the effect of inhibitors and uncouplers of oxidative phosphorylation upon photoaccumulation of *Euglena*. No effect was observed, indicating that in this organism, there is probably no link between respiration and the step-down photophobic response. However, the step-up response is still exhibited by photosynthetically incompetent cells (DIEHN, 1973), thus suggesting that the transduction system for this response is linked to either the glycolytic or respiratory pathways. No experiments have been performed to date to test or refine this hypothesis.

In the green alga *Chlamydomonas reinhardi*, STAVIS (1974) observed specific inhibition of photoaccumulation by sodium azide. Unlike the situation in *Euglena* (DIEHN and TOLLIN, 1967) complete inhibition occurred at azide concentrations (0.35 mM) where mobility of the organisms was not yet affected. Azide is a potent inhibitor of respiration, and in *Euglena* inhibits photoaccumulation indirectly through its effects on flagellar movement by reducing the ATP supply. In *Chlamydomonas*, however, the ATP level in the cell is not affected by azide concentrations of 1–2 mM. STAVIS (1974) points out that the latter observation might simply mean that inhibition of oxydative phosphorylation is masked by an increased ATP supply from glycolysis, but a more interesting alternative explanation for the azide effect is his suggestion that it might act by depolarizing a membrane potential, as it does in frog muscle (HOROWICZ et al., 1962). If this were the case, it would indicate an electric nature of the signal traveling from the (as yet unidentified) photoreceptor to the effector of *Chlamydomonas*.

In summary, there is little evidence thus far for a link between the photosensory and nonphotosynthetic energy-transducing pathways in photomotile microorganisms.

E. Behavioral Mechanisms

I. Strategies Utilized in Accumulation/Dispersal and Taxis

The mechanisms by which motile microorganisms can accumulate or disperse in response to a stimulus have been summarized in the discussion of terminology in the introductory section of this chapter. Depending on the individual organism, kineses, phobic responses, true taxes, or combinations of the above can mediate accumulation or dispersal as a result of photic stimulation. In the following, I will discuss examples of these photomotive mechanisms to the extent that they have been studied and identified unequivocally.

1. Kineses and Phobic Responses

Step-up negative orthokinesis leads to photodispersal in *R. rubrum* (THROM, 1968). These cells swim faster when illuminated than in the dark and thus spend less time in the illuminated region. Step-up negative klinokinesis mediates photoaccumulation in *Navicula peregrina* (NULTSCH and WENDEROTH, 1973); and the cells reverse direction less frequently when illuminated than in the dark.

Paramecium bursaria, by contrast, accumulates as the result of step-down positive klinophobic responses, by changing direction 13%–17% more frequently when swimming down a light gradient than when swimming up (SAJI and OOSAWA, 1974).

Orthophobic responses can also result in accumulation or dispersal. *Chromatium* responds to an increase in light intensity, e.g., at the border of a light trap, with flagellar reversal and then reverse movement. Upon adaptation it remains immobile for up to a second, and then resumes forward swimming (ENGELMANN, 1882). If a passive change in cell orientation has occurred during the immobile phase, the direction of forward movement differs from the original one.

Strophophobic responses differ from orthophobic ones in that a change in orientation, which is random with respect to the ensuing final direction, is brought about by an active turning process that may persist for many revolutions before adaptation occurs. In this type of reaction, the classical phobic response is turning around a lateral axis, and hence forward movement ceases for the duration of the response.

The term positive is superfluous for the strophophobic responses as no example of a negative response is known, and indeed would be hard to imagine, since it would require an organism that turns constantly in the absence of a stimulus, and reduces its rate of turning upon stimulation. Strophophobic responses and klinokinesis are similar in that they involve directional changes. In the former, however, the response occurs only once and within milliseconds after stimulation, whereas in the latter, changes in direction are exhibited in the absence of stimulation and it is their average frequency which is modulated by the stimulus. Thus, the *transduction time* (the time between stimulation and first observed response) can be quite long in klinokinesis.

The step-up and step-down photophobic responses of *Euglena*, which are identical in appearance, are examples of strophophobic responses. The step-up response has been studied quantitatively by DIEHN et al. (1975) via high speed cinemicrography. The response is expressed as a turn toward the dorsal, i.e., stigma-bearing, side of the cell (see Fig. 4) which starts with white light of 0.5 kW/m^2 intensity after a transduction time of $\sim$200 ms. Turning occurs because the normal 180° bend of the flagellum, where it emerges from the anterior end of the cell, straightens to 90° as the result of stimulation. At no time does the flagellum stop beating; it beats continuously, whether the photosensory system is being stimulated or not, and the only consequence of photic stimulation is reorientation of the flagellum toward the ventral side of the cell.

Upon removal of the light stimulus, the response ceases after a delay time that approximates the transduction time. If the light stimulus is left on, the cell's sensory system adapts within approximately 15 s, and the response ceases. Removal of the stimulus after adaptation has no visible effect yet it restores the cell's responsiveness toward illumination, i.e., it "resets" the photosensory system for the step-up photophobic response. The frequency of the flagellar beat changes only slightly during the phobic response, or during resetting of the sensory apparatus by darkening.

The photophobic response of *C. reinhardi* toward 6.4 μs flashes of 12.6 mW/m^2 intensity was studied by BOSCOV (1974). 46% of 392 cells studied showed a turning (strophophobic) reaction in response to a single flash; one-third of these turned toward the light and the remaining two-thirds turned away from it. 42% of the cells exhibited no response to a flash, while 12% showed a stop (negative orthophobic) response. The interpretation of these data is difficult because a short flash does not present a single stimulus to the cell, but rather a step-up stimulus immediately followed by a step-down stimulus. In the case of *Euglena*, for instance, a short flash results in a delayed and attenuated phobic response (DIEHN, 1969a)—a phenomenon that is not observed when pure step-up or step-down stimuli are presented.

The distribution of turning, stopping, and nonresponsive cells upon repetitive (60 flashes/s) stimulation at 10 mW/m^2 was very similar to the distribution found after a single flash. This could mean that in *Chlamydomonas* the response to continuous stimulation closely resembles that to a single flash. (At a flash frequency of 60 Hz, the intervals between flashes are shorter than the transduction times usually observed in microorganisms, and thus one would expect the cell's sensory system to perceive the flashes as continuous illumination.)

Since videomicrographic techniques were used for recording the cells' movement, the resolution obtained was not sufficient for observation of the cell's flagellar responses. High speed cinemicrography of *Chlamydomonas*, which represents a formidable tracking task because of the cell's rapid movement, will be required for attaining a description of its photophobic responses in terms of flagellar configuration.

2. Taxes

Theoretically, two mechanisms are conceivable for determining the vectorial asymmetry of a stimulus (i.e., the direction from which the stimulus originates):

a) The direction of the stimulus is detected by "comparing" its simultaneous intensity in two different regions of the cell. This requires an extended photoreceptor, or two or more localized photoreceptor organelles. This has been called a one-instant mechanism (FEINLEIB, 1975). No such mechanism has yet been proven to be utilized by a cell, although it appears likely that this type of mechanism would operate in an organism such as *Micrasterias*, which has no identifiable receptor organelle and exhibits smooth and slow orientation (NEUSCHELER, 1967).

b) A cell can also determine the stimulus direction by comparing its intensity at two points in time. This type of mechanism determines the stimulus direction with only a point receptor, but whenever the cell moves in a direction that requires a course correction it relies on movement of the organism to differentiate between intensities at the two points in time. The term two-instant has been used for this detection mode, but the term movement-modulated or simply modulated mode would appear more descriptive.

Regardless of the detection mode, tactic orientation with respect to a light stimulus is achieved when the perceived stimulus distribution is symmetrical. In the following, some movement-modulated detection modes will be discussed. Previously I mentioned that FEINLEIB and CURRY's population-method apparatus is generally thought to measure only phototaxis, without interference by phobic responses (Sec. A III 3b). Yet FEINLEIB (1975) observed that photoaccumulation in this setup can occur in response to a single flash of 6.4 µs duration. She proposed that this phenomenon might be explained in terms of an one-instant detection

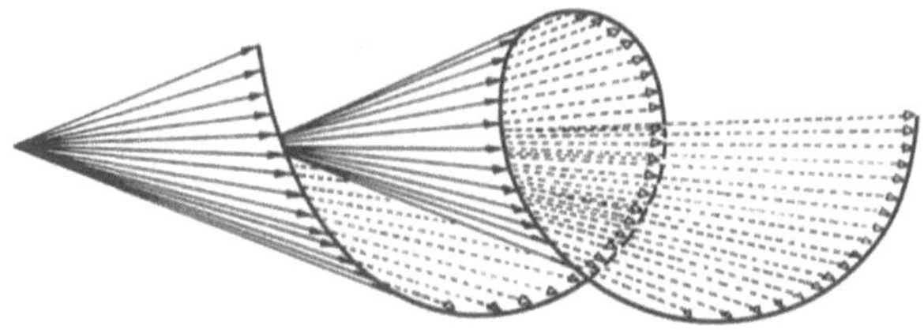

Fig. 14. Forward movement of *Euglena*. *Arrows* indicate successive positions of the anterior end of the cell. [After CHECCUCCI (1976)]

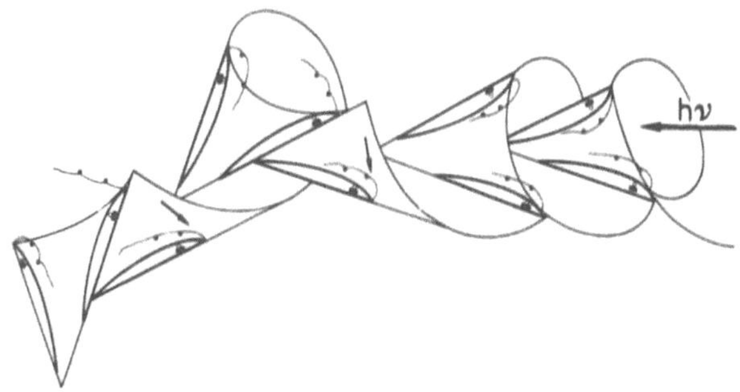

Fig. 15. Phototactic orientation toward light in *Euglena*. *Light arrows* indicate course corrections accomplished in preceding step-down responses induced via periodic shading of the photoreceptor by the stigma. [After CHECCUCCI (1976)]

mode. Other videomicroscope studies (BOSCOV, 1974) suggest an alternative interpretation. Since only about one-half of the cells showed a turning response to a single flash (see above), it may be that the only cells to respond were those in which the photoreceptor [organelle] happened to be in a "correct" position at the time of the flash. If so, then a movement-modulated mechanism may well apply in *Chlamydomonas*. A satisfactory explanation for determination of stimulus direction in this organism awaits more detailed examination of the behavior of single cells.

A reasonable model for the phototactic orientation mechanism in *Euglena* has been proposed in an elementary form by MAST (1911) and was refined and experimentally tested by DIEHN (1969 a) and DIEHN et al. (1975):

As a result of the trailing configuration of the flagellum (Fig. 4) and the torque generated by its helical beat, *Euglena* rotates as it moves forward, and its anterior end progresses along a spiral path, with the flagellum facing the inside of the helix, and the stigma the outside (Fig. 14). If light of an intensity such that its dimming will cause a step-down phobic response (i.e., below 0.2 kW/m^2 of white light) is applied at right angles to the cell's path, then the shadow of the stigma will fall upon the photoreceptor (the paraflagellar swelling) once every revolution, causing a phobic response that ceases when the stigma has turned past the shading position. This results in stepwise course corrections that lead to positive phototactic orientation (Fig. 15). This orientation process is easily recognized as based on a movement-modulated detection mode: the cells correct their course as a result of comparing the intensities perceived during two periods of time that differ with respect to the relative positions of the stigma, photoreceptor, and light source. In a test of this hypothesis, DIEHN (1969a) illuminated a population of *Euglena* in the phototaxigraph with repetitive flashes of light and observed a clear resonance peak of photoaccumulation at a flash frequency that was equal to the rotation frequency of the cells.

At high intensities of lateral light (above 0.2 kW/m^2), *Euglena* exhibits negative phototaxis as well as step-up responses upon application of the high-intensity stimulus. DIEHN (1969a) has proposed that the two phenomena are causally related as follows (see also DIEHN et al., 1975):

If the intensity of lateral illumination is above the threshold for the step-up photophobic response, the cell will commence a turning reaction upon stimulation. Since every shading event stops this response and resets the cell's sensory mechanism for a further step-up response upon reillumination, the only situation in which there will be no further phobic response is one in which the photoreceptor is permanently shaded. Permanent shading of the photoreceptor is best accomplished by the posterior end of the cell, and thus will lead to an orientation of the organism directly away from the light source.

CHECCUCCI (1976) states that DIEHN (1973) lists the above mechanism as mediating dispersal from the actinic zone of the phototaxigraph at high light intensities. This statement is incorrect (see Sec. C,I 3).

An entirely different rationale for the phototactic orientation mechanism of positive phototaxis in *Euglena* has been developed by CREUTZ and DIEHN (1976). As previously mentioned (Sec. CI 3) the photoreceptor molecules of *Euglena* are aligned such that their transition moments, for absorption in the visible wavelength region, are oriented perpendicular to the long axis of the cell. An unpolarized light stimulus is actually composed of light randomly polarized in all directions except in the direction of propagation of the stimulus. If the photoreceptor molecules are aligned as described here, cells moving transversely to the direction of a stimulus will have only part of their photoreceptor molecules parallel to the direction of polarization (Fig. 16, Cells *a* and *c*). These molecules would be in such a position that they could absorb light. A cell moving directly toward the stimulus, however, will have all its photoreceptor molecules parallel to one of the polarizations present in the stimulus. All photoreceptors could then absorb light (Cell *b*). Hence, motion toward the stimulus would occur in preference to motion with a component transverse to the stimulus direction. Any cell that "drifts" from a transverse or negative orientation to a positive one could do so without constraint, since such reorientation will be accompanied by an increase in the amount of light absorbed, but the reverse process, involving a decrease in the amount of light absorbed, would elicit a step-down phobic response and bring the cell back to the positive phototactic orientation. This mechanism would require that a posteriorly shaded photoreceptor perceives less light than a transversely oriented one that is in an orientation of minimum light absorption.

This type of movement-modulated detection mode compares the amount of light absorbed at two points in time at which there is a difference in the orientation of the photoreceptor molecules with respect to the stimulus direction. The polarization-orientation hypothesis is every bit as reasonable as the periodic-

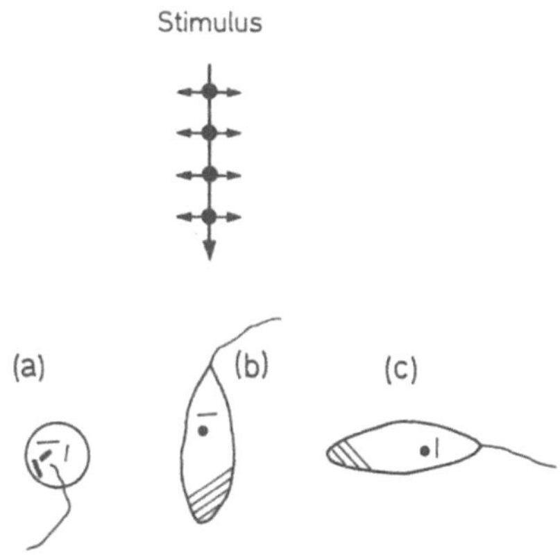

Fig. 16. Relative positions of cells and photoreceptor molecules of *E. gracilis* in various orientations toward nonpolarized light. [After CREUTZ and DIEHN (1976)]

shading hypothesis, and circumvents one vexing problem associated with the latter: no modern researcher has actually observed and recorded the stepwise course correction postulated in Figure 15, in part perhaps because *Euglena* rarely exhibits positive phototaxis. Most probably, both orientation mechanisms can be employed by the cell to achieve phototactic alignement.

Another organism for which a movement-modulated mode for detecting the direction of the stimulus has been proposed is *G. dorsum*. This cell exhibits a negative orthophobic (stop) response upon step-up stimulation with light of moderate intensity. Several seconds later, a strophophobic (turning) response commences that ceases when the cell is oriented toward the light (vide infra). *Gyrodinium* has both a transverse and a longitudinal flagellum. The former, protruding from the girdle (see Fig. 5), causes rotation of the cell during normal locomotion and is immobile throughout the phobic response. The longitudinal flagellum, projecting posteriorly from the sulcus, is reoriented laterally at the beginning of the stop response, and causes the turning response when it resumes beating in this orientation. HAND and SCHMIDT (1975) have proposed that the photoreceptor is located at the sulcus/girdle junction, and suggest that shading of the photoreceptor stops the phobic response when the cell faces the light. Unlike *Euglena*, which accomplishes positive tactic orientation via step-down phobic responses, *Gyrodinium* is thus presumed to utilize step-up responses for achieving positive phototaxis. When the cell deviates from this orientation, light can reach the photoreceptor through the channel formed by the sulcus, and trigger a course correction.

II. Signal Processing: Systems Analytical Approaches

As is evident from the preceeding section, the photoaccumulation of microorganisms involves kineses or phobic responses. Taxes, however, require the active maintenance of the cell's orientation and are mediated solely by phobic responses in all organisms that utilize movement-modulated modes for detecting the stimu-

Table 1. Photophobic responses of *Euglena* as a function of light intensity and preadaptation

Adaptation state	Stimulus regimen	Intensity of white light W/m²		
		800	200	50
Light adapted[a]	Light off	No immediate shock, slight delayed shock	50% shock	Shock
	Light on after 1.0 s dark	—	Stops shock, shocks previously unshocked cells	Stops Shock
	Light on	Shock	50% shock	No shock
Dark-Adapted	Light off after 1.0 s illumination	Stops shock, delayed 2nd shock	Stops shock, shocks previously unshocked cells	Shock

[a] To the intensities shown.

lus direction, i.e., in all cells for which the detection mode has been identified. What makes phobic responses suitable for mediating not only accumulation/dispersal phenomena, but also directional homeostasis, are two characteristics of these responses:

1. The existence of distinct thresholds for the step-up and step-down modes of the same response. Thus, a stimulus can act as step-up or step-down stimulus, depending on its magnitude.

2. The "resetting" phenomenon, i.e., the fact that removal of a stimulus restores responsiveness of the cell's sensory system to this same stimulus by cancelling the system's adaptation. One might list the adaptation phenomenon as a third essential characteristic, but this is a redundant statement since adaptation is what distinguishes phobic responses from kineses in the first place.

The combination of these characteristics allows the cell to utilize the same basic response for achieving sophisticated types of behavior such as positive and negative taxes. It is evident that the way in which a cell accomplishes the internal management of response thresholds, adaptation, and resetting processes is an extremely important function within its sensory chain. As discussed in Sec. B,I2, this is an information-processing function and, for the purposes of facilitating our analysis of its operation, we assume it is handled by a hypothetical processor.

The only experimental studies conducted to date with the aim of elucidating signal processing in unicells have been performed by DIEHN (1972, 1974) with *E. gracilis*. Table 1 summarizes the results of a series of experiments in which the photophobic responses of *Euglena* were observed during sequential stimulation.

The following characteristics of the stimulus/response system of *Euglena* are evident from these data:

1. A single light pulse of low intensity will reset the transduction system for the step-down phobic reaction without inducing a step-up photophobic response. Conversely, a "dark pulse" resets for the step-up phobic response during high-intensity illumination.

2. If the system is adapted to a given light intensity a change in intensity induces a photophobic reaction only when the direction of change is away from an internal reference level (adaptation level) which, for the particular culture studied, corresponded to 200 W/m^2 of white light from a xenon lamp. It appears as though the system were naturally adapted to this light intensity (which, not unexpectedly, corresponds to the saturating intensity for photosynthesis), even though the cells had for many generations been cultivated at much lower levels of illumination. On the assumption that at the output of the photoreceptor an electric signal appears that is a function of the incident light intensity (DIEHN and KINT, 1970), the processor must be capable of comparing the magnitude of this signal with that of an internal reference level and of determining the sign and magnitude of dI/dt in order to activate the effector.

3. A system that has adapted to a low-light intensity will exhibit an immediate step-up photophobic response upon an increase of the intensity beyond the reference level. By contrast, after adaptation to the high intensity, darkening will elicit only a delayed and attentuated step-down reaction.

4. If a step-up photophobic response is induced by high-intensity illumination, darkening after 1 s will not result in an immediate transition to a step-down response. Conversely, we found that transition to a step-up response is not observed upon high-intensity illumination of a cell that has just commenced a strong step-down response.

The behavior of the cell not only indicates that signal processing occurs but also yields information on the dynamic characteristics of the processor. An attempt can now be made to use the techniques of systems analysis in order to elucidate the approximate internal structure of the processor, the "black box" (DIEHN, 1973).

Studies of signal processing are usually approached in terms of electronics. The processor may be treated as an electronic device, the characteristics of which can be considered in terms of interdependent chemical reactions if the electronic approach requires assumptions that are unreasonable within the constraints of a single cell.

An electronic analog of the processor is most easily constructed in the form of a flow sheet. Such a diagram can then be converted into a computer program for analysis of the system's performance with various simulated sensory inputs.

The input of the processor was assumed to consist of an electric signal the magnitude of which is a function of the incident light intensity. The output consists of commands that activate the appropriate effector.

An internal reference level in the processor corresponds to the photoreceptor signal at the adaptation level of light intensity. Figure 17 shows a sequence of processing steps that will generate effector control signals exactly duplicating the response of the actual receptor/effector system of *Euglena* as summarized in Table 1.

Although this diagram appears complex, it describes in essence only the charge or discharge of two storage devices (e.g., membrane capacitances) toward the receptor or reference potential level as shown. If the charge rate exceeds a threshold value, the effector is activated; otherwise the system is considered adapted. The second of the capacitances is utilized only when a change in light

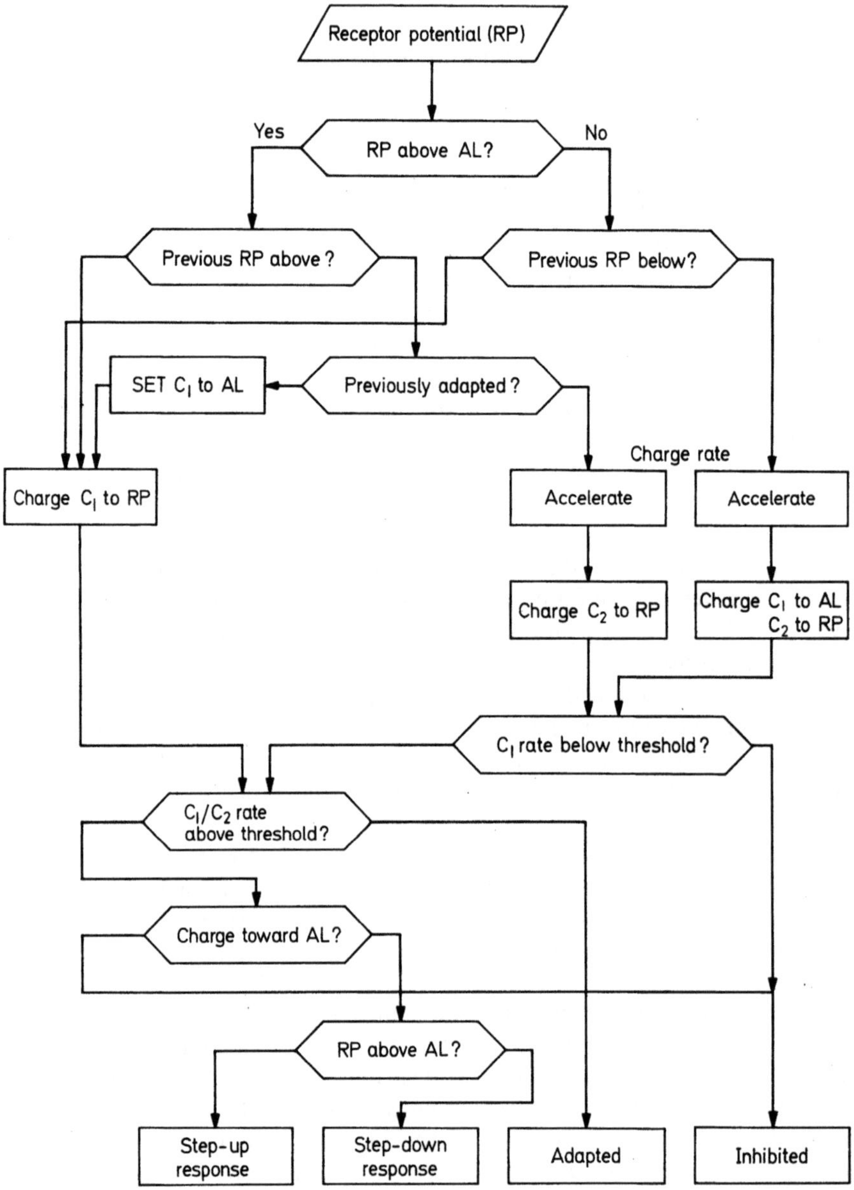

Fig. 17. Computer flow sheet of signal processing steps that are presumed to link photoreception and the expression of photophobic responses in *Euglena*. AL, adaptation level. [After DIEHN (1973)]

intensity from below the adaptation level to above, or vice versa, is imposed upon the system before it has adapted to the previous illumination, i.e., when the first capacitance is incompletely charged or discharged. When both storage devices are being utilized, charge and discharge occur at an accelerated rate and the signal activating the effector is suppressed.

This scheme has been translated into a FOCAL computer program by DIEHN (1973). He used a PDP-8 computer to simulate the behavior of *Euglena* under photostimulation by a variety of stimulus regimes. In all experiments, the computer-generated behavior patterns were qualitatively identical to the actual behavior of *Euglena*, and differed quantitatively only in that the capacitance-charge model predicted a logarithmic dependence of response duration times upon the stimulus intensity, whereas *Euglena* was found to exhibit a linear relationship between these two parameters. A constant-current device in the model would correct this discrepancy.

Computer simulation is most useful when it can be used for predicting behavior of a system under conditions that have not yet been investigated experimentally. With a computer, a large number of varied stimulation sequences can be tested with little effort on the experimenter's part and if an unusual or unexpected behavior is predicted for a given stimulus regime, it can then be studied in the live cell. A number of characteristics of the phototransduction system of *Euglena*, such as "additivity" of the responses, were discovered in this way and subsequently verified in vivo (DIEHN, 1974).

A decision flow sheet such as the one shown in Figure 17 appears rather abstruse in the context of the behavior of a simple unicellular organism. It is, however, fairly easily translated into a scheme of interrelated biochemical reactions with the control functions exercised by, e.g., regulatory enzymes or channel-gating proteins (DIEHN, 1974).

PICCINNI and OMODEO, ignoring the above work, discovered the homeostatic nature of phototaxis in *Euglena* in 1975. Not surprisingly, they propose that the stimulus intensity is compared by the cell with a previously experienced one, and that a phobic response is triggered if a difference is detected. The delay device required to "store" a parameter that is a function of the previous light intensity (in DIEHN's model, a membrane capacitance), is hypothesized to be a synaptic type junction between the membranes of the nonemergent and the locomotor flagella. The model is not elaborated beyond suggesting that the "comparator" (corresponding to DIEHN's processor) might reside in the flagellar bases. An attempt by the same authors to explain the step-up phobic responses of *Euglena* in terms of pigment bleaching is discussed and refuted in a paper by CREUTZ et al. (1978).

F. Interactions of Photosensory and Other Stimuli

Throughout this discussion the photoresponse systems of the cells that have been considered were treated as though they were the only sensory systems present. In actuality, this would seem to be a very improbable supposition. Although in certain organisms there appears to be a preference for exploring the environment via one of the sense modalities, it has long been known that cells such as *Paramecium aurelia* respond to chemical as well as mechanical stimuli. It is clear that the different sensory pathways must converge and probably interact whenever the respective responses are expressed through the same motor apparatus.

Generally, these interactions can be eliminated by varying only the stimulus to be studied; indeed, this is usually accomplished without any conscious effort, and it is often the failure of such unintentional stimulus separation that leads to the discovery of other sensory systems.

Although *Paramecium* would appear to be a suitable organism for studying the interaction of various sensory pathways, no such investigations have been conducted to date. The problem with such work is that at least one of the stimuli must be such that it can be easily manipulated, and this is most easily accomplished if the stimulus is light. However, only *P. bursaria* is photoresponsive, and its responses to mechanical and chemical stimuli have not yet been studied. A very similar situation exists with bacteria.

Recently, however, additional sensory pathways of *Euglena* have been discovered and the initial results of these studies will now be discussed.

I. Mechanical Stimulation

Most scientists working with *Euglena* have seen spontaneous body contractions in the cells while they are being observed under the microscope. MIKOLAJCZYK and DIEHN noticed that these contractions occurred only in cells that were resting on the microscope slide, and never under constant red-light illumination. In a systematic study of the phenomenon (1976a), they obtained the following results:

Body contractions of the cell are triggered by a step-up or step-down stimulus if the cell is simultaneously stimulated mechanically. Light or mechanical stimuli alone do not induce body contractions. Both types of stimuli do, however, induce flagellar phobic responses when acting alone, as shown by observations of the cells' behavior at an interface between low viscosity (1 cP) and high viscosity (4000 cP) medium. In red light, the cells exhibited typical "avoiding" (mechanophobic) responses when touching the interface obliquely. They stopped, executed a fractional turn toward the dorsal side, and continued normal swimming in the new direction. Cells approaching the interface head on entered the high viscosity medium and became immobilized without contraction, maintaining their normal elongated shape.

Illumination with white light of 0.14–1.1 kW/m^2 caused step-up photophobic responses in both free-swimming organisms and in those immersed in the high viscosity medium. In the latter, the response consisted of flagellar reorientation toward the ventral side of the cell, followed by contraction of the cell body. Such a contraction lasts for a few seconds and is followed by more contractions at 5–10 s intervals. No attempt was made to establish whether the phenomenon represents a morphophobic response or a morphokinesis, i.e., whether adaptation to the light occurs. When step-down phobic responses were induced by removal of a 472 nm stimulus of 1.4 W/m^2 intensity, body contractions were also observed.

For cells immersed in media of varying viscosity, the transduction time for light-induced body contractions was found to depend on the viscosity of the medium, with the shortest transduction times observed at the higher viscosities.

From 500 to 2000 cP, when the cells are capable both of body contraction and forward locomotion (only at 4000 cP and above does complete immobilization occur), a logarithmic dependence of both transduction time and forward velocity upon the viscosity was found. Within this limited viscosity range, each doubling of the viscosity resulted in a decrease of the transduction time to 60%, and of the forward velocity to 20% of the preceding value.

Stokes' Law calculations based on these data indicate that the frictional drag on the cell body decreases as the viscosity increases, whereas the stress on the flagellum or flagellar attachment point probably increases. This implicates the flagellum or the flagellar base as the site of mechanoreception, as did the observation of mechanophobic responses upon glancing collisions of the anterior end of the cell with a high viscosity interface.

Although the transduction time for contraction depended on the intensity of the mechanical stimulus, it was found to be independent of the intensity of the light stimulus from threshold (at 0.10 kW/m^2) to 1.13 kW/m^2. This is a very intriguing situation: a Weber–Fechner relationship exists not between the response and the triggering stimulus, but between the response and the mechanical stimulus that couples the body contraction system to the apparatus responsible for flagellar reorientation. Further studies are clearly in order.

Mechanophobic responses have also been observed in *Chlamydomonas* (STAVIS, 1974; RIEDL, 1976) and can lead to a reversal in the direction of phototaxis in that organism (RIEDL, 1976).

II. Chemoresponses

COLOMBETTI and DIEHN observed, while studying photoresponses in thin preparations of *Euglena*, that dynamic ring patterns involving the majority of cells form spontaneously under very dim illumination and in the dark. They proceeded to define the parameters of this motor response (1978).

In layers of 0.1 mm thickness and with concentrations of approximately 10^6 cells/ml, the cells aggregate in one or more regions of the suspension within 2–5 min, apparently via a negative kinesis. 20–30 s after the local cell concentration has increased by a factor of about 10, all cells exhibit a typical (stropho-) phobic response and after adaptation proceed to migrate with a net outward component of movement of 20 µm/s, thus forming a ring pattern that entrains other cells entering its path. Phobic responses are exhibited by cells moving both inward and outward from the ring. The authors present evidence that these responses are chemophobic reactions toward oxygen; they occur in response to a step-down stimulus when moving inward, and to a step-up stimulus when moving outward from the ring. Apparently, the cells possess a chemosensory system that determines the upper and lower thresholds of a chemosensory adaptation level.

What makes these responses interesting in the context of the photosensory system of *Euglena* is that they are dramatically affected by illumination. Red light will cause a rapid inward collapse of the ring pattern, apparently not by acting as a photic stimulus but via stimulation of photosynthetic oxygen evolution of the cells. Blue light illumination, on the other hand, results in unrestricted movement of the cells, and thus destruction of the ring pattern by randomization. It does so by suppressing the chemophobic responses and it is effective only if the cells are capable of step-down photophobic responses. Dark-grown or otherwise etiolated cells, which are not capable of step-down responses, exhibit perfectly normal ring patterns that are, nevertheless, insensitive to blue light. In blue-sensitive cells, the patterns, and hence the ability to respond chemophobically, return after 30–60 s of continuous blue-light illumination.

Since the responses to both light and chemical stimulation are identical, one would indeed expect both sensory pathways to converge; yet these results suggest that they do not simply compete for the motor apparatus, but rather interact in a more complex fashion. For instance, applying low-intensity blue light causes not only resetting of the photosensory stimulus/response system for the step-down response, but also commits the motor apparatus to the photosensory system exclusively, whether or not a photophobic response is going to be executed. Control of the motor apparatus is returned to the chemosensory system after the photic stimulus has acted on the cell for 30–60 s, indicating that the "receptor-switching" mechanism is itself subject to an adaptation phenomenon.

Tracing the molecular basis of the interactions between the photo- and chemosensory systems should provide interesting new insights into the mechanism of stimulus transduction in *Euglena*.

III. Gravity Sensing

CREUTZ and DIEHN (1976) noticed, while studying the responses of individual *Euglena* cells to polarized light, that in the absence of any stimulation the majority of cells assume an otherwise random orientation in the horizontal plane, i.e., perpendicular to the force of gravity. They suggested that *Euglena* may possess a gravity sensor that interacts with the cell's motor apparatus thus causing transverse geotaxis. This observation has not yet been followed up by further work and therefore no information is available on how the hypothetical geosensory system might interact with the other stimulus/response systems of *Euglena*. There is, however, some indication of a competition between the photic and gravity transduction systems for the motor apparatus. As a result of the orientation of the photoreceptor molecules, the cells preferentially assume an orientation perpendicular to the plane of polarization (Sec. C,I 3, Fig. 16) at the same time the receptor-shading mechanism tends to orient the cells toward the light. Therefore an experimental situation could be devised in which orientation in any of the three spatial directions is "forbidden" in terms of one of the light and gravity stimulus parameters (CREUTZ and DIEHN, 1976). Since the cells obviously have to assume some orientation, the resulting distribution of orientations yields information on the relative degree of control exerted upon the motor apparatus by the photosensory vs the geosensory system.

G. Summary and Outlook

This final section is not meant to be a listing of all pertinent work accomplished in the field of photosensory transduction, but rather a summary of the most interesting and promising avenues of future inquiry—a selection understandably biased by this reviewer's scientific philosophy and research interests.

I. Experimental Approaches

Barring the appearance of radically novel techniques, one would expect two general developments: The application to photomovement methodology of techniques developed for other fields of study, and the application of methods already used for the study of photomovement, to additional organisms.

In the first category, one of the most promising techniques is that of fluorescence. This technique has potential far beyond the limited application it has found for the identification of *Euglena*'s photoreceptor pigment (Sec. C,I 3). Fluorescence lifetime and dynamic fluorescence polarization measurements, for instance, can yield data on the kinetics of the primary photoprocess, on the excited-state orientation of the pigment, and on its mode of relaxation to the resting state (SCHULDINER et al., 1975).

Another technique that has not yet been applied to photomovement studies is work with behavioral mutants. Such mutants as exist, e.g., in *Euglena*, are strains originally produced as pigment mutants for studies of photosynthesis, and only incidentally exhibit variant behaviors. Some behavioral mutants of *Chlamydomonas* are known, but have been characterized only in terms of their accumulation or dispersal upon illumination (SMYTH, 1972). Genetic analysis of mutants has been shown to be a very powerful tool for identifying the components of other sensory systems in microorganisms, e.g. the system mediating mechanotransduction in *Paramecium* (KUNG et al., 1975); the application of genetic analysis of mutants to studies of phototransduction is overdue.

In the second category, even an incomplete list of promising investigative methods that have been applied to only one organism, would be long indeed. The reasons for this are understandable to some extent in those cases where specialized and/or expensive equipment is required, but this would, if anything, only appear to be an argument for more extensive collaboration between research groups. Examples of such techniques are electrophysiologic studies, now under way only with *Stentor* and *Euglena;* automated tracking methods; high speed cinemicrography; and combinations thereof.

There are other experimental approaches that would appear to be adaptable to different organisms with little effort, such as investigations with polarized light, and the study of individual step-up and step-down responses. Further examples can easily be identified by the readers of this review.

Fig. 18. Summary and Outlook. [From Brookhaven Lecture Series No. 9; BNL 699 (T-241)]

II. Open Problems

As in most fields of scientific inquiry, it appears that in the study of photosensory transduction much more is unknown than has been satisfatorily explained. Leaving out both that which appears easily done but simply has not yet been attempted (e.g., molecular studies of photokinesis in eucaryotes), and those areas of inquiry for which no experimental approach has yet been conceived (such as the identification of effector organelles and mechanisms), there remains a large number of avenues of inquiry that are both interesting and endowed with some probability of success.

Among these are the isolation of and biochemical studies with photoreceptor organelles, notably the paraflagellar swelling of *Euglena*. This would allow unambiguous pigment identification, and help identify the physical process by which the excitation energy of the photoreceptor molecule is utilized to generate a sensory signal. Another important area of study is the determination of the mode of transmission of the physiologic signal into which the light signal has been transduced, perhaps with electrofluorochromes if not by electrophysiologic methods. In this context, a fascinating complex of questions deals with the modulation of the photosensory transduction system by, for instance, the photosynthetic apparatus. How is a certain state of the photosynthetic energy transduction system signaled to the processor, and how, for that matter, does the processor work? Processor mutants might be formed that could provide a means of answering this question. In all probability new insights will also be derived from studying the interactions between the various sensory systems of the cell. Specific inhibition or chemical modification of selected components of these pathways will undoubtedly be an important tool in such work.

A molecular approach does not seem appropriate to certain areas of research. Strictly behavioral studies will probably lead to the discovery of an instantaneous mode for the detection of stimulus direction, if such exists. And then there is the intriguing possibility that photosensory systems might exhibit habituation and dishabituation upon repetitive stimulation. Such phenomena, which are analogous to learning and memory in higher sensory systems, have, for instance, been studied in the responses of *Spirostomum* to mechanical stimuli (EISENSTEIN et al., 1973).

One may reasonably conclude that significant progress in the knowledge of photosensory transduction processes in unicells is possible in the next few years, and I have summarized the content of this section in the final illustration (Fig. 18).

References

Bamberger, E. S., Black, C. C., Fewson, C. A., Gibbs, M.: Inhibitor studies of carbon dioxide fixation, adenosine-triphosphate formation, and triphosphopyridine nucleotide reduction by spinach chloroplasts. Plant. Physiol. **38**, 483—487 (1963).

Bartlett, C. J., Walne, P. L., Schwarz, J., Brown, D. H.: Large scale isolation and purification of eye-spot granules from *Euglena gracilis* var. *bacillaris*. Plant Physiol. **48**, 881—885 (1972).

Batra, P. P., Tollin, G.: Phototaxis in *Euglena* I. Isolation of the eye-spot granules and identification of the eye-spot pigments. Biochim. Biophys. Acta **79**, 371—378 (1964).

Bendix, S.: Phototaxis. Botan. Rev. **26**, 145—208 (1960).

Benedetti, P. A., Bianchini, G., Checcucci, A., Ferrara, R., Grassi, S.: Spectroscopic properties and related functions of the stigma measured in living cells of *Euglena gracilis*. Arch. Microbiol. **111**, 73—76 (1976 a).

Benedetti, P. A., Bianchini, G., Chitti, G.: Fast-scanning microspectroscopy; an electrodynamic moving-condenser method. Appl. Optics **15**, 2554—2557 (1976 b).

Benedetti, P. A., Checcucci, A.: Paraflagellar body (PFB) pigments studied by fluorescence microscopy in *Euglena gracilis*. Plant Sci. Letters **4**, 47—51 (1975).

Berg, H. C.: How to track bacteria. Rev. Sci. Instrum. **47**, 868—873 (1971).

Berg, H. C., Brown, D. A.: Chemotaxis in *E. coli* analyzed by three-dimensional tracking. Nature **229**, 500—504 (1972).

Berg, S. P., Krogmann, D. W.: Mechanism of KCN inhibition of Photosystem I. J. Biol. Chem. **250**, 8957—8962 (1975).

Böhme, H., Reimer, S., Trebst, A.: The effect of dibromothymoquinone, an antagonist of plastoquinone, on noncyclic and cyclic electron flow systems in isolated chloroplasts. Z. Naturforsch. **26 b**, 341—352 (1971).

Boscov, J. S.: Response of *Chlamydomonas reinhardii* to single flashes of light. M. S. Thesis Tufts University (1974).

Bound, K. E., Tollin, G.: Phototactic response of *Euglena gracilis* to polarized light. Nature **216**, 1042—1044 (1967).

Brinkmann, K.: Keine Geotaxis bei *Euglena*. Z. Pflanzenphysiol. **59**, 12—16 (1968).

Bünning, E., Schneiderhöhn, G.: Über das Aktionsspektrum der phototaktischen Reaktionen von *Euglena*. Arch. Mikrobiol. **32**, 310—321 (1959).

Clayton, R. K.: Primary processes in bacterial photosynthesis. Ann. Rev. Biophys. Bioeng. **2**, 131—156 (1973).

Checcucci, A.: Photomovement methodology. In: Primary molecular events in photobiology. Amsterdam: Elsevier 1973.

Checcucci, A.: Molecular sensory physiology of *Euglena*. Naturwissenschaften **63**, 412—417 (1976).

Checcucci, A., Colombetti, G., Ferrara, R., Lenci, F.: Action spectra for photoaccumulation of green and colorless *Euglena;* evidence for identification of receptor pigments. Photochem. Photobiol. **23**, 51—54 (1976).

Colombetti, G., Diehn, B.: Chemosensory responses toward oxygen in *Euglena gracilis*. J. Protozool. **25**, 211—217 (1978).

Cramer, W. A., Horton, P.: Recent studies on the chloroplast cytochrome b-559. Photochem. Photobiol. **22**, 304—307 (1975).

Creutz, C., Diehn, B.: Motor responses to polarized light and gravity sensing in *Euglena gracilis*. J. Protozool. **23**, 552—556 (1976).

Creutz, C., Colombetti, G., Diehn, B.: Photophobic behavioral responses of *Euglena* in a light intensity gradient, and the kinetics of photoreceptor pigment interconversions. Photochem. Photobiol. **27**, 611—616 (1978).

Davenport, D., Culler, G. J., Greaves, J. O. B., Forward, R. B., Hand, W. G.: The investigation of the behavior of microorganisms by computerized television. IEEE Trans. Biomed. Eng. BME-**17**, 230—237 (1970).

Diehn, B.: Phototactic response of *Euglena* to single and repetitive pulses of actinic light: Orientation time and mechanism. Exp. Cell Res. **56**, 375—381 (1969 a).

Diehn, B.: Action spectra of the phototactic responses in *Euglena*. Biochim. Biophys. Acta **177**, 136—143 (1969 b).

Diehn, B.: Mechanism and computer simulation of the phototactic accumulation of *Euglena* in a beam of light. Photochem. Photobiol. **11**, 407—418 (1970).

Diehn, B., Kint, B.: The flavin nature of the photoreceptor pigment for phototaxis in *Euglena*. Physiol. Chem. Phys. **2**, 483—488 (1970).

Diehn, B.: The receptor-effector system of phototaxis in *Euglena*. Acta Protozool. **11**, 325—332 (1972).

Diehn, B.: Phototaxis and sensory transduction in *Euglena*. Science **181**, 1009—1015 (1973).

Diehn, B.: Cybernetics and the behavior of microorganisms. In: Aneural Organisms in Neurobiology. New York: Plenum 1974.

Diehn, B., Feinleib, M., Haupt, W., Hildebrand, E., Lenci, F., Nultsch, W.: Terminology of behavioral responses in motile microorganisms. Photochem. Photobiol. **26**, 559—560 (1977).

Diehn, B., Fonseca, J. R., Jahn, T. L.: High speed cinemicrography of the direct photophobic response of *Euglena* and the mechanism of negative phototaxis. J. Protozool. **24**, 492—494 (1975).

Diehn, B., Tollin, G.: Phototaxis in *Euglena*. II. Physical factors determining the rate of phototactic response. Photochem. Photobiol. **5**, 839—844 (1966).

Diehn, B., Tollin, G.: Phototaxis in *Euglena*. IV. Effect of inhibitors of oxidative and photophosphorylation on the rate of phototaxis. Arch. Biochem. Biophys. **121**, 169—177 (1967).

Eckert, R., Naitoh, Y.: Bioelectric control of locomotion in the ciliates. J. Protozool. **19**, 237—241 (1972).

Eisenstein, E. M., Osborn, D., Blair, H.: Behaviour modification in protozoa. In: Behavior of microorganisms. London-New York: Plenum 1973.

Engelmann, T. W.: Über die Licht- und Farbenperzeption niederster Organismen. Pflügers Arch. **29**, 387—400 (1882).

Feinleib, M. E.: Phototacic response of *Chlamydomonas* to flashes of light. I. Response of cell populations. Photochem. Photobiol. **21**, 351—354 (1975).

Feinleib, M. E., Curry, G. M.: Methods for measuring phototaxis of cell populations and individual cells. Physiol. Plantarum **20**, 1083—1095 (1967).

Feinleib, M. E., Curry, G. M.: The nature of photoreceptor in phototaxis. In: Loewenstein, W. R. (Ed.): Handbook of Sensory Physiology, Vol. 1. Berlin-Heidelberg-New York: Springer 1971.

Forward, R. B.: Phototaxis by the Dinoflagellate *Gymnodinium splendens* Lebour. J. Protozool. **21**, 312—315 (1974).

Froehlich, O., Diehn, B.: Photoeffects in a flavin-containing lipid bilayer membrane and implications for algal phototaxis. Nature **248**, 802—804 (1974).

Goessel, I.: Über das Aktionsspektrum der Phototaxis chlorophyllfreier Euglenen und über die Absorption des Augenflecks. Arch. Mikrobiol. **27**, 288—305 (1957).

Haeder, D.-P.: Untersuchungen zur Photo-phototactic bei *Phormidium uncinatum*. Dissertation Marburg 1973.

Haeder, D.-P.: The effect of inhibitors on the electron flow triggering photophobic reactions in Cyanophycea. Arch. Microbiol. **103**, 169—174 (1974a).

Haeder, D.-P.: Participation of two photosystems in the photo-phototaxis of *Phormidium uncinatum*. Arch. Microbiol. **96**, 255—266 (1974b).

Haeder, D.-P.: Further evidence for the electron pool hypothesis. The effect of KCN and DSP on the photophobic reaction in the filamentous blue-green alga *Phormidium uncinatum*. Arch. Microbiol. **110**, 301—303 (1976).

Haeder, D.-P., Nultsch, W.: Negative photo-phototactic reactions in *Phormidium uncinatum*. Photochem. Photobiol. **18**, 311—318 (1973).

Hand, W. G., Schmidt, J.: Phototactic orientation by the marine dinoflagellate *Gyrodinium dorsum* Kofoid. II. Flagellar activity and overall response mechanism. J. Protozool. **22**, 494—498 (1975).

Harayama, S., Iino, T.: Phototactic response of aerobically cultivated *Rhodospirillium rubrum*. J. Gen. Microbiol. **94**, 173—179 (1976).

Haupt, W.: Phototaxis in plants. Intern. Rev. Cytol. **19**, 267—299 (1966).

Hildebrand, E., Dencher, N.: Two photosystems controlling behavioural responses of *Halobacterium halobium*. Nature **257**, 46—48 (1975).

Horowicz, P., Caputo, C., Robeson, J. A.: Effects of azide on the electrical properties of frog striated muscle. Fed. Proc. **27**, 702—704 (1962).

Kung, C., Chang, S.-Y., Satow, Y., Van Houten, Y., Hansma, H.: Genetic Dissection of Behavior in *Paramecium*. Science **188**, 898—904 (1975).

Lindes, D., Diehn, B., Tollin, G.: Phototaxigraph; recording instrument for determination of rate of response of phototactic microorganisms to light of controlled intensity and wavelength. Rev. Sci. Instrum. **36**, 1721—1724 (1965).

Macnab, R. M., Koshland, D. E.: The gradient sensing mechanism in bacterial chemotaxis. Proc. Natl. Acad. Sci. (USA) **69**, 2509—2512 (1972).

Mast, S. O.: Light and the behavior of organisms. New York: Wiley 1911.

Mikolajczyk, E., Diehn, B.: The effect of potassium iodide on photophobic responses in *Euglena;* evidence for two photoreceptor pigments. Photochem. Photobiol. **22**, 269—271 (1975).

Mikolajczyk, E., Diehn, B.: Light-induced body movement of *Euglena gracilis* coupled to flagellar photophobic responses by mechanical stimulation. J. Protozool. **23**, 144—147 (1976a).

Mikolajczyk, E., Diehn, B.: Differential effect of the cationic detergent CTAB on the photophobic responses of *Euglena*. Proc. 7th Intl. Congress Photobiol., Rome (1976b).

Møller, K. M.: On the nature of stentorin. C. R. Trav. Lab. Carlsberg **32**, 471—497 (1962).

Naitoh, Y., Eckert, R.: Ionic mechanisms controlling behavioral responses of *Paramecium* to mechanical stimulation. Science **164**, 963—965 (1969).

Neuscheler, W.: Bewegung und Orientierung bei *Micrasterias denticulata* Breb. im Licht. Z. Pflanzenphysiol. **57**, 151—172 (1967).

Nultsch, W.: Photokinesis in *Phormidium uncinatum*. Planta **57**, 613—617 (1962).

Nultsch, W.: Phototaxis and photokinesis. In: Primitive Sensory and Communications Systems; the Taxes and Tropisms of Microorganisms and Cells. London-New York-San Francisco: Academic Press 1975.

Nultsch, W., Wenderoth, K.: Phototactic investigations in single cells of *Navicula peregrina* (Ehrenberg) Kuetzing. Arch. Mikrobiol. **90**, 47—58 (1973).

Pagni, P. G. S., Walne, P. L., Wehry, E. L.: Fluorometric evidence for flavins in isolated eyespots of *Euglena gracilis* var. *bacillaris*. Photochem. Photobiol. **24**, 373—375 (1976).

Piccini, E., Omodeo, P.: Photoreceptors and phototactic programs in protista. Boll. Zool. **42**, 57—79 (1975).

Racker, E., Stoeckenius, W.: Reconstitution of purple membrane vesicles catalyzing light-driven proton uptake and adenosine triphosphate formation. Nature New Biol. **233**, 149—152 (1974).

Radmer, R., Kok, B.: Energy capture in photosynthesis; photosystem. II. Ann. Rev. Biochem. **44**, 409—433 (1975).

Riedl, G.: Änderung der phototaktischen Reaktionsrichtung bei *Chlamydomonas reinhardi* durch freie und thigmophobische Umkehr. Zulassungsarbeit, Universität Erlangen-Nürnberg (1976).

Saji, M., Oosawa, F.: Mechanism of photoaccumulation in *Paramecium bursaria*. J. Protozool. **21**, 556—567 (1974).

Schuldiner, S., Spencer, R. D., Weber, C., Weil, R., Kaback, H. R.: Lifetime and rotational relaxation time of dansylgalactoside bound to the *lac* carrier protein. J. Biol. Chem. **250**, 8893—8896 (1975).

Smyth, R. D.: Genetic control of phototactic aggregation in *Chlamydomonas reinhardtii*. Ph. D. diss. University of California, Los Angeles (1971). Ann Arbor: University Microfilms, No. 72 11899 (1972).

Song, P.-S.: On the basicity of the excited state of flavins. Photochem. Photobiol. **7**, 311—313 (1968).

Sperling, P. G., Walne, P. L., Wehry, E. L.: Fluorometric evidence for flavins in isolated eyespots of *Euglena gracilis* var. *bacillaris*. Photochem. Photobiol. **24**, 373—376 (1976).

Stahl, E.: Über den Einfluß von Richtung und Stärke der Beleuchtung auf einige Bewegungserscheinungen im Pflanzenreiche. Bot. Z. **38**, 297—413 (1880).

Stavis, R. L.: The effect of azide on phototaxis in *Chlamydomonas reinhardtii*. Proc. Natl. Acad. Sci. (USA) **71**, 1824—1827 (1974).

Strother, G. K., Wolken, J. J.: Microspectrophotometry of *Euglena* chloroplast and eyespot. Nature **188**, 601—602 (1960).

Taylor, B. L., Koshland, D. E.: Intrinsic and extrinsic light responses of *Salmonella typhimurium* and *Escherichia coli*. J. Bacteriol. **123**, 557—569 (1975).

Throm, G.: Untersuchungen zum Reaktionsmechanismus von Phototaxis und Kinesis in *Rhodospirillum rubrum*. Arch. Protistenk. **110**, 313—371 (1968).

Tollin, G., Robinson, M. I.: Phototaxis in *Euglena*. VI. Correlations between ATP production by light and phototactic rates. Bioenergetics **1**, 139—145 (1970).

Trebst, A., Burba, M.: Über die Hemmung photosynthetischer Reaktionen in isolierten Chloroplasten und in *Chlorella* durch Disalicylidenpropandiamin. Z. Pflanzenphysiol. **57**, 419—433 (1967).

Tsang, N., Macnab, R., Koshland, D. E.: Common mechanism for repellents and attractants in bacterial chemotaxis. Science **181**, 60—63 (1973).

Walne, P. L., Arnott, H.: The comparative ultrastructure and possible function of eyespots: *Euglena granulata* and *Chlamydomonas eugametos*. Planta **77**, 325—353 (1967).

Wolken, J. J., Shin, E.: Photomotion in *Euglena gracilis*. I. Photokinesis. II. Phototaxis. J. Protozool. **5**, 39—46 (1958).

Wood, D. C.: Action spectrum and electrophysiological responses correlated with the photophobic responses of *Stentor coeruleus*. Photochem. Photobiol. **24**, 261—266 (1976).

Chapter 3

Ocular Optical Filtering

WILLIAM H. MILLER, New Haven, Connecticut (USA)

With 39 Figures

Contents

A. Introduction

Visual perception is initiated by the eye's optics in the sense that the neural aspects of the visual process can only operate on the information that is presented to the nervous system. The varied ways in which this information is presented in the eyes of vertebrates and invertebrates raise questions about the consequences of the function of optical structures for vision. The purpose of this chapter is to review the effects of optical filtering on eye function emphasizing such unresolved issues, for example, as the function of optical filtering in the eyes of animals with high (suprahuman) visual acuity. Aspects of optical filtering will be reviewed commencing with the physical basis of transparency of the ocular media. The effects of interference and absorption filters will be examined. Polarization filters, which are covered by WATERMAN (this volume, Part B), and visual pigments, which have been recently reviewed by several authors in the Handbook of Sensory Physiology Vol. VII/I and HAMDORF (this volume), will be omitted. Filtering that results from the spatial arrangements and the physical properties of the photoreceptor cells will also be discussed. The unifying nature of this topic lends itself to a consideration of these factors independent of phylogenetic classification, and for this reason examples will be drawn freely from both the vertebrates and invertebrates.

B. Transparency

Transparency depends on the absence of absorbing substances and on uniformity of the media's refractive index *(n)*. The presence of particles causing variation in refractive index results in reflection of light that reduces transparency. Reflection by particles smaller than the wavelengths of light is called scattering (Sec. II, below). Absorption is considered first.

I. Absorption

1. Aromatic Amino Acids

Absorption occurs when a quantum of light energy collides with and gives up its energy to an absorbing element. When an absorbing substance is uniformly distributed in an optical medium, absorption is proportional to the concentration of the substance and the path length

$$\log_{10}(I_i/I_t) = \varepsilon_\gamma c \cdot l \tag{1}$$

where I_i is the incident intensity of illumination, I_t is the transmitted intensity, $\log_{10}(I_i/I_t)$ is the optical density, absorbance or extinction of the medium, ε_γ is the extinction coefficient in liters per mole·centimeter that is a characteristic of the substance and the wavelength of radiation, c is concentration in moles per liter, and l is path length in centimeters. Illustrations of absorbing structures in the ocular media with specific filtering functions are given in B,2 and B,3 below. As a rule, transparent parts of the ocular media do not contain substances that absorb strongly in the visible spectrum (but see "yellow filters" B,3 below). Both vertebrate and invertebrate optical media contain protein that absorbs light in the near ultraviolet (UV). The constituents of arthropod cuticle vary depending on its origin, but it is roughly half protein and half chitin (FRAENKEL and RUDALL, 1947). Cuticular chitin is in its α-form which is a polysaccharide chain of chitobiose residues linked by 1-4-β glucoside bonds (CARLSTRÖM, 1966). Radiation absorption centers are associated with the π electrons of unsaturated double bonds. The acetyl glucosamine units of chitobiose are saturated. The main source of absorption would therefore be expected to come from the unsaturated ring structures of the aromatic amino acids that occur in protein. Arthropod cuticle contains tyrosine [79 residues per 1000 for *Bombyx mori* (larval)] with absorption $\lambda_{max} = 278$ nm and phenylalanine (24 residues per 1000) with $\lambda_{max} = 259$ nm (HACKMAN and GOLDBERG, 1971). The corneal cuticle itself has not been analyzed for aromatic amino acids.

Figure 1 *(open circles)* from BERNHARD et al. (1965) illustrates the extinction spectrum of arthropod cornea. The results are similar to those of GOLDSMITH and FERNANDEZ (1968), KOLB et al. (1969), and CARRICABURU and CHARDENOT (1967). The human cornea, an order of magnitude thicker, yet comparably transparent, is also shown on Figure 1 and additional factors that contribute to its transparency are discussed in Section B,II,2 below. There is a marked decrease in transparency in both corneas in the near UV which, because of the location of the peak absorption, is probably caused by the aromatic amino acids in the cuticular protein. The absorbance of arthropod cornea is of the order of 0.01 log units/μm in the near UV. Since $\varepsilon_{278} = 10^3$ liters/mol·cm for tyrosine, the measured absorbance suggests the presence of a concentration of tyrosine on the order of 1 *M*. The concentration of the higher molecular weight visual pigments in vertebrate photoreceptor organelles is 1–4 m*M* (LIEBMAN, 1972). The presence of aromatic amino acids in arthropod cuticle may protect the animal from UV radiation. Nevertheless, tryptophan, which has the highest extinction coefficient of all the

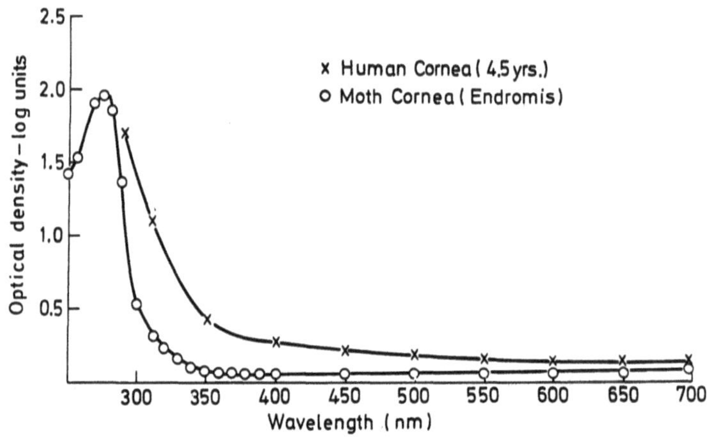

Fig. 1. Optical density of arthropod and vertebrate corneas. Arthropod is the moth, *Endromis* (after BERNHARD et al., 1965); vertebrate is the human. (After Fig. 3, BOETTNER and WOLTER, 1962)

aromatic amino acids at its λ_{max}=279 nm is not present in arthropod cuticle. Many arthropod eyes contain a visual pigment of λ_{max}=350 nm, providing visual sensitivity in the near UV (reviewed in GOLDSMITH, 1972). Because there is little absorption in the arthropod cornea and crystalline cone for wavelengths longer than 310 nm (BERNHARD et al., 1965; KOLB et al., 1969, SKRZIPEK and SKRZIPEK, 1974), the dioptrics are well matched to the visual pigments, whereas the presence of blue-absorbing substances in the vertebrate dioptrics (Sec. B,I,3 below) insures the same degree of match for those eyes.

2. Screening Pigment Pupillary Functions

Stationary and mobile screening pigments composed of melanins, ommochromes, or pteridines function to regulate the access of light to photoreceptor organelles. For reviews of this subject see KUNZE, this volume; GOLDSMITH and BERNARD (1974); FRANCESCHINI (1975); WALCOTT (1975); LANGER (1975); and PARKER (1932) and MAZOKHIN-PORSHNYAKOV (1969). In general, screening pigment in superposition eyes moves parallel to the axis of the ommatidium whereas in apposition eyes the movement is predominantly radially within the ommatidium, perpendicular to the ommatidial long axis. The illustrative examples below deal only with the ideal type of superposition eye. See HORRIDGE (1975) and LAND (1976) for a wide range of examples dealing with eyes that fall short of the ideal.

a) Superposition Compound Eye

α) Sensitivity. The diurnal superposition compound eye, for example that of the lepidopteran skipper, has accessory screening pigment that is stationary, separating the crystalline cones regardless of ambient light levels (DØVING and MILLER, 1969; HORRIDGE, 1975). The accessory pigment in nocturnal superposition eyes migrates proximally to screen the receptors' organelles from bright light (KUNZE,

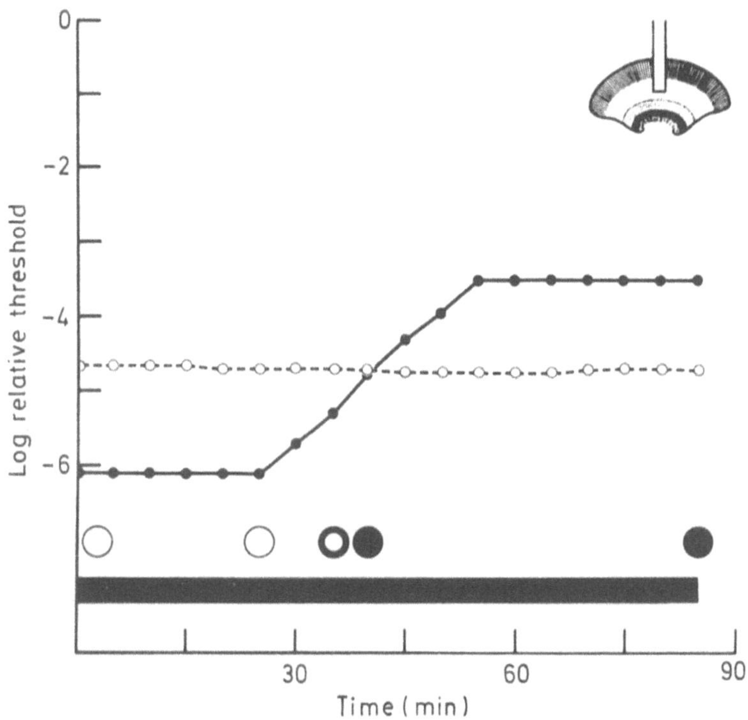

Fig. 2. Illustrating changes in sensitivity of moth *(Deilephila)* receptors caused by accessory cell pigment migration. The eye is kept in darkness during experiment, as indicated by *lower dark line*. Pigment moved spontaneously to light position as indicated by glow disappearance (diagrams just above *lower dark line*). As glow disappeared, threshold of photoreceptors as measured by illumination presented conventionally increased over 2.5 log units *(filled circles)*. When pigment was bypassed by the light stimulus being presented through a 75-μm diameter optical waveguide inserted through the cornea *(upper-right diagram)*, receptor threshold *(open circles)* was constant during the time of full, spontaneous pigment migration. (HÖGLUND, 1966)

this volume). When a light-adapted moth eye is placed in the dark there is a rapid decrease in threshold of 2 log units unrelated to pigment movement. This is followed by a decrease in sensitivity of between 2 and 3 log units, which takes about 30 min to reach full sensitivity. The sensitivity changes are correlated with pigment position and hence the degree of screening (BERNHARD and OTTOSON, 1960a, b, 1964; HÖGLUND and STRUWE, 1971). HÖGLUND (1966) inserted a 75-μm diameter glass fiber through the pigment layer, as shown in Figure 2, which he used to test the sensitivity of the receptors independent of pigment position. As the pigment moved spontaneously to the light position in a moth eye kept continuously in the dark, the glow decreased (see KUNZE, this volume; STAVENGA, this volume) and the sensitivity of the eye, as measured by illumination through the cornea in the normal way, decreased about 2.5 log units. However, when the screening pigment was bypassed so that the receptors were stimulated using the glass fiber, the sensitivity of the receptor cells was found to be unchanged by pigment migration *(open circles)*. Two to three log units of sensitivity are therefore controlled exclusively by screening pigment migration in the moth and perhaps other nocturnal arthropod superposition compound eyes.

β) Acuity. In the superposition eye (EXNER, 1891; KUNZE, this volume; HORRIDGE, 1975; GOLDSMITH and BERNARD, 1974; MILLER, 1976) diffraction-limited resolution is a function of the facet diameter rather than the superposition aperture, which may be ten or more facets in diameter. The individual facets still function as separate apertures with the dioptric apparatus of each ommatidium directing the incoming light to the superposition image plane. When the eye is light adapted, the superposition eye would be expected to function as an apposition eye (EXNER, 1891) and hence have the same theoretical resolution if the facet aperture limits resolution by diffraction. Assuming hexagonal packing of the facets, the highest spatial frequency (ν) that can be reconstructed by the array of facets is 0.58 $\Delta\Phi$, where $\Delta\Phi$ is the angle between ommatidia (SNYDER and MILLER, 1977a; SNYDER, this volume). In the prawn, $\Delta\Phi$ is about 5° (WATERMAN and CAMPBELL, private communication). The diffraction-limited theoretical resolution limit is therefore about 3°, regardless of illumination conditions. HASSENSTEIN (1954) and DE BRUIN and CRISP (1957) find no difference in visual acuity as a function of adaptation conditions. DE BRUIN and CRISP determined that proximal pigment migration caused by light adaptation gives an increase in acuity by reducing scatter from the tapetum. Distal pigment migration, which determines whether the eye functions to give a superposition or apposition image, has no effect on acuity. The maximum acuity is independent of adaptation and is on the order of 4° for a square grating on a revolving cylinder, so the resolving power falls short of the theoretical maximum. See Section D,I below for factors other than iris pigment that influence acuity.

b) Apposition Compound Eye

The optical and physiological effects of screening pigment both in primary pigment cells and retinular cells of apposition eyes are reviewed by STAVENGA (Chapter 7, this volume). Nevertheless, a few examples dealing with the interaction of shielding pigment and rhabdoms will be discussed here because they raise questions concerning the mechanisms of the interaction. Shielding pigment migration has been studied in apposition compound eyes by both histologic and optical techniques (e.g. MILLER, 1958a; BEHRENS and KREBS, 1976; BUTLER, 1971; MENZEL, 1972; KOLB and AUTRUM, 1972, 1974; KIRSCHFELD and FRANCESCHINI, 1969; FRANCESCHINI and KIRSCHFELD, 1976; STAVENGA and KUIPER, 1977; STAVENGA et al., 1977; CHAMBERLAIN and BARLOW, 1977; RIBI, 1978). Figure 3, taken from KIRSCHFELD and FRANCESCHINI (1969), illustrates the phenomenon. Pigment granules that approach the rhabdom or rhabdomere during light adaptation would be expected to raise the n outside the rhabdom, tending to frustrate total internal reflection, and to both scatter and absorb modal energy propagating outside of or bled from the rhabdom. Of these three effects, that of absorption has been shown to be the most important for dipteran rhabdomeres by FRANCESCHINI and KIRSCHFELD (1976). Nevertheless, these experiments also suggest that n of the granules is high because the rhabdomeres are brightly lit in the region of the spectrum where the absorption index of the ommochrome is highest. FRANCES-

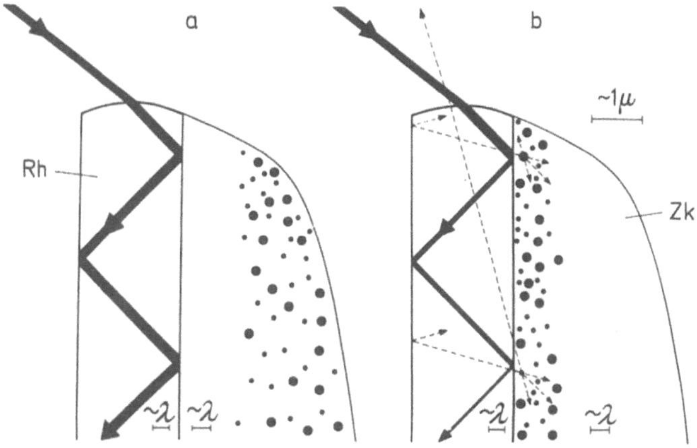

Fig. 3. Mechanism of attenuation of light propagated in rhabdom waveguide by screening pigment. In dark-adapted condition light within the acceptance angle is propagated by total internal reflection (a). In the light-adapted condition (b) pigment granules move near to (less than a wavelength, λ) the rhabdom. A small amount of energy is propagated outside the rhabdom waveguide and this is scattered and absorbed by the pigment, reducing the intensity of the energy guided inside the rhabdom. In addition the screening pigment increases the refractive index of receptor cell *(Zk)* cytoplasm near the rhabdom, reducing the critical and acceptance angles and the intensity of illumination in the rhabdom. (KIRSCHFELD and FRANCESCHINI, 1969)

CHINI and KIRSCHFELD (1976) show that the increased scatter that causes the dipteran rhabdomeres to light up using orthodromic illumination is a result of an increased n caused by anomolous dispersion. The complex n is raised because of the imaginary term containing the high absorption index. In fact there is evidence from another compound eye that in one species the mitochondria rather than pigment cause an increase in n outside the rhabdom, reducing the critical angle and hence the acceptance angle of the rhabdom. The critical angle is the angle whose cos is the ratio of n of the surround to n of the rhabdom; the acceptance angle is twice the critical angle. HORRIDGE (1966) showed that for *Locusta* a watery palisade around the rhabdom is replaced by higher n mitochondria during light adaptation. The half sensitivity angle is reduced from about 6° to 3° and there is a 2 log unit decrease in sensitivity to illumination caused by light adaptation. WILSON'S (1975) more recent data on the same preparation indicates a narrowing from 2.5° to 1.5°. Interestingly, pigment granules seem to have the same effect. BUTLER and HORRIDGE (1973) working with the compound eye of *Periplaneta* and BEERSMA (1978) with several species of dipteran compound eyes both find a narrowing of the acceptance angle with light adaptation. Therefore, the relative contributions caused by changes in n, absorption and scattering are unanswered questions.

A variety of other types of screening pigment movements have been described in apposition type compound eyes. WALCOTT (1975) reviewed pigment migration adjacent to the crystalline cone in several insects and YAGI and KOYAMA (1963) described similar changes in butterflies that change the appearance of the pseudopupils (see STAVENGA, Ch. 7).

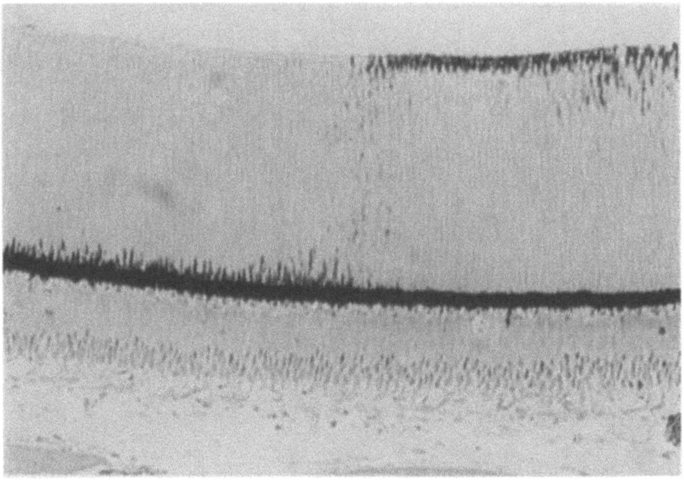

Fig. 4. Screening pigment in the receptor cells of the cephalopod retina moves to the vitreal surface and corresponds to the introduction of a 0.6 log unit neutral density filter. Left half of retina was in dark, right half in light. (DAW and PEARLMAN, 1974)

The physiologic effect of a longitudinal pigment migration located proximal to the basement membrane is uncertain in some butterflies. In the light-adapted condition this pigment migrates away from the optic ganglion and towards the basement membrane. The time course of the movement is in the order of seconds (DEMOLL, 1909). This may be a mechanism to decrease scattered light from wavelengths in the pass band of the tracheolar filters (Sec. C,II,3 below). Wavelengths in the rejection band are of course returned to the rhabdom, but it would be advantageous to suppress the reflection of light within the eye passed by the filters in order to prevent that illumination from reaching the rhabdoms.

c) Cephalopod Retinal Pigment

In addition to a mobile iris pupil the camera eye of the mollusk cephalopod has a well-developed screening pigment movement. Illumination causes pigment granules (mostly within the photoreceptors) to move about 300 μm from the proximal to the distal border of the rhabdomeres where they form a dense screen at the vitreal surface of the retina (COHEN, 1973). DAW and PEARLMAN (1974) found that when 3%–10% of the rhodopsin of the squid eye is isomerized, the screening pigment moves to the vitreal surface with a time course of 5–15 min. In the dark it moves back again with the same time course unless more than 10% of the rhodopsin is isomerized. Figure 4 from their paper shows the pigment position in a section of the squid retina taken at the edge of the adapting light spot. When the screen has migrated to the vitreal surface of the retina, it corresponds to the introduction of a 0.6 log unit neutral density filter.

d) Vertebrate Pigment Epithelium

The eyes of lower vertebrates (teleosts, anurans, urodeles, birds and some reptiles) also have screening pigment movement (WALLS, 1967, p. 150). It is char-

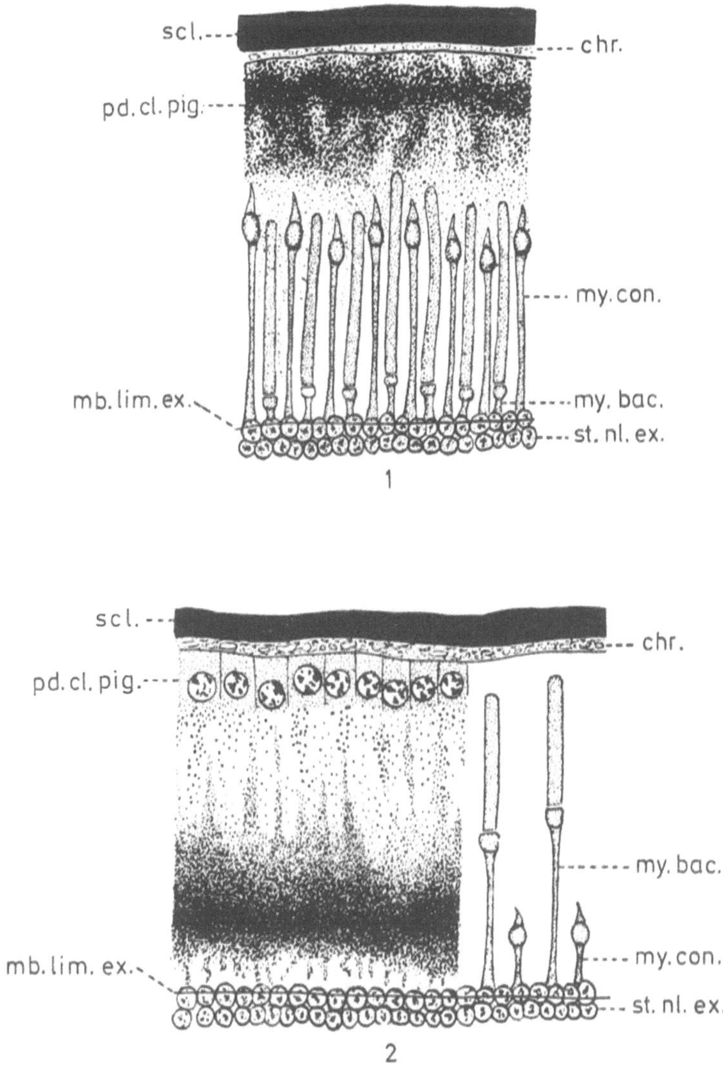

Fig. 5. Illustrating screening pigment position in vertebrate eyes with photomechanical movements. Dark adapted (1), light adapted (2). *chr.*, choroid; *mb.lim.ex.*, external limiting membrane; *my.bac.*, rod myoid; *my.con.*, cone myoid; *pd.cl.pig.*, base of pigment cell; *scl.*, sclera; *st.nl.ex.*, external nuclear layer; *st.pig.*, pigment layer. (AREY, 1919)

acterized by migration of pigment granules within the apical processes of retinal pigment epithelial cells from an extreme sclerad direction in the dark so that in the light-adapted condition the myoid inner segments of the rods and cone outer segments are enveloped by pigment. This envelopment (Fig. 5) would theoretically be expected to decrease the illumination reaching the rod outer segments by "bleeding" the myoids of light (MILLER and SNYDER, 1972, and Sec. E below). GARTEN (1907) found that adrenaline causes the retinal pigment epithelial granules to migrate to the light position in the dark-adapted frog. DODT and JESSEN

(1961) used this technique to determine that the screening pigment raises the threshold of the frog eye by ~1 log unit as measured using the b wave of the ERG. BÄCK et al. (1965) measured the amount of bleaching of rhodopsin in frog rod outer segments and concluded that the screening pigment of the retinal pigment epithelium reduces the effective intensity of illumination reaching the rod outer segments to $^1/_3$ which is equivalent to the introduction of a neutral density filter of 0.5 log units. The critical factor in obtaining the screening effect is whether or not the screening pigment surrounds the rod ellipsoid and myoid.

e) Extinction Spectra

Whereas the spectra of melanin, the vertebrate-shielding pigment (BOWNESS and MORTON, 1952), and the pigments of the butterfly (LANGER, 1975) and *Limulus* compound eyes (WASSERMAN, 1967) are relatively flat, most arthropod-screening pigments have a broad maximum of absorption peaking at about 550 nm, with high absorption in the blue, and are transparent for wavelengths longer than about 680 nm (HÖGLUND et al., 1970; LANGER, 1975; SCHNEIDER, DRAŠLAR, LANGER and SCHLECHT, 1978; NOLTE and BROWN, 1970). GOLDSMITH (1965) called attention to a physiologic effect of the stray red light caused by the imperfect screening. This red light can give rise to physiologic responses that are easily confused with a red-sensitive receptor process. STAVENGA et al. (1973) postulate that this screening defect serves a useful purpose by flooding the eye with long-wavelength illumination that photochemically converts dipteran metarhodopsin to rhodopsin (STAVENGA, this volume; HAMDORF, this volume).

f) Mechanisms

Measurement of the action spectrum of vertebrate retinal pigment epithelial migration indicates that absorption of light by rhodopsin of the pink rods triggers screening pigment movement (LIEBMAN et al., 1969). Likewise, absorption of light by the rhodopsin of the cephalopod retina triggers its screening pigment movement (DAW and PEARLMAN, 1974), and a green sensitive mechanism triggers pigment migration in certain butterfly apposition eyes (BERNARD, 1976). The train of effects after initiation of the signal by rhodopsin is unknown. However, dipteran retinular cell pigment migration occurs when these cells are depolarized by the receptor potential (KIRSCHFELD and FRANCESCHINI, 1969). Therefore STAVENGA et al. (1973) propose an electrophoretic theory of granule movement. Charged granules would move from the electric field force generated by the receptor potential. A mechanism involving much larger granules is discussed by MILLER and CAWTHON (1974) who found radially oriented microtubules in the retinular cells of the compound eye of *Limulus*. Since colchicine caused both the migration of the pigment granules to the light position and changes in the palisade of endoplasmic reticulum near the rhabdom, they postulated that the microtubules may be a link mediating similar light-induced changes in that animal. There is evidence that both microfilaments and microtubules may be involved in position changes of vertebrate pigment epithelial apical processes and melanin granules with light and dark adaptation (MURRAY and DUBIN, 1975; BURNSIDE, 1976).

g) Vertebrate Pupil

The iris of vertebrate eyes is usually a pigmented movable diaphragm that controls the size and shape of the pupil aperture. It is a thick muscular organelle that is relatively opaque even when unpigmented. The size and shape of the diurnal pupil determine the maximum amount of diffraction-limited information in the retinal image (reviewed in WESTHEIMER, 1972a, b). Because of diffraction, the highest spatial frequency in object space that passes into the eye (for incoherent light, regardless of pupil shape for symmetric pupils) is a function only of the pupil diameter and the wavelength of light

$$\nu = \frac{D}{\lambda} \tag{2}$$

where ν is the cut-off frequency of an infinite grating in cycles per radian, D is the pupil diameter and λ is the wavelength expressed in the same units. The cut-off frequency, ν, is frequently expressed as $D/57.3\,\lambda$ cycles/deg of visual field (57.3 is the conversion factor for radians to degrees).

The pupil adjusts approximately to optimize acuity over a wide range of luminances (DENTON, 1956; CAMPBELL and GREGORY, 1960), contracting to the diurnal diffraction limit in bright light and expanding in dim light to the point where aberrations and the intrinsically poor acuity of the scotopic system rather than the pupil diameter are limiting. Nocturnal vertebrates such as the cat tend to have a small diurnal pupil that has the advantage of protecting the retina from bright light to prepare it for subsequent dark adaptation (WOODHOUSE and CAMPBELL, 1975). The optics of the cat eye with dilated pupil (WÄSSLE, 1971; ROBSON and ENROTH-CUGELL, 1978) and its cone mosaic in the area centralis (STEINBERG et al., 1973) can theoretically pass spatial frequencies of the order of 20 cycles/deg. Nevertheless, ν for the visual system as a whole is 6–9 cycles/deg (BERKLEY and WATKINS, 1971; JACOBSON et al., 1976). WOODHOUSE and CAMPBELL (1975) point out that the maximum acuity of the visual system is adjusted to the diurnal pupil which is a 0.2-mm slit that restricts spatial frequencies in the vertical meridian to about 6 cycles/deg.

When the human pupil is 2.4 mm, the theoretical ν [from Eq. (2)] should be 75 cycles/deg. CAMPBELL and GUBISCH (1966) found the best quality image of the human eye occurs with a 2.4-mm pupil at which ν is only 60 cycles/deg. The theoretical potential maximum image quality of the diurnal human pupil is not realized because of demodulation caused by aberrations and scatter (GUBISCH, 1967).

Table 1

	Diurnal Pupil Diameter	ν (Theoretical assuming $\lambda = 555$ nm)	ν (Measured from line spread function)
Man (CAMPBELL and GUBISCH, 1966)	2.4 mm	76 cycles/deg	60
Eagle (SHLAER, 1972)	6.25 mm	197 cycles/deg	120

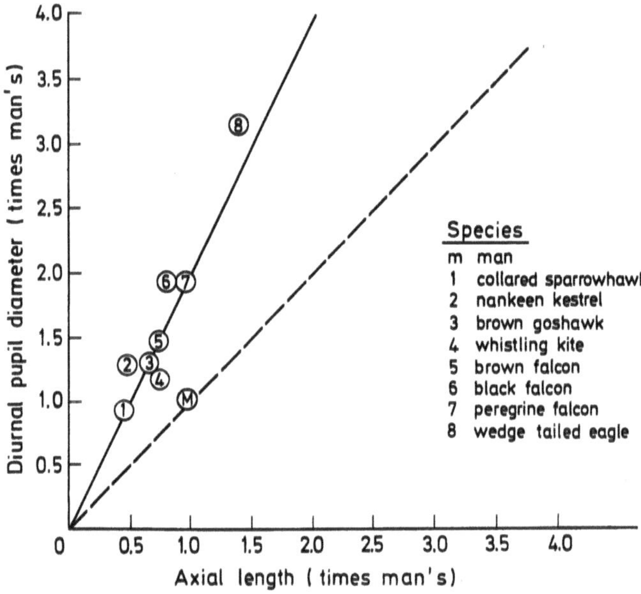

Fig. 6. Axial eye length and diurnal pupil diameter of the eyes of eight birds of prey compared with that of man. Based on the assumption that diurnal pupil diameter determines the spatial cut-off frequency, resolving power is proportional to diurnal pupil diameter. The ratio of the slopes of the *dashed* and *solid curves* implies that the diurnal *f* number of the birds is half that of man. (MILLER and SNYDER, 1979)

Table 1 shows similar data for the eagle indicating that the best quality image in the eagle eye falls short of the theoretical maximum as is the case for man's v. Nevertheless it is interesting that the best quality image in the eagle eye is twice that of man. SNYDER and MILLER (1978) have suggested that v of bird eyes is proportional to the axial length of the eye. In the case of very large bird eyes it may be three or four times that of man's eye. The pupil diameters of eight birds and one human were photographed simultaneously on a cloudless day. Two birds with about the same axial length as man's eye had about twice the pupil diameter. The other bird pupils are shown to fall approximately on the *solid line* of Fig. 6, which has a slope of 2, compared with the *dashed line* passing through the data of man's eye and the origin. An underlying assumption is that the diurnal pupil is the main determinant of image quality in all of these birds, as it has been shown to be in the eagle and man. This assumption is worthy of additional research. A tentative conclusion is that the diurnal pupil determines the maximum amount of diffraction-limited information on the retina. The visual system is assumed to be adjusted to transmit the maximum amount of information passed by the pupil, e.g., 6 cycles/deg for the cat and 120 for SHLAER's (1972) eagle. The significance of the fact that the bird data falls approximately on a line of slope equal to 2 (Fig. 6) is that the "*f*-number" [focal length (F)/pupil diameter (D)] of birds in this study is about half of man's *f* number. The spatial cut-off frequency *in the retinal image plane*, v_r, is

$$v_r = \frac{D}{\lambda F} = \frac{1}{\lambda \cdot f} \tag{3}$$

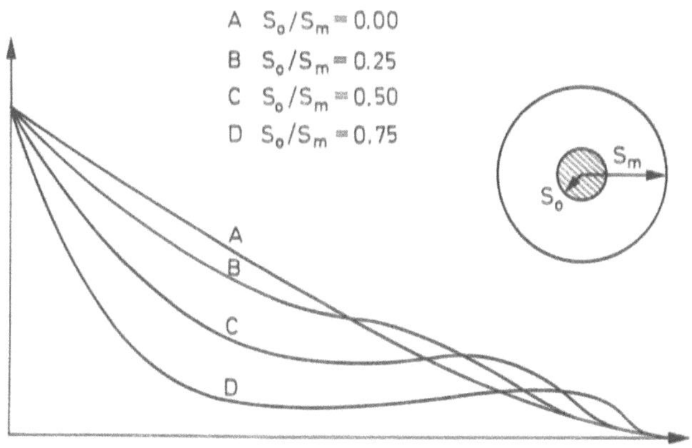

Fig. 7. Modulation transfer function of a circular aperture with no *(A)*, 25% *(B)*, 50% *(C)* and 75% *(D)* central obscuration. *Ordinate* is modulation contrast and *abscissa* is normalized spatial frequency. (From "The International Encyclopedia of Optical, Electro-Optical and Laser Technology", 1976, page E-155)

in cycles per the unit of length in which D, F and λ are expressed. Thus the eyes of a number of birds of prey, independent of axial length, have half the f number of man. The spatial cut-off frequency expressed as cycles per micrometer in the retinal image plane is twice that of man. The problem of how the bird's visual system could recognize *retinal* spatial frequencies that are twice man's is discussed in connection with filtering effects of photoreceptor packing in Section D,1 below.

It is theoretically possible for a nocturnal animal to have a slit pupil that protects the retina and still passes high spatial frequencies using central obscuration (ZUCKERMAN et al., 1979), a technique that is used, for example, in cassegrain telescopes. The catfish *Plecostomus* and the mammal *Procavia* have pupil shapes that maintain the pupil opening with various degrees of central obscuration caused by the expansion of an operculum during light adaptation (WALLS, 1967). When the center of the pupil is obscured, v is the same as without obscuration, (Fig. 7) and is very much higher than the v to be expected from a slit of equivalent area (ZUCKERMAN et al., 1979). Thus, fine detail is maintained in the retinal image by a pupil configuration that has the advantages of a slit pupil for reducing the brightness of the image but with relative enhancement of fine detail. Curves A, B, C and D call attention to the fact that the pupil shape is just as important a determinant of the spatial information for all spatial frequencies as for v. The sucker catfish *Plecostomus* and *Loricaria* of ~10-cm length have a circular pupil with a diameter of ~2 mm that shows no central obscuration in the dark-adapted condition. These animals are night feeders, remaining quiescent during the day. With the onset of illumination, the operculum fills out with a time course of ~3 min to give ~90% central obscuration in bright light (ROTH, 1911). Figure 8 shows the partially expanded operculum of *Plecostomus*.

The question whether the high spatial frequencies are utilized is unresolved The retina is rich in both rods and cones and uses photomechanical responses to switch from one receptor type to the other. In the light-adapted state the cones are densely packed with a center-to-center spacing of about 5 μm in a *Plecostomus* eye

Fig. 8. Central obscuration of the catfish pupil. Living fish photographed in aquarium. Central obscuration of the pupil passes higher spatial frequencies than a slit of equivalent area, but at the expense of frequencies in the middle range. STAVENGA, this volume, has shown that the pseudopupil pattern in insect eyes is "designed" to blend into the body markings for camouflage. Note that in the catfish, the body markings mimic the pupillary pattern, apparently for camouflage. From ZUCKERMAN et al. (1979)

of 7-mm axial length. Because of the small size of the eye, the receptor spacing would have to be about 1 μm to detect the cut-off frequency transmitted by a 1.5-mm diameter centrally obscured pupil. However, in a large adult catfish with an eye axial length of 20 mm it is conceivable that the information could be used (ZUCKERMAN et al., 1979).

The large size of the vertebrate and cephalopod camera eyes appears to be advantageous for absolute scotopic sensitivity under scotopic conditions because light gathering for the retinal image is proportional to pupil area. Nevertheless, as KIRSCHFELD (1974) has shown, this is true only when the object is a point source (stars, for example). When the object is an extended source, the quantal flux at the receptor is proportional to the f number (ratio of focal length to pupil diameter). Nocturnal vertebrate eyes and insect apposition eyes have about the same f number and hence the numbers of quanta reaching the rods and rhabdoms are about the same for extended sources. But KIRSCHFELD (1974) points out that in the optical superposition compound eye the area of the pupil is about 25 times that of the apposition eye (25 facets), and at the same time the capture area of the larger rhabdom of these eyes is at least 36 times greater. Thus for extended sources, the numbers of quanta reaching the receptors are roughly 1000 times that of the compound apposition and nocturnal vertebrate eyes.

3. Yellow Absorption Filters

WALLS and JUDD (1933) (summarized in WALLS, 1967) proposed that yellow corneas, lenses, colored oil droplets, and macular pigment of diurnal vertebrate eyes function as filters. They increase visual acuity by reducing chromatic aberration and glare, by absorbing scattered blue sky light (blue haze) and by increasing color contrast. Modern evidence on this subject has been thoroughly reviewed by MUNTZ (1972) and WOLBARSHT (1976). The main points are summarized:

Dorsally located yellow corneal pigment in fish acts as an eye-shade against the glare source of downwelling light (MORELAND and LYTHGOE, 1968, and Sec. C,II,1 below).

Fish lenses had been believed to be achromatic (PUMPHREY, 1961 a); however SIVAK and BOBIER (1978) have demonstrated a significant chromatic aberration (5–11 diopters) and have shown that yellow coloration in fish corneas functions to reduce chromatic aberration in the way that yellow lenses of diurnal terrestrial animals such as squirrels and snakes probably do. MUNTZ (1972) points out that reduction of blue haze only improves visibility of very distant objects on clear days. This would be most useful for birds that do not have yellow lenses but do have yellow oil droplets.

According to WALLS and JUDD (1933), oil droplets may have first developed in rods in order to concentrate illumination on the outer segment. BAYLOR and FETTIPLACE (1975) confirmed this function for turtle-cone oil droplets. Some fish, amphibians, reptiles, birds, and marsupial mammals have oil droplets in their cones. The system is most highly developed in the colored droplets of turtles and birds. In the retina of the turtle *Pseudemys* cones have either colorless, yellow, orange, or red oil droplets. LIEBMAN (1972) found three cone visual pigments in the retina with peak absorption at 450 nm, 518 nm and 620 nm. He calculated the expected shift in λ_{max} that could be attributed to the oil droplets, which are cut-off filters that decrease the transmitted intensities of the short wavelengths. The 620 nm pigment would have its maximum shifted to 640 nm by the red oil droplet, the 518 pigment would be shifted to 560 nm by the orange oil droplet, and the 450 pigment's peak would not be altered by the colorless droplet. He further found that each pigment may be in receptors containing different oil droplets and he postulated that the turtles may have a five-receptor color-vision system in addition to the other advantages for acuity mentioned above. This idea has been extended by BOWMAKER (1977) and BOWMAKER and KNOWLES (1977) to the chicken and pigeon retinas. BOWMAKER has found five types of oil droplet combined with three types of cone visual pigment in the pigeon to give at least six cone receptors with different peak absorptions and he suggests that the combination of oil droplets and visual pigments produce narrow band sensitivity channels similar to those produced by neural mechanisms in fish and mammals. All the oil droplets cut off blue light to help prevent chromatic aberration. In addition, the red oil droplets are missing in the deep fovea of a number of birds with high visual acuity. However, in the visual streak and area centralis of *Pseudemys*, all four types of oil droplets are present. BAYLOR and HODGKIN (1973) found only three cone receptor types, but because their sample was small they did not rule out the existence of undetected subtypes. The peak sensitivity and response of shorter wavelengths were found to be decreased, as would be expected from the effects of the oil droplets. Yellow carotenoid-like photostable pigments resembling those of the oil droplets and macular pigments of vertebrate eyes have been found in the ellipsosomes of certain fish cones (MAC NICHOL et al., 1978) as well as the invertebrate *Limulus* photoreceptors (BENOLKEN, 1976), *Balanus* photoreceptors

(MINKE and KIRSCHFELD, 1976) and in rhabdomeres number 7 of dipteran compound eyes (KIRSCHFELD et al., 1978). The function of these invertebrate pigments is not yet understood. But it has been shown in the last example that this pigment in rhabdomere 7 causes a shift in the spectral sensitivity of rhabdomere 8 to longer wavelengths. Another (possibly carotenoid) photostable pigment in dipteran rhabdomeres 1–6 absorbs strongly in the UV and has been shown to function as an "antenna" pigment to transfer this absorbed UV energy to the rhabdomere's visual pigment (KIRSCHFELD et al., 1977).

The vertebrate oil droplets act as cut-off filters only down to about 400 nm. For wavelengths shorter than 400 nm, the vertebrate cornea and lens cut off the near UV, except in birds whose ocular media are transparent down to 340 nm (GOVARDOVSKII and ZUEVA, 1977) permitting sensitivity for these animals and possibly some fish, amphibians and reptiles in the near UV (reviewed in KREITHEN and EISNER, 1978). The cut-off function of the cornea and lens of most vertebrates other than birds and of oil droplets in birds, except for the clear ones, may help to prevent confusion that might have resulted from the absorption by the cis-peaks of visual pigments (WOLBARSHT, 1976). The mechanism of the cut-off filtering effect of corneas and lenses (shown in Fig. 1) will be discussed in the next section (B II).

II. Scattering

1. Spherical Scatters

Particles of a different n from their surround, even if much smaller than the wavelengths of light, would be expected to decrease the intensity of transmitted light by scattering. The effect of electromagnetic energy incident on a scattering center is to create a dipole that oscillates synchronously with the incident field, causing the radiation of scattered light. Even though the particles are very small, they will create an observable effect if their number density is high enough or the optical path length long enough. RAYLEIGH (1911) showed that the blue color of the sky results from scattering of sunlight by molecules of air. His classic analysis of scattering is also applicable to the ocular media. Spherical scatterers, small with respect to a wavelength, have a characteristic wavelength dependence. RAYLEIGH (1871) showed, by considering the dimensions alone, that the intensity of scatter is proportional to the inverse fourth power of the wavelength. He assumed that the amplitude of scatter is directly proportional to its relative n, which is dimensionless, and to the volume of the scatters, while it is inversely proportional to both the distance of the observer to the scatterers and the wavelength. It is evident that the wavelength dependence must be inverse squared if the scattered intensity is not to have the dimension of length. Since intensity is amplitude squared, the scattered intensity is porportional to the inverse fourth power of wavelength, giving a relationship for scattering which is similar to Eq. (1), the law for absorption

$$\log (I_s/I_t) = \tau V/d\,\lambda^4 \tag{4}$$

where I_s is the intensity of scattered light, τ is the turbidity coefficient, V is volume in cm^3, d is distance in centimeters, and λ is wavelength in centimeters. Rayleigh

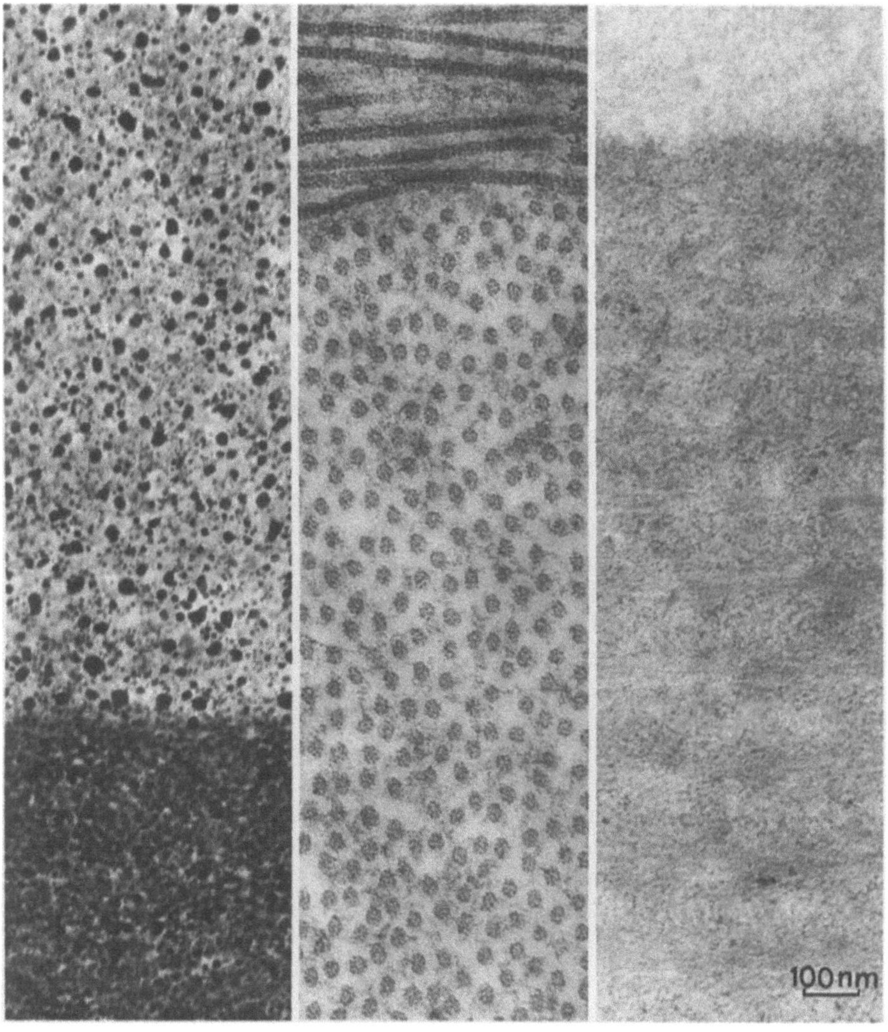

Fig. 9. Illustrating possible sources of scattered radiation. (A) arthropod crystalline cone, cortex at lower part of figure; (B) gull cornea; (C) arthropod cornea. Identical magnification for all three pictures

scattering has not been investigated for the ocular media. Nevertheless there are structures for which the theory could be applied. For example the core of the crystalline cones of many arthropods (FISCHER and HORSTMANN, 1971) contains densely packed spherical granules that are small with respect to the wavelengths of light (Fig. 9A).

2. Cylindrical Scatterers

The vertebrate cornea provides an example of the effects of scattering by cylindrical scatterers in the ocular media. It consists of two layers of cells, the epi- and endothelia and a central stroma. The cornea is 100–800 μm thick depending on the species. The stroma makes up more than 90% of the corneal thickness and is composed of lamellae several micrometers thick, each of which consists of fibrils of collagen packed in a ground substance of mucopolysaccharide. Parts of two lamellae of the herring gull cornea are shown in Fig. 9B. The collagen fibril diameter is about 30 nm and the center-to-center spacing about 50 nm. These figures vary somewhat from species to species and for different parts of the stroma. The long axes of the fibrils lie in planes roughly parallel with the corneal surface. Compare the vertebrate fibrils with those of an arthropod cornea at the same magnification in Fig. 9C.

The vertebrate cornea has been the subject of numerous studies of transparency (reviewed by MAURICE, 1962). MAURICE (1957) assigned a value of $n=1.47$ for the collagen fibrils and $n=1.345$ for the ground substance, based on measurements of the volume fractions of the stroma components and on birefringence studies. The effect of the phase relations and directions of scatter from *every* fibril must be taken into account in order to estimate the intensity of scattered light. MAURICE (1957) calculated that if there were no correlation in phase, i.e., if the fibrils acted as independent scatterers, the cornea would be opaque. For wavelengths near the middle of the spectrum and for a cornea the thickness of that of the human, about 90% of the incident illumination would be scattered.

Actual measurements of corneal transparency indicate that the cornea is very transparent. The *X*s on Figure 1 are data from BOETTNER and WOLTER (1962) of the optical density of human cornea. They made measurements of transmittance, accepting light within only one degree of the direct beam in order to exclude light that was scattered in a forward direction. The increased absorbance in the near UV is probably caused by aromatic amino acids, as is the case for arthropod cuticle. FREEMAN et al. (1968) found that corneal collagen contains tyrosine and phenylalanine. Note that even though the vertebrate cornea is more than an order of magnitude thicker than that of the arthropod, its absorbance in the visible spectrum is about the same. Since the vertebrate cornea is transparent, it is safe to conclude that the collagen fibrils do *not* act as independent scatterers.

a) Theory

Crystalline materials most closely approach the condition of a uniform refractive index necessary to minimize scattering. This explains the virtual absence of scattering in mica, for example, compared with that of glass (see discussion in WOOD, 1967). MAURICE (1957) compared the arrangement of the stromal fibrils to crystalline lattice structure in order to explain transparency. He showed that each row of fibrils with long axes lying perpendicular to normal incidence illumination of the cornea acts as a diffraction grating, and he argued that therefore the whole array of fibrils in the lamellae should act as diffraction gratings to scatter light at normal incidence where the wavelength is less than the interfibrillar spacing. The grating equation is

$$\sin\theta = m\,\lambda/d \tag{5}$$

where θ is the diffraction angle for light of wavelength λ incident perpendicularly on the grating, d is the spacing of the grating lines (or fibrils of the stroma), and m

is an integer. For $m \geqq 1$ the grating equation cannot be satisfied if $d \leqq \lambda$. In other words, when the spacing of the grating lines is smaller than the wavelength of the incident light, there is no diffracted (scattered) light since $\sin \theta > 1$. Only the zero order comprising the central image of the incident beam is transmitted. The scattered light can be considered to be nulled by destructive interference so that all of the energy appears in the central image.

Although the corneal stroma may be likened to a stack of gratings, the more accurate model was shown by BENEDEK (1971) to be a three-dimensional one such as used to describe scattering of x-rays by atomic or molecular planes in crystal lattices. The well-known Bragg relation for scattering by a three-dimensional lattice is

$$\text{Sin}\,\theta = m\lambda/2\,d \qquad (6)$$

The importance of the three-dimensional model, as BENEDEK (1971) pointed out, is that it predicts that only variations in the refractive index of stroma in which the spatial frequency components are longer than half a wavelength of perpendicularly incident light will result in scatter. Strictly speaking, the wavelength must be the wavelength in the medium so that λ is λ/n. In analogy to the grating equation (5), when twice the spacing of the stromal collagen fibrils is less than λ/n, the sine of the angle of scattered light is greater than 1 and only the central image is transmitted. The condition for transparency is that there should be no variations in refractive index over distances of a half a wavelength or greater of light in the medium.

BENEDEK stressed that fibrils of the corneal stroma do not have to be oriented regularly as lines in a grating because the important variable is simply the refractive index *(n)*. Interestingly, Bowman's layer of the corneal stroma is literally a felt-work showing no apparent orientation of the fibrils as shown in Figure 10 from GOLDMAN and BENEDEK (1967). Nevertheless, their analysis of densitometric tracings of slit-lamp photographs indicates that Bowman's layer scatters less than the other parts of the stroma.

It will be recalled that the intensity of scattering by Rayleigh (spherical) scatterers shows an inverse fourth-power dependence on wavelength. Because the collagen fibrils are cylindrical, their scattering properties are different from spheres (RAYLEIGH, 1881). It can be shown from dimensional considerations that the intensity of light scattered from infinite cylinders of small circular cross section is proportional to the fourth power of cylinder radius and the inverse *third* power of wavelength (Summary in KERKER, 1969). FARRELL et al. (1973) measured the scatter at 120° from the rabbit cornea and observed the predicted inverse cubic power dependence on wavelength. Their measurements are from 700 to 375 nm. At shorter wavelengths, transparency decreases more rapidly because of the aromatic amino acids (Fig. 1 and discussion in previous section).

Additional insight into the relation between scattering and transparency has been gained from investigation of loss of transparency in corneas that have become swollen by imbibition of fluid. GOLDMAN et al. (1968) argue that the loss of transparency is caused by areas larger than half a wavelength without fibrils which they call "lakes." FARRELL et al. (1973) observe that the scatter from swollen

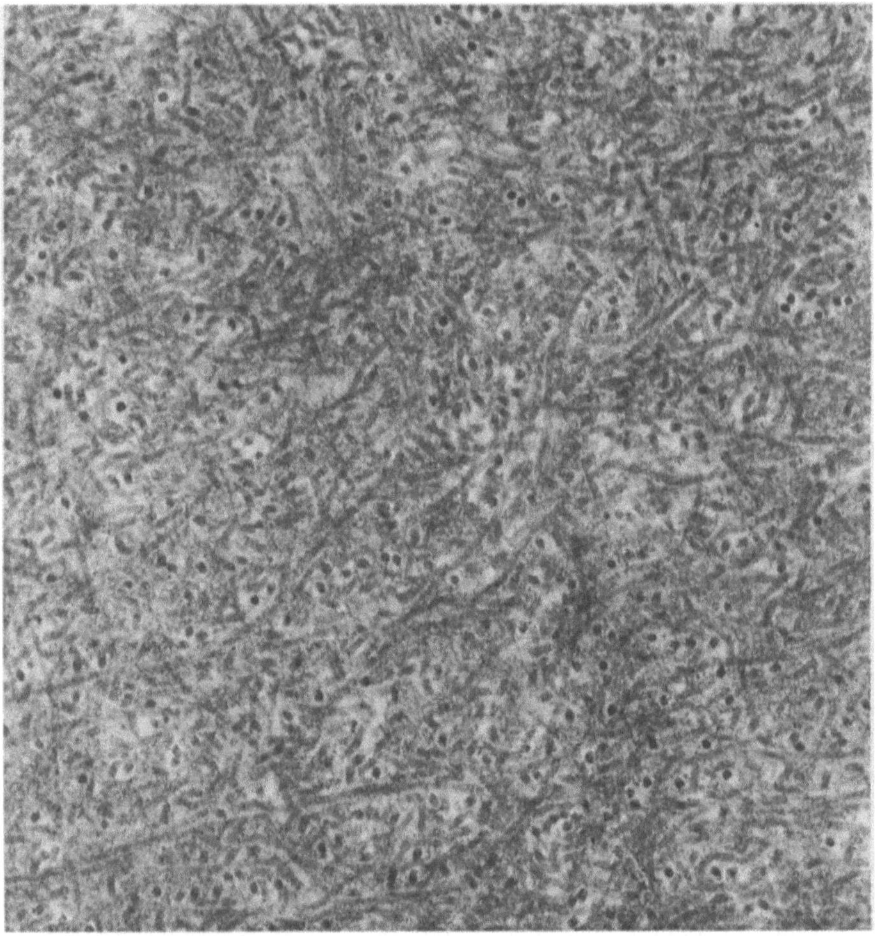

Fig. 10. Feltwork of fibrils in Bowman's layer of cornea showing a lack of regular orientations. Fibril diameters about 300 Å. (GOLDMAN and BENEDEK, 1967)

corneas has a dependence on the inverse square of wavelength. This, they argue, could result from scatter from the lakes. HART and FARRELL (1969) measured the average number of fibrils per unit area in electron micrographs of the fibrils in cross section. They constructed a radial distribution function for the fibrils of rabbit stroma showing that the nearest neighbors are correlated to be at a distance of 60 nm and that there is a short-range ordering going out to about 150 nm. It is not known what relation these findings on fixed embedded and sectioned material have to the fibril distribution in the living state (LYTHGOE, 1976). Even disregarding this, controversy among those making electromagnetic models about the exact effect of the short-range ordering on scattering, and about concepts of the relation of the models to details of the fine structure in the living state still predominate (cf. FARRELL and MCCALLY, 1975; TWERSKY, 1975; BERGER, 1976).

Table 2. Optical density in log units of ocular media at 500 nm

		O.D.
LUDVIGH and MCCARTHY, 1938	Human	0.31
KINSEY, 1948	Rabbit	0.08
WIESINGER et al., 1956	Rabbit	0.03
PITTS, 1959	Cow	0.10
GEERAETS et al., 1960	Rabbit	0.05
GEERAETS et al., 1960	Human	0.09
BOETTNER and WOLTER, 1962	Human—(including forward scatter)	0.12
BOETTNER and WOLTER, 1962	Human—(excluding forward scatter)	0.43

b) Scattering and Visual Function

Transparency of the ocular media is important for the absolute sensitivity function of scotopic receptors as well as for photopic receptors because maximal intensity at the receptor and lack of scatter (glare) are important for contrast detection and visual acuity. An estimate of the optical density of the ocular media caused by absorption and scatter is therefore important for understanding photoreceptor function. HECHT et al. (1942) estimated that at the absolute threshold of human vision a range of between 54 and 248 quanta are incident on the cornea and between 5 and 14 quanta are absorbed by a group of about 500 retinal rods. They used the data of LUDVIGH and MCCARTHY (1938) showing that for blue-green light the absorbance of the ocular media is about 0.3 log units. This literature has recently been reviewed by NORREN and VOS (1974). It is evident that there is a large range of estimates of the absorbance of the ocular media even allowing for the difference in path length of various species (typical values summarized in Table 2).

The explanation for these variations appears to lie in the methodology of the measurements. Note the last two data of Table 2 from BOETTNER and WOLTER (1962). They carefully defined the degree of scatter included in their transmission measurements. The 0.43 log unit optical density for human ocular media was made by excluding all but the light within one degree of the direct beam; the 0.12 figure included all the radiation emerging over a cone of 170°. In future estimates similar to those of HECHT et al. (1942), the contribution of scatter to the rod response should be taken into account.

As mentioned above (Fig. 1), arthropod dioptrics are more transparent than vertebrate dioptrics principally because of their shorter path length, but are substantially less transparent on the basis of optical density per unit thickness of corneas (cf. Fig. 1). The reasons for these differences in transparency have not been analyzed. Nevertheless, the increased transparency of the arthropod dioptrics over the vertebrate combined with the longer rhabdom length gives the arthropod photoreceptor slightly higher quantum efficiencies than those of the vertebrate. Between five and ten quanta incident on the cornea of an ommatidium of the fly *Musca* are sufficient to generate an electrophysiologically detectable response (KIRSCHFELD, 1966), whereas at the threshold of the optomotor response ~ 30 quanta per ommatidium are incident on the cornea.

III. Summary

Transparency of the ocular media is a function of the uniformity of its refractive index and the consequent presence or absence of absorption and scattering centers. Absorption of light in the near UV is principally caused by aromatic amino acids of ocular proteins. Structures in the ocular media with spatial frequency components larger than $\lambda/2n$ contribute to loss of transparency by scattering. The ocular media of arthropods are more transparent than that of vertebrates because of their shorter path length. Sensitivity of the vertebrate and large invertebrate camera eyes for point sources is greater because of their greater light capturing ability, but considerably less than superposition compound eyes for extended sources.

Screening pigments have a pupillary function. In superposition arthropod eyes migrating iris pigment controls sensitivity with apparently little effect on resolving power, which appears to be limited by facet diameter in both the light- and dark-adapted eyes. Migrating pigment in the retinular cells of the apposition eye decreases sensitivity and increases acuity by approaching close to the rhabdom, where it both absorbs light and narrows the rhabdom-acceptance angle. The vertebrate pupil has a protective function for nocturnal animals and the diurnal pupil of nocturnal and diurnal vertebrates determines the amount of diffraction-limited information available to the visual system. In the case of central obscuration, the pupil can subserve the protective function of the slit and still pass high spatial frequencies. Yellow filters in the vertebrate ocular media, the lens, macular pigment, and oil droplets enhance acuity principally by absorbing short wavelengths, thus decreasing the effects of chromatic aberration. For cones with visual pigments that absorb in the red and have minor peak absorption also in the visible spectrum, the oil droplet sharpens the color response. In addition, colored oil droplets appear to offer enhanced color vision by locating the same visual pigment in receptors with different colored oil droplets and thus increasing the number of primary wavelength detectors. In birds where the ocular media are transparent to near ultraviolet, the oil droplets, except for the clear ones, probably also prevent wavelength confusion resulting from absorption of light by the cis-peaks of visual pigments. Photostable pigments resembling the yellow pigments of vertebrate ocular media are also found in some invertebrate rhabdomeres. These include an "antenna" pigment that absorbs energy in the UV and transfers it to a visual pigment and a yellow pigment located in one rhabdomere that acts as a filter to shift the peak response of another rhabdomere to longer wavelengths.

C. Interference Filters

Refracting surfaces are necessary for image formation and are therefore a feature of eyes either of the camera or compound type. The refracting surfaces also give rise to small reflections. Such reflections originate wherever there are sharp transitions in refractive index. The Purkinje images, for example, are reflections from the anterior and posterior surfaces of the cornea and lens in the vertebrate eye. The amplitude of reflection for normal incidence is independent of the polarization of the incident light and is given by the Fresnel formula:

$$r = \frac{n_1 - n_2}{n_1 + n_2} \tag{7}$$

where r is the amplitude of reflection, n_1 is the n of the rare medium and n_2 is that of the dense medium. When $n_2 = 1.5$ and $n_1 = 1.0$, $r = -0.2$. The minus sign signifies a phase shift of π radians that happens to occur at rare to dense reflections. The intensity of reflection is amplitude squared, 0.04 or 4%. The vertebrate cornea's $n \simeq 1.38$ (MAURICE, 1962); its anterior surface is usually covered with a tear film or water of $n = 1.33$. The arthropod cornea's $n \simeq 1.5$ [e.g., VARELA and WIITANEN (1970) find $n = 1.49$ for the honeybee, *Apis*, cornea]. Therefore, 4% is an upper limit for the intensity of reflection from ocular refractive surfaces at normal incidence. A similar order of reflection would be expected from the front and rear surfaces of camera, microscope or other optical lenses. However, the surfaces of better quality lenses today are routinely coated with thin layers designed to reduce this reflection. An adaptation of certain arthropod eye corneas serves the same purpose.

I. Reflection Suppression

1. Corneal Nipple Array Structure

Reflections at refracting surfaces degrade the retinal image by causing glare and ghosts as well as loss of intensity of the image. These effects are small, but the reflections are easily detected and measured in the vertebrate eye as noted above. On the other hand the corneal cuticle of the eyes of certain species of insects is relatively nonreflective. Figure 11 (from BERNHARD et al., 1965) shows living bee and night moth eye photographed side by side in their reflected light. The eyes are in their light-adapted condition in order to minimize the tapetal glow of the moth eye so that the front surface reflections may be compared. There is a bright specular reflection from the bee eye; however the moth eye is relatively nonreflective.

The front surface of arthropod corneas may be observed at high magnification using incident illumination and a metallographic compound microscope. Isolated corneas may be prepared to minimize reflection from the corneal back surface by floating them on india ink, which is then allowed to dry. A grasshopper *(Schistocerca)* cornea (Fig. 12A), a hornworm moth *(Manduca)* cornea (Fig. 12B), and a fly *(Musca)* cornea (Fig. 12C) are photographed by white incident light using this technique. The reflection from the grasshopper and fly corneas is white and corresponds in intensity to the reflection from Cargille index of refraction fluid $n = 1.500$ floating on carbon black (Fig. 12D). This intensity is the same as that from a plane glass surface $n = 1.5$, 4% of the incident illumination. The reflection intensity from the moth cornea is 0.7 log units dimmer and is tinted blue.

Fig. 11. Reflection from corneal front surface of bee (left) and light-adapted moth (right) eyes. (BERNHARD et al., 1965)

The surface structure of the moth cornea that accounts for the decreased reflection consists of an array of cuticular protuberances termed corneal nipples (BERNHARD and MILLER, 1962; BERNHARD et al., 1965; MILLER et al., 1966; BERNHARD et al., 1970). Electron micrographs (EMs) of corneal surfaces or of replicas of these surfaces combined with EMs of sectioned material are needed to visualize this surface structure, the components of which are smaller than the wavelengths of light and therefore not visible in the light microscope. Figure 13 (from MILLER et al., 1966) is an EM of a replica of a grashopper *(Melanoplus)* cornea that reflects light as a smooth glass surface. Figure 14 is a scanning EM of the surface of sulfur butterfly *(Phoebis)* facet that shows a reduced front surface reflection. The corneal nipples are evident. They are shown in greater detail in EM's of the moth, *Prodenia*, cornea that is sectioned normal of the surface in Figure 15A and in a plane parallel to the surface of the cornea in Figure 15B. These pictures (from BERNHARD and MILLER, 1962) of facet surfaces and of sections of the surfaces show that the corneal nipples are congruent conical protuberances of corneal material.

BERNHARD et al. (1970) classified the corneal nipples by height; those 50 nm or less were called group I, those between 50 and 200 nm, group II, and those 200 nm or greater, group III.

Their findings, summarized in Table 3, indicate that the highest nipples tend to occur in insects of the anagenetically most advanced orders (but see PAULUS, 1972). They discuss the significance of this trend but conclude that it is difficult to decide whether there has been a progressive appearance of corneal nipples during

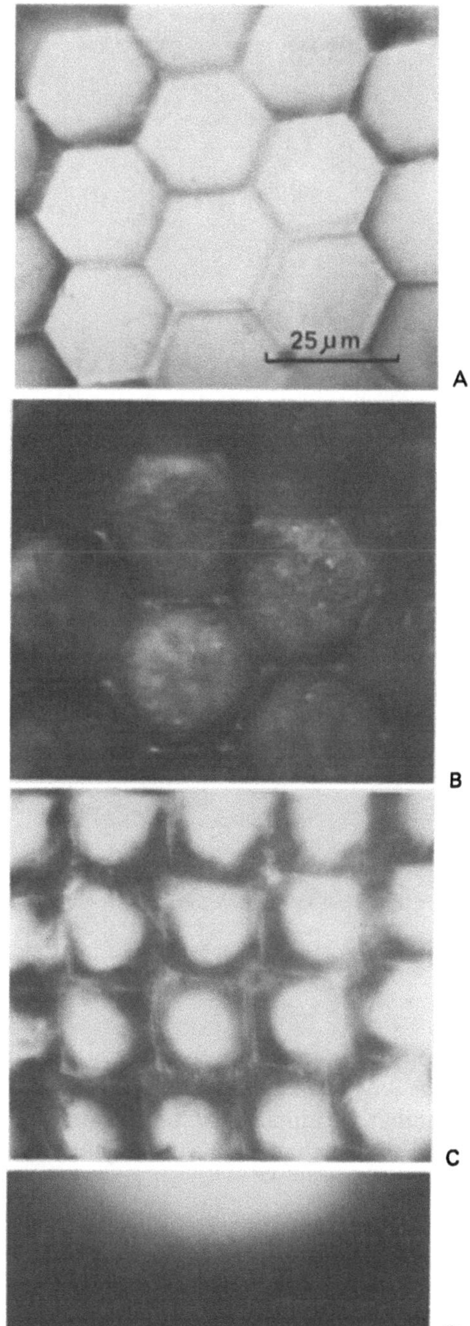

Fig. 12. Reflected light from surfaces of grasshopper (A), moth (B) and fly (C) corneas in comparison with a small drop of fluid of $n = 1.5$. Curvature of the four surfaces is not identical and the comparison simply serves to illustrate the greatly decreased reflection from the front surface of the moth cornea in comparison with the others

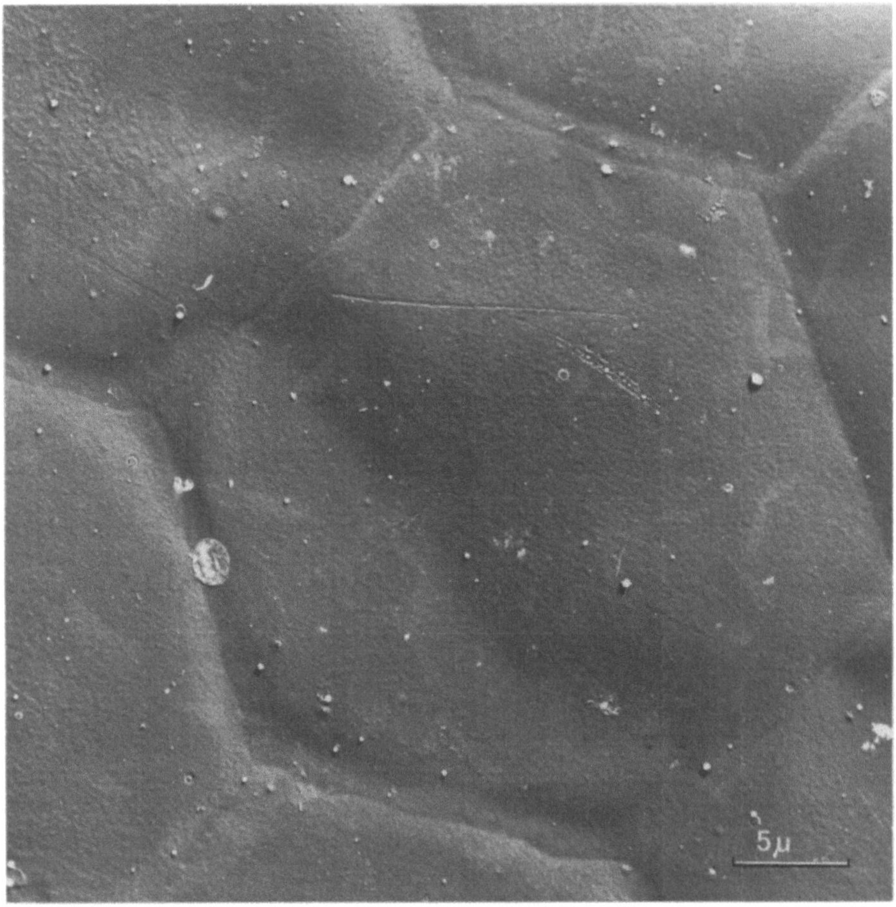

Fig. 13. Transmission EM of replica of a grasshopper corneal front surface. (MILLER et al., 1966)

phylogenesis or whether the finding of lower height nipples in lower orders points to a regression from the fully developed structures in a common ancestor. Alternatively, they point out that the nipples may represent different degrees of expression of an underlying character that occurred many times during insect phylogenesis.

GEMNE (1971) reported that hexagonally distributed microvilli arise from the corneagenous cell surface during ontogeny. He suggests these and other homologous cuticular elements may be the basis for the morphogenesis of cuticular sculpturing on the cornea or elsewhere on the body cuticle. Similar microvilli of epidermal cells of the larva of *Calpodes* generate the extracorneal cuticle in that animal (LOCKE, 1966) and microvilli of the cornea secreting cells are a common feature of compound eyes with smooth surfaces and group I protuberances. The mechanism by which these microvilli determine the nature of the cuticular sculpturing is unknown.

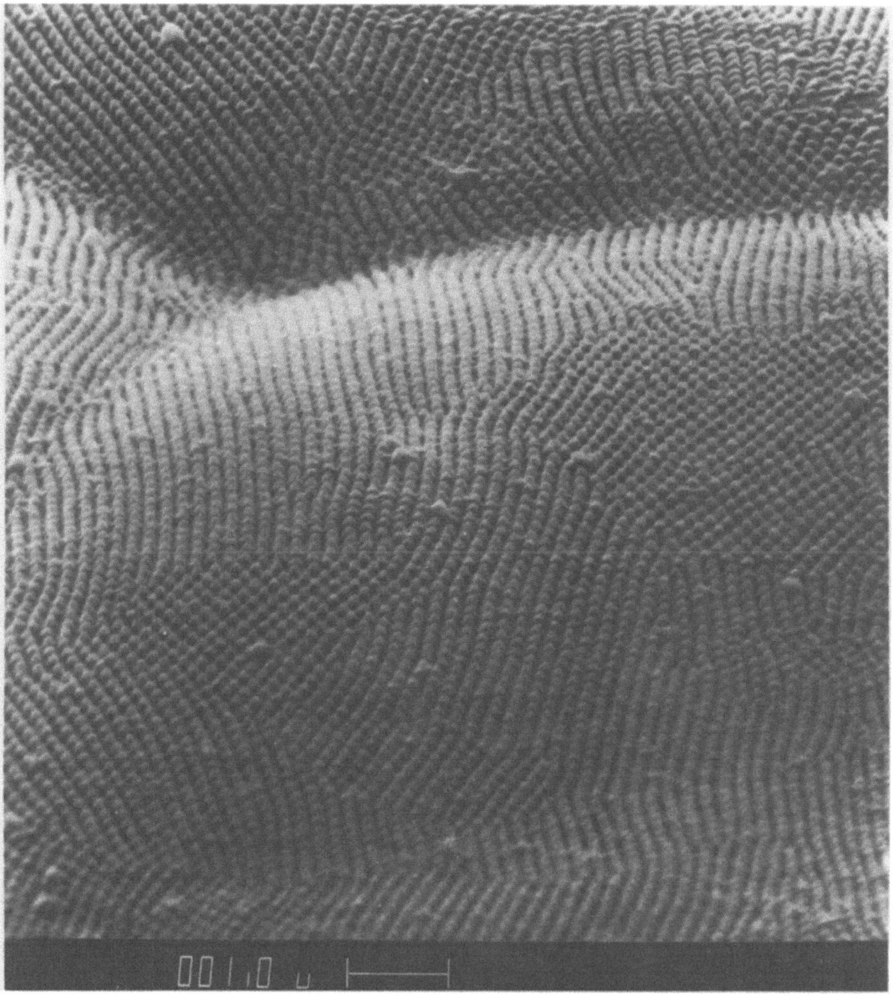

Fig. 14. Scanning EM of the front surface of a sulfur butterfly cornea

BERNHARD et al. (1970) found considerable variation from animal to animal within species and they note a tendency towards irregularity of the nipple pattern in Group I species. PAULUS (1972) found the corneas of different species of Collembola range from smooth to 150 nm-high nipples. The nipples in the facet centers of *Tomocerus* are 80–120 nm high whereas those on the facet sides are 100–150 nm. In *Sminthurides* the nipples are conical with hexagonal rather than circular cross sections and the tips are not pointed but have minute bulb-shaped swellings.

a) Function

BERNHARD et al. (1965) investigated the function of the insect corneal nipple array by measuring the radiation intensity of reflected and transmitted micro-

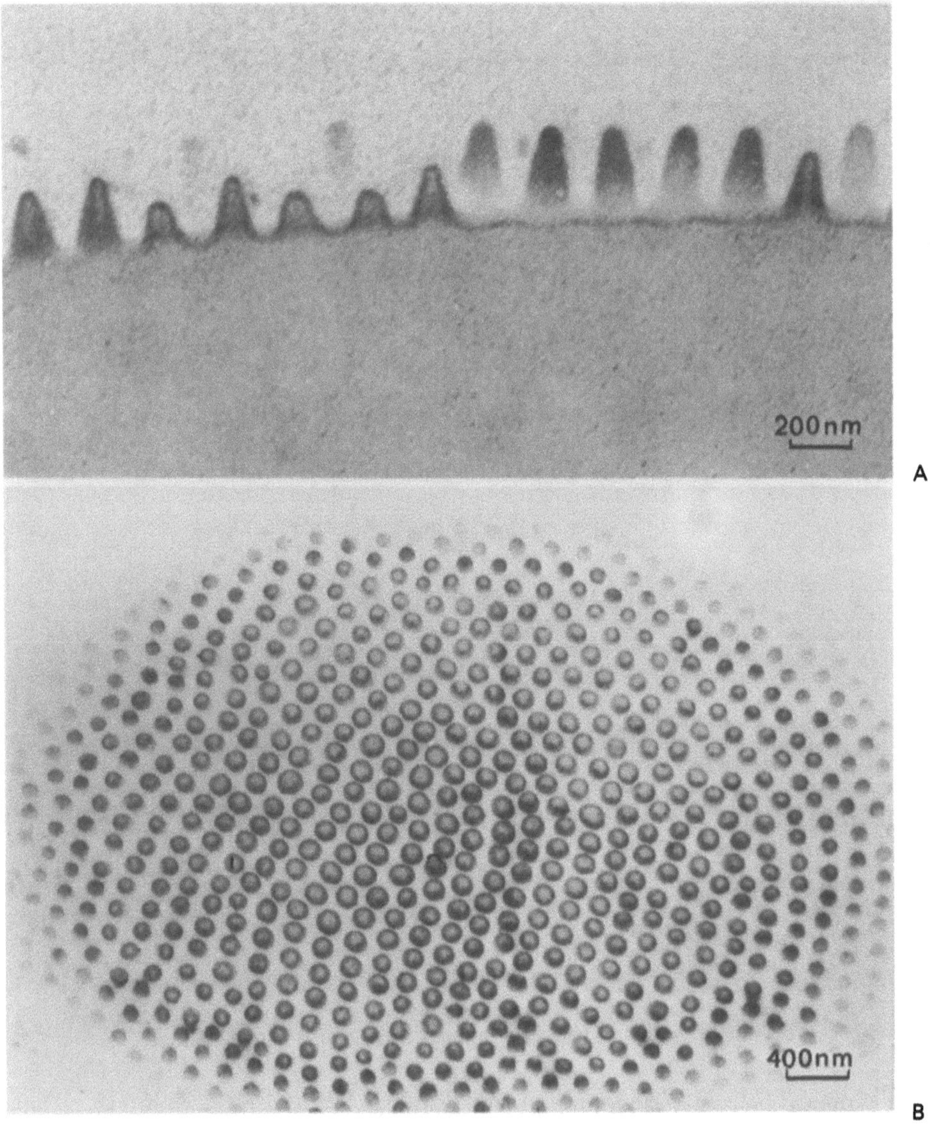

Fig. 15. Transmission EMs of sections normal to (A) and tangent to (B) the surface of a moth cornea. (Bernhard and Miller, 1962)

waves using scaled dielectric models of the cornea, by comparing the intensity of light reflected from corneas that have nipples with those that do not and by calculating the reflection coefficient of a simulated nippled surface. These investigations established that the Group II and III nipple arrays function as thin-film antireflection coatings on the corneal surfaces.

The reflectance properties of a thin film at the corneal surface for normal incidence illumination are summarized in Figure 16 (see Born and Wolf, 1965 or

Table 3. (After BERNHARD et al., 1970)

	Groups		
	I	II	III
Entognatha			
Collembola (springtails)	2	–	–
Collembola (Data of PAULUS, 1972)	1	5	
Ectognatha			
Archaeognatha, Zygentoma (Thysanura; silverfish)	–	4	–
Ephemeroptera (mayflies)	1	–	–
Odonata (dragonflies, damselflies)	2	–	–
Phasmida (walkingsticks)	1	–	–
Saltatoria (grasshoppers)	6	–	–
Dermaptera (earwings)	1	–	–
Blattodea (cockroaches)	4	–	–
Mantodea (mantids)	2	–	–
Plecoptera (stoneflies)	1	–	–
Psocoptera (booklice)	1	–	–
Thysanoptera (thrips)	1	–	–
Homoptera (cicadas, leafhoppers, scale insects, etc.)	2	–	–
Heteroptera (bugs)	6	–	–
Neuroptera (dobsonflies, alderflies, snakeflies, lacewings, mantispids)	10	–	–
Coleoptera (beetles, weevils)	11	–	–
Hymenoptera (sawflies, ants, bees, wasps)	9	–	–
Mecopteroida			
Mecoptera (scorpionflies)	4	–	–
Diptera (flies)	27	7	–
Amphiesmenoptera			
Trichoptera (caddisflies)	5	5	5
Lepidoptera (moths, butterflies)	62	56	126
Total number of species: 367	159	77	131

BREKHOVSKIKH, 1960 for comprehensive treatments). Assume the cornea, $n_3 = 1.5$, is covered with a thin film, n_3, of thickness d, as shown in Fig. 16A. If the n of the film is the same as either air or the cornea, the intensity of reflection would be 4% as given by Eq. (7) and this would be depicted as a *straight line* parallel with the *abscissa* and intersecting the *ordinate* on Figure 16B at 4. If $n_2 = 1.2$ (the geometric mean of n_1 and n_3), reflection intensity is dependent on the optical path length ($n_2 d$) (*dashes*, Fig. 16B). When the optical path length is one quarter and odd multiples of a quarter the wavelength in the medium λ_0 (λ/n_2), the reflection is zero. The peak reflection of 4% occurs at even multiples of $\lambda_0/2$.

The minima result from suppression of reflection by destructive interference. The reason for the destructive interference is that (as mentioned above) the incident light wave going from the rare medium, n_1, to the denser medium, n_2, undergoes a phase shift of π radians or half a cycle on reflection from the interface. The intensity of reflection from the Fresnel formula is 1% for this example. The refracted wave passes the boundary without any change in phase and traverses the medium of n_2. At the n_2, n_3 boundary 1% is reflected, again with a phase change of π radians. The path length of the wave in the medium n_2 is π radians. The difference in phase between the two reflected waves is π radians or one-half cycle. The two reflected waves half a cycle out of phase and of the same intensity thus exactly cancel. For other values of n_2 between 1.0 and 1.5 the intensities of the two reflections will

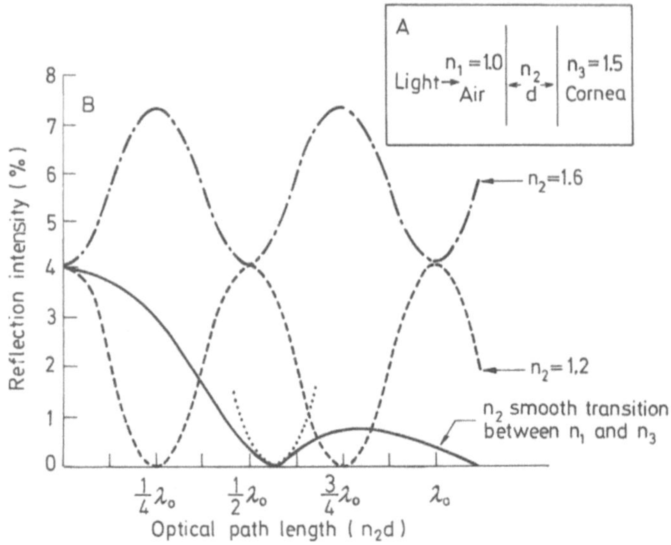

Fig. 16. Reflectance properties of a thin film

be unequal and the minima at $\lambda_0/4$ and odd multiples of $\lambda_0/4$ will accordingly not reach zero. For values of $n_2 > 1.5$ constructive interference occurs because the second reflection is at a dense to rare boundary that undergoes *no* phase change. The reflection is always greater than 4% and maxima occur at $\lambda_0/4$ and odd multiples of $\lambda_0/4$ (*dot–dash* Curve, Fig. 16B).

The corneal nipple array is different from films of uniform *n*. The average *n* increases smoothly from that of air at the tips to that of the cornea at the bases. The thickness of this layer is about 200 nm or $\lambda_0/2$ for light near the middle of the spectrum. It has long been known that if n_2 is a gradual transition between n_1 and n_3 with a thickness $d \simeq \lambda_0/2$, a more broadband antireflection effect is achieved than the quarter wave plate described above (e.g., EPSTEIN, 1930; BREKHOVSKIKH, 1960; MONACO, 1961). The *solid curve* (Fig. 16B) shows the reflection coefficient of a smooth transition by comparison with the quarter wave plate discussed above and may be directly compared with the quarter wave plate *(dots)* that has been shifted so that both minima coincide.

Because the smooth transition layer on insect corneas is composed of the discrete nipples, the effects of increased scatter at the shorter wavelengths must be considered. As mentioned in Section B, II, above, variations in *n* over distances of a half a wavelength or greater of light in the medium give rise to scattered light for the three-dimensional case. However, because of the thinness of the nipple layer, the two-dimensional model that resembles a diffraction grating is more appropriate. Therefore, variations in *n* over distances of a wavelength or greater will give rise to scattered light. The longest periodicity of nipples is about 250 nm. Since the wavelengths of light in the medium (λ/n) will be smaller than 250 nm, this periodicity is well below the shortest wavelengths of the visible spectrum for insects. In conclusion, the nipple array is an effective broadband antireflection coating.

Although the function of the Group II and III corneal nipples is understood in terms of measurements and theory discussed above, the same cannot be said for

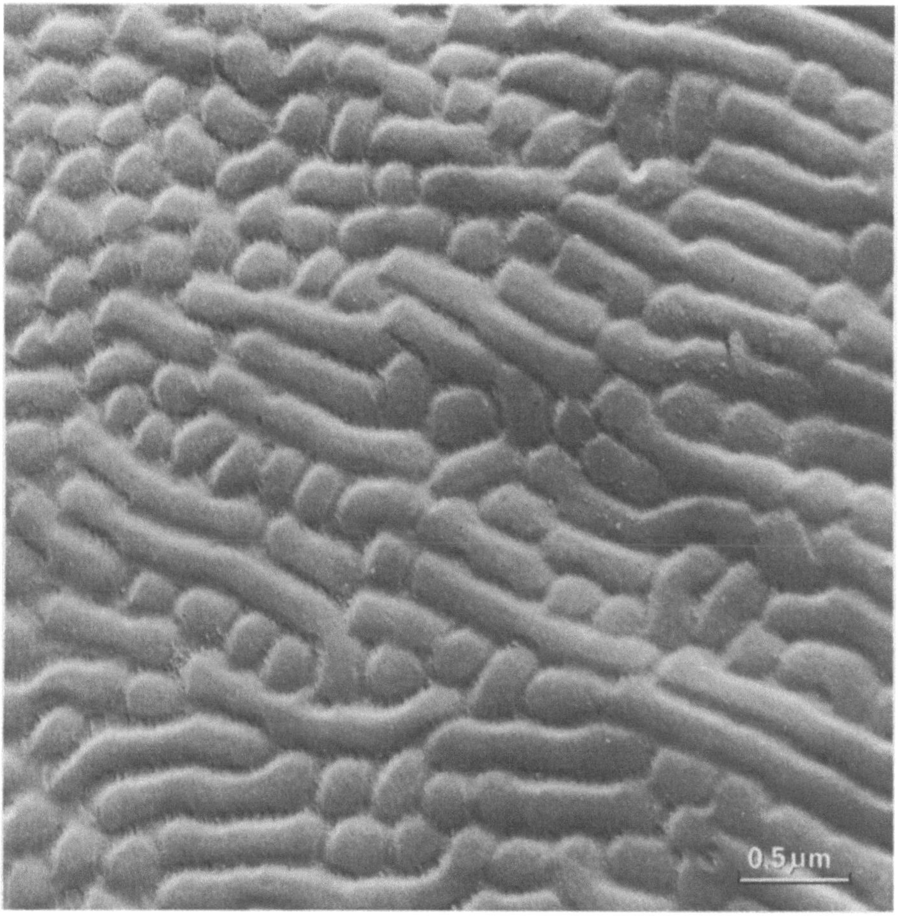

Fig. 17. Group I protuberances. Transmission EM of replica of fly cornea. (MILLER et al., 1966)

the group I corneal protrusions. These appear in surface view to be low-profiled mounds (Fig. 17), or mounds that show a tendency to coalesce. MILLER et al. (1966) find that a densely staining substance is often seen to overlie the corneal protrusions as illustrated in Figure 18A. The actual thickness of the layer containing this substance is not precisely known but appears to be less than 100 nm. The thickness cannot be judged from sections unless the exact plane of the section is known. It is equally difficult to judge thickness from surface views because there is no way of knowing whether the troughs go down to the corneal surface. Note the appearance of this densely staining substance in tangential sections through the facet surface of the dipteran, *Musca*, compound eye cornea (Fig. 18B). One may conclude that the average n of the layer overlying the cornea in group I is not very different from the underlying cornea. If this is correct, there would be no interference effects. Measurements of the intensity of reflection similar to those of Figure 12 suggest that the corneal front surface of *Musca* reflects white light of the same intensity as the grasshopper, *Schistocerca*, the n of which $= 1.5$. Nevertheless, fur-

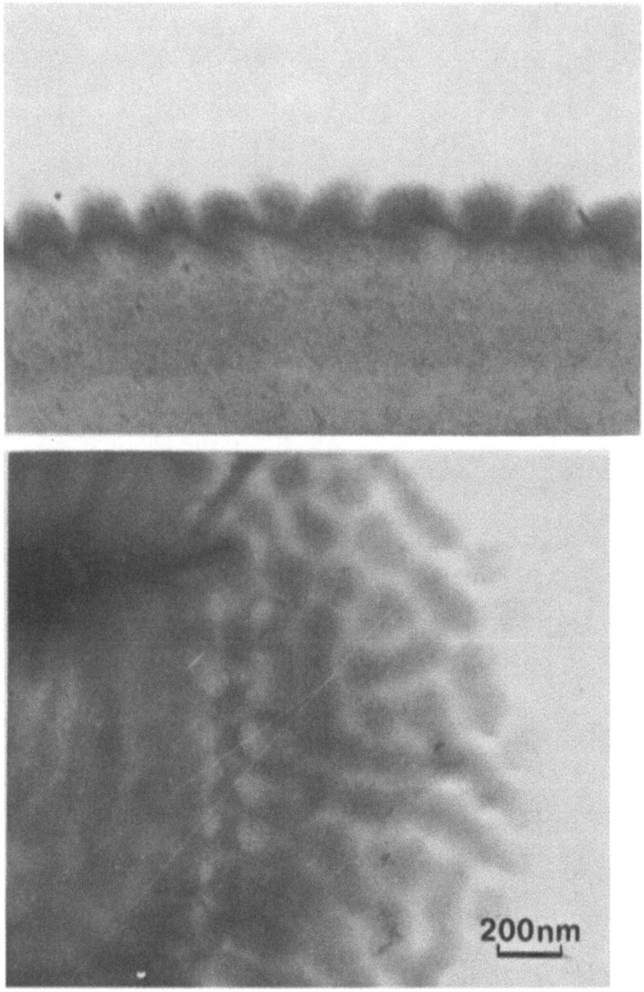

Fig. 18. Group I protuberances. Transmission EM of the cornea of the fly, *Musca*, (A) normal to and (B) oblique to corneal surface

ther studies of the anatomy, development and physical properties of the group I protrusions should be untertaken not only because their properties are poorly understood, but also because of their widespread occurrence.

b) Biologic Significance

The biologic advantages of reduced corneal front surface reflection are discussed by BERNHARD et al. (1965), MILLER et al. (1968) and BERNHARD et al. (1970). Possible advantages are (1) many nocturnal insects depend on camouflage for protection during the day. Since the compound eyes often occupy much of the surface area of the head, the dull appearance of the eyes as contrasted with a specular reflection may aid camouflage. (2) The Group II and III nipples increase the intensity of light transmitted in the ocular media by about 4%. It is argued that this increase provides an advantage particularly for scotopic vision. (3) Group II and III nipple arrays are associated with reflecting tapeta in the eyes of both nocturnal and diurnal Lepidoptera. Reflection of light from the corneal air interface of light that is *emerging* from the eye could give rise to confusing optical signals. The complete absence of the nipple array in the Papilionidae, which seems to be the only family of butterflies that lacks a tapetum, supports this argument.

Although these three common sense conclusions are possible biologic advantages of the corneal nipple array, there is as yet no experimental evidence that offers direct proof.

II. Reflection Enhancement

1. Increased Corneal Reflection

The corneas of certain vertebrates (LYTHGOE, 1976) and invertebrates (BERNARD and MILLER, 1968) show enhanced reflectance at selective wavelengths. Increased reflectance resulting from anatomic structures in the optic axis will be discussed rather than the reflections that result from layering in the facet intersections of many arthropods such as the neuropteran ant lion *Myrmeleon* and lace wing *Chrysopa*. Colored eye patterns have been reported in 72 species belonging to 23 of the 60 dipteran families (STEYSKAL, 1957). For the vertebrates, corneal iridescence has been noted in Tetradontidae (puffer fish), Siganidae (rabbit fish), Scaridae (parrot fish), Gobiidae (gobies), Labridae (wrasse), Cottidae and Serranidae by LYTHGOE, 1971.

a) Gross appearance of eye patterns

Dipteran eye patterns are most common in the family Tabanidae where the horizontally or vertically oriented alternating bright and dim colored stripes or figurings have the metallic appearance of interference colors. The long-legged flies, family Dolichopodidae, have a checkerboard pattern with alternate rows of facets reflecting red and green over part of the eye. Figure 19 shows the eye patterns of a few of the 67 examples of patterns in the genus *Chrysops* depicted by DAECKE (1906). There may be small variations from animal to animal in the same species and usually a greater variation between male and female, particularly in the tabanids where the striped pattern usually does not extend to the dorsal region of the eye in the male. Colored photographs of typical dipteran eye patterns may be seen in BERNARD and MILLER (1968).

b) Quarter-wavelength layering

FRIZA (1928) reviews the earlier literature on the origin of colors in compound eyes. He concludes on the basis of his anatomic studies that dipteran corneas have a highly refractile "enamel layer" at the front surface of the cornea that produces metallic colored reflections by interference (c.f. his Fig. 4 and discussion on pp. 296 and 297). This enamel layer can be clearly visualized now by phase-contrast light

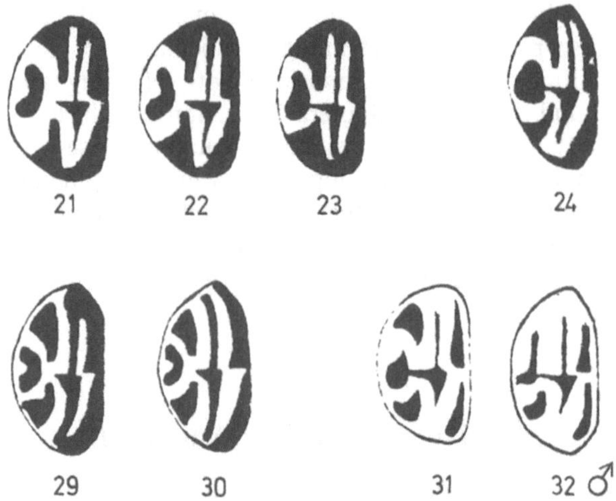

Fig. 19. Illustrating variations in eye maculations from animal to animal of the same species in the genus *Chrysops*. All figures except 32 taken from females. 21—23 *C. dimmocki*, 24 *C. indus*, 29—30 *C. bistellatus*, 30—31 *C. moechus*. (Daecke, 1906)

microscopy that registers differences in refractive index and by electron microscopy. Figure 20 (from Bernard and Miller, 1968) shows the same corneal facet from the horsefly, *Hybomitra lasiophthalma*, by bright field, phase-contrast light, and electron microscopy. The phase-contrast pictures indicate the presence of layering with a periodicity of about 200 nm near the corneal front surface. The eye was prepared for section by infiltration with epoxy resin. Figure 20B shows a 1 μm thick section taken normal to the corneal surface. The degree to which the cornea is permeable to the epoxy resin is unknown. Nevertheless, the phase microscopy suggests an underlying refractive index difference caused by a difference in mass in the six rare and seven dense layers. This difference also appears as a difference in electron density of the layers in Figure 20C.

From Figure 16 we recall that if $n_2 > n_3$, an increase in reflection occurs for all thicknesses of n_2 but the reflection is maximum when the optical path length of the layer is $\lambda_0/4$ and odd multiples of $\lambda_0/4$. From Figure 16 one can also see that if there were a stack of layers of alternating high and low refractive index so that each layer had an optical path length of $\lambda_0/4$, reflection would be suppressed by destructive interference for the rare layers and a large net reflection would result from the stack for the wavelengths nearly four times the optical thickness of the layers. The physics and biology of such quarter-wavelength interference filters have been comprehensively reviewed by Land (1972). The reader should consult this work for a summary of the reflectance properties of animal reflectors and for a lucid exposition of computational methods for predicting reflectance properties. Figure 21A shows Land's theoretical spectral reflectance functions for ideal multilayers (for both dense and rare layers $nd = \lambda_0/4$) with the maximum wavelength for constructive interference chosen as 560 nm. This filter function is for a chitin-water system (n of dense layers, $n_b = 1.56$ and n of rare layers, $n_a = 1.33$). As the number of dense layers (p) in the system increases the peak sharpens, the side

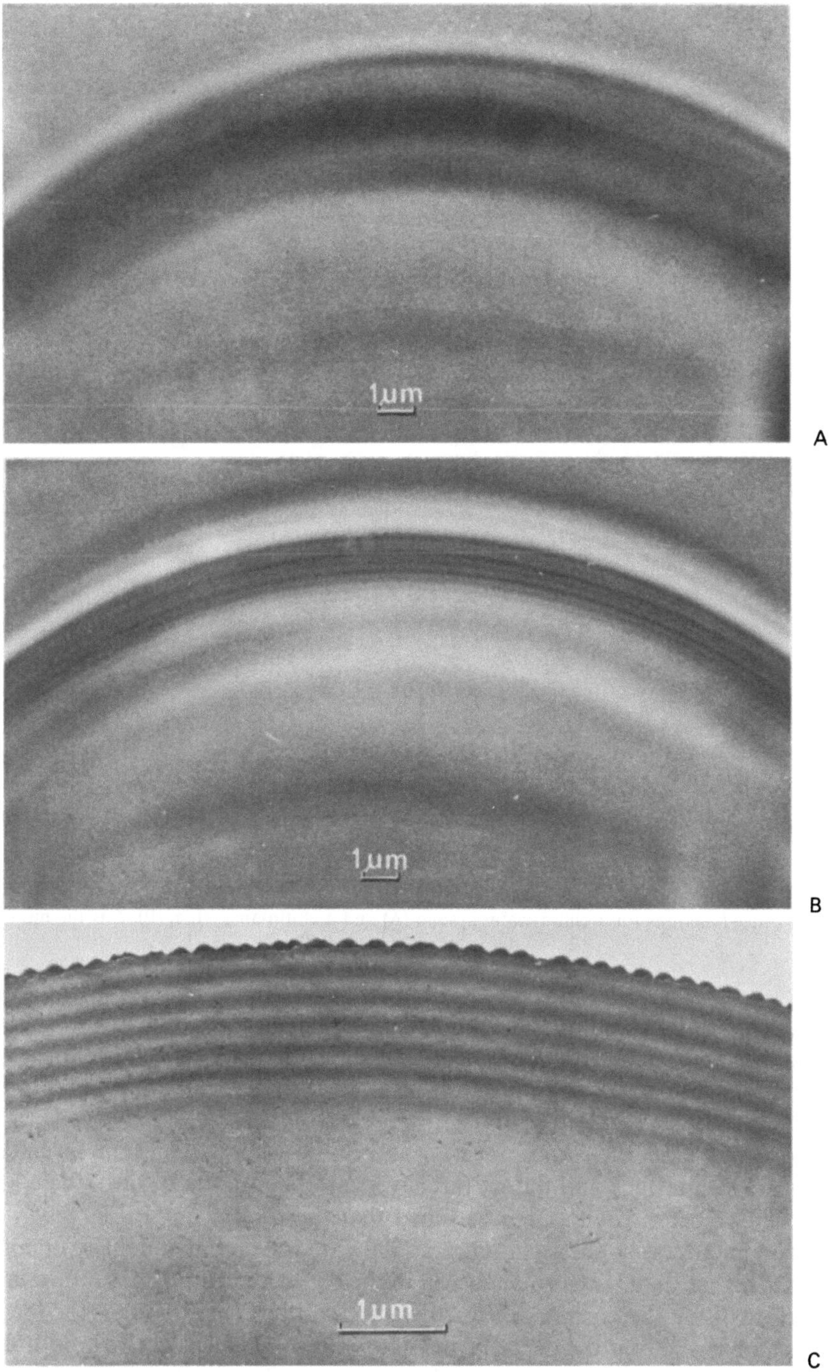

Fig. 20. Horsefly facet by bright field (A), phase contrast (B), and (C) transmission EM. (BERNARD and MILLER, 1968)

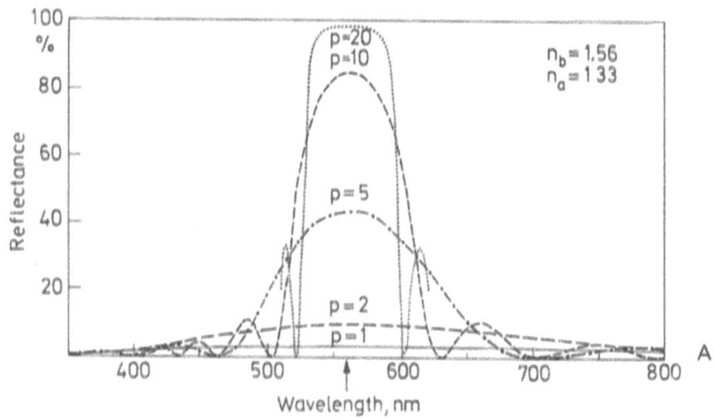

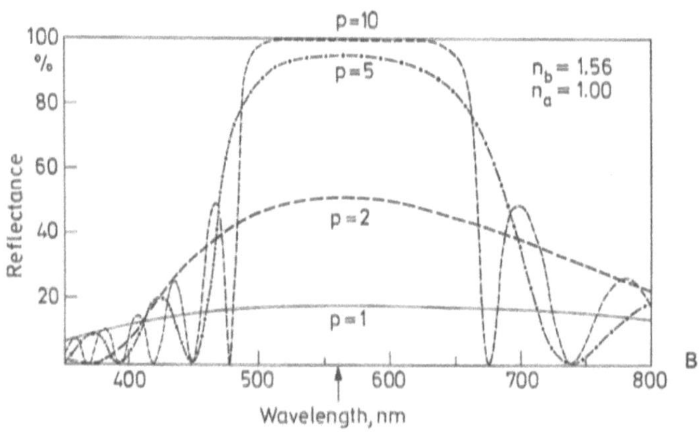

Fig. 21. Filter functions for a chitin-water system (A) and a chitin-air system (B). *P* is the number of dense layers. (LAND, 1972)

lobes increase in amplitude and number; and when p approaches 20, the peak reflectance approaches 100%. As n_b-n_a increases, the peak reflectance for a given number of dense layers and the bandwidth of the main peak increases (Fig. 21B, from LAND, 1972).

We can think of the corneal layering as a chitin-water system. The seven dense layers are 89 nm thick and the six rare layers are 112 nm thick. Since the corneas lose their color on drying, it is assumed that the rare layers are largely water. Electron micrographs such as Figure 22 show that the number density of corneal fibrils is higher than for normal cornea in the dense and lower in the rare layers. Therefore, if for the cornea of Fig. 20 we arbitrarily assign $n = 1.34$ for the rare layers and $n = 1.69$ for the dense layers, the result is an ideal set with both dense and rare layers a quarter wavelength for $\lambda = 600$ nm. The facet shown on Figure 20 appears orange when illuminated with white light. In common with all interference filters, as the angle of incidence of the illumination is increased, the light in the rejection band of the filter shifts to shorter wavelengths.

Fig. 22. Number density of corneal fibrils is smaller in rare than in dense layer. EM of *Culex* cornea

The principle of the mechanism of these filters is proven beyond doubt by the phase microscopy (Fig. 20A) that reveals a periodicity of a half wavelength and the electron microscopy of the same facet showing the filter is constructed of quarter-wavelength layers (Fig. 20B). Nevertheless, the absence of sufficiently accurate measurements of both the thickness of the individual layers and their *ns* in fresh corneas makes it impossible to accurately predict filter spectral properties. A few actual measurements of spectral reflectance are published by BERNARD (1971). These are mostly of the neighboring red- and green-reflecting facets ($p=6$) of the long-legged fly of the genus *Condylostylus.* Figure 23 is from that paper. The green-reflecting facet's peak intensity is 60% and its half-power bandwidth is 70 nm, or 14% of the peak wavelength (BERNARD, personal communication). If this facet had the characteristics of Fig. 21A ($n_b=1.56$, $n_a=1.33$, $p=6$), the peak reflection intensity would be 55% and the half-power bandwidth 19%. If $n_b=1.58$ and $n_a=1.33$, peak reflectance is 60% and half-power bandwidth is 20%. These comparisons are made on the assumption that both the rare and dense layers are each a quarter wavelength as in an "ideal" filter set. It is evident that the measured filter function does not exactly meet these criteria. If n_b-n_a is larger, as suggested in the previous paragraph, the peak reflectance would be greater than 60%, but if at

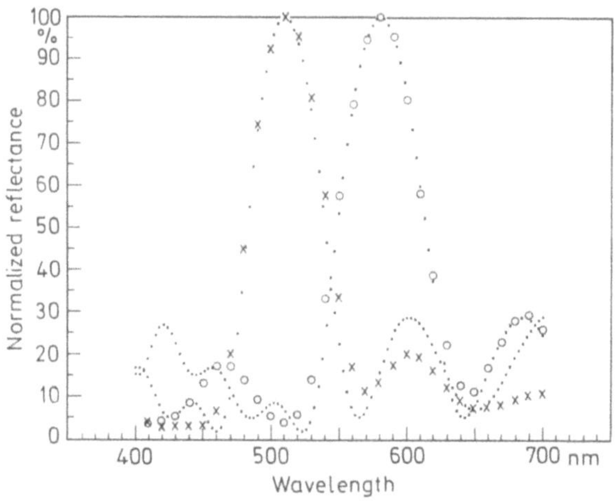

Fig. 23. Measured reflectances of neighboring orange-reflecting *(open circles)* and blue-green reflecting (*X*s) facet surfaces of the fly *Condylostylus*. (BERNARD, 1971). Peak reflectances about 50%. *Dotted curves* are theoretical calculations (G. D. BERNARD, unpublished), assumptions: $n_a = 1.34$, $n_b = 1.65$, the filter contains 12 layers resting on a substrate of $n = 1.473$, except that the anterior layer is a dense layer that is three times the thickness of the other five dense layers; for the blue-green filter $d_a = 28.9$ µm, $d_b = 133.1$ µm; for the orange filter $d_a = 32.9$ µm, $d_b = 151.4$ µm (notation that of LAND, 1972)

the same time the optical path length of a dense plus rare layer are a half wavelength with neither being a quarter wavelength, this is said to be a "nonideal" system (c.f. LAND, 1972, p. 87). The peak reflectance would be decreased and the half-power bandwidth also decreased. These characteristic properties of a "nonideal" set are suggested by these measurements on *Condylostylus* and by measurements on many other dipteran corneal filter systems (BERNARD, personal communication). The bright areas on the corneas of deerflies (genus *Chrysops*, shown in Fig. 19) display one of the brightest reflections observed from dipteran corneas. Electron microscopy shows a filter system with $p = 15$. As with many bright facets of other dipterans, the reflection is red for normal incidence illumination. A major unanswered question is the reflectance of dipteran corneas in the near UV. This is especially important since the published reflectance curves suggest that these are nonideal systems, and in such systems strong peak reflectance occurs at half the wavelength of the first-order peak (LAND, 1972, Fig. 9).

c) Biologic Function

LYTHGOE (1976) has convincingly shown that vertebrate multilayer corneal interference filters function as a sophisticated sun shade to prevent ocular flare from downwelling light. The biologic significance of dipteran corneal filters is less clear. It is argued that these filters may enhance color contrast (BERNARD and MILLER, 1968; BERNARD, 1971). BERNARD (1971) stresses this possibility for *Condylostylus* where he has shown that the reflectance characteristics of neighboring red- and green-reflecting facets fit, in that the wavelength of peak reflectance of one lies at the same wavelength as a minimum of the other (Fig. 23). The presence of bright reflections commonly found in the red in many dipteran corneas is especially puzzling if it is assumed that these filters serve a visual function. The transmission of red light to the rhabdom is diminished and this results in decreased intensities where the animal already has very little sensitivity. The dimly reflecting facets with a small number of multilayers that are equally common cannot be expected to alter transmission sufficiently to mediate a visual affect (BERNARD, personal communication). TRUJILLO-CENÓZ and BERNARD (1972) have shown in another dolichopodid, *Symphycnus lineatus*, that the rhabdomal microvilli of both the superior and inferior central cells are oriented vertically and parallel to one another in the ommatidia with red-reflecting facets, whereas those of the yellow-reflecting facets have orthogonally oriented microvilli in the superior and inferior central cells as in other diptera. This anatomic neural correlate of the corneal filter system in this animal is the strongest evidence that the filters have an optical function that is somehow being used by the nervous system. In spite of the fact that multilayer interference filters are found in the cornea on the optic axis of a number of dipterans, they may function chiefly for display. Evidence for a visual function is scant, with the exception of the Dolichopodidae where the possibility of a visual function should be further investigated.

2. Silvered Crystalline Cones

Two mechanisms have been described for the formation of the arthropod compound eye superposition image by which light from many facets contributes to the image. The most common mechanism occurs in many nocturnal Lepidoptera, Neuroptera and Coleoptera. In these animals the image is focused by a lens-cylinder effect of the crystalline cone mediated by its inhomogeneous n (HORRIDGE, 1975; KUNZE, this volume). Another mechanism is found in the compound eyes of decapod crustaceans where the crystalline cones are homogeneous but are lined with a thin reflecting layer that shows interference film properties. This reflecting layer functions as a mirror to reflect illumination entering the crystalline cones of many facets such that the reflected light forms the superposition image (VOGT, 1975; LAND, 1976 and this volume, Part B).

3. Tapeta

a) Location

A mirror-like reflecting layer called the tapetum lucidum lies beneath the retina causing the familiar eyeshine often observed in nocturnal animals. The topic of tapeta has been reviewed in general by LAND (1972), for vertebrates by WALLS (1967), for vertebrate carnivores by PEDLER (1963) and COLES (1971), and for fish by ZYZNAR (1975a, 1975b) and NICOL (1975).

Light not absorbed on the first pass through the photoreceptors is reflected back for a second pass by the tapetum. It is advantageous for the tapetum to abut the photoreceptor so that the reflected light is restricted to the same photoreceptor. This is important for the tapetum designed to function in daylight so as not to degrade acuity. It has been hypothesized that the inverted retina evolved in order

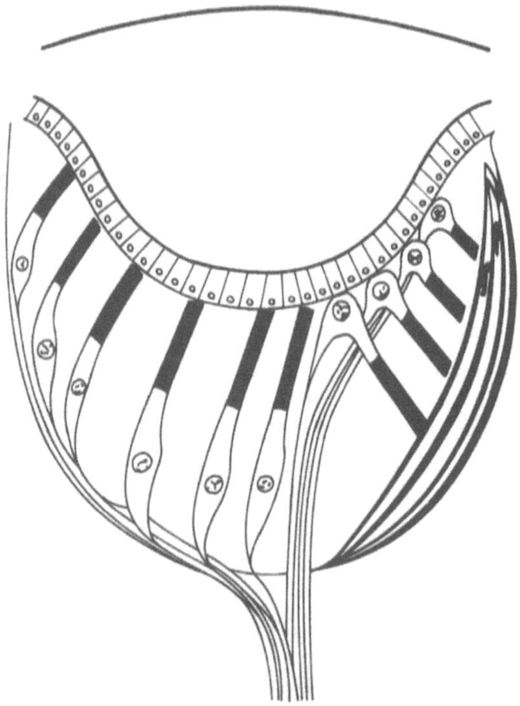

Fig. 24. The same receptor is illuminated by light reflected from the tapetum when the retina is inverted (right half of figure). Conception of VERSLUYS and DEMOLL (1923) of how the inverted retina of the arachnoid eye evolved

to achieve the proximity of tapetum and photoreceptor organelle. For example, Figure 24 (from VERSLUYS and DEMOLL, 1923) shows their conception of how the arachnoid eye evolved from the prototype with a noninverted retina into the eye with an inverted retina in which the photoreceptor organelles end on the tapetum. By analogy one could speculate that the pineal eye of lower vertebrates was the prototype that evolved into the inverted retina of the lateral eye. Nevertheless, one cannot generalize from the development of the arachnoid inverted retina because most of the compound eyes that have tapeta, including those of the neuropterans and lepidopterans, do not have inverted retinas.

b) Mechanism

Most tapeta have in common a mechanism involving constructive interference. Crystals with an optical path length of a quarter wavelength for visible wavelengths have been found in the median ocellus of a copepod crustacean *Macrocyclops* (FAHRENBACH, 1964), the lycosid spider arachnoid *Arctosa* (BACCETTI and BEDINI, 1964), several molluscs including *Pecten* (LAND, 1965, 1966), the vertebrate elasmobranchs (DENTON and LAND, 1967), some teleosts (NICOL and ZYZNAR, 1973) and the mammalian bush baby *Galago* (DARTNALL et al., 1965). In other animals the crystals take a variety of shapes: needles arranged in lattice planes in the cat (COLES, 1971) and spheres in teleosts (NICOL, 1975). Guanine is

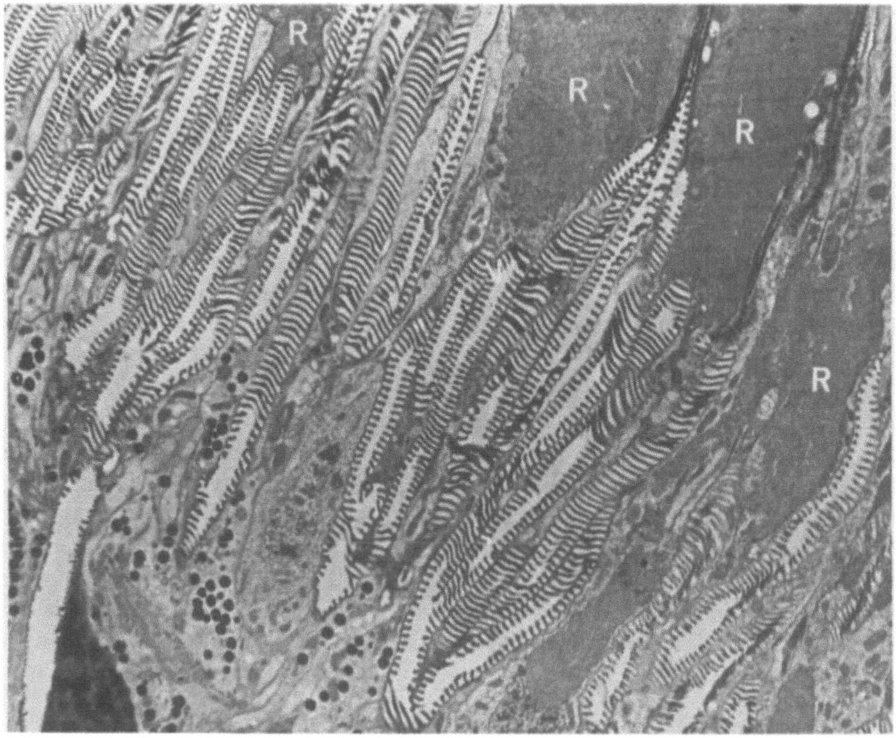

Fig. 25. Tracheolar tapetum of the nocturnal lepidopteran eye. Tracheolar branch penetrates basement membrane *(lower left)*, divides into many branches with prominent taenidial ridges and envelops proximal part of rhabdom *(R)*

thought to be the crystalline substance of many tapeta, though in the bush baby it is riboflavin (PIRIE, 1959), and in teleosts a variety of complex organic substances are implicated (NICOL, 1975).

The spectral reflectance of these tapeta is usually broad and peaked at the wavelengths appropriate for crepuscular, nocturnal or low-level illumination such as encountered by deep-sea fish. Though the individual crystals are sharply tuned for constructive interference at one wavelength, the system is detuned by staggering spacings of the crystals (e.g., COLES, 1971).

Insect tapeta are formed from a specialized tracheolar system (BUGNION and POPOFF, 1914). In the nocturnal lepidopteran compound eye the tracheols branch after penetrating the basement membrane. The tracheolar walls are lined with taenidial ridges that present a periodic ridge-air interface to incoming light (Fig. 25). This tapetum is responsible for the well-known glow of the moth eye (LEYDIG, 1864). The tracheolar tapetal system of diurnal moths, for example, *Composia*, a noctuoid of the subfamily Pericopidae, is essentially of the same construction, but the rhabdom resembles that of butterflies because of its reduced volume. The skippers, family Hesperoidea, and the sphinx moth *Hemaris* are diurnal Lepidoptera with superposition eyes and properties in common with both diurnal butterflies and the nocturnal moths (DØVING and MILLER, 1969; MILLER et al., 1968; HORRIDGE et al., 1972; MILLER, 1976). The rhabdoms are large in volume as in the nocturnal moths. The tracheols form a sheetlike enclosure of the rhabdom so that the rhabdom is surrounded by an axially periodic set of alternating air and chitin taenidial rings with a periodicity of 180 nm. Incoming illumination with wavelengths shorter than 500 nm is diffracted back through the rhabdom and out of the eye, tinting the light reaching the rhabdom and also causing skipper eyeglow (MILLER and BERNARD, 1968a).

One of the most highly developed tapetal systems is that of the diurnal lepidopteran butterfly compound eye. In all of the butterflies with the exception of the Papilionidae (swallowtails), a tracheol forms a multilayered quarter-wavelength tapetum; one private tapetum is fed by the waveguide rhabdom of each ommatidium of the compound eye (MILLER and BERNARD, 1968b). Although the appearance of these filters varies somewhat depending on the species, the plan of construction of the tapetum is basically the same in all butterflies. A single large tracheolar branch penetrates the basement membrane proximal to a rhabdom and subdivides it so that if a cross section is made at any level proximal to the rhabdom, one observes that the taenidial ridges have formed plates perforated by two, three, or four air holes. At the level of the rhabdom just distal to the filter, the tracheol usually splits into eight separate air tubes that travel distally between the retinular cells (MILLER and BERNARD, 1968b, Fig. 5). The tapetal filter typically contains 15–40 quarter-wave plates, each composed of a thin layer of chitin with tracheolar cell cytoplasm inside (see Fig. 27 and 28). Therefore, MILLER and BERNARD (1968b) assign n for the plates at 1.4.

Using an ophthalmoscope, EXNER (1891) discovered the eye shine produced by the butterfly tapetum. Butterfly glow is only visible by means of such a device because of the narrow acceptance angle of the rhabdom that feeds each tapetal filter. The glow is evanescent in many butterflies and disappears in a matter of seconds, as Exner noted. STAVENGA et al. (1977) have shown that the evanescence of the glow is caused by radial migration of pigment granules within the retinular cells similar to that occurring in other insects (Sec. B, 1, b above). The colors of the glow range from pastels to highly saturated. Neighboring facets may have identical or similar glow colors as for instance, the orange and red facets in the central part of the monarch, *Danaus*, eye, or constrasting colors such as the red and green neighboring facets in the dorsal part of the eyes of satyrs (see BERNARD and MILLER, 1970, for color pictures). Overall, in a large number of genera, the glow colors are arranged in bands with reddish hues located ventrally and bluer hues located dorsally as depicted in Figure 26 from MILLER and BERNARD (1968b). There is often considerable variation from animal to animal in the same species similar to the variations in dipteran eye patterns depicted in Figure 19.

SWIHART et al. (1974) suggested butterfly glow may be caused by reflections from the corneal processes, but there is no doubt that the glow actually originates in the tracheolar tapetal filters. MILLER and BERNARD (unpublished observations) found that the rhabdoms can be removed by making a scalp with a sharp razor cut and still leaving the cut surface relatively intact. A faint colored reflection that may originate from the corneal processes remains in the corneal part of the scalp of some animals. The unmistakable glow of individual tapeta is usually easily visible at the exposed cut surface made just distal to the basement membrane.

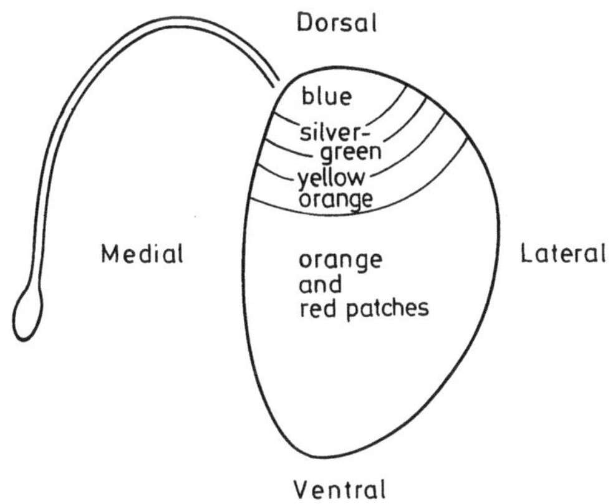

Fig. 26. Typical distribution of butterfly glow hues. (MILLER and BERNARD, 1968 b)

Although it is well established that the tracheolar filter causes the reflection that results in butterfly glow (e.g., STAVENGA, 1975a), the question of how the reflection properties of the filter and the absorbance of the visual pigments and stable photoproducts combine to produce the final glow color is a matter of conjecture (STAVENGA et al., 1977). The following examples are selected to illustrate the range of variations that have been encountered. The sulfurs, family Pieridae, have a red glow throughout the eye, though it is difficult to observe ventrally. At the proximal end of the rhabdom in some genera the tracheols cling to the rhabdom in a manner reminiscent of the skippers (Fig. 27A, and KOLB, 1977, Fig. 9B). The rhabdom feeds into the filter as shown in Figure 27A. It is evident that the periodicity of the taenidial ridges and plates gradually changes over the length of the filter so that near the distal end the rejection wavelength is 470 nm and at the proximal end it is 730 nm. The rejection bandwidth of the filter is fairly broad, yet the glow is a saturated red. On the other hand, the filter of the clouded sulfur *Colias philodice*, shown in Figure 27B, is a very regular nonideal, air cytoplasm set, with its theoretical rejection band peaked at 525 nm. Since the glow is red, it seems likely that most of the green reflection is absorbed by the visual pigment, and the red glow is from a subsidiary peak in the rejection band of the filter (MILLER and BERNARD, 1976). MILLER and BERNARD (1968b) reported that the buckeye *Junonia*, which has a pastel blue glow throughout the eye, has a filter composed of approximately ideal quarter-wavelength sets for incident light of 490 nm. Note the periodicity decreases rapidly going proximally from the rhabdom and settles down to remain constant over most of the filter (Fig. 28). This is consistent with the pastel blue glow. The available evidence indicates that in many, but not all instances, the filter characteristics are the most important determinant of glow color and probably cause the systematic variation in glow color over the eye. The filters, consisting of air-cytoplasm quarter-wave sets of at least 15 plates, reflect 100% of the light in the rejection band. The glow intensity is difficult to measure but is probably in the order of 5% of the incident intensity. The complex question of the interaction of filter and visual pigment and photoproducts absorbance needs additional investigation.

A 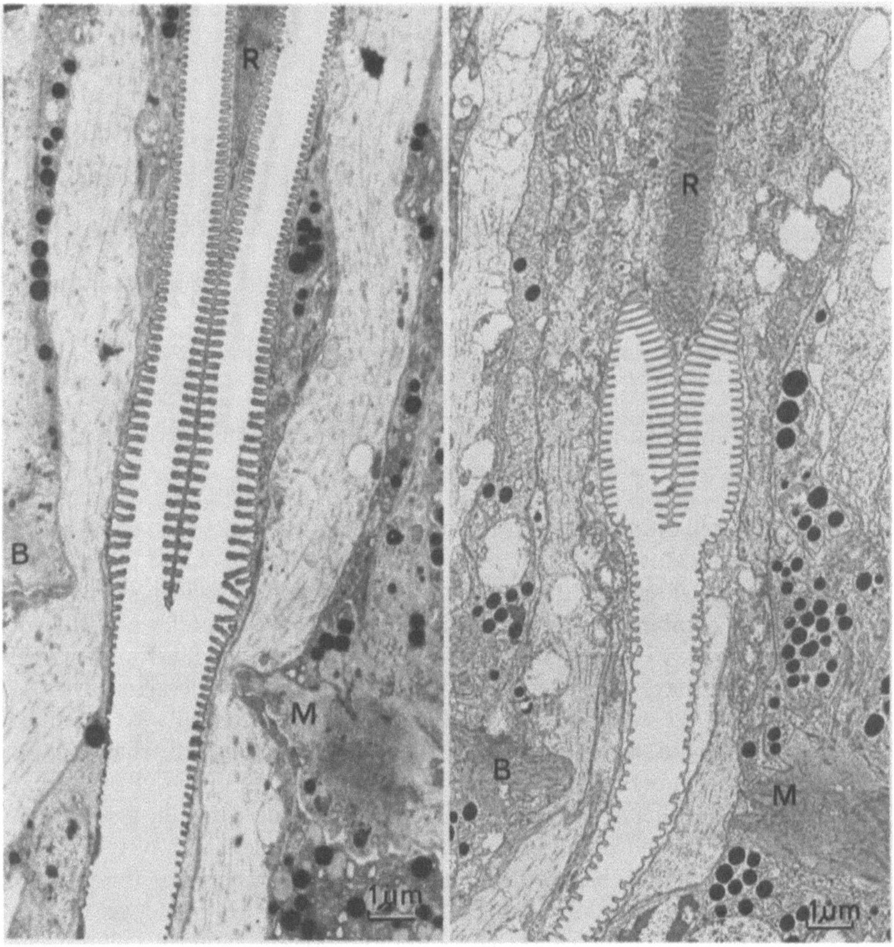 B

Fig. 27. Tapetal filters of different species of sulfur (Fam. Pieridae) that both show a red glow through the eye. Spacing of plates as measured in EMs of (A) predicts a broad range of wavelengths in the rejection band of the filter. Spacing of plates in (B) predicts a rejection band peaked at 525 nm. Rhabdom *(R)*, basement membrane *(BM)*

c) Function

The tapetum lucidum is chiefly an adaptation for nocturnal vision to enhance sensitivity by increasing the intensity of illumination reaching the visual pigment. DENTON and NICOL (1964) found that the length of the rod outer segments of elasmobranchs with a tapetum is half of the outersegment length in teleosts that do not have a tapetum. Since the fraction of light incident on the retina would be the same both with and without a tapetum, DENTON and NICOL (1964) argue that the advantage of the tapetum is to allow the same amount of light absorption with half the concentration of visual pigment. Two possible advantages of less visual pigment they call attention to are (1) the reduction of intrinsic "dark" noise resulting from thermal breakdown of visual pigment and (2) the improvement of signal-to-noise ratio to the extent that the receptor acts as an ideal square law

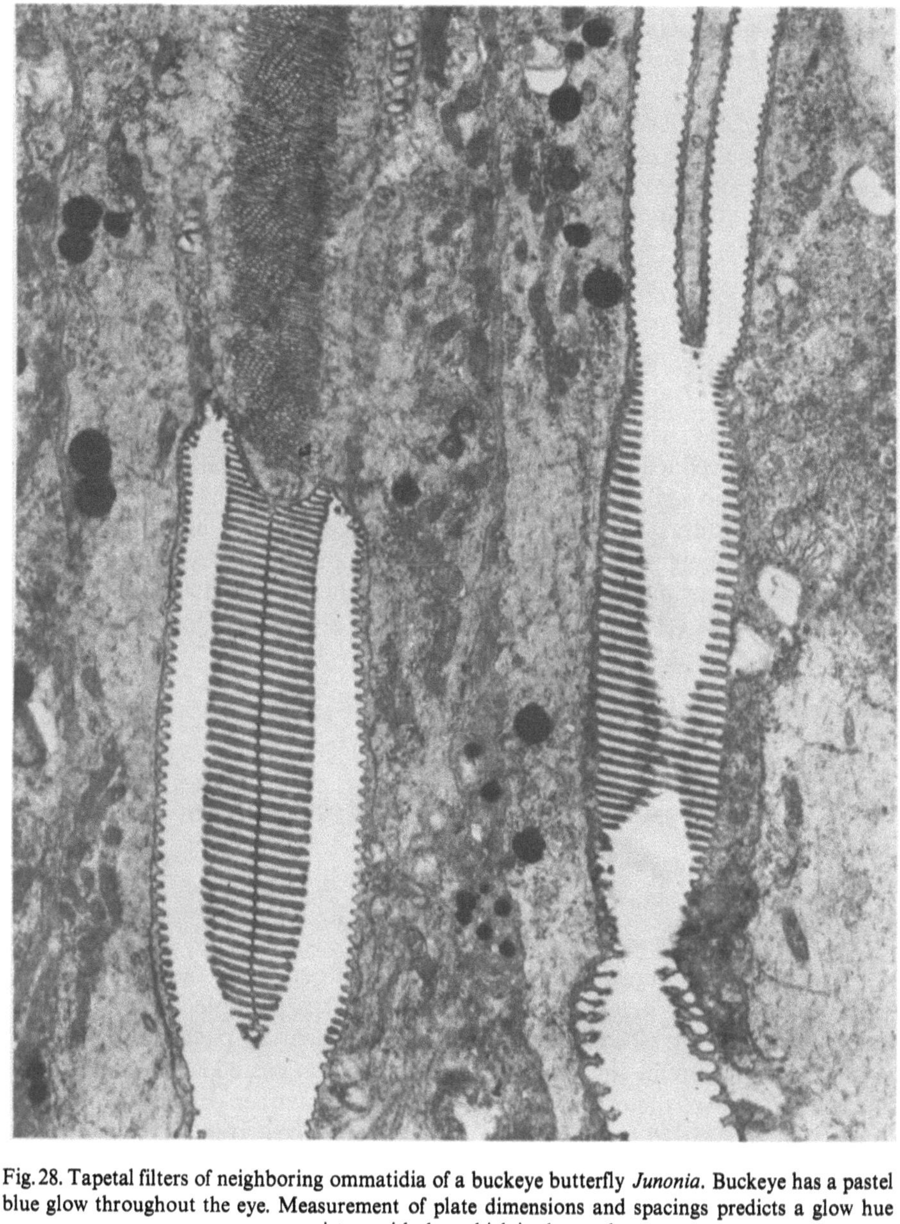

Fig. 28. Tapetal filters of neighboring ommatidia of a buckeye butterfly *Junonia*. Buckeye has a pastel blue glow throughout the eye. Measurement of plate dimensions and spacings predicts a glow hue consistent with that which is observed

detector (ROSE, 1948; BARLOW, 1965). Comparable statements cannot be made for the nocturnal lepidopteran eyes, and perhaps the eyes of the deepsea animals where the emphasis seems to be on increasing *both* light capture and the amount of visual pigment.

The eye of the mollusc scallop *Pecten* is unusual in that the tapetum has a catoptric function of focusing the image onto the distal sense cells of the retina (LAND, 1965; BARBER, EVANS and LAND, 1967; LAND, this volume Part B).

The biologic function of the tapetum in diurnal moths and skippers is unknown, but it may facilitate the photoconversion of metarhodopsin to rhodopsin.

The unique quarter-wavelength multilayer tapeta of the diurnal lepidoptera (butterflies) would be expected to play a role in enhanced color discrimination either by a direct effect of color vision or by enhanced contrast effects (MILLER and BERNARD, 1968b). The exact functioning of the system is still poorly understood in terms of the relations between visual pigments and their photoproducts.

III. Summary

Corneal interference filters function (1) to suppress reflections and thus aid in camouflage and reduction of ghosts (2) to enhance spectral reflections and thus shade the eye from sunlight, (3) create colored patterns for display and (4) possibly enhance color contrasts for vision. Tapetal interference filters increase light capture by the visual pigments of scotopic eyes and some diurnal eyes, function as a mirror to focus an image on photoreceptors in the scallop, *Pecten*, retina and probably aid in wavelength discrimination in the diurnal lepidopteran butterfly eye.

D. Arrangement of Photoreceptors

I. Packing

1. Foveal Receptors

The array of photoreceptors acts as a filter to determine the spatial, wavelength, intensity, and polarization information that is presented to the nervous system. The filtering effect of the spatial arrangement of photoreceptors with regard to the maximum amount of spatial information passed is considered here. The central area of the eye containing the highest number density of photopic receptors is specialized for acuity tasks. This area is located in a depression, the fovea, in some fish, birds, lizards, and primates, but the term fovea is also applied to the region of highest receptor density in the compound eye. The definition of maximum acuity depends on the task (RIGGS, 1966; WESTHEIMER, 1972b, 1975). For this section the task is defined as the resolution of the image of an infinite sinusoidal grating. This has been recognized as a valuable criterion by modern researchers because no retinal image is passed by the optics above the cut-off frequency, ν (Eq. 2 and 3). Detection at the cut-off frequency therefore defines the maximum ability of the visual system to recognize diffraction-limited information by the criterion adopted here.

Starting with the premise that foveas may be "designed" to detect at the diffraction limit, ν, we can ask how the foveal receptors might have been best arranged. HELMHOLTZ (1911) showed that to detect two luminous points, the minimum number of photoreceptors necessary is a receptor at the image of each point and one in between. The same argument may be applied for detecting the image of a sinusoidal grating. Recall that the cut-off spatial frequency, v, from Eq. 2 above is

$$v = \frac{D}{\lambda} \tag{2}$$

where D is the ommatidial or pupil diameter and λ is wavelength. If the receptors are rectangularly packed, the minimum number of receptors necessary for sampling to reconstruct accurately the image of a grating at the spatial cut-off frequency is found by placing a receptor at each node and antinode of the image pattern. The maximum angular separation between neighboring receptors in radians, $\Delta\phi_R$, is (SNYDER and MILLER, 1977)

$$\Delta\phi_R = 1/2v = \lambda/2D\,. \tag{8}$$

When the receptors are hexagonally packed, the maximum angular separation, $\Delta\phi_H$, independent of grating orientation, becomes (SNYDER and MILLER, 1977)

$$\Delta\phi_H = 1/\sqrt{3}\,v = \lambda/D\sqrt{3}\,. \tag{9}$$

Equations (8) and (9) predict the angular spacing of photoreceptors necessary to resolve information on the retina at the diffraction limit, given λ and D. Although it is usually assumed that the foveal cone mosaic is the limiting factor for resolving diffraction-limited image information (WESTHEIMER, 1960; CAMPBELL and GREEN, 1965), certain cautions have been raised regarding a strict anatomic theory of resolution. For example, RIGGS (1966) points out that visual acuity remains high when monochromatic illumination is used and asks how, if all the cones are being stimulated, wavelength discrimination is possible. The very center of the human fovea subtending an angle of 7–8 min is blue blind (WALD, 1967) and the absorption peaks of the green and "red" absorbing cones are separated by only 30 nm (BROWN and WALD, 1963). The center of the human fovea is specialized for acuity at the expense of wavelength discrimination. The fact that the deep foveas of birds contain only one color of oil droplet (WALLS and JUDD, 1933) suggests the same valid conclusion for the bird eye deep fovea, which should therefore have high visual acuity for monochromatic illuminations, but poor wavelength discrimination as does the human.

How well does foveal receptor packing conform to the predictions of equations (8) and (9)? SNYDER (1977 and this volume) predicts that only those animals that move at low angular velocity in bright sunlight will have enough illumination to provide sufficient contrast sensitivity so that they can profit by having their foveal receptors spaced near the diffraction limit. He argues that fast angular motion such as is seen in *Musca* reduces the light available per receptor and is therefore not compatible with high visual acuity. HORRIDGE (1977) and SHERK (1977) confirmed this prediction by showing that certain diurnal locusts, dragonflies, and wasps come close to the diffraction limit. HORRIDGE finds that the receptor spacing in the fovea of the desert wasp, *Bembix*, is 93% of the diffraction limit, compared to ~22% for the fast flying *Musca* (STAVENGA, 1975), which also tends to be active at relatively dim illuminations. These compare with a receptor spacing that is 47% of the diffraction limit in the diurnal skipper, *Epargyrus* (DØVING and MILLER, 1969), 29% in the diurnal spider, *Phidippus* (LAND, 1969) and only 0.008% for the nocturnal spider, *Dinopis* (BLEST and LAND, 1977) where the diffraction limit based on pupil diameter is meaningless because the pupil functions mainly to increase light capture and sensitivity in dim illumination. These last three animals mentioned for comparison do not have foveas.

The above conclusions are based on measurements of ommatidial diameters (D's) and the angles between the optical axes of ommatidial pseudopupils ($\Delta\phi$'s) which are relatively easy to perform with accuracy. Knowledge of vertebrate eye foveas is much less accurate because of the difficulty in measuring $\Delta\phi$. The measurement equivalent to the angle between ommatidia is the center-to-center distance, d_{cc}, between vertebrate foveal cones which differs from $\Delta\phi$ by the factor F, the posterior nodal distance. Assuming hexagonal packing the d_{cc} is

$$d_{CC} = \Delta\phi_H F = \frac{\lambda F}{D\sqrt{3}} = \frac{\lambda f}{\sqrt{3}} \tag{10}$$

The f-number, f, is the ratio of F to D. The d_{cc} may be obtained from histologic sections taken tangential to the retinal surface at the very center of the fovea. The measurements must be corrected for changes caused by the histologic procedures. Figure 29 shows the deep fovea of a female sharp shinned hawk, *Accipiter striatus*, with an eye of 14-mm axial length. The section is slightly vitread to the external limiting membrane, ELM. The receptors at the ELM in the foveal center are shown in Figure 30B and compared with the inner segment region just sclerad to the ELM in the fovea of the monkey, *Macaca fascicularis*, with an eye of 20-mm axial length. The pictures of Figure 30 were divided into 30 boxes of 132.1 μm^2 each and the number of receptors in each box was counted. Correction for histologic processing was made by the method of OGDEN (1975). The maximum number of receptors per box in the sharp shinned fovea (Fig. 30B) is 41 and the range is from 26 to 41. Forty-one receptors correspond to 310000 per mm^2 and a $d_{cc} = 1.9$ µm, assuming hexagonal packing. The maximum number density of receptors at the very center of the monkey *(M. fascicularis)* fovea (Fig. 30A) is half or 151000 per mm^2 and the $d_{cc} = 2.8$ µm.

Although a large number of investigations of fovea d_{cc} have been conducted, no unanimity has been reached even on the value for the human fovea. It is a matter of judgement as to which value to emphasize and no attempt will be made to be comprehensive. ØSTERBERG (1935) gives a value of $d_{cc} = 2.8$ µm for the human fovea which is widely accepted and is included in Table 4 for purposes of comparison with other foveal values. O'BRIEN (1951) believed that ØSTERBERG "missed the very center of the fovea" and gives a range of d_{cc} from 2.0 to 2.3 based on an exhaustive histologic study and claimed to be in agreement with POLYAK (1941). Nevertheless, HARTRIDGE (1950) made a careful study of the published

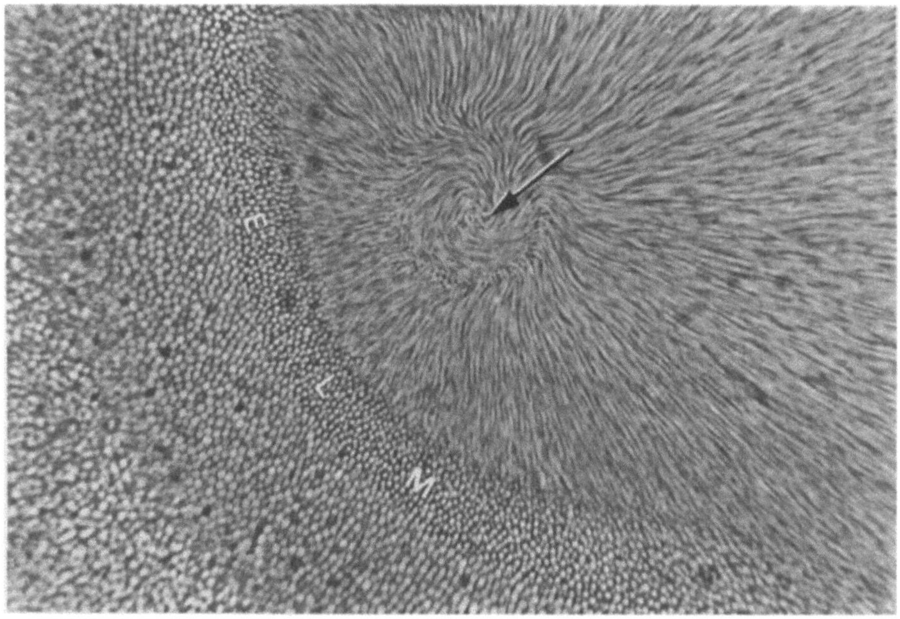

Fig. 29. Center of the deep fovea of *Accipiter striatus (arrow)* in a section just vitread to external limiting membrane (ELM). (MILLER and SNYDER, 1979)

photographs of the foveal center by POLYAK (1941) and found a maximum density of 127000 receptors/mm^2 or a $d_{cc}=3.0$. These last values are consistent with a $d_{cc}=3.0$ (Table 4) that MILLER and SNYDER (1979) found for a 26-year-old female whose eye was removed because of a malignant melanoma and whose vision was 20/20 when corrected with -2.5 diopter spectacles. MILLER and SNYDER (1979) examined the foveal $d_{cc's}$ of a number of birds of prey by the same procedure used on the human and monkey, thus facilitating comparisons (Table 4); their bird data is in agreement with that of FITE and ROSENFIELD-WESSELS (1975) processed by another technique. Although we shall use these values of the primate foveal cone $d_{cc} \simeq 3.0$ and the bird deep foveal cone $d_{cc} \simeq 2.0$ in the following comparisons and arguments for heuristic purposes, it is important to note that a number of recent investigators have obtained different results. WÄSSLE and RIEMANN (1978) find a peak cone density of 300000 per mm^2 ($d_{cc}=2.0$), uncorrected, and a $d_{cc}=2.3$, corrected, in the monkey *Macaca mulatta* and OEHME (1964) obtained a value for two birds of prey *(Falco* and *Buteo)* of $d_{cc}=1.5$.

By combining Eq. (2) and (10) the relation of the measured d_{cc} and the experimentally determined spatial cut-off frequency, v, given here in cycles per degree can be compared:

$$v = \frac{F}{(d_{cc})(\sqrt{3})(57.3)} \tag{11}$$

The agreement between d_{cc} and v can be tested by Eq. (11) to determine whether or not the measured foveal minimum d_{cc} is consistent with the hypothesis that the

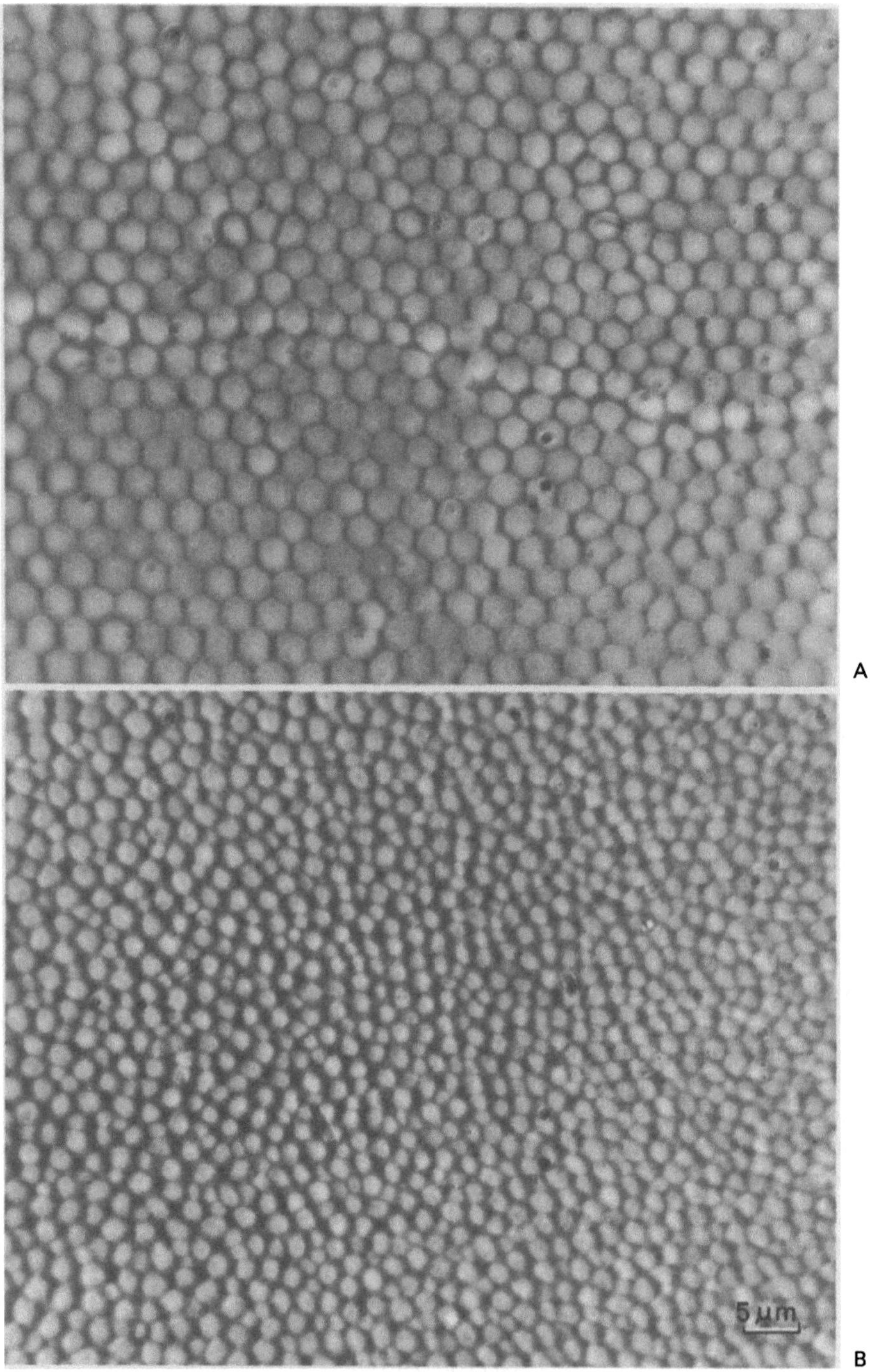

Fig. 30. Foveal cone inner segments at foveal center in the monkey, *Macaca fascicularis* (A). Foveal cone inner segments at the level of ELM, deep fovea of *Accipiter striatus* (B). From MILLER and SNYDER (1979)

vertebrate fovea can detect spatial information at the diffraction limit. This comparison is made in Table 4 for certain primates and birds of prey. For example, v calculated from the line spread function by CAMPBELL and GUBISCH (1966) is slightly greater than 60 cycles per degree when the human eye has a 2.4-mm pupil, a size giving the best image quality. As shown in Table 1, a 2.4-mm pupil would cut off at 76 cycles per degree in the absence of aberrations. The line spread function measurement suggests a marked attenuation above 60 cycles per degree and it is of interest to ask what the human foveal cone mosaic plus retina and brain can detect when the optics are bypassed by directly projecting interference fringes onto the retina. With 100% contrast WESTHEIMER (1960) finds that above ~53 cycles per degree, resolution is no longer achieved. Following the development of the laser, CAMPBELL and GREEN (1966) were able to project high contrast and high frequency fringes onto the retina. Comparison with WESTHEIMER's data is difficult because the mean luminance is not specified. However, near the end point at 99.5% contrast, 63 cycles per degree is resolved. The contrast sensitivity for gratings viewed through a variety of pupil sizes was always lower than that obtained with the interference fringes. Thus the best measurements to date suggest (1) that spatial frequencies passed over 60 cycles per degree by the human pupil are of low contrast on the retina and (2) that the photoreceptor spacing is approximately set to sample a maximum of 55–60 cycles per degree. How then can one explain the reliable reports of detection of gratings of higher spatial frequencies both through the pupil and with the optics bypassed: for example SHLAER (1937), 63 cycles per degree; a 10-year-old boy measured by BERGMANN and cited in HELMHOLTZ (1911, p. 36), 70 cycles per degree; RIGGS (1966, p. 325), 72 cycles per degree; KEESEY (1960), 78 cycles per degree; and BYRAM (1944), 86 cycles per degree with the optics bypassed. The acceptance of a human $d_{cc} \simeq 3$ μm implies that detection of spatial frequencies above 56 cycles per degree either bypassing the optics using interference fringes or by viewing a grating through the optics may not represent an accurate resolution of each line in the pattern. Spatial frequencies passed by the optics at the diffraction limit (e.g., 75 cycles per degree for a 2.4 mm pupil) though demodulated by the optics, might still be detected by receptors spaced to sample at 56 cycles per degree. GOETZ (1956) showed that dipteran eyes can detect spatial frequencies higher than the sampling limit. The demodulation caused by undersampling would prevent the unambiguous reconstruction of an image waveform. Main sources of demodulation other than imperfections of the optics are the finite diameter of the receptors, the finiteness of the photoreceptor array and the degree of undersampling.

DONNER (1951) measured the acuity of a number of small passerine birds with eyes less than half as large as a human's but with acuities about half that of a human's. The comparison including bird eyes in Table 4 are of interest because some birds have been suspected of having superhuman visual acuity. FISCHER (1969) measured a behavioral cut-off frequency for an old-world vulture with an eye about the size of a human's at 138 cycles per degree or more than twice that of the human's. Recall from Table 1 that the best image quality of such an eagle eye derived from the line spread function implies that the highest "usable" frequency passed is 120 cycles per degree for that eye, compared to 60 cycles per degree for humans. Finally FOX et al. (1976) found a behavioral v of 160 cycles per degree for

Table 4. Comparison of maximum foveal photoreceptor densities with the spatial frequencies these densities should theoretically resolve. Note that the human foveal receptor density should resolve spatial frequencies close to the cut-off measured using the line-spread function while the eagle deep-foveal receptor density would only be expected to resolve 64% of the measured cut-off. It is suggested that the deep-foveal pit may have an optical function which has the effect of eliminating the discrepancy between theoretical and measured cut-off frequencies

Species	ν calculated from line spread function (cycles/degree)	D (mm) to give ν[k]	F (mm)	f-number (effective)[j]	Receptors/mm²	d_{cc}(μm)[m]	Theoretical ν from Eq. (11)[l]	$\frac{\nu\ \text{(theoretical)}}{\nu\ \text{(measured)}}$[n]
Human	Slightly greater than 60[a]	1.9	16.7	8.8	147000[b]	2.8	60	1.0
Human					127000[c]	3.0	56	0.93
Human					125000[d]	3.0		
Monkey (*Macaca fascicularis*)		1.6[e]	13.9[e]	8.8	151000[d]	2.8		
African serpent eagle	120[f]	3.8	15.2[g]	4.0	300000[h]	2.0[h]	77	0.64
Wedgetailed eagle					303000[d]	2.0		
Red tailed hawk					333000[i]	1.9		
Sparrow hawk					297000[i]	2.0		
Sharp shinned hawk					310000[d]	1.9		
Goshawk					343000[i]	1.8		

[a] CAMPBELL and GUBISCH (1966).
[b] ÖSTERBERG (1935).
[c] HARTRIDGE (1950).
[d] SNYDER and MILLER (1978).
[e] Based on measured axial length = 20 mm.
[f] SHLAER (1972).
[g] 66% of axial length of 23 mm based on the assumption that the retina "follows a segment of a spherical surface centered on the posterior nodal point" (PUMPHREY, 1961b).
[h] Based on eagle of SNYDER and MILLER (1978).
[i] FITE and ROSENFIELD-WESSELS (1975).
[j] F/D from this Table.
[k] Theoretical pupil diameter for ν calculated from line spread function. See Table 1 for comparison of theoretical ν at diffraction limit with ν calculated from line spread function.
[l] D, F, and d_{cc} from this Table, $\lambda = 555$ nm.
[m] From number density assuming hexagonal packing.
[n] Sampling efficiency, based on measured receptor spacing and ν calculated from line spread function.

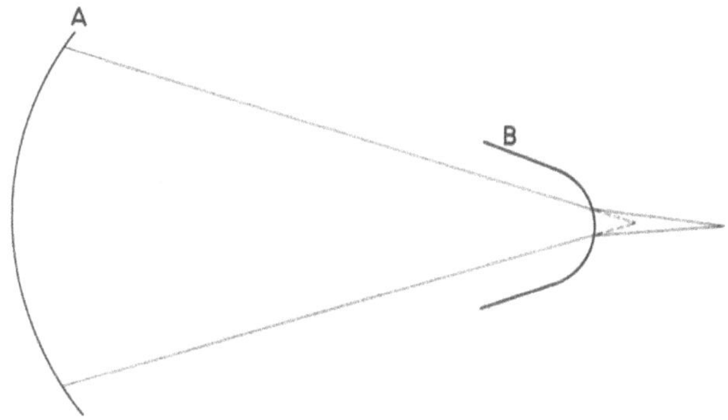

Fig. 31. Schematic of hypothesized diverging lens effect of bottom-most pit of the avian deep fovea. The location of the diverging element very close to the focal plane (at 100 μm) insures that the new focus is within the depth of focus of the system. Refracting surfaces of cornea and lens (A), refracting surface of foveal pit (B)

the kestrel which would require a 5-mm pupil and detection rather than true resolution (based on the assumption of $d_{cc} \simeq 2.0$ μm).

The line-spread function gives perhaps the most objective measure of the v that the diurnal vertebrate eye is "designed" to detect in the sense that it may predict the maximum spatial frequency that is passed at a high enough contrast to be reliably detected. The diurnal *f*-number would then be "designed" to set the image size consistent with the optimum photoreceptor size and spacing, which appear to be limited by the wave and particle nature of light.

While the pupil diameter determines the theoretical maximum acuity, (Eq. 2), the *f*-number alone determines the actual image size on the retina (Eq. 3), assuming λ is constant. The data in Figure 6 suggest that the *f*-number of birds of prey is about half that of the human eye. This means that regardless of the axial length of the bird eye and the maximum spatial frequency passed by the pupil, the maximum frequency in the retinal image is always twice that of the human's. Thus the d_{cc} of the retina of birds of prey should be half that of the human's. However, the bird d_{cc} appears to be too large by a factor of ~ 1.5 (from Table 4).

If the eagle and vulture eyes of the size of a human's have deep foveas that are set to sample accurately at something less than 80 cycles per degree as is indicated by a $d_{cc} = 2.0$ μm, two possibilities should be considered. One possibility is a case of extreme undersampling akin to the example of *Musca*. By this argument, d_{cc} is placed at 2 μm because of insufficient contrast sensitivity to support a closer spacing with its consequent loss of light per receptor and increased probability of loss of contrast sensitivity resulting from cross-talk (Sect. III, E). The minimum *deep* foveal receptoral spacing that we have observed is 2.0 μm. However, the fovea of the white throated sparrow, *Zonotrichia*, has a $d_{cc} = 1.6$ μm. This indicates that a $d_{cc} < 2.0$ μm is functional, but because the *Zonotrichia* fovea is shallow and contains several colors of oil droplets, it is possible that these receptors are primarily specialized for color vision rather than acuity. A second possibility is that the deep fovea may have an optical function that effectively increases the focal length of the eye so that with this same d_{cc} the fovea may sample accurately at 120 cycles per degree so as to take better advantage of the eye's optical quality.

In support of the latter possibility, MILLER and SNYDER (1977) and SNYDER and MILLER (1978) have speculated that the bottom-most pit of the deep fovea acts as the diverging element of a telephoto lens system (Fig. 31). WALLS (1937) had previously speculated that the steep sides of the fovea act as a magnifier. PUMPHREY (1948) proved this analysis wrong, showing that the sides merely distort. He thought the distortion may aid movement detection.

The model of MILLER and SNYDER (1977) and SNYDER and MILLER (1978) postulates a foveal pit of radius of curvature R and index of refraction n_r, compared with that of the vitreous n_v. The floor of the pit is located a distance, S, from the foveal cone oil droplets. Then the magnification factor M, attributed to the foveal pit would be

$$M=1+\frac{S}{R}\left(\frac{n_r-n_v}{n_v}\right) \tag{12}$$

An n difference between retina and vitreous of $\sim 3\%$ is indicated to provide $M=1.5$ along with the observed $S \simeq 100$ µm and $R \simeq 6$ µm for birds of prey. The difference between the retinal and vitreal n's gives rise to the retinal reflex used for retinoscopy (GLICKSTEIN and MILLIDOT, 1970). VALENTIN's measurements from 1879 quoted in WALLS (1940) indicate an n difference $\simeq 0.6\%$ for most retinas, but the foveal pit n was not measured. SNYDER and MILLER (1978) find the foveal pit is formed by Müller cells packed tightly with microfilaments indicating a higher n than for the retina as a whole. Interference micrographs suggest a complex pattern of n changes in the deep fovea of the sharp shinned hawk, *Accipiter*, the main feature of which is the foveal pit. The increased n of the vitreal surface is not evident outside the fovea (Fig. 32). The deep fovea's pit bottom acting as the diverging element of a telephoto lens offers a unifying theory to explain both how the African serpent eagle could sample at 120 cycles per degree and how other birds of prey could sample at spatial frequencies consistent with their large diurnal pupils relative to eye size as shown in Figure 6.

2. Extrafoveal Receptors

The superficial appearance of the compound eye shows that the basic organization of the hexagonal or less efficient rectangular packing of ommatidia is maintained outside the fovea over the entire eye. This organization is accurately maintained at least to the level of the second-order neurons (HORRIDGE and MEINERTZHAGEN, 1970). The spatial, wavelength, and polarization identification of each of the receptor components of every ommatidium must be confirmed for an exact description of the filtering effect of receptor packing. For example, the UV, blue and green receptors of the drone and worker bee ommatidium (first identified by AUTRUM and ZWEHL (1964) can now be exactly localized in each ommatidium (MENZEL and BLAKERS, 1976). The ommatidia with short ninth retinular cells that detect polarized light (MENZEL and SNYDER, 1974) are randomly distributed in the dorsal part of the worker bee eye (WEHNER et al., 1975; see WATERMAN, this volume, Part B). Thus an exact mapping of the filter functions attributable to the receptor mosaic is within reach for the entire compound eye.

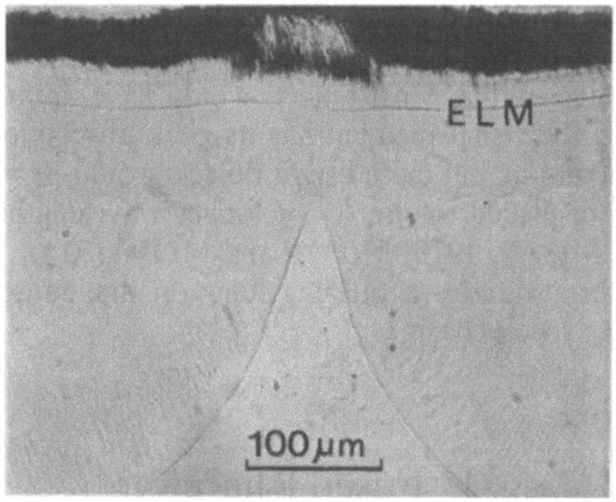

A

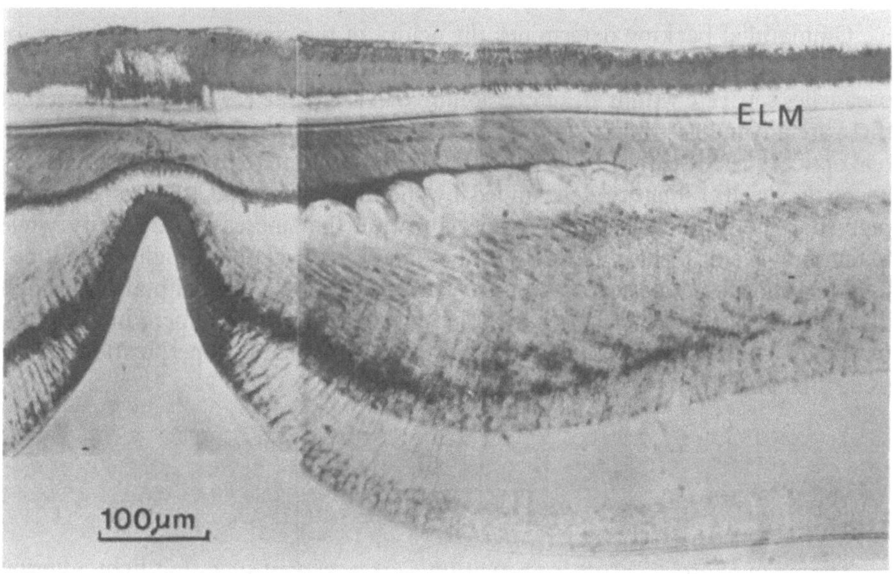

B

Fig. 32. Bright field (A) and interference (B) microscopy of a 10 µm-thick section of the deep foveal region of *Accipiter striatus*. The material was fixed in glutaraldehyde and embedded in epoxy resin. The dark material sclerad to the external limiting membrane (ELM) is shielding pigment. The darkening vitread to the ELM is caused by increased refractive index relative to the lighter appearing retinal material. From SNYDER and MILLER (1978)

The general receptor mosaic also shows a very regular organization in many vertebrate retinas. For example, in teleosts a number of patterns of double and single cones have been reviewed by ENGSTRÖM (1963) and a regular array of receptors has been demonstrated in the chick (MORRIS, 1970). Recent microspectrophotometric studies on outer segments are making available information about the spectral properties of specific morphologic cell types of photoreceptors (STELL and HAROSI, 1976; BOWMAKER and MARTIN, 1978; LOEW and LYTHGOE, 1978). These last investigators have investigated the square array patterns of teleost cones where they find that the binary twin cones and central single cones generally have their absorption peaks matched to wavelengths at which the water is most transparent while the additional single cones, which are located at the corners of the arrays of twin cones, have visual pigments with absorption peaks offset to shorter wavelengths. Most objects would have optimal contrast and be best detected when the visual pigment has its peak absorption where the water transmission is maximum, but close bright objects would be expected to require "offset" pigments, displaced to shorter or longer wavelengths (LYTHGOE, 1972; MCFARLAND and MUNTZ, 1975; MUNTZ and MCFARLAND, 1975). The array of extrafoveal receptors exercises a filtering function that must be considered by retinal location on a species basis.

II. Fused Rhabdom

Ommatidial packing determines the acuity of the compound eye so that for purposes of spatial resolution there is no advantage in keeping the receptor organelles separate within an ommatidium. SNYDER et al. (1973) have shown that there are advantages to having the rhabdomeres so close together that they form one optical structure, the fused rhabdom. Each rhabdomere in the fused rhabdom keeps its spectral identity determined by the absorption spectrum of its visual pigment. The fused rhabdom is a specialization that increases sensitivity while maintaining color discrimination at the ommatidial level and is highly developed in both apposition and superposition eyes. An exception is the adaptation for dim illumination of the dipteran eye in which one separate rhabdomere in each ommatidium of the pseudopupil functions to form a neural superposition image (KIRSCHFELD, 1973).

III. Tiering

Tiering of photoreceptors is an anatomic arrangement such that light entering one type of photoreceptor organelle must first pass through one or more different organelles. Tiering is found in the eyes of many species of arthropods, mollusks and vertebrates and appears to be a strategy to achieve specialization with minimum retinal area.

In some cases the nature of the specialization is known. The ninth retinular cell of the worker bee has a rhabdomere that is proximal to those of the other eight rhabdomeres. This ninth retinular cell, which receives the illumination not absorbed by the other rhabdomeres, is specialized to detect polarized light (MENZEL and SNYDER, 1974; see WATERMAN, this volume, Part B).

Another example of tiering as a specialization for polarization detection (WATERMAN, this volume, Part B SNYDER, 1973) is a long rhabdomere with microvilli of one orientation that interdigitates with rhabdomeres with other microvillar orientations. These rhabdoms with alternating layers of orthogonally oriented microvilli are found in apterygote insects (PAULUS, 1975), Coleoptera (BUTLER et al., 1970), Lepidoptera (MEYER-ROCHOW, 1972) and many crustaceans (EGUCHI, 1965;

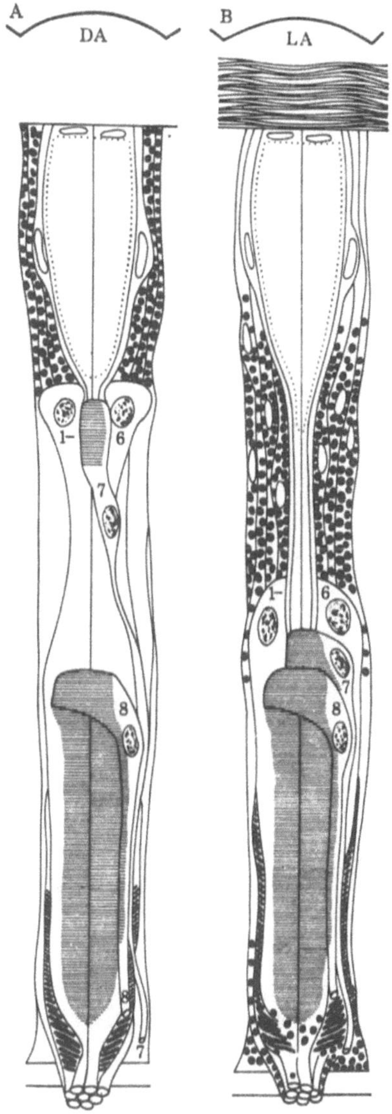

Fig. 33. Tiering in the neuropteran superposition eye of *Chrysopa*. (A) dark adapted, *(DA)*; (B) light adapted *(LA)*. Retinular cells are numbered 1–8. Number 7 is photopic rhabdomere, others are scotopic. (HORRIDGE and HENDERSON, 1976)

EGUCHI and WATERMAN, 1966). The specialization involved in the tiering of dipteran rhabdomeres 7 and 8 is not entirely clear but may involve spectral and/or polarization detection (KIRSCHFELD and FRANCESCHINI, 1977).

The distal sense cell of the tiered duplex retina of the bivalve mollusk, *Pecten*, (LAND, 1965; MILLER, 1958b) responds to the cessation of illumination (HARTLINE, 1938); the proximal sense cell (MILLER, 1960, p. 342) responds to an increase in illumination (HARTLINE, 1938). In other cases the nature of the specialization is

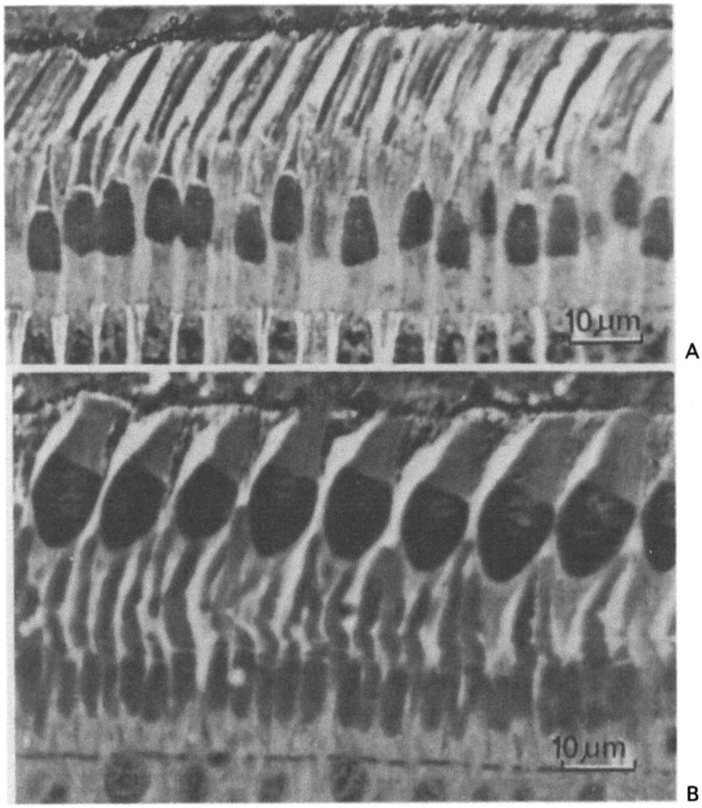

Fig. 34. Tiering in duplex vertebrate retinas that lack photomechanical movements. Pig retina is permanently light adapted (A). Cones take up most of retinal light capture area and light not absorbed by cones is available to rods. Cat-eyed snake *Leptodeira* (B) is in the permanently dark-adapted position. (MILLER and SNYDER, 1977b)

unknown, for example, in arthropod ocelli and in the ommatidia of apterygote insects such as *Entomobrya* and *Lepisma* (PAULUS, 1975).

Tiering in the superposition eye typical of Neuroptera (HORRIDGE and HENDERSON, 1976) and some Coleoptera (HORRIDGE and GIDDINGS, 1971) is illustrated in Figure 33 (from HORRIDGE and HENDERSON, 1976). In both the light- and dark-adapted condition the light must pass through the small, probably photopic, rhabdomere of retinular cell number 7 before reaching the other rhabdomeres. With dark adaptation, extensive cellular movements occur (Sec. B, I, 2, above, and STAVENGA, this volume). Rhabdomere 7 still only receives light from one facet whereas the other rhabdomeres receive light from a large number of facets.

An even more efficient photomechanical repositioning of the photopic and scotopic receptors occurs in a number of vertebrate eyes; the mammals and reptiles are the main exceptions (summarized in WALLS, 1967, p. 150). With light adaptation the photopic receptors migrate so that their inner segments occupy most of the retinal area at the level of the external limiting membrane. With dark adaptation the photopic receptors elongate and move sclerad whereas the scoto-

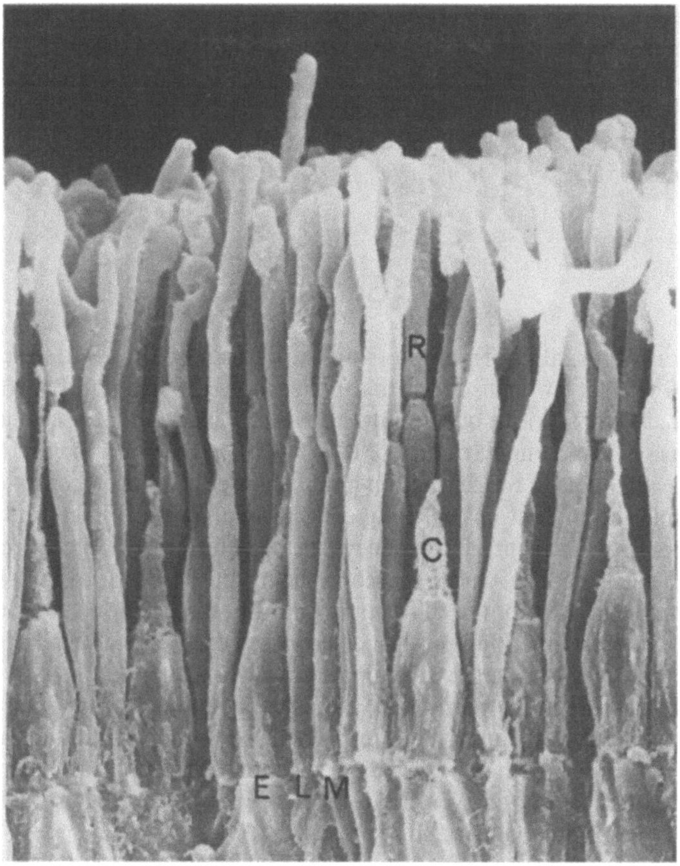

Fig. 35. Scanning EM of receptor layer of pig retina sclerad to ELM. Pigment epithelium is removed but pigment epithelial strand is attached to cone outer segment at left. Cone outer segments *(C)* end vitread to rod outer segments *(R)* in this retina that is permanently in the light-adapted position

pic receptors shorten and occupy most of the retinal light capture area at the level of the external limiting membrane. The peripheral retinas of vertebrates that lack photomechanical movements do not necessarily have to be maximally efficient for both photopic and scotopic functions but at least in some retinas a strategy of tiering favors the efficient use of retinal light-capture area (MILLER and SNYDER, 1973, 1977b; ZYZNAR and ALI, 1975). In these retinas both the rods and cones simultaneously occupy almost the entire retinal capture area so that under both photopic and scotopic conditions the illumination must pass through one type of receptor before reaching the other. The pig retina is permanently light adapted and the light must pass through the cones where some is absorbed by cone visual pigments before reaching the rods (Fig. 34A, Fig. 35). The retinas of nocturnal snakes of the family Homalopsidae lack photomechanical movements and their photoreceptors are in a permanently dark-adapted position (WALLS, 1967; UNDERWOOD, 1966). Light must pass through the rods before the cones in the tiered retina of the cat-eyed snake, *Leptodeira* (Fig. 34B).

The tiering of rods in the retinas of deep sea fish is a distinctive example in which from two to six layers of scotopic receptors are tiered (MUNK, 1966; LOCKET, 1969). Although this extreme example of tiering is evidently an adaptation for increased scotopic sensitivity, the advantages of tiering six rod outer segments of 15-μm length over one of 90 μm are not completely understood (c.f. discussion MUNK, 1966, p. 51).

IV. Summary

The array of foveal photoreceptors appears to sample at the diffraction limit only in the eyes of a few arthropods that hover in bright sunlight. Undersampling by a factor of two or more occurs in the foveas of insects that are "starved" for light because of rapid angular motion. It is suggested that sampling in the foveas of primates falls short of the diffraction limit because the photoreceptor spacing is set to sample the highest spatial frequency passed by the optics that provides sufficient contrast. That cut-off frequency can be determined in practice from the line-spread function. The photoreceptor spacing in the deep foveas of birds of prey not only falls short of the diffraction limit but also of the cut-off determined from the line spread function. This last discrepancy can be resolved if the bottom-most pit of the fovea acts as the diverging element of a telephoto lens system. The packing of receptors in the peripheral retina is functionally capable of providing spectral, polarization and acuity filtering. The fused rhabdom is an adaptation for increasing sensitivity without sacrificing spectral discrimination. Tiering is an adaptation common to both vertebrate and invertebrate retinas to achieve specialization with minimal light capture area.

E. Waveguide Effects

The refractive index of photoreceptor organelles and associated feeding structures such as inner segments, crystalline tracts and rhabdomere caps, n_n, is higher than the refractive index of the surrounding material n_s (SIDMAN, 1957; STAVENGA, 1975b). As a concequence of this difference in refractive indices, such structures will tend to hold light inside and guide it by total internal reflection, provided that the illumination is approximately directed into the structure. This last condition is not realized for the crystalline tracts of superposition eyes (EXNER, 1891; KUNZE, this volume). Consequently crystalline tracts which are otherwise capable of guiding do not function as light guides in the dark-adapted superposition eye. When the illumination is incident on the photoreceptor structure, n_r and n_s basically determine directionality because the acceptance angle is twice the critical angle, ϕ. Cos $\phi = n_s/n_r$. During light and dark adaptation, n_s around

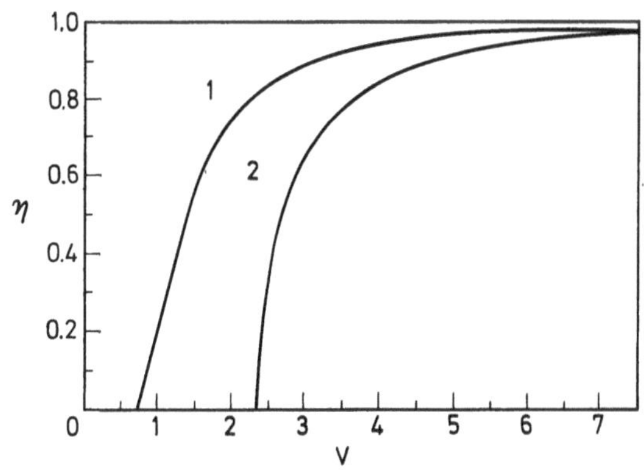

Fig. 36. Fraction of modal energy propagated inside an optical guide (η) as a function of the waveguide parameter (V) for the two lowest order modes. (See SNYDER, this volume)

rhabdomeres may change and give rise to large changes in the acceptance angle as discussed in Section B,2,b above. The acceptance angles of the vertebrate rods and cones may be related to the Stiles-Crawford-I directional sensitivity function as measured psychophysically, but the relation is a complex one (MAKOUS, 1977). The critical angle may be a main determinant of directionality of the rods and cones. The critical angle can be estimated from measurements of the radiation pattern of illumination *emerging* from the myoids. The 5° acceptance angles inferred from "reverse path" illumination of rods and cones by ENOCH (1975) predicts an $n_r - n_s$ of the order of 0.001 for the rod and cone inner segments and their surround. This small difference in *n*s also predicts that as guiding commences, the waveguide parameter, V, (see below) is less than one so that most of the incident light is outside the excited inner segment. As n_r increases because of increased density of cellular organelles (mitochondria), V and the fraction of light guided inside the inner segment also increases; $n_r - n_s$ for the rat rod outer segment is 0.06 (ENOCH and TOBEY, 1978).

The waveguide parameter V (SNYDER, 1975) defines the waveguide properties

$$V = (\pi D/\lambda)\,(n_r^2 - n_s^2)^{1/2} \tag{13}$$

where D is diameter, λ is wavelength and the *n*s are as defined above. The light is guided in modal patterns and the observation of these patterns in retinal photoreceptors is evidence of their waveguide function (ENOCH, 1961; VARELA and WIITANEN, 1970; FRANCESCHINI and KIRSCHFELD, 1971). The waveguide parameter V as a function of the fraction of light propagated inside, η, is shown for the first two modes in Figure 36. When V is greater than about 4, most of the light is propagated inside the guide. When V is less than 2.4, only the first mode is propagated and the fraction of modal energy inside the guide becomes wavelength dependent (from Eq. 13). The consequences of waveguide effects have been reviewed by SNYDER (this volume) and will be briefly summarized here. The difficulty in accurately specifying V results from the difficulty in exactly specifying the *n*s. KIRSCHFELD

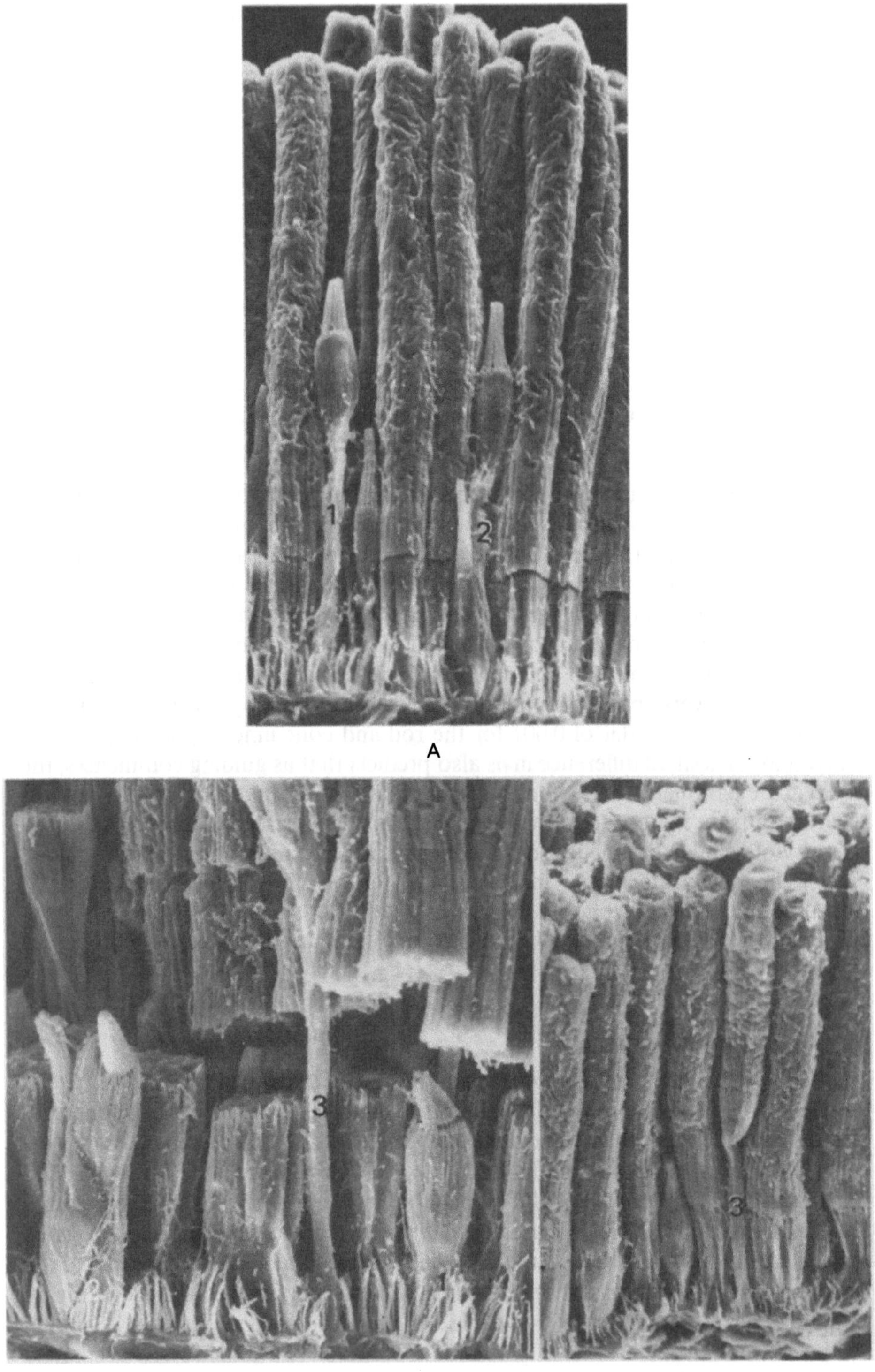

A

B

C

and SNYDER (1976) have surmounted this by estimating *V* of fly rhabdomeres by measuring their birefringence. They conclude that the small rhabdomeres (*R*7 and *R*8, of about 1 μm diameter) have pronounced waveguide effects for wavelengths greater than 400 nm. This would be expected to cause a slight shift in the extinction maximum of a visual pigment to shorter wavelengths and could thus theoretically be a mechanism for creating differing spectral absorptions with the same visual pigment (KIRSCHFELD, 1969; SNYDER and MILLER, 1972; SNYDER and PASK, 1973).

The possibility of filtering resulting from waveguide effects should be examined wherever a small *V* is suspected. Dipteran rhabdomeres 7 and 8 were mentioned above. Other rhabdomeres of small diameter, those of diurnal Lepidoptera and Hymenoptera, may be candidates. The *V* of rhabdomeres may decrease with light adaptation based on the observation of WHITE and LORD (1975) that rhabdomere volume decreases during light adaptation. The inner segments, myoids, of vertebrate photoreceptors are interesting in this regard (MILLER and SNYDER, 1972). The myoids of cones that respond to illumination with photomechanical movements elongate and become thin when dark adapted (Fig. 37A and B). The theoretically small *V* of these myoids suggests an optical cut-off function (MILLER and SNYDER, 1972). This is consistent with the idea that during dark adaptation photomechanical movements have the effect of substituting the rods for the cones to effectively comprise the entire retinal light-capture area.

The green rod myoid may be an example of a structure causing a waveguide spectral filtering effect. About 90% of frog rods are pink and 10% are green when a dark-adapted frog retina is spread on a slide, receptor side up, and transilluminated. KÜHNE (1878) noted: "It is quite different with the really green, that is, strictly grass green rods—which are present in the retina of the frog. These are mostly transparent, they give the Leeuwenhoek images, they bleach somewhat slower than the red, they stand the isolation test of the ocular diaphragm; and when I have occasionally met with them isolated, and standing on end, their colour has been fully developed. I do not doubt that BOLL had these rods particularly in mind in his second communication to the *Acad. d. Lincei* of 7 Jan. 1877, S. 3."

The source of the green color is somewhat of a mystery. DENTON and WYLLIE (1955); DARTNALL (1967); and LIEBMAN and ENTINE (1968) investigated this problem and determined that the green rod outer segment contains a visual pigment with maximal absorption at 430 nm. These investigators postulate a non-bleaching red absorbing pigment, a "structural" color to explain the green appearance. MILLER and SNYDER (1972) postulate that the green rod myoid η is wavelength

---

◀ Fig. 37. Scanning EM's of frog *(Rana catesbeiana)* retinal receptors. (A) partially dark adapted showing various stages of elongation of single cone *(1)* and double cone *(2)* myoids. Thinning of myoids provides an optical and perhaps an electric cut-off function. Note accessory member of double cone with rodlike outer segment stays in the rod dark-adapted position. Dark-adapted single and double cones shown in (B). Rod outer segments became detached from inner segments in (B). Green rod myoid *(3)* is little affected by adaptation. Its normal relation to other cells shown in (C) (MATSUURA et al., 1978)

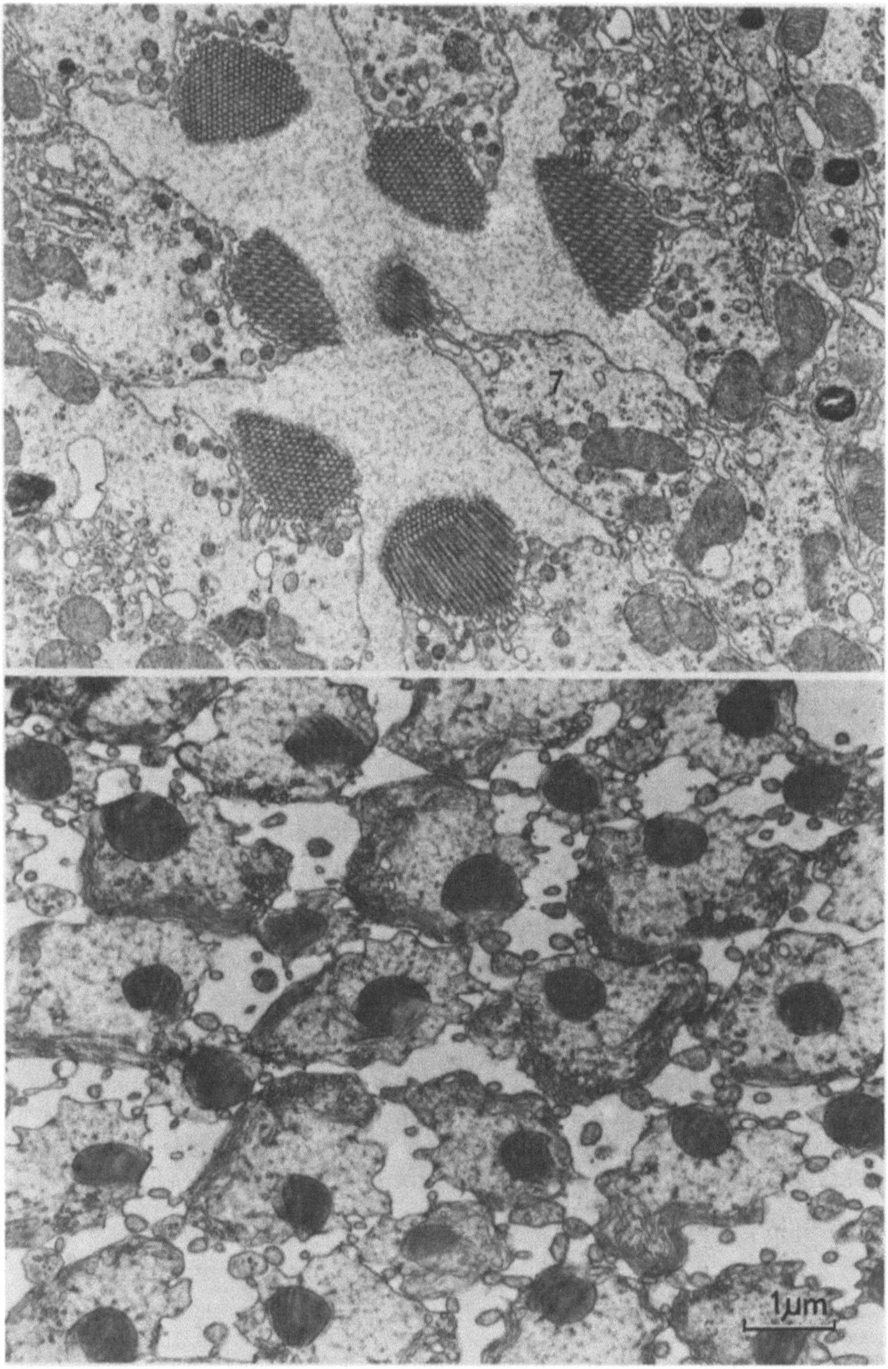

A

B

Fig. 38. Fly *Phaenicia* (A) and falcon foveal (B) photoreceptor organelles at the same magnification. Retinular cell seven *(7)*. The avian foveal cone structure is unusual in that the cone disks that appear as *filled circles* in (B) do not fill the entire outer segment. The consequences, both optical and physiologic, are not known

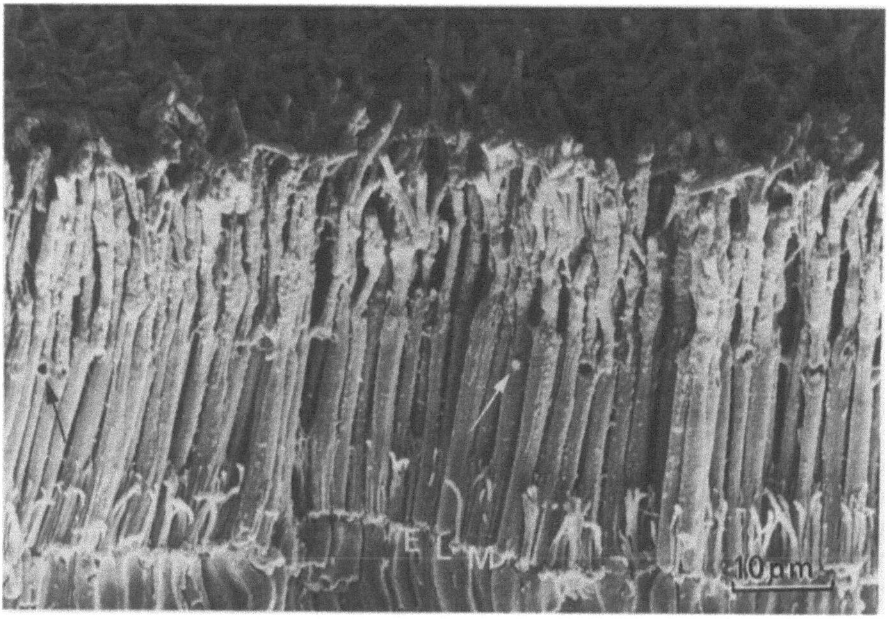

Fig. 39. *Zonotrichia leucophrys*: avian foveal cones have the highest packing density of all foveal receptors. Inner segments do not taper. Oil droplets are seen here in various stages of popping out (*arrows*). Pigment epithelium has been removed

dependent so that red light is not guided into the outer segment. The green rod myoid is long and thin (Fig. 37B and C). Even though the green rod outer segments are about the same diameter as the pink and half the length, they are much dimmer than the pink rods after being fully bleached. This suggests the green rods have a smaller light-capture area than the pink ones. For all of these reasons the hypothesis that the green rod myoid mimics a structural red absorption by not propagating red light is attractive. However, efforts to prove the hypothesis, for example by immersing the myoid in a high n surround to negate waveguide effects, have so far been unsuccessful, and the possibility that the ellipsoid may have the required filtering effect has not been ruled out.

Aside from determination of directionality and the rather minor waveguide spectral effects, the most important function of the waveguide properties of receptors and their associated feeding structures is to maintain optically separate cellular channels. One exception is the fused rhabdom mentioned above. The vertebrate double rods and cones, of unknown function, are another exception (WALLS, 1967, pp. 59–63; TOBEY et al., 1975; Fig. 37C). Optically separate channels are maintained in dipteran ommatidia by a slight bowing of the rhabdomeres which prevents significant cross talk (WIJNGAARD and STAVENGA, 1975). This bowing of the rhabdomeres is an indication that cross talk is avoided in spite of the very small diameter of these photoreceptor organelles. The physical arrangements of very small diameter photoreceptors that would be expected to maintain separate optical channels are of interest. Such separate channels would seem to be less important for vertebrate rods, which may be of small diameter to minimize the latency of response (PAPERMASTER et al., 1978) and more important for dipteran rhabdomeres of one ommatidium and for vertebrate foveal photoreceptors. The rhabdomeres of *Phaenicia* and the foveal cone outer segments of the brown falcon, *F. berigora*, are shown in Figure 38 at the same magnification. Figure 39 is a scanning EM of avian foveal cones. Both rhabdomere 7 and the cone outer segment disks are less than a micrometer in diameter. Note that the avian cone disks usually only touch the plasma membrane of the outer segment at one point and the ratio of outer segment cytoplasm to disk material is higher than in most vertebrate outer segments. It is not known if this is an adaptation for optical isolation. Further research is needed to define whether cross talk is a limiting factor in determining the packing density of the foveal cones of birds where the receptor densities are the highest of any retina.

In summary, visual receptor organelles and associated feeding structures have a higher n than their surrounding material which gives them the potential for functioning as light guides. The main example of potential guides that do not function as such are the crystalline tracts of dark-adapted superposition eyes. The waveguide properties may play a role in determining the spectral response of guiding structures with a V less than 2.4. The directional properties of retinal guiding structures are basically determined by their waveguide properties and a main function of waveguides in retinal tissues appears to be to direct illumination to the visual pigment and in the case of photopic receptors to maintain optically separate channels.

F. Conclusion

An attempt has been made to focus on the unanswered questions about the role of optical filters. One class of questions has to do with mechanisms. For example, what are the mechanisms of pigment migration and photomechanical effects, by what mechanisms are separate optical channels preserved, and how is the remarkably high visual acuity of birds achieved?

A theme of this survey is that from the front surface of the cornea to the photoreceptors themselves, the optics and catoptrics of eyes can be characterized in terms of the information passed or rejected. The functional significance of these filtering processes is that the spatial, polarization and spectral information that is passed is the only sensory information available to the visual system and must fill the needs of each specific system in terms of its evolutionary development. These optical filtering properties of eyes absolutely determine the information available to the nervous system. This is a perspective from which the biologic importance of the optical filters can be understood so that the significance of the unanswered questions concerning transparency, interference filters, arrangement of photoreceptors and waveguide properties can eventually be appreciated.

G. Acknowledgements

I appreciate the great help of Benjamin Jones for his research assistance. The preparation of this chapter was supported in part by National Institutes of Health grants EY00089 and EY00785.

References

Arey, L. B.: A retinal mechanism of efficient vision. J. comp. Neurol. **30**, 343—349 (1919).

Autrum, H., Zwehl, V. von: Spektrale Empfindlichkeit einzelner Sehzellen des Bienenauges. Z. vergl. Physiol. **48**, 357—384 (1964).

Baccetti, B., Bedini, C.: Research on the structure and physiology of the eyes of a lycosid spider. Arch. ital. Biol. **102**, 97—122 (1964).

Bäck, I., Donner, K. O., Reuter, T.: The screening effect of the pigment epithelium on the retinal rods in the frog. Vision Res. **5**, 101—111 (1965).

Barber, V. C., Evans, E. M., Land, M. F.: The fine structure of the eye of the mollusc *Pecten maximus*. Z. Zellforsch. **76**, 295—312 (1967).

Barlow, H. B.: Optic nerve impulses and Weber's law. In: Cold Spring Harbor symposia on quantitative biology, Vol. XXX, Sensory receptors. New York: Cold Spring Harbor, L.I. 1965.

Baylor, D. A., Fettiplace, R.: Light path and photon capture in turtle photoreceptors. J. Physiol. (Lond.) **248**, 433—464 (1975).

Baylor, D. A., Hodgkin, A. L.: Detection and resolution of visual stimuli by turtle photoreceptors. J. Physiol. (Lond.) **234**, 163—198 (1973).

Beersma, D. G. M., Stavenga, D. G., Kuiper, J. W.: Organization of visual axes in the compound eye of the fly *Musca domestica* L. and behavioural consequences. J. comp. Physiol. **102**, 305—320 (1975).

Beersma, D. G. M.: Spatial sensitivity and adaptation in fly visual sense cells. Poster Session No 17, Visual Sensitivity and Adaptation Meeting, The University of Surrey, Guildford, England (1978).

Behrens, M., Krebs, W.: The effect of light-dark adaptation on the ultrastructure of *Limulus* lateral eye retinular cells. J. comp. Physiol. **107**, 77—96 (1976).

Benedek, G. B.: Theory of transparency of the eye. Appl. Opt. **10**, 459—473 (1971).

Benolken, R. M.: Photopigments of the lateral eye of *Limulus*. J. comp. Physiol. **107**, 339—347 (1976).

Berger, S. B.: Computations for Twersky's theory of corneal transparency based on Hart-Farrell's swelling pressure. J. Opt. Soc. Amer. **66**, 864—865 (1976).

Berkley, M. A., Watkins, D. W.: Visual acuity of the cat estimated from evoked cerebral potentials. Nature (Lond.) New Biol. **234**, 91—92 (1971).

Bernard, G. D.: Evidence for visual function of corneal interference filters. J. Insect Physiol. **17**, 2287—2300 (1971).

Bernard, G. D.: Spectral sensitivity of the dark-adapted butterfly pupil. ARVO abstract, Sarasota meeting (1976).

Bernard, G. D., Miller, W. H.: Interference filters in the corneas of diptera. Invest. Ophthal. **7**, 416—434 (1968).

Bernard, G. D., Miller, W. H.: What does antenna engineering have to do with insect eyes? IEEE Student J. **8**, 2—8 (1970).

Bernhard, C. G., Gemne, G., Sällström, J.: Comparative ultrastructure of corneal surface topography in insects with aspects on phylogenesis and function. Z. vergl. Physiol. **67**, 1—25 (1970).

Bernhard, C. G., Miller, W. H., Møller, A. R.: The insect corneal nipple array. Acta physiol. scand. **63**, Suppl. 243, 1—79 (1965).

Bernhard, C. G., Ottoson, D.: Comparative studies on dark adaptation in the compound eyes of nocturnal and diurnal lepidoptera. J. Gen. Physiol. **44**, 195—203 (1960a).

Bernhard, C. G., Ottoson, D.: Studies on the relation between the pigment migration and the sensitivity changes during dark adaptation in diurnal and nocturnal lepidoptera. J. Gen. Physiol. **44**, 205—215 (1960b).

Bernhard, C. G., Ottoson, D.: Quantitative studies on pigment migration and light sensitivity in the compound eyes at different light intensities. J. Gen. Physiol. **47**, 465—478 (1964).
Blest, A. D., Land, M. F.: The physiological optics of *Dinopis subrufus* L. Koch: a fish-lens in a spider. Proc. roy. Soc. B **196**, 197—222 (1977).
Boettner, E. A., Wolter, J. R.: Transmission of the ocular media. Invest. Ophthal. **1**, 776—783 (1962).
Born, M., Wolf, E.: Principles of optics. New York: Pergamon Press 1965.
Bowmaker, J. K.: The visual pigments, oil droplets and spectral sensitivity of the pigeon. Vision Res. **17**, 1129—1138 (1977).
Bowmaker, J. K., Knowles, A.: The visual pigments and oil droplets of the chicken retina. Vision Res. **17**, 755—764 (1977).
Bowmaker, J. K., Martin, G. R.: Visual pigments and colour vision in a nocturnal bird, *Strix aluco* (tawny owl). Vision Res. **18**, 1125—1130 (1978).
Bowness, J. M., Morton, R. A.: Distribution of copper and zinc in the eyes of fresh-water fishes and frogs. Occurence of metals in melanin fractions from eye tissues. Biochem. J. **51**, 530—535 (1952).
Brekhovskikh, L. M.: Waves in Layered Media. New York-London: Academic Press, 1960.
Brown, P. K., Wald, G.: Visual pigments in human and monkey retinas. Nature (Lond.) **200**, 37—43 (1963).
Bugnion, E., Popoff, N.: Les yeux des insectes nocturnes. Arch. Anat. Micr. Morph. Exp. **16**, 18—304 (1914).
Burnside, M. B.: Possible roles of microtubules and actin filaments in retinal pigmented epithelium. Exp. Eye Res. **23**, 257—275 (1976).
Butler, L., Roppel, R., Zeigler, J.: Post-emergence maturation of the eye of the adult Black Carpet Beetle *Attagenus megatoma* (Fab.)—an EM study. J. Morph. **130**, 103—127 (1970).
Butler, R.: The identification and mapping of spectral cell types in the retina of *Periplaneta americana*. Z. vergl. Physiol. **72**, 67—80 (1971).
Butler, R., Horridge, G. A.: The electrophysiology of the retina of *Periplaneta americana*. J. comp. Physiol. **83**, 263—278 (1973).
Byram, G. M.: Physical and photochemical basis of visual resolving power. J. Opt. Soc. Amer. **34**, 718—738 (1944).
Campbell, F. W., Green, D. G.: Optical and retinal factors affecting visual resolution. J. Physiol. (Lond.) **181**, 576—593 (1965).
Campbell, F. W., Gregory, A. H.: Effect of size of pupil on visual acuity. Nature (Lond.) **187**, 1120—1122 (1960).
Campbell, F. W., Gubisch, R. W.: Optical quality of the human eye. J. Physiol. (Lond.) **186**, 558—578 (1966).
Carlström, D.: The structure of chitin. In: Bernhard, C.G. (Ed.): The functional organization of the compound eye. Oxford-New York: Pergamon Press 1966.
Carricaburu, P., Chardenot, P.: Spectres d'absorption de la cornée de quelques arthropodes. Vision Res. **7**, 43—50 (1967).
Chamberlain, S. C., Barlow, R. B.: Morphological correlates of efferent circadian activity and light adaptation in the *Limulus* lateral eye. Biol. Bull. **153**, 418—419 (1977).
Cohen, A. I.: An ultrastructure analysis of the photoreceptors of the squid and their synaptic connections I. J. comp. Neurol. **147**, 351—378 (1973).
Coles, J. A.: Some reflective properties of the tapetum lucidum of the cat's eye. J. Physiol. (Lond.) **212**, 393—409 (1971).
Daecke, E.: On the eye-coloration of the Genus *Chrysops*. Entomol. News **17**, 39—42 (1906).
Dartnall, H. J. A.: The visual pigment of the green rods. Vision Res. **7**, 1—16 (1967).
Dartnall, H. J. A., Arden, G. B., Ikeda, H., Luck, C. P., Rosenberg, M. E., Pedler, C. M. H., Tansley, K.: Anatomical, electrophysiological and pigmentary aspects of vision in the bush baby: an interpretative study. Vision Res. **5**, 399—424 (1965).
Daw, N. W., Pearlman, A. L.: Pigment migration and adaptation in the eye of the Squid, *Loligo pealei*. J. Gen. Physiol. **63**, 22—36 (1974).
DeBruin, G. H. P., Crisp, D. J.: The influence of pigment migration on vision of higher crustacea. J. Exp. Biol. **34**, 447—464 (1957).
Demoll, R.: Über eine lichtzersetzliche Substanz im Facettenauge, sowie über eine Pigmentwanderung im Appositionsauge. Pflügers Arch. Physiol. **129**, 461—475 (1909).
Denton, E. J.: The responses of the pupil of gekko to external light stimulus. J. Gen. Physiol. **40**, 201—218 (1956).

Denton, E. J., Land, M. F.: Optical properties of the lamellae causing interference colours in animal reflectors. J. Physiol. (Lond.) **191**, 23—24 (1967).
Denton, E. J., Nicol, J. A. C.: The choroidal tapeta of some cartilaginous fishes (Chondrichthyes). J. mar. biol. Ass. U.K. **44**, 219—258 (1964).
Denton, E. J., Wyllie, J. H.: Study of the photosensitive pigments in the pink and green rods of the frog. J. Physiol. (Lond.) **127**, 81—89 (1955).
Dodt, E., Jessen, K. H.: Das adaptive Verhalten der Froschnetzhaut, untersucht mit der Methode der konstanten elektrischen Antwort. Vision Res. **1**, 228—243 (1961).
Donner, K. O.: The visual acuity of some passerine birds. Acta Zool. Fenn. **66**, 3—40 (1951).
Døving, K. B., Miller, W. H.: Function of insect compound eyes containing crystalline tracts. J. Gen. Physiol. **54**, 250—267 (1969).
Eguchi, E.: Rhabdom structure and receptor potentials in single crayfish retinular cells. J. cell. comp. Physiol. **66**, 411—430 (1965).
Eguchi, E., Waterman, T. H.: Fine structure patterns in crustacean rhabdoms. In: Bernhard, C. G. (Ed.): The functional organization of the compound eye. Oxford-New York: Pergamon Press, 1966.
Engström, K.: Cone types and cone arrangements in teleost retinae. Acta Zool. **44**, 179—243 (1963).
Enoch, J. M.: Nature of the transmission of energy in the retinal receptors. J. Opt. Soc. Amer. **51**, 1122—1126 (1961).
Enoch, J. M.: Vertebrate rod receptors are directionally sensitive. In: A. W. Snyder, Menzel, R., (Eds.): Photoreceptor optics. Berlin-Heidelberg-New York: Springer, 1975.
Enoch, J. M., Tobey, F. L. Jr.: Use of the waveguide parameter v to determine the difference in the index of refraction between the rat rod outer segment and the interstitial matrix. J. Opt. Soc. Am. **68**, 1130—1134 (1978).
Epstein, P. S.: Reflection of waves in an inhomogeneous absorbing medium. Proc. nat. Acad. Sci. (Wash.) **16**, 627—637 (1930).
Exner, S.: Die Physiologie der Facettirten Augen von Krebsen und Insecten. Leipzig-Wien: Franz Deuticke, 1891
Fahrenbach, W. H.: The fine structure of a nauplius eye. Z. Zellforsch. **62**, 182—197 (1964).
Farrell, R. A., McCally, R. L.: On corneal transparency and its loss with swelling. J. Opt. Soc. Amer. **66**, 342—345 (1975).
Farrell, R. A., McCally, R. L., Tatham, P. E. R.: Wave-length dependencies of light scattering in normal and cold swollen rabbit corneas and their structural implications. J. Physiol. (Lond.) **233**, 589—612 (1973).
Fischer, A.: Laboruntersuchungen und Freilandbeobachtungen zum Sehvermögen und Verhalten von Altweltgeiern. Zool. Jb. Syst. **96**, 81—132 (1969).
Fischer, A., Horstmann, G.: Der Feinbau des Auges der Mehlmotte, *Ephestia kuehniella* Zeller (Lepidoptera, Pyralididae). Z. Zellforsch. **116**, 275—304 (1971).
Fite, K. V., Rosenfield-Wessels, S.: A comparative study of deep avian foveas. Brain Behav. Evol. **12**, 97—115 (1975).
Fox, R., Lehmkuhle, S. W., Westendorf, D. H.: Falcon visual acuity. Science **192**, 263—265 (1976).
Fraenkel, G., Rudall, K. M.: The structure of insect cuticles. Proc. roy. Soc. B **134**, 111—143 (1947).
Franceschini, N.: Sampling of the visual environment by the compound eye of the fly: fundamentals and applications. In: Snyder, A. W., Menzel, R. (Eds.): Photoreceptor Optics, pp. 98—125. Berlin-Heidelberg-New York: Springer 1975.
Franceschini, N., Kirschfeld, K.: Étude optique in vivo des éléments photorecepteurs dans l'oeil composé de *Drosophila*. Kybernetik **8**, 1—13 (1971).
Franceschini, N., Kirschfeld, K.: Le contrôle automatique du flux lumineux dans l'oeil composé des Diptères. Biol. Cybern. **21**, 181—203 (1976).
Freeman, I. L., Stevens, F. S., Jackson, D. S.: Isolation and amino acid composition of bovine corneal polymeric collagens. Biochem. biophys. Acta **154**, 252—254 (1968).
Friza, F.: Zur Frage der Färbung und Zeichnung des Facettierten Insektenauges. Z. vergl. Physiol. **8**, 290—336 (1928).
Garten, S.: Die Veränderungen der Netzhaut durch Licht. Von Graefe-Saemisch, Handb. Augenheilk., Bd. 3, S. 1—250. 1907.
Geeraets, W. J., Williams, R. C., Chan, G., Ham Jr., W. T., Guerry III, D., Schmidt, F. H.: The loss of light energy in retina and choroid. Arch. Ophthal. **64**, 606—615 (1960).
Gemne, G.: Ontogenesis of corneal surface ultrastructure in nocturnal Lepidoptera. Phil. Trans. B **262**, 343—363 (1971).

Glickstein, M., Millodot, M.: Retinoscopy and eye size. Science **168**, 605—606 (1970).
Goetz, K. G.: Behavioral analysis of the visual system of the fruitfly *Drosophila*. In: Proceedings of the symposium on information processing in light sensory systems. Pasadena: California Institute of Technology 1965.
Goldman, J. N., Benedek, G. B.: The relationship between morphology and transparency in the non-swelling corneal stroma of the shark. Invest. Ophthal. **6**, 574—600 (1967).
Goldman, J. N., Benedek, G. B., Dohlman, C. H., Kravitt, B.: Structural alterations affecting transparency in swollen human corneas. Invest. Ophthal. **7**, 501—519 (1968).
Goldsmith, T. H.: Do flies have a red receptor? J. Gen. Physiol. **49**, 265—287 (1965).
Goldsmith, T. H.: The natural history of invertebrate visual pigments. In: Dartnall, H. J. (Ed.): Handbook of sensory physiology, Vol. VII/I, pp. 685—719. Berlin-Heidelberg-New York: Springer, 1972.
Goldsmith, T. H., Bernard, G. D.: The visual system of insects. In: Rockstein, M. (Ed.): Physiology of Insects. p. 165—272, 2nd Ed., Vol. II, pp. 165—272. New York-London: Academic Press 1974.
Goldsmith, T. H., Fernandez, H. R.: The sensitivity of housefly photoreceptors in the mid-ultraviolet and the limits of the visible spectrum. J. exp. Biol. **49**, 669—677 (1968).
Govardovskii, V. J., Zueva, L. V.: Visual pigments of chicken and pigeon. Vision Res. **17**, 537—543 (1977).
Gubisch, R. W.: Optical performance of the human eye. J. Opt. Soc. Amer. **57**, 407—415 (1967).
Hackman, R. H., Goldberg, M.: Studies on the hardening and darkening of insect cuticles. J. Insect Physiol. **17**, 335—347 (1971).
Hamdorf, K.: This volume
Hart, R. W., Farrell, R. A.: Light scattering in the cornea. J. Opt. Soc. Amer. **59**, 766—774 (1969).
Hartline, H. K.: The discharge of impulses in the optic nerve of *Pecten* in response to illumination of the eye. J. cell. comp. Physiol. **11**, 465—478 (1938).
Hartridge, H.: Recent Advances in the Physiology of Vision. Philadelphia: Blakiston Company 1950.
Hassenstein, B.: Studi sulla *Lysmata seticaudata* Risso. VII. Pubbl. Staz. Zool. Napoli **25**, 1—8 (1954).
Hecht, S., Shlaer, S., Pirenne, M. H.: Energy, quanta, and vision. J. Gen. Physiol. **25**, 819—840 (1942).
Helmholtz, H.: Helmholtz's treatise on physiological optics (1911). (Translated from the Third German Edition, Vol. II). New York: Dover Publications, Inc. 1962.
Höglund, G.: Pigment migration, light screening and receptor sensitivity in the compound eye of nocturnal lepidoptera. Acta physiol. scand. **69**, Suppl. **282**, 1—56 (1966).
Höglund, G., Struwe, G.: Pigment migration and illumination of single photoreceptors in a moth. Z. vergl. Physiologie **74**, 336—339 (1971).
Höglund, G., Langer, J., Struwe, G., Thorell, B.: Spectral absorption by screening pigment granules in the compound eyes of a moth and a wasp. Z. vergl. Physiol. **67**, 238—242 (1970).
Horridge, G. A.: The retina of the locust. In: Bernhard, C. G. (Ed.): The Functional Organization of the Compound Eye, pp. 513—541. Oxford-New York: Pergamon Press 1966.
Horridge, G. A.: Optical mechanisms of clear-zone eyes. In: Horridge, G. A. (Ed.): The Compound Eye and Vision of Insects, pp. 255—298. Oxford: Clarendon Press 1975.
Horridge, G. A.: The compound eye of insects. Sci. Amer. **237**, 108—120 (1977).
Horridge, G. A., Giddings, C.: Movement on dark-light adaptation in beetle eyes of the neuropteran type. Proc. Roy. Soc. B **179**, 73—85 (1971).
Horridge, G. A., Henderson, I.: The ommatidium of the lacewing *Chrysopa* (Neuroptera). Proc. Roy. Soc. Lond. B **192**, 259—271 (1976).
Horridge, G. A., Giddings, C., Stange, G.: The superposition eye of skipper butterflies. Proc. Roy. Soc. B **182**, 457—495 (1972).
Horridge, G. A., Meinertzhagen, I. A.: The accuracy of the patterns of connexions of the first- and second-order neurons of the visual system of *Calliphora*. Proc. Roy. Soc. B **175**, 69—82 (1970).
Jacobson, S. G., Franklin, K. B. J., McDonald, W. I.: Visual acuity of the cat. Vision Res. **16**, 1141—1143 (1976).
Keesey, U. T.: Effects of involuntary eye movements on visual acuity. J. Opt. Soc. Amer. **50**, 769—774 (1960).
Kerker, M.: The scattering of light and other electromagnetic radiation. New York-London: Academic Press 1969.
Kinsey, V. E.: Spectral transmission of the eye to ultraviolet radiations. Arch. Ophthal. **39**, 508—513 (1948).

Kirschfeld, K.: Discrete and graded receptor potentials in the compound eye of the fly *(Musca)*. In: Bernhard, C. G. (Ed.): The functional Organization of the Compound Eye, Vol. 7, pp. 291—307. New York: Pergamon Press 1966.

Kirschfeld, K.: Absorption properties of photopigments in single rods, cones and rhabdomeres. In: Processing of Optical Data by Organisms and by Machines. New York and London: Academic Press 1969.

Kirschfeld, K.: Das neurale Superpositionsauge. Fortschr. Zool. **21**, 230—257 (1973).

Kirschfeld, K.: The absolute sensitivity of lens and compound eyes. Z. Naturforsch. **29c**, 592—596 (1974).

Kirschfeld, K., Franceschini, N.: Ein Mechanismus zur Steuerung des Lichtflusses in den Rhabdomeren des Komplexauges von *Musca*. Kybernetik **6**, 13—22 (1969).

Kirschfeld, K., Franceschini, N.: Photostable pigments within the membrane of photoreceptors and their possible role. Biophys. Struct. Mech. **3**, 191—194 (1977).

Kirschfeld, K., Snyder, A. W.: Measurement of a photoreceptor's characteristic waveguide parameter. Vision Res. **16**, 775—778 (1976).

Kirschfeld, K., Franceschini, N., Minke, B.: Evidence for a sensitising pigment in fly photoreceptors. Nature (Lond.) **269**, 386—390 (1977).

Kirschfeld, K., Feiler, R., Franceschini, N.: A photostable pigment within the rhabdomere of fly photoreceptors no. 7. J. comp. Physiol. **125**, 275—284 (1978).

Kolb, G.: The structure of the eye of *Pieris brassicae* L. (Lepidoptera). Zoomorphologie **87**, 123—146 (1977).

Kolb, G., Autrum, H.: Die Feinstruktur im Auge der Biene bei Hell- und Dunkeladaptation. J. comp. Physiol. **77**, 113—125 (1972).

Kolb, G., Autrum, H.: Selektive Adaptation und Pigmentwanderung in den Sehzellen des Bienenauges. J. comp. Physiol. **94**, 1—6 (1974).

Kolb, G., Autrum, H., Eguchi, E.: Die spektrale Transmission des dioptrischen Apparates von *Aeschna cyanea* Müll. Z. vergl. Physiol. **63**, 434—439 (1969).

Kreithen, M. L., Eisner, T.: Ultraviolet light detection by the homing pigeon. Nature **272**, 347—348 (1978).

Kühne, W.: The photochemistry of the retina and on visual purple. (Translated from the german). London: MacMillan and Co. 1878.

Land, M. F.: Image formation by a concave reflector in the eye of the scallop, *Pecten maximus*. J. Physiol. (Lond.) **179**, 138—153 (1965).

Land, M. F.: A multilayer interference reflector in the eye of the scallop, *Pecten maximus*. J. exp. Biol. **45**, 433—447 (1966).

Land, M. F.: Structure of the retinae of the principal eyes of jumping spiders (Salticidae: Dendryphantinae) in relation of visual optics. J. exp. Biol. **51**, 443—470 (1969).

Land, M. F.: The physics and biology of animal reflectors. Progr. Biophys. **24**, 77—105 (1972).

Land, M. F.: Superposition images are formed by reflection in the eyes of some oceanic decapod crustacea. Nature (Lond.) **263**, 764—765 (1976).

Langer, H.: Properties and functions of screening pigments in insect eyes. In: Snyder, A. W., Menzel, R. (Eds.): Photoreceptor Optics, pp. 429—455. Berlin-Heidelberg-New York: Springer 1975.

Leydig, F.: Das Auge der Gliederthiere. Tübingen: Verlag der H. Laupp'schen Buchhandlung 1864.

Liebman, P. A.: Microspectrophotometry of photoreceptors. In: Dartnall, H. J. A. (Ed.): Handbook of sensory Physiology, Vol. VII/1, pp. 481—528. Berlin-Heidelberg-New York: Springer 1972.

Liebman, P. A., Entine, G.: Visual pigments of frog and tadpole *(Rana pipiens)*. Vision Res. **8**, 761—775 (1968).

Liebman, P. A., Carroll, S., Laties, A.: Spectral sensitivity of retinal screening pigment migration in the frog. Vision Res. **9**, 377—384 (1969).

Locke, M.: The structure and formation of the cuticulin layer in the epicuticle of an insect, *Calpodes ethlius* (Lepidoptera, Hesperiidae). J. Morph. **118**, 461—494 (1966).

Locket, N. A.: The retina of *Poromitra nigrofulvus* (Garman)—an optical and electron microscope study. Exp. Eye Res. **8**, 265—275 (1969).

Locket, N. A.: Adaptations to the deep-sea environment. In: Crescitelli, F. (Ed.): Handbook of Sensory Physiology VII/5, pp. 67—192. Berlin-Heidelberg-New York: Springer 1977.

Loew, E. R., Lythgoe, J. N.: The ecology of cone pigments in teleost fishes. Vision Res. **18**, 715—722 (1978).

Ludvigh, E., McCarthy, E. F.: Absorption of visible light by the refractive media of the human eye. Arch. Ophthal. **20**, 37—51 (1938).
Lythgoe, J. N.: Iridescent corneas in fishes. Nature (Lond.) **233**, 205—207 (1971).
Lythgoe, J. N.: The adaptation of visual pigments to the photic environment. In: Dartnall, H. J. A. (Ed.): Handbook of Sensory Physiology VII/1, pp. 566—603. Berlin: Springer 1972.
Lythgoe, J. N.: The arrangement of collagen fibrils in the iridescent cornea of the scorpion fish, *Taurulus (Cottus) bubalis*, and the transparency of vertebrate corneal stroma. J. Physiol. (Lond.) **262**, 1—13 (1976).
MacNichol, E. F. Jr., Kunz, Y. W., Levine, J. S., Harosi, F. I., Collins, B. A.: Ellipsosomes: organelles containing a cytochrome-like pigment in the retinal cones of certain fishes. Science **200**, 549—552 (1978).
Makous, W.: Some functional properties of visual receptors and their optical implications. Opt. Soc. Amer. **67**, 1362 (1977).
Matsuura, T., Miller, W. H., Tomita, T.: Cone-specific c-wave in the turtle retina. Vision Res. **18**, 767—775 (1978).
Maurice, D. M.: The structure and transparency of the cornea. J. Physiol. (Lond.) **136**, 263—286 (1957).
Maurice, D. M.: The cornea and sclera. In: Davson, H. (Ed.): The Eye, Vol. 1, pp. 289—368. New York-London: Academic Press 1962.
Mazokhin-Porshnyakov, G. A.: Insect Vision. New York: Plenum Press 1969.
McFarland, W. N., Munz, F. W.: The photic environment of clear tropical seas during the day. Vision Res. **15**, 1063—1070 (1975).
Menzel, R.: The fine structure of the compound eye of *Formica polyctena*—functional morphology of a hymenopteran eye. In: Wehner, R. (Ed.): Information Processing in the Visual Systems of Arthropods, pp. 37—47. Berlin-Heidelberg-New York: Springer 1972.
Menzel, R., Blakers, M.: Colour receptors in the bee eye—morphology and spectral sensitivity. J. comp. Physiol. **108**, 11—33 (1976).
Menzel, R., Snyder, A. W.: Polarised light detection in the bee, *Apis mellifera*. J. comp. Physiol. **88**, 247—270 (1974).
Meyer-Rochow, V. B.: A crustacean-like organization of insect-rhabdoms. Cytobiologie **4**, 241—249 (1971).
Miller, W. H.: Fine structure of some invertebrate photoreceptors. Ann. N.Y. Acad. Sci. **74**, 204—209 (1958a).
Miller, W. H.: Derivatives of cilia in the distal sense cells of the retina of *Pecten*. J. biophys. biochem. Cytol. **4**, 227—228 (1958b).
Miller, W. H.: Visual photoreceptor structures. In: Brachet, J., Mirsky, A. E. (Eds.): The Cell, pp. 325—364. New York-London: Academic Press 1960.
Miller, W. H.: Optical guiding in photoreceptor cells. Fed. Proc. **35**, 37—43 (1976).
Miller, W. H., Bernard, G. D.: Skipper glow. Quarterly Progress Report No. 88, Research Laboratory of Electronics, Massachusetts Institute of Technology, 114—119 (1968a).
Miller, W. H., Bernard, G. D.: Butterfly glow. J. Ultrastruct. Res. **24**, 286—294 (1968b).
Miller, W. H., Bernard, G. D.: Red glow of sulfur butterfly is caused by green tapetum. ARVO Regional Meeting, Atlantic Section, Yale Univ., New Haven (1976).
Miller, W. H., Bernard, G. D., Allen, J. L.: The optics of insect compound eyes. Science **162**, 760—768 (1968).
Miller, W. H., Cawthon, D. F.: Pigment granule movement in *Limulus* photoreceptors. Invest. Ophthal. **13**, 401—405 (1974).
Miller, W. H., Snyder, A. W.: Optical function of myoids. Vision Res. **12**, 1841—1848 (1972).
Miller, W. H., Snyder, A. W.: Optical function of human peripheral cones. Vision Res. **13**, 2185—2194 (1973).
Miller, W. H., Snyder, A. W.: Deep fovea of birds functions as telephoto lens. Invest. Ophthal. Vis. Sci. Suppl. p. 105 (1977a).
Miller, W. H., Snyder, A. W.: The tiered vertebrate retina. Vision Res. **17**, 239—255 (1977b).
Miller, W. H., Snyder, A. W.: In preparation (1979).
Miller, W. H., Møller, A. R., Bernhard, C. G.: The corneal nipple array. In: Bernhard, C. G. (Ed.): Functional Organization of the Compound Eye, pp. 21—33. Oxford-New York: Pergamon Press 1966.
Minke, B., Kirschfeld, K.: Microspectrophotometric evidence for two photointerconvertible states of visual pigment in the barnacle lateral eye. J. gen. Physiol. **71**, 37—45 (1978).

Monaco, S. F.: Reflectance of an inhomogeneous thin film. J. Opt. Soc. Amer. **51**, 280—282 (1961).
Moreland, J. D., Lythgoe, J. N.: Yellow corneas in fishes. Vision Res. **8**, 1377—1380 (1968).
Morris, V. B.: Symmetry in a receptor mosaic demonstrated in the chick from the frequencies, spacing and arrangement of the types of retinal receptor. J. comp. Neurol. **140**, 359—398 (1970).
Munk, O.: Ocular anatomy of some deep-sea teleosts. Dana report no. 70. Copenhagen: Carlsberg Foundation 1966.
Muntz, W. R. A.: Inert absorbing and reflecting pigments. In: Dartnall, H. J. A. (Ed.): Handbook of Sensory Physiology. Vol. VII/1, pp. 529—565. Berlin-Heidelberg-New York: Springer 1972.
Muntz, F. W., McFarland, W. N.: Presumptive cone pigments extracted from tropical marine fishes. Vision Res. **15**, 1045—1062 (1975).
Murray, R. L., Dubin, M. W.: The occurrence of actinlike filaments in association with migrating pigment granules in frog retinal pigment epithelium. J. Cell Biol. **64**, 705—710 (1975).
Nicol, J. A. C.: Studies on the eyes of fishes: structure and ultrastructure. In: Ali, M. A. (Ed.): Vision in Fishes, pp. 579—608. New York-London: Plenum Press 1975.
Nicol, J. A. C., Zyznar, E. S.: The tapetum lucidum in the eye of the big-eye *Priacanthus arenatus* Cuvier. J. Fish. Biol. **5**, 519—522 (1973).
Nolte, J., Brown, J. E.: The spectral sensitivities of single receptor cells in the lateral, median, and ventral eyes of normal and white-eyed *Limulus*. J. gen. Physiol. **55**, 787—801 (1970).
Norren, D. V., Vos, J. J.: Spectral transmission of the human ocular media. Vision Res. **14**, 1237—1244 (1974).
O'Brien, B.: Vision and resolution in the central retina. J. Opt. Soc. Amer. **41**, 882—894 (1951).
Oehme, H.: Vergleichende Untersuchungen an Greifvogelaugen. Z. Morph. Ökol. Tiere **53**, 618—635 (1964).
Ogden, T. E.: The receptor mosaic of *Aotes trivirgatus:* Distribution of rods and cones. J. Comp. Neurol. **163**, 193—202 (1975).
Østerberg, G.: Topography of the layer of rods and cones in the human retina. Acta Ophthal. **13** Suppl. **6** (1935).
Papermaster, D. S., Schneider, B. G., Zorn, M. A., Kraehenbuhl, J. P.: Immunocytochemical localization of a large intrinsic membrane protein to the incisures and margins of frog rod outer segment disks. J. Cell Biol. In press
Parker, G. H.: The movements of the retinal pigment. Ergebn. Biol. **9**, 239—291 (1932).
Paulus, H. F.: Zum Feinbau der Komplexaugen einiger Collembolen. Eine vergleichend-anatomische Untersuchung (Insecta, Apterygota). Zool. Jahrb. Anat. **89**, 1—116 (1972).
Paulus, H. F.: The compound eyes of apterygote insects. In: Horridge, G. A. (Ed.): The Compound Eye and Vision of Insects, pp. 3—19. Oxford: Clarendon Press 1975.
Pedler, C.: The fine structure of the tapetum cellulosum. Exp. Eye Res. **2**, 189—195 (1963)
Pirie, A.: Crystals of riboflavin making up the tapetum lucidum in the eyes of a lemur. Nature (Lond.) **183**, 985—986 (1959).
Pitts, D. G.: Transmission of the visible spectrum through the ocular media of the bovine eye. Amer. J. Opt. **36**, 289—298 (1959).
Polyak, S. L.: The retina. Chicago: The University of Chicago Press 1941.
Pumphrey, R. J.: The theory of the fovea. J. exp. Biol. **25**, 299—310 (1948).
Pumphrey, R. J.: Concerning vision. In: Ramsay, J. A., Wigglesworth, V. B. (Eds.): The Cell and the Organism. Cambridge: University Press 1961 a.
Pumphrey, R. J.: Sensory organs: vision. In: Biology and Comparative Physiology of Birds. New York: Academic Press 1961 b.
Rayleigh, Lord: On the light from the sky, its polarization and colour. Philos. Mag. **41**, 107—120, 274—279 (1871).
Rayleigh, Lord: On the electromagnetic theory of light. Philos. Mag. **12**, 81—101 (1881).
Rayleigh, Lord: Sky. In: Encyclopedia Britannica, Vol. XXV, 11th Ed. 1911.
Ribi, W. A.: Ultrastructure and migration of screening pigments in the retina of *Pieris rapae* L. (Lepidoptera, Pieridae). Cell Tiss. Res. **191**, 57—73 (1978).
Riggs, L. A.: Visual acuity. In: Graham, C. H. (Ed.): Vision and Visual Perception. New York-London-Sydney: John Wiley 1966.
Robson, J. G., Enroth-Cugell, C.: Light distribution in the cat's retinal image. Vision Res. **18**, 159—173 (1978).
Rose, A.: The sensitivity performance of the human eye on an absolute scale. J. Opt. Soc. Amer. **38**, 196—208 (1948).

Roth, W.: Ein paar interessante Aquarienfische. Dtsch. Fischerei Korrespondenz **15**, 107—109 (1911).
Schneider, L., Gogala, M., Drašlar, K., Langer, H., Schlecht, P.: Feinstruktur und Schirmpigment-Eigenschaften der Ommatidien des Doppelauges von *Ascalaphus* (Insecta, Neuroptera). Cytobiologie **16**, 274—307 (1978).
Scholes, J.: The electrical responses of the retinal receptors and the lamina in the visual system of the fly *Musca*. Kybernetik **6**, 149—162 (1969).
Sherk, T. E.: Development of the compound eyes of dragonflies (Odonata). J. exp. Zool. **201**, 391—416 (1977).
Sherk, T. E.: Development of the compound eyes of dragonflies (Odonata). III. Adult compound eyes. J. exp. Zool. **203**, 61—80 (1978).
Shlaer, R.: An eagle's eye: quality of the retinal image. Science **176**, 920—922 (1972).
Shlaer, S.: The relation between visual acuity and illumination. J. gen. Physiol. **21**, 165—188 (1937).
Sidman, R.: The structure and concentration of solids in photoreceptor cells studied by refractometry and interference microscopy. J. biophys. biochem. Cytol. **3**, 15—30 (1957).
Sivak, J. G., Bobier, W. R.: Chromatic aberration of the fish eye and its effect on refractive state. Vision Res. **18**, 453—455 (1978).
Skrzipek, K.-H., Skrzipek, H.: Die spektrale Transmission und die optische Aktivität des dioptrischen Apparates der Honigbiene (*Apis mellifica* L.). Experientia **30**, 314—315 (1974).
Snyder, A. W.: Polarization sensitivity of individual retinula cells. J. comp. Physiol. **83**, 331—360 (1973).
Snyder, A. W.: Photoreceptor optics—theoretical principles. In: Snyder, A. W., Menzel, R. (Eds.): Photoreceptor Optics, pp. 38—55. Berlin-Heidelberg-New York: Springer 1975.
Snyder, A. W.: Acuity of compound eyes: physical limitations and design. J. comp. Physiol. **116**, 161—182 (1977).
Snyder, A. W., Menzel, R., Laughlin, S. B.: Structure and function of the fused rhabdom. J. comp. Physiol. **87**, 99—135 (1973).
Snyder, A. W., Miller, W. H.: Fly colour vision. Vision Res. **12**, 1389—1396 (1972).
Snyder, A. W., Miller, W. H.: Photoreceptor diameter and spacing for highest resolving power. J. Opt. Soc. Amer. **67**, 696—698 (1977a).
Snyder, A. W., Miller, W. H.: Telephoto lens system of falconiform eyes. Nature **275**, 127—129 (1978)
Snyder, A. W., Pask, C.: Spectral sensitivity of dipteran retinula cells. J. comp. Physiol. **84**, 59—76 (1973).
Stavenga, D. G.: Visual adaptation in butterflies. Nature (Lond.) **254**, 435—437 (1975a).
Stavenga, D. G.: Optical qualities of the fly eye—an approach from the side of geometrical, physical, and waveguide optics. In: Photoreceptor optics. Berlin-Heidelberg-New York: Springer 1975.
Stavenga, D. G.: Waveguide modes and refractive index in photoreceptors of invertebrates. Vision Res. **15**, 323—330 (1975b).
Stavenga, D. G., Kuiper, J. W.: Insect pupil mechanisms. I. J. comp. Physiol. **113**, 55—72 (1977).
Stavenga, D. G., Numan, J. A. J., Tinbergen, J., Kuiper, J. W.: Insect pupil mechanisms. II. J. comp. Physiol. **113**, 73—93 (1977).
Stavenga, D. G., Zantema, A., Kuiper, J. W.: Rhodopsin processes and the function of the pupil mechanism in flies. In: Langer, H. (Ed.): Biochemistry and Physiology of Visual Pigments, pp. 175—180. Berlin-Heidelberg-New York: Springer 1973.
Steinberg, R. H., Reid, M., Lacy, P. L.: The distribution of rods and cones in the retina of the cat *(Felis domesticus)*. J. comp. Neurol. **148**, 229—248 (1973).
Stell, W. K., Harosi, F. I.: Cone structure and visual pigment content in the retina of the goldfish. Vision Res. **16**, 647—658 (1976).
Steyskal, G. C.: Notes on color and pattern of eye in diptera. II. Bull. Brooklyn Entomol. Soc. **52**, 89—94 (1957).
Swihart, S. L., Gordon, W. C., Machwart, R. J.: Reflections on the eyes of butterflies. J. Insect Physiol. **20**, 359—381 (1974).
Tobey, F. L. Jr., Enoch, J. M., Scandrett, J. H.: Experimentally determined optical properties of goldfish cones and rods. Invest. Ophthal. **14**, 7—23 (1975).
Trujillo-Cenóz, O., Bernard, G. D.: Some aspects of the retinal organization of *Sympycnus lineatus* Loew (Diptera, Dolichopodidae). J. Ultrastruc. Res. **38**, 149—160 (1972).
Twersky, V.: Transparency of pair-correlated, random distributions of small scatterers, with applications to the cornea. J. Opt. soc. Amer. **65**, 524—530 (1975).
Underwood, G.: On the visual-cell pattern of a homalopsine snake. J. Anat. **100**, 571—575 (1966).

Varela, F. G., Wiitanen, W.: The optics of the compound eye of the honeybee *(Apis mellifera)*. J. Gen. Physiol. **55**, 336—358 (1970).
Versluys, J., Demoll, R.: Das *Limulus*-Problem. Ergeb. Fortschr. Zool. **5**, 67—388 (1923).
Vogt, K.: Zur Optik des Flußkrebsauges (Optics of the crayfish eye). Z. Naturforsch. **30c**, 691 (1975).
Wässle, H.: Optical quality of the cat eye. Vision Res. **11**, 995—1006 (1971).
Wässle, H., Riemann, H. J.: The mosaic of nerve cells in the mammalian retina. Proc. Roy. Soc. B **200**, 441—461 (1978).
Walcott, B.: Anatomical changes during light-adaptation in insect compound eyes. In: Horridge, G. A. (Ed.): The Compound Eye and Vision of Insects, pp. 20—36. Oxford: Clarendon Press, 1975
Wald, G.: Blue-blindness in the normal fovea. J. Opt. Soc. Amer. **57**, 1289—1301 (1967).
Walls, G. L.: Significance of the foveal depression. Arch. Ophthal. **18**, 912—919 (1937).
Walls, G. L.: Postscript on image expansion by the foveal clivus. Arch. Ophthal. **23**, 831—832 (1940).
Walls, G. L.: The Vertebrate Eye and its Adaptive Radiation. New York-London: Hafner Publ. Comp. 1967.
Walls, G. L., Judd, H. D.: The intra-ocular colour-filters of vertebrates. Br. J. Ophthalmol., Nov. 1933, pp. 641—675; Dec. 1933, pp. 705—725 (1933).
Wasserman, G. S.: Density spectrum of *Limulus* screening pigment. J. Gen. Physiol. **50**, 1075—1077 (1967).
Wehner, R., Bernard, G. D., Geiger, E.: Twisted and non-twisted rhabdoms and their significance for polarization detection in the bee. J. comp. Physiol. **104**, 225—245 (1975).
Westheimer, G.: Modulation thresholds for sinusoidal light distributions on the retina. J. Physiol. (Lond.) **152**, 67—74 (1960).
Westheimer, G.: Optical properties of vertebrate eyes. In: Fuortes, M. G. F. (Ed.): Handbook of Sensory Physiology, VII/2, pp. 449—482. Berlin-Heidelberg-New York: Springer, 1972a.
Westheimer, G.: Visual acuity and spatial modulation thresholds. In: Jameson, D., Hurvich, L. M. (Eds.): Handbook of Sensory Physiology VII/4, pp. 170—187. Berlin-Heidelberg-New York: Springer 1972b
Westheimer, G.: Visual acuity and hyperacuity. Invest. Ophthal. **14**, 570—572 (1975).
White, R. H., Lord, E.: Diminution and enlargement of the mosquito rhabdom in light and darkness. J. gen. Physiol. **65**, 583—598 (1975).
Wiesinger, H., Schmidt, F. H., Williams, R. C., Tiller, C. O., Ruffin, R. S., Guerry III, D., Ham Jr., W. T.: The transmission of light through the ocular media of the rabbit eye. Amer. J. Ophthal. **42**, 907—910 (1956).
Wijngaard, W., Stavenga, D. G.: On optical crosstalk between fly rhabdomeres. Biol. Cybern. **18**, 61—67 (1975).
Wilson, M.: Angular sensitivity of light and dark adapted locust retinula cells. J. comp. Physiol. **97**, 323—328 (1975).
Wolbarsht, M. L.: The function of intraocular color filters. Fed. Proc. **35**, 44—50 (1976).
Wood, R. W.: Physical Optics. (See p. 427). New York: Dover Publications, Inc., 1967.
Woodhouse, J. M., Campbell, F. W.: The role of the pupil light reflex in aiding adaptation to the dark. Vision Res. **15**, 649—653 (1975).
Yagi, N., Koyama, N.: The Compound Eye of Lepidoptera. Approach from Organic Evolution. Tokyo: Shinkyo-Press & Co., Ltd., 1963.
Zuckerman, J., Srinivasan, M., Miller, W. H.: Central obscuration of the vertebrate pupil. In preparation 1979.
Zyznar, E. S.: Tapeta lucida and the organisation of visual cells in teleosts. In: Vision in Eyes. New York-London: Plenum Press 1975a.
Zyznar, E. S.: Theoretical considerations about tapeta lucida. In: Vision in Eyes. New York-London: Plenum Press, 1975b.
Zyznar, E. S., Ali, M. A.: An interpretative study of the organization of the visual cells and tapetum lucidum of *Stizostedion*. Can. J. Zool. **53**, 180—196 (1975).

Chapter 4

The Physiology of Invertebrate Visual Pigments

K. HAMDORF, Bochum, Germany

With 39 Figures

Contents

A. Introduction

The visual pigments of all invertebrates investigated so far are membrane-coupled chromoproteids having similar spectral absorption characteristics, the chromophore of which has been, in certain species, shown to be the same as that of the vertebrate rhodopsin (11-*cis*-retinal). In this study rhodopsin is symbolized by P and its photoproduct metarhodopsin by M, and where wavelengths of maximal absorption (λ_{max}) for these are known, then this is also denoted, e.g., P490, or M580.

In the in vitro vertebrate retina, as well as in extraction experiments, quantum absorption leads to the bleaching of the visual pigment, i.e., to the separation of the chromophore, the all-*trans* retinal from the protein opsin (Fig. 1), and the hypothesis of the universality of the correlation of receptor excitation and adaptation to the bleaching reaction was advanced for many years, until the thermostable visual pigments were first discovered in the cephalopods (BROWN and BROWN, 1958; HUBBARD and ST. GEORGE, 1958; see also review by GOLDSMITH, 1972).

In the cephalopod visual pigment the light-induced reaction chain was found to stop with the metarhodopsin intermediates, the chromophore remaining attached to the opsin. Both tautomeric forms of metarhodopsin (the acid, Ms, and the basic, Mb) then reconverted to rhodopsin by further quantal absorption, although it was later shown for the cephalopod *Eledone* (SCHWEMER, 1969), that long-term dark regeneration of rhodopsin also took place. In this species then, the regeneration system is a mixed one, consisting of a fast photoregeneration and a much slower, metabolic-dependent dark reaction.

Later, in some other invertebrates having rhabdomeric structures, the same principle of photo-regeneration as found in the cephalopods was discovered and a generalization in this respect to all other invertebrates quickly followed. For example, in preparing slices from cephalopod retinas for microspectrophotometric measurements, glutaraldehyde was used as a fixative and the preparation seemed to behave as in vivo. This finding was extended to other invertebrates and the irreversible bleaching which in some species occurred in vitro was thought to be the naturally occurring phenomenon. There were, however, problems with such extensions of findings. The *Ascalaphus* visual pigment, for instance, exhibits the same in vitro thermostability as do the cephalopods' pigment; with *Deilephila*, extraction measurements show irreversible bleaching, although the system is really thermostable in nature. It is important therefore, when making in vitro measurements from either 1) isolated whole membranes, or 2) pigment extracts, that care is applied to the subjected ionic milieu, the pH, the nature of the extracting agents, the fixatives used, etc.

The most salient differences between the invertebrate and the vertebrate visual pigment systems are then the following (see Fig. 1). As far as has been detected, the number of the visual pigment intermediates are fewer in the invertebrates (in *Ascalaphus* even Mb is naturally excluded from the sequence of events). Also, while in vertebrates the all-*trans* retinal detaches from the opsin and is utilized, and new 11-*cis* retinal must be provided to combine with opsin in the dark regenerating

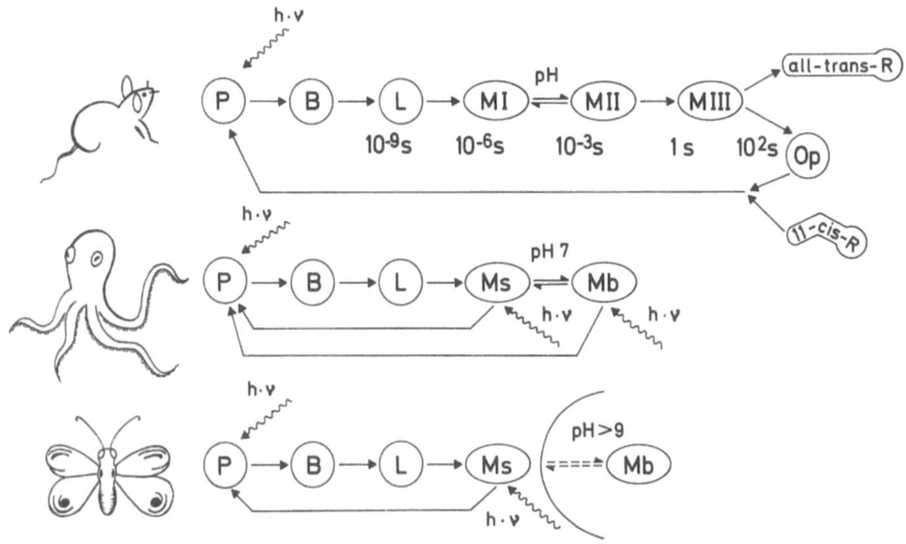

Fig. 1. The reaction scheme after quantum absorption in a vertebrate, a cephalopod and in an insect (*Ascalaphus*). *Wavy arrows* indicate light reactions; *straight arrows* the temperature-dependent dark reactions. *P*, rhodopsin; *B*, prelumirhodopsin (a collective term for both batho- and hypsorhodopsin, see YOSHIZAWA and HORIUCHI, 1973); *L*, lumirhodopsin; *MI*, *MII*, and *MIII*, metarhodopsin of vertebrates; *Ms* and *Mb*, acid and basic metarhodopsins of invertebrates; *Op*, the protein opsin. MI⇌MII and Ms⇌Mb are pH-dependent reactions. N.B., in cephalopods the pK of the Ms⇌Mb transition varies between 6.3 (in *Eledone*, SCHWEMER, 1969) and 9.1 (in *Todarodes pacificus*, HARA and HARA, 1972), indicating that perhaps in some species under natural conditions, existence, and therefore contribution of Mb may be extremely small. In insects same pK range seems to be much narrower, favoring the predominance of the acid metarhodopsin, *Ms*

process, in invertebrates the chromophore is not separated from opsin and the regeneration of rhodopsin is effected by further quantal absorption. This principle of pigment photoregeneration (Fig. 2) seems to be widely distributed in invertebrate receptors with rhabdomeric light-absorbing structures (see review by EAKIN, 1972).

This photochemical "flip-flop" system of excitation and regeneration of the invertebrate visual pigment will be seen to be of great physiologic importance. For example, the naturally existing light conditions in the biotope (the spectral composition of the skylight) so affects this mechanism that in such light-adapted receptors the rhodopsin content is shifted to quite high values at the expense of the concentration of the intermediates.

Also the absorption spectrum of the irreversible thermostable system of rhodopsin and its metarhodopsin covers a wide spectral region (Fig. 2), and the absorption maxima of these two photointerconvertible states are either shifted from one to another or almost overlapping, and are seemingly not predictable by a common rule. It will be shown later how these shifts have distinct advantages for the individual species in relation to their natural habitats.

With the possible exception of the cephalopods, other invertebrate visual systems exhibit a high correspondence between the electric response and the visual

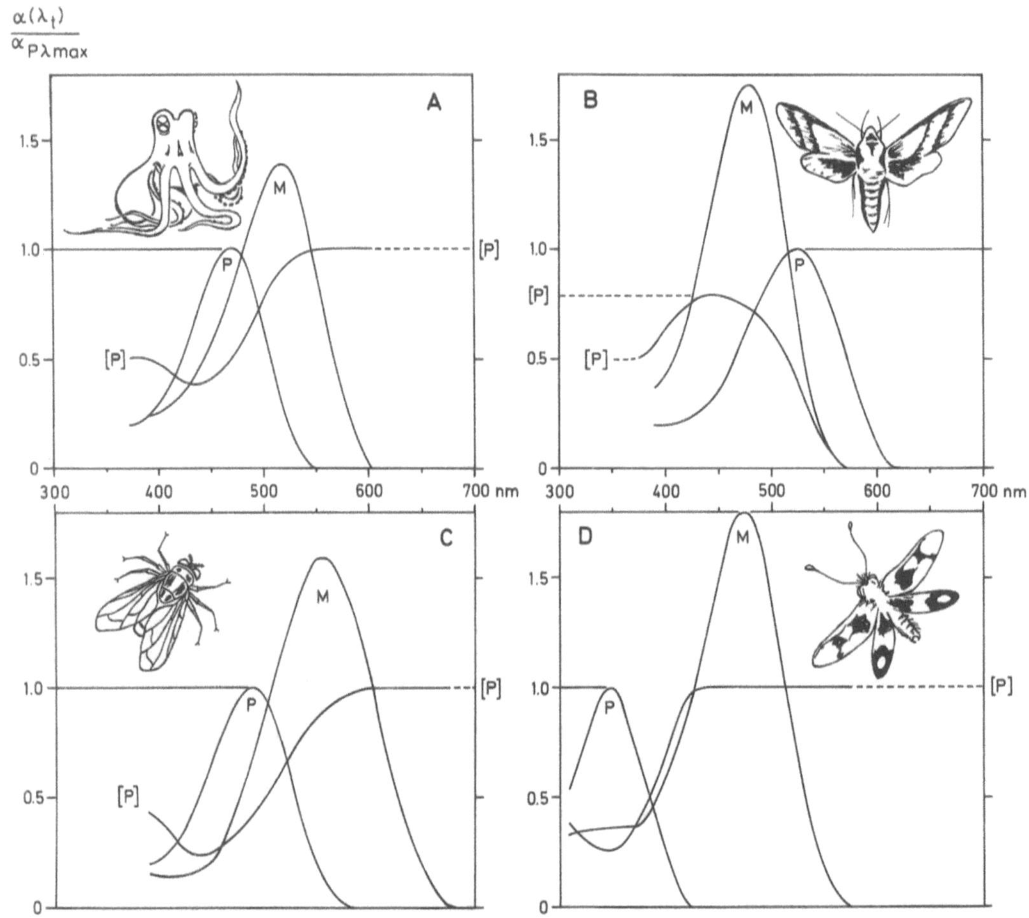

Fig. 2. Visual pigment system of cephalopod, *Eledone* (A), green receptor of sphingid moth, *Deilephila* (B), receptors R1–6 of blowfly (C), and UV receptors of owlfly, *Ascalaphus* (D). *P* and *M* refer to spectral absorptions of rhodopsin and its metarhodopsin respectively. N.B., in all four cases M peaks are higher than P peaks. Figure shows inter-species individuality of shifts in λ_{max} of P and M and their positions in the spectrum. In cephalopod peaks are not far apart; in *Loligo* and *Sepia* (not shown) they coincide almost exactly. Curves marked [*P*] show the relative rhodopsin content at equilibria as a function of the wavelengths of adaptation (see eqn. 3)

pigment content. The recently employed method of altering the total visual pigment concentrations thus opens a new experimental approach to the problems of the primary events of the membrane photoexcitation and the secondary adaptation mechanisms.

Lastly, the revelation of the thermostable visual pigment molecules in the microvillus membrane by electron microscopy, together with data from microspectrophotometry, electrophysiology and biochemistry indicate the microvillus structure to be the possible functional entity of species with rhabdomeric structures.

B. Physiology and Experimental Approaches

I. Basic Light and Dark Reactions

1. Receptors of Cephalopods

Since the biochemical scheme of the cephalopod visual pigments has been recently reviewed (GOLDSMITH, 1972; HARA and HARA, 1972), we may thus comment only on its functional aspects. If an acid rhodopsin extract of *Eledone moschata* (a species of octopus active only in dim light) is irradiated with monochromatic light, then after a certain time a photoequilibrium will be established between the P→Ms and the Ms→P transitions (Fig. 3), resulting in apparently fixed P and Ms concentrations. If the wavelength of the isosbestic point[1] is used for adaptation, a mixture of precisely 50%P and 50% Ms is found in the extract. If a wavelength is used which is more strongly absorbed by the visual pigment than by its Ms, i.e., 440 nm, the photoequilibrium is shifted to greater Ms content.

The use of a wavelength more strongly absorbed by Ms than by P shifts the equilibrium towards greater P concentration (90%P, 10% Ms; Fig. 3a). From Fig. 3b it may be seen that there is good agreement between the experimentally measured pigment concentrations and the calculated $C_{P_{eq}}$ values. This indicates that the photoequilibria are dominantly determined by the ratio of the absorption coefficients of both states, P and Ms [see Eqs. (3) and (4) in Sec. C.I], and this suggests that under physiological light conditions the visual pigment concentration $C_{P_{eq}}$ in the cephalopod receptor will never sink below 60%.

Apparently, in the living retina of cephalopods the metarhodopsins are reconverted also in the dark to visual pigments, i.e., after saturating monochromatic light adaptation, the pigment content increases from the photoequilibrium concentration by about 10% within 10 min (at 20° C), and after 2 hrs in the dark about 95% of the original P content is reached (SCHWEMER, 1969).

This slow dark reaction may be attributed to the retinochrome pigment system which has been found accumulated in the receptor cell somata, incorporated possibly in myeloid bodies nearby the basic membrane. These pigments most likely contain the all-*trans* retinal as the stable chromophoric group which is then photoconverted into 11-*cis* retinal. This 11-*cis* retinal intermediate is thermolabile in digitonin extracts. This may suggest that the retinochromes may supply the 11-*cis* retinal for visual pigment regeneration in the dark. Relevant enzymes catalyzing the suggested reaction are unknown (HARA and HARA, 1972, 1973).

Presently, only a few experimental data concerning the physiologic relevance of this pigment-regenerating cycle are available. In *Eledone*, bright light adaptation results in a long lasting sensitivity loss, depending on intensity, duration, and the color of the adapting light, and from Fig. 4 it can be seen that the slope of the potential increase in the dark depends dominantly on the number of quanta absorbed by the receptor during light adaptation. Together with biochemical data

[1] This term refers, on the spectrum scale, to wavelengths for which the interconvertible states of a visual pigment have equal absorption coefficients ($\alpha_P = \alpha_M$), see Fig. 3.

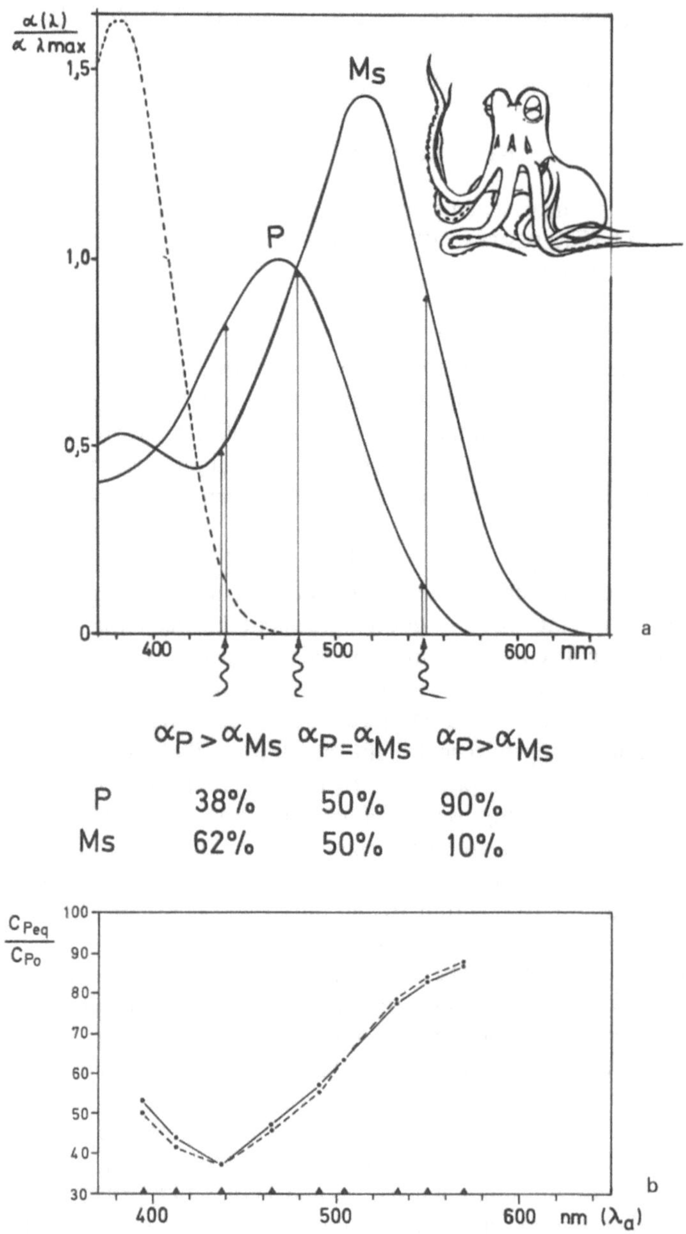

Fig. 3. (*a*) The visual pigment *P* and its acid (*Ms*) and basic (*Mb*) metarhodopsins of the cephalopod *Eledone moschata* and the concentrations of P and Ms in photoequilibrium state (C_{Peq}, C_{Mseq}) after irradiation with wavelengths 440 nm, 480 nm, and 550 nm. *Eledone* pigment is specially favored for such experiments, as absorption peaks of P and Ms are more spectrally separated than in most other cephalopod visual pigments (*Mb*, dashed curve). (*b*) Concentration of visual pigment in photoequilibrium state, depending on wavelength of adaptation ($C_{Peq}[\lambda_a]$). Experimental values: *dots* connected by *dashed line*. Calculated values, assuming relative quantum efficiency $\gamma_{M/P}=1$, connected by *continuous line* (after SCHWEMER, 1969)

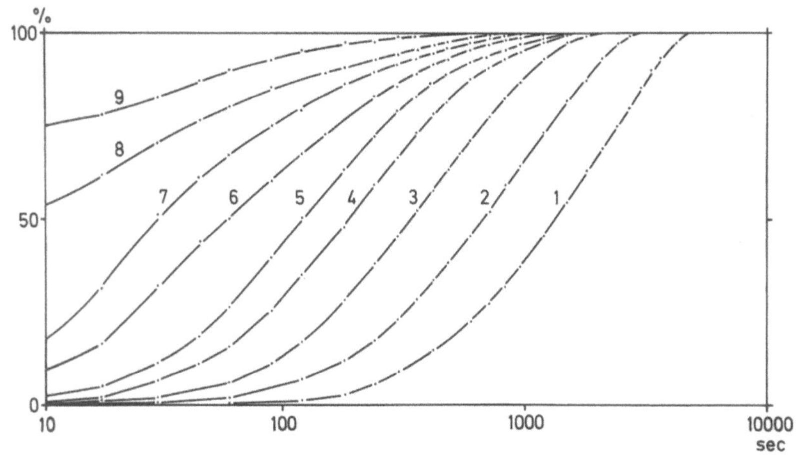

Fig. 4. Time course of dark adaptation after light adaptation by monochromatic flashes (508 nm, $\simeq 2$ ms) of different intensities. Numbers 1–9 on adaptation curves correspond to intensity steps of 0.4 log units. Almost parallel curves 1–5 show that duration of response increase depends strongly on intensity of adapting flash

on the time course of the P regeneration in the dark, this implies that the time course of the sensitivity increase in this cephalopod is determined presumably by the renewal of the visual pigment in the dark.

Specific color-dependent effects, however, have also been shown to exist. After quanta-equal monochromatic adaptation, the time course of the dark adaptation reflects the λ-specific photoequilibrium (created by the adapting light) and not the spectral sensitivity curve of the *Eledone* receptors. The most effective wavelength (440 nm) to produce the lowest pigment concentration induces the longest dark adaptation, whereas after 500 nm (which is almost equal to 440 nm in the response amplitude) the duration of dark adaptation is almost half as long (HAMDORF, 1970).

In conclusion, it is seen that the regeneration of the visual pigment in cephalopods is a mixed process, a photo back-reaction and a slower dark one, a system which lends itself to the mathematic formalism presented in the theoretic section, and it will be shown later that this dark regeneration process in insects is either entirely absent or greatly diminished (Secs. B.II, V, VI).

2. A Note on Crustacea

The visual system of the Crustacea has been recently reviewed (GOLDSMITH, 1972), and will not be dealt with here in detail except for the following. Microspectrophotometric measurements have shown (HAYS and GOLDSMITH, 1969) that in some species (e.g., the spider crab, *Libinia emarginata*) the metarhodopsins are thermostable, as is the case with the cephalopods, and it was shown moreover, that the thermostability of the metarhodopsin as well as the rhodopsin was fixation dependent (GOLDSMITH, 1972). And again in the cases of the spider crab, the blue crab (*Callinectes*), and the green crab (*Carcinus*), where the spectra of

P and M overlap, the small amount of bleaching observed has suggested a photoequilibrium state (GOLDSMITH, 1973), similar to that in the cephalopods.

As with the cephalopods, immediately after light adaptation a long-lasting loss in sensitivity has also been observed in crabs and lobsters by WALD (1968). It was electrophysiologically determined that after a fast initial rise in sensitivity there followed a much slower sensitivity increase of 1–1.5 logarithmic units over a period of 2–3 hrs. This latter slower increase has been interpreted (WALD, 1968) as an indication of a second, temperature-dependent regeneration of rhodopsin.

II. *Ascalaphus*, a Model of Insect Vision

Since the original work of K. v. FRISCH, demonstrating through behavioral experiments that worker bees perceived UV radiation, UV receptors have been identified in the eyes of numerous insects, mainly by electrophysiologic techniques.

In most insect eyes so far investigated, UV receptors are smaller cells and rare in number (see Sec. B.V), while in some species, for example in the neuropter *Ascalaphus macaronius* Skop., which is active in bright sunshine only, UV receptors are the most frequently found cells. The superposition eye of this species is divided into two morphologically distinguishable spheres: the upper and larger (the so called frontal eye) which is exclusively sensitive to UV radiation, whereas the lateral eye, built of smaller facets, responds also to blue-green light but with less sensitivity (about 1%) than to UV (GOGALA, 1967). For detailed analysis of eye structure see SCHNEIDER et al., 1978.

This UV-sensitive eye was the key to a better understanding of insect vision, and it was through experiments with the frontal eye of *Ascalaphus* that most of the photometric, the electrophysiologic, and the analysis of photoconvertible visual pigment systems came to be developed (GOGALA et al., 1970; HAMDORF et al., 1971; SCHWEMER et al., 1971; PAULSEN and SCHWEMER, 1972; HAMDORF and GOGALA, 1973; HAMDORF et al., 1973a, b, 1974, 1975).

Absorption of UV quanta by a digitonin extracted visual pigment, P345, causes the formation of the photoproduct Ms475 of high molar absorbance in the blue range of the visible spectrum. Blue, green or yellow illumination completely reconverts this thermostable Ms475 into the UV visual pigment. This is shown by the difference spectra in Fig. 5b, adaptation with 354 nm (Curve *1*) causes a maximal decrease of extinction at 345 nm and a maximal increase at 475 nm; subsequent illumination with 571 nm totally inverts this difference spectrum (Curve *2*). For a collection of visual pigment molecules, of course, UV irradiation leads to a statistical distribution of states, a photoequilibrium between P345 and Ms475. Like in cephalopods, these equilibria are directly proportional to the ratio of the absorption coefficients of both pigments at the wavelength of adaptation, and thus the concentrations of P345 and Ms475 (C_{Peq}, C_{Meq}) in the photoequilibrium can be calculated by Eq. (3) (Fig. 5a, Curve *3*), and there is good agreement between such calculations and the experimentally determined concentrations (by microspectrophotometry) in digitonin extracts, as well as in situ (Fig. 5b, Curve *3*).

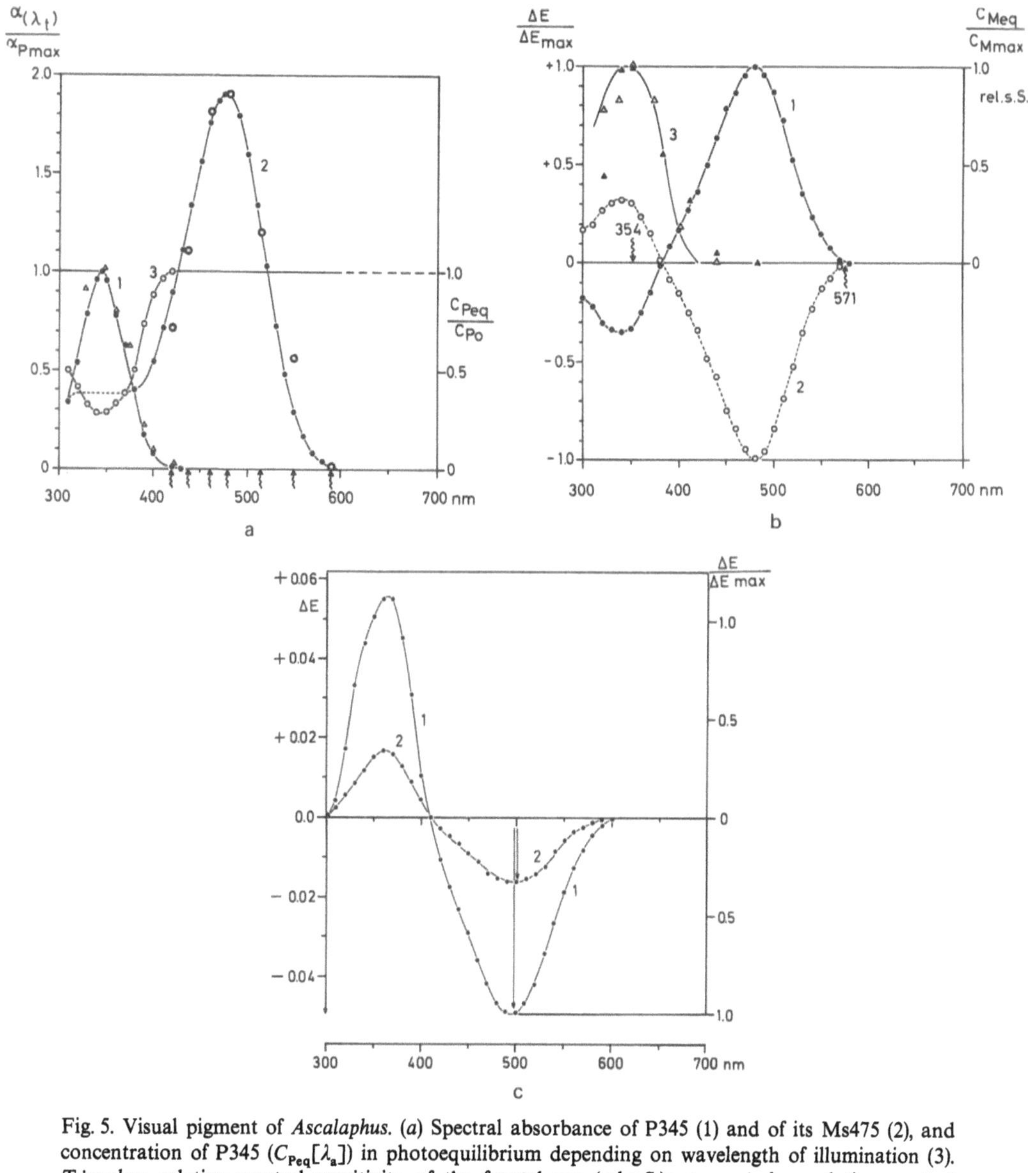

Fig. 5. Visual pigment of *Ascalaphus*. (*a*) Spectral absorbance of P345 (1) and of its Ms475 (2), and concentration of P345 ($C_{Peq}[\lambda_a]$) in photoequilibrium depending on wavelength of illumination (3). *Triangles*: relative spectral sensitivity of the frontal eye (rel. s.S.); *open circles*: relative spectral sensitivity of photoregeneration based on electrophysiologic data similar to that in Fig. 7c, normalized for comparison with peak of Ms475 (rel. s.S.). (*b*) Difference spectra of P345 in digitonin extract after irradiation at 354 nm (1), and after subsequent 571 nm irradiation (2). Amplitudes of these difference spectra measured at 475 nm are proportional to the Ms content in photoequilibrium, which depends on wavelengths of UV irradiation; $C_{Mseq}(\lambda_a)$ is calculated by Eq. (3) (Sec. C.I.1) and normalized to the most effective irradiation (3). *Filled triangles*: experimental data from extract; *open triangles*: experimental data from retina. (*c*) Cattle rhodopsin synthesized from chromophore of UV pigment of *Ascalaphus*. Difference spectra show bleaching of synthesized cattle rhodopsin in presence of NH_2OH. Cattle rhodopsin formed from denatured P345 (1). Cattle rhodopsin formed from UV (354 nm) irradiated P345 (2). In second experiment only 1/3 of previously formed cattle rhodopsin is synthesized (after GOGALA, 1967; GOGALA et al., 1970; HAMDORF et al., 1971, 1973; SCHWEMER et al., 1971; PAULSEN and SCHWEMER, 1972)

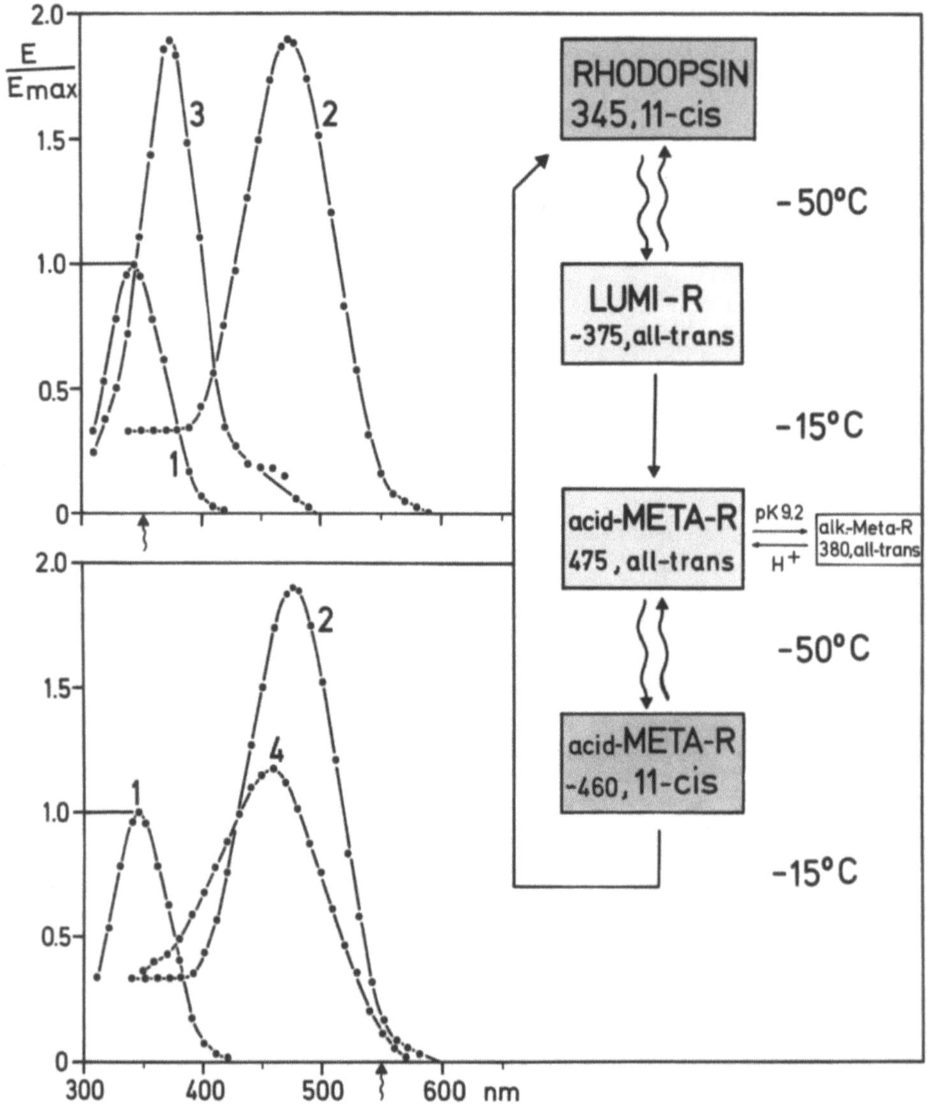

Fig. 6. Light and dark reactions of P345 of *Ascalaphus*. *Left*: Spectral absorbance of P345 (1), acid metarhodopsin Ms475 (2), lumirhodopsin (3), and 11-*cis* acid metarhodopsin (4). Spectra of intermediates (3 and 4) are calculated from difference spectra. *Wavy arrows*: experimentally used wavelengths (λ_a) to form intermediates (3 and 4). *Right*: Scheme of photoreactions (*wavy arrows*) and dark reactions (*straight arrows*). Alkaline metarhodopsin does not occur in measurable amounts at physiologically existing pH values, and its spectral absorbance is very similar to curve (3) (HAMDORF et al., 1973)

The UV visual pigment is an 11-*cis* retinal proteid (PAULSEN and SCHWEMER, 1972), the chromophore of which can be separated from the protein and combined with cattle opsin. Bleaching of this synthesized "cattle" rhodopsin in the presence of NH_2OH produces a difference spectrum of high amplitude shown in Fig. 5c,

Curve *1*. If the separation experiment is carried out on a UV-irradiated pigment sample, the "cattle" rhodopsin formed is only about one-third of that obtained in a control experiment without previous UV irradiation. This value corresponds to the content of 11-*cis* retinal which is to be expected at a photoequilibrium between the 11-*cis* pigment P345 and its all-*trans* photoproduct Ms475.

Under UV illumination and at $-50°$ C, P345 becomes converted into an all-*trans* pigment (HAMDORF et al., 1973), absorbing strongly in the UV region (Fig. 6). This intermediate, which is probably comparable to the lumirhodopsin of cephalopods, when warmed turns into Ms475, the all-*trans* acid metarhodopsin. The latter, illuminated at 550 nm and at $-50°$ C turns into another intermediate which does not absorb as strongly as Ms475 and differs in λ_{max} by a small shift. While the protein configuration of this compound should be analogous to that of the all-*trans* acid Ms475, its chromophore is 11-*cis* retinal, because on warming, and in darkness, it is converted back to P345. This latter intermediate, therefore, is designated as 11-*cis* acid metarhodopsin. A comparable 11-*cis* Ms occurs in *Eledone* during photoreconversion and here the alkaline metarhodopsin (all-*trans* Mb) is also converted to the corresponding 11-*cis* Mb (HAMDORF et al., 1973). Corresponding intermediates of the regeneration chain have been also described in the squid and in the octopus (SUZUKI et al., 1972, 1973; AZUMA et al., 1975). In the visual cycle of *Ascalaphus*, however, Mb is found only in negligible amounts, as the pK (9.2) of the indicator reaction Ms⇌Mb is out of the physiologic range (SCHWEMER et al., 1971).

The rate of the photoregeneration depends on the intensity of illumination, as shown by microspectrophotometric measurements on the retina (Fig. 7a). The time courses of transmission changes at 475 nm were recorded after saturating UV illumination. The rate increases with the intensity of the 475 nm measuring beam, and Fig. 7b shows a direct proportionality between the reciprocal intensity of the regenerating light and the rate of photoreconversion, indicating a first order reaction. This relationship has also been electrophysiologically demonstrated (HAMDORF et al., 1971).

Exposure of the eye of *Ascalaphus* to UV light causes a considerable decrease in the response amplitude (Fig. 7c), and the amplitude increases only slightly after cessation of the adapting light. During exposure to blue light (480 nm), however, the amplitude increases to the original value. The rate of this sensitivity increase is directly proportional to the intensity of the blue light and varies with the wavelength of the regenerating light.

Using equal quanta illumination in the range of 400 nm to 589 nm, the rate of regeneration was found to be maximal at 475 nm. Comparing all other rates $t_{50}(\lambda_{a_{var}})$ to the maximal value $[t_{50}(\lambda_{a_{max}})/t_{50}(\lambda_{a_{var}})]$, the relative sensitivity curve for photoregeneration is obtained, which is directly proportional to the absorbance spectrum of Ms475 (Fig. 5a, *open circles*). This would be expected if a proportionality exists between the response amplitude of the receptor and its P345 concentration. The decay of C_M during irradiation of intensity I and wavelength λ_a is given by:

$$C_{Ms}(\lambda_a, I, t) = C_{Ms0} \cdot \exp - \alpha_M(\lambda_a)\gamma I t$$

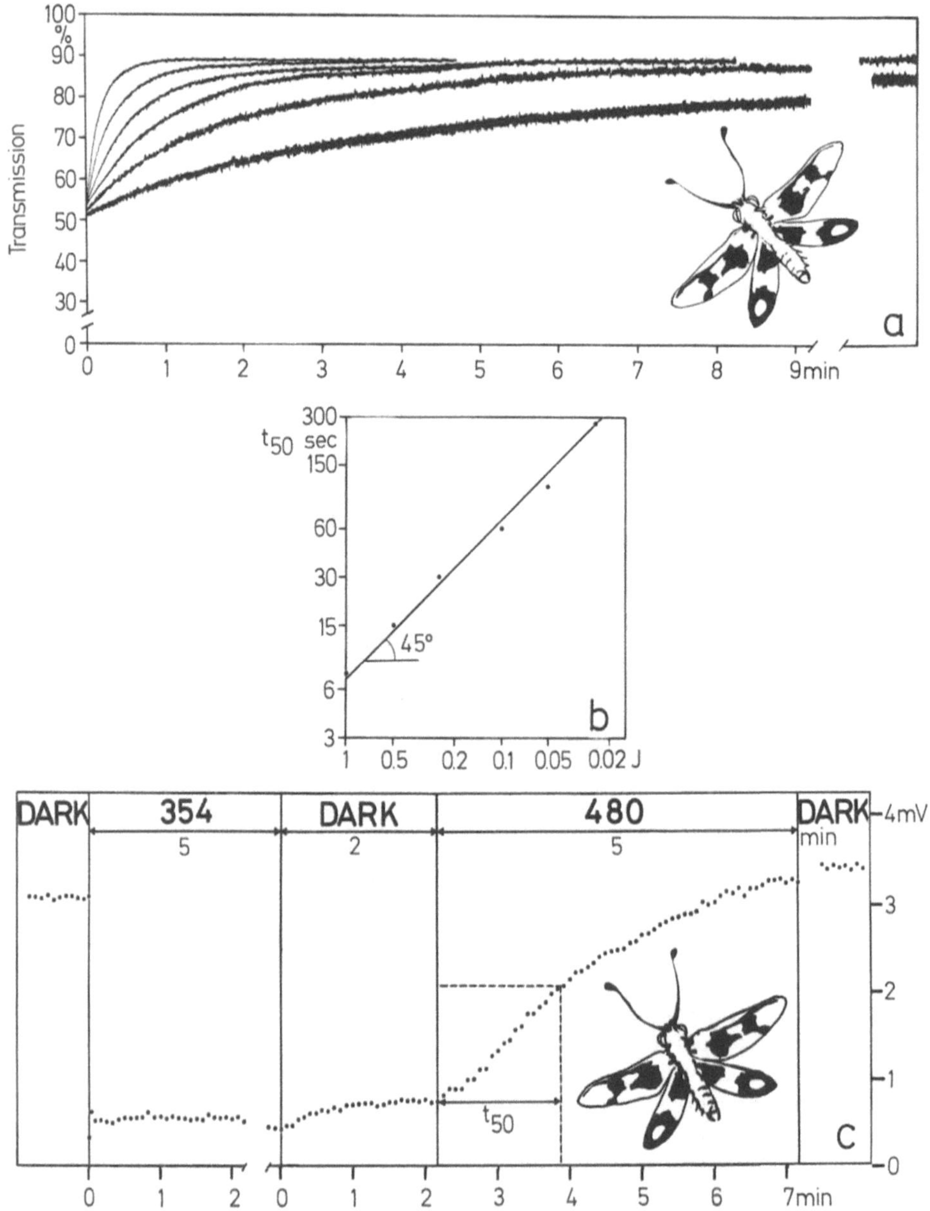

Fig. 7. Photoreconversion of Ms475 in the retina of *Ascalaphus* by blue light of various intensities, and the corresponding regeneration of the sensitivity of the retina. (*a*) Time course of transmission change during 475 nm illumination after photoequilibration by UV at 354 nm, as a function of measuring intensities (0.5, 0.25, 0.13, 0.06, and 0.03, respectively, relative to the next most intense beam; the uppermost curve). The width of these 6 original recordings indicate increasing electric noise on decreasing intensity of 475 nm measuring beam. (*b*) Demonstration of direct proportionality between rate of photoreconversion and intensity of regenerating light. (*c*) The amplitude of extracellularly, recorded receptor response (*dots*) to low intensity UV test flash in darkness, during bright UV adaptation (354 nm), and during illumination with blue light (480 nm). Rate of response increase (t_{50}) depends on wavelength, using light of equal quanta, as shown in Fig. 5a (*open circles*) (SCHWEMER et al., 1971; HAMDORF et al., 1971)

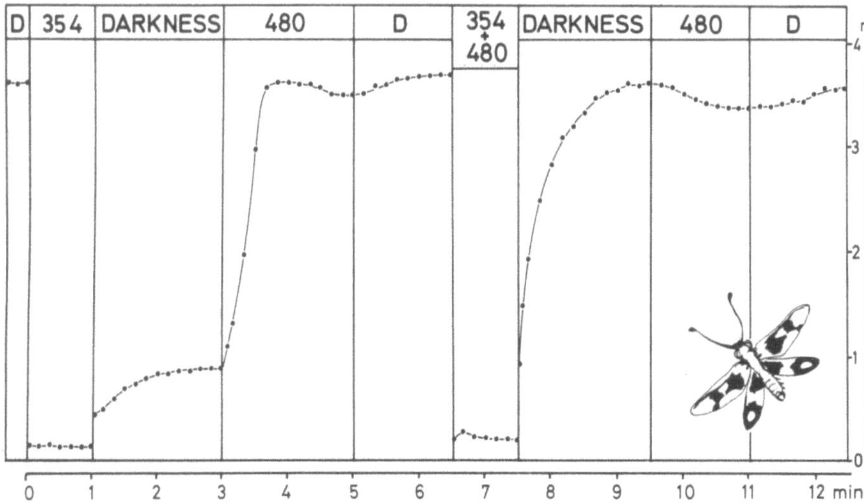

a

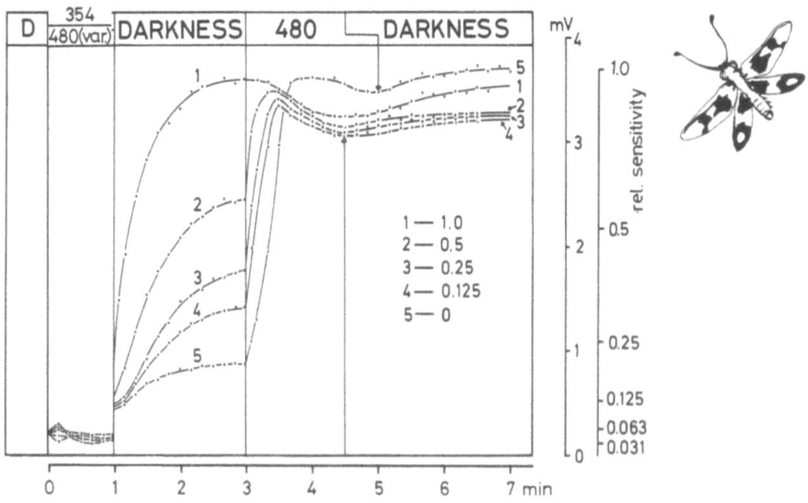

b

Fig. 8. Regeneration of retinal sensitivity by blue light in *Ascalaphus*. (*a*) Amplitude of extracellularly recorded receptor response (*dots*) to low-intensity UV test flash in darkness, during bright UV adaptation (354 nm), during illumination by blue light (480 nm), and by combination of UV and blue light (354 nm + 480 nm). (*b*) Regeneration of sensitivity depending on ratio I_{blue}/I_{UV}, with I_{UV} held constant. Almost equal number of quanta of UV and blue light (1) and ratios of quantal flux $I_{blue}/I_{UV}=0.5$, 0.25, 0.125, and 0.00 (2–5). *Right ordinate*: rel. sensitivity. This scale corresponds to response amplitudes at reduced intensities of the test flash (0.5, 0.25, ... 0.03), making possible transformation of amplitudes to sensitivity (HAMDORF and GOGALA, 1973)

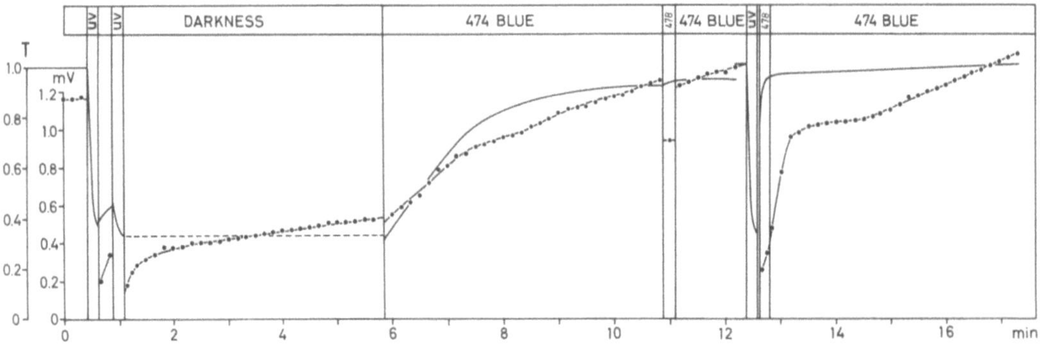

Fig. 9. Simultaneous measurement of response amplitude (*curve with dots*) and Ms concentration (*solid curve*) in the UV sensitive retina of *Ascalaphus*. Transmission T of retina was measured at 474 nm, indicating the Ms475 content. During the minutes 1–6 the 474 nm measuring beam was switched off (*broken line*), to demonstrate that during slight increase of receptor response Ms content is unchanged. If immediately after UV adaptation visual pigment is completely regenerated by bright blue light (478 nm; minutes 12–13), then time course becomes much faster, and independent of C_P, indicating "electric" adaptation mechanisms (GOGALA et al., 1975; HAMDORF and SCHWEMER, 1975)

if quanta-equal radiations ($I_1 = I_2$) at all wavelengths ($\lambda_{a_1}, \lambda_{a_2}, \ldots$) are used, then for equal concentrations $[C_{Ms}(\lambda_{a_1}, t_1) = C_{Ms}(\lambda_{a_2}, t_2)]$ one obtains:

$$\alpha_{M_1}/\alpha_{M_2} = t_2/t_1$$

and from this it follows that:

$$\alpha_M(\lambda_a)/\alpha_M(\lambda_{a_{max}}) = t_{50}(\lambda_{a_{max}})/t_{50}(\lambda_a)$$

(HAMDORF et al., 1971).

The first information concerning the sensitivity regulation of UV receptors in the biotope of *Ascalaphus* was obtained by simultaneous exposure to equal quanta of blue and UV light. As shown in Fig. 8 (HAMDORF and GOGALA, 1973), after the cessation of the UV adaptation the response amplitude increases only slightly, whereas after simultaneous UV-blue adaptation the response amplitude increases to the maximal values. But the increase in amplitude, and thus the sensitivity increase, depends on the ratio of the blue intensity to the UV intensity (I_{bl}/I_{UV}) of the adapting light (Fig. 8b). This result is in harmony with the theory of photoequilibration when using simultaneously diverse wavelengths for adaptation (Eq. 11, Sec. C.I.3, HAMDORF and GOGALA, 1973). The spectral content of skylight has two maxima, one in the blue and the other in the UV region. The high content of blue light, and the high absorption of Ms475 in this spectral region causes almost all of the latter to be reconverted into P345, and calculations according to Eq. 9a, b (Sec. C.I.3) show that the P345 content is close to 90%. Furthermore, it was shown that at such physiologically existing high P345 levels, the sensitivity of the UV receptor is maximal, whereas at 100% P345 content, which could presumably be achieved only experimentally, the sensitivity is somewhat lower, indicating a perfect phylogenetic adaptation of the UV receptors to the naturally available spectrum of the mediterranean biotope, and very recent histologic data corroborate this early hypothesis (LANGER, 1975; SCHNEIDER et al., 1978), showing that the dioptric apparatus is screened by two types of pigment cells, and the primary cells which determine the aperture of the facet contain exclusively a pale yellow pigment that screens only the UV and the violet radiation, and that the aperture for the receptor excited by UV radiation is much smaller than the aperture for the regenerating blue light. At present little is known about the metabolic pathway of UV pigment regeneration.

It was mentioned above that in *Ascalaphus*, after the cessation of the UV adaptation the response amplitude increases only slightly. The supposition that this slight further increase may be due to a slow thermal regeneration of P345 from Ms475 has not been confirmed by simultaneous electrophysiologic and microspectrophotometric measurements. As shown in Fig. 9, after UV adaptation, the transmission due to Ms475 content is unchanged during increase of the receptor response (2–6 min). Thus it seems to be more likely that the slow part of the adaptation curve is affected by other properties of the receptor system, i.e., by the ion transport mechanism (HAMDORF and SCHWEMER, 1975).

III. Other Electrophysiologic Findings Yielding Information on Pigment Function

The response of invertebrate photoreceptors has been divided into three distinct components. The measure of sensitivity has been the so-called late receptor response (LRP), the amplitude of which as a valid index of the P:M state has been clearly demonstrated by parallel electrophysiologic and spectrophotometric measurements (see Sec. B.II and V, C.II.3). This component of the response, however, is preceded by another, the early receptor potential (ERP), and followed by the prolonged depolarizing afterpotential (PDA), or in certain cases by a prolonged afterhyperpolarization (AHP). We shall now consider the findings concerning these other components of the receptor response.

1. Early Receptor Potential

Since the discovery of the early receptor potential (BROWN and MURIKAMI, 1964) the direct correlation of this fast photovoltage response (FPV) to quanta absorption by the visual pigment and its intermediates has been the subject of numerous investigations in vertebrate and invertebrate photoreceptors (see review by CONE and PAK, 1971). The polarity and the time course of the FPV responses suggest that they are generated by reversible alterations of the charge distribution in the pigment (or an intermediate) molecule itself (CONE, 1967; HAGINS and MCGAUGHY, 1967).

Briefly, the findings are that in vertebrate receptors bright flashes elicit diphasic ERPs, reflecting the time course of the transitions from P⇝MI→MII→MIII, and also in the reverse direction. Component analysis of this complex response has shown that quanta hits to the visual pigment cause a positive signal, whereas quanta hits to MI induce negative-going and to MII, again, positive-going potentials.

Comparable results have been obtained in squid receptors by extracellular recordings. P responds with a positive potential wave, Ms with a negative and Mb with a positive wave (HAGINS and MCGAUGHY, 1967). ERPs have been also intracellularly recorded in the receptors of the *Limulus* lateral eye (SMITH and BROWN, 1966).

Incidentally, analysis of these intracellular ERPs by using the double flash technique and rapid microphotometry gave first the evidence that within the receptor membrane of *Limulus* the photopigment of the green receptor is thermostable (FEIN and CONE, 1973), whereas in digitonin extracts (HUBBARD and WALD, 1960), P530 is irreversibly bleached. The interpretation (FEIN and CONE, 1973), however, that the time course and lifetime of the photoproduct B (Ms?), which possibly triggers the excitation of the membrane, is identical with the photometrically observed transmission change seems doubtful, as the wavelength used for the measurements (463 nm) speedily reconverts Ms back into P530.

A clearer connection between the various components, the polarity of the ERP and the different pigment states was demonstrated by comparable investigation on green receptors of the lateral eye of *Limulus* and on those of two barnacle species: *Balanus amphitrite* and *B. eburneus* (HILLMAN et al., 1973; HOCHSTEIN et al., 1973; MINKE et al., 1973). In all three cases, the polarity of the intracellularly recorded ERPs depends on the color of preadaptation. After blue, only negative potentials have been recorded and after red preadaptation only positive responses (Fig. 10). Furthermore, the spectral sensitivity curves of both responses are apparently different in λ_{max} and can be approximated by Dartnall nomograms of λ_{max} at 532 nm and 495 nm. As the spectrum of the 532 pigment coincides with the spectral sensitivity of the late receptor response, it can be postulated that the negative ERP originates from excitation of the visual pigment while the positive ERP from a thermostable intermediate, presumably metarhodopsin M495. Thus the resulting amplitude and polarity of the ERP could be used as a valid index of the P:M states present in the receptor membrane (Fig. 10), which has been strongly supported recently by parallel ERP and microspectrophotometric measurements (MINKE and KIRSCHFELD, 1977). Furthermore, this study has convincingly demonstrated that the barnacle possesses a P532 and M492 interconvertible system (with a ratio of 1:1.6), similar to that found in the green receptors of the butterfly *Deilephila* (see Sec. B.VI).

2. Prolonged Depolarizing Afterpotential

The common scheme of the UV receptor mechanism has been strongly supported by electrophysiologic investigations on *Limulus* receptors (NOLTE and BROWN, 1972a, 1972b). In the median ocellus of *Limulus* two types of photoreceptors are found, one maximally sensitive to UV (λ_{max} = 360 nm) and the other to green light (λ_{max} = 525 nm). Biphasic receptor potentials consisting of a small

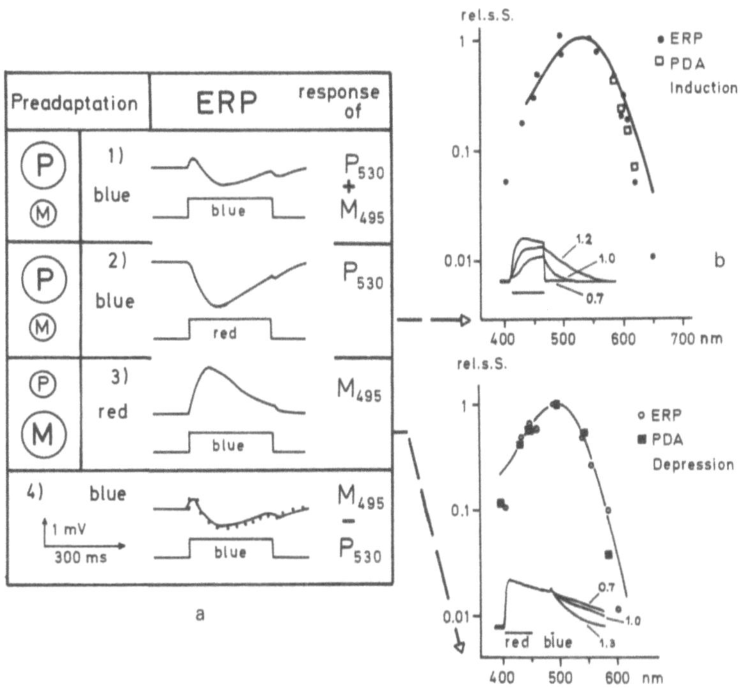

Fig. 10. Early receptor potential (ERP) of *Balanus eburneus* and *B. amphitrite* and its correlation to PDA induction and PDA depression. (*a*) ERP responses depending on the color of preadaptation and stimulus; trace 1) and 2) after blue preadaptation; trace 3) after red-adaptation, trace 4) simulation of trace 1 by subtraction of P530 and M495 responses, corresponding to trace (2) and (3). The sizes of P and M in the left column indicate the expected concentrations of P and M in the photoequilibrium produced by the wavelength of preadaptation (*B. eburneus*). (*b*) Spectral sensitivity of the negative component of ERP after saturating blue preadaptation (447 nm, 10 s, followed by 2 min dark) in comparison to the spectral sensitivity of PDA induction (*open squares*) after corresponding blue preadaptation. The insert shows the criterion response for PDA induction (middle trace) chosen for the measurements. N.B., 0.2 log units stronger light induces upper PDA, whereas 0.3 log units weaker light does not elicit PDA (*B. eburneus*). (*c*) Spectral sensitivity of positive component of the ERP after saturating red adaptation (650 nm) in comparison to spectral sensitivity of PDA depression (*black squares*) after PDA induction by 620 nm, 15 s stimulus. Insert shows the criterion response for PDA depression chosen in the measurements (middle trace). Note that 0.3 log units stronger blue light induces lower trace, whereas 0.3 log units weaker blue light causes upper trace (*B. amphitrite*). The comparisons show that spectral data of PDA induction coincide with the P530 spectrum and spectral data of PDA depression with M495 spectrum (DARTNALL nomogram; after HOCHSTEIN et al., 1973; MINKE et al., 1973; HILLMAN et al., 1976)

initial hyperpolarizing phase (the ERP) followed by a large depolarizing phase (the LRP) can be recorded from both receptors. After the cessation of bright UV adaptation, however, the UV cells *remain depolarized* (Fig. 11). This prolonged depolarizing afterpotential (PDA) is abolished by long wavelength stimuli, i.e., visible light repolarizes the membrane to rest potential (Fig. 11b). Moreover, long wavelength stimuli superimposed on bright UV irradiation elicit sustained repolarizing responses indicating competition of membrane depolarizing (UV) and repolarizing (visible) effects (Fig. 11c).

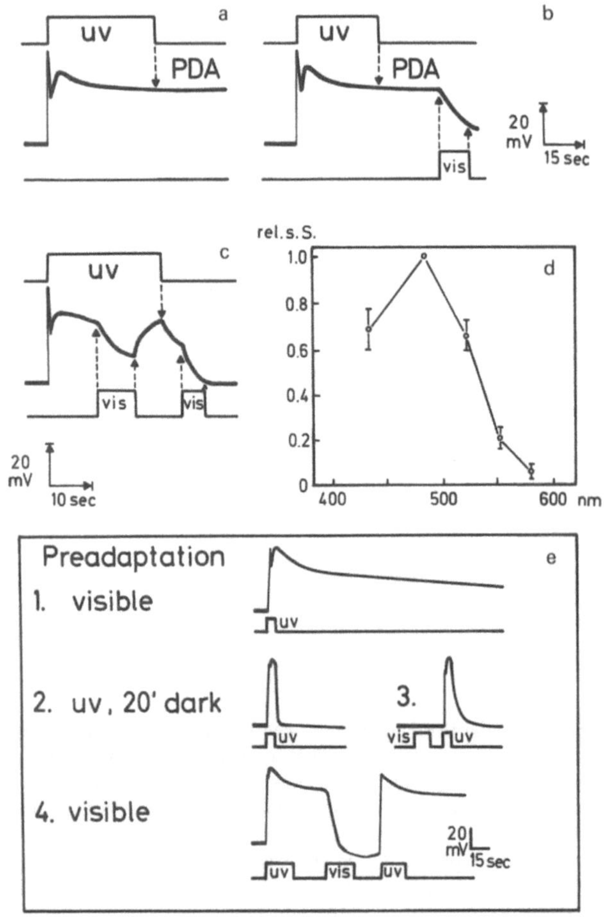

Fig. 11. Photoreconversion in UV receptor of the median ocellus of *Limulus* (a–d after NOLTE and BROWN, 1972b; e after MINKE et al., 1973). (*a*) Bright UV stimulus elicits a depolarizing receptor potential, which remains nearly unchanged after cessation of UV stimulus. (*b*) This PDA, however, is abolished by visible light stimulus (vis.). (*c*) Also during UV adaptation the superposition of visible light reduces the amplitude of receptor response. This decrease depends on the proportion of the intensity of UV to visible light. (*d*) Spectral sensitivity of repolarizing stimulus. N.B., most effective wavelength is 480 nm, indicating a similar M480 as in the UV receptors of *Ascalaphus*. (*e*) Dependence of PDA on the state of adaptation. Trace 1: PDA induced after visible preadaptation; trace 2: receptor response after the decay of the PDA; trace 3: as in trace 2 but with short preadapting visible light, showing a small PDA; trace 4: PDA initiated as in trace 1, followed by suppression by visible light followed immediately by another UV stimulus resulting in response and the subsequent PDA with the same long lifetime

The spectral sensitivity curve of the repolarizing stimulus (Fig. 11d) is approximated by an Ms480 spectrum comparable to that for *Ascalaphus* (Fig. 6). Furthermore, it has been shown through current injection techniques that the reversal potential of the repolarizing and the depolarizing responses are the same and both depend in the same way on the external sodium concentration (NOLTE and BROWN, 1972b). These results support the hypothesis that the amount of

M480 in the rhabdomeric membrane determines the permeability of the receptor membrane to sodium ions, and hence the PDA will last as long as M480 is present in the membrane. Furthermore, this hypothesis implies that during the decay of the PDA, M480 is either thermally regenerated to the visual pigment or inactivated by a still unknown process. Through a few striking experiments, however, it has been doubted that the M480 content in the membrane is responsible for the lifetime of the PDA (Fig. 11e; MINKE et al., 1973b). For only if the receptor has been exposed to visible light, which maximally photoreconverts M480 into P360, can a normal PDA be elicited by bright UV test stimuli. In contrast, if the receptors have been UV preadapted, i.e., most of the P360 has been converted to M480, and if the PDA caused by the UV preadaptation has declined to the rest potential within several minutes, responses of normal height are induced by the UV test stimulus but without the PDA (Trace 2). Then, after photoregeneration, up to 5 min in the dark are necessary for eliciting again a normal PDA by UV stimulus (compare Trace 3 with Trace 1). In contrast, if the PDA is suppressed by visible light (Trace 4), then immediately after the membrane repolarization normal PDAs can be reinduced. From these experiments it has been concluded that:

a) PDA induction depends on the number of P→M transitions, and the time course of its decline does not directly reflect either the destruction of M480 or its thermal reconversion to P360, since the concentration of M480 remains constant in the membrane until it is photoreconverted to P360.

b) The fact that after the decay of PDA and subsequent photoregeneration about a further 5 min are necessary for reinducing normal PDAs indicates a PDA inhibitory process of relatively long time constants for activation and inactivation.

c) Since after the decay of the PDA no response to visible light is observed, it is evident that quanta absorption by M480 does not induce a substantial voltage response (repolarizing response).

d) "The PDA must be an extra-pigmental process, and the suppression of the PDA must involve another, independent and antagonistic, extra-pigmental process. The excitation is activated by the P360→M480 transition and the suppression by the reversal transition; both transitions can be induced only by light" (MINKE et al., 1973).

The conclusion arrived in (1) stands in contradiction to the findings (NOLTE and BROWN, 1973b) that PDA is directly proportional to the concentration of M480. The explanation sought for this and presented in conclusion (4) necessitates the addition of new "extra-pigmental processes".

In the barnacle (MINKE et al., 1973; HOCHSTEIN et al., 1973; BROWN and CORNWALL, 1975; HILLMAN et al., 1976) the findings are as follows. As has already been mentioned, this species has a P532–M492 system, and the PDA can be induced only when rhodopsin is by red light photoconverted to metarhodopsin, the maximal PDAs thus produced having life times of several minutes. The spectral sensitivity of its induction corresponds well with the Dartnall nomogram of P532 (Fig. 10b), and its extent is a supra-linear function of the amount of the P→M transition, where this extra-linear dependence is believed to be a cooperative effect between the PDA-inducing agents. As in *Limulus*, in this species it was also shown that the PDA and the late receptor potential are both correlated with an increase of membrane conductance to sodium ions (BROWN and CORNWALL, 1975).

In this animal the PDA is abolished by blue light which reconverts M492 to P532, and the spectral sensitivity of its abolition agrees well with the M495 spectrum (Fig. 10c). When the decline of a PDA is interrupted in its early stages by an M→P transition (by blue light), a further PDA can be soon after elicited by red light, whereas when a PDA is allowed to decline to the resting potential (lasting several minutes in the dark), an immediate P→M transition will not yield a PDA; some time must elapse before such stimuli begin to cause gradually greater PDAs, during which period it is believed that an "anti-PDA" agent is slowly destroyed.

A further observation was that when the PDA was soon abolished by M→P transitions, after the lapse of some time, a P→M inducing red light of reduced intensity caused no PDA, while a further similar stimulus led to a large PDA. This apparent "facilitation" effect was suggested to be due to either another cooperative effect, or the same which was seen to cause the previously mentioned supra-linearity. Such findings have led to the following model of an excitor-inhibitor mechanism, originally proposed by HOCHSTEIN et al. (1973) and recently summarized by HILLMAN et al. (1977):

a) each P→M transition produces one unit of an excitor and each M→P transition one unit of an inhibitor;

b) there is a positive cooperative effect among the excitors, which is non-local and extends to several 100 Å;

c) excitors open membrane ionic channels; the inhibitors do nothing to the membrane, however, the excitors and inhibitors mutually annihilate each other, 1:1, and, in addition, if a finite time is assumed for the excitor-inhibitor mutual annihilation, the late receptor response to a mixed light (one that causes no *net* pigment transfer) is thus predicted by the model, but other properties of this response are not;

d) if not annihilated, excitors and inhibitors "die slow natural deaths".

In flies the PDA phenomenon has been studied in *Drosophila* (COSENS and BRISCOE, 1972; STARK, 1972; MINKE et al., 1975; COSENS and WRIGHT, 1975; HARRIS et al., 1976; STARK and ZITZMANN, 1976; WRIGHT and COSENS, 1977), in *Calliphora* (MUIJSER et al., 1975; RAZMJOO and HAMDORF, 1976; HAMDORF and RAZMJOO, 1977, 1979), and in *Eristalis* (TSUKAHARA et al., 1977). The receptors R1–6 of these species possess visual pigments similar to one another (ranges in absorption maxima being: P460–494, M550–580, $\lambda_{P_{max}}:\lambda_{M_{max}} \simeq 1:1.6$). Blue light shifts the equilibrium in the P→M direction, while orange to red light causes the regeneration (M→P) of the pigment (see Fig. 2, also Sec. B.V and C.II.3). In all these flies intense blue light creates the PDA which can be abolished by orange-red light.

In *Calliphora* and *Eristalis*, direct intracellular recordings show maximally induced PDAs to decline to the resting potential within a few minutes (less than 5 min, HAMDORF and RAZMJOO, 1977; TSUKAHARA et al., 1977), and the extrapolations of the few interrupted intracellular records in *Drosophila* (MINKE et al., 1975; MINKE, 1977) suggest similar decay times. Most of the data for *Drosophila*, however, are from extracellular (ERG) measurements, which suggest decay times of 3–6 hrs or more (COSENS and BRISCOE, 1972; STARK, 1972; MINKE et al., 1975; WRIGHT and COSENS, 1977). However, in these cases, the criterion of the end of

PDA has not been the return of the potential to the resting level (as with intracellular measurements in *Calliphora* and *Eristalis*), since the true level becomes masked by extensive drifting of the potential (WRIGHT and COSENS, 1977; TSUKAHARA et al., 1977), but the monitoring of the late receptor response to continuous test stimuli. Therefore, such sensitivity losses (which are really due to reduced P concentration as a result of P→M transitions, see Sec. B.II.3) have been considered as a measure of the extent of the PDA, and the time of the increase of this response to the original value taken as its lifetime.

If the PDA did last for as long as it is suggested in *Drosophila*, only then could the hypothesis concerning its mechanism (WRIGHT and COSENS, 1977) be considered as plausible, i.e., that M580 content depolarizes the membrane, and this state persists for as long as M580 concentration remains unchanged. The lengthy decline of the PDA is thus attributed to the dark, metabolic reconversion of metarhodopsin to rhodopsin which, based on ERP measurements, is reported to take about 6 hrs (PAK and LIDDINGTON, 1974).

Sensitivity measurements in vitamine A-deprived and enriched *Drosophila* after blue and red adaptation (Fig. 26) have led to the proposal of another model (STARK and ZITZMANN, 1976). This considers finite P→M transitions to release "activated transmitters" which then excite a corresponding number of membrane sites (the latter a fixed number irrespective of the animal's diet), so that in the case of vitamine A-deprived animals, the maximum number of activated transmitters produced by an intense blue stimulus does not exceed the number of membrane sites. But when in vitamine A-enriched animals the now greater number of M molecules (created by blue light) produce a greater number of activated transmitters, exceeding the number of membrane sites, these extra transmitters must then take their turn in depolarizing the membrane, until they too are exhausted; and so the production and abolition of the PDA (see also Sec. B.VII).

However, the excitor-inhibitor mechanism as suggested for *Limulus* and *Balanus* has not been conclusively demonstrated for any of the flies so far studied. In the case of the dronefly *Eristalis* (TSUKAHARA et al., 1977), a red and dark adapted cell (in that order, producing a maximally sensitive P-loaded state) is stimulated by an intense blue light, producing PDA. In the same sensitive state, however, a red stimulus produces an opposite effect: an afterhyperpolarization (AHP), with a time course similar to the PDA. A plot of the afterpotentials (the PDA and the AHP) as a function of quanta-equal wavelengths (TSUKAHARA et al., 1977) shows a maximum and a minimum, very similar to the difference spectrum of the P460–M550 system (Fig. 36a, b), and the wavelength at which the PDA and the AHP cancel one another (490 nm) is very nearly the isosbestic wavelength of this system. The same similarity is found in the spectral efficiency function for the *abolition* of the PDA.

It has also been shown that for maximal PDA production, the most effective wavelength is intensity-dependent; at low intensities this is 450 nm (identical to the peak of the spectral efficiency curve as determined from the late receptor response), but shifts at higher intensities to 430 nm, where the ratio $\alpha_P/\alpha_M = \max$ (see Sec. C.I.1) indicates that the amount of P which is transferred during a stimulus is important in producing the PDA. From these measurements (TSUKAHARA et al., 1977), it was concluded that the change in the concentration of P was responsible

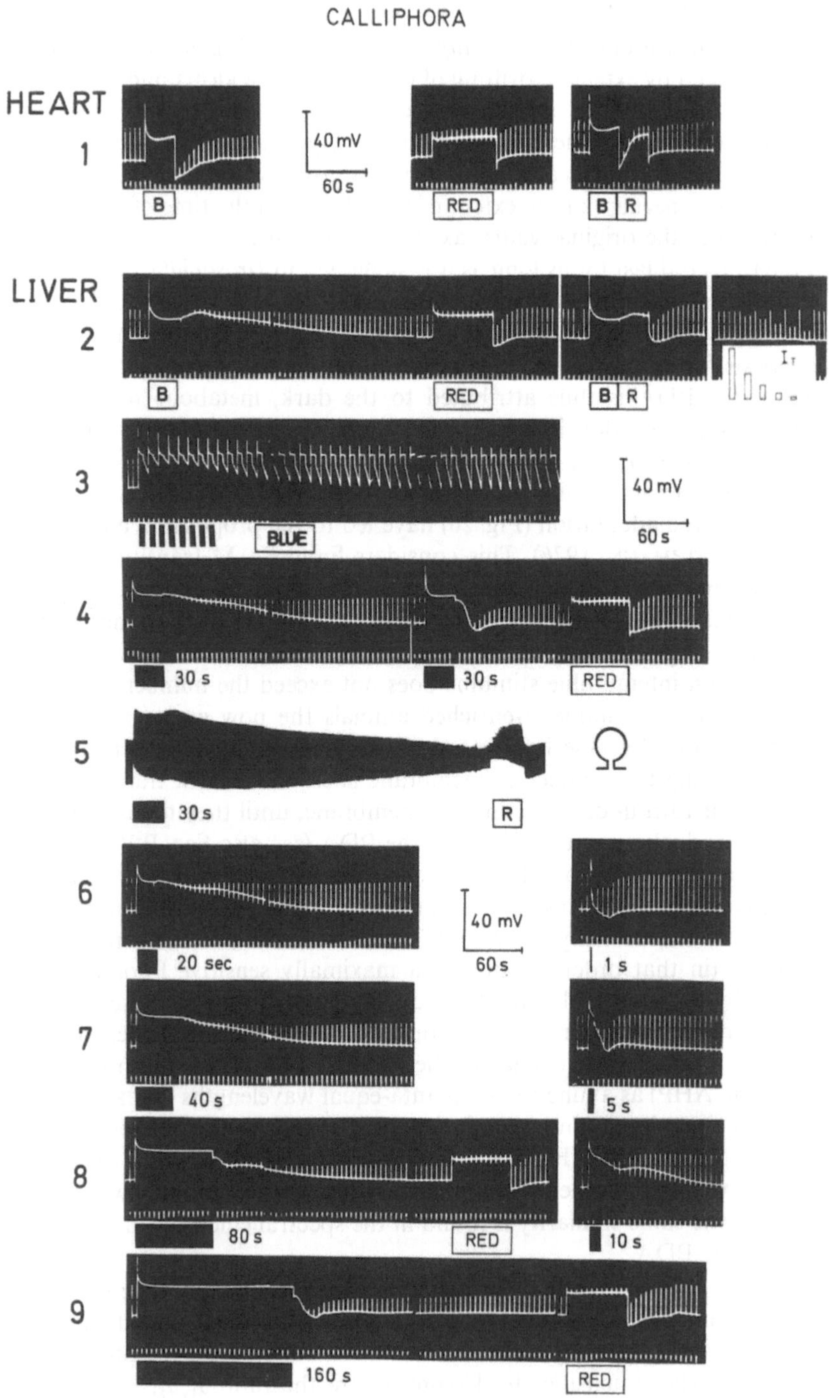
CALLIPHORA
HEART
1
40 mV
60 s
B
RED
B R
LIVER
2
B
RED
B R
3
BLUE
40 mV
60 s
4
30 s
30 s
RED
5
30 s
R
Ω
6
20 sec
40 mV
60 s
1 s
7
40 s
5 s
8
80 s
RED
10 s
9
160 s
RED

for the production and the abolition of the PDA, but that the experiments could not decide whether the PDA was the result of simply a concentration build-up of M or the *rate* of its build-up.

The excitor-inhibitor theory finds further contradiction (TSUKAHARA et al., 1977) when it is observed that after blue adaptation and the subsequent decline of the PDA, a red stimulus, which according to the excitor-inhibitor hypothesis should produce inhibitors in amounts more than 100 times greater than excitors ($M \rightarrow P/P \rightarrow M > 100$), preventing even the late receptor response, does not do so. Since such a response is observed, it is concluded that: "... there can not be excitor and inhibitor substances having unequally large and opposite effects on the membrane potential."

In *Calliphora* the dependence of the production and abolition of the PDA on pigment transitions were shown by MUIJSER et al. (1975), and in the comparison of vitamine A-rich and deficient flies (see Fig. 12, rows 1 and 2) this effect was shown to be present only in pigment-rich membranes, where the density of the pigment molecules varied in the range of less than 100 to a maximum of 2000 per microvillus (SCHWEMER, unpublished results; RAZMJOO and HAMDORF, 1976; HAMDORF and RAZMJOO, 1977). Figure 12 also shows the induction of PDA by a fixed intensity blue light to be a function of time, and maximum PDAs at this particular intensity are seen to be produced after about 20–40 s of illumination (with a mean of 30 s, Fig. 12, rows 6–9). More prolonged illumination periods can be seen to be masking the true rate of decline of the maximally induced PDA by a steady state potential. To illustrate the true effect of more prolonged illuminations without this masking effect, trace 3 shows the same stimulus when it was

◄ Fig. 12. Intracellular recordings of prolonged depolarizing afterpotential (PDA) from photoreceptors R1–6 of white-eyed (*chalky*) mutant of *Calliphora erythrocephala* Meig. Stimuli were: blue (473 nm), red (615 nm), and the test stimuli given all through the experiments (500 nm, pulses of 100 ms every 5 s) were more than 3 orders of magnitude weaker than the intense blue and red. *Heart*; trace 1: intense blue illumination in vitamine A-deprived flies (bred on ox heart) exhibit an afterhyperpolarization (AHP) instead of a PDA, which is rather slowly neutralized when blue light is followed immediately by red. *Liver* (traces 2–9); trace 2: in vitamine A-enriched flies (bred on ox liver), the same blue light causes a PDA of maximal amplitude and duration, which is abolished by red light. The last figure in this trace shows the reduction of the receptor response as the intensity of the test stimulus is decreased in 50% steps; trace 3: receptor response to alternating 5 s blue and 5 s dark periods, showing the accumulation and decline of the PDA. The PDA reaches a maximum after 4 blue exposures; traces 4–5: membrane potential and conductance change during a maximal PDA, measured in the same cell, showing the corresponding decline of membrane conductance during the return of the potential to resting value. In trace 4 notice that after the exhaustion of the first PDA, the second is much smaller; traces 6–9: PDA as a function of stimulus duration at constant intensity of blue light. Before each blue stimulus the receptor was red and dark adapted for 1 and 3 min, respectively, and in that order. The comparison of the exposures to 80 s and 160 s with that at 40 s shows that in the latter two the decline of the PDA is masked by the steady state response. In all these traces note that after PDA decline to the resting potential, the response to test stimuli (shown here only shortly after, but for measurements taken after 5 min of dark adaptation) is reduced when compared to those immediately preceding exposures to blue light, this reduction being dependent on the P content of the receptor (see Eq. 44, Sec. C.II.3). The analyses of these amplitudes show that after maximal PDAs ($\simeq 30$ s at this particular intensity, wavelength, and this particular fly!), only 50% of the photoequilibrium is reached; an indication that PDA is elicited by quanta absorption by dark adapted P molecules. (From HAMDORF and RAZMJOO, 1977) cumulatively presented, allowing a maximal PDA to fall back (during alternating dark periods) to the true level during its decline, and it can be seen that after 4 exposures (totalling about 20 s) the maximal PDA is reached, the shape and rate of the decline of which is very similar to the one produced with 30 s of blue light (trace 4). That the PDA may be compared to the effect of a diminishing background light (which would lead to a gradually decreasing membrane conductance) may be seen from Fig. 12, trace 5, where it is shown that the rate of recovery of the membrane resistance change following a maximally induced PDA is identical with the rate of the membrane repolarization.

The late receptor response to weak test stimuli ($<10^{-3}$ times the PDA inducing light) presented all through the experiments show two effects:

a) As the PDA declines, the amplitude of the late receptor response increases, an effect similar to when such test stimuli are superimposed on a slowly diminishing background illumination (not shown here). The similarity of the steady state response of a receptor during such sustained stimuli to the PDA suggests possibly identical underlying mechanisms;

b) The amplitude of the reduced response to the test stimuli, measured after the complete decline of the PDA (and always checked after a lapse of 5 min in the dark), is directly correlated with the reduced P content as a result of P→M transitions which have caused the PDAs (also see Fig. 25).

From such data it was concluded that PDAs are induced by primary quantum captures by dark adapted rhodopsin molecules, and that during further illumination (which leads to photoequilibrium between the P and the M populations), additional conductance change is due *only* to the fraction of photoregenerated P molecules which can survive for a *certain period* before quantum absorption, and the amplitude of the steady state response during sustained illumination may well be the excitation due to such dark-adapted P molecules.

The excitor-inhibitor model proposed for *Limulus* and *Balanus* (with certain admitted shortcomings) does not explain the findings in flies in the following respects:

a) If the production of a maximal PDA is due to the existence of a large number of excitors which keep the membrane depolarized for correspondingly long periods, then how can during this period, much smaller numbers of excitors produced by the weaker test stimuli have an effect in preference to the PDA-causing excitors which are still extant? Unless it is proposed that there are two different excitors; one causing the late receptor response and the other the PDA (the membrane being somehow more responsive to the former), with the need for two different inhibitors as well.

b) It is maintained that after the decay of a PDA when no more excitors persist, an M→P shifting stimulus creates inhibitors which should cause no membrane depolarization. But in *Eristalis* and *Calliphora* a weak red stimulus shows not only the late receptor response, but as long as the stimulus is maintained, a steady state response too (TSUKAHARA et al., 1977).

c) After the decay of a PDA, when no more excitors exist, an intense M→P transferring stimulus should create a great number of inhibitors, which (it is postulated) have life-times in the same dimension as that of the excitors, and were the latter stimulus to be followed very shortly after by another PDA-inducing one, the resulting PDA should be as brief as that induced without regeneration. This is not the case in *Calliphora*, where the second PDA following the regenerating stimulus is much larger than when the regenerating stimulus is omitted. Thus the inhibitor, if at all present in *Calliphora*, must have a life-time much shorter than that of the excitor (HAMDORF and RAZMJOO, 1977; HAMDORF and RAZMJOO, in HILLMAN et al., 1977).

d) The model holds that P→M-shifting stimuli produce only excitors, which then *depolarize* the membrane, and that the inhibitors, when produced by the reverse transition, apart from annihilating the excitors, do not affect the membrane. If it be conceded that an overproduction of inhibitors may even *hyperpolarize* the membrane to some extent, how does this explain the afterhyperpolarization due to *P→M* transitions in vitamine A-deprived *Calliphora* (see Fig. 12, trace 1)?

IV. A Model for the Function of the Photopigment System

When a phenomenon manifests itself consistently in many species, it becomes tempting to construe an explanation which applies universally. We shall now consider a model which appears to explain the facts as they have so far presented themselves.

In all invertebrate species, *regenerated* and *dark adapted* P molecules; $\boxed{\text{P}}$ (see footnote[2]), initiate membrane excitation through the capture of such photons that can cause P→M transitions (see Fig. 13). The *newly created* metarhodopsin $\boxed{\text{M}}$ releases energy $\textcircled{\scriptsize E}$ (corresponding to the direct release of excitors in the previous model), which is utilized by the membrane sites $\textcircled{\scriptsize S}$ for excitation, following which it is no longer effective in exciting the membrane: $\boxed{\text{P}} \rightsquigarrow \boxed{\text{M}} \xrightarrow{t_1} \textcircled{\scriptsize M} + \textcircled{\scriptsize E}$. Photon captures which can cause the back-reaction $\boxed{\text{M}} \rightsquigarrow \boxed{\text{P}}$, if they occur before $\boxed{\text{M}}$ can release its energy (e.g., as in the case of multiple hits), will cause no excitation; but if there is a lapse of time, the irreversible loss of energy in the above reaction (given for excitation to the membrane) prevents the back-reaction. Next, since photon capture (long after such dark adaptation) does apparently cause the regeneration of rhodopsin, this *freshly light-regenerated* form of the pigment, $\textcircled{\scriptsize P}$, differs from $\boxed{\text{P}}$ by the amount of lost energy, and can be transformed to the latter active form by the uptake of metabolic energy (a dark process), the rate of which, t_5, must be dictated by the particular metabolic rate of the species (e.g., this effect can be observed for *Ascalaphus*, Fig. 9). But before such uptake occurs, appropriate photon capture will lead to the reverse reaction: $\textcircled{\scriptsize P} \rightsquigarrow \textcircled{\scriptsize M}$, without exciting the membrane.

[2] The symbols which follow are *not* new identities with different absorption spectra (since such spectra reveal information only about the direct environment of the chromophore), but light absorption may lead to reactions involving the pigment with other neighboring structures, so that the symbolism here is meant to represent perhaps different "energetic" states of the visual pigment as the result of such interactions.

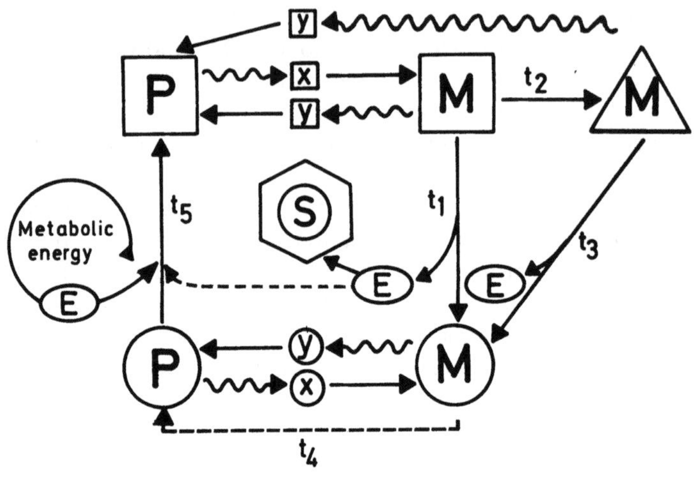

Fig. 13. A scheme of pigment reactions in the microvillus membrane of invertebrate rhabdomeric photoreceptors. Appropriate quantum absorption by dark-adapted visual pigment $\boxed{\text{P}}$ converts this to energy-rich metarhodopsin $\boxed{\text{M}}$ which can activate membrane sites Ⓢ by the release of energy Ⓔ, through which conversion it becomes inactivated, Ⓜ. Further appropriate quantum absorption can change the latter to Ⓟ, which needs metabolic energy to resume the activated form $\boxed{\text{P}}$. This cycle is dominant during normal physiological light levels, but intense, extraphysiologic illuminations increase the $\boxed{\text{P}} \rightsquigarrow \boxed{\text{M}}$ transitions, and the rate t_1, limited by the rate of the membrane utilization of the energy, forces some $\boxed{\text{M}}$ molecules into a storing form $\triangle\!\!\!\text{M}$, the latter releasing energy through a much slower decay to Ⓜ. The intermediate states of the forward and the backward reactions are symbolized by $\boxed{\text{X}}$ and $\boxed{\text{Y}}$; *dashed arrows* denote the possibility of energy by-pass from $\boxed{\text{M}}$ and $\triangle\!\!\!\text{M}$ directly to Ⓟ; $t_1 - t_5$ are thermally controlled rates of the dark reactions, and t_4 indicates the possibility of a very slow metabolic dark regeneration of the visual pigment (STAVENGA, 1974), so that $t_4 \ll t_5$, $t_3 \ll t_1$, but see footnote 7 page 184

The above, then, is the state of affairs for normal physiologic light levels of excitation and regeneration. But in all species discussed here, under experimental conditions of intense illumination, the anomalous phenomenon of the PDA can mean the production of $\boxed{\text{M}}$ from $\boxed{\text{P}}$ and thus energy in greater extent than can be utilized by the membrane. The resultant blockage of $\boxed{\text{M}} \rightarrow$ Ⓜ may then lead to the transformation of $\boxed{\text{M}}$ to an energy-storing form, $\triangle\!\!\!\text{M}$, which could, either directly (but much more slowly) release its energy for membrane excitation: $\triangle\!\!\!\text{M} \xrightarrow{t_3}$ Ⓜ + Ⓔ $(t_3 \ll t_1)$, or that $\triangle\!\!\!\text{M}$ build-up from $\boxed{\text{M}}$ is an equilibrium process $(\boxed{\text{M}} \rightleftharpoons \triangle\!\!\!\text{M})$ having a slower backward reaction rate, of the same dimension as the direct energy-releasing pathway, t_3; but the degrees of these formalistic alternatives do not manifest themselves in the experimental outcome. The result, however, is the same; a slow release of energy (perhaps less influenced by the membrane rate of utilization) manifested as the PDA. It would seem plausible then, that if during the PDA decline further small $\boxed{\text{P}} \rightsquigarrow \boxed{\text{M}}$ transitions occur (by weak stimuli, which, even after maximal PDA production cause further $\boxed{\text{P}} \rightsquigarrow \boxed{\text{M}}$ conversions, see further on), then excitation will be through the faster t_1 pathway, eliciting the normal late receptor response superimposed on the declining level of the PDA.

One critical test of such a model could be the following: Each unit of E produced through the t_1 pathway should lead to a corresponding membrane conductance change. Now if each E unit for the t_3 path produces the same conductance change, we can postulate that the PDA, measured, e.g., as the integral of the conductance change over the expansion interval of the time of its duration, must equal the concentration of $\triangle\!\!\!\!M$, which in turn owes its origin to the total pigment molecules and nothing else ($\triangle\!\!\!\!M \leqq \boxed{P}$). Background illumination experiments in *Calliphora* show that such simulated PDAs are directly correlated with the amount of $P \rightarrow M$ transitions, and that as low as 20% of the total visual pigment is enough to create a maximum PDA. But were the PDA to last for hours (as is claimed—but not clearly demonstrated—to occur in some species, see Sec. B.III), the energy release through $P \rightarrow M$ conversion could not explain such hours-long sustained conductance changes without introducing additional assumptions into the model[3].

This model then, as it stands, explains the various manifestations of the PDA phenomenon in the following way:

a) PDA is due to an overproduction of $\boxed{M}$ molecules, in that its *concentration* as well as the *rate* of its production are decisive:

$$d[\boxed{M}]/dt > d[\textcircled{M}]/dt;$$

b) PDA abolition is by metarhodopsin photon capture in *all* its various "energetic" states:

$$(\boxed{M}, \triangle\!\!\!\!M, \textcircled{M}) \rightsquigarrow (\boxed{P}, \textcircled{P});$$

c) PDA may be prevented if the $\boxed{P} \rightsquigarrow \boxed{M}$ transition is *cancelled* by an equal $\textcircled{M} \rightsquigarrow \textcircled{P}$ transition (an effect which is attributed to the "inhibitor" in the model for *Limulus* and *Balanus*), when $\textcircled{P}$ would compete with the membrane sites for the energy released through the t_1 reaction path. The extent of this "transaction" between $\boxed{M}$ and $\textcircled{P}$ may be a function of the proximity of the two[4];

d) the "anti-PDA" effect of the excitor-inhibitor model is represented here by the time necessary for the t_5 conversion pathway; metabolic dependent and thus species-specific.

The model can be used to explain the findings in *Limulus* (Fig. 10): In a completely regenerated membrane (fully loaded with $\boxed{P}$, Fig. 14a), an UV stimulus will produce a photoequilibrium, dominant in $\boxed{M}$ concentration, consequently resulting in not only the immediate membrane response (the late receptor response), but because of the overflow of energy, in a sustained membrane response, the PDA. After the decay of the PDA, however, (trace b), the UV stimulus will establish a photoequilibrium between $\boxed{P}$ and $\boxed{M}$, as well as between $\textcircled{M}$ and $\textcircled{P}$, resulting in $\boxed{M}$ and $\textcircled{P}$ existing in equal concentrations, and since $\textcircled{P}$ will compete with the membrane sites for this energy, only the late receptor response but no PDA is the outcome. Even when after the gradual decline of the PDA (trace c) the membrane is photoregenerated by visible light, producing high amounts of $\textcircled{P}$, the UV stimulus will again establish the same concentration of $\boxed{M}$ and $\textcircled{P}$ as described in the previous example (trace b), with an identical outcome.

---

[3] However, see HAMDORF and RAZMJOO, 1979.

[4] There are some indications that even the late receptor response may be affected by this proximity; however, the effect is so small in this case that only certain experiments can reveal it (see Fig. 25a).

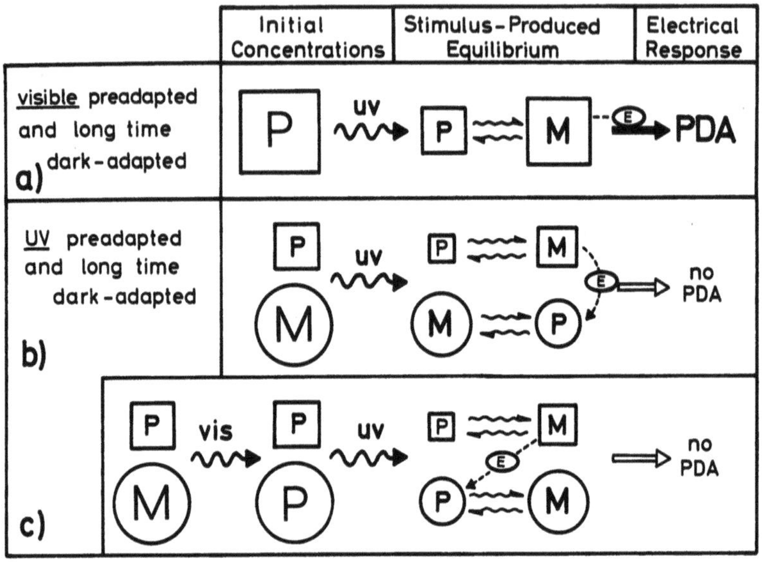

Fig. 14. Supposed concentrations of active and inactive states of UV visual pigment [P] and (P), and its metarhodopsin [M] and (M), in the rhabdomere of *Limulus* during the experimental conditions corresponding to those of Fig. 10e, traces 1, 2, 3. a) Size of squares and circles correspond to concentrations present. PDA is produced by overflow of [M] after visible preadaptation and long-lasting dark adaptation. b) and c) No PDA can be observed either after decay of PDA or after visible stimulus shortly before UV stimulation. Note that in this figure both the energy rich forms of metarhodopsin (△M, [M]) are collectively shown by [M]

This model then, through its ability to predict the various sensitivity states of an invertebrate photoreceptor, seems to obviate the apparent differences of observations which have led to the idea of species-specific mechanisms.

V. Visual Pigments of Fly Receptors

The ommatidia of the neuronal superposition eyes of diptera, *Drosophila*, *Calliphora* and *Musca* are built uniformly of eight receptor cells. The arrangement of the rhabdomeres, as well as the orientation of the microvilli are precisely determined, and the arrangement of the patterns of the rhabdomeres of the upper and the lower half spheres of the compound eye are mirrored about its "equator" (BRAITENBERG, 1967; KIRSCHFELD, 1967; recent reviews: FRANCESCHINI, 1975; STAVENGA, 1975). There is good evidence through electrophysiologic, photometric, and microphotometric data, that the receptors R1–6 contain the same visual pigments, whereas the receptors R7–8, which jointly form the central rhabdomere No. 7 may have different absorption properties.

Early electrophysiological work (BURKHARDT, 1961) suggested that the ommatidia of the blowfly has two (or three) different types of receptors with double peak sensitivities: a very frequent "green" receptor with a smaller sensitivity peak in the UV (≃480–350 nm), and a less frequent UV receptor with a smaller sensitivity peak in the "blue" (350–450 nm). Through marking experiments, MCCANN and

ARNETT (1972) demonstrated that the most numerous "green" receptors are identical with retinula cells R1–6. Optical and behavioral studies in *Musca* (KIRSCHFELD and FRANCESCHINI, 1968; KIRSCHFELD, 1969) indicated that R1–6 in each ommatidium form a low-acuity, high-sensitivity system (HSS), while the central receptors R7–8 form a low-sensitivity, high-acuity system (HAS). Spectral sensitivity measurements based on the optomotor reaction (ECKERT, 1971) showed that in dim light mainly the green receptors (R1–6) are activated, while at high intensities blue-sensitive receptors of higher acuity respond. The conclusion, however, that the spectral sensitivities of the optomotoric response represent the spectral absorption properties of the peripheral and the central receptors seems doubtful, for it greatly contradicts the following electrophysiologic findings. In mass recordings in the blowfly (the *chalky* mutant), as well as in the white-eyed mutants of *Drosophila*, during light adaptation, the double-peaked spectral efficiency curve alters specifically with the wavelength of the adapting light (ROSNER, 1975; STARK, 1975). In *Calliphora* the UV peak is selectively reduced by UV at 348 nm but not by violet at 411 nm, suggesting a UV-sensitive visual pigment of small bandwidth, comparable to that of *Ascalaphus*. Furthermore, adaptation through increasing intensity at 553 nm causes a continuous shift of the second sensitivity peak from 490 nm to less than 460 nm, while adaptation at 450–500 nm depresses nonspecifically the sensitivity in the blue-green range, indicating selective adaptation of two different visual pigments P490 and P440–460, in the blue-green. Intracellular recordings of R1–6 are in agreement with the hypothesis that these peripheral receptors contain uniformly two pigments P345 and P490 (DÖRRSCHEIDT-KÄFER, 1972; ROSNER, 1975; HORRIDGE and MIMURA, 1975; SMOLA and MEFFERT, 1975), and there is some evidence through spectral polarization sensitivity measurements that UV and green-sensitive pigments are spatially separated in the rhabdomere (HORRIDGE and MIMURA, 1975).

A comparable double-peaked spectral sensitivity was found for R1–6 of the dronefly *Eristalis* (HORRIDGE et al., 1975), as was also a UV-blue sensitive receptor, presumably R7 (BISHOP, 1974). Very recent intraaxonal recordings from R7 and R8, identified by staining with Procion yellow, indicate three classes of central receptors: single-peaked UV receptors, double-peaked UV-blue, and single-peaked blue receptors (SMOLA and MEFFERT, 1975, 1976; MEFFERT and SMOLA, 1976). Up to the present two types of cells have been identified clearly as UV-blue sensitive R7 receptors and one as a blue sensitive R8 receptor. The authors, however, suppose that there exist two subclasses of R7, exclusively either UV or UV-blue sensitive, and two subclasses of R8, which are sensitive to either blue or to UV-blue. This idea is supported by microspectrophotometric measurements on *Calliphora* and *Drosophila*, in both of which the existence of two corresponding subclasses of R8 have also been reported by other workers (KIRSCHFELD and FRANCESCHINI, 1975).[5]

Summarizing the data so far available on *Calliphora*, the peripheral receptors R1–6 contain mainly P490, and P345 in smaller amount, while the central rhabdomeres contain P345 and P445 in variable amounts.[6]

In the ommatidia of the fruit fly *Drosophila melanogaster*, a similar receptor pattern has been elucidated by genetic dissection of the photoreceptor system (STARK and WASSERMAN, 1974; STARK, 1974, 1975; HARRIS et al., 1976), as well as by electrophysiologic methods (STARK, 1972, 1973; MINKE et al., 1975). The genetic analysis is based on mutant separation methods developed by BENZER (1967) using behavioral reactions (for detailed description of automatic separation techniques in use see HEISENBERG and GÖTZ, 1975; see also Ch. 12 of this volume).

[5] Recent microspectrophotometric measurements on R7/8 have provided good evidence that ≃2/3 of the R7 cells (the 'yellow', $R7_y$) contain an additional photostable pigment (presumably β-carotene), which selectively filters the spectral input to the R8 photoreceptors, causing the latter to exhibit a shift in spectral sensitivity from 490 nm towards 530 nm, close in fact to the peak sensitivity actually measured intracellularly in R8 (Hardie, 1977).

[6] The greater UV sensitivity of R1–6 is still a matter of debate. There is evidence that this may be caused by either 1) energy transfer from a photostable UV-absorbing pigment to the P490 (the 'antenna' pigment, KIRSCHFELD and FRANCESCHINI, 1977); 2) by the pronounced β-band (≃350 nm) due to the orientation of the rhodopsin molecule in the microvillus membrane (SCHWEMER, unpublished data).

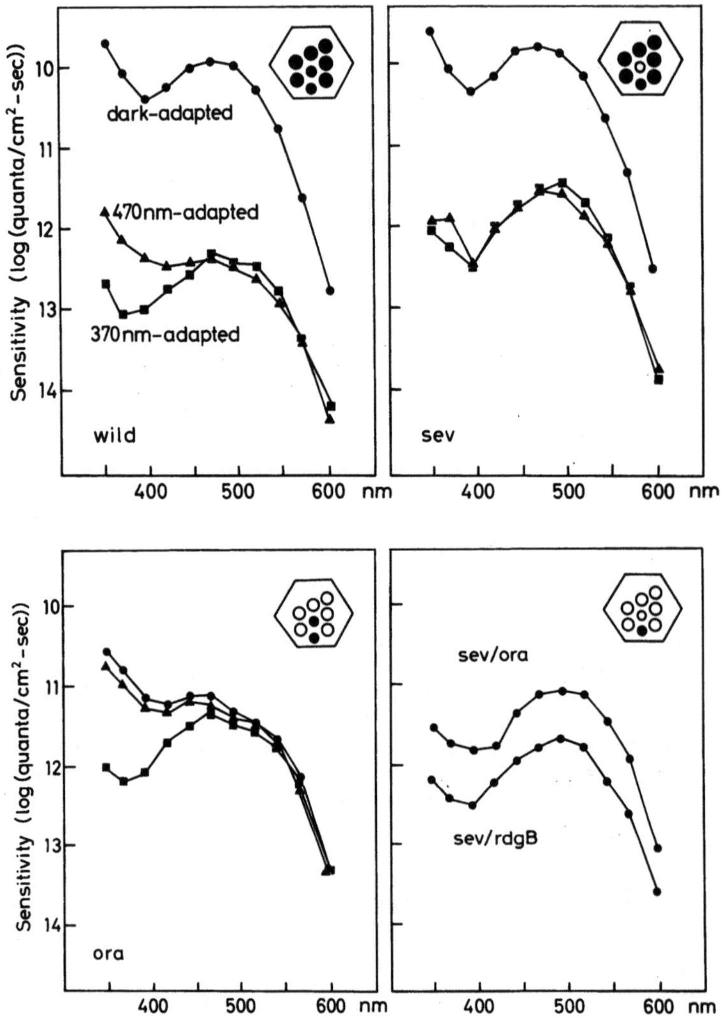

Fig. 15. Spectral sensitivities of wild type and mutants of *Drosophila melanogaster* (after HARRIS et al., 1976; STARK, 1975). Curves of the dark-adapted eyes of wild type and of mutant *sev*, which lacks only R8, are almost equal, and in both cases light adaptation causes high loss in sensitivity, showing breakdown of high sensitivity system (HSS) of R1–6, but note that in contrast to wild type, in *sev*, 470 nm and 370 nm have equal adaptation effects. Dark and light adapted curves of *ora* are in agreement with curves of light adapted *wild* type. The double mutants *ora/sev* and *rdgB/sev*, possessing only receptor R8 of high acuity system (HAS; KIRSCHFELD and FRANCESCHINI, 1968) agree with the light adapted *sev* mutant

Until now, three mutants of *Drosophila* with elimination of specific receptor types have been characterized through spectral sensitivity measurements (mass recordings), by photometry of pigment conversion and by behavioral investigations. Two mutants *rdgB* and *ora* delete the peripheral receptors R1–6, and one mutant *sev* eliminates the central rhabdomere R7. From these single mutants, white-eyed double mutants *sev/rdgB* and *sev/ora* were bred, in which only R8 is present (Fig. 15). The comparison of the spectral sensitivities in the light-adapted states (470 nm, 370 nm) show that the curves of the wild type agree with those of the mutant *ora*, and the curves of *sev* with those of *ora/sev* (*rdgB/sev*). Thus the lack of a specific effect to UV adaptation in *sev*, single or double mutants, indicates that R8 is a blue receptor ($\lambda_{max} = 470$ nm), and consequently R7 an UV receptor with $\lambda_{max} \simeq 350$ nm (HARRIS et al., 1976). The similarity of the dark adapted wild type to the *sev* eyes suggests that R1–6 have two sensitivity peaks at 350 nm and 470 nm, as is shown in *Calliphora*. These results agree in principle with the data of MINKE et al. (1975), but some discrepancies exist concerning λ_{max} of R1–6 = 485 nm, and of R8 = 460 nm.

The photoconversion of the visual pigment in the blowfly retina was first demonstrated by in vivo photometry (HAMDORF and LANGER, 1965), and early microspectrophotometric measurements on single rhabdomeres gave the first evidence that peripheral and central rhabdomeres harbor two different visual pigments comparable to the rhodopsins of vertebrates (LANGER and THORELL, 1966; see Fig. 12, Handb. Sens. Physiol. Vol. VII/3, A, p. 673). Conclusive evidence for the existence of photoreconvertible pigments were presented simultaneously by two laboratories, using different measuring techniques (STAVENGA et al., 1973; HAMDORF et al., 1973; HAMDORF and ROSNER, 1973).

For in vivo microphotometry of insect eyes a special technique has been developed by STAVENGA et al. (1973; STAVENGA, 1975). The compound eye is illuminated antidromically from the brain side by quartz waveguide optics, and the pattern of the deep pseudopupil corresponding to the rhabdomere pattern (FRANCESCHINI and KIRSCHFELD, 1971; FRANCESCHINI, 1975) is observed through the microscope, and the light flux originating from single rhabdomeres can be diaphragmed and measured by a photomultiplier, the observed receptors being light-adapted orthodromically via the microscope. By using this technique, photoreconvertible difference spectra were obtained, indicating uniform pigment systems in blowfly receptors R1–6 (P490, M580), and an additional diverse UV pigment comparable to that of *Ascalaphus* in the jointly measured R7 and R8. By the same technique the visual pigment of the peripheral receptors in the dronefly *Eristalis tenax*, (P460, Ms550), was recently analyzed (STAVENGA, 1976; Fig. 36). These results of in vivo experiments agree in principle with the photoreconvertible difference spectra automatically recorded between monochromatically light adapted and dark adapted dissected eyes (Fig. 16), and also with the corresponding calculations of the P and M spectra (data of SCHWEMER, in HAMDORF and ROSNER, 1973; HAMDORF et al., 1973). Difference spectra of more recent spectrophotometry can be approximated optimally by P494 and Ms580 at a ratio of $\lambda_{Pmax}/\lambda_{Mmax} = 1/1.6$ (see Sec. C.II.1a; Fig. 35), where the relative quantum efficiency of photoreconversion was calculated to be close to 1 (Fig. 37). Also the digitonin extracted visual pigment of R1–6 is thermostable (at 0° C) and photointerconvertible, and the wavelength-dependent concentrations at photoequilibria, $C_{Peq}(\lambda_a)$, agree with the results obtained from receptor photometry (SCHWEMER, unpublished data).

Visual pigments with very similar properties (P480, M580) have been found in two other fly species: in *Drosophila melanogaster* (OSTROY et al., 1974, by pigment extracts; KIRSCHFELD and FRANCESCHINI, 1975, by microspectrophotometry), and in the housefly *Musca domestica* (KIRSCHFELD and FRANCESCHINI, 1975). On the

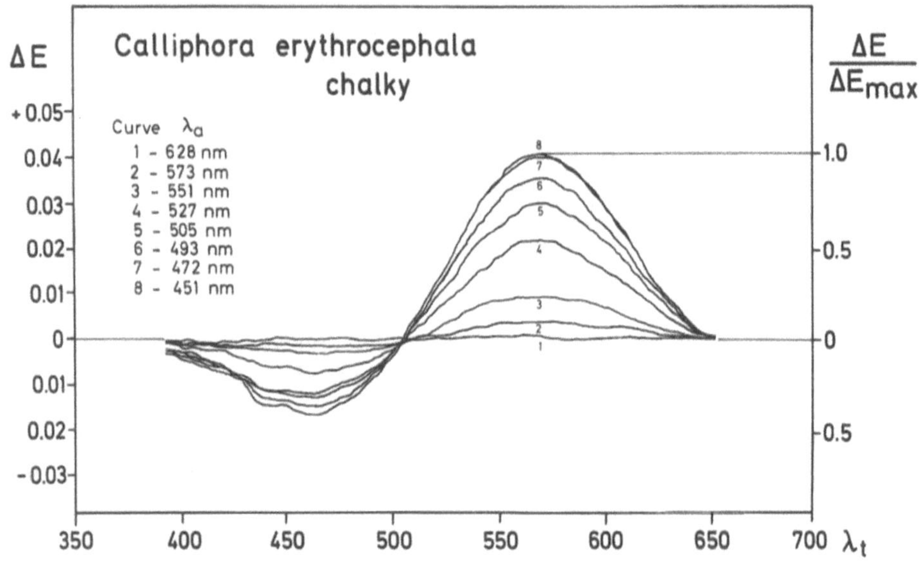

Fig. 16. Photoequilibria in eye of blowfly. Difference spectra obtained by dark adapting one eye (after 628 nm preadaptation) and light adapting other eye for 6 min each to: 628, 573, 551, 527, 505, 493, 472 and 451 nm (curves 1–8). The amplitudes are specific to each λ_a-created equilibrium. For computer analysis of spectra see also Fig. 35; Sec. C.II.1a

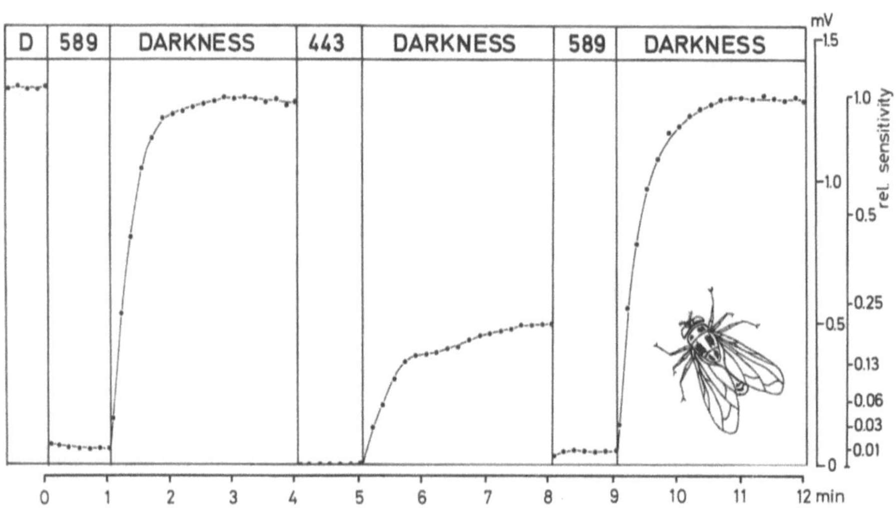

Fig. 17. Retinal response of *chalky* mutant of the blowfly, during and after monochromatic adaptation (589 nm, 443 nm). Dots represent amplitude of sum of potentials of receptors R1–6 to a constant test flash (450 nm, 60 ms, every 10 second). After 589 nm adaptation, receptor response increases rapidly to a maximum value, while after 443 nm adaptation amplitudes are reduced, and further adaptation by 589 nm is necessary to again reach maximum sensitivity. From scale on right (relative sensitivity), it can be seen that response amplitude after 3 min in dark (after 443 adaptation) corresponds to a rel. c.S. = 0.2 (Sec. C.II.3, Eq. 44, HAMDORF and ROSNER, 1973)

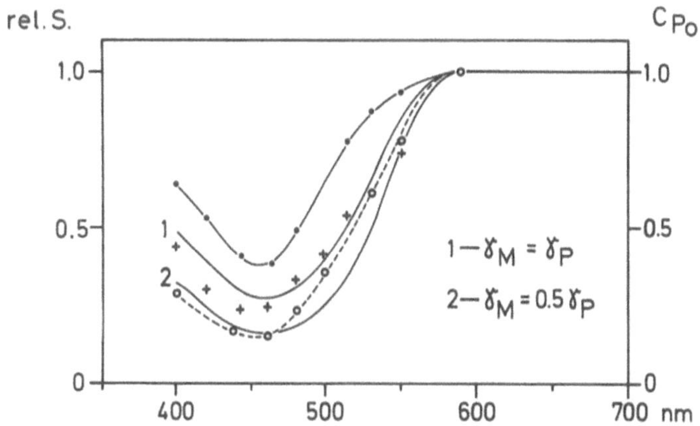

Fig. 18. Relative sensitivity of blowfly receptors 3 min after monochromatic light adaptation, as a function of the wavelength of quanta-equal illumination, and the visual pigment concentration $C_{Peq}(\lambda_a)$, calculated for quantum yields $\gamma_{M/P} = 1.0$ (*Curve 1*) and $\gamma_{M/P} = 0.5$ (*Curve 2*). *Dots* represent mean values of 8 extracellular measurements, *crosses* are from one single extracellular experiment, and *circles* are mean values of 6 intracellular recordings (HAMDORF and ROSNER, 1973)

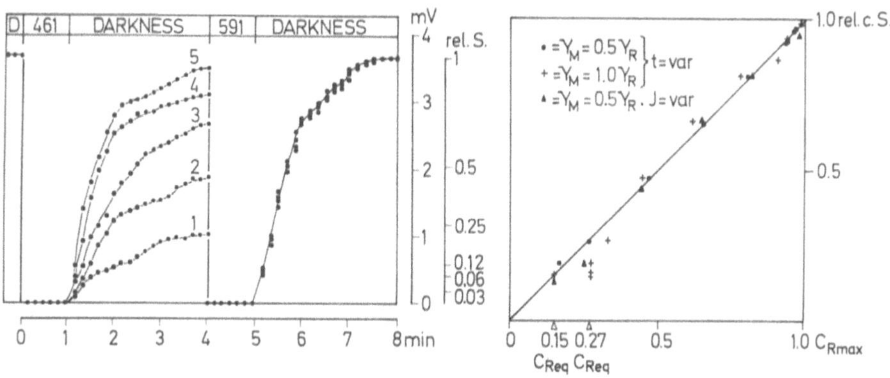

Fig. 19. *a*) The dependence of photoequilibration on intensity of adapting light (461 nm) in blowfly. *Curves* 1–5 show increase of response to constant intensity weak test stimuli (*dots*) after 1 min light adaptation with 461 nm of decreasing relative intensities: 1.0, 0.5, 0.25, 0.125, 0.063 and 0.032. Right hand scale shows response amplitudes to reduced test stimulus intensities after red adaptation. Receptor response is regained after 1 min illumination with 591 nm in all five experiments (*superimposed dots*). Comparable results are also obtained if intensity of adapting light is held constant while illumination period is varied. With increasing exposure time (approaching equilibration), response amplitude (recorded 3 min after light adaptation) becomes depressed continuously until a constant value (corresponding to C_{Peq}) is reached. *b*) Relationship between rel. c. S. and visual pigment content (C_P/C_{P0}), calculated by Eq. 1 (see Sec. C.I.1) for $\gamma_{M/P} = 1$ and 0.5. ● and +, electrophysiologically determined sensitivities 3 min after adaptation with 461 nm, with I being constant and t variable. Graph demonstrates *an almost direct* proportionality between C_P and receptor sensitivity (ROSNER, 1975)

other hand, in the larval mosquito *Aedes aegypti*, a green-absorbing pigment has been described which changes into a hypochromic-shifted blue-absorbing M (BROWN and WHITE, 1972). This implies that possibly principal differences exist between the visual pigments of Nematocera (mosquitos) and of Brachycera (flies).

The physiological relevance of photoregeneration in flies has been shown by extracellular and intracellular electrophysiologic recordings from R1–6. After red adaptation (589 nm, Fig. 17), which reconverts Ms maximally into P, the receptor response increases during a 2-min period to maximal values, whereas after blue adaptation (443 nm) maximal values are attained during the min 5–8 (Fig. 17), and thus after such photoequilibration ($C_{P_{eq}} = 0.2\, C_{P_0}$, Eq. 7 Sec. C.I.2), the response increase is reduced, the red light then reactivates the receptor to maximal sensitivity (9th–12th min). By using different quanta-equal wavelengths for adaptation, an almost direct proportionality has been demonstrated to exist between pigment concentration at photoequilibria and sensitivity of photoreceptors after light adaptation (Fig. 18, HAMDORF and ROSNER, 1973). A detailed analysis of this correlation based on intracellular recordings has been given by ROSNER (1975), (Fig. 19). Very recently, STARK and ZITZMANN (1976) reported similar results in vitamin A-deprived *Drosophila*. Applying adaptation and recording techniques comparable to those for *Calliphora*, in fruitflies of low P content, an almost direct proportionality with $C_{P_{eq}}(\lambda_a)$ was confirmed.

VI. *Deilephila*: an Example of a Trichromatic Visual System

Behavioral observations have shown that insects (i.e., the bee, rev. v. FRISCH, 1965) and some butterflies (CRANE, 1955) are able to distinguish between wavelengths and colors of more complex spectral distribution (rev., AUTRUM and THOMAS, 1973). The discrimination of wavelengths by a visual system requires at least two types of receptors with different spectral sensitivities. Based on training experiments for the worker-bee, a trichromatic visual system with three color receptors having maximal sensitivity at UV (350 nm), blue (444 nm) and green (520 nm) has been postulated (DAUMER, 1956). These three receptors have actually been identified by intracellular spectral sensitivity measurements (AUTRUM and v. ZWEHL, 1964; MENZEL and BLAKERS, 1976). A comparable triple-receptor system has been found in the sphingid moth *Deilephila elpenor* (HÖGLUND and STRUWE, 1970; HÖGLUND et al., 1973a, 1973b), with indications of trichromatic color vision; it has also been demonstrated through behavioral experiments for the closely related *D. livornica* (KNOLL, 1925). At present, that of *Deilephila* is the only triple-receptor system of insects of which all three visual pigments have been identified in extracts as well as by microspectrophotometry. The three color receptors analyzed by the selective adaptation technique show sensitivity peaks at $\simeq 350$ nm, $\simeq 440$ nm, and $\simeq 525$ nm. Moreover, the spectral sensitivity, $S(\lambda_i)$, agrees rather well with the nomograms (DARTNALL, 1953) for rhodopsins absorbing maximally at 350 nm, 440 nm, and 525 nm (see Fig. 20). Through microphotometry, the thermostability of the three metarhodopsins and also the photointerconversion of the green and UV visual pigments have been demonstrated (HAMDORF et al., 1972; HÖGLUND et al., 1973).

Very recently the position of these receptors in the ommatidia (composed of 9 receptor cells) were identified by microspectrophotometry (SCHLECHT et al., 1977 submitted) and electron microscopy (WELSCH, 1977). The very distal part of the fused rhabdome is formed by the rhabdomeres of two UV

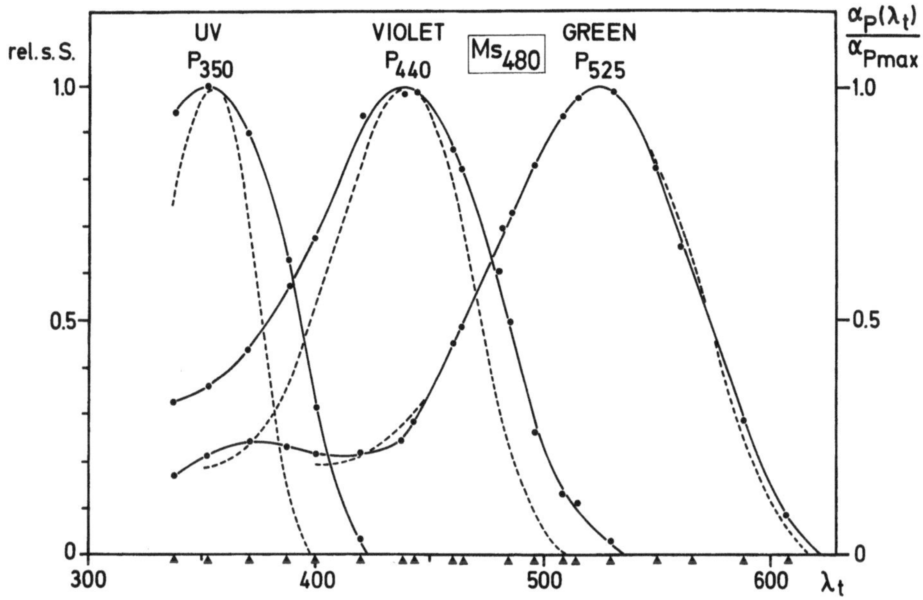

Fig. 20. The three color receptors of *Deilephila*, determined by selective adaptation technique. *Broken lines*, nomograms for rhodopsins absorbing maximally at 350, 440, and 525 nm (HÖGLUND et al., 1973)

receptors, the middle portion of the rhabdome by the microvilli of four green receptors and the proximal part by only a single large receptor cell, presumably the violet receptor. This arrangement of the color receptors over one another is of physiologic importance, as is pointed out below.

In digitonin extracts, however, all three pigments of *Deilephila* are bleached at room temperature after irradiation. This thermal instability of the metarhodopsins has been used to advantage for analyzing the combined extracted pigments, as green, blue and UV absorbing pigments can be selectively bleached, starting with longer wavelengths as the bleaching light. All three pigments result in metarhodopsins of very similar absorption properties with $\lambda_{max}=470$–480 nm. At $-15°$ C these acid metarhodopsins are thermostable and photoreconvertible to their parent pigments as in the membrane. For comparison, Fig. 21 shows difference spectra of the green pigment in extract at $-15°$ C, and in the membrane at room temperature. It is clear that both spectra are nearly identical and can be explained by a simple P525↔M480 transition ($\alpha_{P_{max}}/\alpha_{M_{max}}=1/1.75$).

The fact that in *Deilephila* the metarhodopsins of all three visual pigments have almost the same spectra implies a very similar interaction between the all-*trans*-retinal and the corresponding protein parts, whereas the interaction between the 11-*cis*-retinal and the protein part has to be specific for each pigment. This specificity may be explained by model calculations (ABRAHAMSON and WIESENFELD, 1972) on pigments with 11-*cis*-retiniledene iminium ion as the chromophore. Comparable shifts in λ_{max}, as observed in *Deilephila* pigments might be induced by a negative point charge close to carbon atom 6 (P520) and to carbon 11 (P440) of the chromophore or close to the nitrogen of the Schiff base linkage (P345) (SCHWEMER and PAULSEN, 1973).

Furthermore, it was shown for the example of the green receptors (Fig. 22), that as in *Ascalaphus* and in the blowfly membrane, the photoequilibrium is determined by the spectral absorption ratio $\alpha_{M490}(\lambda_a)/\alpha_{P525}(\lambda_a)$. This λ_a-specific alteration of pigment concentration induces comparable alterations of the ab-

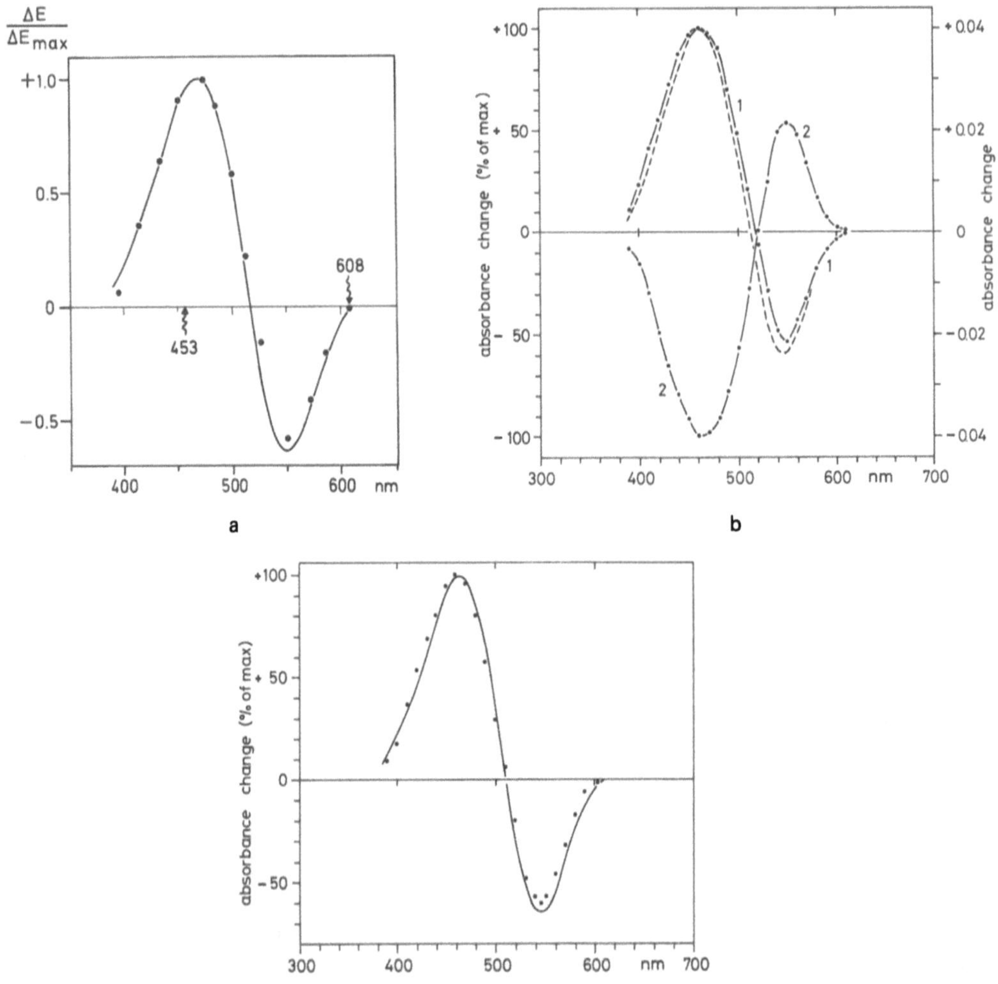

Fig. 21. Difference spectra of the green receptor pigment of *Deilephila.* (*a*) MSP-difference spectra of retina slices induced by alternating short (453 nm) and long (608 nm) adaptation. Continuous curve: difference spectrum calculated from nomograms of vertebrate pigments with λ_{max} at 525 nm and 480 nm, and with a ratio of absorbancies at λ_{max} of 1:1.75 (HAMDORF et al., 1973). (*b*) Difference spectra of a dissected retina illuminated with 628 nm (*Curve 1*) and after subsequent adaptation with 474 nm (*Curve 2*), showing complete reversibility of photokinetics. For comparison, difference spectrum of *c* is shown as dashed line (SCHWEMER and PAULSEN, 1973). (*c*) Difference spectrum of a digitonin extract after 628 nm irradiation measured at −15° C. Empirical data (dots) are seen to fit the theoretical difference spectrum (P520, M480) (SCHWEMER and PAULSEN, 1973)

solute sensitivity of the green receptors. Yellow light which reduces the P525 concentration causes a decrease in sensitivity, whereas blue light which favors the P525 concentration increase, results in increased sensitivity (HÖGLUND et al., 1973). This correlation is shown in Fig. 23.

Comparable results have been obtained for another sphingid species, *Manduca sexta.* The pigments of the UV-receptor and of the more abundant green receptor

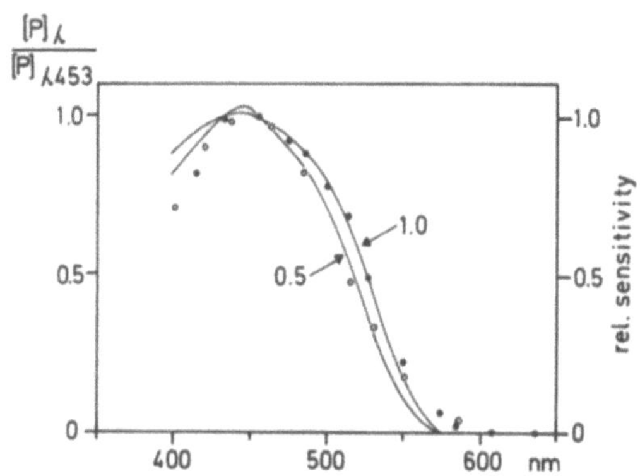

Fig. 22. Relation between sensitivity and concentration of visual pigment in green receptor of *Deilephila*. Continuous lines: calculated relative concentrations of P525 at photoequilibrium for quantum efficiency 1 and 0.5. Dots: photometrically recorded concentrations of P525. Circles: electrophysiologically recorded sensitivity level reached 4 min after monochromatic light adaptation (HÖGLUND et al., 1973; HAMDORF et al., 1973)

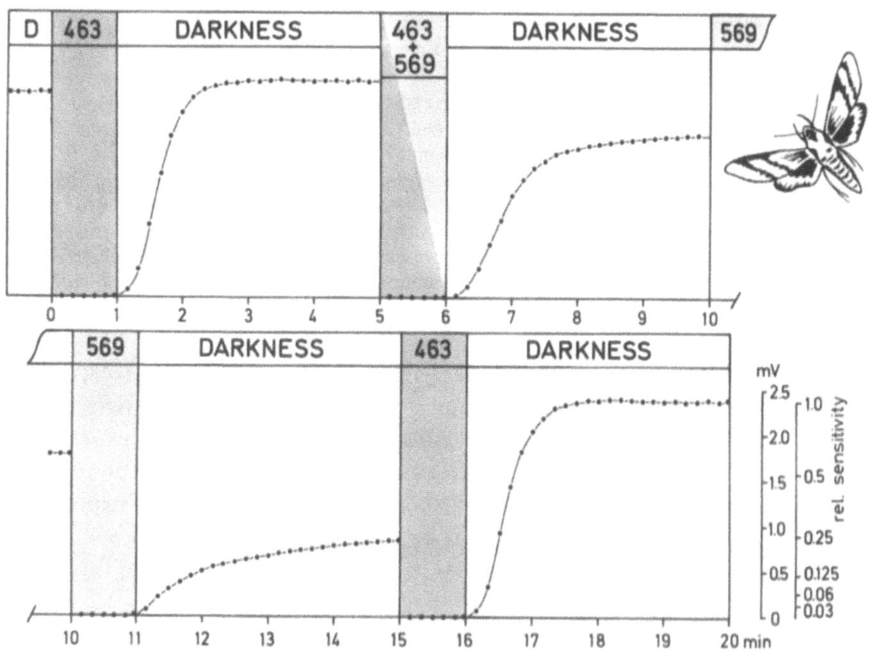

Fig. 23. Dark adaptation of retinal response after monochromatic light adaptation (D = darkness, 463 and 569 = adapting lights). The dots represent response amplitudes to constant test flashes of 590 nm, which stimulated only green receptors. N.B., the sensitivity, 4 min after cessation of 463 nm illumination (minutes 5 and 20) is equal to sensitivity of dark-adapted retina (1.0), while sensitivity after exposure to 569 nm (15th min) is lower (0.2). Sensitivity after combination of 463 and 569 nm (10th min) is 0.65. Quantal content of both adapting lights were almost equal (HÖGLUND et al., 1973)

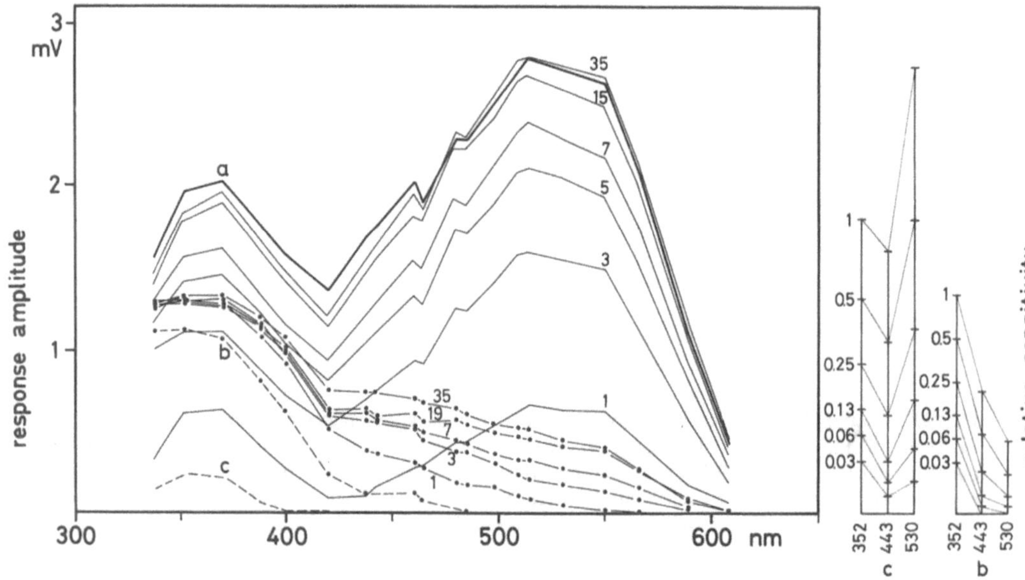

Fig. 24. Spectral efficiency of an isolated retina of *Manduca* during and after light adaptation to 590 and 450 nm. *Curve a*: spectral efficiency before start of experiment. *Curves b* and *c*: recorded during adaptation to 590 and 450 nm. *Curves* with *dots* and *numbers 1, 3, 7, 19* and *35* show change of spectral efficiency in dark, after adaptation at 590 nm. *Solid curves* with numbers *1, 3, 5, 7, 15,* and *35* show increase of spectral efficiency after adaptation with 450 nm. Each curve was recorded in a 40 s period, employing 20 test flashes of equal quanta. After light adaptation, *Curves 1, 3, 5, 7, 15, 19,* and *35* were recorded after elapse of 20, 100, 180, 260, 580, 740, and 1380 s, respectively, in dark. Right hand scales show response amplitudes at wavelengths 352, 443, and 530 nm to relative intensities (1.0, 0.5, 0.25, etc.) about 20 min after light adaptation to 590 nm (scale *b*) and to 450 nm (scale *c*). It can be seen that after 590 nm adaptation, depressing the P525 content, the sensitivity of green receptors are reduced for a long time, while sensitivity of UV receptors (and possibly also of violet receptors) is unchanged. After such P525 depression, blue photoregenerative light is necessary to regain sensitivity of green receptors. Note that increase in UV response after 450 nm adaptation is due to additional β-band response of the green receptors (HAMDORF and SCHWEMER, 1975)

are converted into thermostable blue-absorbing metarhodopsins (Ms480), as in *Deilephila*. This was shown through microspectrophotometry (SCHWEMER and BROWN, unpublished data) and by electrophysiologic adaptation experiments (HAMDORF and SCHWEMER, 1975). At present, the violet visual pigment and its Ms transition has not been precisely analyzed, but a similar Ms is to be expected.

Earlier microspectrophotometric measurements pointed to four pigments with λ_{max} at 350 nm, 450 nm, 480 nm and 530 nm (CARLSON and PHILIPSON, 1972). The former two peaks match the pigments found in *Deilephila*, but the reported P480 as well as the bleaching sequence are doubtful, and it is more likely that this pigment is identical with the metarhodopsins of all three pigments, and that the bleaching was caused by the experimental treatments, i.e., freezing of the tissue and use of isopentane and glycerol, which could reduce the stability of the pigment molecules in the membrane.

The position of all three metarhodopsins within the blue range of the spectrum seems to be of physiologic importance (HAMDORF et al., 1972, 1975; HÖGLUND et al., 1973; SCHWEMER and PAULSEN, 1973). According to Eq. 8, 9, Sec. C.I.3, the concentrations of the visual pigments in photoequilibrium (set by the spectral distribution of the skylight) are calculated to be 90% visual pigment in the UV receptor, and about 70% in the green and violet receptors. Thus the coincidence of the spectral Ms absorption with the radiation spectrum of the sky keeps the concentrations in all three receptors at high constant levels, corresponding to almost equal sensitivity levels. This relationship has been demonstrated electrophysiologically (Fig. 24). Adaptation by yellow, orange, or red lights, which reduce the concentration of the green receptor pigment, results in a long-lasting depression of the green receptor response, whereas the response amplitude to UV stimuli is only slightly reduced. However, after subsequent blue irradiation, which photoregenerates the pigments, the sensitivity of the green receptors is re-established after a few minutes (note: the simultaneously observed increase of the UV response is due to the contribution of the green receptors to the sum potential via the β-band of P525). Corresponding experiments on the compound eyes of *Deilephila* and of the honeybee have provided comparable results.

This mechanism must be of great importance for color discrimination and color constancy. A spectral color (or mixed colors) can be distinguished very easily if the ratio of response amplitudes (R) of three receptor types A, B, and C is constant at a wavelength λ_t (or at a given wavelength distribution). The assumption necessary for a constant response ratio ($R_{A\lambda_t}:R_{B\lambda_t}:R_{C\lambda_t}=\text{const.}$) is that the absorption probability (W) in all three receptors is also held constant ($W_{A\lambda_t}:W_{B\lambda_t}:W_{C\lambda_t}=\text{const.}$). This assumption is actually fulfilled as the visual pigments of all three receptor types are photoregenerated under natural light conditions to constant levels. If the ratio of pigment concentrations were variable, then the ratio of absorption probabilities would alter as well. This would result in an uncontrolled color shift.

VII. Carotenoids and Receptor Sensitivity

The first biochemical evidence that retinaldehyde was also the chromophore in the visual pigments of insects was reported by GOLDSMITH (1958). Lipid extracts of the heads of dark-adapted honeybees contain 0.22 µg retinaldehyde/g of fresh weight, with only traces of retinol, whereas in the light adapted state the amount of retinal decreases and is replaced by retinol (GOLDSMITH and WARNER, 1964). A corresponding NADH-dependent enzyme capable of reducing retinal to retinol has been detected, indicating participation of vitamin A in the visual cycle (GOLDSMITH, 1972; Handb. Sens. Physiol., Vol. VII/1, p. 704). A direct proof that 11-*cis* retinal is the prosthetic group of insect visual pigments (as in vertebrates) has been given by PAULSEN and SCHWEMER (1972).

In insects, carotenoid-deficient diets cause a marked decrease in photoreceptor sensitivity. If flies (*Musca domestica*) are reared for several generations on carotenoid-free diets, the sensitivity threshold of the compound eye is elevated by at least 10^4. This rise can be reversed by the addition of β-carotenoid to the larval diet (GOLDSMITH and FERNANDEZ, 1966). Comparable results were obtained for white-eyed strains of *Drosophila* reared on vitamin-enriched and vitamin-deprived larval diets (STARK and ZITZMANN, 1976), as well as for the *chalky* mutant of *Calliphora* reared on liver or heart meat (RAZMJOO and HAMDORF, 1976). The observed increase in the threshold was $2 \cdot 10^2$ in *Drosophila* (extracellular measurements), and $2.5 \cdot 10^1$ in *Calliphora* (intracellular measurements). Moreover, in *Calliphora* spectrophotometric measurements showed direct proportionality between visual pigment content in the rhabdomeres R1–6 and the sensitivity of the photoreceptors. Freeze-fracture preparations of eyes of *Calliphora* showed a corresponding thinning out of pigment particles in the inner layer of the microvillus membrane (BOSCHEK and HAMDORF, 1976).

Minimal distances of 12 nm (close to the theoretical value of the most densely-packed molecules in a microvillus) were determined in liver-fed animals, corresponding to about 1000 pigment molecules in a microvillus of 1 μm length. The decrease of pigment content to 1/23, as found in experiment of Fig. 25 for heart-fed animals means a reduction to less than 50 molecules/microvillus and an increase in molecular distances to about 70 nm. In extreme cases, no particles can be detected in the microvilli (BOSCHEK and HAMDORF, 1976). Extending this recent data to the early observations in *Musca* (GOLDSMITH and FERNANDEZ, 1966) it may be concluded that among the third generation of flies reared on carotenoid-free diets, only one out of ten microvilli had a pigment molecule. Consequently the sensitivity decrease observed in *Drosophila* should be due to about 5 molecules/microvillus (STARK and ZITZMANN, 1976). The fact that microvilli of vitamin A-deprived animals have plain, particle-free membranes is in agreement with recent experimental data concerning the rate of pigment synthesis in blowfly receptors (SCHWEMER, 1976). If vitamin A-deprived adults receive vitamin A-enriched diets (vitamin A acetate or palmitate), the pigment content increases continuously within about one week to maximal values, indicating a renewal of about half of the pigment molecules within about 3 days. Moreover, the argument that the metabolic pathway, starting from retinal uptake to its incorporation, may limit the time course is dispelled by directly injecting 11-*cis* retinal into one eye. For such introduction does not initiate a rapid increase of pigment content as may be expected if the binding protein was present in the microvillus membrane (the injected retinal is obviously completely stored inside the injected eye, for no pigment increase in the control eye can be observed).

Thus, it is rather more likely that the presence of retinal stimulates the synthesis of the visual pigment and its subsequent incorporation into the microvillus membrane.

At present it is still debatable whether metarhodopsin in the rhabdomeres of the blowfly can be reconverted to P490 via metabolic-dependent pathways. Neither difference spectrophotometry on dissected eyes nor intracellular recordings from fly heads have demonstrated such a metabolic reconversion (HAMDORF et al., 1973; ROSNER, 1975), but in vivo photometry gives some indications of a slow regeneration process, with a time constant of about 25 min (STAVENGA et al., 1973).[7]

Comparable results have also been reported with the white-eyed mutant *Drosophila*, norp A^{P512} (PAK and LIDDINGTON, 1974). This mutant also responds to quanta absorption by metarhodopsin, with a "fast transient" component (presumably comparable to the "on effect" of *Calliphora* lamina neurons). This Ms-specific signal is elicited only by using flashes of high energy ($1.9 \cdot 10^{14}$–10^{15} photons $\cdot$ cm^{-2}), indicating that all Ms molecules of the rhabdomere have to absorb one, two or even more photons to produce the maximal response. In this connection it is of importance to note that this M response decays very slowly in darkness, corresponding to a time constant of more than 6 hrs, indicating a slow metabolic regeneration of Ms.

[7] It has recently been shown, by *in vivo* experiments, that for up to an hour or more after the maximum production of metarhodopsin, there is no measurable dark regeneration of rhodopsin (SCHWEMER, personal communication), and thus the thermal regeneration Ⓜ $\xrightarrow{t_4}$ Ⓟ of figure 13 is non-existent in this mutant and can be omitted from the scheme.

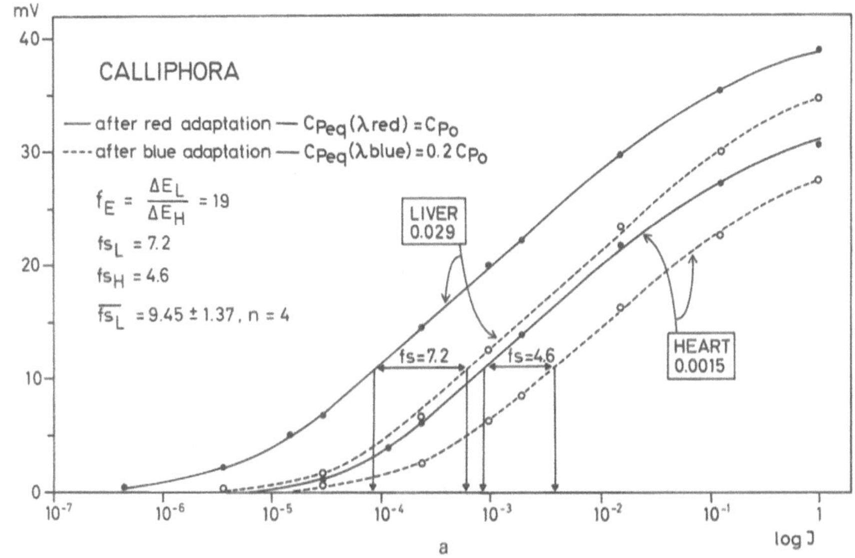

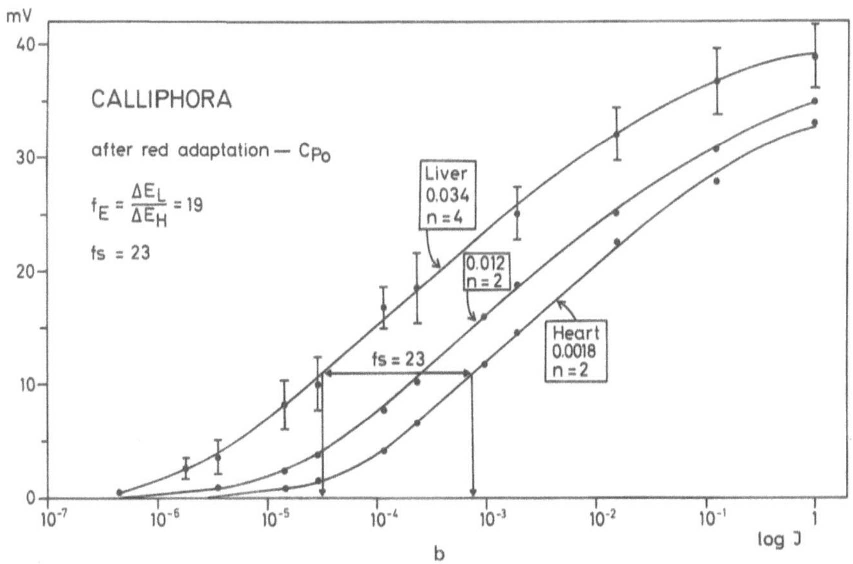

Fig. 25. *I–A* functions of blowfly photoreceptors depending on the content of visual pigment in the rhabdomeres and on the color of adaptation. (*a*) *I–A* functions of receptors from liver-fed (L) and heart-fed (H) fly. N.B., after blue adaptation sensitivity loss ($f_s = 4.6$) in H fly agrees well with expected direct proportionality between sensitivity and pigment concentration in rhabdomere, whereas in the L fly f_s increases to a mean of 9.45, approximately twice that of the H fly. (*b*) *I–A* receptor functions of L and H flies after red adaptation. N.B., with increasing total P content in receptor membrane (corresponding to increasing extinction values of 0.0018, 0.012, 0.034), functions shift to lower stimulus intensities. Sensitivity factor f_s is approximately equal to the extinction factor f_E, indicating direct proportionality between pigment content and sensitivity in red-adapted state. (RAZMJOO and HAMDORF, 1976)

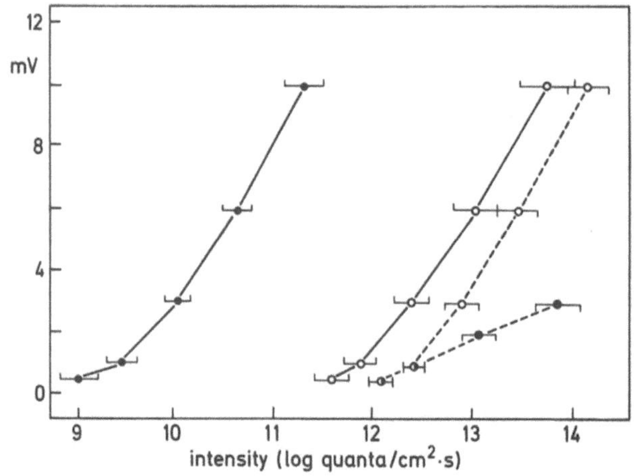

Fig. 26. Extracellularly recorded *I–A* functions of *Drosophila* (*cn bw*), vitamin A-enriched (*curves with dots*) and vitamin A-deprived (*open circles*), after monochromatic adaptation with 570 nm (*solid lines*) and 470 nm (dashed lines). Data are averaged for 8 preparations and *bars* represent S.E. Shape of curves of vitamin A-deprived animals are almost similar and show parallel shift by factor of only 0.5 log unit, while curves of vitamine A-enriched animals differ in shape and distance, showing a great loss of sensitivity by about 3 log units after 470 nm adaptation (after STARK and ZITZMANN, 1976)

At present, the literature suggests different adaptation mechanisms among flies. Whereas after blue adaptation in *Calliphora* (setting $C_{P_{eq}} = 0.2C_{P0}$), sensitivity is only slightly reduced (HAMDORF and ROSNER, 1973; MUIJSER et al., 1975; ROSNER, 1975; RAZMJOO and HAMDORF, 1976), the sensitivity of *Drosophila* receptors drops by several logarithmic units in the light-adapted state. This "inactivation" of the HSS receptors depending on the Ms content after blue adaptation has been reported by three laboratories (STARK, 1975; MINKE et al., 1975; COSENS and WRIGHT, 1975). But in vitamin A-deprived strains of *Drosophila* only a small sensitivity loss was found, as was in the light-adapted *Calliphora* receptors (Fig. 26) (HARRIS et al., 1976). Thus the observed differences seem to be of a quantitative rather than a qualitative nature. To overcome these discrepancies two essentially diverse hypotheses have been advanced.

1. The Transmitter-Membrane-Site Hypothesis. This hypothesis proposes that each dark-adapted receptor contains a certain amount of inactive transmitter molecules, of which a certain number become activated by a P – M transition. These activated transmitters combine with membrane sites, eliciting the receptor potential. Thus the number of transmitters activated by a constant intensity test flash must depend on the concentration of P molecules in the rhabdomere. If the density of P molecules is low, as in the vitamin A-deprived populations, then the number of activated transmitters is less than in rhabdomeres with high pigment content. Now, in cases of high P concentration and a relatively smaller, but finite, number of membrane sites for which the activated transmitter must then compete, one might expect a prolonged depolarizing afterpotential, and (as is actually observed after the cessation of the intense adapting light) a depletion of the inactive transmitter store. Consequently the rate of transmitter inactivation or of transmitter regeneration must limit the recovery of the receptor sensitivity in the dark. Photoreconversion, however, repolarizes the receptor membrane (MUIJSER et al., 1975). To reconcile these facts, the hypothesis has to be further expanded. For example,

a) M itself may activate the transmitter, then the PDA will last as long as the total M is reconverted to P by enzyme-dependent processes (11-*cis*→all-*trans* isomerization). This is rather

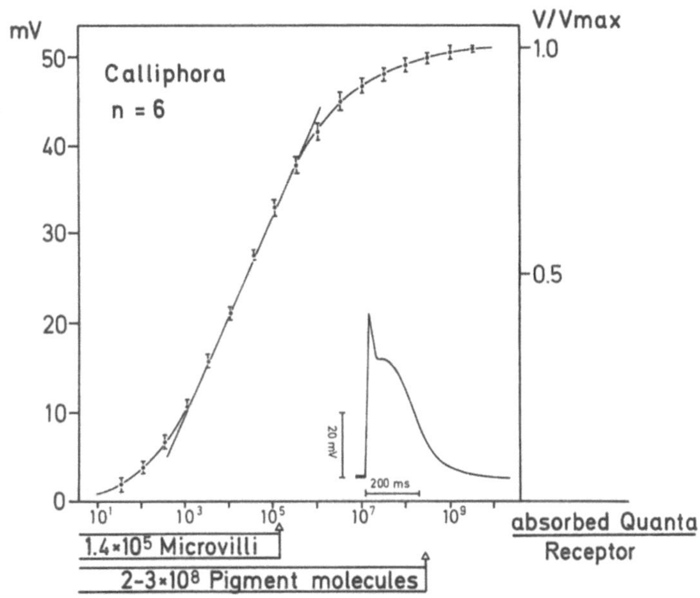

Fig. 27. *I–A* function of blowfly photoreceptors (mean of 6 cells) determined by "white" xenon flashes of ≃1 ms duration (DÖRRSCHEIDT-KÄFER, 1972). N.B., (*a*) the scale of the abscissa (absorbed quanta/receptor) allows a direct comparison with the number of structure elements: pigment molecules and microvilli, and (*b*) end of constant exponential part of curve (*straight line*) almost coincides with the number of microvilli in receptor, and after this the curve becomes slowly saturated until all pigment molecules have absorbed one or more quanta

unlikely, for the decay time of PDA is much faster than the rate of P regeneration, if any P is regenerated at all.

b) P molecules may favor the inactivation of the activated transmitter.

c) An antagonistic transmitter system (inhibitory or competetive) may also exist, depending on C_M, regulating the exciting effect of the excitatory transmitter and activated by the P – M transition. Of these the latter two seem more likely, if at all (see Sec. C.IV).

2. *The Microvillus Hypothesis.* This hypothesis proposes that the microvilli themselves act as the functional units. This assumption is supported by the relationship between the number of absorbed quanta and the response amplitude (Fig. 27). The $I-A$ function of blowfly receptors R1–6 show three phases: up to about 50 absorbed quanta the response amplitude increases linearly, then follows an exponential phase (straight line over 3–4 logarithmic units), and lastly a section with retarded rate of increase, up to the maximum amplitude of about 50 mV.

Can these changes in the $I-A$ function be correlated with some morphologic functional units, and if so, with which—the pigment molecules (1000–2000 in a microvillus) or the microvilli themselves ($1.4 \cdot 10^5$ in a rhabdomere)? The absorption of quanta by the microvilli obeys Lambert-Beer's law along the entire length of the rhabdomere. But the probability of absorption in each "layer" (*dx*, corresponding to either a single layer of microvilli or a layer of pigment molecules) follows a Poisson distribution (HAMDORF et al., 1968), which differs from one layer to the next, for the intensity is reduced by each layer according to Lambert-Beer's law. By translating the values along the intensity axis (Fig. 27) to the number of quanta of light, the point of absorption of $1.4 \cdot 10^5$ quanta (assuming the $1.4 \cdot 10^5$ microvilli to be the absorbing units) intercepts the second phase of the curve at the end of its linear part. The Poisson distribution for this one to one (quantum to microvillus) relationship indicates that 63% of all microvilli absorb at least one photon at this part of the curve. Yet as little as four times this intensity will involve about 100% of the microvilli in single or multiple hits, intercepting the curve further to the right at the beginning of the third phase. Thus it becomes apparent that: a) the second phase coincides with the range of single quanta hits to the microvilli, and that, b) the third phase entails more than one hit to each microvillus, that is, the interactions of quanta with several visual pigment molecules in each microvillus.

As the receptor response increases until all pigment molecules in a microvillus have absorbed one or more quanta, it can be assumed that each P molecule contributes to the electric response by its own share of reduction of membrane resistance. But note that the contribution to the production of the electric signal is much greater in the range of *single* quantal hits to the microvilli. Thus the microvilli might be light-excitable units, determining the depolarizing influx of cations (Na^+) into the receptor cell (HAMDORF, 1975), and the excitation might be mediated by intracellular transmitters acting on membrane sites, or directly caused by the light-excited P molecules, or by adjacent molecular structures. This cannot be decided with the data now available. In any event, the idea of the microvillus as a functional unit would demand that light and dark adaptation be influenced by interactions, or "interference", of P and M molecules statistically distributed inside the microvillar unit (RAZMJOO and HAMDORF, 1976).

VIII. Structure and Function of the Microvillus

Visual pigments are structural elements of photoreceptor membranes and are freed only by destruction of the membrane architecture. As the visual pigments of both the vertebrates and the invertebrates are retinal chromoproteids of comparable molecular weights, one may assume similar molecular structures for the disc and the microvillus membrane.

The disc membrane of vertebrate photoreceptors is a highly fluid phospholipid bilayer in which the pigment molecules are free to rotate and to diffuse, as has been revealed by the technique of rapid microphotometry (see review by LIEBMAN, 1975). Freeze-etch studies have shown that only particles of suitable size are incorporated into the cytoplasmic lipid layer of the disc membrane (CLARK and BRENTON, 1968; LEESON, 1971; MASON et al., 1974) and there is good evidence through immunolabeling experiments that these particles are identical with single or aggregate pigment molecules (JAN and REVEL, 1974). Furthermore, biochemical studies on frog and cattle photoreceptors have demonstrated that about 80% of the total protein content of rod outer segments (ROS) is opsin (HALL et al., 1969; HEITZMANN, 1972; ROBINSON et al., 1972) (of the remaining 20% little is known about their nature and their morphological or functional contribution to the membrane assembly). About 50% of the total dry weight of ROS's are lipids (NIELSON et al., 1970), according to a molar lipid to rhodopsin ratio of $\simeq 68:1$ (DAEMEN, 1973).

The diameter (4.2 nm) and the mean separation of rhodopsin molecules in discs (7.0 nm) have been determined by low-angle x-ray diffraction studies (BLASIE et al., 1969; BLASIE, 1972a, 1972b), and comparable molecular center-to-center distances were calculated from microspectrophotometric (LIEBMAN, 1962; LIEBMAN and ENTINE, 1968; LIEBMAN, 1975), and from biochemical data (DAEMEN, 1973; DEGRIP et al., 1972). Each rhodopsin molecule forms part (13.8 nm^2) of a membrane area of 44.6 nm^2 (DAEMEN, 1973), and the 68 lipid molecules have to cover a bilayer area of 75.4 nm^2 (inner lipid layer, 30.8 nm^2, plus outer layer, 44.6 nm^2), corresponding to 1.11 nm^2 membrane area for each phospholipid molecule. This calculation agrees well with results of monolayer experiments, where a single lecithin molecule (C_{11}:0–C_{22}:6–PC) covers an area of 1.25 nm^2 in the expanded form and 0.70 nm^2 in the compressed form, close to the collapse pressure of the monolayer (DEMEL et al., 1972). This indicates that each pigment molecule will be surrounded by only $\simeq 28$ phospholipids, and will be covered at the extracellular side by $\simeq 40$ lipid molecules.

Comparable data of membrane composition in rhabdomeric photoreceptors are rare, but the small amount of data for the cephalopods is in line with the disc model.

For the squid receptors a phospholipid to pigment ratio of about 86:1 can be calculated (from the data of MASON et al., 1973), and for carefully purified rhabdomeres of the octopus *Eledone* a ratio of 60:1 has been found (PAULSEN and ZINKLER, 1976), and moreover, the phospholipid composition of the rhabdomeres is similar to that of the disc membrane (DAEMEN, 1973; BENOLKEN et al., 1975). In the sphingid moth, *Deilephila*, and in the octopus *Eledone*, similar proportions of phospholipids were found (PE:PC:PS—$\simeq$4:$\simeq$4:$\simeq$1, ZINKLER, 1974; PAULSEN and ZINKLER, 1976). Divergent proportions were found for the rhabdomeres of *Calliphora* ($\simeq$5:$\simeq$2:$\simeq$1, ZINKLER, 1975) and for *Limulus* ($\simeq$2:$\simeq$4:$\simeq$1, BENOLKEN et al., 1975). The cholesterol content of the microvillus membrane, however, is about two to three times greater than in the disc membrane, indicating a greater rigidity and molecular interaction in the rhabdomeric membrane (MASON et al., 1973). In *Eledone* a cholesterol-to-total-phospholipid ratio of 2:1 has been determined (PAULSEN and ZINKLER, 1976), suggesting either a direct coupling between molecular pairs of the phospholipid bilayer, or a one-to-one interaction with phospholipids of the external, or of the cytoplasmic layer. Also one would expect strong molecular interactions to exist between the phospholipids and the visual pigments which are incorporated into the lipid bilayer. The amino acid composition of squid rhodopsin shows an extremely high content of nonpolar residues, especially proline ($\simeq$10%, HAGINS, 1972). Vertebrate opsins have similar characteristics (ABRAHAMSON and WIESENFELD, 1972). This high content of hydrophobic groups is consistent with the insolubility of rhodopsin in aqueous media except in the presence of detergents. There are also indications through proteolytic digestion experiments that the protein portion, bearing the chromophore, is the high hydrophobic region of the pigment molecule which makes contact with the phospholipids (HAGINS, 1973). This contact may stabilize the geometric structure of the pigment molecule, or it may directly prevent the disruption of the Schiff base linkage between the retinal and the protein as a result of bleaching. Although in general the membrane-coupled metarhodopsins of insects and cephalopods are thermostable and photoreconvertible, they react in distinctively different ways, depending on the nature of the detergent extract.

Some metarhodopsins irreversibly bleach when extracted (i.e., those of the visual pigments of *Deilephila*, SCHWEMER and PAULSEN, 1973), while others react similarly to the membrane-coupled metarhodopsins (i.e., those of the visual pigments of *Ascalaphus*, *Calliphora*, and *Eledone*). But the metarhodopsins of the latter group of pigments also become unstable when extracted by other detergents than digitonin (CTAB, Triton X 100), and even the robust visual pigment P470 of *Eledone* becomes unstable in CTAB (PAULSEN and ROMERO, unpublished data).

Similar effects have been observed in the lobster (*Homarus americanus*). While the extracted pigment is bleached by irradiation (WALD and HUBBARD, 1957), the membrane-coupled pigment is converted into the photoreconvertible metarhodopsin (GOLDSMITH and BRUNO, 1973). Also, this lobster pigment is sensitive to glutaraldehyde fixation, whereas other pigments, i.e., that of *Eledone*, remain thermostable even in the fixed state (WATERMAN et al., 1969; HAYS and GOLDSMITH, 1969). In summary, these reactions indicate that membrane components that stabilize the geometric structure of the pigment molecule may more or less be disrupted by the detergent micelles, and that therefore the stabilizing factor probably lies with the phospholipids (PAULSEN and SCHWEMER, 1973).

Based on these data, a structure similar to the disc membrane has been proposed for the microvillus membrane (HAMDORF and SCHWEMER, 1975; LAUGHLIN et al., 1975). Assuming a lipid-to-pigment ratio of 60:1 in *Eledone*, or 86:1 for the squid, and comparable expansions of phospholipids as in the disc membrane, pigment distances of 7.0–7.6 μm are calculated, when "viewed" from inside of the microvillus membrane. Taking into account the bending of the microvillus and the thickness of the inner membrane layer, these values correspond to particle distances of 8.4–9.2 μm when viewed from the outside (looking towards the inner membrane layer).

Microvilli are membrane tubes of very constant diameter, and only relatively small variations (45–80 μm) among various species of arthropods and cephalopods

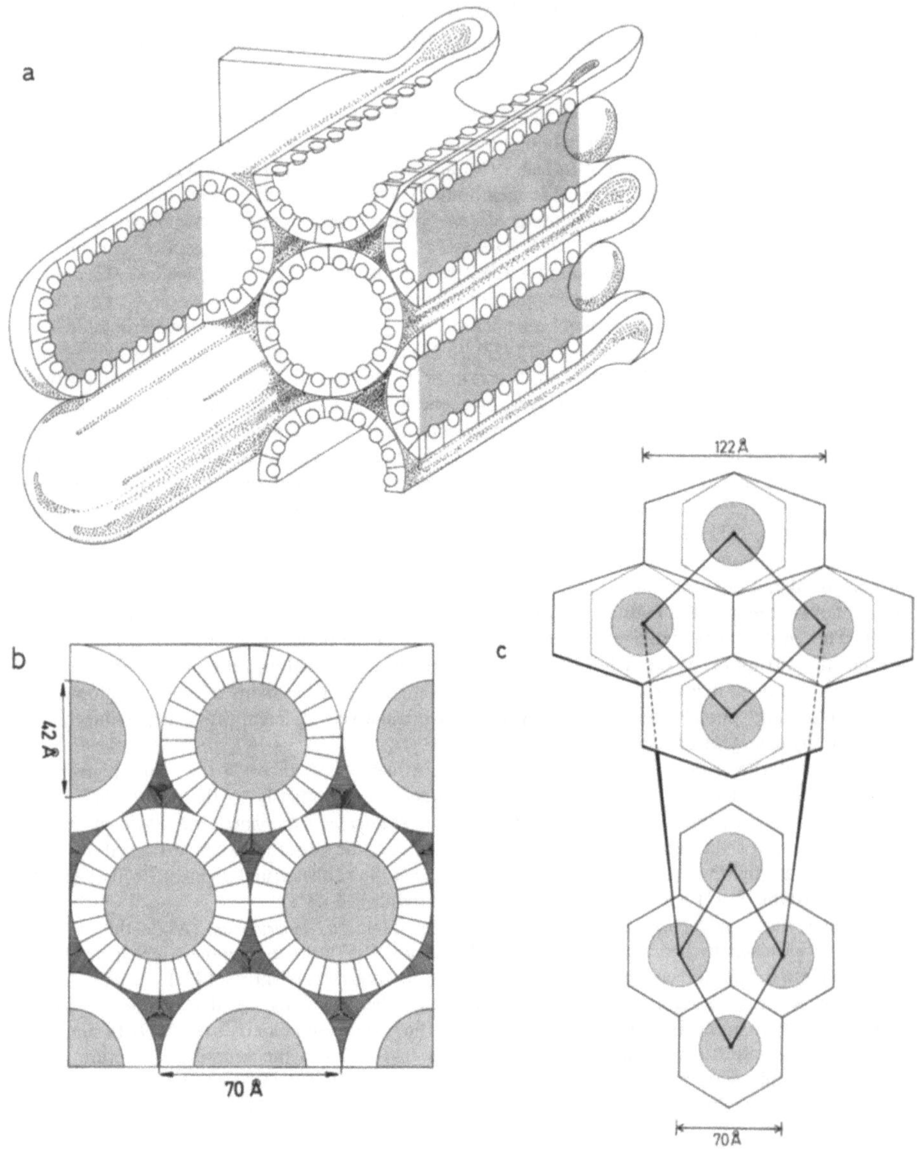

Fig. 28. Molecular model of microvillus structure. (*a*) Three dimensional scheme of the microvillus structure. In a dense molecular packing, about 16 pigment molecules form circumference of tubular microvillus. (*b*) View from within the microvillus membrane. Ratio of phospholipids to pigment molecules suggests each pigment (*grey field*) to be surrounded by a single circle of 28 phospholipid molecules, while outer surface is completely covered by phospholipids. This is based on the data from BLASIE et al. (1969). However, it is more probable that opsin and other adjacent structural proteins actually penetrate the entire thickness of the membrane, and thus c) as a consequence of relatively small diameter and fixed length of microvillus, regular hexagonal array of inner membrane layer (monolayer), having a rhombic center-to-center pattern, has to change into a square array (HAMDORF and SCHWEMER, 1975)

have been found. This constancy and accuracy in tube structure implies that the microvillus membrane is a more or less crystalline cylinder built of tightly packed membrane "bricks". Patterning the microvillus structure on the model of the disc (Fig. 28), it is conceivable that the pigment molecules dip into the membrane bilayer from the cytoplasmic side, each surrounded by a single ring of phospholipids, forming a hexagonal pattern (HAMDORF and SCHWEMER, 1975). This model is supported by recent freeze-fracture studies of crayfish rhabdomeres (FERNANDEZ and NICKEL, 1976). The A-face (the external face of the cytoplasmic layer, corresponding to the P-face in other literature) contains particles of about 8 µm. These "A" particles tend to aggregate into clusters of hexagonal array oriented diagonally to the microvillus axis. The particle distances are about 8.5 µm, i.e., close to the predicted values mentioned above (8.4–9.2 µm). Particles of comparable size were found in the rhabdomeres of the honeybee (PERRELET et al., 1972) and of the bulldog ant *Myrmecia gulosa* (NICKEL and MENZEL, 1976). There is good evidence that these "A" particles may really be the visual pigment molecules, for in the blowfly receptors the density of the "A" particles is in agreement with the pigment concentration in the rhabdomeres, as determined by spectrophotometry (BOSCHEK and HAMDORF, 1976).

The diameter of the microvillus is found to be constant along its entire length, and since the number of structural "bricks" covering the outer membrane (with a greater surface area) must be the same as those constituting the inner layer (with a smaller area), the hexagonal array of the pigment molecules in the inner lipid layer must undergo dimensional changes as the outer lipid layer is approached (Fig. 28). Since the transition from the inner to the outer layer experiences a dimensional expansion not in the direction of the microvillus axis (as this length is fixed), but only in the circumference of the microvillus cross-section, the rearrangement of the "bricks" occurs only along the latter dimension. Thus the rhombic shape of the "unit cell" (in this case the joining of the centers of four hexagons found in the inner layer) must gradually change into a square "unit cell" (see Fig. 28c) with the approach of the outer layer. Considering the dimensions actually found for the outer diameter of 50 µm, 36 µm for the inner diameter, and 7 µm for the thickness of the bilayer (in the blowfly), precisely such a change from the hexagonal to a square pattern is observed.

Some experimental data from the bulldog ant (NICKEL and MENZEL, 1976), where particles were found in both layers under certain experimental conditions, indicate a similar transformation in membrane pattern. The clusters of particles in the inner layer showed a more irregular hexagonal arrangement, whereas the particles of the outer layer (E-face, the same as the B-face) were arranged in rows, aligned with the axis of the microvillus. Looking at the outside layer along the microvillus axis, if the particles of two adjoining rows alternate with one another, then in the inner layer the long axis of the rhombus of the "unit cell" lies parallel to the microvillus axis, and if in the outer layer the particles of two adjoining rows are arranged side by side, then in the inner layer the long axis of the rhombus assumes a position diagonal to the microvillus axis.

The chromophore of visual pigments is a polyene and thus a dipole with vectorial absorption characteristics, and this can be used to define the spatial orientation of the molecule, except in pigment extracts, where the absorption vectors are randomized, and therefore no dichroism can be observed. A similar situation exists in ROS's examined axially, for their absorption is also independent of the polarization plane (***E*** vector) of the measuring beam. ROS's, however, viewed from the side (transversely with respect to the receptor axis), are highly dichroic (SCHMIDT, 1938; LIEBMAN, 1962; HAROSI and MACNICHOL, 1974). This can only be explained if the (dominant) absorption vector of the total number of pigment molecules is oriented parallel to the disc plane, i.e., restricting one degree of freedom, namely the pinwheel rotation perpendicular to the plane, the molecules being allowed to rotate only around their vertical axis and to diffuse laterally within the lamellar plane (CONE, 1972; BROWN, 1972; LIEBMAN and ENTINE, 1974; POO and CONE, 1973; LIEBMAN, 1975). During the transition from P to MII and MIII, the dichroic ratio in the spectral range of the α band decreases only slightly from 5 to 4. This small change corresponds to a tipping movement of the chromophore away from the membrane plane by about 15° (see Sec. C.I.4; Fig. 33, $\Delta\Psi = 15°$; HAROSI, 1971; HAROSI and MACNICHOL, 1974) and furthermore the mobility of the intermediates seems to be unaltered. A drastic change in dichroism is expected, however, when the chromophore is split off from opsin, for the dipoles retinal, retinal oxim and possibly retinol, are aligned vertically to the plane of the membrane. In this case the molecular orientation is determined directly by the polar and hydrophobic forces in the phospholipid bilayer, placing the oxygen atom of retinal into contact with the surface water, while the hydrocarbon part is interdigitated between the membrane lipids (DENTON, 1959; LIEBMAN, 1969, 1972, 1975).

The speed of molecular rotation in the disc membrane has been determined by rapid flash photolysis. The momentary dichroism, induced by bleaching with linearly polarized light, is totally randomized within about $\simeq 60\,\mu s$, indicating a frequency of >1000 cycles/s (CONE, 1972). If the membrane is fixed with glutaraldehyde, rotation is prevented, and thus a high dichroism can be easily established by selective bleaching of the fixed dipoles oriented in the direction of the $\boldsymbol{E}$ vector (BROWN, 1972). The speed of lateral diffusion has been measured through local bleaching experiments in cones. A bleached membrane area is filled with pigment molecules from the adjacent non-illuminated areas during a period of less than one minute. This lateral diffusion is also blocked by glutaraldehyde fixation. From both the rotation and diffusion measurements, the viscosity calculated for the disc membrane is comparable to that of olive oil (POO and CONE, 1974; LIEBMAN and ENTINE, 1974). Hence for the disc membrane the SINGER-NICHOLSON model of a fluid mosaic membrane seems to be more plausible than a crystalline membrane structure (SINGER and NICHOLSON, 1972; WORTHINGTON, 1974).

By transferring the disc model to the microvillus membrane, particular consequences concerning the dichroic properties and photoequilibration have to be expected (LAUGHLIN et al., 1975; SCHLECHT and TÄUBER, 1975). Assuming a statistical distribution of ideal dipoles (free to rotate and to diffuse) in the bent microvillus membrane, a dichroic value of 2 has been postulated (MOODY and PARRISS, 1961; for restrictions see LAUGHLIN et al., 1975). Microspectrophotometric measurements (LANGER, 1965; KIRSCHFELD, 1969) as well as electrophysiologic data from fly receptors (NAKA and KUWABARA, 1959; BURKHARDT, 1962; AUTRUM and v. ZWEHL, 1962; SCHOLES, 1969; MCCANN and ARNETT, 1972) support this hypothesis. The electrophysiologically determined polarization sensitivities (PS) of different crayfish receptors (*Carcinus*, *Callinectes*, *Astacus*, *Procambarus*) however, vary between 3 and 13, and thus exceed the critical value of 2 (SHAW, 1969; WATERMAN and FERNANDEZ, 1970; MULLER, 1973; MOTE, 1974). These results are supported by recent microspectrophotometric measurements on *Orconectes* rhabdomes, showing dichroic ratios of $\simeq 3$ (GOLDSMITH, 1975).

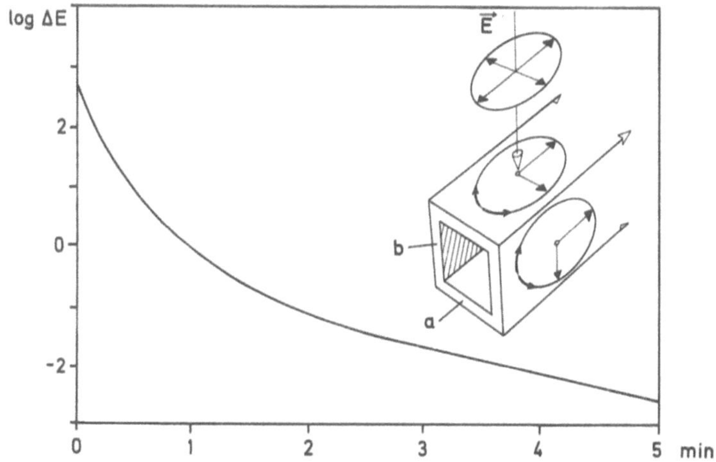

Fig. 29. Theoretically calculated time course of photoequilibration in a microvillus, where pigment molecules are free to rotate. It is apparent from inserted scheme that the kinetics of molecules of *a* walls has to be faster than the kinetics of *b* molecules due to greater absorption probability of *a* molecules. N.B., this plot [$\log \Delta\varepsilon(t)$] would be a straight line for first order kinetics

In these cases therefore, the dipoles cannot be simply statistically distributed in the membrane, but must be oriented more or less in alignment with the direction of the microvillus axis. Such an alignment of dipoles in the cylindric membrane may be caused by Brownian motion if asymmetries in molecular shape exist. As in a cylinder, the molecular motion in the direction of the microvillus axis will be favored in comparison with tangential movements, thus a statistical alignment parallel to the microvillar axis is to be expected. This alignment, caused by both molecular asymmetries and bending of the microvillus, is almost independent of the viscosity of the membrane, and still present at high viscosities. Also in crystalline microvillous structures, where lateral diffusion is axially as well as tangentially severely restricted, as presumed above, the probability of molecular alignment will depend mainly on asymmetries in rotational friction. This preference for statistical alignment in the *z*-axis of the microvillus (Fig. 33) can be described generally by a distribution function according to BOLTZMANN (SCHLECHT and TÄUBER, 1975; see Sec. C.I.4).

At present, few data concerning the dichroism of the intermediates in the microvillus membrane are available. However, the high dichroic ratio found for the thermostable and photoreconvertible acid metarhodopsin of the cephalopod *Eledone* is similarly high for its parent rhodopsin (TÄUBER, 1973; TÄUBER, 1975; SCHLECHT and TÄUBER, 1975). Through numerical calculations of the rhabdomeric dichroism—depending on λ_t, λ_a, the molecular orientation, and the fraction of P and Ms in the membrane (see Sec. C.I.4)—these authors estimated for the transition from P to Ms a tilting of the absorption vector, perpendicular to the membrane surface ($\Delta\Psi$), of about 15°–20°, but it can not be determined whether the absorption vector tilts centrifugally or centripetally. These results are also in agreement with measurements on rod outer segments (HAROSI, 1971), where during transition from rhodopsin to MII only a slight change in the dichroic ratio has been observed. The change from $N=5$ to $N=4$ corresponds to a tilting of the chromophoric group centrifugal to the membrane surface, of $\Delta\Psi=15°$.

There is no doubt that P and Ms molecules can rotate in the living microvillus membrane, but lateral diffusion may possibly be hindered. Such a conclusion could be drawn from photokinetic experiments, but unfortunately until now corresponding measurements have been carried out only on fixed rhabdomeres. However, in those glutaraldehyde-fixed receptors in which both rotation and diffusion are hampered, it has been shown that the time course of photo-equilibration is greatly influenced by the direction factors of the pigment molecules in the bent membrane (see Sec. C.I.4). From Fig. 29 it can be seen that the molecules of the a-walls of the square tube model have a greater probability of quantum absorption than the molecules of the vertical b-walls, even if the molecules rotate. Thus the photokinetics of the a-molecules has to be faster, and consequently, the photoequilibrium will be reached sooner than in the fraction of b-molecules.

When in microphotometry both kinetics are measured simultaneously, a continuous change of the exponential coefficient has to be expected, starting with the coefficient of the fast a-kinetic, and ending with that of the slower b-kinetic.

C. Mathematical Treatment and Calculations

I. Photoequilibrium of Photo-Interconvertible Visual Pigments

1. Basic Photoreactions

In contrast to the bleaching sequence and the dark-state reconversion of visual pigments in vertebrates, in many invertebrate pigments the subsequent dark reaction ends with the indicator pigment metarhodopsin (M), both the acid and the alkaline tautomers of which can then be reconverted by irradiation to rhodopsin (P). This photoregeneration of a visual pigment was first described in the octopus (HUBBARD and ST. GEORGE, 1958; BROWN and BROWN, 1958). More detailed data on the mechanism of photoregeneration has been obtained from experiments on *Eledone moschata*, a species of octopus active in dim light (HAMDORF et al., 1968; TÄUBER, 1967; SCHWEMER, 1969). These investigations have demonstrated that any specific monochromatic irradiation leads to a particular state of photoequilibrium between the visual pigment P and its metarhodopsin M, in digitonin extracts as well as in the natural receptor membrane. By systematic production of different states of photoequilibria, it was discerned that the concentrations of P and M in a photoequilibrium is determined mainly by the ratio of their absorption coefficients at the wavelength of irradiation: $\alpha_M(\lambda_a)/\alpha_P(\lambda_a)$. The greatest concentration of P, i.e., $C_{P_{eq}}(\lambda_a)$, is produced when the ratio is maximal, and the lowest when the ratio is minimal (see Fig. 3). Further parameters affecting the photoequilibrium are: the quantum efficiencies of the forward and the backward reactions, intensity of the monochromatic incident light $I(\lambda_a)$, the temperature and the pH. The extent of the influence of these factors is greatly dependent on each specific step along the bleaching sequence. For the simplest case, in the visual pigment of the blowfly photoreceptors R1–6, for example, the photoequilibrium can be described by the formalism of a reversible first-order reaction: $P \underset{k_2}{\overset{k_1}{\rightleftharpoons}} M$, where:

$$k_1 = \alpha_P(\lambda_a)\cdot\gamma_P(\lambda_a)\cdot I \quad \text{and} \quad k_2 = \alpha_M(\lambda_a)\cdot\gamma_M(\lambda_a)\cdot I.$$

Integration of the differential rate equations:

$$\frac{dC_P}{dt} = -C_P k_1 + C_M k_2 \quad \text{and} \quad \frac{dC_M}{dt} = -C_M k_2 + C_P k_1$$

leads to:

$$C_P(\lambda_a, I, t) = C_{PO}\,[k_2 + k_1 \cdot \exp - (k_1 + k_2)\cdot t]/(k_1 + k_2) \tag{1}$$

and to:

$$C_M(\lambda_a, I, t) = C_{PO} - C_P(\lambda_a, I, t). \tag{2}$$

As the equilibrium is approached, i.e., $I\cdot t \rightarrow \infty$, the concentrations of P and M can be derived from:

$$C_{Peq}(\lambda_a) = C_{PO}\cdot\frac{k_2}{k_1 + k_2} \quad \text{and} \quad C_{Meq}(\lambda_a) = C_{PO}\cdot\frac{k_1}{k_1 + k_2} \tag{3}$$

the ratio of the concentrations being given by:

$$\frac{C_{Peq}}{C_{Meq}}(\lambda_a) = \frac{\alpha_M}{\alpha_P}(\lambda_a)\cdot\frac{\gamma_M}{\gamma_P} \tag{4}$$

the lowest values of C_{Peq} occurring at wavelengths (λ_a) where α_M/α_P reaches a minimum.

2. The Common Scheme of Photoreactions

More complicated than the case of the blowfly is the formalism of the common scheme tentatively suggested for cephalopod pigments (HAMDORF et al., 1968) (Fig. 30). This scheme takes into account the intermediate products of short life-times, where X stands for prelumirhodopsin and lumirhodopsin, Y for 11-cis acid metarhodopsin, and Z for the presumed entity, the 11-cis alkaline meta-rhodopsin.

Whereas the rate of the photoreactions (k_1, k_2, k_3) are dependent on the number of incident light quanta per unit time, the rate of the dark reactions (t_1, t_2, t_3, t_p, t_s, t_b) are temperature dependent. Therefore at a constant illumination intensity,

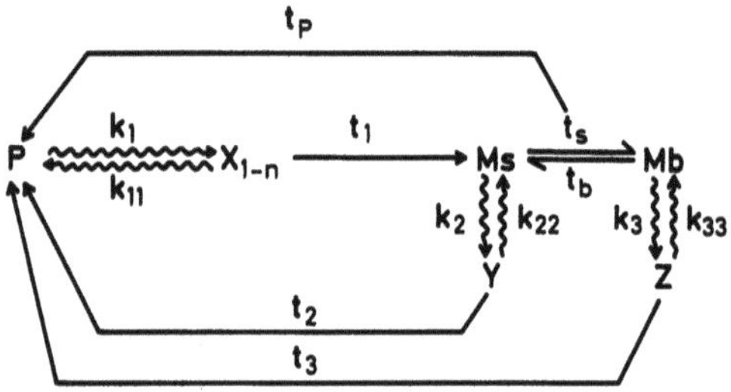

Fig. 30. Common scheme of interconvertible visual pigment. *P*, visual pigment; *Ms* and *Mb*, acid and alkaline metarhodopsin, respectively; *X*, *Y*, *Z*, pigment intermediates of short lifetimes; *wavy arrows*, photoreactions; *straight arrows*, dark reactions; $t_{1,2,3,b,s,p}$, rates of thermal reactions; $k_{1,2,3}$, rates of photochemical forward reactions and $k_{11,22,33}$ are rates of photochemical backward reactions

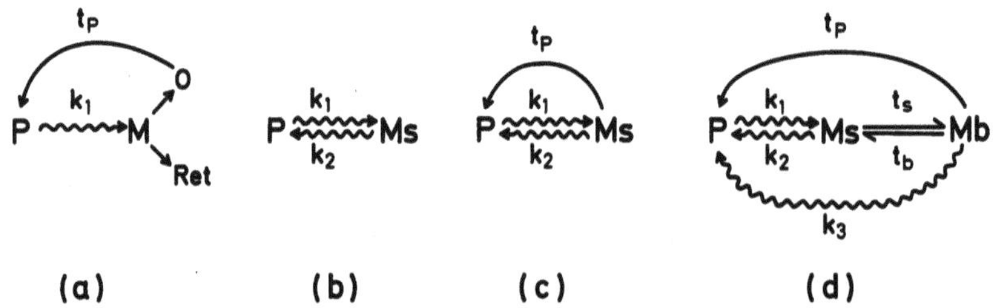

Fig. 31. Schemes of different visual pigment systems: a) vertebrate; b) owl-fly *Ascalaphus*; c) honey-bee *Apis*; d) cephalopod *Eledone*; photochemical reactions, *wavy arrows*; dark reactions, *straight arrows*; *P*, visual pigment; *Ms*, *Mb*, acid and alkaline metarhodopsin, respectively; *O*, opsin; *Ret*, all-trans retinaldehyde; t_P, rate constant of thermal regeneration; t_s and t_b, rate constants of pH-dependent Ms–Mb equilibrium; k_1, k_2, k_3, rates of photoreactions, where $k_1 = \alpha_P \cdot \gamma_P \cdot I$, $k_2 = \alpha_{Ms} \cdot \gamma_{Ms} \cdot I$, and $k_3 = \alpha_{Mb} \cdot \gamma_{Mb} \cdot I$, at wavelength of illumination

depression of the temperature will strongly influence the photoequilibrium, since by the blocking or retarding the steps of the dark reactions, the probability of other accumulated intermediates (*X*, *Y*, *Z*) absorbing quanta and participating in establishing the photoequilibrium increase accordingly. On the other hand, at constant temperatures, increasing the illumination intensity causes the photoreactions to overtake the temperature-dependent dark reactions. Thus the number and the quantity of intermediates detectable in the photoequilibria depend on intensity as well as on the ratios α_X/α_P, α_M/α_P, etc. (for specific irradiation wavelengths), and also on the quantum yields γ_P, γ_X, γ_M, etc. However, in the range of light intensities of physiological norm, this intensity-dependent effect of the photoequilibria is negligible, since all dark reactions [except the thermal regeneration of the visual pigment (t_P)] are much faster than the photoreactions. Since under physiological conditions the common reaction scheme can be simplified as shown in schemes b, c and d (Fig. 31), a generalization including the scheme for the vertebrate pigments (a) can be brought together and collectively described by the scheme c. Conditions for the other schemes are as follows: scheme a: $k_2 \simeq 0$, if the bleaching reaction $M \rightarrow 0 + Ret$ is faster than the photoreactions; scheme b: $t_P = 0$, because no thermal regeneration can be measured; scheme d: at a constant pH, if t_s, $t_b > k_1$, k_2, t_p. Moreover, if $\gamma_{Ms} = \gamma_{Mb}$, Ms and Mb can be represented by a pigment Mg, and the rate of photoreactions k_2 and k_3 by $\bar{k}_2$. The concentration C_{Mg}, and the absorption coefficient α_{Mg} can be derived from:

$$C_{Mg} = C_{Mb} + C_{Ms} \tag{5}$$

$$\alpha_{Mg} = (t_b \cdot \alpha_{Ms} + t_s \cdot \alpha_{Mb})/(t_b + t_s) \tag{5a}$$

from which:
if:

$$\gamma_{Ms} = \gamma_{Mb}$$

then:

$$\bar{k}_2 = \alpha_{Mg} \cdot \gamma_{Mg} \cdot I$$

and if:

$$\gamma_{Ms} \neq \gamma_{Mb}$$

then:

$$\bar{k}_2 = I(\alpha_{Ms} \cdot t_b \cdot \gamma_{Ms} + \alpha_{Mb} \cdot t_s \cdot \gamma_{Mb})/(t_b + t_s). \tag{5b}$$

$C_P(\lambda_a, I, t)$ can then, based on the initial pigment concentration C_{P0}, be written as:

$$C_P(\lambda_a, I, t) = C_{P0}(t_p + k_2 + k_1 \cdot \exp - (k_1 + k_2 + t_p) \cdot t)/(t_p + k_1 + k_2), \tag{6}$$

and the assumption of $I \cdot t \to \infty$, changes this equation to:

$$C_{Peq}(\lambda_a) = C_{P0}(t_p + k_2)/(t_p + k_1 + k_2). \tag{7}$$

This means that, at low intensities—$(k_1, k_2 < t_p)$—, C_{Peq} is dominated by the thermal back-reaction and is approximately equal to C_{P0}. However, with increasing intensities—$(k_1, k_2 \geqq t_p)$—, C_{Peq} tends to values determined dominantly by the photoreactions (Fig. 32), i.e., within a range of only about 2 log units the pigment concentration changes from pure C_{P0} to a particular photoequilibrium independent of t_p.

3. Experimental Methods for Testing Photoequilibria

On the basis of these considerations then, the visual pigment concentration in the tissue, as well as in an extract, can be experimentally altered in the desired direction in the following ways:

a) By the production of photochemical steady states with bright monochromatic illumination (Eq. 7). In cases where the absorbance spectra of Ms are shifted to shorter wavelengths—with respect to the spectra of rhodopsin (i.e., hypochromic shifts in the green receptors of *Manduca, Deilephila*, and *Apis*)—C_{Peq} can be reduced to approximately zero by using yellow-to-red illumination that only affects the P population, and in the opposite direction, where the Ms spectra are shifted bathochromically (i.e., in *Octopus, Eledone, Ascalaphus, Calliphora, Musca, Drosophila, Eristalis*), where minimal C_{Peq} values are obtained at wavelengths where the ratios of $\alpha_{Ms}/\alpha_P(\lambda_a)$ reach a minimum.

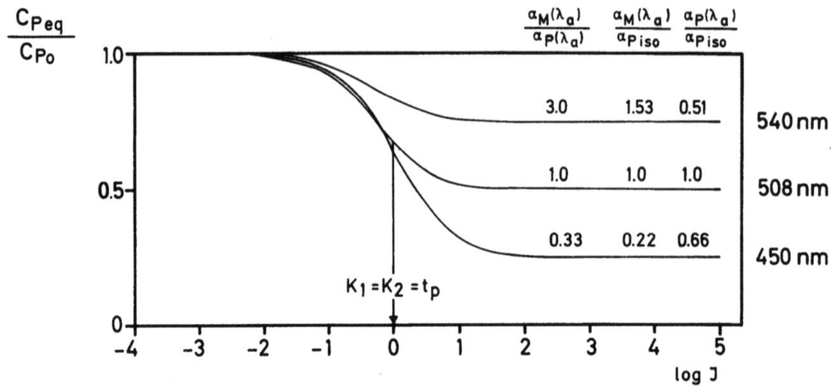

Fig. 32. Curves of equilibria of a photointerconvertible visual pigment of the cephalopod type, depending on number of incident quanta at three different wavelengths (450, 508, and 540 nm). Logarithm of relative I has been normalized to $k_1 = k_2 = t_P$ at wavelength of isosbestic point. $\alpha_{\mathrm{Pmax}} = 494$ nm, $\alpha_{\mathrm{Mmax}} = 562$ nm

b) by varying the intensity, I, at constant t or vice versa by varying the duration of illumination, t, at constant I. To choose the most appropriate wavelength for illumination, points in section a) (above) should be born in mind, and also reference to Figs. 29, 32, 38.

c) If several wavelengths ($\lambda_{a(1 \ldots n)}$) are used simultaneously to set up photo-equilibria, or light sources of special spectral distribution, the photochemical rates of the forward and the backward reactions are given by their integrals over the range $\lambda_{a(1 \ldots n)}$:

$$\bar{k}_{\mathrm{p}} = \int_{\lambda_1}^{\lambda_n} \alpha_{\mathrm{P}}(\lambda) \cdot \gamma_{\mathrm{P}}(\lambda) \cdot I(\lambda) \cdot d\lambda_{\mathrm{a}} \tag{8a}$$

$$\bar{k}_{\mathrm{M}} = \int_{\lambda_1}^{\lambda_n} \alpha_{\mathrm{M}}(\lambda) \cdot \gamma_{\mathrm{M}}(\lambda) \cdot I(\lambda) \cdot d\lambda_{\mathrm{a}}. \tag{8b}$$

Then the pigment concentrations are given by:

$$C_{\mathrm{P}}(\lambda_n, I, t) = C_{\mathrm{PO}} \cdot (t_{\mathrm{P}} + \bar{k}_{\mathrm{M}} + \bar{k}_{\mathrm{P}} \cdot \exp - f \cdot t)/f \tag{9a}$$

$$C_{\mathrm{Peq}}(\lambda_n, I) = C_{\mathrm{PO}} \cdot (t_{\mathrm{P}} + \bar{k}_{\mathrm{M}})/f \tag{9b}$$

where

$$f = t_{\mathrm{p}} + \bar{k}_{\mathrm{M}} + \bar{k}_{\mathrm{P}}.$$

Using only two monochromatic beams λ_1 and λ_2, of intensities I_1 and I_2, respectively, Eqs. (9a) and (9b) can then be simplified as follows:

$$C_{\mathrm{P}}(\lambda_{1,2}, I, t) = C_{\mathrm{PO}} \cdot (t_{\mathrm{p}} + k_{\mathrm{M}_1} + k_{\mathrm{M}_2} + (k_{\mathrm{P}_1} + k_{\mathrm{P}_2}) \cdot \exp - G \cdot t)/G \tag{10a}$$

$$C_{\mathrm{Peq}}(\lambda_{1,2}, I) = C_{\mathrm{PO}} \cdot (t_{\mathrm{p}} + k_{\mathrm{M}_1} + k_{\mathrm{M}_2})/G \tag{10b}$$

where:

$$G=(t_p+k_{M_1}+k_{M_2}+k_{P_1}+k_{P_2}).$$

Substituting for k_M and k_P, we obtain:

$$C_{Peq}(\lambda_{1,2}, I)=C_{P0}\cdot\frac{t_P+(\alpha_{M_1}\cdot I_1+\alpha_{M_2}\cdot I_2)\gamma_M}{t_P+(\alpha_{M_1}I_1+\alpha_{M_2}I_2)\gamma_M+(\alpha_{P_1}I_1+\alpha_{P_2}I_2)\gamma_P} \tag{11}$$

an indication that C_{Peq} becomes strongly dependent on the ratio I_1/I_2. Equations (10a), (10b), and (11) were used to calculate pigment concentrations in the receptors of *Deilephila*, *Ascalaphus* and *Calliphora* under physiological light conditions (HÖGLUND et al., 1973; HAMDORF and GOGALA, 1973).

4. Photoequilibria and the Orientation of the Chromophore in the Microvillus Membrane

It should be noted that these calculations of photoequilibria suppose a random distribution of the P and the M dipoles, as is the case in solutions. However, within the microvillus membrane there is a reduction in the degrees of freedom of a pigment molecule. Thus it should be taken into account that a different orientation of P and M dipoles will influence $C_{Peq}(\lambda_a)$ as well as the time course needed to reach a photoequilibrium. If the direction and the magnitude of an absorbing dipole be symbolized by the vector $\boldsymbol{a}$, and those of the electric field vector by $\boldsymbol{E}$, then the absorption of the pigment molecule A is given by the scalar product:

$$A=(\boldsymbol{a}\cdot\boldsymbol{E})^2=a^2\cdot E^2\cdot\cos^2(\sphericalangle\,\boldsymbol{a},\boldsymbol{E}). \tag{12}$$

The direction of the absorbing chromophore inside the microvillus membrane (Fig. 33) is described by the angles Ψ, Φ and Θ, and thus in the Cartesian coordinate system the components of the vector $\boldsymbol{a}$ depending on these angles are:

$$\begin{aligned} a_x&=a(-\cos\Psi\sin\Phi\sin\Theta+\sin\Psi\cos\Theta)\\ a_y&=a(\cos\Psi\sin\Phi\cos\Theta+\sin\Psi\sin\Theta)\\ a_z&=a\cos\Psi\cos\Phi. \end{aligned} \tag{13a}$$

Linearly polarized light is characterized by the directions of incidence (β), and of the $\boldsymbol{E}$ vector (α). Thus the components of $\boldsymbol{E}$ are obtained by:

$$E_x=E\sin\beta\cos\alpha;\ E_y=E\sin\alpha;\ E_z=E\cos\beta\cos\alpha. \tag{13b}$$

Then for the absorption of one molecule, the result is:

$$A=(E_xa_x+E_ya_y+E_za_z)^2=E^2a^2r(\alpha,\beta,\Psi,\Phi,\Theta) \tag{14}$$

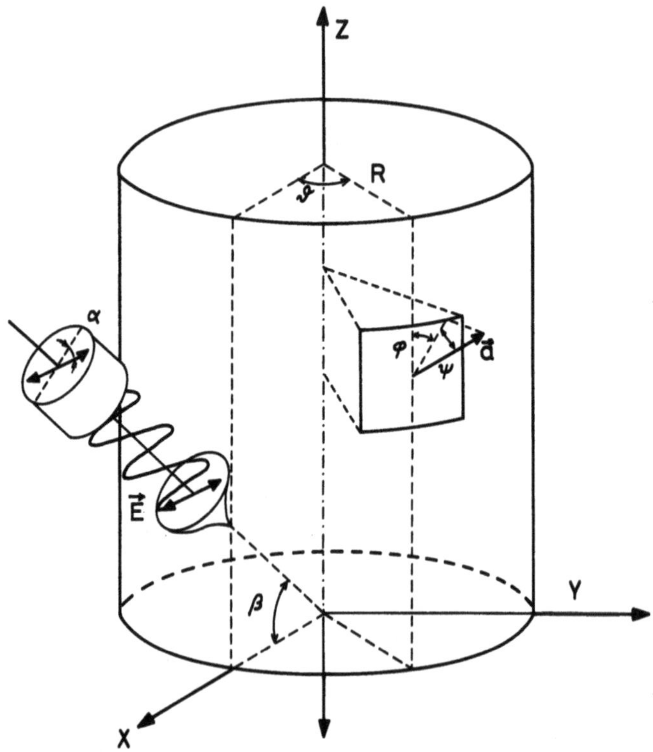

Fig. 33. Direction of incidence and electric vector $\boldsymbol{E}$ of linearly polarized light, and position and orientation of visual pigment molecule on surface of cylindrical microvillus. For further explanation see text

where r is the direction factor that includes all dependencies on the various angles:

$$r(\alpha, \beta, \Psi, \Phi, \Theta) = [\sin\beta \cos\alpha (-\cos\Psi \sin\Phi \sin\Theta + \sin\Psi \cos\Theta)$$
$$+ \sin\alpha (\cos\Psi \sin\Phi \cos\Theta + \sin\Psi \sin\Theta + \cos\beta \cos\alpha \cos\Psi \cos\Phi]^2 . \qquad (15)$$

In the disc membrane of vertebrate receptors, the chromophoric group of the visual pigment is oriented parallel to the disc plane (DENTON, 1959; WALD et al., 1962; HAROSI, 1971), and the molecules are free to rotate in the plane and to diffuse in the fluid phospholipid bilayer (CONE, 1972; POO and CONE, 1973, 1974). Suggesting an analogous orientation of the pigment molecules in the microvillus membrane, it would be reasonable to assume that Ψ of all pigment molecules is small (or zero) and identical with each other (LAUGHLIN et al., 1975; GRIBAKIN and GOVARDOVSKII, 1975; HAMDORF and SCHWEMER, 1975). One may also assume that the value of Ψ could be different for the visual pigment and its metarhodopsin Ψ_P, Ψ_M (SCHLECHT and TÄUBER, 1975), and further, that all possible values of Φ have equal probabilities of existence. There are some indications, however, that in the rhabdomere the molecules are preferentially oriented in the direction of the microvillus axis (SNYDER, 1973; LAUGHLIN et al., 1975). This is presumably caused by energy differences between molecular orientation parallel and perpendicular to the microvillus axis, possibly due to an asymmetry of the pigment molecule and to the bending of the microvillus membrane. Then the distribution of the molecular orientation will be related to $\sin^2\Phi$, and according to BOLTZMANN, this can be described by the distribution function:

$$H(\Phi) = (\exp - \Delta U \sin^2\Phi/kT) \Big/ \int_0^{2\pi} \exp - \Delta U \sin^2 d\Phi/kT. \tag{16}$$

Noticeable deviations from a uniform distribution are obtained if ΔU is equal to or larger than kT.

Assuming a uniform density of pigment molecules σ in small membrane elements of the area $R \cdot l \cdot d\Theta$, the absorption by an entire microvillus is given by:

$$A = \sigma \cdot R \cdot l \cdot E^2 \cdot a^2 \cdot \int_0^{2\pi} \int_0^{2\pi} H(\Phi) \cdot r(\alpha, \beta, \Psi, \Phi, \Theta) \cdot d\Phi \cdot d\Theta. \tag{17}$$

Solving the integrals, the direction factor R, of the whole microvillus is obtained, which is dependent only on fixed parameters. Representing the number of P molecules in a microvillus by Z ($=2\pi\sigma Rl$), the intensity $I(=E^2)$ and the extinction coefficient $\varepsilon(=2a/3)$ defined for pigment solutions, the result is:

$$A = (3/2\pi) \cdot Z \cdot I \cdot \varepsilon \cdot R(\alpha, \beta, \Psi). \tag{18}$$

For physiological angles of incidence, i.e., perpendicular to the microvillus axis ($\beta = 0$), and in the case of a statistically uniform distribution of the molecules with respect to Φ, the direction factor is:

$$R(\alpha, \Psi) = (1/4)\sin^2\alpha(1 + \sin^2\Psi) + (1/2)\cos^2\alpha \cdot \cos^2\Psi \tag{19}$$

and the dichroic ratio N (SHAW, 1969, corresponding to the dichroic sensitivity of SNYDER, 1973), is determined by:

$$N = A(\alpha = 0°)/A(\alpha = 90°) = 2/(1 + 2\,\mathrm{tg}^2\Psi). \tag{20}$$

In the case of total alignment of all molecules ($\Phi = 0°$), the results are:

$$R(\alpha, \Psi) = (1/2)\sin^2\alpha \sin^2\Psi + \cos^2\alpha \cos^2\Psi; N = 2\,\mathrm{cotg}^2\Psi. \tag{21a, b}$$

Figure 34 shows the R functions depending on the angles of the polarized light for various angles of inclination, where it can be seen that these curves are proportional to the absorption. In case *a*, of uniform distribution, N attains a maximal value of 2 (Fig. 34a, $\Psi = 0°$), and in *b*, that of complete alignment, N becomes infinite at $\Psi = 0°$, and in the third case *c*, of incomplete alignment (i.e., preferential orientation computed for $\Delta U = 3kT$, Fig. 34c), N becomes 7.9. Common to all three cases is that, at certain angles of inclination, namely at $\Psi = 35°$, 55°, and 50°, the absorption becomes independent of α, i.e., $N = 1$. Further increase in Ψ causes N values to become less than 1, which produces an inversion

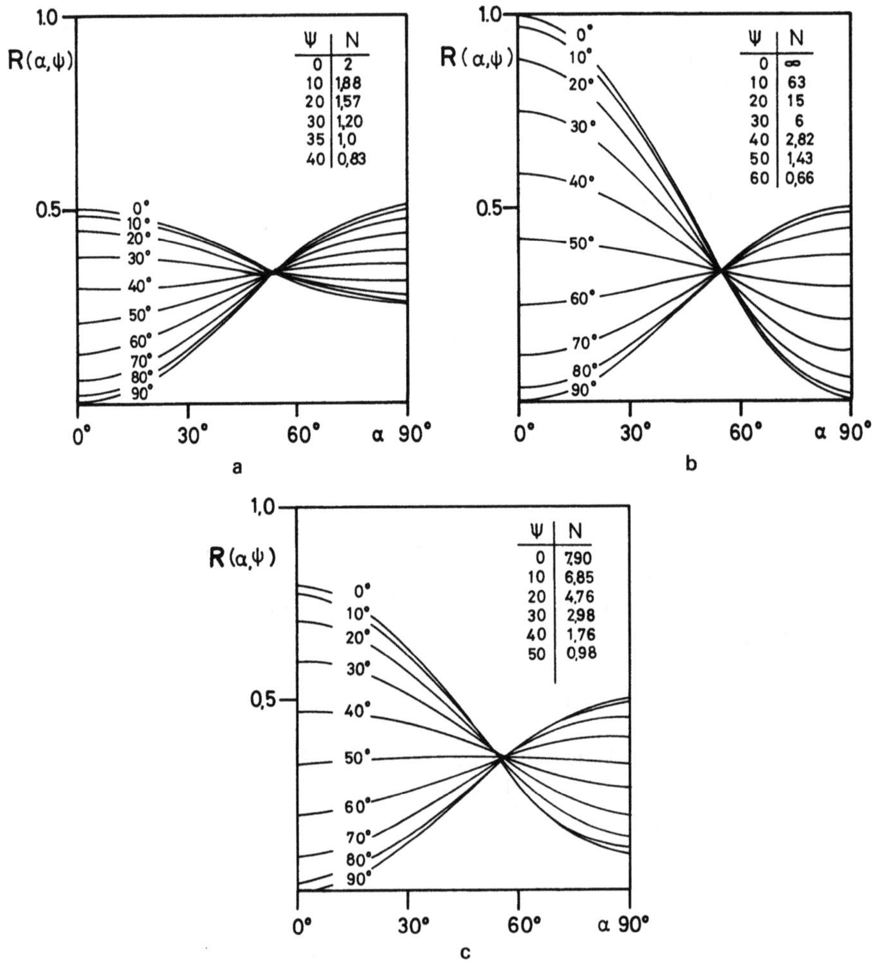

Fig. 34. Dependence of direction factor of tubule containing a homogeneous visual pigment (for example rhodopsin) on parameters α and Ψ; a) for uniform distribution over Φ; b) for the case of complete alignment of molecules in direction of axis of tubule ($\Phi=0$), c) for a preferential distribution of molecules around $\Phi=0$

of dichroism, then polarized light, incident perpendicularly to the microvillus axis, will be absorbed more strongly than polarized light incident at 0°. All the curves intersect at the same α value of 55°, which corresponds to the absorption of pigment in solution $[R(\alpha, \Psi)=1/3]$.

In summary, the inclination, as well as the degree of alignment $H(\Phi)$ of the dipole, influence the absorption characteristics and the dichroic ratio of the microvillus. Thus anomalous dichroic values of less than 1, as observed in fly receptor R8 (KIRSCHFELD, 1975), can possibly be explained by larger values of Ψ and/or of Φ.

The two states, that is, the visual pigment P and its metarhodopsin M(Ms, Mb, Mg), may have different conformations, and therefore the orientation of their dipoles, characterized by Ψ and Φ, may also differ. Thus the transition from P to

M may be accompanied by a change in dipole orientation concerning $\Phi(\Phi_P - \Phi_M)$ as well as $\Psi(\Psi_P - \Psi_M)$. If the degree of alignment varies, the difference $\Delta\Phi = \Phi_P - \Phi_M$ is only of importance and therefore one can equate $\Phi_P = \Phi$, and $\Phi_M = \Phi + \Delta\Phi$. According to Eq. 18 the absorption probability of a molecule is given by the term $3 \cdot I \cdot \varepsilon \cdot r(\alpha, \beta, \Psi, \Phi, \Theta)$, which differs for P and M. Then the absorption probability of a molecule is given by:

$$dW_P/dt = -3I \cdot \gamma \cdot \varepsilon_P \cdot r_P \cdot W_P + 3I \cdot \gamma \cdot \varepsilon_M \cdot r_M \cdot W_M, \tag{22}$$

where W_P and W_M, the absorption probabilities of P and M, are equivalent to the concentrations in macroscopic samples C_P and C_M. The time course for equilibration is obtained by the solution of the differential Eq. 22:

$$W_P(t) = W_P(\infty) + (W_P(0) - W_P(\infty)) \cdot \exp - (3I\gamma(\varepsilon_P \cdot r_P + \varepsilon_M \cdot r_M) \cdot t), \tag{23}$$

and since $W_P + W_M = 1$, the photoequilibrium at $(t = \infty)$ becomes:

$$W_{P_{eq}} = r_M \cdot \varepsilon_M / (r_M \cdot \varepsilon_M + r_P \cdot \varepsilon_P), \tag{24}$$

which differs from equilibrium in solution only if $r_P = r_M$.

If Ms is transformed into Mb in considerable amounts (i.e., in cephalopods), Eq. 22 must be extended by a third term, $+3I\gamma \cdot \varepsilon_{Mb} \cdot r_{Mb} \cdot W_{Mb}$. However, as Mb absorbs in the UV-range only, this term is reduced to zero in the visible range, and Eq. 23 remains valid.

The photoequilibrium will strongly depend on whether the molecules are fixed in the membrane or free to rotate and–or to diffuse; thus three cases are to be distinguished:

a) P and M are fixed in the membrane. The $W_{P_{eq}}$ of each molecule will be different and will depend on the complete direction factors r_P, r_M, as described by Eq. 15, as well as on the wavelength of the adapting light λ_a:

$$W_{P_{eq}} = W_P(\Phi, \Theta, \alpha_a, \beta, \lambda_a, \Psi_P, \Psi_M, d\Phi). \tag{25}$$

b) P and M are free to rotate, whereby some directions may be preferred, as described by the distribution function $H(\Phi)$. Then, for the computation of the equilibration, a mean direction factor $\bar{r}$, averaged over all directions of Φ, can be used, and $W_{P_{eq}}$ becomes independent of Φ:

$$\bar{r}(0) = \int_0^{2\pi} H(\Phi) \cdot r(\Phi, \Theta)\, d\Phi; \; W_{P_{eq}} = W_P(\Theta, \alpha_a, \beta, \lambda_a, \Psi_P, \Psi_M). \tag{26}$$

c) P and M are free to rotate and also free to diffuse. Assuming that diffusion is much faster than the probability of photochemical conversion, the averaging has to be done over Φ as well as over all Θ. Then $W_{P_{eq}}$ becomes independent of both Φ and Θ:

$$r = (1/2) \int_0^{2\pi} \int_0^{2\pi} H(\Phi) \cdot r(\Phi, \Theta) \cdot d\Phi \cdot d\Theta; \; W_{P_{eq}} = W_P(\alpha_a, \beta, \lambda_a, \Psi_P, \Psi_M). \tag{27}$$

To compute the absorption of a microvillus, Eq. 18 must be expanded and the contributions of P and M taken into account:

$$A=(3/2\pi)\cdot Z\cdot I\cdot \int_0^{2\pi}\int_0^{2\pi} H(\Phi)\cdot(\varepsilon_P\cdot r_P\cdot W_P+\varepsilon_M\cdot r_M\cdot W_M)\cdot d\Phi\cdot d\Theta\,. \tag{28}$$

Now, absorption depends on wavelength $\lambda_t(\varepsilon_{P,M}(\lambda))$ and the angle of polarization $\alpha_t(r_{P,M}(\alpha_t, \Psi_P, \Psi_M, \Delta\Phi, \Phi, \Theta))$ of the test beam, and on the wavelength λ_a and angle α_a of the adapting light (via $W_{P,M}$). It also depends on the direction of incidence β, which is assumed to be identical for the test and adapting beams of the experimental and the physiological conditions. Lastly, it depends on the molecular parameters Ψ_P, Ψ_M, and specially in case (1) on $\Delta\Phi$. To find $W_{P,M}$, the specific values of one of the above-mentioned three cases has to be inserted. In most cases, however, an explicit solution of the integral (Eq. 28) is impossible, and thus one must resort to numerical methods of solution.

II. Calculations of Spectra and of Quantum Efficiency

1. Calculations of P and M Spectra by Data from Difference Spectrophotometry

The absorption spectra of the components P and M of the photointerconvertible visual pigments can be either estimated (HAMDORF et al., 1973), or calculated (STAVENGA, 1974, 1976), from their difference spectra.

a) Method Based on Dartnall's Nomogram

If the assumption is made that the spectra of the family of frequency-dependent visual pigments are identical (Dartnall nomogram, 1962), and that the resonance spectra of their metarhodopsins are similar in shape to rhodopsin spectra, then $\alpha_P(\lambda)$ and $\alpha_M(\lambda)$ can be estimated. Few data are necessary for such an estimation, i.e., one difference spectrum of two photoequilibria. But it should also be verified that the family of normalized difference spectra of these photoequilibria produced by different wavelengths (λ_a) are equal in shape and have the same isosbestic wavelength (Figs. 16, 36), for such spectra are to be expected if only two photoreconvertible components determine the photoequilibrium.

If λ_{max} of the visual pigment is known from electrophysiologic determinations of spectral sensitivity, the difference spectra can easily be made to fit by varying the λ_{max} of the metarhodopsin and its relative absorbance ($\alpha_{M\lambda/max}/\alpha_{P\lambda/max}$) within the estimated range of wavelengths. This technique was first applied to the UV visual pigment of *Ascalaphus* (HAMDORF et al., 1971), to the visual pigments of the moth *Deilephila elpenor* (HAMDORF, HÖGLUND and LANGER, 1973), and in the blowfly pigments (HAMDORF, PAULSEN and SCHWEMER, 1973; HAMDORF and ROSNER, 1973). More precise approximations are possible by computer calculations (SCHLECHT, unpublished data). As the normalized difference spectra are invariant,

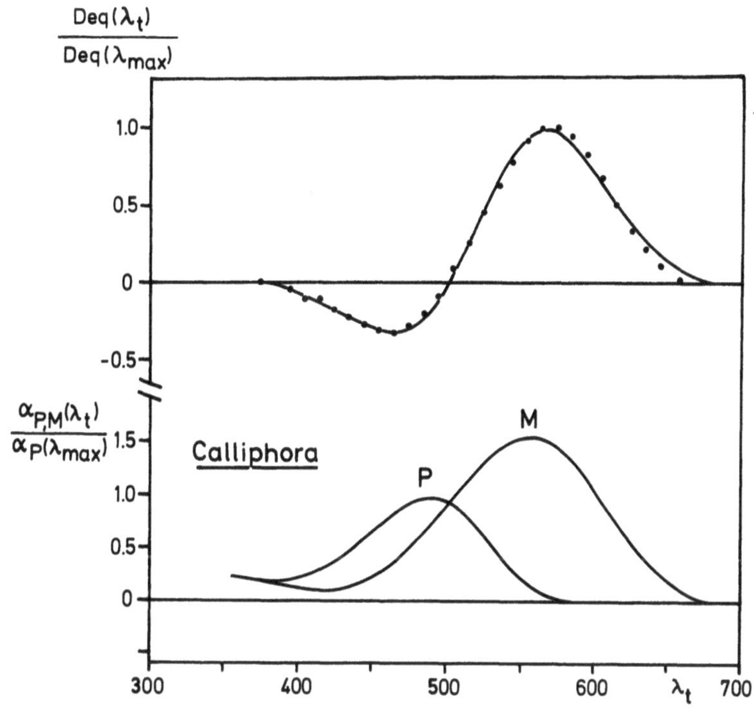

Fig. 35. Visual pigment of blowfly analyzed by computer program. a) Mean value of normalized difference spectra of 10 flies in comparison to calculated difference spectrum. b) P and M spectra calculated by data of (a). (SCHLECHT, unpublished data)

the best fit, i.e., the least square of the standard deviations can be calculated automatically by varying λ_{max} of the P and M as well as the ratio $\alpha_{M\lambda/max}/\alpha_{P\lambda/max}$. Figure 35 shows a computer analysis of a difference spectrum of the blowfly measured in situ, compared to an in vitro pigment analysis.

b) Method Independent of Dartnall's Nomogram

The second method of analyzing the difference spectra was introduced and developed by STAVENGA (1974, 1976). The advantage of this method is that no assumptions have to be made as to the shapes of the pigment spectra. Therefore it can also be applied when the nature of the pigment is unknown, and furthermore, the theory also considers the waveguide properties of rhabdomeres as determined by microspectrophotometry. The transmittance $T(\lambda_t)$ of a rhabdomere which contains P and M is given by:

$$T(\lambda_t)=\exp\left\{-\int_0^l \eta(\lambda_t, x)\,[f_P\alpha_P(\lambda_t)+f_M\alpha_M(\lambda_t)]\,C(x)\cdot dx\right\} \tag{29}$$

where η is the factor taking into account the waveguide properties of a rhabdomere and equals the light power transmitted in the boundary of the waveguide (SNYDER, 1975), x is the coordinate, l the total length of the rhabdomere, $C(x)$ is the concentration of the visual pigment, where $C(x)=C_P(x)+C_M(x)$, f_P and f_M are the fractions of pigment in the P or M state, and α_P and α_M are the molar absorption coefficients (extinction coefficients).

If a photoequilibrium is established by a monochromatic adapting light λ_a, then the fractions of the visual pigment and its metarhodopsin are related by:

$$f_{Peq}=1-f_{Meq}, \quad f_{Meq}(\lambda_a)=1\bigg/\left(1+\gamma_{M/P}\frac{\alpha_M(\lambda_a)}{\alpha_P(\lambda_a)}\right), \tag{30}$$

where $\gamma_{M/P}$ is the relative quantum efficiency α_M/α_P of the photoconversion. Providing the metarhodopsins do have bathochromic shifts, as is the case with the pigments of the blowfly and *Eledone*, it can be expected that at long wavelengths λ_0, $\alpha_P/\alpha_M=0$. Then at the photoequilibrium:

$$f_{Meq}(\lambda_0)=0, \quad f_{Peq}(\lambda_0)=1,$$

and the difference in absorbance, D_{eq}, between the photoequilibria formed at wavelengths λ_0 and λ_a is given by:

$$D_{eq}(\lambda_t, \lambda_a)=\log_{10}\frac{T_{eq}(\lambda_t, \lambda_0)}{T_{eq}(\lambda_t, \lambda_a)}=f_{Meq}(\lambda_a)\,N(\lambda_t)\,[\alpha_M(\lambda_t)-\alpha_P(\lambda_t)], \tag{31}$$

where:

$$N(\lambda_t)=0{,}4343\int_0^l \eta(\lambda_t, x)\cdot C(x)\cdot dx,$$

and the establishment of a photoequilibrium at the isosbestic wavelengths, λ_{iso} (i.e., $\alpha_P(\lambda_{iso})=\alpha_M(\lambda_{iso})$), from the application of Eq. 30 leads to:

$$f_{Meq}(\lambda_{iso})=1/(1+\gamma_{M/P}) \tag{32}$$

and with the application of Eq. 31 to:

$$D_{eq}(\lambda_t, \lambda_{iso})=D_{\lambda iso}(\lambda_t)=1/(1+\gamma_{M/P})\cdot N(\lambda_t)\,[\alpha_M(\lambda_t)-\alpha_P(\lambda_t)]. \tag{33}$$

Relating all other difference spectra to this particular one leads to:

$$Q(\lambda_a)=\frac{D_{eq}(\lambda_t, \lambda_a)}{D_{eq}(\lambda_t, \lambda_{iso})}=\frac{f_{Meq}(\lambda_a)}{f_{Meq}(\lambda_{iso})}=(1+\gamma_{M/P})f_{Meq}(\lambda_a) \tag{34a}$$

$$Q(\lambda_a)=(1+\gamma_{M/P})\bigg/\left(1+\gamma_{M/P}\frac{\alpha_M(\lambda_a)}{\alpha_P(\lambda_a)}\right) \tag{34b}$$

By dropping the suffixes in Eqs. 33 and 34, the absorption spectra of P and M can then be obtained:

$$\alpha_P(\lambda)=\frac{D_{iso}(\lambda)}{N(\lambda)}\frac{\gamma_{M/P}Q(\lambda)}{1-Q(\lambda)}, \quad \alpha_M(\lambda)=\frac{D_{iso}(\lambda)}{N(\lambda)}\frac{1+\gamma_{M/P}-Q(\lambda)}{1-Q(\lambda)}. \tag{35}$$

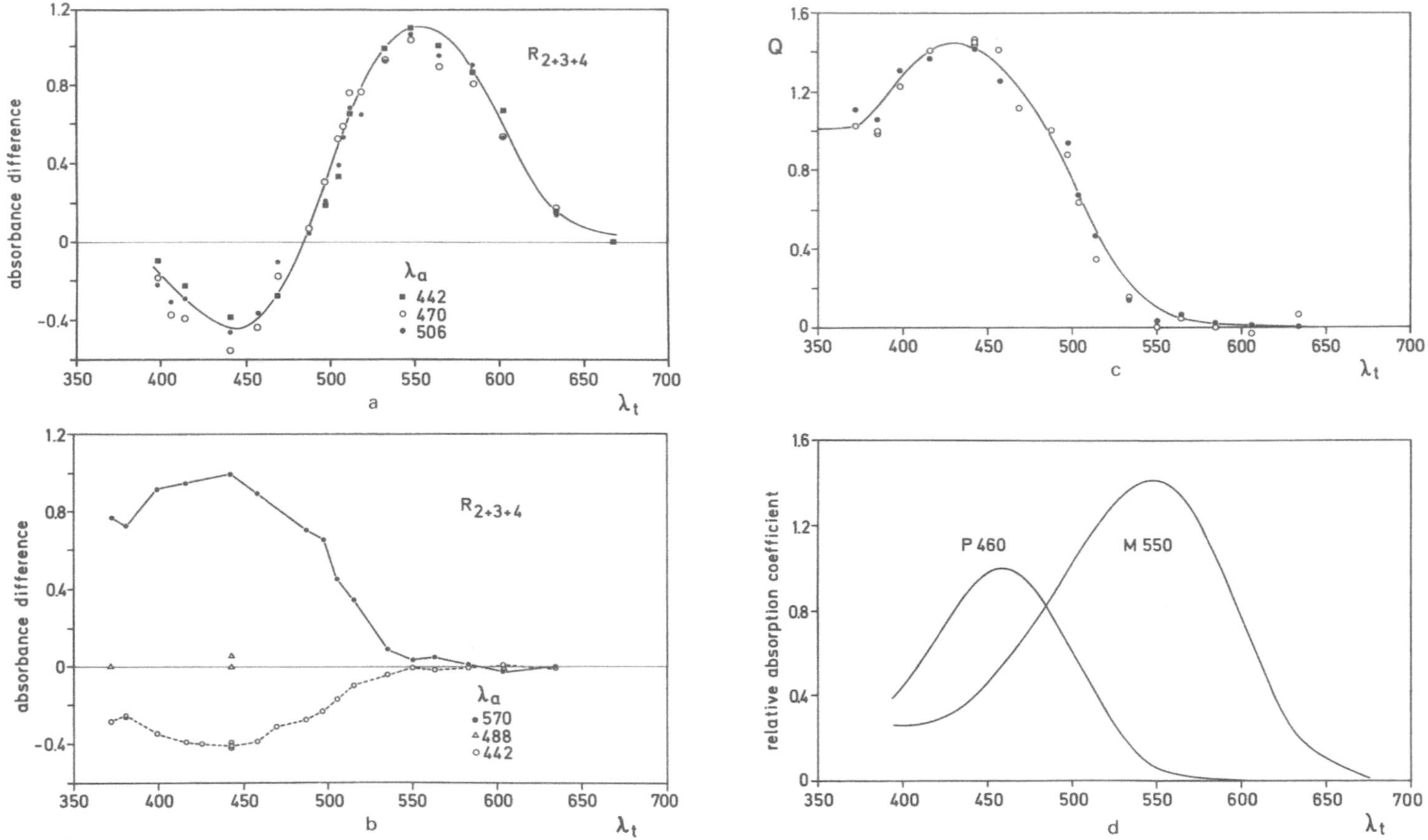

Fig. 36. Visual pigment of Dronefly receptors R2, R3, and R4 (*Eristalis tenax*). a) Standardized difference spectra between red-adapted state ($\lambda_a = 635$ nm, $f_{Meq} = 0$) and other photoequilibria created by wavelengths 442, 470 and 506 nm. b) Amplitudes of difference spectra measured at wavelengths $\lambda_t = 570$, 488, and 442 nm, depending on monochromatic light of adaptation (λ_a). c) The function $Q(\lambda_a)$ of Eq. 34 and curves of (b). d) Spectra of P and M calculated from Eq. 35. Inaccuracy in λ_{max} is estimated to be ± 10 nm, and that for the ratio of $\alpha_{Mmax}/\alpha_{Pmax}$ to be 1.4 ± 0.2 (after STAVENGA, 1976)

Figure 36 shows a recent in vivo analysis of the visual pigment system of dronefly receptors R2, R3 and R4 (STAVENGA, 1976). The normalized difference spectra (Fig. 34a) are almost equal in the ratio Maximum/Minimum and in the isosbestic point (485 nm), a strong argument for the existence of only two photo-interconvertible states of the visual pigment which set up the photoequilibrium. The amplitudes of the difference spectra depending on the adapting wavelengths (λ_a), (Fig. 36b) can be transformed into the $Q(\lambda_a)$ function (Fig. 36c), which is directly proportional to the fraction of $M_{eq}(\lambda_a)$ (Eq. 30). Combining the data of Figs. 36a and c, the relative absorption coefficients of P450 and M550 can be calculated (Fig. 36d).

2. Calculations of Quantum Efficiency

The accuracy of the analysis of difference spectra and consequently of the calculation of the P and the M spectra depend mainly on the accuracy in the determination of the relative quantum efficiency $\gamma_{M/P}$. This problem is overcome by numerical analysis made by the computer. Since no assumptions concerning $\gamma_{M/P}$ have been made in the programm, Eq. 34b can be used to determine the relative quantum efficiency:

$$\gamma_{M/P} = \frac{1 - Q(\lambda_a)}{Q(\lambda_a)\dfrac{\alpha_M(\lambda_a)}{\alpha_P(\lambda_a)} - 1}. \qquad (36)$$

Figure 16 shows difference spectra of the blowfly retina corresponding to the photoequilibria through different monochromatic adaptations. The amplitudes of $D_{eq}(\lambda_a)$ measured at 560 nm are normalized in Fig. 37 to the most effective wavelength ($\lambda_{a\,max} = 450$ nm). The relation $D_{eq}(\lambda_a)/D_{eq}(\lambda_{a\,max})$ can be drawn to the scale of $Q(\lambda_a)$ and the ratio $\alpha_M(\lambda_a)/\alpha_P(\lambda_a)$ can be taken from the computer data of Fig. 35. Then $\gamma_{M/P}$ can be calculated at each equilibrium. This determination of $\gamma_{M/P}$ by Eq. 36 is inaccurate (see values in brackets), only at the lower right part of the Q curve, and also close to the isosbestic point, i.e., in the blowfly, at 508 nm. In the latter species the mean of $\gamma_{M/P}$ is close to 1, as was first shown for the pigment of the cephalopod *Eledone* (SCHWEMER, 1969).

This method is also applicable if the acid metarhodopsin (Ms) changes into the alkaline state (Mb) as a function of pH. At a fixed pH the fractions of Ms and Mb ($f_{Ms} + f_{Mb} = 1$) are constant, and therefore $\alpha_{Ms}(\lambda_t)$ and $\alpha_{Mb}(\lambda_t)$ can be replaced by a mean absorption coefficient:

$$\alpha_{Mg}(\lambda) = f_{Ms}\alpha_{Ms}(\lambda) + f_{Mb}\alpha_{Mb}(\lambda)$$

and Eq. 36 will yield values of $\gamma_{M/P}$, provided that $\gamma_{Ms} = \gamma_{Mb}$.

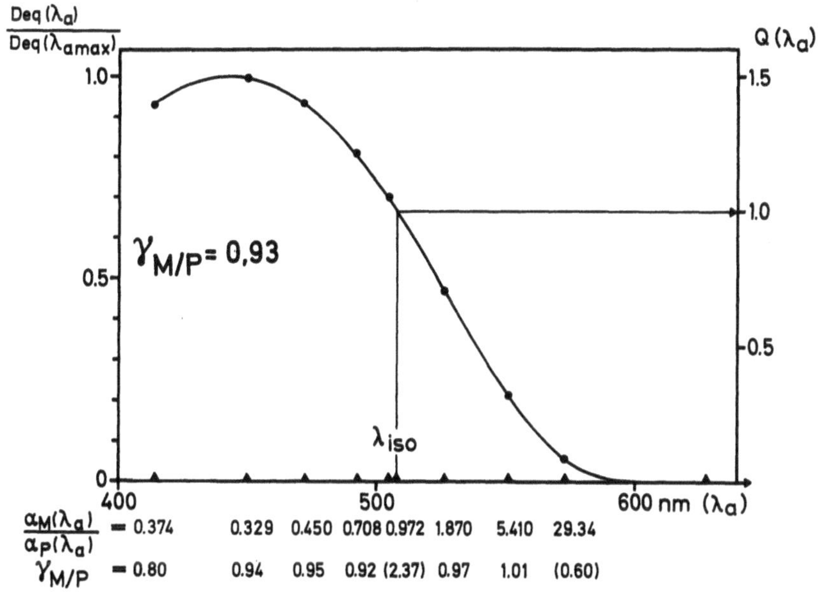

Fig. 37. Amplitudes of difference spectra of *Calliphora* (measured at 580 nm) as a function of wavelength of adaptation (λ_a), normalized to the maximum amplitude [*left scale*; Eq. (31)], and to $Q(\lambda_a)$ [*right scale*; Eq. (34)]. Values $\alpha_M(\lambda_a)/\alpha_P(\lambda_a)$ used for calculation of relative quantum efficiency ($\gamma_{M/P}$) were taken from computer analysis (Fig. 35). In obtaining the average $\gamma_{M/P}=0.93$, values in parentheses (2.37, 0.60) were excluded

3. Calculation of P and M Spectra from Electrophysiologic Data

a) Assumptions of Necessary Conditions

For the data from spectral sensitivities to lead directly to knowledge about the pigment spectra, the following assumptions have to be understood.

α) The diameter of a rhabdomere, and also of a fused rhabdome, has to be large enough so that the properties of small waveguide optics can be ignored, and Lambert-Beer's law holds. The critical diameter, down to which the waveguide effects do not geometrically interfere in appreciable amounts with the spectral absorption, is approximately 1.5 µm. In smaller rhabdomeres the characteristic waveguide parameter V has to be known:

$$V=\frac{\pi d}{\lambda}!\sqrt{n_1^2-n_2^2}$$

where n_1, n_2 are the refractive indices of the rhabdomere and its surround, respectively, d is the diameter of the rhabdomere and λ is the wavelength of incident light in vacuum (Snyder, 1975). The geometric optics are applicable down to values of $V>2.4$.

β) Screening pigments may act as spectral color or edge filters (Langer, 1975; Stavenga et al., 1973; Schneider et al., 1977) and may therefore falsify the spectral sensitivity curves. In superposition eyes this error can be excluded by either inserting glass or quartz rods into the cornea, down to the rhabdomere layer (HÖGLUND, 1966), or by dissection of the complete dioptric apparatus, including the pigment cells (HAMDORF and SCHWEMER, 1975; HAMDORF and HÖGLUND, 1976).

γ) The self-shielding effects inside a rhabdomere can be ignored only if the concentration of the visual pigment and its metarhodopsin is suitably low, i.e., the absorption along the length of a rhabdomere has to be less than 0.5 at the wavelengths of maximal absorption of P and M (see below).

δ) In fused rhabdomes that contain different visual pigments such as UV, blue, and green-absorbing pigments, the spectral sensitivity of the receptors will be markedly altered. This is also the case in the Hymenoptera species, *Apis*, *Vespa* and *Cataglyphis*, where longitudinal spectral filter effects are to be expected (SNYDER et al., 1973; MENZEL, 1974), and also in sphingid moths, where three layers of color receptors act as spectral filters in an axial arrangement (the distal layer being the UV receptor; SCHLECHT et al., unpublished data).

The relative spectral sensitivity (rel. s.S.), as well as the course of sensitivity alterations during and after light adaptation, are usually derived from the amplitude-intensity functions (*A–I* functions) of the receptor or retina potentials. The rel. s.S. is defined as the reciprocal of the factor f, by which the intensity I (the number of incident light quanta) of a test flash at wavelength λ_t has to be increased to evoke a response of the same amplitude (A) as is obtained by the most efficient wavelength $\lambda_{t_{max}}$:

$$\text{rel. s.S.} = 1/f(\lambda_t) = I\lambda_{t_{max}}/I\lambda_t, \tag{37}$$

if: $A(\lambda_t) = A(\lambda_{t_{max}})$.

Thus the spectral sensitivity is simply a descriptive function that allows one to draw conclusions only concerning the spectral absorption properties of the visual pigment if the above-mentioned assumptions are fulfilled.

b) *Relative Spectral Sensitivity and the P Spectrum*

The light I, which is absorbed in a receptor of total length l by the visual pigment of molar absorption $\alpha_P(\lambda_t)$ and concentration C_P, is given by:

$$I(\lambda_t, C_P) = I_{0\lambda t} \cdot (1 - \exp - \alpha_P(\lambda_t) \cdot C_P \cdot l). \tag{38}$$

If equal response amplitudes are always derived from the same number of light quanta absorbed per unit time, i.e., $I_a(\lambda_{t_1}) = I_a(\lambda_{t_n})$, (irrespective of the energy content of the photons and the locale of absorption inside the rhabdomere[8], then

[8] The two assumptions implicit in Eq. (39) are based on the following experimental results: i) psychophysiological determinations of rel.s.S. in lensless human eyes, as well as electrophysiologic measurements in invertebrate photoreceptors have shown that visibility is determined by I, and not by the energy content of the absorbed quanta, ii) the point inside the rhabdomere where the absorption takes place influences neither the formation nor the amplitude of the response. Spot stimulation along the long axis of a fly rhabdomere elicits equal response amplitudes and thus equal absolute sensitivities over the whole length of the rhabdomere (HAMDORF, 1970). Furthermore, it has been shown that the gradient of absorbed quanta over the length of the rhabdomere does not affect the response amplitude (HAMDORF et al., 1968).

rel. s.S. is defined by combining Eqs. 37 and 38 to give:

$$\text{rel.s.S.} = 1/f(\lambda_t) = \frac{I_{0\lambda_{t\max}}}{I_{0\lambda_t}} = \frac{1-\exp-\alpha_P(\lambda_t)\cdot C_P\cdot l}{1-\exp-\alpha_P(\lambda_{t\max})\cdot C_P\cdot l}. \tag{39}$$

If the assumptions made in Eq. (39) are true, one would expect the $A-I$ functions to be identical in shape when the I values are plotted on a logarithmic scale, that is, the curves should remain parallel to that obtained for the most efficient wavelength λ_{max}, but shifted to greater light intensities, these shifts corresponding to $f(\lambda_t)$ in Eq. (39). If it can be shown that $A-\log I$ functions at all wavelengths are identical, $f(\lambda_t)$ for each response amplitude at any test wavelength can be taken from the $A-\log I$ function determined for the most efficient wavelength λ_{max}, provided that the spectral test stimuli used have equal quanta.

As shown in Fig. 39, the curves of rel.s.S. and of $\alpha_P(\lambda_t)$ coincide if the extinction value $(\alpha_{P\lambda_{max}}\cdot C_P\cdot l)$ is less than 0.5. But in many receptors of animals active in dim light, i.e., in species of octopus or crayfish, the extinction exceeds 1, corresponding to an extinction of 0.0055 per each 1 µm length of the rhabdomere for unpolarized light and of 0.0073/µm for polarized light. In the latter case the rel.s.S. differs markedly from $\alpha_P(\lambda_t)$, and the determination λ_{max} of the visual pigment is possible only if the fraction of metarhodopsin present in the rhabdomere is small.

c) *Determination of P Concentration in the Light-Adapted Receptor*

For the eyes of insects (*Ascalaphus*, *Deilephila*, *Manduca*, *Apis*, *Calliphora*, *Musca*, and *Eristalis*) it has been shown that a few minutes after creating photoequilibria by bright monochromatic adaptation, the amplitude of the receptor response elicited by a constant test stimulus is proportional to the visual pigment concentration in the rhabdomere. This relationship is not self-evident and occurs only under certain circumstances. $I_{P,M}$, the number of quanta of a test stimulus I_0 that are absorbed together by the visual pigment and its metarhodopsin inside a rhabdomere of length l is given by:

$$I_{P,M}(C_P,\lambda_t) = I_0\{1-\exp-[\alpha_P(\lambda_t)\cdot C_P + \alpha_{Mg}(\lambda_t)\cdot(C_{P0}-C_P)]\cdot l\} \tag{40}$$

Since only the quanta absorbed by the visual pigment lead to depolarization of the receptor membrane (as has been shown for *Ascalaphus*; HAMDORF et al., 1971), only I_P is of interest and can be obtained from:

$$I_P(C_P,\lambda_t) = \{C_P\alpha_P(\lambda_t)/[C_P\alpha_P(\lambda_t)+(C_{P0}-C_P)\alpha_{Mg}(\lambda_t)]\}\cdot I_{P,M}(C_P,\lambda_t). \tag{41}$$

If in addition to the assumption of Eq. (39), the hypothesis is correct that the content of M $(C_{P0}-C_P)$ in the receptor does not influence the formation of the response amplitude, then the relative sensitivity (at test wavelength λ_t) depending on C_P and related to the highest possible concentration of visual pigment $C_{P_{max}}$, can be written as:

$$\text{rel.s.S.}_{\lambda_t} = 1/f(C_P) = \alpha_P C_P/(\alpha_P C_P + \alpha_{Mg}(C_{PO} - C_P)) \qquad \text{I}$$
$$\cdot [\alpha_P C_{Pmax} + \alpha_{Mg}(C_{PO} - C_{Pmax})]/\alpha_P C_{Pmax} \qquad \text{II}$$
$$\cdot \frac{1 - \exp - [\alpha_P C_P + \alpha_{Mg}(C_{PO} - C_P)] \cdot l}{1 - \exp - [\alpha_P C_{Pmax} + \alpha_{Mg}(C_{PO} - C_{Pmax})] \cdot l} \qquad \text{III.} \quad (42)$$

In cases where metarhodopsin is shifted to longer wavelengths (*Eledone, Ascalaphus, Calliphora, Eristalis*), long wavelength illumination leads to $C_{Pmax} = C_{PO}$, i.e., term II assumes the value of 1. But if the metarhodopsin is shifted to shorter wavelengths (i.e., in the "green receptor" of *Manduca, Deilephila*, and *Apis*), a maximal concentration C_{Pmax} is obtained by illumination with wavelengths where the ratio $\alpha_{Mg}(\lambda_a)/\alpha_P(\lambda_a)$ attains a maximum.

The rel.s.S. can be experimentally verified on the basis of the following considerations (see Fig. 38).

α) If system a (Fig. 31a) is valid, Eq. (42) simplifies to:

$$\text{rel.s.S.}_{\lambda_t} = (1 - \exp - \alpha_P \cdot C_P \cdot l)/(1 - \exp - \alpha_P \cdot C_{PO} \cdot l). \qquad (43)$$

Assuming an extinction value of 0.2 ($\varepsilon_{P\lambda_t} \cdot C_{PO} \cdot l$) for the receptor at the wavelength of the test flash (λ_t), the rel.s.S.$_{\lambda_t}$ is directly proportional to the concentration of the visual pigment C_P:

$$\text{rel.c.S.}_{\lambda_t} = 1/f(C) \simeq C_P/C_{PO}. \qquad (44)$$

This simple relationship holds also for receptors of higher extinction values, $(\varepsilon_{\lambda P_{max}} \cdot C_P \cdot l) \geqq 1$, if one chooses an appropriate test flash λ_t at which the extinction becomes <0.2.

β) If systems b, c, or d are valid, Eq. (42) is simplified by using the wavelength of the isosbestic point λ_{iso} for the test stimulus, and in this case, where $\alpha_{P\lambda_{iso}} = \alpha_{Mg\lambda_{iso}}$, the concentration is given by:

$$\text{rel.c.S.}_{\lambda_{iso}} = 1/f(C) = C_P/C_{Pmax} \qquad (45)$$

that is, using λ_{iso} as the test flash, the total amount of quanta absorbed inside the receptor, $I_{P,M}(\lambda_{iso})$, remains constant, and so does the exponential decrease of intensity along the length l of the rhabdomere. Thus the screening effect of metarhodopsin causes a constant probability of photon absorption for each visual pigment molecule, independent of C_P. This holds for all high extinction values of the receptor, and therefore λ_{iso} is an ideal test stimulus to prove the relationship.

If wavelengths other than λ_{iso} have to be chosen for the test flash, one has to take into account the kind of P–M system present; if metarhodopsin is shifted towards shorter wavelengths, a λ_t can be chosen that is only absorbed by the visual pigment, where again Eq. (44) can be applied. In contrast, if metarhodopsin is shifted towards longer wavelengths, it is more convenient to choose a λ_t at which $\alpha_P > \alpha_{Mg}$, rather than a λ_t at which $\alpha_P \ll \alpha_{Mg}$, for such test lights influence the

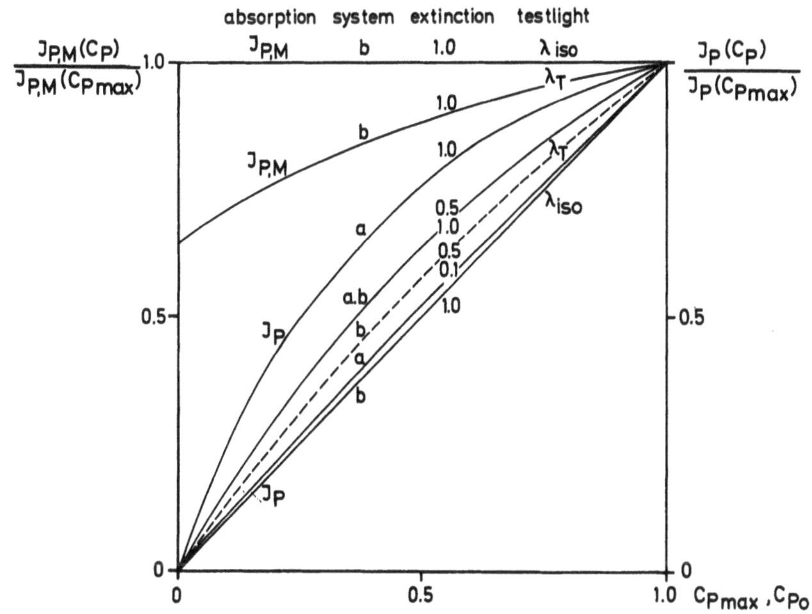

Fig. 38. Number of absorbed photons by both visual pigment and its metarhodopsin, $I_{P,M}$, in relation to concentration of visual pigment C_P. Ordinates are normalized to quantal absorption at maximal pigment concentration $C_{P\max}$ or C_{PO}. Curves (a) are valid if M is bleached, curves (b) are valid if M is thermostable. Curves are calculated for extinctions 1.0, 0.5, and 0.1 at wavelength of test light λ_t. For calculating curves of system (b), a λ_t was chosen at which $\alpha_{P\lambda_t}/\alpha_{M\lambda_t}=2.6$ (corresponding to $\lambda_t=460$ nm in blowfly receptors). If λ_{iso} is used as test flash, total amount of quanta absorbed inside the rhabdomere is constant $[I_{P,M}(C_P)/I_{P,M}(C_{P\max})=1]$, and amount of quanta absorbed by visual pigment is directly proportional to its relative concentration $C_P/C_{P\max}$. In system (a), $I_P(C_P)$ becomes approx. linear only at low extinctions of about 0.1. In system (b) (or c or d), the self-shielding effect of thermostable M favors direct proportionality (curves b, 1.0, and 0.5, dashed line)

concentrations of P and M in the receptor less than do test lights of longer wavelengths, and furthermore, the presence of M in the receptor improves the proportionality between I_P and C_P (compared with system a). This is shown in Fig. 38.

d) Dependence of Spectral Sensitivity on M Content in the Rhabdomere

Since according to Eqs. (40), (41), I_P is a function of C_P as well as of λ_t, a change in C_P in receptors of systems b, c, or d is accompanied by an alteration in the relative spectral absorption $I_P(\lambda_t)/I_P(\lambda_{t\max})$, and will therefore affect the spectral sensitivity, which is defined, with reference to Eq. (39) by:

$$\begin{aligned}\text{rel.s.S.} &= 1/f(\lambda_t, C_P) = I_{0\lambda_{t\max}}/I_{0\lambda_t} = I_{0\lambda_{t_2}}/I_{0\lambda_{t_1}}\\ &= \frac{\alpha_{P_1}(\alpha_{P_2}C_P+\alpha_{Mg_2}(C_{PO}-C_P))}{\alpha_{P_2}(\alpha_{P_1}C_P+\alpha_{Mg_1}(C_{PO}-C_P))}\\ &\quad\cdot\frac{1-\exp-[\alpha_{P_1}C_P+\alpha_{Mg_1}(C_{PO}-C_P)]\cdot l}{1-\exp-[\alpha_{P_2}C_P+\alpha_{Mg_2}(C_{PO}-C_P)]\cdot l}\,. \end{aligned} \tag{46}$$

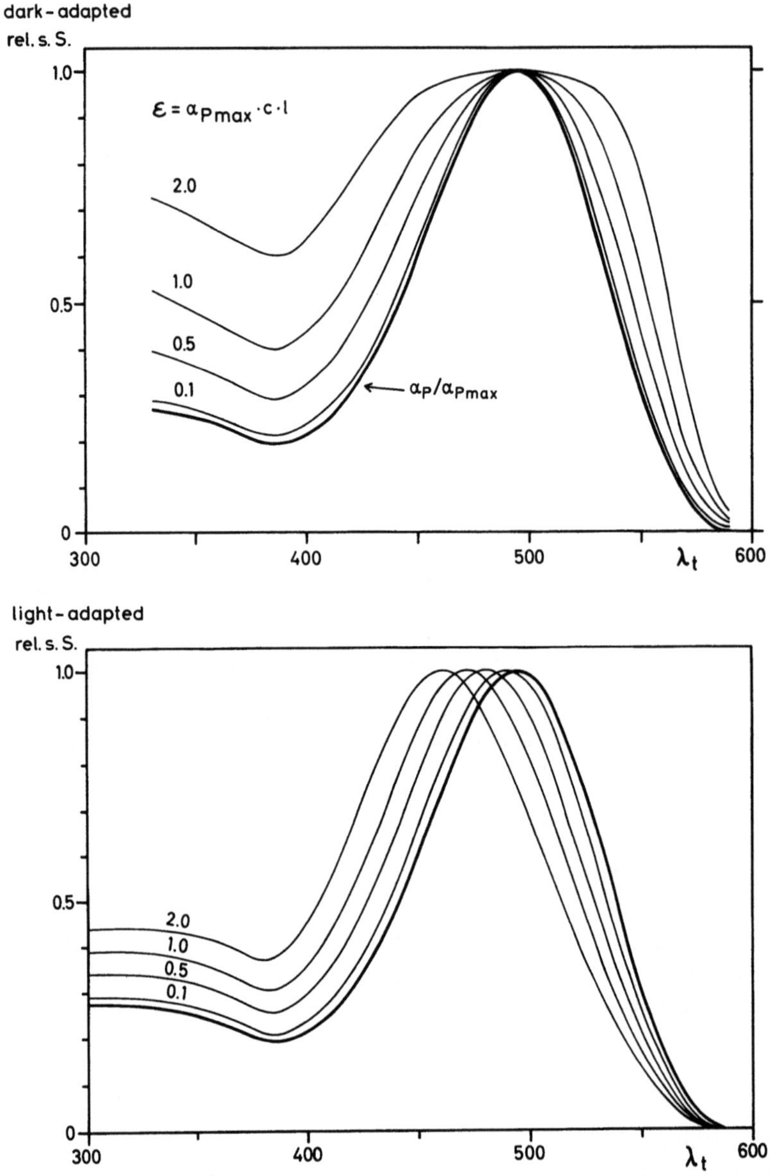

Fig. 39. Postulated modification of rel. s.S. caused by metarhodopsin content in light-adapted photoreceptor of blowfly. Spectral curves of dark-adapted (rel. s.S.$_{DA}$, $C_P = C_{P0}$) and light-adapted (rel. s.S.$_{LA}$, $C_P = 0.22\, C_{P0}$) photoreceptors were calculated for extinction 2.0 and 0.1 at $\alpha_{P\lambda_{t_{max}}}$

From this equation it follows that the spectral sensitivity of light-adapted receptors (rel.s.S.$_{LA}$) must differ from that of a dark-adapted one (rel. s.S.$_{DA}$). This is shown in Fig. 39 for the visual pigment system of *Calliphora* (P492, Ms564) under light- and dark-adapted conditions (calculated for extinctions ($\varepsilon_{P\lambda_{max}} \cdot C_{P0} \cdot l$) equal to 2.0, 1.0, 0.5, and 0.1). In the dark-adapted state, the functions are similar to $\alpha_P(\lambda_t)/\alpha_P(\lambda_{t_{max}})$, but in the light-adapted state, if metarhodopsin acts as a color filter, two phenomena are to be expected: 1) a shift of the sensitivity maximum $\lambda_{t_{max}}$ to a shorter wavelength (in Fig. 39 by 20 nm at extinction 1.0, and by 10 nm at extinction 0.5, respectively); 2) a narrowing of the function. In cases where M is shifted to shorter wavelengths (i.e., in the "green receptor" of *Deilephila* with P525 and M480), an increase of M content causes the opposite effect, that is, the maximal sensitivity $\lambda_{t_{max}}$ is shifted towards longer wavelengths.

In summary, the shape as well as the maxima of spectral sensitivity curves are strongly influenced by the color filter effect of the metarhodopsins, and thus one may only draw conclusions about the spectral absorption of the visual pigment from the rel. s.S. of the dark-adapted receptor.

e) Calculation of M Spectra

The P content in a rhabdomere can be altered specifically by saturating monochromatic adaptation. According to Eq. (30), at photoequilibrium, C_{Peq} which depends on the wavelength of adaptation λ_a, is described by:

$$f_{Peq}(\lambda_a) = \frac{C_{Peq}}{C_{PO}}(\lambda_a) = \alpha_M(\lambda_a) \cdot \gamma_{M/P} / (\alpha_P(\lambda_a) + \alpha_M(\lambda_a) \cdot \gamma_{M/P}) \tag{47}$$

where the relative quantum efficiency is $\gamma_{M/P} = \gamma_M/\gamma_P$. Thus, if $\alpha_P(\lambda_a)$ is known by rel. s.S.$_{DA}$, α_M can be derived from:

$$\alpha_M(\lambda_a) = \frac{\alpha_P(\lambda_a) \cdot (\text{rel. c.S.})}{\gamma_{M/P}(1 - \text{rel. c.S.})}. \tag{48}$$

In cases where M is shifted to longer wavelengths this equation holds in the range of the superposition of the P and the M spectra. Assumptions have to be made only with reference to $\gamma_{M/P}$ (Hamdorf et al., 1973; Hamdorf and Rosner, 1973; Rosner, 1975). In cases, however, where M is shifted towards shorter wavelengths and hence $C_{Peq}(\lambda_a)$ attains a maximum (C_{Pmax}) where $\alpha_M/\alpha_P(\lambda_a)$ becomes maximal, for calculation of $\alpha_M(\lambda_a)$ an additional assumption has to be made concerning C_{PO}/C_{Pmax}, yielding:

$$\alpha_M(\lambda_a) = \frac{\alpha_P(\lambda_a) \cdot (\text{rel. c.S.})}{\gamma_{M/P}[(C_{PO}/C_{Pmax}) - \text{rel. c.S.}]}. \tag{49}$$

D. Appendix: List of Symbols and Abbreviations

P490	Visual pigment of λ_{max} 490 nm
Ms, Mb, Mg	Acid metarhodopsin, alkaline metarhodopsin, and the sum of both
X, Y, Z	Pigment intermediates of short lifetime
C_P, C_{Ms}, C_{Mb}, C_{Mg}	Concentrations of visual pigment and its metarhodopsins

C_{P0}	Concentrations of visual pigment if $C_{Mg}=0$
C_{Peq}, C_{Pmax}	Concentrations of visual pigment in photoequilibrium and in equilibrium caused by the most effective wavelength of reconversion
$f_{Peq}=C_{Peq}/C_{P0}$, $f_{Meq}=C_{Meq}/C_{P0}$	Fractions of concentrations in equilibrium
W_P, W_M	Absorption probabilities of P and M in membrane
α_P, α_{Ms}, α_{Mb}, α_{Mg}	Absorption coefficients of pigment and its photoproducts
ε_P, ε_M	Molar extinctions of P and M
γ_P, γ_{Ms}, γ_{Mb}	Quantum efficiencies
$\gamma_{M/P}$	Relative quantum efficiency
λ_a, λ_t	Wavelength of adaptation and test flash, respectively
t	Time of reaction
k_1, k_2, k_n	Rate factors of photochemical reactions
t_P	Rate constant of pigment regeneration dependent on metabolism
t_s, t_b	pH dependent rate factors of Ms↔Mb transitions
$\boldsymbol{a}$, $\boldsymbol{E}$	Direction vectors of absorbing dipoles and of electric field
r	Direction factor
Ψ, Φ, Θ	Angles describing orientation of chromophore in membrane (see Fig. 33)
α, β	Angles of $\boldsymbol{E}$ vector of polarized light with respect to microvillus axis (see Fig. 33)
I	Intensity of incident or absorbed light
A	Absorption
T	Transmission
N	Dichroic ratio
D_{eq}	Difference in absorbance in photoequilibrium
$Q(\lambda_a)$	Difference in spectral absorbance normalized to difference spectrum caused by wavelength of the isosbestic point
rel. s.S.	Relative spectral sensitivity
rel. c.S.	Relative concentration sensitivity
$I-A$ function	Relationship between intensity and response amplitude
HSS	High-sensitivity system
HAS	High-acuity system

E. Acknowledgement

Parts A.III and IV were written in collaboration with SHAHRDAD RAZMJOO, to whom I offer my gratitude also for many discussions and the critical reading of the entire manuscript; and to Mrs. ELLEN HALFMANN for the preparation of all the illustrations.

References

Abrahamson, E. W., Wiesenfeld, J. R.: The structure, spectra and reactivity of visual pigments. In: Handbook of sensory physiology, Jung, R. (ed.), Vol. VII/1, pp. 69—121. Berlin-Heidelberg-New York: Springer 1972.

Autrum, H., Burkhardt, D.: Spectral sensitivity of single visual cells. Nature (Lond.) **190**, 639 (1961).

Autrum, H., Thomas, I.: Comparative physiology of colour vision in animals. In: Handbook of sensory physiology, Jung, R. (ed.), Vol. VII/3. Berlin-Heidelberg-New York: Springer 1973.

Autrum, H., Von Zwehl, V.: Zur spektralen Empfindlichkeit einzelner Sehzellen der Drohne *Apis mellifica*. Z. vergl. Physiol. **46**, 8—12 (1962).

Autrum, H., Von Zwehl, V.: Die spektrale Empfindlichkeit einzelner Sehzellen des Bienenauges. Z. vergl. Physiol. **48**, 357—384 (1964).

Azuma, K., Azuma, M., Suzuki, T.: Circular dichroism of cephalopod rhodopsin and its intermediates in the bleaching and photoregeneration process. Biochim. biophys. Acta (Amst.) **393**, 520—530 (1975)

Baumann, F., Hadjilazaro, B.: Afterpotential in retinula cells of the drone. Vision Res. **11**, 1198 (1971).

Baumann, F., Hadjilazaro, B.: A depolarizing effect of intense light in the drone visual receptor. Vision Res. **12**, 17—31 (1972).

Benolken, R. M., Anderson, R. E., Mande, M. B.: Lipid composition of *Limulus* photoreceptor membranes. Biochim. biophys. Acta (Amst.) **413**, 234—242 (1975)

Benzer, S.: Behavioural mutants of *Drosophila* isolated by countercurrent distribution. Proc. nat. Acad. Sci. (Wash.) **58**, 1112—1119 (1967)

Bishop, L. G.: An ultraviolet photoreceptor in a dipteran compound eye. J. comp. Physiol. **91**, 267—275 (1974).

Blasie, J. K.: The location of photopigment molecules in the cross-section of frog-retinal receptor disc membranes. Biophys. J. **12**, 191—209 (1972a).

Blasie, J. K.: Net electric charge on photopigment molecules and frog retinal receptor disc membrane structure. Biophys. J. **12**, 205—213 (1972b)

Blasie, J. K., Worthington, C. R.: Planar liquid-like arrangement of photopigment molecules in frog retinal receptor disc membranes. J. molec. Biol. **39**, 417—439 (1969).

Blasie, J. K., Worthington, C. R., Dewey, M. M.: Molecular localization of frog retinal receptor photopigment by electron microscopy and low-angle X-ray diffraction. J. molec. Biol. **39**, 407—416 (1969)

Boschek, B. C., Hamdorf, K.: Rhodopsin particles in the photoreceptor membrane of an insect. Z. Naturforsch. **31c**, 763 (1976)

Braitenberg, V.: Patterns of projection in the visual system of the fly. I. Retina-lamina-projections. Exp. Brain Res. **3**, 271—298 (1967).

Brown, H. M., Cornwall, M. C.: Spectral correlation of a quasi-stable depolarization in barnacle photoreceptors following red light. J. Physiol. (Lond.) **248**, 555—578 (1975).

Brown, H. M., Cornwall, M. C.: Ionic mechanism of a quasi-stable depolarization in barnacle photoreceptor following red light. J. Physiol. (Lond.) **248**, 579—593 (1975).

Brown, H. M., Hagiwara, S., Koike, H., Meech, R. M.: Membrane properties of a barnacle photoreceptor examined by the voltage clamp technique. J. Physiol. (Lond.) **208**, 385—413 (1970).

Brown, K. T., Murikami, M.: A new receptor potential of the monkey retina with no detectable latency. Nature (Lond.) **201**, 622—628 (1964).

Brown, P. K., Brown, P. S.: Visual pigments of the *Octopus* and Cuttlefish. Nature (Lond.) **182**, 1288—1290 (1958).

Brown, P. K., White, R. H.: Rhodopsin of the larval mosquito. J. gen. Physiol. **59**, 401—414 (1972)

Brown, P. K.: Rhodopsin rotates in the visual receptor membrane. Nature (Lond.) New Biol. **236**, 35—38 (1972).

Burkhardt, D.: Allgemeine Sinnesphysiologie und Elektrophysiologie der Rezeptoren. Fortschr. Zool. **13**, 146—189 (1961).

Burkhardt, D.: Spectral sensitivity and other response characteristics of single visual cells in the arthropod eye. Symp. Soc. exp. Biol. **16**, 86 (1962).

Carlson, S. D., Philipson, B.: Microspectrophotometry of the dioptric apparatus and compound rhabdom of the moth *(Manduca sexta)* eye. J. Insect Physiol. **18**, 1721—1731 (1972).

Clark, E. W., Branton, D.: Fracture faces in frozen outer segments from guinea pig retina. Z. Zellforsch. Mikr. Anat. **91**, 586—603 (1968)

Cone, R. A.: Early receptor potential: Photoreversible charge displacement in rhodopsin. Science **155**, 1128—1131 (1967).

Cone, R. A.: The early receptor potential. Proc. Int. School Physics Enrico Fermi **43**, 187—200 (1969).

Cone, R. A.: Rotational diffusion of rhodopsin in the visual receptor membrane. Nature (Lond.) New Biol. **236**, 39—43 (1972).

Cone, R. A., Cobbs, W. H.: Rhodopsin cycle in the living eye of the rat. Nature (Lond.) **221**, 820—822 (1969).

Cone, R. A., Pak, W. L.: The early receptor potential. In: Handbook of sensory physiology, Loewenstein, W. R. (ed.), Vol. I, pp. 345—365. Berlin-Heidelberg-New York: Springer 1971.

Cosens, D., Briscoe, D.: A switch phenomenon in the compound eye of the white-eyed mutant of *Drosophila melanogaster*. J. comp. Physiol. **18**, 627—632 (1972).

Cosens, D., Wright, R.: Light elicited isolation of the complementary visual input systems in white-eye *Drosophila*. J. Insect Physiol. **21**, 1111—1120 (1975).

Crane, J.: Imaginal behaviour of a Trinidad butterfly, *Heliconius erato hydara* Hewitson, with special references to the social use of colour. Zoologica **40**, 167—196 (1955).

Daemen, F. J. M.: Vertebrate rod outer segment membranes. Biochem. biophys. Acta **300**, 255—288 (1973).

Dartnall, H. J. A.: The photobiology of visual processes. In: Davson, H. (ed.): The Eye, Vol. 2. New York-London: Academic Press 1962.

Daumer, K.: Reizmetrische Untersuchungen des Farbensehens der Bienen. Z. vergl. Physiol. **38**, 413—478 (1956).

Demel, R. A., Guerts van Kessel, W. S. M., van Deenen, L. L. M.: The properties of polyunsaturated lecithins in monolayers and liposomes acid. The interactions of these lecithins with cholesterol. Biochem. biophys. Acta **266**, 26—40 (1972).

Denton, E. J.: The contribution of the oriented photosensitive and other molecules to the absorption of whole retina. Proc. roy. Soc. B **150**, 78—89 (1959).

Dörrscheidt-Käfer, M.: Die Empfindlichkeit einzelner Photorezeptoren im Komplexauge von *Calliphora erythrocephala*. J. comp. Physiol. **81**, 309—340 (1972).

Eakin, R. M.: Structure of invertebrate photoreceptors. In: Dartnall, H. J. A. (ed.): Handbook of Sensory Physiology, Vol. VII/1, pp. 625—684. Berlin-Heidelberg-New York: Springer 1972.

Eckert, H.: Die spektrale Empfindlichkeit des Komplexauges von *Musca* (Bestimmung aus Messungen der optomotorischen Reaktion). Kybernetik **9**, 145—156 (1971).

Fein, A., Cone, R. A.: *Limulus* rhodopsin: Rapid return of transient intermediates to the thermally stable state. Science **182**, 495—497 (1973).

Fein, A., Lisman, J.: Localized desensitization of *Limulus* photoreceptors produced by light or intracellular calcium ion injection. Science **187**, 1094—1095 (1975).

Fein, A., deVoe, R. D.: Adaptation in the ventral eye of *Limulus* is functionally independent of the photochemical cycle, membrane potential, and membrane resistance. J. gen. Physiol. **61**, 273—289 (1973).

Fernandez, H. R., Nickel, E. E.: Ultrastructural and molecular characteristics of crayfish photoreceptor membranes. J. Cell Biol. **69**, 721—732 (1976).

Franceschini, N.: Sampling of the visual environment by the compound eye of the fly: Fundamentals and applications. In: Photoreceptor Optics, eds. A. W. Snyder, R. Menzel. Berlin-Heidelberg- New York: Springer 1975

Franceschini, N., Kirschfeld, K.: Etude optique in vivo des éléments photorécepteurs dans l'œil composé de *Drosophila*. Kybernetik **8**, 1—13 (1971).

Franceschini, N., Kirschfeld, K.: Le contrôle automatique du flux lumineux dans l'œil composé des Diptères. Biol. Cybernetics **21**, 181—203 (1976).

Frisch, K., von: Tanzsprache und Orientierung der Bienen, 578 pp. Berlin-Heidelberg-New York: Springer 1965

Gogala, M.: Die spektrale Empfindlichkeit der Doppelaugen von *Ascalaphus macaronius* Scop. (Neuroptera, Ascalaphidae). Z. vergl. Physiol. **57**, 232—243 (1967)

Gogala, M., Hamdorf, K., Schwemer, J.: UV-Sehfarbstoff bei Insekten. Z. vergl. Physiol. **70**, 410—413 (1970)

Gogala, M., Michieli, S.: Das Komplexauge von *Ascalaphus*, ein spezialisiertes Sinnesorgan für kurzwelliges Licht. Naturwissenschaften **9**, 217—218 (1965)

Goldsmith, T. H.: The visual system of the honey bee. Proc. nat. Acad. Sci. (Wash.) **44**, 123—126 (1958).

Goldsmith,T.H.: The natural history of invertebrate visual pigments. In: Handbook of Sensory Physiology, ed. H.J.A.Dartnall, Vol. VII/1, 685—719. Berlin-Heidelberg-New York: Springer 1972

Goldsmith,T.H.: The polarization sensitivity-dichroic absorption paradox in arthropod photoreceptors. In: Photoreceptor Optics, eds. A.W.Snyder, R.Menzel. Berlin-Heidelberg-New York: Springer 1975.

Goldsmith,T.H., Barker,R.J., Cohen,C.F.: Sensitivity of visual receptors of carotenoid-depleted flies: A Vitamin A deficiency in an invertebrate. Science **146**, 65—67 (1964).

Goldsmith,T.H., Bruno,M.S.: Behaviour of rhodopsin and metarhodopsin in isolated rhabdoms of crabs and lobster. In: Biochemistry and Physiology of Visual Pigments, ed. H.Langer. Berlin-Heidelberg-New York: Springer 1973

Goldsmith,T.H., Fernandez,H.R.: Some photochemical and physiological aspects of visual excitations in compound eyes. In: The Functional Organization of the Compound Eye, ed. C.G.Bernhard. New York: Pergamon Press 1966

Goldsmith,T.H., Warner,L.T.: Vitamin A in the vision of insects. J. gen. Physiol. **47**, 433—441 (1964).

Gribakin,F.G., Govardovskii,V.I.: The role of the photoreceptor membrane in photoreceptor optics. In: Photoreceptor Optics, eds. A.W.Snyder, R.Menzel, pp. 215—236. Berlin-Heidelberg-New York: Springer 1975.

Grip,W.J.,de, Daemen,F.J.M., Bonting,S.L.: Enrichment of rhodopsin in rod outer segment membrane preparations. Biochemical aspects of the visual process XVIII. BBA **17**, 1667—1707 (1972).

Hagins,F.M.: Purification and partial characterization of the protein component of squid rhodopsin. J. biol. Chem. **248**, 3298—3304 (1973).

Hagins,W.A.: The visual process: Excitatory mechanisms in the primary receptor cells. Ann. Rev. Biophys. Bioengin. **1**, 131—158 (1972).

Hagins,W.A., McGaughy,R.E.: Molecular and thermal origins of fast photoelectric effects in the squid retina. Science **157**, 813—816 (1967).

Hall,M.O., Bok,D., Bacharachi,A.D.E.: Biosynthesis and assembly of the rod outer segment membrane system. Formation and fate of visual pigment in the frog retina. J. molec. Biol. **45**, 397—406 (1969).

Hamdorf,K.: Korrelation zwischen Sehfarbstoff und Empfindlichkeit bei Photorezeptoren. Verh. dtsch. Zool. Ges. **64**, 148—158 (1970).

Hamdorf,K.: Die Dauer der Dunkeladaptation beim Fliegenauge nach Belichtung mit heterochromatischen Blitzen. Z. vergl. Physiol. **75**, 200—206 (1971).

Hamdorf,K.: Primärprozesse beim Sehen der Wirbellosen. Rhein. Westf. Akad. Wiss. N253, Opladen: Westd. Verlag 1975.

Hamdorf,K., Gogala,M.: Photoregeneration und Bereichseinstellung beim UV-Rezeptor. J. comp. Physiol. **86**, 231—245 (1973).

Hamdorf,K., Gogala,M., Schwemer,J.: Beschleunigung der "Dunkeladaptation" eines UV-Rezeptors durch sichtbare Strahlung. Z. vergl. Physiol. **75**, 189—199 (1971).

Hamdorf,K., Höglund,G., Langer,H.: Photoregeneration of visual pigments in a moth. J. comp. Physiol. **86**, 247—263 (1973).

Hamdorf,K., Langer,H.: Veränderung der Lichtabsorption im Facettenauge bei Belichtung. Z. vergl. Physiol. **51**, 172—184 (1965).

Hamdorf,K., Paulsen,R., Schwemer,J.: Photoregeneration and sensitivity control of photoreceptors of invertebrates. In: Biochemistry and Physiology of Visual Pigments, ed. H.Langer. Berlin-Heidelberg-New York: Springer 1973.

Hamdorf,K., Paulsen,R., Schwemer,J., Täuber,U.: Biochemistry of Visual Pigments. 1. Photoreconversion of invertebrate visual pigments. In: Information Processing in the Visual System of Arthropods, ed. R.Wehner. Berlin-Heidelberg-New York: Springer 1972.

Hamdorf,K., Razmjoo,S.: The prolonged depolarization after potential and its contribution to the understanding of photoreceptor function. Biophys. Struct. Mechanism **3**, 163—170 (1977).

Hamdorf,K., Razmjoo,S.: Photoconvertible pigment states and excitation in *Calliphora*; the induction and properties of the prolonged depolarising afterpotential. Biophys. Struct. Mech. in press, (1979)

Hamdorf,K., Rosner,G.: Adaptation und Photoregeneration im Fliegenauge. J. comp. Physiol. **86**, 281—292 (1973).

Hamdorf,K., Schwemer,J.: Photoregeneration and the adaptation process in insect photoreceptors. In: Photoreceptor Optics, eds. A.W.Snyder, R.Menzel. Berlin-Heidelberg-New York: Springer 1975.

Hamdorf,K., Schwemer,J., Gogala,M.: Insect visual pigment sensitive to ultraviolet light. Nature (Lond.) **231**, 458—459 (1971).

Hamdorf, K., Schwemer, J., Täuber, U.: Der Sehfarbstoff, die Absorption der Rezeptoren und die spektrale Empfindlichkeit der Retina von *Eledone moschata*. Z. vergl. Physiol. **60**, 375—415 (1968).

Hara, T., Hara, R.: Cephalopode retinochrome. In: Handbook of Sensory Physiology, ed. H.J.A. Dartnall, Vol. VII/1. Berlin-Heidelberg-New York: Springer 1972.

Hara, T., Hara, R.: Biochemical properties of retinochrome. In: Biochemistry and Physiology of Visual Pigments, ed. H. Langer, pp. 181—191. Berlin-Heidelberg-New York: Springer 1973.

Hardie, R.C.: Electrophysiological properties of R7 and R8 in Dipteran retina. Z. Naturforsch. **22c**, 887—889 (1977)

Harosi, J.I.: Frog rhodopsin in situ: Orientational and spectral changes in the chromophores of isolated retinal rod cells. Ph. D. Thesis, Johns Hopkins Univ., Baltimore U.S.A. 1971.

Harosi, F.I., MacNichol, jr., E.F.: Dichroic microspectrophotometer: A computer assisted, rapid, wavelength-scanning photometer for measuring the linear dichroism of single cells. J. Opt. Soc. Amer. **64**, 903—918 (1974).

Harris, W., Stark, W.: Light-sensitive hereditary photoreceptor degeneration in *Drosophila*. Abstract of paper presentation at Spring 1975 meeting of the Association for Research in Vision and Ophthalmol., Sarasota, 1975.

Harris, W.A., Stark, W.S., Walker, J.A.: Genetic dissection of the photoreceptor system in the compound eye of *Drosophila melanogaster*. J. Physiol. (Lond.) **256**, 415—439 (1976).

Hays, D., Goldsmith, T.H.: Microspectrophotometry of the visual pigment of the spider crab *Libinia emarginata*. Z. vergl. Physiol. **65**, 218—232 (1969).

Heisenberg, M., Götz, K.G.: The use of mutation for the partial degradation of vision in *Drosophila melanogaster*. J. comp. Physiol. **98**, 217—241 (1975).

Heitzmann, H.: Rhodopsin is the predominant protein of rod outer segment membranes. Nature (Lond.) **235**, 114 (1972).

Hillman, P., Dodge, F.A., Hochstein, S., Knight, B.W., Minke, B.: Rapid dark recovery of the invertebrate early receptor potential. J. gen. Physiol. **62**, 77—86 (1973).

Hillman, P., Hochstein, S., Minke, B.: A visual pigment with two physiologically active stable states. Science **175**, 1486—1488 (1972).

Hillman, P., Hochstein, S., Minke, B.: Nonlocal interaction in the photoreceptor transduction process. J. gen. Physiol. **68**, 227—245 (1976).

Hillman, P., Keen, M.E., Winterhager, J.: V. Discussion of selected topics about the transduction mechanism in photoreceptors. Biophys. Struct. Mechanism **3**, 183—189 (1977).

Hochstein, S., Minke, B., Hillman, P.: Antagonistic components of the late receptor potential in the barnacle photoreceptor arising from different stages of the pigment process. J. gen. Physiol. **62**, 105—128 (1973).

Höglund, G.: Pigment migration, light screening and receptor sensitivity in the compound eye of Nocturnal Lepidoptera. Acta physiol. scand. **69**, 5—56 (1966).

Höglund, G., Hamdorf, K., Langer, H., Paulsen, R., Schwemer, J.: The photopigments in an insect retina. In: Biochemistry and Physiology of Visual Pigments, ed. H. Langer. Berlin-Heidelberg-New York: Springer 1973.

Höglund, G., Hamdorf, K., Rosner, G.: Trichromatic visual system in an insect and its sensitivity control by blue light. J. comp. Physiol. **86**, 265—279 (1973).

Höglund, G., Struwe, G.: Pigment migration and spectral sensitivity in the compound eye of moth. Z. vergl. Physiol. **67**, 229—237 (1970).

Horridge, G.A.: Arthropod receptor optics. In: Photoreceptor Optics, eds. A.W. Snyder, R. Menzel. Springer 1975.

Horridge, G.A., Mimura, K.: Physical separation of two visual pigments in *Calliphora* retinula cells 1—6. Proc. roy. Soc. B **190**, 211—224 (1975).

Horridge, G.A., Mimura, K., Tsukahara, Y.: Fly photoreceptors II. Spectral and polarized light sensitivity in the dronefly *Eristalis*. Proc. roy. Soc. L. B **190**, 225—237 (1975).

Hubbard, R., St. George, R.C.C.: The rhodopsin system of the squid. J. gen. Physiol. **41**, 501—528 (1958).

Hubbard, R., Wald, G.: Visual pigments of the horsehoe crab *Limulus polyphemus*. Nature (Lond.) **18**, 212 (1960).

Jan, L.Y., Revel, J.P.: Ultrastructural localization of rhodopsin in the vertebrate retina. J. Cell Biol. **62**, 257—273 (1974).

Kirschfeld, K.: Die Projektion der optischen Umwelt auf das Raster der Rhabdomere im Komplexauge von *Musca*. Exp. Brain Res. **3**, 248—270 (1967).

Kirschfeld, K.: Absorption properties of photopigments in single rods, cones and rhabdomeres. In: Processing of Optical Data by Organisms and Machines, ed. W. Reichardt, pp. 116—136. New York-London: Academic Press 1969.
Kirschfeld, K.: Optics of the compound eye. In: Processing of Optical Data by Organisms and Machines, ed. W. Reichardt, pp. 144—166. New York-London: Academic Press 1969.
Kirschfeld, K.: Problems of menotactic orientation according to the polarized light of the sky. Z. Naturforsch. **30c**, 88—90 (1975).
Kirschfeld, K., Feiler, R., Franceschini, N.: A photostable pigment within the rhabdomere of fly photoreceptors No. 7. J. comp. Physiol. **125**, 275—284 (1978)
Kirschfeld, K., Franceschini, N.: Optische Eigenschaften der Ommatidien im Komplexauge von *Musca*. Kybernetik **5**, 47—52 (1968).
Kirschfeld, K., Franceschini, N.: Microspectrophotometry of fly rhabdomeres. Lecture of conference on Visual Physiology, Günzburg (Germany) 1975.
Kirschfeld, K., Snyder, A. L.: Waveguide mode effects, birefringence and dichroism in fly photoreceptor. In: Photoreceptor Optics, eds. A. W. Snyder, R. Menzel. Berlin-Heidelberg-New York: Springer 1975.
Knoll, F.: Lichtsinn und Blütenbesuch des Falters von *Deilephila livornica*. Z. vergl. Physiol. **2**, 329—380 (1924).
Langer, H.: Nachweis dichroitischer Absorption des Sehfarbstoffes in den Rhabdomeren des Insektenauges. Z. vergl. Physiol. **51**, 258—263 (1965).
Langer, H.: Metarhodopsin in single rhabdomeres of the fly, *Calliphora erythrocephala*. In: Information Processing in the Visual Systems of Arthropods, ed. R. Wehner. Berlin-Heidelberg-New York: Springer 1972.
Langer, H.: Properties and functions of screening pigments in insect eyes. In: Photoreceptor Optics, eds. A. W. Snyder, R. Menzel. Berlin-Heidelberg-New York: Springer 1975.
Langer, H., Thorell, B.: Microspectrophotometry of single rhabdomeres in the insect eye. Exp. Cell Res. **41**, 673—677 (1966).
Laughlin, S. B., Menzel, R., Snyder, A. W.: Membranes, dichroism and receptor sensitivity. In: Photoreceptor Optics, eds. A. W. Snyder, R. Menzel. Berlin-Heidelberg-New York: Springer 1975.
Leeson, T. S.: Freeze-etch studies of rabbit eye. II. Outer segments of retinal photoreceptors. J. Anat. **108**, 147—157 (1971).
Liebman, P. A.: In situ microspectrophotometric studies on the pigments of single retinal rods. Biophys. J. **2**, 161—178 (1962).
Liebman, P. A.: Microspectrophotometry of retinal cells. Ann. N.Y. Acad. Sci. **157**, 250—264 (1969).
Liebman, P. A.: Microspectrophotometry of photoreceptors. In: Handbook of Sensory Physiology, ed. H. J. A. Dartnall, Vol. VII/1. Berlin-Heidelberg-New York: Springer 1972.
Liebman, P. A.: Birefringence, dichroism and rod outer segment structure. In: Photoreceptor Optics, eds. A. W. Snyder, R. Menzel. Berlin-Heidelberg-New York: Springer 1975.
Liebman, P. A., Entine, G.: Sensitive low-light-level microspectrophotometer detection of photosensitive pigments of retinal cones. J. Opt. Soc. Amer. **54**, 1451—1459 (1964).
Liebman, P. A., Entine, G.: Visual pigments of frog and tadpole. Vision Res. **8**, 761—775 (1968).
Liebman, P. A., Entine, G.: Lateral diffusion of visual pigment in photoreceptor disc membranes. Science **185**, 457—459 (1974).
Lisman, J. E., Brown, J. E.: The effects of intracellular iontophoretic injection of calcium and sodium on the light response of *Limulus* ventral photoreceptors. J. gen. Physiol. **59**, 701—719 (1972).
McCann, G. D., Arnett, D. W.: Spectral and polarization sensitivity of the dipteran visual system. J. gen. Physiol. **59**, 534 (1972).
Mason, W. T., Fager, R. S., Abrahamson, E. W.: Characterization of the lipid composition of squid rhabdome outer segments. Biochim. biophys. Acta (Amst.) **306**, 67—73 (1973).
Mason, W. T., Fager, R. S., Abrahamson, E. W.: Structural response of vertebrate photoreceptor membranes to light. Nature (Lond.) **247**, 188—191 (1974).
Meffert, P., Smola, U.: Electrophysiological measurements of spectral sensitivity of central visual cells in eye of blowfly. Nature (Lond.) **260**, 342—344 (1976).
Menzel, R.: Colour receptors in insects. In: The Compound Eye and Vision of Insects, ed. G. A. Horridge. Oxford: Clarendon Press 1974.
Menzel, R., Blakers, M.: Colour receptors in the bee eye: Morphology and spectral sensitivity. J. comp. Physiol. **108**, 11—33 (1976).
Minke, B.: *Drosophila* Mutant with a Transducer Defect. Biophys. Struct. Mechanism **3**, 59—64 (1977).
Minke, B., Hochstein, S., Hillman, P.: Antagonistic process as source of visible-light suppression of afterpotential in *Limulus* UV photoreceptors. J. gen. Physiol. **62**, 787—791 (1973).

Minke,B., Hochstein,S., Hillman,P.: Early receptor potential evidence for the existence of two thermally stable states in the barnacle visual pigment. J. gen. Physiol. **62**, 87—104 (1973).

Minke,B., Kirschfeld,K.: Microspectrophotometric evidence for two photointerconvertible states of visual pigment in the barnacle lateral eye. J. gen. Physiol., in press (1977—1978).

Minke,B., Wu,C.F., Pak,W.L.: Isolation of light-induced response of the central retinula cells from the electroretinogram of *Drosophila.* J. comp. Physiol. **98**, 345—355 (1975).

Moody,M.F., Parriss,J.R.: The discrimination of polarized light by *Octopus*: A behavioural and morphological study. Z. vergl. Physiol. **44**, 268—291 (1961).

Mote,M.I.: Polarization sensitivity: A phenomenon independent of stimulus intensity or state adaptation in retinular cells of crabs *Carcinus* and *Callinectes.* J. comp. Physiol. **90**, 389—403 (1974).

Muijser,H., Leutscher-Hazelhoff,J.T., Stavenga,O.G., Kuiper,J.W.: Photopigment conversion expressed in receptor potential and membrane resistance of blowfly visual sense cells. Nature (Lond.) **254**, 520 (1975).

Muller,K.J.: Photoreceptors in the crayfish compound eye: Electrical interactions between cells as related to polarized light sensitivity. J. Physiol. (Lond.) **232**, 573—595 (1973).

Naka,K., Kuwabara,M.: Two components from the compound eye of the crayfish. J. exp. Biol. **36**, 51 (1959).

Nickel,E., Menzel,R.: Insect UV- and green-photoreceptor membranes studied by the freeze-fracture technique. Cell Tiss. Res. **175**, 357—368 (1976).

Nielson,N.C., Fleischer,S., McConnell,D.G.: Lipid composition of bovine retinal outer segment fragments. Biochem. biophys. Acta **211**, 10—19 (1970).

Nolte,J., Brown,J.E.: Electrophysiological properties of cells in the median ocellus of *Limulus.* J. gen. Physiol. **59**, 167 (1972a).

Nolte,J., Brown,J.E.: UV-induced sensitivity to visible light in UV receptors of *Limulus.* J. gen. Physiol. **59**, 186—200 (1972b).

Nolte,J., Brown,J.E., Smith,T.G.: A hyperpolarizing component of the receptor potential in the median ocellus of *Limulus.* Science **162**, 677—679 (1968).

Ostroy,S.E., Wilson,M., Pak,W.L.: *Drosophila* rhodopsin: Photochemistry, extraction and differences in the *norp* A^{P12} phototransduction mutant. Biochem. Biophys. Res. Commun. **59**, 960—966 (1974).

Pak,W.L., Lidington,K.J.: Fast electrical potential from a long-lived, long-wavelength photoproduct of the fly visual pigment. J. gen. Physiol. **63**, 740—756 (1974).

Paulsen,R., Schwemer,J.: Studies on the insect visual pigment sensitive to ultraviolet light: Retinal as the chromophoric group. Biochim. biophys. Acta (Amst.) **283**, 520—529 (1972).

Paulsen,R., Schwemer,J.: Proteins of invertebrate photoreceptor membranes. Europ. Biochem. **40**, 577—583 (1973).

Paulsen,R., Zinkler,D.: Lipid and protein composition of Cephalopod photoreceptor membranes. Europ. J. Biochem. (1978)—submitted—

Perrelet,A., Bauer,H., Fryder,V.: Fracture faces of an insect rhabdomere. J. Microscop. **13**, 97—106 (1972).

Poo,M.M., Cone,R.A.: Lateral diffusion of rhodopsin in *Necturus* rods. Exp. Eye Res. **17**, 503—510 (1973).

Poo,M.M., Cone,R.A.: Lateral diffusion of rhodopsin in the photoreceptor membrane. Nature (Lond.) **247**, 438—441 (1974).

Razmjoo,S., Hamdorf,K.: Visual sensitivity and the variation of total photopigment content in the blowfly photoreceptor membrane. J. comp. Physiol. **105**, 279—286 (1976).

Robinson,W.E., Gordon-Walker,A., Bownds,D.: Molecular weight of frog rhodopsin. Nature (Lond.) **235**, 112—114 (1972).

Rosner,G.: Adaptation und Photoregeneration im Fliegenauge. J. comp. Physiol. **102**, 269—295 (1975).

Salem,L.: Theory of photochemical reactions. Science **191**, 822—830 (1976).

Schlecht,P., Hamdorf,K., Langer,H.: The arrangement of colour receptors in a fused rhabdom of an insect. J. comp. Physiol. **123**, 239—243 (1978)

Schlecht,P., Täuber,U.: The photochemical equilibrium in rhabdomeres of *Eledone* and its effect on dichroic absorption. In: Photoreceptor Optics, eds. A.W.Snyder, R.Menzel. Berlin-Heidelberg-New York: Springer 1975.

Schmidt,W.J.: Polarisationsoptische Analyse eines Eiweiß-Lipoid Systems, erläutert am Außenglied der Sehzellen. Kolloid-Z. **85**, 137 (1938).

Schneider, L., Drašlar, K., Langer, H., Gogala, M., Schlecht, P.: Feinstruktur und Schirmpigment-Eigenschaften der Ommatidien des Doppelauges von *Ascalaphus* (Insecta, Neuroptera). Cytobiologie, **16**, 274—307 (1978).
Scholes, J.: The electrical responses of the retinal receptors and the lamina in the visual system of the fly *Musca*. Kybernetik **6**, 149—162 (1969).
Schwemer, J.: Der Sehfarbstoff von *Eledone moschata* und seine Umsetzung in der lebenden Netzhaut. Z. vergl. Physiol. **62**, 121—152 (1969).
Schwemer, J., Gogala, M., Hamdorf, K.: Der UV-Sehfarbstoff der Insekten: Photochemie in vitro und in vivo. Z. vergl. Physiol. **75**, 174—188 (1971).
Schwemer, J., Paulsen, R.: Three visual pigments in *Deilephila elpenor* (Lepidoptera, Sphingidae). J. comp. Physiol. **86**, 215—229 (1973).
Seldin, E. B., White, R. H., Brown, P. K.: Spectral sensitivity of larval mosquito ocelli. J. gen. Physiol. **59**, 415—420 (1972).
Shaw, S. R.: Sense cell structure and interspecies comparisons of polarized light absorption in arthropod compound eye. Vision Res. **9**, 1031 (1969a).
Shaw, S. R.: Interreceptor coupling in ommatidia of drone honey bee and locust compound eyes. Vision Res. **9**, 199 (1969b).
Shichi, H.: Phospholipid requirement and opsin conformation for regeneration of bovine rhodopsin. J. biol. Chem. **246**, 6178—6182 (1971).
Singer, S. J., Nicholson, G. L.: The fluid mosaic model of the structure of cell membranes. Science **175**, 720—731 (1972).
Smith, T. G., Brown, J. E.: A photoelectric potential in invertebrate cells. Nature (Lond.) **212**, 1217—1219 (1966).
Smola, U., Meffert, P.: A single-peaked UV-receptor in the eye of *Calliphora erythrocephala*. J. comp. Physiol. **103**, 353—357 (1975).
Smola, U., Meffert, P.: Spectral sensitivity of central visual cells in the eye of blowfly *Calliphora erythrocephala*. Verh. dtsch. Zool. Ges. Hamburg **30**, 282 (1976)
Snyder, A. W.: Polarization sensitivity of individual retinula cells. J. comp. Physiol. **83**, 331—360 (1973).
Snyder, A. W.: Photoreceptor optics—theoretical principles. In: Photoreceptor Optics, eds. A. W. Snyder, R. Menzel. Berlin-Heidelberg-New York: Springer 1975.
Snyder, A. W., Menzel, R., Laughlin, S. B.: Structure and function of the fused rhabdom. J. comp. Physiol. **87**, 99—135 (1973).
Snyder, A. W., Pask, C.: Spectral sensitivity of dipteran retinula cell. J. comp. Physiol. **84**, 59—76 (1973).
Stark, W. S.: Light-elicited wavelength-specific shifts in the ERG visual sensitivity of *Drosophila*. Drosoph. Inf. Serv. **49**, 69 (1972).
Stark, W. S.: The effect of eye colour pigments on the action spectrum of *Drosophila*. J. Insect Physiol. **19**, 999—1006 (1973).
Stark, W. S.: Spectral selectivity of visual response alterations mediated by interconversions of native and intermediate photopigments in *Drosophila*. J. comp. Physiol. **96**, 343—356 (1975).
Stark, W. S., Prunsky, G. S., Zitzmann, W. G.: Separation of photopigment from membrane adaptation by vitamin A deprivation in the *Drosophila* retina. In: Society for Neuroscience Program and Abstracts (1975) (abstract).
Stark, W. S., Wassermann, G. S.: Transient and receptor potentials in the electroretinogram of *Drosophila*. Vision Res. **12**, 1771—1775 (1972).
Stark, W. S., Wassermann, G. S.: Wavelength-specific ERG characteristics of pigment and white-eyed strains of *Drosophila*. J. comp. Physiol. **91**, 427—441 (1974).
Stark, W. S., Zitzmann, W. G.: Isolation of adaptation mechanisms and photopigment spectra by Vitamin A deprivation. J. comp. Physiol. **105**, 15—27 (1976).
Stavenga, D. G.: Visual receptor optics, rhodopsin and pupil in fly retinula cells. Diss. Groningen 1974.
Stavenga, D. G.: Dark regeneration of invertebrate visual pigments. In: Photoreceptor Optics, eds. A. W. Snyder, R. Menzel. Berlin-Heidelberg-New York: Springer 1975.
Stavenga, D. G.: The neural superposition eye and its optical demands. J. comp. Physiol. **102**, 297—304 (1975).
Stavenga, D. G.: Optical qualities of the fly eye—an approach from the side of geometrical, physical and waveguide optics. In: Photoreceptor Optics, eds. A. W. Snyder, R. Menzel. Berlin-Heidelberg-New York: Springer 1975.
Stavenga, D. G.: Derivation of photochrome absorption spectra from absorbance difference measurements. Photochem. Photobiol. **21**, 105—110 (1975).
Stavenga, D. G.: Fly visual pigments. Difference in visual pigments of blowfly and dronefly peripheral retinula cells. J. comp. Physiol. **111**, 137–152 (1976).

Stavenga, D. G., Zantema, A., Kuiper, J. W.: Rhodopsin process and the function of the pupil mechanism in flies. In: Biochemistry and Physiology of Visual Pigments, ed. H. Langer. Berlin-Heidelberg-New York: Springer 1973.

Suzuki, T., Sugahara, M., Azuma, K., Azuma, M., Saimi, Y., Kito, Y.: Studies on cephalopod rhodopsin conformational changes in chromophore and protein during the photoregeneration process. Biochim. biophys. Acta (Amst.) **333**, 149—160 (1973).

Suzuki, T., Sugahara, M., Kito, Y.: An intermediate in the photoregeneration of squid rhodopsin. Biochim. biophys. Acta (Amst.) **275**, 260—270 (1972).

Täuber, U.: Mikrospektrophotometrische Untersuchungen an Sehrezeptoren von *Eledone moschata*. Dipl.-Arbeit München 1967.

Täuber, U.: *Octopus* rhodopsin in situ: Microphotometric measurements of orientational and spectral changes. In: Biochemistry and Physiology of Visual Pigments, ed. H. Langer. Berlin-Heidelberg-New York: Springer 1973.

Täuber, U.: Photokinetics and dichroism of visual pigments in the photoreceptors of *Eledone (Ozoena) moschata*. In: Photoreceptor Optics, eds. A. W. Snyder, R. Menzel. Berlin-Heidelberg-New York: Springer 1975.

Tsukahara, Y., Horridge, G. A.: Visual pigment spectra from sensitivity measurements after chromatic adaptation of single fly retinula cells. J. comp. Physiol. **114**, 233—251 (1977).

Tsukahara, Y., Horridge, G. A., Stavenga, D. G.: Afterpotentials in dronefly retinula cells. J. comp. Physiol. **114**, 253—266 (1977).

Wald, G.: Visual pigments of the fresh water crayfish. Federation Proc. **21**, 344 (1962).

Wald, G.: Single and multiple visual systems in arthropods. J. gen. Physiol. **51**, 125—156 (1968).

Wald, G.: The molecular basis of visual excitation. Nature (Lond.) **219**, 800—807 (1968).

Wald, G.: Visual pigments and photoreceptors—review and outlook. Exp. Eye Res. **18**, 333—343 (1974).

Wald, G., Hubbard, R.: Visual pigment of a decapod crustacean: The lobster. Nature (Lond.) **180**, 278—280 (1957).

Wald, G., Brown, P. K., Gibbons, I. R.: Visual excitation: A chemo-anatomical study. Symp. Soc. exp. Biol. **16**, 32—57 (1962).

Walker, J., Stark, W., Harris, W.: Genetic dissection of the *Drosophila* retina. Genetics **80**, 582—583 (1975).

Waterman, T. H., Fernandez, H. R.: *E*-vector and wavelength discrimination by retinular cells of the crayfish *Procambarus*. Z. vergl. Physiol. **68**, 154—174 (1970).

Waterman, T. H., Fernandez, H. R., Goldsmith, T. H.: Dichroism of photosensitive pigment in rhabdomes of the crayfish *Orconectes*. J. gen. Physiol. **54**, 415—432 (1969).

Welsch, B.: Ultrastruktur und funktionelle Morphologie der Augen des Nachtfalters *Deilephila elpenor* (Lepidoptera, Sphingidae). Cytobiologie **14**, 378—400 (1977).

Wolken, J. J.: Photoprocesses, photoreceptors, and evolution. New York-San Francisco-London: Academic Press 1975.

Worthington, C. R.: Structure of photoreceptor membranes. Ann. Rev. Biophys. Bioengin. **3**, 53—81 (1974).

Wright, A. K.: Ellipsoid models for rotational diffusion of rhodopsin in a digitonin micelle and in the visual receptor membrane. Biophys. J. **14**, 243—245 (1974).

Wright, R., Cosens, D.: Blue-adaptation and orange-adaptation in white-eyed *Drosophila*: Evidence that the prolonged afterpotential is correlated with the amount of M 580 in R_{1-6} J. comp. Physiol. **113**, 105—128 (1977).

Wright, W. E., Brown, P. K., Wald, G.: The orientation of rhodopsin and other pigments in dry films. J. gen. Physiol. **59**, 201—212 (1972).

Wright, W. E., Brown, P. K., Wald, G.: Orientation of intermediates in the bleaching of shear-oriented rhodopsin. J. gen. Physiol. **62**, 509—522 (1973).

Yoshizawa, T., Horiuchi, S.: Studies on intermediates of visual pigments by absorption spectra at liquid helium temperature and circular dichroism at low temperatures. In: Biochemistry and Physiology of Visual Pigments, pp. 69—81, ed. H. Langer. Berlin-Heidelberg-New York: Springer 1973.

Zinkler, D.: Charakterisierung und Stoffwechsel der Lipide des Komplexauges von Insekten. Thesis, Bochum 1974.

Zinkler, D.: Zum Lipidmuster der Photorezeptoren von Insekten. Verh. dtsch. Zool. Ges. Bochum 1975.

Chapter 5

Physics of Vision in Compound Eyes

ALLAN W. SNYDER, Canberra, Australia

With 36 Figures

Contents

A. Introduction

Invertebrates are the class *par excellence* where one finds all manner of variations in the structure and arrangement of photoreceptors and their associated light-gathering and processing structures. We are compelled to search for a physiologic explanation of these unique forms. This article offers a highly personalized view of the physics suited to the task.

Our emphasis is on the more highly evolved compound eyes, i.e., those with the most sophisticated visual apparatus. Consequently, we commence by clarifying how animals with compound eyes reconstruct their environment, as this is presumed to be a primary objective. We learn from the outset that the environmental intensity is a crucial determinate of visual acuity and that an eye must, therefore, optimize the capture of all available photons. Thus, one of the major themes of our development is the appreciation of those adaptations that increase absolute sensitivity.

An apology is due to readers who expect a historical development, for we have instead attempted to present a self-consistent, unifying treatment. Two important topics have been omitted completely, spectral sensitivity and the coding of information by the neurons, but these are expertly dealt with by MENZEL (this volume) and LAUGHLIN (this volume part B), respectively. We recommend the excellent review article by KIRSCHFELD (1969) on the optics of the compound eye to those readers unfamiliar with the subject. In addition, the review article by GOLDSMITH and BERNARD (1974) contains a wealth of material that also serves as a useful background to many topics considered here.

B. Visual Acuity

In this section, we consider the capacity of a compound eye to resolve details of its spatial environment. We begin in Section A by considering the factors that limit an animal's resolving power and then in Section III use them to determine the design of a compound eye. Figure 1 taken in part from KIRSCHFELD (1969), is presented to illustrate the three types of compound eyes referred to in this chapter. The symbols most frequently referred to are summarized in Table 1.

I. Physical Limitations to Spatial Resolving Power

1. Receptor Grain (Limitation of the Finite Angular Spacing of Ommatidia)

Because there is only a finite number of ommatidia, the image at the retina is a quantized version of the object. The smaller the interommatidial angle $\Delta\phi$, the greater the *potential* of the eye to resolve fine detail. In general, compound eyes are

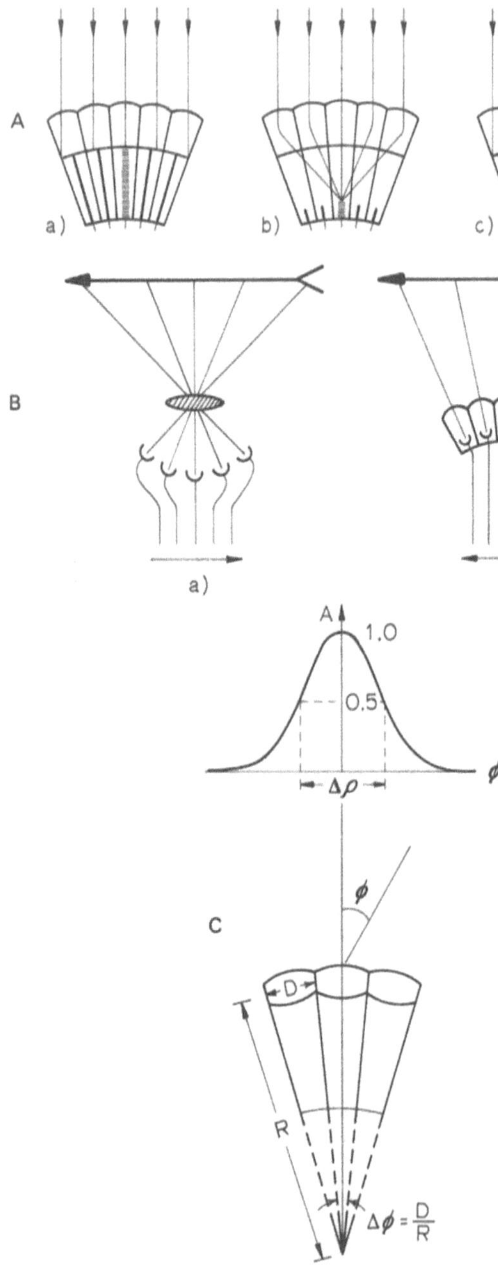

Fig. 1. (A) Schematic representation of three different types of compound eyes: *a* Apposition eye, *b* optical superposition eye, and *c* neural superposition eye. Each eye is illuminated by a distant point source and the illuminated photoreceptor (rhabdom) is indicated by the dashed line. (B) Schematic representation of image formation in the vertebrate lens eye *a* and in the compound eye *b*. (A) and (B) are from KIRSCHFELD (1969). (C) Cross-section of several ommatidia, where *D* is the facet diameter, *R* the eye radius, and $\Delta\phi$ the interommatidial angle. The acceptance function of a retinula cell is also shown, where $\Delta\varrho$ is the width of the function at 50% sensitivity and ϕ the inclination to the axis of the ommatidium

Table 1. List of important symbols

p	=	eye parameter $=D\Delta\phi=D^2/R=R(\Delta\phi)^2$; [μm]. At the diffraction limit, $p=\lambda/2$ (square lattice); $p=\lambda/\sqrt{3}$ (hexagonal lattice)
$\Delta\phi$	=	interommatidial angle $=D/R$; [radians] in formulae, [degrees] in figures
D	=	facet diameter, assumed to be equal to the entrance pupil diameter; [μm]
R	=	(local) eye radius; [μm]
f	=	distance from distal tips of rhabdom to posterior nodal point
$\Delta\varrho_l$	=	width of lens blur function at 50% height. When diffraction is limited, $\Delta\varrho_l=\lambda/D$. The angular diameter $\Delta\varrho_A$ of the Airy disc is $\Delta\varrho_A=2.44\,\Delta\varrho_l$
λ	=	wavelength of light in vacuum
$\Delta\varrho_r$	=	effective angular diameter of the rhabdom [radians]. Includes additional light gathering due to crystalline cone, which when not present $\Delta\varrho_r=d_{Rh}/f$, where d_{Rh} is the rhabdom diameter
$\Delta\varrho$	=	width of the angular acceptance function of the photoreceptor retinula cell at 50% sensitivity; [radians]. $\Delta\varrho^2=(\lambda/D)^2+(\Delta\varrho_r)^2$
ν_s	=	sampling frequency, i.e., highest spatial frequency that can be reconstructed by the array of ommatidia; [radians] $\nu_s=1/2\,\Delta\phi$ (square lattice); $\nu_s=1/\sqrt{3}\,\Delta\phi$ (hexagonal lattice)
ν_{co}	=	cutoff frequency, i.e., highest spatial frequency passed by the optics of an individual ommatidium. $\nu_{co}=D/\lambda$, in case of diffraction limit; [radians]$^{-1}$
v	=	angular velocity; [radians] [s]$^{-1}$
SNR	=	signal to noise ratio
m	=	modulation (or contrast) of sinusoidal grating
$\tilde{m}$	=	m/SNR
$\bar{N}$	=	mean number of photons absorbed by each photoreceptor of an array, per integration time of the eye, due to a uniform source, infinite in extent 1) $\bar{N}=\hat{I}(D\Delta\varrho_r)^2$ 2) $\hat{I}$ = intensity parameter $=0.89\,\varepsilon\Delta t\bar{I}$ 3) $\bar{I}$ = mean number of photons per second entering the entrance pupil per square μm per steradian of field; [sr]$^{-1}$ [μm]$^{-2}$ [s]$^{-1}$ 4) ε = quantum efficiency, i.e., the fraction of photons entering the pupil that are counted by the photoreceptors 5) Δt = integration time (effective shutter time) of the eye; [s]
M_l	=	modulation transfer function (MTF) of lens pupil $=[\exp\{-3.56(\nu\Delta\varrho_l)^2\}]$
M_r	=	MTF of rhabdom $=[\exp\{-3.56(\nu\Delta\varrho_r)^2\}]$; $M=M_lM_r=\exp\{-3.56(\nu\Delta\varrho)^2\}$

neither spheric nor are they composed of the same size facets (for review, see MAZOKHIN-PORSHNYAKOV, 1969). Consequently, the interommatidial angle $\Delta\phi$ must also depend on the region of the eye. From Fig. 1

$$\Delta\phi=D/R\,, \tag{B.1}$$

where R is the local eye radius and D the facet diameter. Of these three parameters, R usually displays the greatest variation over one eye as well as between different eyes. Thus, the relatively flat eye regions, where R is large, provide the highest acuity. The size of the head ultimately limits the *fraction* of total surface area that can be devoted to high resolving power. In many animals (e.g., bee and dragonfly) the interommatidial angles $\Delta\phi$ tend to be smallest in the central (or flatter) region of the eye (PORTILLO, 1933), which has led MAZOKHIN-PORSHNYAKOV (1969),

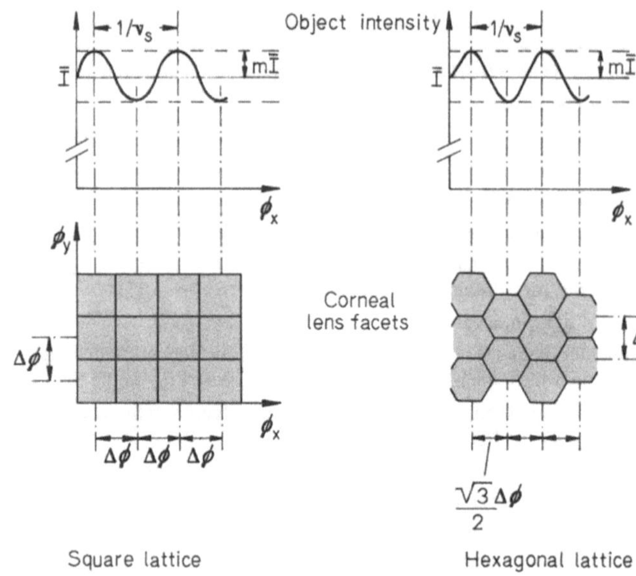

Fig. 2. Highest spatial frequency ν_s that can be reconstructed *independent of the grating orientation*, by a square and hexagonal lattice of ommatidia visual axes. Note that there are *special* orientations of the grating for which a larger ν can be theoretically reconstructed, e.g., $\nu = \Delta\phi/\sqrt{2}$ for the square array and $\Delta\phi/2$ for the hexagonal array. A consequence of nonspheric eyes is that facet shape is not in general related to the lattice of visual axes (STAVENGA, 1975). However, for simplicity, we have illustrated the situation for a spheric eye in which the shape of lens facets coincidentally represents the lattice geometry for the visual axes

BARROS-PITA and MALDONADO (1970), and HORRIDGE (1977) to note the analogy with the vertebrate fovea.

The eye parameter p where

$$p = D\Delta\phi = R(\Delta\phi)^2 = D^2/R \tag{B.2}$$

is nearly constant over large portions of many compound eyes, e.g., the honey bee (KUIPER and LEUTSCHER-HAZELHOFF, 1965), the fly *Musca* (STAVENGA, 1975), and other arthropods (HORRIDGE, 1976). Thus, this is a useful parameter to characterize compound eyes.

a) Highest Resolvable Spatial Frequency

The highest angular spatial frequency that can be reconstructed by the array of ommatidia (called the ommatidia sampling frequency ν_s) is determined from the geometry of Fig. 2 to be (SNYDER, 1977)

$$\nu_s = \text{sampling frequency} \tag{B.3a}$$

$$= 1/2\,\Delta\phi, \quad \text{square lattice} \tag{B.3b}$$

$$= 1/\sqrt{3}\,\Delta\phi, \quad \text{hexagonal lattice} \tag{B.3c}$$

assuming the array is regular (FRENCH et al., 1977). Both lattice arrangements are found in compound eyes (HORRIDGE, 1976). It is convenient to view ν_s as the anatomic resolving power of the compound eye; however, ν_s is in general greater than the physiologic resolving power unless certain conditions are satisfied. These conditions include a sufficient luminance and a restricted angular velocity. Furthermore, because of aliasing (Sec. E.III), animals in theory can detect frequencies in excess of $\nu = \nu_s$ but they are percieved as lower frequencies.

2. Optical Limitations (Diffraction and Finite Rhabdom Diameter)

Here we consider the limitation to spatial resolving power due to diffraction and the finite diameter of a rhabdom.

a) Angular Sensitivity Function

Because of imperfect dioptrics, ultimately limited by the wave nature of light (pupil diffraction), it is impossible to specify the precise location of a distant point source. This spatial uncertainty is characterized, at the level of the photoreceptor, by the angular sensitivity or acceptance function A of a retinula cell (Fig. 1c) where $\Delta\varrho$ is the width of the function at 50% sensitivity. The minimum $\Delta\varrho$ is set by diffraction (BORN and WOLF, 1970)

$$\Delta\varrho_{\min} \cong \lambda/D\,, \tag{B.4}$$

where λ is the wavelength in vacuum and D the diameter of the entrance pupil, which is approximately the facet diameter. In practice, a Gaussian function is a good fit to the measured angular sensitivity function of a retinula cell (GOETZ, 1965; TUNSTALL and HORRIDGE, 1967) so that

$$A = \mathrm{e}^{-2.77(\phi/\Delta\varrho)^2}\,, \tag{B.5}$$

where A is normalized to unity (Fig. 1c) at maximum and ϕ is the inclination to the axis of the ommatidia. We next express $\Delta\varrho$ in terms of the physical parameters of the compound eye. Ignoring lens-pupil diffraction and aberrations and assuming that the distal tips of the rhabdomeres are close to the focal plane (as they are in *Drosophila*, *Musca*, and worker bee; for review, see FRANCESCHINI, 1975), $\Delta\varrho$ is simply the angular projection of the rhabdom in object space, i.e.,

$$\Delta\varrho = d_{\mathrm{Rh}}/f\,, \tag{B.6}$$

where d_{Rh} is the diameter of the rhabdom and f is the distance from the distal tips of the rhabdomeres to the posterior nodal point (for review, see STAVENGA, 1975). To account for the possibility of additional light gathering or funnelling by the crystalline cone, we replace d_{Rh}/f by the effective angular diameter of the rhabdom $\Delta\varrho_r$.

We next investigate how imperfections of the lens-pupil modify our geometric optics result that $\Delta\varrho = \Delta\varrho_r$. Because of diffraction (but possibly aberrations in the

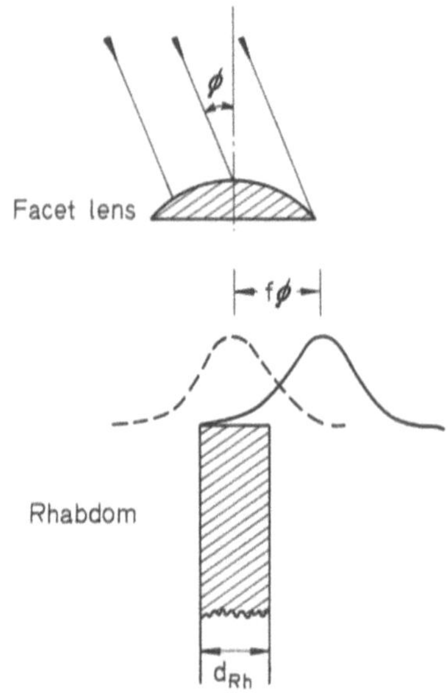

Fig. 3. Displacement of the lens pupil intensity function as a distant point source is moved an angular amount ϕ off axis; f is the distance from the distal tips of the rhabdomeres to the posterior nodal point

optical superposition eyes of nocturnal animals), a distant point source appears as a distribution of light intensity at the distal tips of the rhabdom. Moving the point source a distance ϕ from the axis of the ommatidium as shown in Fig. 3 displaces the center of the distribution by an amount $f\phi$. The angular sensitivity A of a retinula cell is the result of viewing the intensity pattern by a rhabdom of finite width, as the distant point source moves across the animal's visual field. Such a process is described mathematically by convolving (BRACEWELL, 1965) the intensity pattern with the aperture of the rhabdom. The exact analysis first given by KUIPER (1966) for the fly is rather complicated and must be computed numerically. However, approximating both the angular intensity distribution A as well as the rhabdom angular acceptance function A_r by Gaussian functions, we derive a simple result for $\Delta\varrho$ with insignificant error (Fig. 4). The process of convolving the Gaussian function representation of A_l and A_r results in the angular sensitivity A of a retinula cell, given by Eq. (B.5), of half width (SNYDER, 1977)

$$(\Delta\varrho)^2 = (\Delta\varrho_l)^2 + (\Delta\varrho_r)^2 \tag{B.7a}$$

also shown in Fig. 4, where

$$\Delta\varrho_l = \lambda/D \tag{B.7b}$$

when the lens pupil is diffraction limited and

$$\Delta\varrho_r = d_{Rh}/f \tag{B.7c}$$

when there is no additional light gathering due to a crystalline cone. Thus, the geometric optics expression $\Delta\varrho = \Delta\varrho_r$ applies, to an accuracy within 10%, when

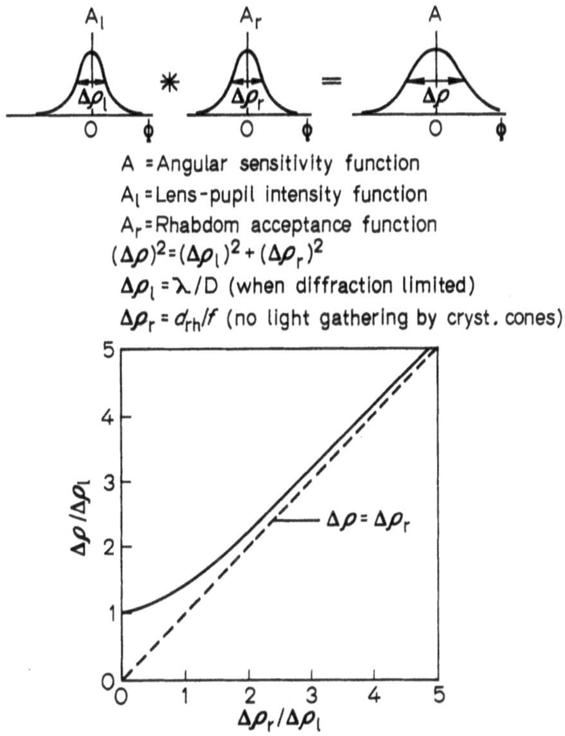

Fig. 4. The angular sensitivity function A of a photoreceptor is determined by convolving the lens pupil function A_l shown in Fig. 3 with the rhabdom acceptance function A_r for the case when both are assumed to be Gaussian functions. The asterisk signifies convolution. For convenience, we note that the angular diameter of the Airy disc $\Delta\varrho_{\text{Airy}} = 2.44\,\Delta\varrho_l$. We have normalized all functions to unity at $\phi = 0$, but in general only $A_r = 1$ at $\phi = 0$. The function A_l at $\phi = 0$ is proportional to the pupil area $\pi D^2/4$ times the fraction of photons entering the rhabdom $(\Delta\varrho_r/\Delta\varrho)^2$. The height of A at $\phi = 0$ is found by the process of convolving the two-dimensional Gaussian functions, leading to a constant times $(D\Delta\varrho_r/\Delta\varrho)^2$ (see SNYDER, 1977)

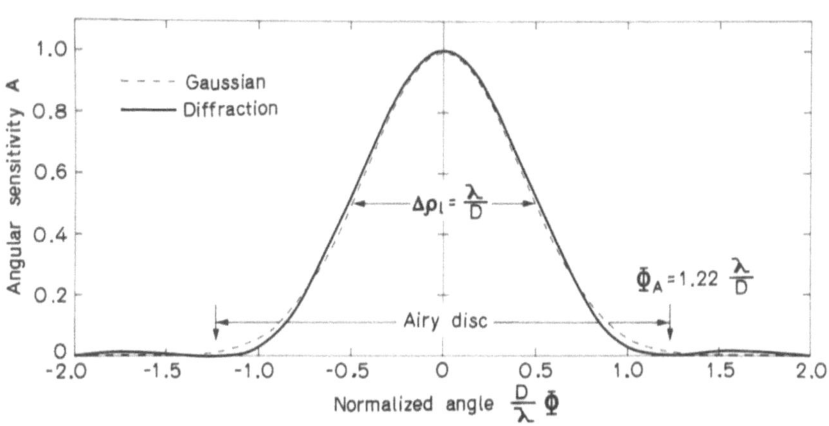

Fig. 5. Angular distribution of intensity pattern A due to diffraction by a circular aperture; ϕ is the inclination of a distant point source to the axis of the ommatidium (see Fig. 1). The angular diameter of the Airy disc is $2.44\,\lambda/D$, where λ is the wavelength in vacuum and D the pupil diameter. The dashed line is given by Eq. (B.5) while the solid curve is given by $A = \{2J_1(x)/x\}^2$; $x = \pi\nu_{co}\phi = \pi D\phi/\lambda$

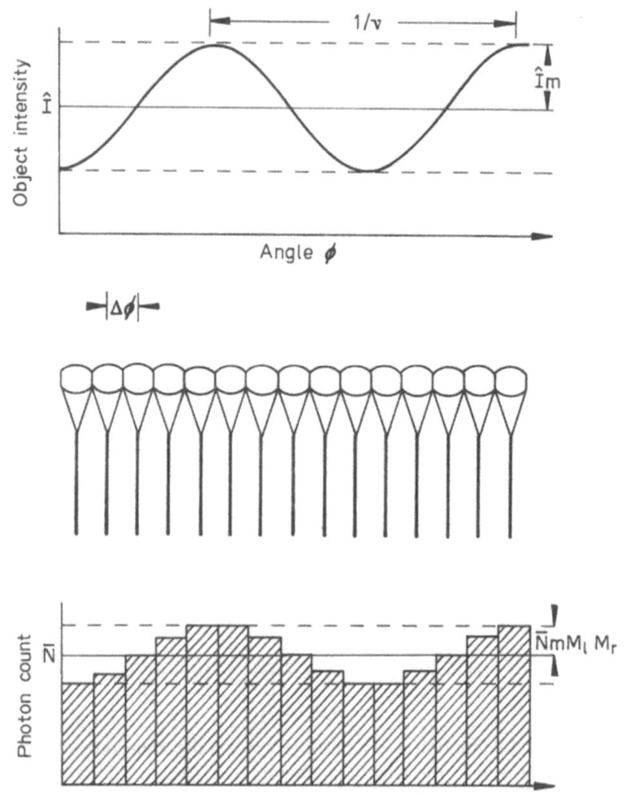

Fig. 6. The effect on object reconstruction of the lens pupil optics and the finite angular diameter of the rhabdom. The object intensity is due to a sinusoidal grating of mean intensity $\bar{I}$ and modulation m. A quantized version of the sinusoid is reconstructed by the array of photoreceptors. The amplitude of the sinusoid is $\bar{N}mM_lM_r = \bar{N}mM$, where $\bar{N}$ is the mean number of absorbed photons in one integration time, M_l and M_r are the modulation transfer functions of the lens pupil and finite rhabdom diameter, respectively

$\Delta\varrho_l < 0.45\,\Delta\varrho_r$. For diffraction-limited optics and without light-gathering effects of the crystalline cone, this requires $d_{Rh} > 2.22\,\lambda f/D$, i.e., d_{Rh} greater than the Airy disc of the diffraction pattern (Fig. 5). Equation (B.7) together with Eq. (B.5) allow for a very simple description of the angular sensitivity of a retinula cell. Equation (B.7) is not significantly altered by even the most pronounced wave guide effects (see Appendix A) and therefore has general applicability.

b) Modulation or Contrast Transfer Functions (MTF)

The most convenient way to view the effect of spatial uncertainty caused by imperfect optics and a finite rhabdom acceptance angle is via the amount of demodulation that a spatial frequency ν experiences when passing through the system (Fig. 6). The modulation transfer function (MTF) at the retinula cell level is obtained by the Fourier transform of the angular sensitivity function, normalized to unity at zero spatial frequency (GOODMAN, 1968). Fourier transforming Eq. (B.5) leads to

$$M(\nu)=M_l(\nu)M_r(\nu)=e^{-3.56(\nu\Delta\varrho)^2}, \quad \text{(B.8a)}$$

$$M_l(\nu)=e^{-3.56(\nu\Delta\varrho_l)^2}, \quad \text{(B.8b)}$$

$$M_r(\nu)=e^{-3.56(\nu\Delta\varrho_r)^2}, \quad \text{(B.8c)}$$

where $\Delta\varrho$ is given by Eq. (B.7).

c) Highest Spatial Frequency Passed by Optics

The highest spatial frequency or *cutoff frequency* ν_{co} passed by the optics of an individual ommatidium is defined by

$$M_l(\nu_{co})=0. \quad \text{(B.9a)}$$

However, because of our Gaussian representation of M, there is no finite value of ν satisfying Eq. (B.9a). An excellent approximation to the cutoff frequency is given by

$$\nu_{co}=\text{cutoff frequency}\simeq 1/\Delta\varrho_l, \quad \text{(B.9b)}$$

where $M_l(\nu_{co})\simeq 0.028$. If the lens pupil is diffraction limited, the cutoff frequency as defined by Eq. (B.9a), is given exactly (GOODMAN, 1968) by

$$\nu_{co}=D/\lambda \quad \text{(B.9c)}$$

as shown in Fig. 7.

If the array of retinula cells is to reconstruct the highest spatial frequency ν_{co} *passed by the optics, then by Fig. 2 and Eq.* (B.3), $\nu_s \geqq \nu_{co}$, *i.e., the interommatidial angle must not be greater than*

$$\Delta\phi \simeq \Delta\varrho_l/2\ ; \text{ square lattice} \quad \text{(B.10a)}$$

$$\simeq \Delta\varrho_l/\sqrt{3}\ ; \text{ hexagonal lattice}. \quad \text{(B.10b)}$$

Since $\Delta\varrho_{min}=\lambda/D$, *there is no advantage in having the interommatidial angle smaller than* $\Delta\phi=\lambda/2D$ *for resolving the highest spatial frequency passed by the optics.*

Optimum $\Delta\phi$ for Diffraction Limited Eye

Assuming diffraction is the only limit to resolving power, what is the optimum interommatidial angle $\Delta\phi$ for a spherical eye? The smaller $\Delta\phi$, the greater the anatomical resolving power or sampling frequency ν_s of Eq. (B.3). However, the smaller $\Delta\phi$, the poorer the image quality, i.e. the lower the cutoff frequency $\nu_{co}=D/\lambda$. This is because the facet diameter D is reduced. The optimum $\Delta\phi$ occurs when the sampling frequency equals the cutoff frequency, i.e. $\nu_s=\nu_{co}$ so that $\Delta\phi=\lambda/\sqrt{3}D$ or $p=\lambda/\sqrt{3}$ assuming hexagonally packed ommatidia, where p is the eye parameter of Eq. (B.2). For a spherical eye of radius R, Eq. (B.1) leads to $\Delta\phi=(\lambda/R\sqrt{3})^{1/2}$. However, as we show in section (B.III.6), only animals that are active in the brightest sunlight and are relatively stationary when examining their prey have a region of their eye fitting this description. To explain $\Delta\phi$ in other animal requires appreciating the limitations to resolving power discussed below.

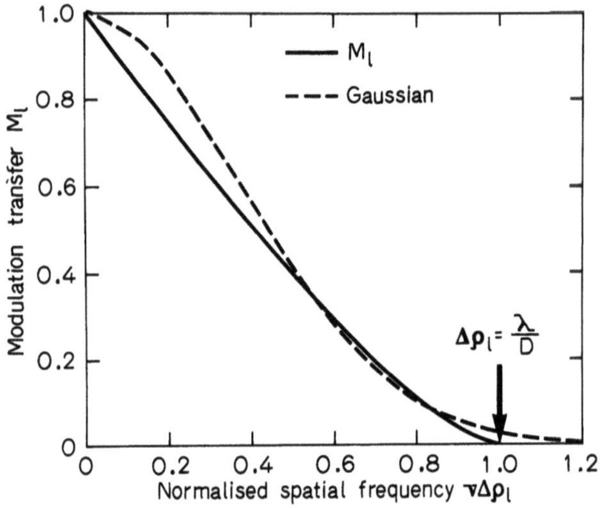

Fig. 7. The modulation transfer function for a circular pupil M_l compared with the Gaussian approximation Eq. (B.8b). The solid curve is given by

$$M_l = \left(\frac{2}{\pi}\right)\{\cos^{-1}x - x(1-x^2)^{1/2}\} \quad \text{for} \quad |x|<| \quad \text{where} \quad x = \nu\Delta\varrho_l = \nu/\nu_{co}$$

We have been discussing grating resolution. Wave theory does not set a limit on $\Delta\phi$ for two point resolution, since the image of two distant point sources appears different from one alone. The Rayleigh (1879) criterion arbitrarily sets the minimum at $\Delta\phi = 0.61\,\lambda/D$ for diffraction limited optics which is comparable to $\Delta\phi = 0.58\,\lambda/D$ we found above for gratings.

3. Motion (Limitation to Spatial Resolving Power of Finite Integration Time)

Because neither transduction nor neutral transmission is instantaneous, it is impossible to specify the precise time of an event, just as it is impossible to specify the precise location of a distant point source. Thus, there is a temporal uncertainty of duration Δt in exact analogy with the spatial uncertainty of angular extent $\Delta\varrho$ discussed in Section 2a, where Δt is the integration or exposure time of the eye. If a distant point object moves at an angular velocity v, then it is displaced by an angular distance $v\Delta t$ across the retina in the integration time Δt. Consequently, the effect of movement on spatial resolution is analogous to the process of viewing a stationary scene through a receptor of finite angular diameter. Thus, the *additional* spatial uncertainty introduced by movement is found by convolving the acceptance function of the retinula cell A with the temporal response function of the retinula cell. If we assume that the temporal response function is Gaussian of half width $v\Delta t$, then the spatial uncertainty function in the presence of movement is characterized by half width $\Delta\varrho$ where (SNYDER, 1977)

$$(\Delta\varrho)^2 = (\Delta\varrho_l)^2 + (\Delta\varrho_r)^2 + (v\Delta t)^2 \tag{B.11}$$

in analogy with Eq. (B.7a). This intuitive approach can be verified by the formalism of SRINIVASAN and BERNARD (1975) who also provide a more complete discussion of the degrading effects of motion on spatial resolving power.

Thus, when there is motion, we must replace $\Delta\varrho$ *in the MTF function Eq.* (B.8) *by* $\Delta\varrho$ *of Eq.* (B.11) *so that* $M = M_1 M_r M_m$ *where* $M_m = \exp\{-3.56(\nu v \Delta t)^2\}$. *Consequently, the highest frequency passed by the optics is then* $\nu_{co} = 1/\Delta\varrho$ *with* $\Delta\varrho$ *from Eq.* (B.11), *and there is no advantage in the interommatidial angle* $\Delta\phi$ *being smaller than*

$$\Delta\phi \cong \{(\Delta\varrho_1)^2 + (\Delta\varrho_r)^2 + (v\Delta t)^2\}^{1/2}/2 \tag{B.12}$$

for resolving the highest sinusoidal frequency passed by the optics. The greater the angular velocity v, *the greater the spatial uncertainty and the larger we can make* $\Delta\phi$. Thus, for an animal that flies predominantly in linear motion, along its longitudinal axis, we might anticipate larger ommatidial angles along the horizontal axis than the vertical axis, as is the case in several species including bees, muscoid flies, dragonflies, and locusts (Review see MAZOKHIN-PORSHNYAKOV, 1969).

4. Light Intensity (Limitation of Photon Noise to Contrast Sensitivity)

If it were not for the presence of noise, ultimately due to the quantum nature of light, there would be no limit to the contrast sensitivity of the compound eye, and the MTF would be sufficient to determine the highest spatial frequency reconstructed by the array of retinula cells. However, because of the random character of photon absorption, a uniform light source appears irregular even to an array of ideal detectors, i.e., those without intrinsic noise (Fig. 8). The standard deviation σ_{noise} in photon counts appearing across an infinite array of photoreceptors is

$$\sigma_{\text{noise}} = \sqrt{\bar{N}}, \tag{B.13}$$

where $\bar{N}$ is the mean number of photons captured by the individual photoreceptors per integration time or shutter time of the eye. This result follows from the fact that photon arrival obeys a Poisson distribution (for review, see ROSE, 1973). Because there is an uncertainty of $\sqrt{\bar{N}}$ in the meaning of a photoreceptor count, two light sources can be reliably distinguished only if they differ in intensity sufficiently to exceed $\sqrt{\bar{N}}$ at the photoreceptors. Although the standard deviation in photon counts increases with increasing $\bar{N}$, its effect on a measurement is relatively less. Thus, *the greater the intensity of light, the greater the contrast sensitivity of ideal photoreceptors.* Since FERMI and REICHARDT (1963) have provided evidence that the photoreceptors of the fly *Musca* are close to ideal, the theory presented here may be more general than simply an upper limit to performance.

From the measurements of LILLYWHITE and LAUGHLIN (1978), we now know that the intrinsic noise of the photoreceptor also obeys (B.13).

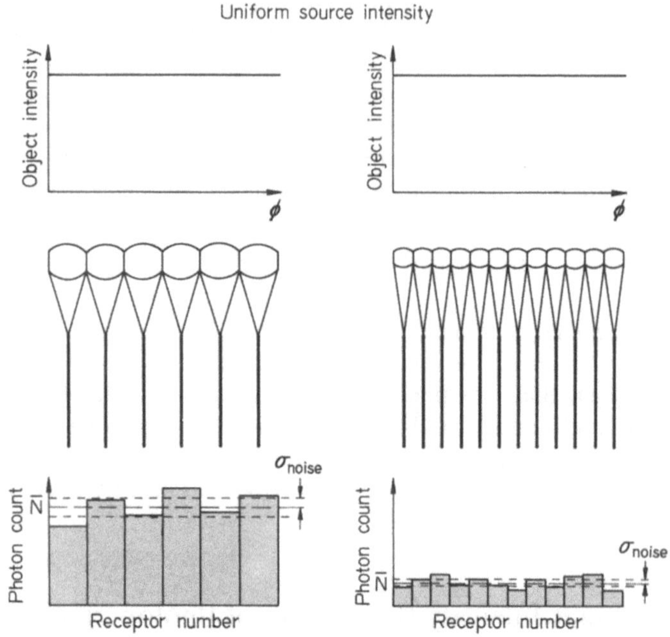

Fig. 8. The fluctuation in photon counts across the photoreceptors (in one integration time) due only to the random arrival of photons (photon noise). The object is a two-dimensional uniform source, infinite in extent. Ommatidia with larger facets accept more photons than those with smaller facets (*assuming nearly equal f/D ratios*) so that their signal-to-noise ratio is greater

a) Signal-to-Noise Ratio (SNR) of a Sinusoidal Grating at the Retinula Cell Level

We now combine the concepts of modulation transfer function with photon fluctuation to determine the SNR of a sinusoidal grating reconstructed by the retinula cells (Fig. 6). The amplitude σ_{sig} of the fluctuation in photon counts about the mean $\bar{N}$, due to a sinusoidal grating, is shown in Fig. 6 to be

$$\sigma_{\text{sig}} = \bar{N} m M(\nu) \tag{B.14}$$

if we ignore the fluctuation (photon noise) due to the random arrival of photons. The most likely amplitude in photon noise σ_{noise} is $\sqrt{\bar{N}}$, so that the SNR is

$$\text{SNR} = \sqrt{\bar{N}}\, m M(\nu) \tag{B.15}$$

at the retinula cell level.

1) Mean Number $\bar{N}$ of Photons Captured per Integration Time of the Eye Due to a Uniform Source, Infinite in Extent

If we let $\bar{I}$ be the mean number of photons entering the eye per steradian of object field per cm² per second, ε the fraction of this number that is absorbed in the rhabdom, Δt the sampling time of the eye, then $N = \bar{I}\varepsilon\Delta t$ times the pupil area

$D^2\pi/4$, multiplied by the solid angle that the rhabdom subtents in object space. We find this solid angle by integrating the assumed Gaussian field of view A_r of the rhabdom as shown in Fig. 4, so that

$$\bar{N} = (\pi/4)\, D^2 \bar{I} \varepsilon \Delta t \left\{ 2\pi \int_0^{\pi/2} A_r \sin\phi \, d\theta d\phi \right\} \tag{B.16a}$$

$$\cong (\pi/4)\, D^2 \bar{I} \varepsilon \Delta t \left\{ 2\pi \int_0^{\infty} \phi A_r d\phi \right\} \tag{B.16b}$$

$$= \hat{I} (D \Delta \varrho_r)^2 , \tag{B.16c}$$

where the intensity parameter $\hat{I}$ is defined in Table 1. It is convenient to express the intensity parameter $\hat{I}$ in terms of the field luminance L (in units of candelas $[m]^{-2}$) (WYSZECKI and STILES, 1967) so that

$$\hat{I} = 1.31 \times 10^3 (\eta \Delta t)\, L , \tag{B.17}$$

where Δt is in seconds. As a representative example, we take $\varepsilon = 0.5$ and $\Delta t = 20$ ms, so that $\hat{I} = 13.1\ L$ or

$$\log \hat{I} = 1.1 + \log L . \tag{B.18}$$

The range in log L of interest is: 3–5 for daylight, 0–3 for inferior lighting, and -4 to -1 for night (LE GRAND, 1968).

2) SNR Expressed in Terms of $\hat{I}$

It is convenient for later use to express Eq. (B.15) in terms of $\hat{I}$. From Eqs. (B.8), (B.15), and (B.16b), we have

$$\mathrm{SNR} = D \Delta \varrho_r (m^2 \hat{I})^{1/2} \, e^{-3.56 (\nu \Delta \varrho)^2} , \tag{B.19}$$

where $\Delta \varrho$ is given by Eq. (B.7).

5. Improvement of Signal to Noise Ratio by Neural Processing

By appropriate neural averaging, the signal to noise ratio (SNR) can in theory be improved at higher levels over that at the level of the photoreceptors. Allowing for parallel processing, this is achieved without rendering the retina more functionally course, i.e. without reducing the anatomical resolving power of some higher level from that of the photoreceptor matrix. There are numerous possibilities depending on the plasticity of the brain in adapting to different stimuli. For example, for a repetitive image, like a grating, then the greater the number of photoreceptors participating in the resolution task, the better. This leads to a SNR proportional to $\sqrt{N_S}$ rather than $\sqrt{N}$, assuming the retina is limited by receptor Poisson noise, where N_S is the summed photon capture of the retina.

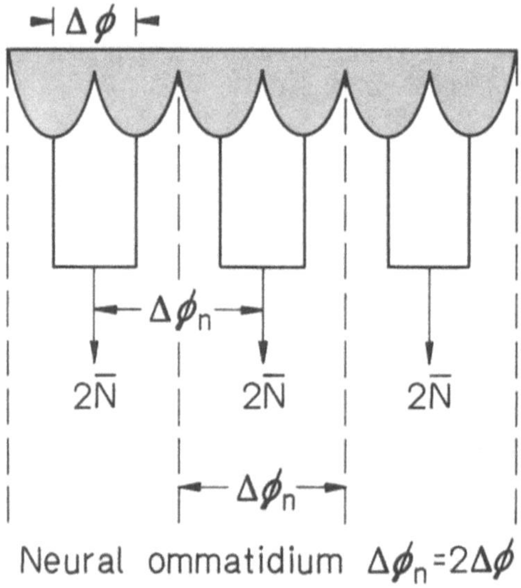

Fig. 9. Example of neural ommatidia

Neural Ommatidia

The example of processing discussed above requires rather sophisticated neural machinery. A less demanding scheme results from a simple combination of ommatidia at higher neural levels and running these summed neurons in parallel with the original receptor neurons. These new functional units are called neural ommatidia and allow the retina to be simultaneously course and fine. For example, suppose that ommatidia, with different fields of view, converge to form a neural ommatidium (e.g. Fig. 9) that has an effective facet diameter $D_n = R\Delta\phi_n$, where $\Delta\phi_n$ is the angular separation of adjacent neural ommatidia. The highest spatial frequency that can be constructed by these neural ommatidia is $v_s = 1/\sqrt{3}\Delta\phi_n$. The functional difference between this scheme and the example above is most noticeable at spatial frequencies $v \cong v_s$ for which the present strategy offers no improvement whereas the former gives an improvement proportional to the number of ommatidia involved in the resolution task.

Neural Isolation and the Receptive Field

As we have stressed, neural isolation is not a requisite for preserving the photoreceptor sampling frequency $v_s = 1/\sqrt{3}\Delta\phi$ at higher neural levels. In the presence of lateral interconnectivity, possibly manifested by a center surround receptive field at some higher order neuron as with the ganglion cells of the vertebrate retina, the sampling frequency may still be reconstructed across the array of higher order neurons. For example, consider a hypothetic two-dimensional version of the $\sin x/x$ type of receptive field: $J_1(\psi\phi)/(\psi\phi)$, where $\psi = 2\pi v_s$ and ϕ is the angular displacement in object space from the center of the field. The angular diameter of this receptive field (measured between zero crossings) is $2.1\Delta\phi$, i.e., about two interommatidial angles; nevertheless, by FOURIER transforming this function, we find that the transfer function of the eye (up to the array of neurons) with this receptive field is *flat* from $\nu=0$ to ν_s and zero for greater values of ν. Far from limiting resolving power, neural interconnectivity can *in theory* compensate for optical blur. Whether or not this is advantageous depends on a number of factors discussed in Section B.V.2.

II. Minimum Number of Photons Necessary for Threshold Resolution of a Sinusoidal Grating

To emphasize the need for photon capture in order to unmask the signal from the random fluctuation in photon arrival (photon noise), we determine the minimum, mean number of photons $\bar{N}$ necessary for threshold resolution of a sinusoidal grating by a *hypothetic eye*. The eye is hypothetic not only because of ideal photoreceptors but also because we assume that all ommatidia have equal light capture [i.e., $D\Delta\varrho_r$ is the same for all ommatidia, which is nearly so in *Musca* (STAVENGA, 1975)] and that dark adaptation mechanisms, such as the increase in the acceptance angle $\Delta\varrho$, are inactive. For our example, we assume a square array of lens facets.

To determine the *threshold* frequency ν of a *given* eye for a specified mean number $\bar{N}$ of absorbed photons per integration time, we reexpress Eq. (B.19) in the form

$$\nu\Delta\varrho = 0.57\,\{\log \tilde{m}^2 \bar{N}\}^{1/2} \tag{B.20}$$

which is plotted in Fig. 10. The value of $\tilde{m}=m/\mathrm{SNR}$ depends on the grating modulation m and the *threshold* SNR for the animal, which depends on the number of ommatidia that participate in viewing the grating plus the reliability or decision criterion adopted by the animal (ROSE, 1973). It is convenient to take the threshold $\mathrm{SNR} \cong 1$ as we have done in our figures.

The $\nu\Delta\varrho$ scale has a maximum of unity, since the highest frequency ν_{co} passed by the optics is shown in Section B.III to be $\nu_{co} \cong 1/\Delta\varrho$. However, the highest frequency that can be reconstructed by the array of retinula cells is limited by the ommatidial sampling frequency $\nu_s = 1/2\Delta\phi$ so that $\nu\Delta\varrho$ can never be greater than $\Delta\varrho/2\Delta\phi$ for any given eye. We have *arbitrarily* selected $\Delta\varrho/2\Delta\phi = 0.66$ for the example of Fig. 10, but one should realize that the position of $\Delta\varrho/2\Delta\phi$ and hence the horizontal portion of the curve can be located anywhere from zero to any position along the dashed curve. We are again reminded that ν is the *highest* spatial frequency that can be resolved for a given number of absorbed photons $\bar{N}$ per integration time of the eye.

Several important facts emerge from a perusal of Fig. 10. If an animal is to reconstruct the highest frequency passed by its optics ($\nu_{co}=\nu_s$ or $\Delta\varrho/2\Delta\phi=1$), then from Fig. 10, $\log \tilde{m}^2\bar{N}=3.09$, i.e., $\bar{N}=1230$ photons must be captured per integration time if the grating has unity modulation and assuming the threshold $\mathrm{SNR}=1$. Taking *Musca* as an example, in which case $\Delta\varrho \cong \Delta\phi$ for retinula cells 1–6 (SNYDER, 1976), it is then necessary to have about $\bar{N}=6$ photons to reconstruct

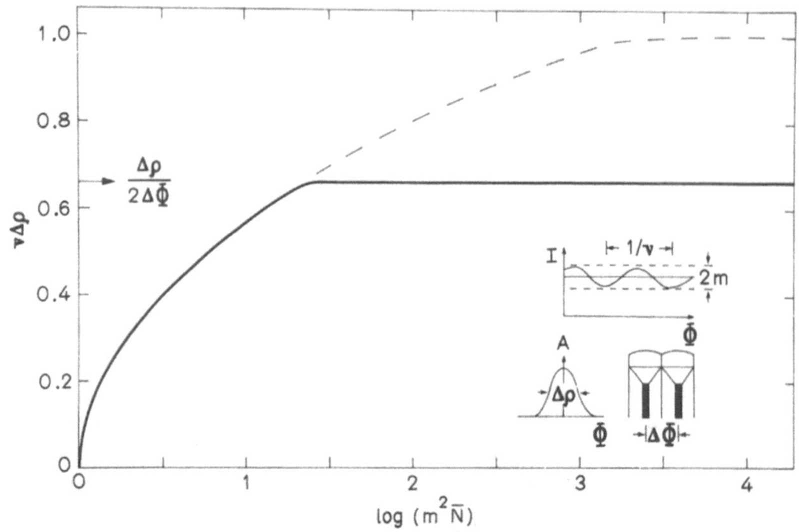

Fig. 10. Highest resolvable spatial frequency for an eye of given $\Delta\phi$ and $\Delta\varrho$. $\bar{N}$ is the mean number of photons absorbed per integration time per receptor cell

the ommatidia sampling frequency ($\Delta\varrho/2\Delta\phi=0.5$). Because of the neural summation in *Musca* (KIRSCHFELD and FRANCESCHINI, 1968), rhabdomeres 1–6 each require only one photon per integration time. If $\bar{N}$ is insufficient to reach the break point of the curve where $\nu\Delta\varrho=\Delta\varrho/2\Delta\phi$, then the acuity of the eye is less than optimum. In other words, as the light intensity falls below a certain minimum, the acuity also falls as shown in Fig. 10. The fall in acuity is due to photon noise masking the image of the grating at the retinula cell level. We are reminded from Eq. B.11 that $\Delta\varrho$ depends on the angular velocity v of the grating, so that the greater v, the greater $\Delta\varrho$ and hence the lower the spatial frequency ν that can be resolved. This is because $\nu\Delta\varrho$ must be less than or equal to unity.

To compare Fig. 10 with the measured performance of an animal, we must know the conversion from N to intensity (Sec. B.IV) in addition to the threshold SNR of the animal. Thus, there is a margin of error in deciding how well an eye agrees with the hypothetic eye results of Fig. 10. This difficulty can be in part avoided if we determine the slope of acuity fall-off on a log ν vs. log (intensity) graph, since the slope is independent of the value of SNR at threshold as well as the conversions from N to intensity. Taking the log of Eq. (B.21) and noting that $\bar{N}$ is proportional to intensity, we have $\log\nu=1.2\log$ (intensity). Thus, a slope of about 45° *throughout* the region of acuity fall-off indicates that the animal is limited only by photon noise *and obeys the hypothetic eye criteria discussed at the beginning of this section.*

III. Design of Compound Eyes

Having discussed the fundamental limitations to spatial acuity, we now consider the logic behind the design of compound eyes. We assume that the actual resolving power of a diurnal animal is specified by its interommatidial angle $\Delta\phi$. This means, among other limitations discussed above, that the rhabdoms receive sufficient light for the eye to have adequate contrast sensitivity at the sampling frequency $\nu=\nu_s$, where ν_s of Eq. (B.3) is the highest spatial frequency that can be faithfully reconstructed by the ommatidia. Thus, *animals with the smallest interommatidial angles have the highest resolving power.* Now it turns out, as we show below, that the principal determinate of high resolving power is the size of the animal's head. High resolving power requires a large head in order to accommodate a long focal length and large facet diameters. We remind the reader that, unless the head is spheric, the radius of curvature R of Fig. 1 is in general independent of head size, e.g., flat portions of the head have $R=\infty$. Because focal length f is defined in Eq. (B.7c) via the angular projection of the rhabdom into object space $\Delta\varrho_r$, we first determine the $\Delta\varrho_r$ necessary for maximum contrast sensitivity at the sampling frequency $\nu=\nu_s$.

1. Angular Rhabdom Diameter for Maximum Contrast Sensitivity at the Sampling Frequency $\nu=\nu_s$

By Eq. (B.16c), the greater the angular rhabdom diameter $\Delta\varrho_r$, the greater the photon capture and hence the greater the potential for higher contrast sensitivity. However, by Eq. (B.8c), the greater $\Delta\varrho_r$, the smaller the modulation of the image at the level of the photoreceptors and hence the smaller the potential contrast sensitivity. Thus, there is an optimum $\Delta\varrho_r$ for maximum contrast sensitivity. Assuming the eye is limited by Poisson noise at the photoreceptor level, the optimum $\Delta\varrho_r$ is determined by maximizing the SNR [Eq. (B.19)] leading to $\Delta\varrho_r=0.375/\nu$. But the highest spatial frequency that the animal can faithfully reconstruct is the sampling frequency $\nu=\nu_s$, where ν_s is defined by Eq. (B.3). Consequently, the angular rhabdom diameter for maximum contrast sensitivity at the animal's sampling frequency is

$$\Delta\varrho_r=0.87\,\Delta\phi=d_r/f \tag{B.21}$$

for ommatidia packed in a hexagonal array and $\Delta\varrho_r=0.75\,\Delta\phi$ for a square array, where d_r is the rhabdom diameter and f the focal length of the optics. In other words, the acceptance angle of the rhabdom $\Delta\varrho_r$ is slightly less than the interommatidial angle $\Delta\phi$. This assumes the eye is adapted to bright light. As a consequence of Eq. (B.21) we see from Eq. (B.16c) that the number of photons absorbed per rhabdom is proportional to p^2, where p is the eye parameter defined by Eq. (B.2).

2. Influence of Focal Length on Resolving Power

Here we determine the focal length of the optics f, which is approximately the distance from the distal tips of the rhabdomeres to the corneal facets. This gives an indication of the axial length of an ommatidium. From Eq. (B.21) we observe that f is proportional to the animal's resolving power $(\Delta\phi)^{-1}$ and thus proportional to ν_s. Therefore, the greater the animal's resolving power, the greater the axial length of ommatidium and hence the larger the animal's head. Since it is obviously advantageous for the eye to be as small as possible, particularly for flying insects, we now consider the limitations of having a smaller focal length f while retaining the animal's resolving power $(\Delta\phi)^{-1}$. From Eq. (B.21), f can be reduced arbitrarily provided the rhabdom diameter d_r is also reduced equivalently. However, because of wave effects which are discussed in Section VII, when the diameter of a rhabdom of fixed length is reduced it becomes a poorer absorber of light, and below about 0.5 μm it fails to act as an efficient absorber of photons. For open rhabdoms, e.g., *Musca*, there is the additional possibility of optical cross-talk, which reduces contrast sensitivity to a negligible amount for a rhabdomere center to center spacing less than 1 μm. Thus, the *minimum* focal length possible is seen from Eq. (B.21) to occur when $f \simeq (\Delta\phi)^{-1}$ μm, with $\Delta\phi$ in radians. For example, $f \simeq 57$ μm when $\Delta\phi = 1°$ and about 115 μm when $\Delta\phi = 0.5°$. In practice, as expected, the rhabdom diameter is greater than the minimum value so that f is larger than $(\Delta\phi)^{-1}$ and can be determined from Eq. (B.21). Another advantage of a longer focal length is a greater depth of focus because focal depth depends on f/D. In conclusion, for a fixed sending power the minimum focal length, and hence head size is limited by restrictions on the rhabdom diameter.

By equating $\Delta\phi$ of Eqs. (B.21) and (B.1), we obtain the relation between facet diameter, local eye radius R, rhabdom diameter d_r and focal length f (Fig. 1) for a given interommatidial angle $\Delta\phi$

$$R d_r = 0.87\, D f\,. \tag{B.22}$$

This can also be expressed in terms of the eye parameter p of Eq. (B.2) and the F number f/D as

$$p = 1.15 \frac{d_r}{F}\,. \tag{B.23}$$

These relations hold approximately in many compound eyes adapted to diurnal conditions.

3. Increasing Resolving Power by Isomorphic Scaling

In Fig. 11 we show how the resolving power of the compound eye is doubled by isomorphic scaling. In particular, the facet diameter D must be doubled to double the highest spatial frequency [ν_{co} of Eq. (B.9c)] passed by the optics while the interommatidial angle $\Delta\phi$ must be halved to sample this higher frequency. If large and small eyes are to have equal SNRs at the photoreceptor level, then the angular rhabdom diameter $\Delta\varrho_r$ must be halved. This is because $M_r(\nu_s)$ of Eq. (B.8c) depends on the ratio $\Delta\varrho_r/\Delta\phi$. If we assume the small eye is designed with the smallest focal length f possible, i.e., smallest rhabdom diameter possible, then to halve $\Delta\varrho_r$ we must double f. Because the eye parameter $D\Delta\phi$ is equal in both eyes, $\Delta\phi^{-1}$ is proportional to $\sqrt{R}$. Accordingly, the focal eye curvature must be four times larger to double the resolving power. By doubling resolving power, there are four times as many ommatidia per field of view. Furthermore, because $D\Delta\phi$ is the same in both eyes, they both have the same depth of focus and the rhabdoms of each have equal photon capture. These scaling laws should apply when comparing

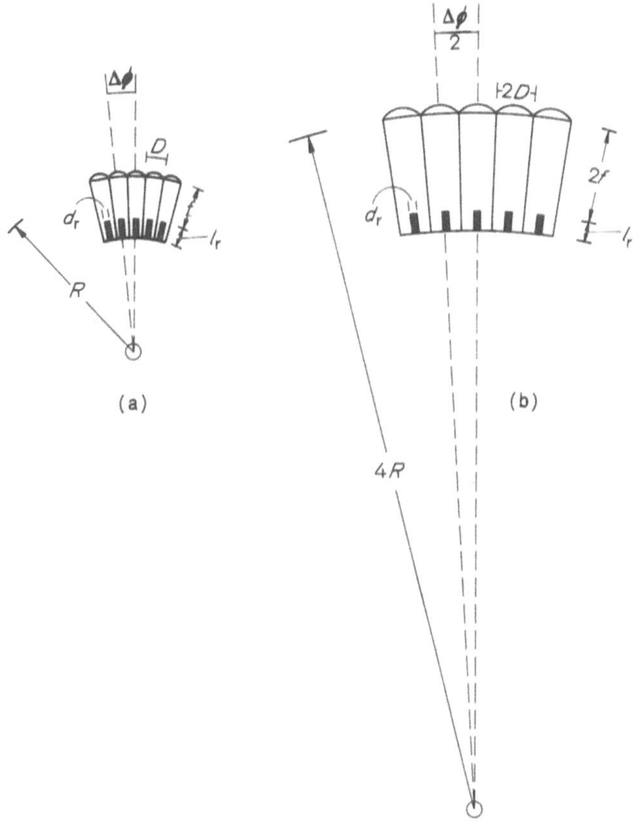

Fig. 11a and b. Doubling resolving power of an eye by isomorphic scaling. R is the radius of curvature and only equals the radius of the animal's head for a spherical compound eye. The eye in (b) has twice the resolving power of (a)

animals with similar life styles and employing identical optical and neural strategies.

Larger eyes have a potential for enhanced contrast sensitivity because of their greater photon capture. This assumes the appropriate neural processing (section B.5). It could also be that larger eyes can afford more space to specialized regions, e.g. a larger fovea.

4. Degree to Which a Compound Eye Takes Advantage of its Potential (Diffraction-Limited) Resolving Power

Here we distinguish between the actual resolving power of an animal as specified by the interommatidial angle $\Delta\phi$ (or equivalently by the sampling frequency ν_s) on one hand and the degree to which an animal realizes its potential resolving power $\nu = \nu_{co}$ as set by the pupil diameter on the other hand. In other words, is the interommatidial angle sufficiently small to sample the highest spatial frequency $\nu_{co} = D/\lambda$ passed by the optics, i.e., is $\nu_s = \nu_{co}$ [see Eqs. (B.3) and (B.9c)]?

When the eye does take advantage of all spatial frequencies passed by the optics, the interommatidial angle is

$$\Delta\phi = 1/\nu_{co}\sqrt{3} = \lambda/D\sqrt{3} \tag{B.24}$$

for hexagonal packing, which is the smallest value needed to reconstruct all frequencies $\nu \leqq \nu_{co}$. It is convenient to express the degree to which eyes utilize their full diffraction-limited potential by the eye parameter p of Eq. (B.2). When diffraction limited

$$p = D\Delta\phi = \lambda/\sqrt{3}\,. \tag{B.25}$$

To take an example, $p = 0.29$ µm in diffraction-limited eyes at $\lambda = 500$ mm. The legitimate application of Eq. (B.25) presumes knowledge of the wavelength at which the animal desires its highest resolving power.

Eyes that satisfy $\nu_s = \nu_{co}$ or equivalently Eq. (B.25) have two interesting properties. Firstly, from Eqs. (B.16c) and (B.21), the number of photons captured by a rhabdom N is proportional to p^2 and by Eq. (B.25) is independent of any physical eye parameter such as pupil diameter or head size. Furthermore, from Eqs. (B.21) and (B.25), the F number f/D of an eye obeys the simple relation $F \simeq 2d_r/\lambda$.

Although no known compound eye has an eye parameter p as small as 0.29 µm, many foveas of bright-light locusts, dragonflies, and wasps closely approach it, i.e., animals that are active in brilliant sunlight and are nearly stationary while examining their prey. For example, in the central fovea of the Australian sand wasp *Bembix*, $p \simeq 0.31$ µm (HORRIDGE, 1977) which satisfies Eq. (B.25) exactly at $\lambda \simeq 540$ nm, while $p \simeq 0.42$ µm in the fovea of hoverfly (COLLETT and LAND, 1975).

These eyes are rather exceptional. The minimum p values for many diurnal animals is greater than 0.5 µm, e.g., mantids (HORRIDGE, 1977). In other words, the interommatidial angle $\Delta\phi$ is too large to sample all available information passed by the optics. Why is it that so many eyes undersample, i.e., have $\nu_s < \nu_{co}$? SNYDER (1977) and SNYDER et al. (1977) have provided an explanation. Since the modulation transfer function of the optics $M_1(\nu)$ of Fig. 7 is zero at $\nu = \nu_{co}$, the contrast sensitivity of the animal is zero at $\nu = \nu_{co}$. Furthermore, unless the animal is in the brightest light and stationary relative to its prey, it has nearly negligible contrast sensitivity for high frequencies near $\nu \simeq \nu_{co}$. This is because, from Fig. 7, $M_1 \ll 1$. Thus, unless the animal can receive sufficient light (within the integration time of the eye) to have adequate contrast sensitivity at high spatial frequencies, it need not sample at the diffraction limit. This means the ommatidia can be shorter (f shorter) and the eye less flat in the region of high acuity or the facet diameter can be larger allowing for greater light capture per ommatidia. The greater portion of most eyes have 0.5 µm $< p <$ 1 µm except in the small locally flat region of certain bright-light animals where p approaches 0.3 µm. Eyes that must be spheric, e.g., for image formation reasons as in the skipper optical superposition eyes, must therefore be comparatively large to accommodate low p values.

We extend our discussion of eyes with p greater than the diffraction limit in Section B.III.6 after noting an interesting consequence of undersampling.

5. Aliasing (Effects of Undersampling)

Eyes with $p > \lambda/\sqrt{3}$ undersample, i.e., the highest spatial frequency they can faithfully reconstruct ν_s is less than ν_{co}. However, spatial frequencies in the range $\nu_s \leqq \nu \leqq \nu_{co}$ appear as lower spatial frequencies to the animal (BRACEWELL, 1965, p. 197):

$$\nu_{app} = \nu_s - (\nu - \nu_s) = 2\nu_s - \nu, \tag{B.26}$$

where ν_{app} is the apparent spatial frequency. In other words, the animal can *detect* frequencies higher than ν_s up to ν_{co} but these frequencies appear lower than $\nu = \nu_s$ to the animal. GOETZ (1965) has shown that *Drosophila* has aliasing effects by measuring the turning response above $\nu = \nu_s$. One would expect that under normal environmental conditions for an animal that aliasing is a small effect; otherwise, the animal could have made use of its diffraction-limited optics.

Lastly, we note that this deleterious effect of undersampling at the retinular cell level can be removed at a higher level by a neural filter mechanism producing a flat transfer function up to $\nu = \nu_s$ where it then cuts off. Such a neuron would have a sin x/x type of receptive field of the form $2J_1(\pi\nu_s\psi)/\pi\nu_s\psi$ where ψ is the angular distance from the center of the receptive field.

6. Determination of Eye Parameter p

We noted in Section B.IV.4 that only those animals that are active in the brightest sunlight and remain relatively stationary while examining their prey have an eye parameter p near that given by Eq. (B.25), i.e., have interommatidial angles $\Delta\phi$ small enough to sample the highest spatial frequency passed by the pupil. These animals receive sufficient light (within the integration time of the eye) to have adequate contrast sensitivity near the sampling frequency $\nu_s = \nu_{co}$ where $M_1(\nu_{co}) = 0$. All other animals have increased $\Delta\phi$ and hence p to some minimum value necessary for adequate contrast sensitivity at their sampling frequency. In this Section, we follow the analysis of SNYDER (1977) to determine this minimum p value assuming that the visual system is limited only by Poisson noise at the photoreceptor level.

Here we consider the SNR across the array of photoreceptors in which case increasing p increases the SNR. If we assume no further neural processing, then this SNR is proportional to the perceived contrast sensitivity of the animal. Other possibilities, discussed in section B.1.5 render SNR independent of individual receptor noise but this is not considered in the following analysis. We instead assume that the noise $\sqrt{N}$ of the individual photoreceptor is limiting.

Contrast Sensitivity. The contrast sensitivity (CS) of an animal is the reciprocal of the minimum grating modulation m necessary for threshold detection of a grating at frequency ν. If it were not for noise in the visual pathways, m^{-1} would be infinite. However, because of photoreceptor noise, the CS at the photoreceptor level is found from Eq. (B.15) to be

$$\mathrm{CS}(\nu) = m^{-1}(\nu) = \sqrt{\bar{N}}M(\nu), \tag{B.27}$$

where $\bar{N}$ from Eq. (B.16) is the number of photons captured per rhabdom and from Eq. (B.8) M is the modulation transfer function of the optics and rhabdom, assuming threshold detection occurs when the ratio SNR is unity. The greater the spatial frequency ν, the lower the animal's contrast sensitivity reaching a minimum value of zero at $\nu = \nu_{co}$. It is observed from Eq. (B.15) that the contrast sensitivity of an array of photoreceptors is closely related to the SNR of fluctuations across the array.

Our purpose is to determine the eye parameter p of Eq. (B.2) and the angular rhabdom diameter $\Delta\varrho_r$ of Eq. (B.7c) that maximize the contrast sensitivity at the sampling frequency ν_s of the ommatidia as defined by Eq. (B.3). We have already shown in Section II.C.1 above that $\Delta\varrho_r = 0.87\Delta\phi$ maximizes the contrast sensitivity. Now we determine the value of p that will provide adequate contrast sensitivity at the highest spatial frequency sampled by the ommatidia $\nu = \nu_s$ assuming the animal is in the brightest light of its usual environment.

From Eqs. (B.16c), (B.21), (B.8a), (B.3b), and (B.11), we can rewrite the expression for contrast sensitivity Eq. (B.27) at the sampling frequency

$$CS(\nu_s) = 0.45p\sqrt{\bar{I}}\, e^{-0.89\left[\left(\frac{\lambda}{p}\right)^2 + \left(\frac{v\Delta t}{\Delta\phi}\right)^2\right]} \tag{B.28}$$

for square packing. For hexagonal packing, we replace $\Delta\phi$ by $\sqrt{3}\,\Delta\phi/2$ so that p in Eq. (B.28) equals $D\sqrt{3}\,\Delta\phi/2$.

We begin by assuming the animal and object are stationary, relative to one another so that $v = 0$ in Eq. (B.28). From Eq. (B.28), it is clear that the greater p, the greater the contrast sensitivity $CS(\nu_s)$ at the sampling frequency. Of course, the greater p, the more inefficient the animal is in sampling its diffraction-limited optics. Therefore, we want the minimum value of p that provides a tolerable contrast sensitivity for the animal in its brightest environment. Lets suppose that the minimum value is $CS = 2$ so that the animal can detect gratings at $\nu = \nu_s$ of 50% contrast or greater. Since in nature, contrasts above 50% are rare, $CS = 2$ is a reasonable minimum value for tolerable contrast sensitivity.

In Fig. 12 we show the optimum eye parameter p vs. log luminance (upper scale) for $\lambda = 500$ nm and a minimum contrast sensitivity of $CS = 2$. The results are found from Eq. (B.28) with $v = 0$. The luminance scale is taken from the example of Section II.D with $(\Delta\varrho/\Delta\phi)^2 = (\lambda/p)^2 + 0.56$. It is clear from this figure that the appropriate eye parameter p depends on the luminance, assumed to be the largest value in the animal's typical environment. In particular, from Fig. 12 for hexagonal packing, the diffraction-limited eye with $p = 0.29$ μm requires the brightest natural light for the barely adequate contrast sensitivity of $CS = 2$. In Appendix E, we show that $\log m^2\hat{I} \geqq 5$ for a diffraction-limited eye to be the optimum design independent of the packing geometry.

For those animals that are *not* active in the brightest light, there is no loss in having p greater than the diffraction limit and in fact there is a gain because the focal length can be shorter, the eye less flat, or photon capture increased by increasing D. Note from Fig. 12 that p need only be slightly larger than the diffraction-limited value to be optimum for an environment 2–3 log units less than the brightest sunlight. For example, an eye with a hexagonal lattice of ommatidia

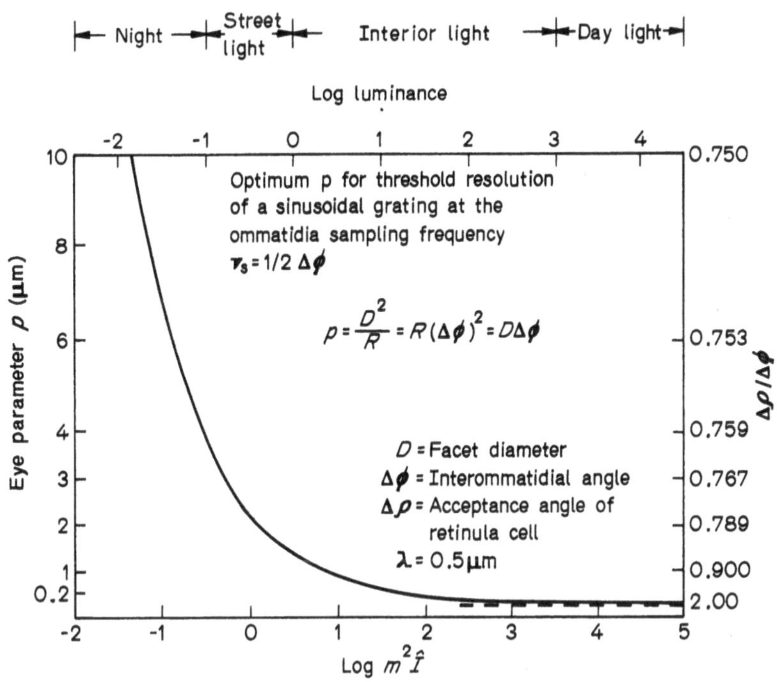

Fig. 12. Optimum eye parameter p for a given intensity parameter $\hat{I}$ and contrast or modulation m. The luminance scale is in candelas per square meter for the case of $m = 0.5$. The results are for square facets. For hexagonal facets, we replace $p = D\Delta\phi$ by $p = D\sqrt{3}\,\Delta\phi/2$. In both cases, the dashed line is the asymptote $p = 0.25$ μm

and $p = 0.5$ μm is the optimum design for an environment 3 log units less intense than necessary for the diffraction-limited eye where $p = 0.29$ μm. To use Fig. 12 for an hexagonal lattice, define p as $D\Delta\phi(\sqrt{3}/2)$. In Fig. 12 we also show the ratio $\Delta\varrho/\Delta\phi$ corresponding to each p value, where $\Delta\varrho$ is the acceptance angle of a retinular cell and $\Delta\phi$ the interommatidial angle.

Comparison with Measured p Values. HORRIDGE (1977) has compared the predicted p values of Fig. 11 to those measured in various animals. The foveas of bright-light locusts, dragonflies, and wasps have p between 0.31 μm and about 0.45 μm. The foveas of mantids and a large portion of the eye of bright-light insects have $0.45 \leqq p \leqq 1$; eye edge of many diurnal insects, $1 \leqq p \leqq 2$; eye of nocturnal dragonfly, $2 \leqq p \leqq 3$; eyes of deep-sea crustacean, $3 \leqq p \leqq 6$. The trend is clear. The dimmer the animal's environment, the less it takes advantage of its diffraction-limited optics.

Since there is an optimum eye parameter p for each intensity, the compound eye ideally would have a range of values for p across its head if it is to perform *optimally* over a range of values for intensity. For example, if p varies from 0.3 to 0.5 μm (representative of the dragonfly *Hemicordula*, HORRIDGE, 1976), this allows for a range of 3 log units in luminance for which the eye design is optimum. If p

varies from 0.3 to 1 μm, then the eye can accommodate 4 log units change in luminance while remaining at optimum.

Despite the apparent advantage of having a substantial range in p across the eye, in many animals p appears rather constant, e.g., as in *Musca* (STAVENGA, 1975). It remains possible that, because of neural summation, p of the neural ommatidia vary significantly. For example, because of convergence in the human peripheral retina, there is an enormous variation in the receptor spacing at the neural level while there is none at the rod photoreceptor level (SNYDER et al., 1977). Furthermore, the variation follows a curve like that of Fig. 12. Thus, although the facet topology may not display large values of p, the neural ommatidia quite possible do (Sec. V).

While there would appear to be no alternative strategies for optimizing to different contrast levels, there are other possibilities for dealing with reductions in intensity. One is by the method of optical superposition, which effectively increases the light-gathering aperture for a rhabdom (HORRIDGE, 1975; KIRSCHFELD, 1975). The other is to increase the acceptance angle $\Delta\varrho$ during dark adaptation (Sec. IV).

Although the theory presented here suggests why many animals have p above the diffraction limit, it does not answer why some animals have comparatively large p values even though they are in bright light. For example, *Musca* has $p \simeq 1.3$ μm (STAVENGA, 1975), i.e., about 4.5 times greater than the diffraction limit. One possible explanation is that such animals have a high angular velocity. We consider this possibility next.

7. Influence of Angular Motion on Eye Design

It is apparent from Eq. (B.28) that, when the animal is undergoing angular velocity v, the effect is equivalent to a reduction in intensity from $\hat{I}$ to $\hat{I}_t$, where

$$\hat{I}_t = \hat{I}\, e^{-1.78\left(\frac{v\Delta t}{\Delta\phi}\right)^2} \qquad \text{(B.29)}$$

so that, to include angular motion in Fig. 12, we simply replace $\log \hat{I}$ by $\log(\hat{I}) - 0.78\,(v\Delta t/\Delta\phi)^2$. The quantity $v\Delta t$ is the amount the animal turns in one integration time. We see, for example, that the effect of $v\Delta t = 2\Delta\phi$ is equivalent to a reduction in 3 log units of intensity. Taking $v\Delta t = 3°$, $\Delta\phi = 1.5°$, $\Delta t = 20$ ms, we find that $v\Delta t = 2\Delta\phi$ is equivalent to the animal turning 180°/s. This is less than the *average* chasing behavior of the housefly (COLLETT and LAND, 1974). Thus, if an animal, like *Musca*, is active in the midrange of daylight (say, log luminance $\simeq 3.5$) but, say, typically turns at 180° in 1 s, Fig. 11 shows that $D\Delta\phi = 1$ as in *Musca* rather than $D\Delta\phi \simeq 0.33$, when $v = 0$.

From the above logic, it is also easy to understand the advantage of increasing $\Delta\phi$ from the center of the eye to the side (along the horizontal axis) of an animal with a direct flight pattern, because increasing v is compensated for by increasing $\Delta\phi$ in the ratio $v\Delta t/\Delta\phi$ discussed above. Assuming that $D\Delta\phi$ is constant around the eye, this is consistent with an eye having its greatest eye radius in front and smallest on the side (PORTILLO, 1936; MAZOKHIN-PORSHNYAKOV, 1969).

8. Diurnal Eyes Employing Optical Superposition: Skipper Butterflies

Our discussions so far have been primarily for apposition eyes, although they apply fairly generally to neural superposition eyes as well. However, animals employing optical superposition require special consideration. Here we study the skipper eye which provides a unique example of a diurnal animal with high resolving power while employing optical superposition. Optical superposition is a strategy generally restricted to the eyes of nocturnal moths. Neural superposition eyes are discussed in Section D (see Fig. 1).

The skipper eye is hemispheric (DØVING and MILLER, 1969; HORRIDGE et al., 1972), probably as a consequence of its image formation by optical superposition. Accordingly, facet diameter D and interommatidial angle $\Delta\phi$ are interrelated by the radius R of the animal's head: $D = R\Delta\phi$. Furthermore, D, $\Delta\phi$, and hence the eye parameter p are constant everywhere on the skipper eye. By virtue of the image-forming characteristic of superposition optics, $\Delta\phi$ is also related to the center-to-center distance d_{cc} between rhabdoms: $d_{cc} = f\Delta\phi$, where f is the focal length (or posterior nodal distance) of an individual lens facet. Because the eye is spheric, f is also the distance from the center of the eye to the distal tips of the rhabdoms. By equating our two expressions for $\Delta\phi$, we have a fundamental relation for optical superposition eyes

$$Rd_{cc} = fD \tag{B.30}$$

which also holds for neural superposition eyes (Sec. D) and is quite likely at least approximately valid for apposition eyes.

The light-gathering capacity of a rhabdom is given by Eq. (B.16) for superposition and neural superposition eyes. For optical superposition, many facets are recruited in the task of gathering light for one rhabdom. From the work of HORRIDGE et al. (1972), it appears that the angular diameter of the summed participating facets $\phi_\Sigma \simeq 30°$, independent of the animal's size or the diameter of the individual facets. For angles ϕ_Σ greater than 30°, it is probably difficult for a crystalline cone to sufficiently bend the light rays for accurate image formation. The mean number of photons counted by a rhabdom due to a uniform extended source $\bar{N}$ is proportional to $(R\phi\Sigma\,\Delta\varrho_r)^2$ where $\Delta\varrho_r$ is the angular rhabdom diameter. Since the rhabdoms are surrounded by tracheal walls that in turn are closely packed with their neighbors (HORRIDGE et al., 1972), and assuming that all light falling within the tracheal walls is absorbed, we then have $\Delta\varrho_r \simeq \Delta\phi$. Consequently, each rhabdom captures an amount of light proportional to $(D\phi_\Sigma)^2$ due to a uniform source infinite in extent. Assuming that the angular diameter of the participating facets is nearly constant in all animals ($\phi_\Sigma = \text{constant} \simeq 30°$), then the *photon capture of each rhabdom depends only on the diameter of the individual facet D*. The optical superposition eye thus has high contrast sensitivity, i.e., an amount $(\phi_\Sigma/\Delta\phi)^2$ greater than an apposition eye of the same size.

The image quality is limited by the diffraction of the *individual* facet diameter D; however, we cannot exclude the possibility of some aberrations due to the superposition optics. Nevertheless, intracellular measurements of the acceptance angle in *Epargyreus clarus* by DØVING and MILLER (1969) indicate that aberrations are not very significant. In particular, they find that $\Delta\varrho \simeq 2°$ (see Fig. 1C for nomenclature). To see how this compares with the theoretic expression for $\Delta\varrho$ when only diffraction is limiting, we use the theory of Section B.2.a. From DØVING and MILLER (1969), $D = 27\,\mu m$, $\Delta\phi = 1.5°$, and $\lambda \simeq 520\,nm$. Recalling from our discussion above that $\Delta\varrho_r = \Delta\phi = 1.5°$ together with $\Delta\varrho_l = (\lambda/D) = 1.11°$, then from Eq. (B.7) $\Delta\varrho \simeq 1.9°$. This is sufficiently close to the measured value $\Delta\varrho \simeq 2°$ to show that the optical quality of skipper is nearly diffraction limited even though many facets participate in the image formation.

Recall our discussion of Section B.III.4 where we showed that the eye parameter $p = D\Delta\phi$ gives the efficiency of the animal in sampling the highest frequency passed by the optics. For *Epargyreus clarus*, $p \simeq 0.7\,\mu m$, which according to the logic of Fig. 12 is the correct design for an animal not normally in the brightest sunlight. However, skippers usually favor bright conditions and we are compelled to explain the discrepancy. There are two possibilities. Skippers have a rapid chasing behavior and accordingly, like *Musca*, receive only a small amount of light per integration time of the eye (Sec. 3 above). Since skipper eyes are unable to be locally flat, only those animals with large heads can have comparatively small p.

Because there are many skipper eyes with D nearly independent of head size R, it follows in these animals that resolving power is proportional to R, i.e., $\Delta\phi$ and hence p are proportional to R^{-1}. Remember also that the light capture of a rhabdom is proportional to $(\phi_\Sigma)^2$ for extended sources and thus, in this instance, the same for large and small eyes. For point sources, the greater ϕ_Σ the better. This observation is consistent with the second explanation above, i.e., only big eyes can accommodate small $\Delta\phi$ and have large absolute resolving power.

One question remains unanswered. Because skipper undersamples, (varying between 0.5 and 1.2 µm depending on eye size), why should it not increase light capture (i.e., increase ϕ_Σ) an additional amount until aberrations limit the highest spatial frequency passed to the sampling frequency $\nu_s = 1/\sqrt{3}\,\Delta\phi$? Perhaps the superposition image deteriorates rapidly beyond 30° or possibly the animal can make use of its comparatively good image quality in movement detection by aliasing.

9. Summary

We have considered the physical factors that influence the performance and design of compound eyes. In this section, eye design is characterized by the facet topology through the parameter $p = D\Delta\phi$ (facet diameter multiplied by the interommatidial angle) and by the acceptance angle $\Delta\varrho$ of the ommatidium. The most suitable value of these parameters is shown to depend critically on the intensity of light as well as on the angular velocity of the animal. This assumes that pupil diffraction and Poisson noise of the individual rhabdomeres are limiting. For example, it is advantageous for an animal to have its commatidia spaced at the limit set by pupil diffraction, i.e., $p \cong 0.3\,\mu m$, only if it is active in bright light and relatively stationary while examining objects, otherwise p should be larger (Fig. 11).

Since an increase in the angular velocity of an animal was shown to be equivalent to a decrease in light intensity, the design of *Musca* ($p \cong 1$ μm) is consistent with an animal performing high-speed aerial acrobatics at high light intensities (LAND and COLLETT, 1974).

The *absolute* (or anatomic) resolving power of an animal with compound eyes is set by the interommatidial angle $\Delta\phi$ and not by p, e.g., when comparing eyes with foveas that are designed at the limit set by lens diffraction ($p \cong 0.58\lambda$), we typically find that the larger eyes have the smaller $\Delta\phi$. In this regard, we note that the highest resolving power achieved in vertebrate eyes (say three times that of man), e.g., eagles and falcons (FOX et al., 1976), is about 100-fold that in those compound eyes with the highest anatomic resolving power, e.g., the dragonfly *Aeschna palmata* with $\Delta\phi = 0.24°$ (SHERK, 1976). The biologic reason for this significant difference is probably embodied in the concept of subjective resolution (KIRSCHFELD, 1976).

Finally, it is necessary to mention that our investigation was limited to extended objects as the acuity task. An insect that is contrasted by the clear sky is a good example of a nonextended object (e.g., two *Musca* separated by more than 70 cm appear like distant point sources to each other; SNYDER, 1977). Further discussions on resolution of isolated points are given in Appendix D.

IV. Strategies for Dark Adaptation

In this Section, we examine various strategies for design that allow an animal to have sufficient resolving power over a range of contrast and intensity when the eye parameter $p = D\Delta\phi$ is constant over the eye. Except when stated otherwise, the eye is assumed to be spheric with ommatidia that have identical facet diameters D and interommatidial angles $\Delta\phi$. The theoretic calculations are given by SNYDER (1977).

1. Eyes Without Adaptation Mechanisms

In Section II we showed that, because of photon noise, the resolving power of an eye decreases as the luminance decreases. The highest spatial frequency ν that can be resolved at a particular luminance is given by Curve *1* of Fig. 13. This figure depicts a Curve *1* for two different eyes, one with $p = 0.25$ μm (p at the minimum value set by diffraction), the other with $p = 1.0$ μm. It is convenient to plot $\nu\Delta\phi$ rather than ν because the curves then apply to an eye of arbitrary interommatidial angle $\Delta\phi$. The results are found from Eq. (B.19) by determining the threshold ν at each intensity, assuming SNR = 1.

The results of Fig. 13 assume that Poisson noise of *individual* rhabdomeres is limiting. However, by the appropriate neural averaging discussed in section 1.5, the SNR can be greatly enhanced at higher levels over that at the photoreceptor array. This is because of the repetitive or coherent nature of the grating stimulus. Consequently, if say 100 ommatidia are involved in the resolution task, then the luminance scale for curve 1 of Fig. 18 is shifted to the right by two log units, i.e. the animal performs as if the environment were two log units brighter.

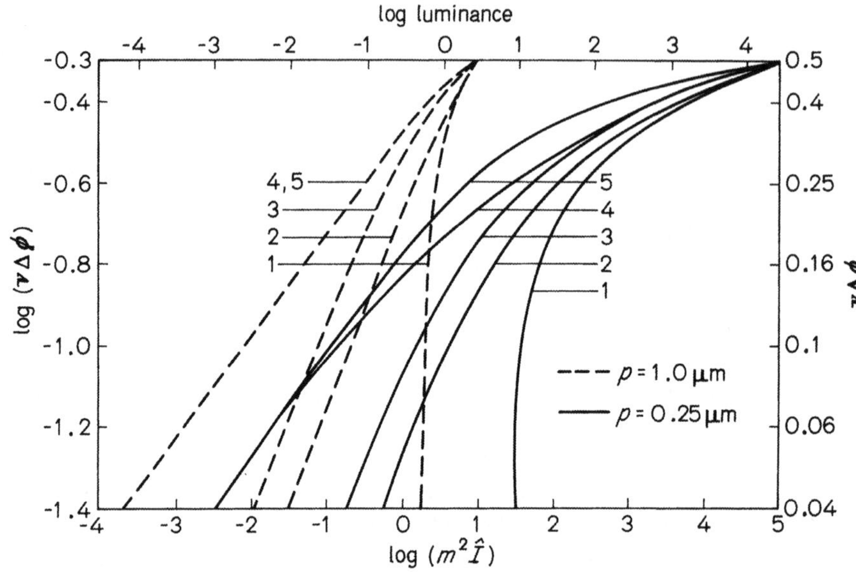

Fig. 13. The highest resolvable (or threshold) spatial frequency ν vs. luminance in candelas per square meter for various strategies of dark adaptation. The luminance scale is for a sinusoidal grating of 50% modulation m or contrast (the mean of all possible contrasts 0 to 1) and $\lambda = 500$ nm. Two different eyes are considered, both spheric, one with $p = 1.0\,\mu$m and the other with $p = 0.25\,\mu$m. Curve *1* is for an eye with no dark adaptation mechanism, Curve *2* for widening the acceptance angle, Curve *3* for neural pooling, Curve *4* for neural pooling together with widening the acceptance angle, and Curve *5* for a range in facet diameters over the eye. Simple expressions for these curves can be easily found for the low intensity limit from Eq. (B.19) because $\nu \ll \nu_{co}$ so that $M_1 = 1$. Accordingly, for Curve *1*: $\nu\Delta\phi = 0.5$ provided $m^2\hat{I} \geqq 1.8p^{-2}$; for Curve *2*: $\nu\Delta\phi = 0.23p(m^2\hat{I})^{1/2}$; for Curve *3*: $\nu\Delta\phi = 0.38p(m^2\hat{I})^{1/2}$; and for Curves *4* and *5*: $\nu\Delta\phi = 0.34p^{1/2}(m^2\hat{I})^{1/4}$. Curve *2* is less than Curve *3* because $M_r = 0.61$ instead of unity

2. Increase in Acceptance Angle with Decrease in Luminance

Suppose that the acceptance angle $\Delta\varrho$ of the ommatidia increases as the luminance decreases. This occurs in a number of animals (for review, see Walcott, 1975). The advantage of increasing $\Delta\varrho$ has been given qualitatively by Laughlin (1975). As the luminance decreases, the contrast sensitivity at the high spatial frequencies becomes inadequate. By increasing $\Delta\varrho$ the contrast sensitivity of these high spatial frequencies is further reduced but that of lower spatial frequencies is enhanced. Since the highest potential resolving power is irretrievably reduced, the slight loss to acuity by increasing $\Delta\varrho$ a certain amount can be offset by the gain in photon capture allowing for resolution of lower spatial frequencies.

Figure 14 demonstrates that there is an optimum acceptance angle $\Delta\varrho$ at each luminance. When the eye is designed at optimum ($\Delta\varrho_r = 0.75\Delta\phi$) shown as the dashed line, we see that the sampling $\nu_s = 1/2\Delta\phi$ can be resolved but only if log $m^2\hat{I} \cong 5$. However, as log $m^2\hat{I}$ decreases below 1, not even extended uniform objects ($\nu \cong 0$) can be seen reliably. Had we taken $\Delta\varrho_r = 2\Delta\phi$, the eye would be unable to resolve the sampling frequency $\nu_s = 1/2\Delta\phi$, but it would be able to resolve low frequencies at luminances when the eye with $\Delta\varrho_r = 0.75\Delta\phi$ is unable to resolve anything. In other words, the greater $\Delta\varrho_r$, the better the detection of low frequency

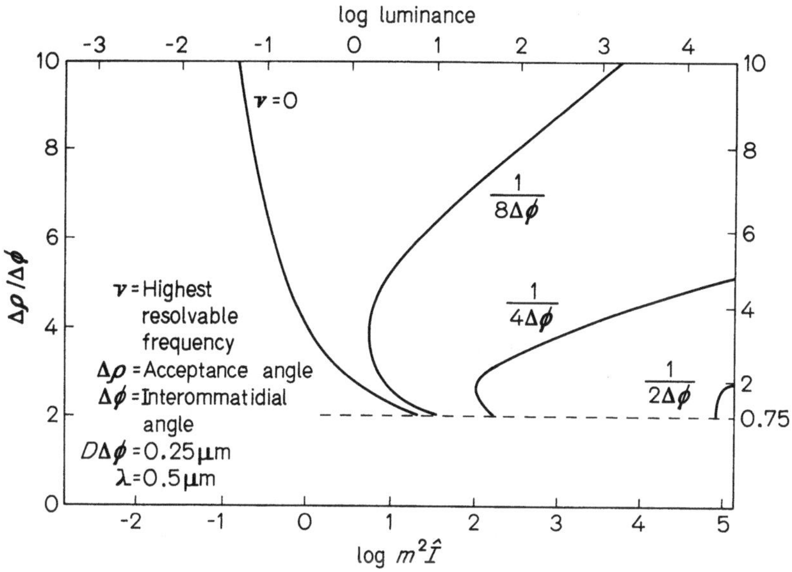

Fig. 14. Highest resolvable frequency for an animal with a given $\Delta\phi$ and $\Delta\varrho$. The luminance scale is in candelas per square meter for the case of 50% object modulation

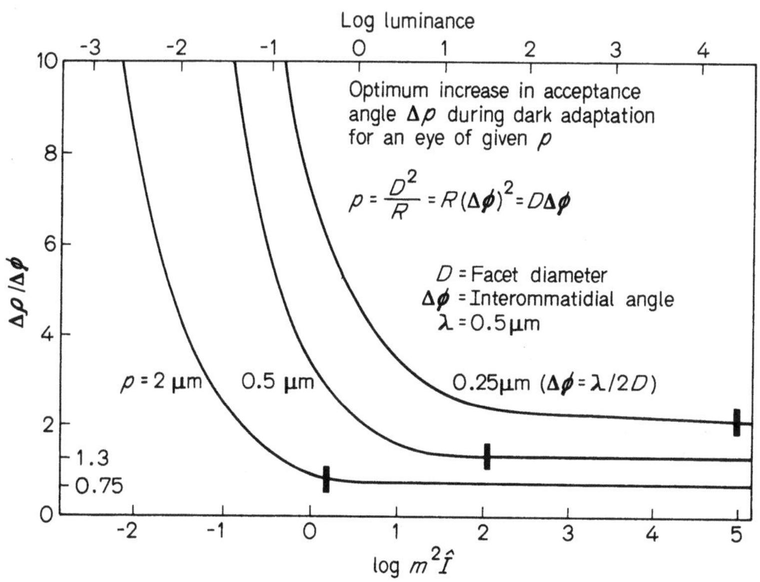

Fig. 15. Optimum acceptance angle $\Delta\varrho$ for each intensity during dark adaptation for an eye of given $\Delta\phi$ and p. The luminance scale is in candelas per square meter for 50% object modulation

gratings at low luminances. At each value of $m^2\hat{I}$, there is an optimum value of $\Delta\varrho_r$ given by the turning points on Fig. 14, e.g., $\Delta\varrho_r = 1.5\Delta\phi = 0.375/v$ when $\log m^2\hat{I} \cong 2$. This optimum $\Delta\varrho$ is shown in Fig. 15 for three different eyes with $p = 2$ µm, 0.5 µm, and 0.25 µm, respectively. By plotting $\Delta\varrho/\Delta\phi$, we have generalized our curve to an eye of arbitrary interommatidial angle. It should be noted that a comparatively small change in $\Delta\varrho$ is required to accommodate large changes in luminance. WALCOTT (1975) has described anatomic changes during light adaptation that can easily accommodate the theoretic optimum suggested here, e.g., $\Delta\varrho$ of *Lethocerus* changes from 3.5° to greater than 9° when dark adapted. Curve *2* of Fig. 13 shows the highest spatial frequency ν vs. luminance for an eye that increases its acceptance angle in the optimum manner discussed above. Because there can be a range in environmental contrasts at each intensity, it is theoretically advantageous to have a range in $\Delta\varrho$ across the eye.

3. Neural Pooling

If the luminance is reduced below that necessary for sufficient contrast sensitivity at the sampling frequency, it is advantageous to form a larger neural ommatidium by neural convergence and gain in contrast sensitivity at low spatial frequencies. This neural convergence could be hard wired, i.e., exist in parallel with through running neurons or it could depend on intensity. In both cases, the effect is to form an effective larger ommatidium. The results of this strategy are shown in Fig. 13, Curve *3*.

Now suppose, together with neural pooling, that the acceptance angle $\Delta\varrho$ changes such that $\Delta\varrho_r$ is proportional to the neural interommatidial angle $\Delta\phi_n$ so that $D\Delta\phi_n = D_n\Delta\phi_n$, where D_n is the effective pupil diameter of the neural ommatidium. The gain in contrast sensitivity at lower spatial frequencies equals $p_n(I)/p$, where $p_n(I)$ is the neural eye parameter $D_n\Delta\phi_n$ and is set according to Fig. 12 while $p = D\Delta\phi$. The highest spatial frequency resolved at any given luminance I equals the neural sampling frequency $\nu_{ns}(I) = \{p/p_n(I)\}^{1/2}\nu_s$. The result of this strategy is given in Fig. 13, Curve *4*.

4. Comparisons and Discussion

It is clear from Fig. 13 that all of the strategies considered offer an improvement in spatial resolution at low intensities compared to that of Curve *1* for no adaptation mechanism. DVORAK and SNYDER (1978) have measured the variation in grating resolving power by monitoring the extracellular response of direction-sensitive, motion-detecting neurons and find that the results indicate a dark adaptation strategy similar to that of Curve *4* in Fig. 13.

It is of interest to compare these results to a *spheric* eye where the facet diameter D changes (i.e., p changes) over the animal's head, but the eye radius remains as in the above examples. We show in Appendix B that Curve 5 of Fig. 13 applies, where $\Delta\phi$ and p now refer to the minimum values on the eye. Although an eye with a range in D is theoretically better than one with neural pooling (large facet diameters have less pupil diffraction), this assumes the radius is sufficiently large to accommodate a substantial number of facets with a large range of diameters. In comparison, an eye with versatile neural pooling can function nearly as well and yet be considerably smaller, because in theory the whole head can act as one neural facet at low intensities while at high intensities the individual ommatidia do the sampling. Moreover, for eyes with a range in the parameter p, each portion of the head is best suited to a different contrast luminance level which need not be the case for a neural pooling strategy. Obviously, it is advantageous to combine neural pooling with an eye that has a range in p. Lastly, we stress that only apposition eyes have been considered. Optical superposition eyes (for review, see HORRIDGE, 1975; KIRSCHFELD, 1975) can theoretically maintain their highest potential acuity for a change of log q units in luminance, where q is the number of recruited facets making up the superposition image. See Section III.8.

V. Neural Processing

In this section, we consider the role of neural processing in regard to spatial information as well as possible strategies to achieve this role. We begin by considering two important tasks that may involve neural processing: 1) the improvement of contrast sensitivity and 2) the transformation of spatial information from one form to another.

1. Enhancement of Contrast Sensitivity by Neural Processing

In the above section, we showed that photon noise of an individual photoreceptor limited the contrast sensitivity or equivalently the SNR across the photoreceptor array. Now it should be rather obvious that, by appropriate weighted averaging and correlating of the individual photoreceptor signals, the SNR appearing across an array of higher order neurons can be improved upon that appearing across the receptor array assuming the averaging procedure itself is noise-free. The most elementary example is for a uniform extended source that has an SNR proportional to $\sqrt{\bar{N}}$ at the photoreceptor level but a potential SNR of $\sqrt{i\bar{N}}$ at some higher level, where $\bar{N}$ is the mean number of photons captured by an individual photoreceptor and i is the number of photoreceptors. This result holds more generally with slight modification. In particular, the SNR or contrast sensitivity to sinusoidal gratings of frequency ν is $\sqrt{i\bar{N}}\,M(\nu)$, where i is the number of photoreceptors per angular area participating in the resolution measurement and M is the modulation transfer function of the optics and rhabdom. *With this form of processing, the contrast sensitivity is independent of the diameter of an individual photoreceptor (except weakly through M_r) and depends instead upon the total angular retinal area $i(\Delta\varrho_r)^2$ devoted to the detection process.* Thus, the gain in contrast sensitivity is proportional to $\sqrt{i}$. Whether or not the nervous system has such versatility remains unknown. The great advantage in contrast sensitivity discussed above is due mostly to the coherent or repe nature of the signal, i.e., a sinusoidal grating. For the incoherent type of object world that occurs naturally, there is no corresponding increase in spatial discrimination by the weighted, correlated averaging discussed above. Recall that improving contrast sensitivity for gratings at each frequency, e.g., by parallel Fourier channels, need not improve spatial discrimination of broad spectrum signals, even if the system is linear. This is because, due to noise, the subtle fluctuations of the image in the frequency domain are no easier to detect than subtle fluctuations in the space domain, in addition to the added problem of phase correlation. Thus, there is may be no compelling reason for the nervous system to adopt such a strategy. However, if the animal had one or several preferred acuity tasks, e.g., the recognition of a characteristic pattern, then the nervous system could be designed to respond only to this pattern with an enormous contrast sensitivity. Furthermore, the pooling of ommatidia as discussed in Section BI.5 and IV would appear advantageous to the animal as luminance decreases.

2. Transformation of Spatial Information

The ommatidia provide a two-dimensional sampled version of the object world, which the nervous system must subsequently analyze. We can view each level in the visual pathway as a generalized transform of the visual input. Each transformation results in a form more convenient for decoding. For example, one direct transformation is via a pointwise measure of visual intensity (ROBSON, 1975). Another is by a spatial frequency transform. For example, if the nervous system had the capability of performing a Fourier transformation, e.g., as manifested by parallel Fourier channels of narrow band width, then the contrast sensitivity to *sinusoidal gratings* is proportional to $D^2 M^2(\nu)/(\Delta\nu)^2$, where D is the pupil diameter, M the modulation transfer function of the optics and rhabdom, and $\Delta\nu$ the band width of each channel, assuming only photon noise at the photoreceptor level is limiting. The minimum bandwidth $(\Delta\nu)^2$ is proportional to the inverse angular area of the eye devoted to the resolving task. Thus, maximum contrast sensitivity is proportional to the number of photons captured within the angular region devoted to the resolution task, as mentioned in Section 1 above. However, the evidence for effective Fourier channels in the nervous system is lacking (ROBSON, 1975, p. 112).

a) Neural Compensations for Optical Blur and Photoreceptor Noise

SNYDER and SRINIVASAN (1979) have suggested a transformation for the visual pathways in the human retina that may also have some relevance for compound eyes. They propose that neural processing compensates for both optical degradation as well as noise contamination at the photoreceptor level—a hypothesis consistent with many psychophysical measurements. We now present the argument as originally applied to the human retina.

It is clearly impossible for neural processing to restore those high frequencies already lost at the photoreceptor level, nor can it filter the noise from the image without simultaneously filtering out some of the object information as well. In other words, subsequent neural processing is unable to recover object information that is already lost at the photoreceptor level, but it can transform this information from one form to another analogous to translating from one language to another. We now determine the transfer properties of the neural processing, assuming its only task is to transform the image at the photoreceptor level to a cortical image that looks like the object. This requires that the nervous system compensates for optical blur as well as for noise contamination. Our discussion commences with the limiting case of extreme suprathreshold conditions.

b) Extreme Suprathreshold Conditions (Noiseless Blurred Image)

When the photoreceptor image is comparatively free of noise, presumed to be the case in extreme suprathreshold conditions, then theoretically all photoreceptor image distortion is due only to the blurring effect of the eyes' optics, i.e., due only to the nonuniform properties of the transfer function for the dioptrics M_l. The neural image will then appear most like the object when the transfer function of the neural apparatus M_n is the inverse of the optical transfer of M_l, i.e.,

$$M_n = M_l^{-1}. \tag{B.32}$$

Taking the data of CAMPBELL and GUBISCH (1966) for the human dioptrics shown in Fig. 16a, then M_n becomes that shown in Fig. 16b. The product of the two curves gives Fig. 16c. In other words, because of this neural compensation for optical blur, the overall transfer function of the visual system M is uniform with spatial frequency as shown in Fig. 16c. Deblurring is made physically realizable by having the nervous system increase its sensitivity (gain) at those spatial frequencies that are attenuated most by the optics, i.e., neural processing amplifies the high spatial frequencies relative to the lows as shown in Fig. 16b.

One significant consequence of a uniform transfer function M for the entire visual system is that gratings of equal contrast are perceived to have identical contrast, independent of their spatial frequency. Indeed, brightness-matching measurements show this to be approximately true, in photopic conditions, provided the grating contrast is sufficiently high (WATANABE et al., 1968; BLAKEMORE et al., 1973; GEORGESON and SULLIVAN, 1975). Consequently, contrast constancy at suprathreshold conditions is consistent with the hypothesis that the neural system transforms the photoreceptor image to a cortical image more nearly resembling the object. The receptive field of a hypothetic neuron at the site of this cortical image would have the center surround form discussed in Section B.II.5 above, i.e., by FOURIER transforming Fig. 16c.

c) Threshold Conditions (Noisy Blurred Image)

Now we consider the other extreme when, in addition to optical blur, the image at the photoreceptor level suffers from significant noise contamination. This situation is most dramatic at threshold conditions. What properties must then be manifested by neural processing for the neural image to look most like the object? When photoreceptor noise is significant, amplifying the signal also amplifies the noise. Furthermore, high frequencies are comparatively noisier than lower spatial frequencies because of the modulation transfer function of the optics shown in Fig. 17a. It is intuitive that the neural image will then appear more like the object if the nervous system is less sensitive to these noisy or unreliable frequencies compared to the more reliable ones. Frequencies which are 100% reliable (i.e., $v=0$) should be reproduced in their entirety (i.e., transmitted without any attenuation or amplifi-

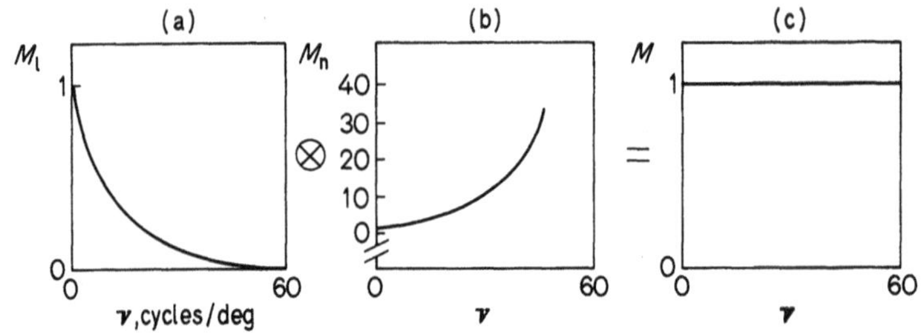

Fig. 16. The modulation function for the visual system $M(\nu)$ determined as a product (denoted by the symbol $\otimes$) of the optical transfer function $M_l(\nu)$ with the neural transfer function $M_n(\nu)$. The case of extreme suprathreshold conditions is shown when the neural modulation transfer function equals the inverse of the optical transfer function, i.e., $M_n = M_l^{-1}$. The optical modulation transfer function M_l is for a pupil diameter of about 3 mm and is found by interpolating the data of CAMPBELL and GUBISCH (1966) for the human eye

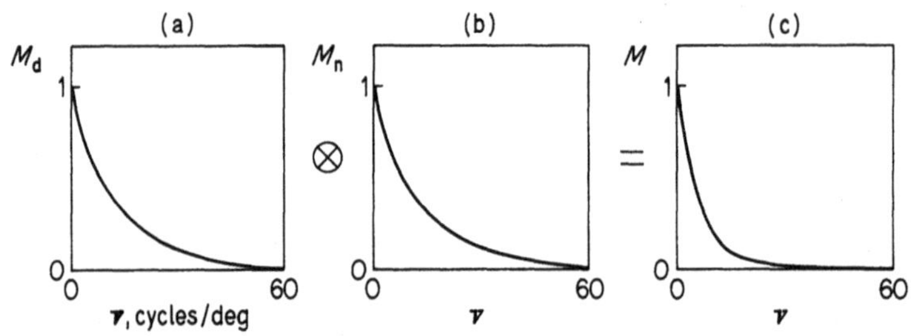

Fig. 17a and b. The same as for Fig. 16 but now for threshold conditions. (a) is the same as Fig. 16a but (b) is given by $M_n = M_l$ so that $M = M_l^2$

cation). On the other hand, frequencies that are 0% reliable, i.e., frequencies which contain only noise and no signal like $\nu = 60$ cpd in Fig. 17a should be completely excluded because they contain no object information and will only contaminate the image with noise. Frequencies that are partially reliable should be attenuated to an extent that depends upon their reliability. At threshold, the reliability of the image is proportional to M_l^2. Consequently, the neural compensatory mechanism must fulfill two competing roles: deblurring on the one hand, characterized by $M_n = M_l$. It turns out that (SNYDER and SRINIVASAN, 1979) the transfer function of the neural apparatus M_n is given by the product of these two competing effects

$$M_n = M_l^{-1} M_l^2 = M_l \tag{B.33}$$

i.e., when the nervous system deblurs the image in a weighted manner, comparatively suppressing the relatively noisier frequencies. The transfer function of the entire visual system, $M = M_l M_n$, then becomes

$$M = M_l^2 . \tag{B.34}$$

In other words, if the neural image is to look like the object in contrast threshold conditions, then the transfer function for the visual system depends only upon the square of the transfer function of the optics. Taking the data of CAMPBELL and GUBISCH (1966) for the transfer function of the human dioptrics as shown in Fig. 17a and multiplying it by the transfer function for the nervous system $M_n = M_l$ shown in Fig. 17b leads to the transfer function of the visual system $M = M_l^2$ shown by Fig. 17c. Indeed, above about three cycles per degree, the shape of the well-known human threshold contrast sensitivity function (e.g., CAMPBELL and ROBSON, 1968) is nearly identical to that of Fig. 17c when plotted on a linear frequency scale as we show here. Thus, the threshold contrast sensitivity function (above three cycles per degree) as well as the extreme suprathreshold sensitivity function are consistent with the hypothesis that neural processing transforms the photoreceptor image to a neural image that better resembles the object. The receptive field of a hypothetic neuron at the site of the cortical image at threshold looks like that at suprathreshold but without the surround.

In summary then, the essential concept for noise compensation is that the neural network must be least sensitive to those spatial frequencies that are comparatively unreliable, i.e., those spatial frequencies with a low SNR appearing across the photoreceptor array. Because of attenuation by the optics, the high spatial frequencies are more unreliable so the neural network is comparatively insensitive to them. The essential concept for optical compensation is that the neural networks must be most sensitive at those spatial frequencies attenuated most by the optics. In extreme suprathreshold conditions, a wide range of spatial frequencies are reliable so that only optical compensation is necessary. At threshold conditions, all spatial frequencies are comparatively unreliable so that both optical and noise compensation is necessary.

VI. Spatial Information Capacity of Eyes

In this Section we derive a measure of an animal's capacity for spatial acuity. We begin with a strictly mathematic definition of the spatial information capacity of an eye H and then show how H can be interpreted as an animal's picture-making ability. We assume that the eye is limited by photoreceptor noise. From information theory (e.g., GOLDMAN, 1953), the information capacity of an eye per visual angle is

$$H = \int_0^{\nu_s} \nu d\nu \ln\{1 + \sigma_s^2(\nu)/\sigma_n^2(\nu)\}\,, \tag{B.34}$$

where $\nu_s = 1/\sqrt{3}\Delta\phi$ is the sampling frequency of the photoreceptors, Eq. (B.3), σ_s^2, σ_n^2 the power spectra (BRACEWELL, 1965) of the signal and noise at the level of the photoreceptor, respectively.

The power spectrum σ is related to the total power P by the integral

$$P = 2\pi \int_0^{\infty} \sigma^2 \nu d\nu\,, \tag{B.35}$$

where P is the variance for a random process. The noise power is $\bar{N}$ assuming photon shot noise while an example of signal power is $N^2M^2(\nu)$ for an object spectrum of a single frequency, i.e., a delta function.

This definition aids in specifying σ_n^2 and σ_s^2. For example, since the noise of the photoreceptors is statistically independent, the noise power spectrum is approximately flat up to the sampling frequency so that from the above equation

$$N = 2\pi\sigma_n^2 \int_0^{\nu_s} \nu d\nu \tag{B.36}$$

and $\sigma_n^2 = \bar{N}/\pi\nu_s^2$. This is the noise power spectrum due to a uniform extended object.

Since all spatial frequencies are equally likely within the object world and because the eye is assumed to have no prior knowledge of the visual scene, we assume the object world has objects of random size and shape or equivalently the power spectrum of the object world is flat. Therefore σ_s^2 is proportional to $\{\hat{I}D^2(\Delta\varrho_r)^2\}^2(CM)^2$ where all qualities are defined in Table 2 while C is the mean contrast of the object world. Assuming C is small, then the noise power spectrum σ_n^2 is proportional to $\hat{I}D^2(\Delta\varrho_r)^2(\Delta\phi)^2$ so that

$$\frac{\sigma_s^2(\nu)}{\sigma_n^2(\nu)} = k\hat{I}(CD)^2\left(\frac{\Delta\varrho_r}{\Delta\phi}\right)^2 M^2(\nu), \tag{B.37}$$

where $k=1$, with units $(\Delta\phi)^2$. Replacing this equation into the integral expression for H gives the information capacity of the eye from which the optimum eye parameters can be determined. In the limit of low luminance ($\hat{I}\to 0$), using the fact that $\ln(1+\varepsilon)\simeq\varepsilon$ when $\varepsilon \ll 1$, Eq. (B.34) can be expressed by a simple analytic form. We then find that for an eye of fixed facet diameter, the ratio $\Delta\phi_r/\Delta\phi$ should be as large as possible. However, our primary concern is for the high intensity limit because nocturnal animals generally employ optical superposition (Sec. B.III.8 above). When $\sigma_s \gg \sigma_n$, Eq. (B.34) has the asymptotic solution (SNYDER et al., 1977).

$$H \simeq (\nu_s^2/4)\left\{\ln 1 + k\hat{I}(CD)^2\left(\frac{\Delta\varrho_r}{\Delta\phi}\right)^2 M(\nu_s)\right\}^{1/2} \tag{B.38}$$

assuming that M can be approximated by a Gaussian as given by Eq. (B.8), i.e., $M(\nu_s) = \exp\{-3.56(\nu_s\Delta\varrho)^2\}$. Equation (B.1) is accurate over a wide range of intensities but becomes inaccurate as $\hat{I}\to 0$. Furthermore, Eq. (B.38) can be shown to equal the number of different pictures an eye can reconstruct (SNYDER et al., 1977).

1. Concept of an Animal's Spatial Information (Picture-Reconstructing) Capacity

An eye must reconstruct its spatial environment from an array of intensity measurements, each measurement provided by an individual ommatidium. It is convenient to view this spatial quantization as a two-dimensional mosaic or checkerboard, i.e., a picture constructed from many smaller elements. Thus, the number of ommatidia, per field of view, sets the capacity of the eye for fine detail.

The fine detail of a picture is lost if there is inadequate contrast between the various elements. Accordingly, the capacity of the eye for contrast sensitivity is determined by the number of *different* intensity levels that can be discriminated by an array of ommatidia. At first, it might be thought that an infinite number of different intensity levels can be distinguished; however, photon noise sets a lower limit to the number that are reliable. The more photons captured by the ommatidia, the greater the number of intensity levels and thus the greater the contrast sensitivity of the animal. In other words, both space and intensity are quantized by the eye.

Now it is clear that *for a spheric eye*, as the number of ommatidia per field of view increases, the potential of the eye for resolving fine detail also increases, but there must be fewer photons available for each of the ommatidia and hence a smaller number of recognizable intensity levels. Because of this unavoidable competition between the capacity of the eye for fine detail on the one hand and the capacity for contrast sensitivity on the other, what is the most appropriate number of ommatidia per field of view? The answer clearly depends on the number of available photons, but some measure of acuity performance is required. A natural metric is the number of *different* pictures that can be reconstructed by the mosaic of ommatidia, i.e., by the many elements that make up each picture. Assuming that there are n_p ommatidia per field of view, each one with *one* of n_i possible intensity levels, the *maximum* number of pictures that can be reconstructed by the ommatidia is $n_i^{(n_p)}$. Now it follows from the classic arguments of information theory (GOLDMAN, 1953; PIERCE, 1961) that the logarithm of the *maximum* number of *different* pictures that can be reconstructed, per field of view, by the ommatidia is the *spatial information capacity of the eye*, denoted here as H.

$$H = \ln n_i^{(n_p)} = n_p \ln n_i \tag{B.39}$$

Spatial Information Capacity = (No. of Ommatidia per Field of View)⊗ln (No. of Different Intensity Levels),

where ln is to the base e. By using the logarithm of the number of pictures, we preserve the intuitive notion that doubling the number of ommatidia per field of view n_p doubles the spatial information capacity H of an eye. Having established the essential concept, we next relate n_p and n_i to the physical parameters of the compound eye. The number of ommatidia per square radian of object space n_p is proportional to $(\Delta\phi)^{-2}$ which in turn is proportional to v_s^2. The number of intensity levels n_i that can be reliably distinguished by the ommatidia is limited by noise, ultimately due to the quantum nature of light, i.e., photon noise, and by imperfect optics, ultimately due to the wave nature of light, i.e., by diffraction.

Because of the random nature of photon absorption, a uniform light source (infinite in extent) appears nonuniform to an array of ommatidia and σ_n is the standard deviation in photon counts across an infinite array of ommatidia. Because of this σ_n degree of uncertainty in the interpretation of a photon count, there is only a finite number of intensity levels that can be distinguished with certainty by the ommatidia.

We suppose here that the intensity levels need be separated by $2\sigma_n$ intervals to be reliably discriminated, i.e., the standard deviation of one level just meets the standard deviation of the neighboring level. This is the usual reliability criterion for communication systems (e.g., CARLSON, 1975) and is analogous to assuming that the threshold SNR is unity (see SNYDER et al., 1977). *Thus, n_i is found by determining the number of intervals of width $2\sigma_n$ that can fit into a given range of mean object intensity.* This intuitive procedure gives the maximum number of intensity levels that can be reliably distinguished (CARLSON, 1975).

We are reminded that our expression for spatial information capacity H, given by Eq. (B.34), equals the *maximum* number of pictures that ommatidia can reconstruct. It is convenient to know what distribution of object intensities would produce this maximum. From information theory (e.g., PIERCE, 1961), we learn that a random scene fulfills the requirement, i.e., a scene containing objects of random contrast and random size (or identical size at random distances from the eye). Such a scene is the epitome of the unexpected since every spatial frequency has equal importance. The spatial information capacity H of an eye is, therefore, equivalent to the amount of information that it can extract from a random scene. *Accordingly, we determine the number of possible contrast levels that exist when the object intensity is random.*

In the absence of noise, the standard deviation in photon counts, due to the random scene is σ_s. In the presence of noise, twice the standard deviation in photon counts is given by $2(\sigma_{\text{sig}}^2 + \sigma_{\text{noise}}^2)^{1/2}$, remembering that variances and not standard deviations must be summed (GOLDMAN, 1953). Dividing this expression by $2\sigma_n$ gives the number n_i of possible intensity levels

$$n_i = \{1 + \sigma_s^2/\sigma_n^2\}^{1/2}. \tag{B.40}$$

From our discussion, it is clear that Eq. (B.38) has the form of Eq. (B.39) (with an effective σ_s/σ_n) so that the asymptotic form for H can be associated with the picture-making capacity of the ommatidia.

2. Optimum Eye Parameters

We now determine those eye parameters that maximize the information capacity of the eye. From Eq. (B.38), H is maximized when $\Delta\varrho_r = 1.06\Delta\phi$ so that Eq. (B.38) becomes

$$H = (v_s^2/4) \ln \{1 + 0.26 k\hat{I}(CD)^2 M_1(v_s)\}^{1/2}. \tag{B.41}$$

Recalling that ν_s is proportional to $\Delta\phi^{-1}$ and that $M_1(\nu_s)$ depends only on the eye parameter $D\Delta\phi$, we see that *for a given facet diameter* H is maximized when the eye samples at the diffraction limit $p = D\Delta\phi = \lambda/\sqrt{3}$ or equivalently when $\nu_s = \nu_{co}$. Furthermore, there is no loss by oversampling, i.e., ν_s can exceed ν_{co} without lowering H. In other words, if the eye can be flat in some small region, then according to information theory, it pays to have $\Delta\phi$ at the diffraction limit.

Now, if we consider the optimum eye parameter p for an eye of *fixed radius R*, then an entirely different result emerges. In this case $D = R\Delta\phi$ so that image quality and photon capture must be sacrificed by decreasing $\Delta\phi$. Consequently, there exists a trade-off between fine detail and contrast for which information theory via Eq. (B.41) gives an optimum p. The diffraction-limited eye is the optimum design only when the luminance $\hat{I}$ is sufficient. From Eq. (B.41) this luminance is determined by the equation

$$\ln\{0.26 k\hat{I} RC^2 p\, e^{-2.67(\lambda/p)^2}\} = 1 \tag{B.42}$$

with $p = \lambda\sqrt{3}$. Thus, at $\lambda = 500$ nm, $\log(C^2 R\hat{I}) \simeq 5$ with R in units μm and $\hat{I}$ in the units of Table 2. The greater the eye radius R, the smaller the luminance necessary for the animal to utilize its diffraction-limited optics. For example, with $R = 10^3$ μm, $\log C^2\hat{I} = 2$, which from Fig. 12 is seen to require moderate inferior lighting if a diffraction-limited eye is to resolve a grating of 50% contrast.

Summary. If we design an eye to maximize its information capacity, then $\Delta\varrho_r \simeq \Delta\phi$ and, for a given facet diameter, the $\Delta\phi$ should be at the diffraction limit $\Delta\phi = \lambda/\sqrt{3}\, D$ provided the eye can be locally flat. If on the other hand, the eye is spheric and *limited to radius R*, then a minimum luminance is required for the diffraction limit to be the design that maximizes information capacity. The greater the head size, the smaller the required luminance. Or stated differently, larger eyes can make more efficient use of their optics.

These conclusions differ from those of Section C.III.6, where we determined eye design by finding the minimum parameter p necessary for adequate contrast sensitivity across the photoreceptor array at the ommatidial sampling frequency $\nu = \nu_s = 1/\sqrt{3}\Delta\phi$. The results of Section C.III.6 maximize the SNR across the rhabdoms. In contrast, the results of information theory are in general independent on the SNR at the photoreceptor level because it is implicitly assumed that weighted averaging can be performed neurally to achieve the high SNR of Eq. (B.37) at some higher level. The equivalent expression for signal to noise power at the photoreceptor level is proportional to $\bar{N}M^2(\nu)$. In other words, information theory assumes the theoretically best neural processing possible and therefore the amount of light captured per ommatidium is deemphasized.

Eyes that can be locally flat do not in general have their ommatidial angles at the diffraction limit, just as in the vertebrate lens eye cones are not in general set by the diffraction limit. Consequently, the prediction of information theory is not confirmed in either case, while that of Section III seems rather broadly applicable.

The results of the original presentation on information theory by SNYDER et al. (1977a) and SNYDER et al. (1977b) differ from those presented here due to an error in their expression for the ratio of the signal to noise power spectrum. The correct expression is presented in Eq. (B.38) which also applies to the vertebrate eye when $\Delta\varrho_r$ is set equal to $\Delta\phi$.

VII. Comparison of the Compound Eye with the Vertebrate Lens or "Camera Eye"

We are now in a position to consider differences between the compound and camera eye. This is presented by way of a brief review of the vertebrate eye.

1. Cutoff Frequency of Optics (Theoretic Limit to Resolving Power)

For all practical purposes, the *highest* spatial frequency passed by the vertebrate optics ν_{co} is given by diffraction theory

$$\nu_{co} = D/\lambda \tag{B.43}$$

in cycles per radian, where D is the average entrance or apparent pupil diameter in bright light (e.g., about 2.3 mm in man; CRAWFORD, 1936) and λ the wavelength at peak photopic sensitivity (about 555 nm in man; LEGRAND, 1968, p. 84). Thus, at $\lambda = 555$ nm, $\nu_{co} \simeq 10\pi D$ cycles per degree, where D is in mm, so that $\nu_{co} \simeq 72$ cpd in man.

While it is true that Eq. (B.34) also applies in dimmer light when D is larger and hence so is ν_{co}, nevertheless because of aberrations, the modulation transfer function of the optics is extremely small at frequencies significantly less than ν_{co} (CAMPBELL and GUBISCH, 1966). Consequently, due to the noise of any measuring device, the highest observable spatial frequency passed by the optics rarely exceeds that predicted by Eq. (B.1) with D in bright light.

Because the theoretic limit to resolving power is given by ν_{co}, the ratio of a single facet diameter of a compound eye to the pupil diameter of a vertebrate in bright light gives the relative potential resolving power of the two animals. For example, the smallest humming birds have a resolving power more than ten times greater than the largest dragonflies. The larger an eye, the greater its diurnal pupil diameter D, and hence the greater its capacity for high resolving power. However, the realization of this capacity demands a sufficient number of photoreceptors per visual angle as we discuss next.

2. Anatomic Resolving Power (Photoreceptor Sampling)

The cutoff frequency ν_{co} sets the limit to resolving power as imposed by pupil diffraction. However, the highest spatial frequency that can be faithfully reconstructed by the retina, or anatomic sampling frequency ν_s, is limited by the angular center-to-center spacing of photoreceptor $\Delta\phi$ (SNYDER and MILLER, 1977).

$$\nu_s = 1/\Delta\phi\sqrt{3} \tag{B.44}$$

for hexagonal packing. Only when $\nu_s = \nu_{co}$ can the photoreceptor reconstruct frequencies up to ν_{co}, requiring a *maximum* angular photoreceptor spacing of $\Delta\phi = 1/\nu_{co}\sqrt{3}$ necessary for $\nu_s = \nu_{co}$.

Note that it is possible to oversample in a vertebrate retina of fixed size by reducing the diameter of photoreceptors. However, in a compound eye of fixed size and shape, reducing $\Delta\phi$ also reduces D so that the resolving power is lowered.

The logic for selecting $\Delta\phi$ is the same as for selecting the optimum compound eye parameter $p = D\Delta\phi$ discussed in Section B.III.4. In particular, $D\Delta\phi$ is the minimum value for sufficient contrast sensitivity at the sampling frequency $\nu = \nu_s$ at the highest luminance of the animal's environment. Thus, those animals that are active in the brightest light and are comparatively stationary relative to their prey are expected to have the smallest $D\Delta\phi$ values. These animals can take the greatest advantage of their potential optics. Curiously, desert hawks, eagles, and probably man take significantly less advantage of their potential optics ($D\Delta\phi \simeq 0.5\,\mu m$) (see MILLER, this volume) than wasps and dragonflies ($D\Delta\phi \simeq 0.31\,\mu m$). One possible explanation for this pronounced undersampling is that the modulation transfer function for vertebrate optics is significantly smaller near $\nu = \nu_{co}$ than for the compound eye. In other words, the compound eye, with a pupil diameter 10^{-2} that of birds, is more nearly diffraction limited. Furthermore, because of their small diameter, the cones in the eagle may be inefficient light absorbers and, because of their close packing, not optically well isolated. Both of these effects cause reduced contrast sensitivity so that it would be advantageous for $D\Delta\phi$ in the eagle to be comparatively large for adequate contrast sensitivity at the anatomic sampling frequency $\nu = \nu_s$.

As in the compound eye (Sec. B.III.5), because of aliasing, an animal with $\nu_s < \nu_{co}$ can detect spatial frequencies within the interval $\nu_s < \nu < \nu_{co}$, but these appear to the animal as frequencies lower than $\nu = \nu_s$. We expect that aliasing is difficult to observe under natural environmental conditions; otherwise, the animal would gain more information with a smaller $D\Delta\phi$, i.e., a larger ν_s.

3. Contrast Sensitivity

The contrast sensitivity of the photoreceptor array is given by Eq. (B.27). From this expression, we can determine the angular photoreceptor diameter $\Delta\varrho_r$ and eye parameter $D\Delta\phi$ that maximize the contrast sensitivity at $\nu = \nu_s$ in the highest luminance of the animal's natural environment. The photoreceptor inner segments should touch $\Delta\varrho_r = \Delta\phi$ for maximum contrast sensitivity (SNYDER and MILLER, 1977). Thus, the photon capture of a photoreceptor is proportional to $(D\Delta\phi)^2$ and is consequently independent of eye size and absolute resolving power when comparing eyes with the same $D\Delta\phi$.

4. Limitation to Minimum Eye Length

One would imagine that it is advantageous to have the shortest eye possible. Weighing against this are decided disadvantages. The shorter the eye, the greater the possibility for chromatic aberration, the smaller the depth of focus and the more narrow the photoreceptors necessary for holding $\Delta\phi$ constant. A short depth of focus is correctable by a rapid accommodation response while chromatic aberration is counteracted by having monochromatic vision in the retina region for highest resolving power. Consequently, the minimum photoreceptor diameter limits the axial length of the eye. The smaller the photoreceptor diameter, the more poorly it absorbs light and the more it becomes optically coupled to its neighbors. This effect is intolerable for a photoreceptor center-to-center spacing d_{cc} of less than $1\,\mu m$

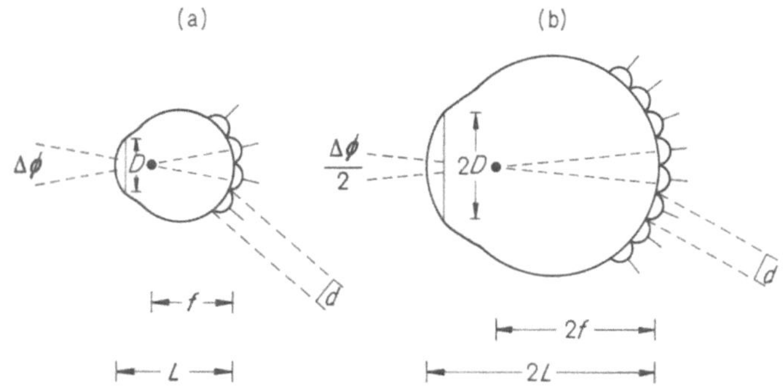

Fig. 18a and b. Doubling resolving power by isomorphic scaling. The resolving power of eye (a) is half that of (b), but the center-to-center cone spacing is the same in both cases

because the contrast sensitivity at $v = v_s$ is negligible. Since $\Delta\phi = d_{cc}/f$, where f is the focal length or more accurately the posterior nodal distance (about 16.7 mm in man), $f = v_s d_{cc}\sqrt{3}$ and the minimum focal length possible must certainly exceed $f = v_s\sqrt{3}$ μm.

5. Change in Resolving Power by Scaling

It is reasonable to suppose that the axial length of an eye is as small as possible or equivalent that the density of photoreceptors is at the maximum tolerable following the logic of the above arguments. Consequently, within any one strategy, say birds of prey, resolving power is determined by head size. In other words, both pupil diameter D and focal length are linearly proportional to the axial length of the eye. In such cases, the resolving power of the eye is doubled by doubling the eye size. This isomorphic scaling procedure is shown in Fig. 18. Note that the F number f/D, the photon capture per receptor $(D\Delta\phi)^2$, and d_{cc} are unchanged by isomorphic scaling, but four times as many photoreceptors are necessary. However, allowing for appropriate neural processing (Section 1.5) larger eyes have potentially greater contrast sensitivity because they capture more photons.

The strategy of Fig. 16 applies to birds of prey (see Miller, this volume) and is analogous to that discussed for the compound eye in Section B.III.3. In these birds, $f \simeq 0.66\,L$ while $D \simeq 0.25\,L$, where L is the axial eye length. In the centermost region of the deep fovea, f may even exceed L due to the negative lens action of the foveal pit (Snyder and Miller, 1978).

6. Dark Adaptation Mechanisms in the Fovea

While it is theoretically possible for each rhabdom to open its acceptance angle to 180° and capture all light entering the ommatidium, each cone is limited to an acceptance angle of $\Delta\phi$. Cones like rhabdoms can be pooled neurally to form larger receptive units as intensity diminishes. Thus, in theory, the gain in contrast sensitivity by neural pooling is the same for vertebrate and compound eyes. However, the highest resolvable frequency at any given luminance is $v_s\,p/p_n(I)$ for the vertebrate instead of $v_s\{p/p_n(I)\}^{1/2}$ as for the compound eye, where $p = D\Delta\phi$, $p_n = D_n\Delta\phi_n$ for compound eye, $D\Delta\phi_n$ for vertebrate eye with the subscript indicating neural quantities. *In other words, comparing a compound eye with a vertebrate eye of the same p value, the vertebrate eye suffers a greater loss in its high frequency resolving power for a given decrease in luminance*, i.e., the vertebrate eye must sacrifice more by neural pooling than the compound eye. This is a consequence of an adjustable rhabdom acceptance angle ($0 < \Delta\varrho < 180°$).

7. Comparative Advantages of Compound and Camera Eye

In this section, we compare the lens and compound eyes with the same resolving power. *For a given resolving power*, the pupil diameter equals the facet diameter while the focal length of the ommatidium and vertebrate eye must be approximately equal. Thus, the depth of both eyes is approximately the same, assuming photoreceptors of equal length. However, the compound eye covers a significantly greater surface area and consequently occupies a significantly greater volume. The increase in surface area of the eye has the potential advantage of a greater field of view in addition to a higher photon capture for the eye. This additional photon capture, together with the advantage of a changeable acceptance angle and the possibility of optical superposition confer an enormous advantage in contrast sensitivity on the compound eye, particularly at spatial frequencies lower than the anatomic sampling frequency $\nu_s = 1/\sqrt{3}\,\Delta\phi$. However, if the animal can cover its head with ommatidia, it could equally cover its head with many lens eyes or several with large pupil diameters, each with the same axial length as an ommatidium. Such a visual system lacks the advantages of a variable acceptance angle at the photoreceptor level, and so at best remains slightly inferior in contrast sensitivity to the theoretic best compound eye design. *Thus, in theory, an animal with a compound eye is better in contrast sensitivity than one with many vertebrate lens eyes each having the same axial length as an ommatidium.* However, the lens eye has a greater number of receptors per 180° of visual angle and thus the advantage of specialized regions of the retina. If man were to have a compound eye with as many facets as cones, he would require an enormous head (KIRSCHFELD, 1976). Many compound eyes only occupy a small region of the head, e.g., mantids, which could be better replaced by two lens eyes of equal size. The smallest hummingbirds have a head size equal to the head of large dragonflies, but their resolving power is ten times greater. Such observations make one wonder if the option for camera eyes was available to these animals. Finally, some invertebrates have lens eyes, e.g., spiders.

C. Absolute Sensitivity

In this section, we are concerned with adaptations that maximize the capture and absorption of light. Because of the inherent particle nature of electromagnetic radiation, light arrives at the eye in irregular bursts of energy or photons. Consequently, all visual information must be extracted from photons, so that a rhabdom is ideally a photon counter.

We have already emphasized (Sec. B.I.4) that the random arrival of photons introduces a fundamental uncertainty in the interpretation of an array of photon counts and that the greater the number of absorbed photons the better the reliability. It is also true that the more sophisticated the animals visual system, the greater the need for efficient photon capture. For example, the finer the animal measures space and time, the more accurately it can determine position and movement. However, an accurate assessment of motion precludes protracted integration in time, just as a precise determination of position precludes extended integrations in space (BARLOW, 1964). Consequently, the animal is forced to optimize its often meagre allocation of photons. We can now begin to appreciate that animals are starved for light, no matter how bright the environmental intensity, and can anticipate many adaptations for maximizing photon capture. Our intent in this section is to provide the theoretic basis for evaluating the absolute sensitivity of an ommatidium. In Section B we considered the effects of optical and neural superposition on absolute sensitivity, so they are not mentioned again here.

I. Light-Gathering Capacity of an Ommatidium

In this section, we consider the capacity of an ommatidium to collect light from a uniform source of angular diameter $\Delta\psi$. The number N of photons absorbed by a rhabdom in the integration time of the eye is easily derived (Appendix B) leading to

$$N = \bar{N}\{1 - e^{-0.69(\Delta\psi/\Delta\varrho)^2}\}, \tag{C.1}$$

where $\bar{N}$ is the number of photons when the source is infinite in extent (Sec. B.I.4) and $\Delta\varrho$ is the width of the acceptance angle at 50% sensitivity

$$\bar{N} = \hat{I}(D\Delta\varrho_r)^2, \tag{C.2}$$

$$(\Delta\varrho)^2 = (\lambda/D)^2 + (\Delta\varrho_r)^2. \tag{C.3}$$

All symbols are defined in Table 1 and described in Section B. We find that, to an accuracy of 90%, the source appears infinite in extent to a photoreceptor when its angular diameter has a value $\Delta\psi > 1.8\,\Delta\varrho$, i.e., $N \cong \bar{N}$, and appears like a distant point source when $\Delta\psi < 0.4\,\Delta\varrho$, i.e., $N \cong 0.69\,\bar{N}(\Delta\psi/\Delta\varrho)^2$.

RODIECK (1973) and KIRSCHFELD (1974) have noted that the pupil diameter D of a human is of the order 10^5 greater than most ommatidia but that the F numbers (f/D) of ommatidia can be equal or greater than the human eye. They concluded that compound eyes do worse than human eyes for detecting distant point sources but as good or better for detecting extended sources. To see this, we neglect any additional light gathering of the crystalline cone, so that $\Delta\varrho_r = d_{Rh}/f$, where d_{Rh} is the diameter of the rhabdom. Thus, N for extended sources is proportional to $(d_{Rh}/F)^2$.

For nonextended sources (Appendix B), i.e., those with $\Delta\psi < 0.4\,\Delta\varrho$, the light-gathering capacity is proportional to $(d_{Rh}/F\Delta\varrho)^2$. Two extreme cases now emerge.

When the acceptance angle is limited by diffraction, $\Delta\varrho \cong \lambda/D$, the light-gathering capacity of the ommatidia is proportional to $(d_{Rh}/F)^2$. Whereas, when $\Delta\varrho \cong d_{Rh}/f$, as it is for *Musca* rhabdomeres 1–6, then the light gathering is proportional to D^2. Allowing for appropriate neural processing, however the greater the summed photon capture of an eye, the greater its potential for detecting both point and extended source. Consequently larger eyes are theoretically superior.

II. Influence of Length, Diameter, and Shape of Rhabdoms on Absorption

The intensity parameter $\hat{I}$ in Eq. (C.2) is proportional to the efficiency ε of photons captured by the rhabdom (SNYDER and PASK, 1973) (see Eq. B.16)

$$\varepsilon = \varepsilon_0 (1 - e^{-\eta \alpha l}), \tag{C.4}$$

where ε_0 is the fraction of photons that have not been diverted by scattering or absorption prior to the rhabdom, α is the length of the rhabdom, l is the absorption coefficient (specific optical density of the visual pigment), and η the fraction of the light that travels inside the rhabdom (SNYDER, 1975 and Sec. G).

If we ignore waveguide effects, $\eta = 1$ and ε is independent of the diameter or shape of the rhabdom. [The diameter d_{Rh} of the rhabdom appears directly in Eq. (C.2)]. Thus, for economy of photopigment, the rhabdom could taper without loss in absorption. Note that absorption increases the longer the rhabdom or the more densely packed the membrane (larger α).

Waveguide effects become important when both $(l/d_{Rh}) \gg 1$ and the waveguide parameter V is less than 2, where (SNYDER, 1975)

$$V = \pi d_{Rh} (n_{Rh}^2 - n_s^2)^{1/2} / \lambda . \tag{C.5}$$

n_{Rh} and n_s are the refractive indexes of the rhabdom and its surround, respectively. In Fig. 19, we present η for both a uniform and atapered rhabdom. KIRSCHFELD and SNYDER (1975) have shown that rhabdomeres 7 and 8 of *Musca* have $1 < V < 2$ for 600 nm $< \lambda <$ 300 nm. Rhabdomeres 1–6 have twice these V values but, unlike 7 and 8, they are tapered with $d_{Rh}/d_0 = 2$.

III. Influence of Rhabdomeric Dichroism on Absorption

Rhabdomeric (microvilli) membrane is dichroic, i.e., its absorption depends on the direction of the electric vector **E** of linearly polarized light (MOODY, 1964; BORN and WOLF, 1970). Measurements of dichroism indicate the degree to which the absorbing dipoles (chromophore group) are aligned within the visual membrane. In Section F this subject is discussed more fully. Our purpose here is to modify Eq. (C.4) for rhabdomeric photoreceptors, i.e., to provide a *macroscopic* description of the effect of dipole alignment within the visual membrane.

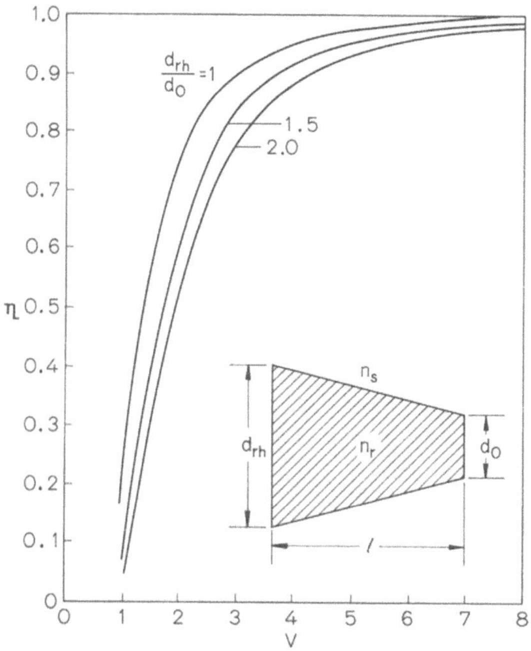

Fig. 19. The fraction η of light power within a photoreceptor of length l

At first sight, it would appear that the only importance of dichroism is to provide a photoreceptor with the potential for polarization sensitivity. This is misleading. Our analysis of dipole alignment within microvilli membrane in Section IV reveals that dichroism is in fact an important determinate of the *absolute* sensitivity of a photoreceptor to *unpolarized light*. However, we must first present the macroscopic description before we can appreciate this last fact.

The dichroic ratio δ of an infinitesimal section of microvilli membrane is defined as

$$\delta = \alpha^x / \alpha^y \tag{C.6}$$

where α^x is the absorption coefficient for **E** parallel to the microvilli membrane and α is the absorption coefficient for **E** perpendicular to the microvilli membrane.

1. Open Rhabdoms

Taking account of dichroism for the rhabdomeres of Fig. 20, Eq. (C.4) becomes (SNYDER, 1975)

$$\varepsilon = \varepsilon_0 \left\{ 1 - \frac{(e^{-\eta\alpha^x l} + e^{-\eta\alpha^y l})}{2} \right\} \qquad \text{(C.7)}^1$$

for *unpolarized light*.

[1] Transcription errors appear in Eqs. (20)–(22) of SNYDER (1975) which are corrected by replacing $\eta/2$ by η and noting a necessary sign change within the brackets of Eq. (C.21) to conform with the result given here.

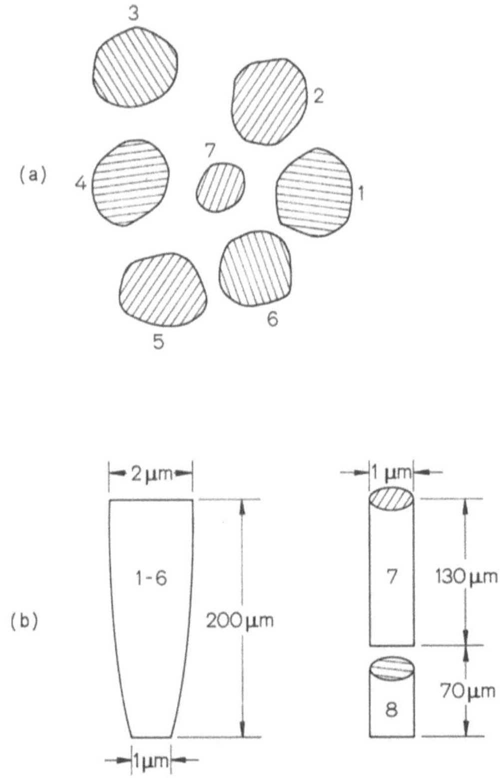

Fig. 20a and b. Rhabdomeres of fly *Musca* eye. (a) is a schema of a cross-section in the distal end. The parallel dark lines represent the microvilli. (b) represents a schema of the rhabdomeres in a longitudinal section. Rhabdomeres 1–6 are tapered with the largest diameter at the distal end. The drawings are from the electron micrographs of MELAMED and TRUJILLO-CENÓZ (1968) and BOSCHEK (1972)

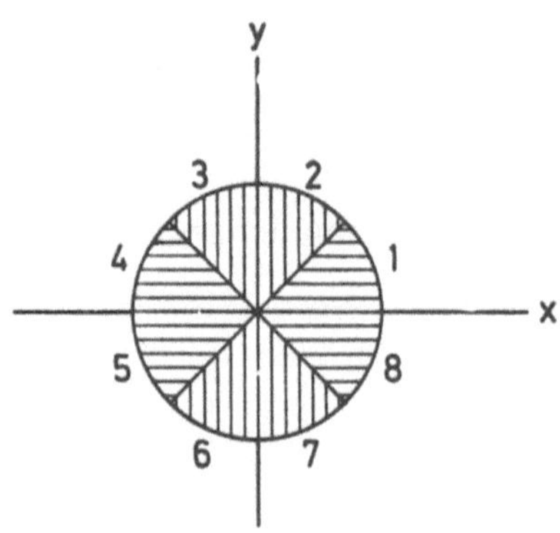

Fig. 21. Hypothetic fused rhabdom, with eight rhabdomeres each with equal cross-sectional area. The dark lines represent the microvilli, which are either parallel or perpendicular to the x and y axes

2. Fused Rhabdoms

The most common photoreceptors of those animals with compound eyes have all rhabdomeres fused into a cylindric light guide called a fused rhabdom (Fig. 21). The participating rhabdomeres are optically (and possibly electrically) coupled to each other and share the total light accepted by the rhabdom. A full treatment of the fused rhabdom is presented by SNYDER (1973) and SNYDER et al. (1973) so that we need only borrow some of their results for the remaining discussion.

For simplicity, we limit our discussion to rhabdoms with the special property that the summed light adsorbed by all rhabdomeres is independent of the direction of **E** (SNYDER, 1973). This is exactly fulfilled by crustacean rhabdoms but is approximated by the rhabdoms of many animals (see also Appendix G). Then Eq. (C.4), for rhabdomere *i*, becomes

$$\varepsilon_i = \varepsilon_0 \left(\frac{A_i}{A}\right) \left(\frac{\alpha_i^x + \alpha_i^y}{2\alpha}\right) (1 - e^{-\eta\alpha l}) \qquad \text{(C.8a)}$$

for *unpolarized light*, where α_i is the absorption coefficient and A_i the area in cross-section of rhabdomere *i*

$$\alpha = (\alpha_1^x A_1 + \alpha_2^x A_2 \ldots \alpha_8^x A_8)/A \qquad \text{(C.8b)}$$

$$= (\alpha_1^y A_1 + \alpha_2^y A_2 \ldots \alpha_8^y A_8)/A\,, \qquad \text{(C.8c)}$$

$$A = A_1 + A_2 + A_3 \ldots A_8\,. \qquad \text{(C.8d)}$$

The superscripts *x* and *y* are for **E** parallel and perpendicular to the microvilli of rhabdomere *i*, respectively. The general situation is discussed by SNYDER and SAMMUT (1973).

If we want the number of photons absorbed by the entire rhabdom, i.e., the summed absorption of all rhabdomeres, Eq. (C.4) becomes

$$\varepsilon = \varepsilon_0 (1 - e^{-\eta\alpha l})\,, \qquad \text{(C.9)}$$

where α is given by Eq. (C.8b). For the special case of a symmetric rhabdom, with eight rhabdomeres, all with equal area and perpendicular microvilli $\alpha = (\alpha^x + \alpha^y)/2$ with α^x and α^y defined as in Eq. (C.6).

IV. Dependence of Photon Capture on Chromophore Alignment

The light-absorbing molecule, the chromophore group retinal, is a dipole and absorbs maximally when **E** is parallel to the π-electron cloud of its conjugated double bonds. Retinal is embedded in the protein molecule (opsin) such that the dipole is parallel to the membrane surface (LIEBMAN, 1975; TÄUBER, 1975).

Now it is well-known from dichroic measurements that the dipoles are arranged differently in the photoreceptor membrane of various animals. For example, the dipoles are arranged at random within the plane of the vertebrate outer segment disc membrane (SCHMIDT, 1938; LIEBMAN, 1975). They are also arranged at random within the plane of the microvillus membrane of fly rhabdomeres 1–6 (KIRSCHFELD and SNYDER, 1975). In sharp contrast, the dipoles appear to be highly aligned along the axis of the microvillus membrane in the crustacean rhabdom (SNYDER, 1973; MOTE, 1974; GOLDSMITH, 1975) and relatively highly aligned in the central rhabdome of the fly (KIRSCHFELD and FRANCESCHINI, 1976). We are compelled to search for a functional basis to explain the diversity in dipole orientation. Why is the visual membrane of one animal more dichroic than that of another? SNYDER and LAUGHLIN (1975) have provided a possible answer to this question. Only a précis of their argument is given here, since an intuitive interpretation of their mathematics is also available (LAUGHLIN et al., 1975).

As is usual in physics, it is advantageous to simplify the problem, preserving only features that contain the essential physics. SNYDER and LAUGHLIN (1976) accomplished this by approximating the cylindric microvillus with a square microvillus. The results are nearly identical to those obtained by the relatively complicated mathematics necessary for an exact analysis of a microvillus (ISRAELACHVILI et al., 1976). However, the square microvillus model provides the intuitive insight helpful for solving the problem (see also Sec. VI).

For simplicity, we neglect the form dichroism and assume perfect dipoles all lying parallel to the microvillus membrane. Such dipoles have no absorption when **E** is perpendicular to their axis. Our conclusions do not require these assumptions, but the necessary mathematics and intuition is simplified. It is rather obvious from Fig. 22 that dipoles should be oriented at random in the plane of the vertebrate disc membrane to have maximal absorption to unpolarized light. In rhabdomeres also, absorption depends on the degree to which the dipoles are aligned with the microvillus axis. Taking this into account, the absorption coefficients for a fly rhabdomere have the form (SNYDER and LAUGHLIN, 1975)

$$\alpha^x = 4\alpha_{\parallel} \cos^2 \phi \,, \tag{C.10a}$$

$$\alpha^y = 2\alpha_{\parallel} \sin^2 \phi \,, \tag{C.10b}$$

where ϕ is the inclination of the dipole axis with the microvillus axis and $\alpha_{\parallel}$ is the absorption of a dipole when **E** is parallel to its axis. Substituting this result into Eq. (C.7) we can calculate the alignment angle ϕ necessary for maximum absorption for any given length absorption ratio $\alpha_{\parallel} l$. The results are shown in Fig. 22.

It is clear that long and short fly rhabdomeres should have their dipoles oriented differently for optimum absorption of unpolarized light. Long rhabdomeres like 1–6 should have $\phi \cong 45°$ which is mathematically equivalent to a random orientation in the membrane. Short photoreceptors like rhabdomere 7 and particularly rhabdomere 8 should have dipoles comparatively more aligned. The membrane dichroic ratio Δ associated with these values of ϕ is also determined by SNYDER and LAUGHLIN (1975). Rhabdomeres 1–6 should have a Δ slightly less than 2:1 whereas rhabdomere 8 could be larger than 8:2 (Fig. 7 of SNYDER and LAUGHLIN, 1975).

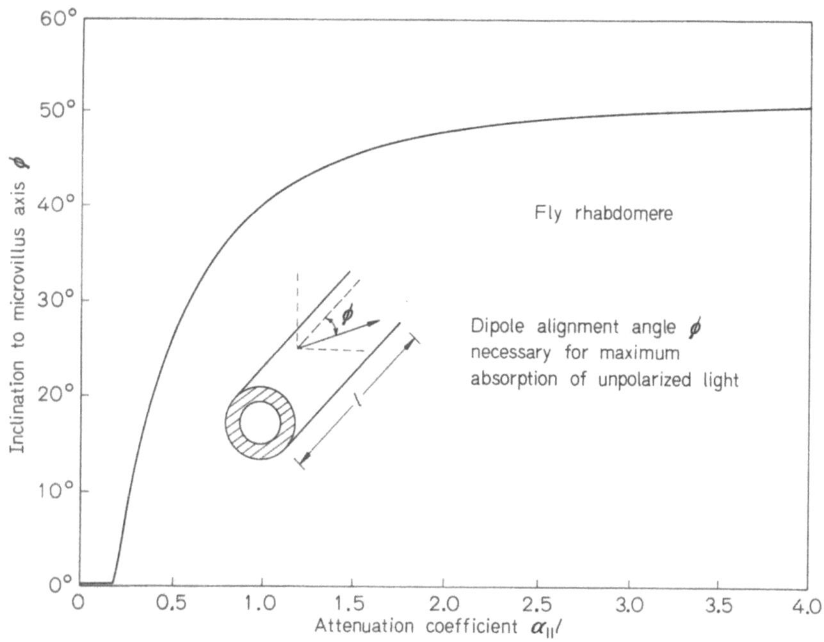

Fig. 22. The dipole alignment angle ϕ necessary for maximum absorption of unpolarized light. $\alpha_{\parallel}$ is the absorption coefficient for **E** parallel to the axis of the microvilli of Fig. 20

The physical principle required to understand Fig. 22 is that, in a long receptor, dipoles at the top act as filters in front of those at the bottom. If all dipoles are perpendicular to the direction of incident light and parallel to each other when unpolarized light is introduced, only the component with **E** parallel to the microvilli is absorbed. The perpendicular component passes through unabsorbed (perfect dipoles). The dipoles at the top absorb most of the light while the dipoles at the bottom are ineffective because they are illuminated by light with **E** perpendicular to the dipoles. Thus, if the rhabdomere is long enough, it is advantageous to randomize dipole orientation.

To determine the dipole alignment necessary for optimum photon capture in fused rhabdoms, we substitute Eq. (C.10) into Eq. (C.9), where $\alpha = \alpha^x + \alpha^y$. When α is maximum, photon capture is maximum. This occurs when $\phi = 0°$ *independent of the length* of the fused rhabdom. Thus, in fused rhabdoms of arbitrary length, like in crustacea, the dipoles need be highly aligned for optimum photon absorption when the light is unpolarized. The dichroic ratio δ was shown by SNYDER and LAUGHLIN (1975) to be theoretically as high as 20 which is not unrealistically large.

In conclusion, the above theory and its consistency with dichroic measurements provide strong evidence for the view that *dipoles are oriented in the plane of visual membrane to provide photoreceptors with the maximum absorption to unpolarized light.* This is consistent with the view that a photoreceptor's primary objective is to maximize the effectiveness of every available rhodopsin molecule and that membrane dichroism is only a secondary consequence of this primary objective.

The physical basis for the high dichroic ratio of only certain rhabdomeric membranes is still open, but recent studies by KIRSCHFELD and FRANCESCHINI (1976) suggest that the presence of a photostable pigment, possibly a carotene, may change the fluidity of the microvillus membrane in some of the No. 7 rhabdomeres of *Musca* and hence affect the alignment of dipoles.

D. Application of Theory to the Fly *Musca*: A Neural Superposition Eye

Having considered the physical limitations to resolving power in addition to the absolute sensitivity of eyes, we now apply our concepts to a real compound eye. This requires *precise* knowledge of $\Delta\phi$, D, and $\Delta\varrho_r$ in the various regions of the eye. Of all compound eyes, only the neural superposition variety has been investigated sufficiently for an in-depth theoretic interpretation. Accordingly, much of our knowledge is derived from the brilliant studies of KIRSCHFELD (1967), KIRSCHFELD and FRANCESCHINI (1968, 1969), FRANCESCHINI and KIRSCHFELD (1971a, b), FRANCESCHINI (1975), and STAVENGA (1975).

I. Basic Principles of Neural Superposition

Musca has an open rhabdom (Fig. 20), i.e., geometrically isolated rhabdomeres with separate fields of view (AUTRUM and WIEDEMANN, 1962; BOSCHEK, 1971). Only rhabdomeres 7 and 8 view the world centered along the axis of their ommatidium. Each of the other six rhabdomeres look in the direction of one of the neighboring ommatidia (KIRSCHFELD, 1967; FRANCESCHINI, 1975; STAVENGA, 1975). Consequently, the interommatidial angles $\Delta\phi$ equal the angular spacing of rhabdomeres, i.e., the center to center spacing d_{cc} between the neighboring rhabdomeres of Fig. 23 obeys the relation

$$\frac{d_{cc}}{f} = \frac{D}{R} = \Delta\phi\,, \tag{D.1}$$

where f is the distance from the distal tips of the rhabdomeres to the posterior nodal point (KIRSCHFELD and FRANCESCHINI, 1968; STAVENGA, 1975). Although the values of f and $\Delta\phi$ depend on the region of the eye (the range of f and $\Delta\phi$ is given in Table 2), the quantity f/D or F number of the ommatidia is nearly the same in all regions of *Musca* studied by STAVENGA (1975) and BEERSMA et al. (1975)

$$F = f/D \cong 1.9\,, \tag{D.2}$$

where D is the diameter of the entrance pupil.

Table 2. Anatomy and theory for *Musca*

D	=	diameter of the entrance pupil which is approximately the facet diameter
f	=	distance from the distal tips of the rhabdomeres to the posterior nodal point which is approximately the focal length of the lens
λ	=	wavelength in vacuum
$\Delta\varrho$	=	width of retinula cell acceptance function at 50% sensitivity
$\Delta\varrho_r$	=	angular diameter of rhabdomere (d_{Rh}/f)

Rhabdomere diameter (distal end) $d_{Rh} \cong 2\ \mu m$; (R1–6) $\cong 1\ \mu m$; (R7 and 8)	*$\Delta\varrho$ for retinula cells 1–6* $\Delta\varrho \cong 0.9\ \Delta\phi$ $4.6° > \Delta\varrho > 2.3°$
Center to center spacing of rhabdomeres $d_{cc} \cong 2.5\ \mu m$	$\frac{\Delta\varrho_r}{\Delta\phi} = \frac{d_{Rh}}{d_{cc}} = 0.8$
F number $= f/D \cong 1.9$ $16\ \mu m < D < 32\ \mu m$ $30\ \mu m < f < 61\ \mu m$	*$\Delta\varrho$ for retinula cells 7 and 8* $\Delta\varrho \cong \{0.59\ \lambda^2 + 0.64\}^{1/2}\ \Delta\phi$ $2.6° > \Delta\varrho > 1.3°$; ($\lambda = 0.49\ \mu m$) $2.4° > \Delta\varrho > 1.2°$; ($\lambda = 0.36\ \mu m$)
Interommatidial angle $\Delta\phi = d_{cc}/f$ $4.8° > \Delta\phi > 2.4°$	*Angular diameter of airy disc $\Delta\varrho_A$* $4.3° > \Delta\varrho_A > 2.2°$; ($\lambda = 0.49\ \mu m$) $3.2° > \Delta\varrho_A > 1.6°$; ($\lambda = 0.36\ \mu m$)
Eye radius $R = D/\Delta\phi$ $189\ \mu m < R < 781\ \mu m$	*Eye parameter p* $p = D\Delta\phi = \frac{D^2}{R} = \frac{d_{cc}}{F}$ $\cong 1.3\ \mu m$

The values used in this table are taken from BOSCHEK (1971), FRANCESCHINI (1975), and STAVENGA (1975).

If $\Delta\phi$ depends on the region of the eye, so must the eye radius R. Since $D = R\Delta\phi$, from Eqs. (D.1) and (D.2), $R = f^2/F d_{cc}$, i.e., R is a constant times f^2. Assuming d_{cc} is constant in all regions of the eye, Table 2 shows that R changes by about a factor of 4 in *Musca*.

Depth of Focus

It is necessary for the rhabdomeres of *Musca* to function as light guides if they are to be optically isolated. This is because the depth of focus $\delta f \cong 2\lambda F^2 \cong 4\ \mu m$ at $\lambda = 500$ nm is significantly less than the rhabdomere length. However, the focal depth of fly optics is sufficient to neglect the defocus caused by chromatic aberrations, which have been measured by MCINTIRE and KIRSCHFELD (1978) to be about 3.3 μm from $\lambda = 300$ nm to 580 nm.

1. Eye Parameter p

From these anatomic facts, we can determine that the eye parameter $p = D\Delta\phi$ given by Eq. (B.2) is nearly constant over the eye and that

$$p = D\Delta\phi = d_{cc}/F \cong 1.3\ \mu m \qquad (D.3)$$

since $d_{cc} \cong 2.5\ \mu m$ (e.g., BOSCHEK, 1971). Thus, *the sampling frequency ν_s of Musca is 5.2 times greater than that necessary to reconstruct the highest frequency* $\nu_{co} = \lambda/D$, *passed by the lens pupil, assuming $\lambda = 500$ nm and square facets or about 4.5 times greater for hexagonal facets (see Sec. B.I).*

The theoretic results of Sections B.II and B.IV suggest that $p \cong 1.3\ \mu m$ is optimum for an animal most active in the late afternoon or at dusk. However, when we account for the high angular motion of *Musca*, which conservatively is 180° turn per second (LAND and COLLETT, 1974), the reasoning of Section B.I.3 shows that this is *equivalent to a reduction of about 3 log units in intensity*. Thus, although *Musca* is active in daylight, its high angular velocity requires the same eye design as an animal in a darker environment.

II. Half Width Acceptance Angle $\Delta\varrho$

Since there is no additional light gathering due to a crystalline cone in *Musca*, the acceptance angle $\Delta\varrho_r$ is simply the angular diameter of the rhabdomeres,

$$\Delta\varrho_r = d_{Rh}/f, \tag{D.4}$$

where d_{Rh} is the diameter of the rhabdomeres. Thus, the acceptance angle of the fly is taken directly from Eq. (B.7) or Fig. 2.4 for the diffraction-limited lens pupil. Using Eq. (D.1), we can express $\Delta\varrho_r$ given by Eq. (4.4) as $\Delta\varrho_r = d_{Rh}\Delta\phi/d_{cc}$.

We showed in Section B.II that the optimum $\Delta\varrho_r$ is $0.75\,\Delta\phi$. Taking $d_{Rh} \cong 2\ \mu m$ for rhabdomeres 1–6 (BOSCHEK, 1971), we find that $\Delta\varrho_r \cong 0.8\,\Delta\phi$ in *Musca* which is very close to the theoretic prediction. For rhabdomeres 7 and 8, $\Delta\varrho_r \cong 0.4\,\Delta\phi$, which is far from optimum.

From Eq. (B.7) we find that the half width acceptance angle $\Delta\varrho$ of *retinula cells 1–6* is

$$\Delta\varrho = \{0.59\,\lambda^2 + 0.64\}^{1/2}\,\Delta\phi, \tag{D.5}$$

and is virtually insensitive to changes in λ, between 350–650 nm, having an average value of about $\Delta\varrho \cong 0.9\,\Delta\phi$. Thus, $\Delta\varrho$ depends on the region of the eye, ranging between $4.6° > \Delta\varrho > 2.3°$ (Table 2). We note that HORRIDGE et al. (1976) found that $\Delta\varrho$ for cells 1–6 in *Calliphora* is independent of λ.

From Eq. (B.7) we find that the half width acceptance angle $\Delta\varrho$ of retinula cells 7 and 8 is

$$\Delta\varrho = \{0.59\,\lambda^2 + 0.15\}^{1/2}\,\Delta\phi, \tag{D.6}$$

which is also rather insensitive to changes in λ (Table 2).

Finally, we have shown in Appendix A that the presence of the pronounced waveguide effects in rhabdomeres 7 and 8 (KIRSCHFELD and SNYDER, 1976) does not significantly alter $\Delta\varrho$, but the shape of the angular sensitivity function is more Gaussian than in cells 1–6.

III. Absolute Sensitivity of Retinula Cells

In this section, we consider the absolute sensitivity of *Musca* retinula cells to extended and point sources. Before beginning, it is necessary to clarify the criterion to be a point or an extended source. It is shown in Appendix B that, as far as one ommatidium is concerned, an extended source is one with an angular diameter that exceeds 1.8 $\Delta\varrho$, while a point source is one with an angular diameter less than 0.4 $\Delta\varrho$. To make these numbers more meaningful, we calculate the separation between two *Musca* necessary for one to appear as an extended or point source to the other. The value of $\Delta\varrho$ depends on the region of the eye and is different for rhabdomeres 7 and 8 and 1–6 (Table 2). We take $\Delta\varrho = 3°$ as a representative value and the length of *Musca* to be 1 cm. Thus, *to appear like extended sources to each other, two Musca must be separated by less than 25 cm, while a separation of more than 100 cm is necessary to appear like point sources to each other.*

1. Sensitivity of *Musca* to Extended Sources

From Section C we know that the absolute sensitivity of a rhabdomere to an extended source is proportional to $(D\Delta\varrho_r)^2$, which in *Musca* is proportional to p^2. Thus, each ommatidium of *Musca* is equally sensitive to an extended source.

2. Sensitivity of *Musca* to Point Sources

From Section C and Appendix B, we learn that the absolute sensitivity of a rhabdomere to a point source is proportional to $D^2(\Delta\varrho_r/\Delta\varrho)^2$. The ratio $\Delta\varrho_r/\Delta\varrho$ depends only on the parameters p and λ. Thus, the sensitivity to point sources depends on D^2. Since the facet diameter D changes by a factor of two over the eye, a point source appears four times brighter in the region of the eye with highest acuity (smallest $\Delta\phi$) than in the region of poorest acuity (largest $\Delta\phi$).

E. Polarization Sensitivity of Retinula Cells

The sensitivity of a retinula cell to the direction of the electric vector **E** of linear polarized light depends not only on the dichroism of the microvilli but also on the gross morphology of the rhabdom in addition to the optical properties of cells above it. Since polarization sensitivity is fully treated by WATERMAN (Part B, this volume), we only consider some of the more theoretic concepts necessary for a quantitative analysis.

I. Origin of Polarization Sensitivity

Nearly all retinula cells show some sensitivity to the direction of **E**. The origin of this polarization sensitivity is the dichroic property of the individual microvillus membrane as discussed in Section F. Nevertheless, evidence is accumulating in favor of the view that in most cases the dichroism of a microvillus is only a by-product of an adaptation to maximize the *absolute* sensitivity of the cell to *unpolarized* light (see Secs. C.IV and F). Recent evidence has in fact shown that the potentially high polarization sensitivity of retinula cells is often deliberately decreased by a twisting of the rhabdom (MENZEL, 1975; SNYDER and MCINTYRE, 1975; WEHNER et al., 1975).

When a retinula cell does show a high polarization sensitivity, its rhabdomere is usually associated with a very specialized arrangement or structure. The best known examples include the crustacean rhabdom with interdigitating orthogonal microvilli and the ninth cell of the bee which is quite short and proximal to the other eight retinula cells (SNYDER, 1973; MENZEL and SNYDER, 1974). Here we give a brief account of polarization sensitivity in several rhabdom types and discuss the effect of twisting.

II. Retinula Cells of Diptera

SHAW (1969b) and SNYDER (1973) have shown that, because of self-screening, the longer a dipteran rhabdomere or the greater the concentration of photopigment within it, the lower its sensitivity to polarized light. The measured polarization sensitivity ratio (PS) of cells 1–6 is between 1 and 2 (KIRSCHFELD and SNYDER, 1975) which is consistent with the view that the dipoles are aligned to optimize absorption of *unpolarized* light. To see this we assume that the absorption coefficient of fly microvilli is between 0.5 and 1.0% per μm and find from Fig. 22 that the dipoles should lie near 45° (or at random) with the microvillus axis.

Because rhabdomeres 7 and 8 are shorter (Fig. 20), they have less self-screening and therefore potentially greater polarization sensitivity. Furthermore, since rhabdom 8 lies beneath rhabdomere 7 and because their microvilli are perpendicular, the polarization sensitivity of cell 8 can be amplified (SNYDER, 1973; GRIBAKIN and GOVARDOVSKII, 1975). Lastly, we note from Fig. 22 that it is advantageous from the point of view of absolute sensitivity to *unpolarized* light for the microvillus dichroism to be greater (higher alignment of dipoles with microvillus axis) in short than long cells. All of these facts taken together lead to the theoretic possibility of a high polarization sensitivity in cell 8. Most of this argument can also be applied to the ninth cell of the bee (SNYDER, 1973; MENZEL and SNYDER, 1974).

III. Retinula Cells of Crustacean Rhabdoms

A theoretic analysis of crustacean rhabdoms shows that the polarization sensitivity of each retinula cell is independent of its absolute sensitivity and exactly equal to the dichroic ratio of the microvilli (SNYDER, 1973). The high polarization sensitivity measured intracellularly (e.g., SHAW, 1969b; GOLDSMITH, 1975) demonstrates that the dichroism of crustacean microvilli is also high. This is an advantage both for polarization sensitivity and absolute sensitivity to unpolarized light, since, for optimum absolute sensitivity, dipoles should be highly aligned in fused rhabdom structures (Sec. C.IV).

IV. Retinula Cells in Fused Rhabdoms

A simple analysis for determining polarization sensitivity of retinula cells that form a fused rhabdom is given by SNYDER (1973). It should be noted that the direction of **E** for maximum retinula cell response need not in general be parallel or perpendicular to the microvilli of the cell (SNYDER and SAMMUT, 1973). When the absorption of the *entire* rhabdom, i.e., the summed absorption of all rhabdomeres, is independent of the direction of **E**, the polarization sensitivity of each retinula cell equals that of its microvilli dichroism (SNYDER, 1973).

Symmetric Fused Rhabdoms

There is a class of fused rhabdoms that is particularly simple to analyze and also have an important physical property. In these rhabdoms, the summed number of photons absorbed by all of the rhabdomeres is independent of the direction of **E**, i.e., the *whole* or macroscopic structure is isotropic (see Appendix G) and *the polarization sensitivity ratio of each retinula cell equals that of the dichroic ratio of the microvilli* (SNYDER, 1973). Many fused rhabdoms approximate this situation. An example of this type of rhabdom is one that has eight rhabdomeres, each with an equal cross-sectional area, equal microvilli dichroism, and equal absorption where four of the rhabdomeres have microvilli oriented perpendicular to the direction of the microvilli of the other four rhabdomeres (Fig. 21).

V. Twisted Fused Rhabdoms

The fused rhabdoms of a number of hymenopterans have a significant structural twist (MENZEL, 1975; WEHNER et al., 1975; MENZEL and BLAKERS, 1976; WEHNER, 1976). As first shown by SNYDER and MCINTYRE (1975), the twist reduces the polarization sensitivity (PS) of the retinula cells. The greater the amount of twist, the greater the reduction in PS. For example, taking a twist of 180° as found by WEHNER et al. (1975) in the worker bee, the analysis of SNYDER and MCINTYRE (1975) shows that the PS is negligible, independent of the amount of microvilli dichroism or birefringence. WEHNER (1976) gives a particularly lucid account of the functional role of twisted rhabdoms in the worker bee.

In Appendix G, we derive a simple analytic expression for the polarization sensitivity ratio PS of a retinula cell within a twisted fused rhabdom. The results are shown in Fig. 23 for a total twist angle of 52°, assuming the rhabdom undergoes a uniform twist rate throughout its length L. The results are for $\alpha_{\|}L=2$, e.g., a 1% absorption per μm and $L=200$ μm, but they are changed insignificantly for values

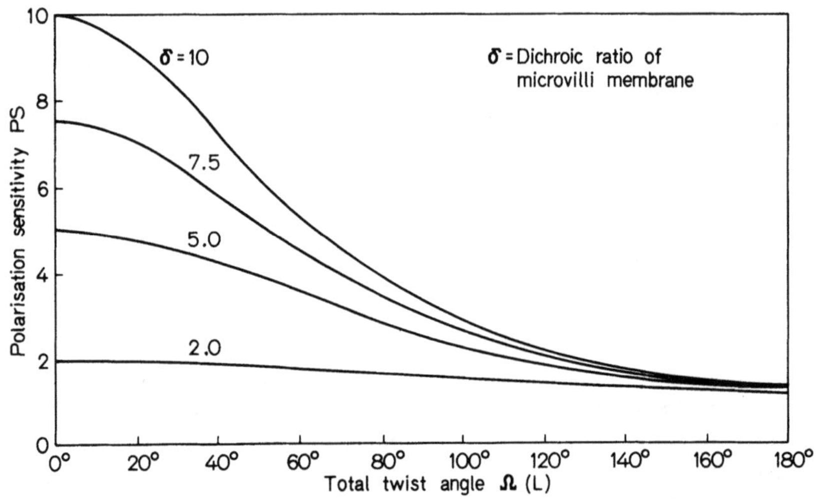

Fig. 23. The polarization sensitivity ratio PS for a retinula cell of a fused rhabdom that is twisted an amount $\Omega(L)$ degrees in its total length L. The twist rate is uniform along the length of the rhabdom. We have taken $\alpha_{\|}L=2$, where $\alpha_{\|}$ is the absorption coefficient of the rhabdomeric membrane when **E** is perpendicular to the microvilli, but the results are changed insignificantly for values of $\alpha_{\|}L$ between 1 and 3

between $\alpha_{\|}L$ equal to 1 and 3, corresponding to 40 and 78% total absorption by the whole rhabdom. It is clear from the figure that 180° twist obliterates PS. This result is intuitive when one realizes that for a 180° twist, an average of half of the length of the rhabdom has microvilli that are perpendicular to the other half. Thus, unless the absorption is so large as to take place principally in the top half, the effect is like crossed polarized lenses.

F. Optical Properties of Rhabdomeric Membrane

The optical properties of a material arise from the various ways that electromagnetic energy interacts with its atoms or molecules. Since every structure is distinguished by a unique arrangement of molecules, its optical properties represent a characteristic "fingerprint", i.e., a positive means of identification. We are concerned here mainly with birefringence and dichroism. Birefringence is the dependence of the refractive index on the direction of the electric vector **E** of linearly polarized light. It indicates the degree of alignment of the membrane substructure, i.e., its crystallinity. Dichroism is the dependence of absorption on the direction of **E** and indicates the degree of alignment of the absorbing dipoles within the membrane. Our purpose in this section is to show how measurements of birefringence and dichroism provide specific information on the properties of visual membrane. Because we must introduce some simplifications in our model of visual membranes, it is advisable to remind ourselves of the complexity of a photoreceptor cell (LAUGHLIN et al., 1975).

I. Review of Photoreceptor Structure

Photoreceptors are specialized cells evolved for high sensitivity to light. The light-absorbing molecule is a dipole embedded in a protein molecule. This chromophore-protein complex, the rhodopsin molecule, is part of the cell membrane, where it is free to undergo lateral and rotational diffusion. The high quantum-capture property of highly evolved photoreceptors is the result of several molecular, fine and gross structural mechanisms: 1. The concentration of rhodopsin molecules within the membrane is extremely high. 2. The membranes holding the rhodopsin molecules are organized in closely packed stacks of discs (vertebrate photoreceptors) or dense packages of tubes (rhabdomeric invertebrate photoreceptors). 3. Light is contained within the photoreceptor structure as the result of the high optical density of these membrane stacks (light guide) (LAUGHLIN et al., 1975; LIEBMAN, 1975). Besides these facts, the quantum-capture property of the whole light-absorbing structure of a photoreceptor is very sensitive to the orientation of the dipole molecule relative to the light path. In general, the dipole must be perpendicular to the light path for highest absorption. This is indeed found in the membrane stacks of the vertebrate rod outer segment (ROS). However, the dipoles are in general not randomly orientated in the plane perpendicular to the light path for maximal absorption of unpolarized light (Sec. C.IV).

1. The Binding of Rhodopsin in Vertebrate Membrane

Direct microspectrophotometric (MSP) measurements show that rhodopsin and its associated photoproducts are localized within the outer segments of rods (LIEBMAN, 1972). Electron microscopy demonstrates that each outer segment is a stack of membrane discs. Each disc consists of a single contiguous membrane enclosing a broad flattened intracellular vacuole (Fig. 24). Rhodopsin is essentially a membrane lipoprotein (WALD, 1973); it can only be extracted using detergents and requires a proportion of bound lipids to show its normal activity of the ROS disc membrane; on bleaching, they withdraw into the hydrophobic phase almost completely. These findings have recently been confirmed by freeze-fracture and freeze-etching techniques, although there is still real doubt as to whether the rhodopsin is bound to the inside or the outside surface of the disc (MASON et al., 1974; cf. RAUBACH et al., 1974). Thus, all evidence points toward rhodopsin conforming to the Singer-Nicholson model of a membrane-bound lipoprotein. It appears to be an approximately spheric molecule of about 4 nm diameter bound to the fluid lipid phase of the disc membrane by strong hydrophobic forces. The diffusion kinetics of rhodopsin in vivo more than confirm this view (see below).

2. Chromophore Orientation in Vertebrate Membrane

The structure of the opsin molecule and in particular the distribution of external hydrophilic and hydrophobic groups must determine the orientation at which rhodopsin sits in the lipid matrix. It also determines the region of the molecule to which the dichroic chromophore, retinal, is bound. Thus, there should be a constant relationship between the position of the hydrophobic/hydrophilic

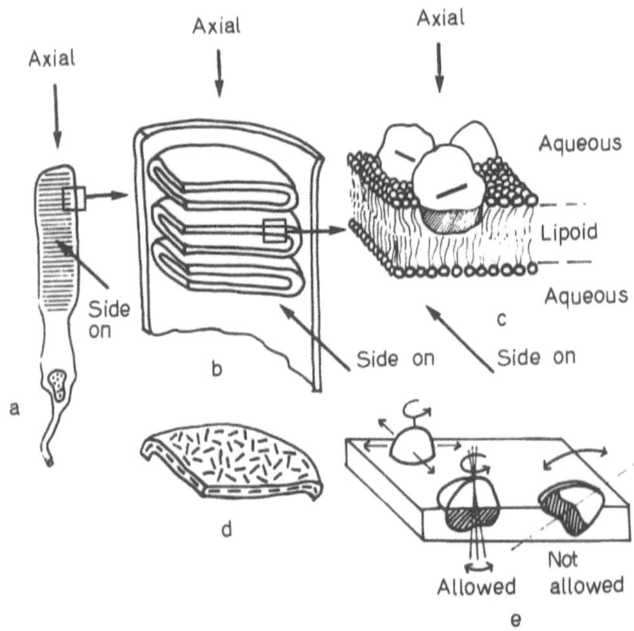

Fig. 24a–e. Structural and molecular organization of the photoreceptive membrane in rod outer segments. (a) A vertebrate rod photoreceptor. The arrows give the two important directions of light paths discussed in the text: axial light travels through the outer segment parallel to its long axis (under physiologic conditions, light reaches the outer segment from the inner segment); with *side-on* illumination, light travels along the short axis perpendicular to the long axis of the outer segment. (b) The disc membrane envelopes of the ROS. Axial light falls normally on the flat disc membrane; side-on light travels parallel to the disc membrane. (c) A fluid membrane model of the disc membrane. Rhodopsin molecules are globular proteins partly embedded in the lipid double layer. The hydrophilic part of the protein (light area) sticks out into the aqueous phase; the hydrophobic part (shadowed) dips into the membrane. The dipole chromophore (black line) is nearly parallel to the membrane surface. (d) A schematic illustration of the randomly oriented dipole molecules in the disc membrane. (e) Rhodopsin molecules are free to move laterally and rotationally within the fluid membrane. Wobbling, however, is strongly reduced by hydrophilic-hydrophobic forces

interface of the membrane as a whole and the position of the chromophore itself. In a wide, flat sea of membrane such as the disc surface, the chromophores should all have the same orientation relative to the plane of the membrane surface itself. This is indeed the case. When the stack of discs constituting the ROS is viewed from the side ("side-on"), the ROS is seen to be highly dichroic. When unbleached, light with an **E** vector parallel to the discs and thus perpendicular to the long axis of the rod is preferentially absorbed relative to light with the **E** vector parallel to the rod long axis (SCHMIDT, 1938). Comparison of the absorption spectrum for perpendicular and parallel polarized light shows that it is the chromophore of rhodopsin that is dichroic (LIEBMAN, 1962; HAROSI and MACNICHOL, 1974). Moreover, the photoproduct metarhodopsin II has the same direction of dichroism, but the free vitamin A molecule released from opsin at the close of bleaching has an orientation perpendicular to the membrane surface (DENTON, 1959; HAROSI and MACNICHOL, 1974). It seems that vitamin A dissolves in the lipid bilayer, the long hydrocarbon chain coming to lie parallel with the lipid chains. In this position, it is parallel to the direction of physiologically incident light and presumably has negligible physiologic absorption.

It is clear that when viewed from the side, the ROS membrane is a dichroic structure. However, the measured dichroism of this membrane is small compared with the measured arthropod polarization sensitivity (PS) values. The most accurate determination of side-on dichroism gives a value of 4–5:1, but careful controls show that the measured value is underestimated as a consequence of scattering of the measuring beam (HAROSI and MACNICHOL, 1974). The magnitude of this error cannot be assessed. For this reason, the upper value of 5 is used for membrane dichroism throughout this work. Note that this value of 5 appears to be uncorrected for form dichroism (MOODY, 1964; see also below).

It is important to emphasize that the measured dichroism of the disc membrane is relatively small. One can postulate two principal reasons as to why dichroism might be reduced: 1. Although the chromophore is perfectly aligned with the membrane, it is not a perfect dipole. Thus, the measured dichroism in vertebrate rods represents the greates possible dichroism that the retinal-opsin system can achieve. 2. The chromophore is an extremely effective dipole but is not perfectly aligned with the surface of the membrane; rather it is tilted. A slight tilt could arise from wobbling of the rhodopsin axes even though this type of rotation is opposed by the strong hydrophobic-hydrophilic bonding (Fig. 24). It is at present impossible to assess the relative importance of these two effects in reducing the measured dichroism. In particular, if the intrinsic dichroism of the chromophore were maximal, then it would be possible to calculate the tilt factor. Unfortunately, to our knowledge no such estimate of rhodopsin's intrinsic dichroism exists. The highest measured dichroism is 5:1, measured in sheared films (WRIGHT, 1973). In this situation, it cannot be assumed that the chromophore dipoles are perfectly orientated.

3. Motion of Rhodopsin Molecules in Vertebrate Membrane

The evidence given above suggests that the rhodopsin chromophore is aligned approximately parallel to the surface of the membrane. However, under normal physiologic illumination, the rod shows no dichroism with respect to light propagating along its optic (long) axis. It is now known that results from a random distribution of dipoles about the axis perpendicular to the membrane disc surface. This random distribution results from the ability of rhodopsin molecules to rotate within their lipid matrix.

When rods are illuminated along their long axis with linearly polarized light, it should be possible to selectively bleach those chromophores whose long axes run predominantly in the same direction as the **E** vector. The bleached photoreceptor should then be dichroic with an absorption maximum at right angles to that of the bleaching light. In practice it is extremely difficult to induce dichroism by this method in the intact cell. Extremely rapid flash photolysis experiments show that this is because the rhodopsin molecules are continually rotating in the membrane. The speed of rotation is so great that within 60 μs the momentary dichroism induced by bleaching with linearly polarized light is totally randomized (CONE, 1972). When the same membrane is fixed with glutaraldehyde, the rhodopsin molecules are loosely bound together, rotation is prevented, and a high dichroism can easily be induced by selective bleaching (BROWN, 1972). From the kinetics and temperature dependence of the rotation effect together with estimates of the diameter of the rhodopsin molecule, it is possible to estimate that the lipid matrix within which rhodopsin is embedded has the viscosity of olive oil (CONE, 1972).

Two further series of elegant selective bleaching experiments show that rhodopsin is not just free to rotate in the lipid matrix; it can diffuse laterally across the matrix as well. When a large frog rod is viewed side-on and a thin longitudinal strip parallel to the optical axis is bleached, there is an initial large decrease in the absorbance measured at the peak wavelength of rhodopsin. This decrease is confined to the area bleached. However, within 40 s, the absorbance increases within the bleached strip which is accompanied by a *decrease* in absorbance in the surrounding unbleached region. Controls show that this effect is a redistribution of bleached and unbleached chromophores resulting from lateral diffusion of molecules in the membrane and that it is abolished by fixation in glutaraldehyde. Again it is possible to calculate the diffusion coefficient and from it derive the viscosity. Surprisingly, it is the same as the viscosity for rotational diffusion (POO and CONE, 1974; LIEBMAN and ENTINE, 1974).

The studies of rhodopsin distribution across disc membrane, using the chromophore's known absorption properties as a "label", provide compelling evidence that the ROS disc membrane conforms to the Singer-Nicholson model of a fluid mosaic membrane. This evidence is supported by the X-ray diffraction data which show that the lipid matrix is a bilayer, the rhodopsin molecules float in this matrix, and that the rhodopsin molecules are not arranged across the surface of the membrane in a definite pattern, but rather seem to be distributed at random (WORTHINGTON, 1974).

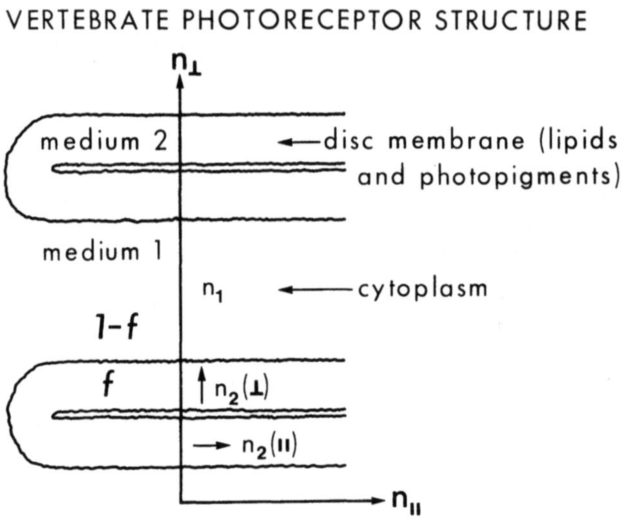

Fig. 25. Schematic vertebrate photoreceptor. The aqueous cytoplasm (medium 1) is isotropic, nonabsorbing, and of refractive index n_1. The disc membrane (medium w) is anisotropic with two intrinsic refractive index components $n_2(\|)$ parallel and $n_2(\perp)$ perpendicular to the discs and two intrinsic absorption coefficients $\alpha(\|)$ and $\alpha(\perp)$. The membrane volume fraction is f

II. Dichroic Property of Photoreceptor Membrane

In this section, we analyze the dependence of absorption on the direction of the electric vector **E** of linearly polarized light. For comparison, we begin with vertebrate visual membrane. The analysis follows that given by SNYDER and LAUGHLIN (1975) and ISRAELACHVILI et al. (1976).

1. Vertebrate Disc Membrane Dichroism

In this section, we discuss the properties of a stack of absorbing slabs as a model for ROS membrane (Fig. 25). The amount of light absorbed when the polarization is parallel to the discs is generally found to be greater than when it is perpendicular to the discs.

The dichroic properties of the photoreceptor are related to the two absorption coefficients $\alpha_{\|}$ and $\alpha_{\perp}$ (Fig. 25) for light polarized parallel and perpendicular to the disc planes. The absorption coefficient α is defined as the fraction of light intensity absorbed per infinitesimal unit length of material; the fraction of light absorbed by a material of finite length l being proportional to $(1-e^{-\alpha l})$. α is related to the molar extinction coefficient ε and concentration c by $\alpha = 2.3\,\varepsilon c$. The dichroic ratio $\alpha_{\|}/\alpha_{\perp}$ of a photoreceptor may be separated into an *intrinsic* and a *form* component. *Intrinsic dichroism* arises when the absorbing photopigments are anisotropically oriented in the disc membranes.

Form dichroism arises from the overall photoreceptor structure and the different refractive indices of the membrane and aqueous regions. This difference in the refractive indices results in a different electric field intensity inside the disc membranes when the electric field polarization of the incoming light is parallel to the discs from that when the polarization is perpendicular to the discs. When the mean photoreceptor field **E** is parallel to the discs, the internal fields are continuous across each boundary (Fig. 25) and therefore the same in both media 1 and 2 and equal to **E**, so that $\mathbf{E}_2 = \mathbf{E}$. But when the mean field **E** is perpendicular to the discs, the electric displacements **D** are now continuous across each boundary and equal to $\mathbf{D} = \varepsilon_\perp \mathbf{E}$, so that $\varepsilon_2(\perp)\mathbf{E}_2 = \mathbf{D} = \varepsilon_\perp \mathbf{E}$. Thus, the ratio of the fields in the absorbing medium 2 when **E** is parallel so that it is perpendicular to the discs is $\varepsilon_2(\perp)/\varepsilon_\perp = (n_2(\perp)/n_\perp)^2$.

Since the rate of absorption of light energy is proportional to the square of the electric field, the ratio of the absorption rates for light polarized parallel and perpendicular to the discs gives us an expression for the form dichroic ratio δ_F due only to the existence of absorbing *isotropic* plates (SNYDER and LAUGHLIN, 1975; ISRAELACHVILI et al., 1976).

$$\text{Form dichroism} = \delta_F \cong (n_2/n_1)^4\,, \tag{F.1}$$

where n_2 is the refractive index of the disc membrane and n_1 of the surrounding cytoplasm. Thus, even when the chromophores are randomly oriented (in three dimensions) within the disc membrane, the dichroic ratio will always be greater than unity, since $n_2 > n_1$. The precise value for n_2/n_1 is uncertain, but ISRAELACHVILI et al. (1976) argue that its most likely value is 1.53/1.34 so that $\delta \cong 1.7$. They also point out that the actual absorbing part of the membrane is the retinal group or chromophore attached to the protein opsin; thus, it is not the refractive index of the membrane that should appear in Eq. (F.1) but that of the retinal. Since the refractive index around the retinal is likely to be close to that of the protein, the form dichroism could well be as high as $(1.60/1.34)^4 \approx 2.0$. Equation (F.1) is an approximation for $f \ll 1$ in Fig. 25. The exact expression is given by ISRAELACHVILI et al. (1976) and shown in Fig. 26.

The intrinsic dichroic ratio δ_I in Fig. 25 is given as $\alpha_{||}/\alpha_\perp$ so that the dichroic ratio δ of the total photoreceptor is

$$\text{Total dichroism} = \delta = \delta_F \delta_I \cong (n_2/n_1)^4 (\alpha_{||}/\alpha_\perp)\,. \tag{F.2}$$

In vertebrate rod outer segments, the form factor δ_F is probably greater than 1.5 and represents a purely structural amplification of the intrinsic dichroism. The most recent measurements of HAROSI and MACNICHOL (1974) show that the maximum $\delta \cong 5$ in frog rods. Correcting for form dichroism, this gives a maximum value of 3 for the intrinsic dichroism of the disc membrane.

2. Dichroism of Rhabdomeric Membrane

Our most sophisticated model for rhabdomeric photoreceptors is presented in Fig. 27, but it is both mathematically and intuitively easier to deal with the box model of the microvillus as introduced by SNYDER and LAUGHLIN (1975). There is

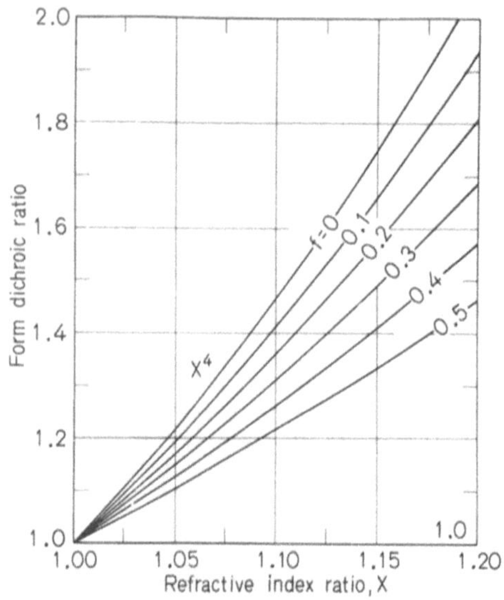

Fig. 26. Form dichroic ratio δ_F or relative absorption $\alpha_{\|}/\alpha_{\perp}$ of a vertebrate photoreceptor (Fig. 25) plotted for no intrinsic birefringence or dichroism. The total or net dichroic ratio is given by multiplying the plotted value by $\alpha(\|)/\alpha(\perp)$. The refractive index ratio $x = n_2/n_1$

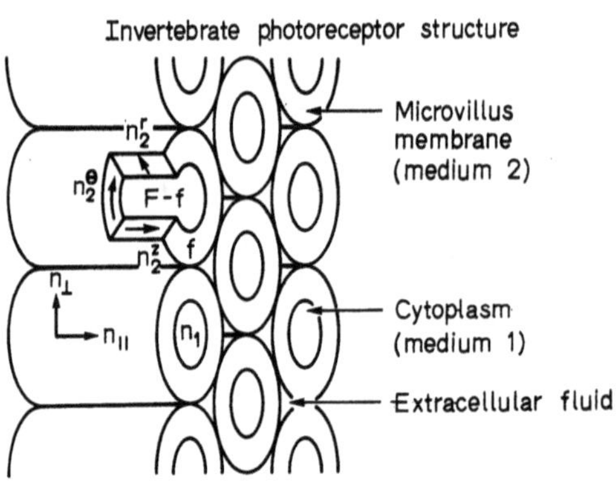

Fig. 27. Schematic rhabdomere-type invertebrate photoreceptor composed of microvilli tubules. The microvilli have a volume fraction F, and each is bounded by the photomembrane (medium 2) of volume fraction f, with three intrinsic refractive index components n_2^r, n_2^θ, and n_2^z, and intrinsic absorption coefficients α^r, α^θ, and α^z. The aqueous cytoplasm and extracellular fluid are assumed to have similar properties (medium 1), being isotropic and nonabsorbing. The bulk photoreceptor refractive indices are $n_{\|}$ and $n_\perp$ and the bulk absorption coefficients are $\alpha_{\|}$ and $\alpha_\perp$. For close packed microvilli $F = \pi/2\sqrt{3} = 0.907$

only about a 1% difference between the results based on either model (ISRAELACHVILI et al., 1976).

We found in Section I above that the disc structure of the vertebrate photoreceptor amplified any intrinsic dichroism. Thus, we would again anticipate some influence by the rhabdomeric structure of Fig. 27. However, it is clear from Fig. 28 that the effect of form dichroism δ_F is comparatively degraded or neutralized

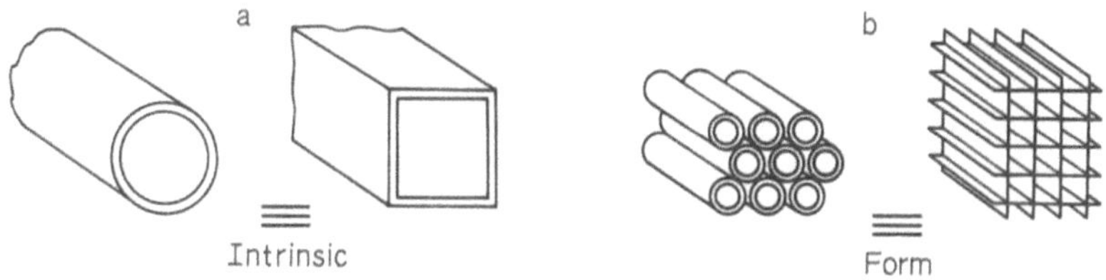

Fig. 28. (a) The individual microvillus as four vertebrate disc outer segments giving rise to the intrinsic component of dichroism. (b) Microvilli (rhabdomeric medium) as crossed sheets of disc membrane giving rise to the form component of dichroism. The fact that a vertebrate disc is formed from two membrane layers separated by an intradisc cytoplasmic phase does not alter the validity of our simple presentation

in rhabdoms because of the presence of crossed plates. The dichroic ratio $\delta = a_{\|}/\alpha_{\perp}$ of rhabdomeric membrane is then (SNYDER and LAUGHLIN, 1975)

$$\delta = \frac{2\alpha^z \delta_F}{\alpha^r + \alpha^\theta \delta_F}, \tag{F.3}$$

$$\delta_F = n_2/n_1, \tag{F.4}$$

where α^r, α^z, α^θ are the absorption coefficients for **E** parallel to the radial, axial, and azimuthal directions of a microvillus. When the individual microvillus lacks intrinsic dichroism, $\alpha^r = \alpha^\theta = \alpha^z$, and $\delta = 2\delta_F/(1+\delta_F)$ or $\delta \cong 1.2$ when $\delta_F = 1.5$. *Thus, the rhabdomeric gross structure has a comparatively small effect on the dichroic ratio of rhabdomere membrane.*

It is rather difficult to estimate the dichroic ratio of rhabdomeric material from Eq. (F.3) without making assumptions about the absorption coefficients α^z, α^r, α^θ, i.e., assumptions about the orientation of the chromophore. If, as did MOODY and PARRISS (1961), we assume ideal dipoles lying flat in the plane of the membrane, i.e., $\alpha^r = 0$ and $\alpha^z = \alpha^\theta$, we find from Eq. (F.3) that the dichroic ratio 8 of rhabdomeric material is 2. *This result is equivalent to assuming an infinite dichroic ratio (measured side-on) for vertebrate disc membrane because, if $\alpha^r = 0$ and the dipoles are ideal, $\alpha_{\|}/\alpha_{\perp}$ is infinite.* Thus, if the dipoles are random in the plane of the microvillus as they are in vertebrate discs, then the largest possible dichroic ratio is 2 for rhabdomeric membrane compared to infinity for vertebrate side-on dichroic measurements.

If we assume that the microvillus membrane is identical to rod disc membrane, we then have $\alpha^z = \alpha^\theta = \alpha_{\|}$ and $\alpha^r = \alpha_{\perp}$ so that from Eq. (F.3)

$$\delta = \frac{2\delta_{\mathrm{vert}}}{1+\delta_{\mathrm{vert}}}, \tag{F.5}$$

where δ_{vert} is the dichroic ratio for disc membrane given by Eq. (F.2). Taking the maximum measured value of $\delta_{\mathrm{vert}} = 5$, the value for the dichroic ratio of rhabdomeric material is 1.67. We note that the dichroic ratio of rhabdomeric membrane of *Musca* rhabdomeres 1–6 ranges from 1 to 2 (KIRSCHFELD and SNYDER, 1976).

Whenever the dichroic ratio of rhabdomeric material exceeds about 1.7, there must be some alignment of the dipoles with the microvillus axis. SNYDER and LAUGHLIN (1975), showed that a mean alignment of about 170 with the microvillus axis gives a dichroic ratio of about 10, while perfect alignment gives about 20. This last value is not excessively greater than the largest dichroic ratio of 13 measured intracellularly in Crustacea by SHAW (1969b).

The *absolute sensitivity* of a photoreceptor to *unpolarized* light depends on the orientation of the dipoles within the membrane and also the gross structure of the photoreceptor. In both vertebrate discs and fly rhabdomeres 1–6, the optimum dipole orientation is a random alignment within the plane of the membrane, but in fused rhabdoms, characteristic of Crustacea, the absolute sensitivity is enhanced by high alignment of dipoles with the microvillus axis (SNYDER and LAUGHLIN, 1975). This is consistent with the measurements and supports the notion that dichroism is only a secondary consequence of the photoreceptor's primary objective-maximizing absolute sensitivity to unpolarized light. Animals that make use of polarization information could appear to do so with highly specialized modifications to the rhabdomeric structure, such as produced by tiered rhabdoms (WATERMAN, this volume, Part B). In fact, animals seem to counteract the effect of the dichroic material by twisting the rhabdom (MENZEL, 1975; SNYDER and MCINTYRE, 1975; WEHNER et al., 1975).

III. Birefringence of Photoreceptor Membrane

In this section, we analyze the dependence of the refractive index on the direction of the electric vector **E** of linearly polarized light. For comparison, we again begin with vertebrate disc membrane. The development follows that presented by ISRAELACHVILI et al. (1976). We emphasize from the outset that sign conventions are *not* consistent throughout the literature. Parallel here is taken to mean parallel to the disc membrane or parallel to the microvillus axis.

1. Birefringence of Vertebrate Disc Membrane

We again refer to Fig. 25. The total birefringence of the vertebrate photoreceptor has a form component Δ_F and an intrinsic component Δ_I.

$$\underset{\substack{\text{total}\\ \text{birefringence}}}{(n_{\|}-n_{\perp})/n_1} = \underset{\substack{\text{form}\\ \text{birefringence}}}{\Delta_F} + \underset{\substack{\text{intrinsic}\\ \text{birefringence}}}{\Delta_I} \tag{F.6}$$

where *we have normalized by dividing with* n_1 and

$$\Delta_F \cong \frac{1}{2} f(1-f)\left(x-\frac{1}{x}\right)^2, \tag{F.7a}$$

$$\Delta_I \cong -\frac{1}{2} f(1+f)\Delta, \tag{F.7b}$$

$$\Delta = \{n_2(\|)-n_2(\perp)\}/n_1, \tag{F.7c}$$

$$X = n_2(\perp)/n_1, \tag{F.7d}$$

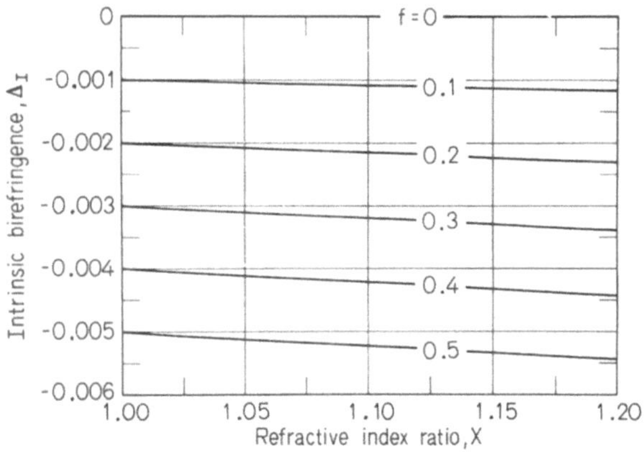

Fig. 29. Intrinsic birefringence Δ_I of vertebrate and invertebrate photoreceptors, plotted for $\Delta = +0.01$ for vertebrates (Fig. 25) and for $(2\delta - \Delta) = +0.02$, $F = \pi/2\sqrt{3}$, for invertebrates (Fig. 27) $x = m_2(\perp)/m_1$. The desired intrinsic birefringence Δ_I is obtained by linearly scaling the plotted value to the real value of Δ or $(2\delta - \Delta)$, i.e., by multiplying the plotted value by $\Delta/+0.01$ or $(2\delta-\Delta)/+0.02$. The curves are reliable to 1%. The total birefringence $(n_{\parallel} - n_{\perp})/n_1$ of the photoreceptor is given by adding the intrinsic birefringence Δ_I to the form birefringence Δ_F as given by Fig. 30 (for vertebrates) and Fig. 31 (for invertebrates)

and f is the volume fraction of the disc membrane (Fig. 25). *We differentiate between Δ_I, the intrinsic birefringence of the bulk structure, and Δ the intrinsic birefringence of the disc membrane.*

Reasonable limits for the variables are:

$$1.08 < X < 1.2\,, \tag{F.8a}$$

$$0.2 < f < 0.5\,, \tag{F.8b}$$

$$-0.02 < \Delta < +0.02\,. \tag{F.8c}$$

The variation of Δ_I with x at different values of f is shown in Fig. 29 plotted for $\Delta = +0.01$. Thus, for any value of Δ different from $\Delta = +0.01$, the value of Δ_I that goes into Eq. (F.6) is given by Δ_I as obtained from Fig. 29 multiplied by $\Delta/0.01$. For example, if $f = 0.3$, $x = 1.1$, and $\Delta = -0.005$, Δ_I is -0.0032 $(-0.005/0.01) = +0.0016$. Figure 30 is included as an example to show how the total birefringence varies with x for $f=0.2$ and $f=0.4$. The solid lines are for form birefringence only ($\Delta = 0$) while the dashed lines are for $\Delta = 0.005$.

Using our sign convention, LIEBMAN and ENTINE (1974) found that $n_{\parallel} - n_{\perp} = +0.0010$ for the total birefringence of frog rods and estimated the form birefringence $n_1\Delta_F = +0.0040$ and the intrinsic birefringence $n_1\Delta_I = -0.0050$. Putting their values for $f=0.37$ and $x = 1.475/1.365$ into our Eq. (F.7a) leads to $n_1\Delta_F = +0.0039$ which is consistent with their estimation. From Eq. (F.7b) we obtain a value for the intrinsic *membrane* birefringence of frog rods of $n_1\Delta = +0.013$. *This last value indicates the degree of crystallinity for disc membrane.*

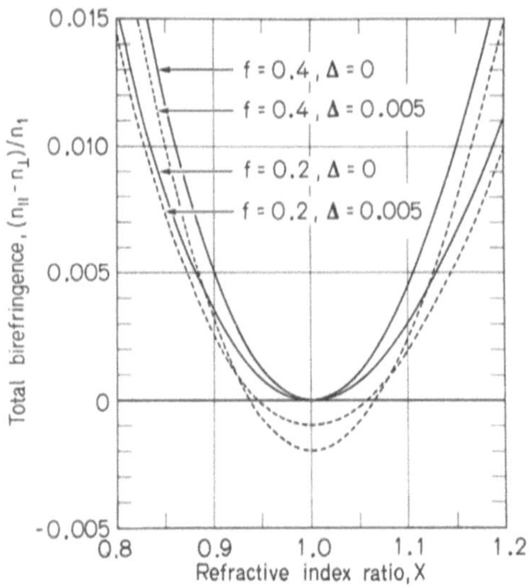

Fig. 30. Total or net birefringence of a vertebrate photoreceptor, illustrating its dependence on x, f, and Δ

2. Birefringence of Rhabdomeric Membrane

We now refer to Fig. 27. The microvillus core is filled with cytoplasm, whereas the external microvillus spaces are filled with extracellular fluid. The intracellular and extracellular regions will be assumed to have identical optical properties, characteristic of aqueous cytoplasm, i.e., isotropic and nonabsorbing and of refractive index n_1. The membrane (medium 2) must now be assumed to be highly anisotropic with three refractive index components: n_2^r (for radial field direction), r_2^θ (for azimuthal field direction), and n_2^z (for axial field direction). We can again use Fig. 28 to gain intuition and derive a simple expression for birefringence of rhabdomeric membrane. For example, it is intuitive that the presence of crossed plates reduces the regularity of the membrane and we can anticipate a comparatively lower bulk form and intrinsic birefringence than for parallel plates. We again have Eq. (F.6) but with

$$\Delta_F \cong \frac{1}{4} f(1-f)\left(x-\frac{1}{x}\right)^2, \tag{F.9a}$$

$$\Delta_I \cong -\frac{1}{4} f(1+x)(2k-\Delta), \tag{F.9b}$$

$$x = n_2^r/n_1, \tag{F.9c}$$

where the *two* intrinsic *membrane* birefringence components are

$$\Delta = (n_2^r - n_2^\theta)/n_1, \tag{F.9d}$$

$$\bar{k} = (n_2^r - n_2^z)/n_1. \tag{F.9e}$$

If the membrane is isotropic within its plane, $n_2^z = n_2^\theta$ and $\bar{k} = \Delta$. Then both form and intrinsic birefringence, and hence the total birefringence, are about half that of the vertebrate rod, assuming equal f, x, and Δ.

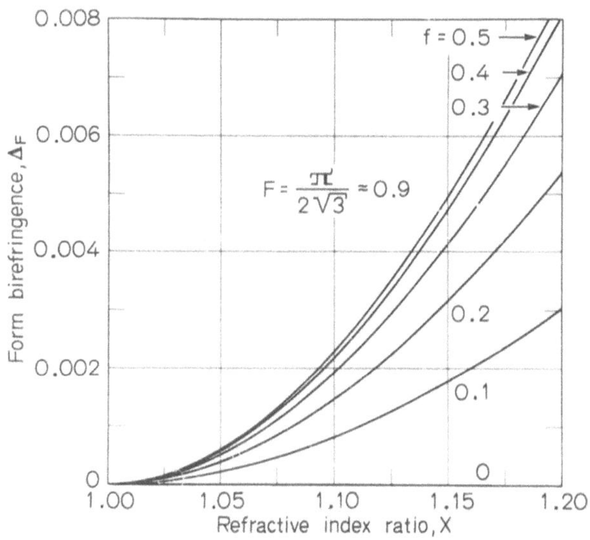

Fig. 31. Form birefringence $\Delta_F = (n_{\|} - n_{\perp})/n_1$ of invertebrate photoreceptors, plotted for $F = \pi/2\sqrt{3}$ (i.e., closely packed microvilli) and no intrinsic membrane birefringence ($\bar{k} = \Delta = 0$). The intrinsic birefringence component Δ_I is given in Fig. 29

The microvilli tubules in invertebrate photoreceptors are usually closely packed (LASANSKY, 1970) so that $F \approx \pi 2\sqrt{3} \approx 0.9069$, although values for F as low as 0.6 are possible (TRUJILLO-CENÓZ, 1972). The membrane volume fraction f is related to F and to the ratio of the inner and outer microvilli radii (a/b) by

$$f = F\left(1 - \frac{a^2}{b^2}\right). \tag{F.10}$$

Figure 31 shows the variation of the form component $\Delta_F \cos x$ at different values for $F = 0.9069$ and $\delta = \Delta = 0$. There is negligible difference between the birefringence for a hexagonal arrangement of microvilli ($F = \pi/2\sqrt{3}$) and a square array ($F = \pi/4$).

Recent measurements of birefringence on the rhabdomeric material of *Musca* (KIRSCHFELD and SNYDER, 1975) show that the large refractive index is obtained when **E** is *perpendicular* to the microvilli of rhabdomeres 1–6 and a total birefringence $n_{\|} - n_{\perp} = -(0.003$ to $0.005)$. By assuming that the density of rhabdomeric membrane is half that of rod disc membrane $f \cong 0.2$, and that the membrane refractive index ratio x is the same for both vertebrate and rhabdomeric material, KIRSCHFELD and SNYDER (1975) found the intrinsic component $n_1 \Delta_I = -(0.0046$ to $0.0066)$, i.e., about the same as measured by LIEBMAN et al. (1974) for disc membrane.

If we also assume $f = 0.2$ and $x \cong 1.08$ as for the vertebrate case, we find from Eq. (F.9b) that the intrinsic *membrane* birefringence $n(2\delta - \Delta) \cong 0.05$, i.e., about four times that of frog rods. *We conclude, therefore, that microvillus membrane of Musca is significantly more crystalline (ordered) than frog outer segment disc membrane.*

G. Waveguide Properties of Visual Photoreceptors

Many animals with highly evolved visual systems have long, narrow photoreceptive structures (outer segments or rhabdomeres) formed by a dense membrane packing with a greater refractive index than the surrounding medium (STOCKHAMMER, 1956; SIDMAN, 1957; SEITZ, 1968; STAVENGA, 1974: KIRSCHFELD and SNYDER, 1975). Consequently, these structures act as light guides, i.e. light is transmitted within them by the mechanism of total internal reflection. Light guides offer the potential for improved optical isolation from neighboring photoreceptors. However, their main advantage may be economy of photopigment for a given capture, because photon capture depends mainly on the length of the photoreceptor (outer segment) being rather insensitive to its diameter or shape (section C.II). This provides one explanation for why many animals have narrow or tapered outer segments. They would probably be even more narrow if it were not for the fact that below a certain diameter wave effects render the structure an inefficient absorber. Thus, the waveguide effects discussed below are most likely a by-product of nature's attempt to minimise the volume of photopigment per photon capture.

Light has been observed to propagate along outer segments as well as rhabdomere cross-sections in patterns known as waveguide modes (ENOCH, 1963; VARELA and WIITANEN, 1970; FRANCESCHINI and KIRSCHFELD, 1971a, b). The occurrence of these patterns is basically a consequence of the small diameter of these structures. The patterns formed by modes most probably have no role in vision, although in theory the fused rhabdom photoreceptors of some arthropods could distinguish between different modes (SNYDER and PASK, 1972; BERNARD, 1975). Nevertheless, the observation of modes serves to emphasize that the photoreceptor is a dielectric or optical waveguide (SNYDER and LOVE, 1980; SNYDER, 1975). Optical waveguides exhibit two properties that might also play a functionally significant role in vision or at least set the limiting performance of an photoreceptors:

1. Only a fraction of a mode's light energy is transmitted within the photoreceptor, the remainder travels along but outside of the photoreceptor. This phenomenon is related to the evanescent field associated with total internal reflection and is strongly wavelength dependent.

2. Because light energy travels outside of the photoreceptor, it is weakly coupled to neighboring photoreceptors and hence cross-talk is unavoidable.

As we shall show, it is necessary to understand the first property particularly when measuring refractive index or birefringence in intact animals.

Waveguide mode effects are specified when the parameter V is known, where (SNYDER, 1975; SNYDER and LOVE, 1980)

$$V = \frac{\pi d_{\mathrm{Rh}}}{\lambda}(n_{\mathrm{r}}^2 - n_{\mathrm{s}}^2)^{1/2}, \tag{G.1}$$

where d_{Rh} is the diameter of the rhabdom, n_{r} and n_{s} are the refractive indexes of the rhabdom and its surround respectively, and λ is the wavelength in vacuum. For

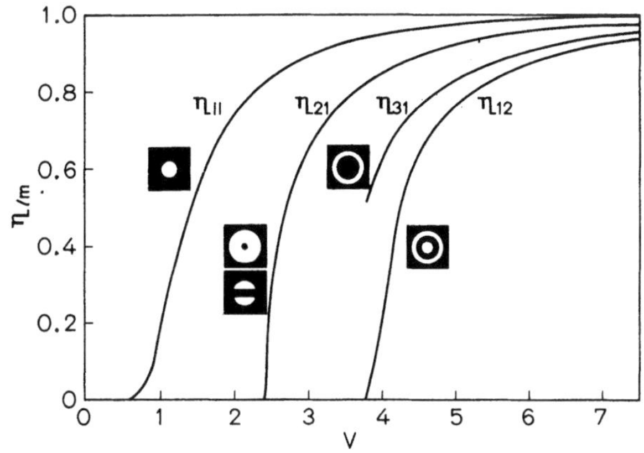

Fig. 32. The fraction η_{lm} of light power within a photoreceptor as a function of V for several low order modes designated by the *lm* notation (SNYDER, 1975). η for mode number 31 does not terminate abruptly as shown, but below $V=3.832$ η depends on the length of the rhabdom (SAMMUT and SNYDER, 1976)

example, the fraction η of light energy within the photoreceptor is shown as a function of V in Fig. 32. *It is assumed here that the length of the rhabdom is greater than ten times its diameter, otherwise waveguide effects are decreased* (SAMMUT and SNYDER, 1976).

It is nearly impossible to determine V using Eq. (G.1) because of the inability to obtain sufficiently accurate values of n_r and n_s representative of in situ conditions. Since $n_r \simeq n_s$, small errors in a measurement of the refractive indices will lead to a larger error in $n_r^2 - n_s^2$ and hence in V. Nor is it possible to determine V from the "cut-off" properties of modes as suggested by ENOCH (1963) because we now know that modes do not exhibit a pronounced cut-off phenomenon (SNYDER, 1974; SAMMUT and SNYDER, 1976). KIRSCHFELD and SNYDER (1975, 1976) circumvented these difficulties by measuring the effect of waveguide modes on birefringence in nearly intact *Musca* rhabdomeres. It is of use to describe the technique of measurement here, since it demonstrates how an appreciation of waveguide mode effects can be used to advantage in obtaining the optical properties of rhabdomeres.

I. Determination of V for Fly Rhabdomeres

The rhabdomeres of *Musca*, shown in Fig. 33, are ideal for measuring waveguide effects. The central rhabdomere is about 1 μm in diameter and has been shown theoretically by SNYDER and MILLER (1972) and SNYDER and PASK (1973) to have pronounced waveguide effects compared to rhabdomeres 1–6, which have a diameter of 2 μm. Consequently, we have a "control" situation within one animal.

Unlike vertebrate photoreceptors, fly photoreceptors are birefringent to light propagating from the physiologic direction along the long axis of the receptor (STOCKHAMMER, 1956; SEITZ, 1969; KIRSCHFELD and SNYDER, 1975), i.e., their

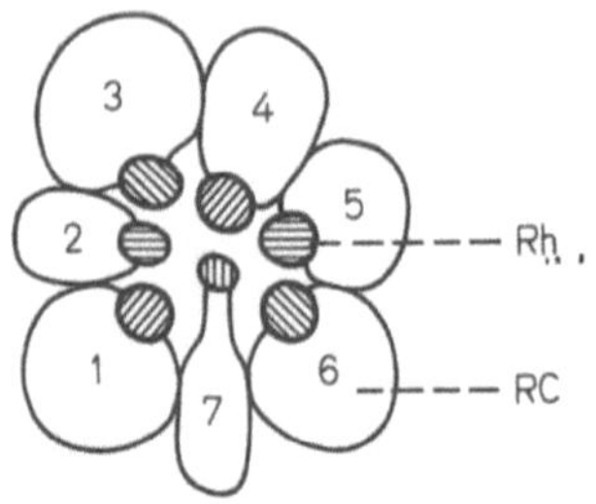

Fig. 33. Cross-section of the photoreceptor cells RC (Nos. 1–7) of the ommatidium of a fly. Rh: rhabdomeres, hatching indicates the direction of the microvilli. Rhabdomeres of cells 1–6 are approximately 2 μm in diameter at the distal end while cell 7 is approximately 1 μm in diameter

refractive index depends on the orientation of the **E** vector of the light. Furthermore, it is possible to accurately measure the birefringence of fly rhabdomeres in nearly undisturbed physiologic conditions.

It is easy to see that the measured (effective or waveguide) birefringence $\langle \Delta n \rangle$ of the rhabdomeres must change with the wavelength of light and their diameter as well, if there are significant waveguide effects. This follows from the fact that only a fraction η of the total energy of a mode is transmitted within the photoreceptor, the remainder $1-\eta$ travels along but outside the structure (SNYDER, 1975). η is highly dependent on V and therefore on the diameter of the structure and on the wavelength of light whenever waveguide effects are significant. Since the measured refractive index is smaller than the actual refractive index of the material for $\eta<1$, the measured birefringence must also depend on η. If $\eta \simeq 1$, the measured birefringence $\langle \Delta n \rangle$ is approximately that of the photoreceptor material Δn, since all the light travels inside the lightguide. When $\eta<1$, a portion of the light travels outside the receptor and $\langle \Delta n \rangle$ is smaller than Δn, since the two characteristic refractive indices of the rhabdomeres approach n_s for $\eta \to 0$ (SNYDER, 1975).

It is implicit from the above discussion that the measured birefringence $\langle \Delta n \rangle$, which includes waveguide effects, equals the birefringence Δn of the photoreceptor material in the absence of waveguide effects multiplied by the fraction η of a mode's light power within the photoreceptor. Thus,

$$\langle \Delta n \rangle = \eta \Delta n, \tag{G.2}$$

where $\langle \Delta n \rangle$ and Δn refer to birefringence due to light from the physiologic direction. This intuitive expression can be derived rigorously from two different theoretic methods (Appendix C). The important point is that the greater the wavelength, the smaller η and hence the smaller the *measured* birefringence. Thus, by measuring birefringence with light *from the physiologic direction*, we obtain information about η and Δn and can by use of Fig. 32 calculate V. The procedure is now described (KIRSCHFELD and SNYDER, 1976).

In order to measure the birefringence, portions of compound eyes of living flies (females of white eye mutant of *Musca*) were cut away and mounted on coverslips in a wet chamber in such a way that the cross-section of the rhabdomeres could be observed by means of a polarization microscope.

Undisturbed sections (rhabdomere lengths 30–50 μm) which are essential for the measurement may be obtained by using a loudspeaker-driven oscillating razor blade for sectioning (frequency 700 Hz, amplitude 50 μm; KIRSCHFELD, 1967). The measurements were performed on white eye mutants of *Musca*, which lack all screening pigments since in the wild type fly pigment granules within the receptor cells move due to illumination of the receptors and change the refractive index of the medium surrounding the rhabdomere (KIRSCHFELD and FRANCESCHINI, 1969). The wavelengths used were $\lambda = 589$ and 656 nm. To our knowledge, the visual pigment in the rhabdomeres at these wavelengths is almost completely photoisomerized into rhodopsin, which by itself shows only negligible absorption at 589 and 656 nm (HAMDORF and ROSNER, 1973; STAVENGA, 1974). Hence, any interference of the visual pigment absorption with the measurements can be avoided. A second reason for selecting these long wavelengths of light is the fact that waveguide effects which are the topic of the investigation are expected to be more prominent and hence more easily detectable in the long wavelength part of the spectrum according to Eq. (G.1); the larger λ, the smaller V and the smaller η (see Fig. 32).

Between crossed polarizers, the birefringent rhabdomeres appear bright if their optical axes (the direction of the microvilli) are 45° to the analyzer orientation. Observed in white light, the rhabdomeres of large diameter (Nos. 1–6 in Fig. 33) appear white, whereas the small rhabdomeres (No. 7 in Fig. 33) appear green. This color is not due to absorption of light; it disappears if analyzer or polarizer are eliminated from the beam path. The effect, however, is exactly what we expect from waveguide theory. The rhabdomeres are more birefringent (and hence appear brighter) at short wavelengths where η is greater and less in the long wavelength part of the spectrum, as can be shown directly by measurements using monochromatic light.

If we know the birefringence Δn of the rhabdomere material in the absence of waveguide effects, we are able to determine η from the measured effective birefringence $\langle \Delta n \rangle$ [Eq. (G.2)]. Since we know η theoretically as a function of V (Fig. 32), we know as well the waveguide parameter of the rhabdomeres. Δn cannot be measured directly, however, since no rhabdomeres of a sufficiently large diameter (allowing a measurement without any waveguide influence) are available in the fly's eye. Nevertheless, we are able to determine V, if we measure the effective birefringence of the rhabdomeres at not one but at two different values of V. This allows us to eliminate Δn from Eq. (G.2). The two different V values necessary can be chosen arbitrarily by either using two different wavelengths of light (measuring at one and the same rhabdomere) or by measuring two rhabdomeres of different diameter in one and the same ommatidium (e.g., rhabdomeres 1 and 7, cf. Fig. 33) keeping the wavelength fixed. In the first case, we assume that Δn is constant within the spectral range used. This seems reasonable since the dispersion of n is the same as that of water with an additional constant (KIRSCHFELD and SNYDER, 1975). In the second case, we assume that Δn is the same for rhabdomeres 1–6 and 7. With these assumptions, the ratio of the two V numbers which we will use for the calculation is known from Eq. (G.1), since the ratios of the wavelength's respective diameters of the rhabdomeres are known.

Using Eq. (G.2), we can then form the ratio of the two measurements

$$\frac{\langle \Delta n_a \rangle}{\langle \Delta n_b \rangle} = \frac{\eta_a}{\eta_b}, \tag{G.3}$$

where a and b indicate numbers for the two different V values. For the known ratio of two V numbers, we are able to calculate the ratio of η_a/η_b as a function of V_a and V_b and we can, therefore, determine V_a and V_b from our measured data and Eq. (G.1).

The ratio of the effective birefringence of five No. 7 rhabdomeres, measured at the wavelengths 656 and 589 nm, was 0.55 ± 0.05 (SDM). We then calculate V at 656 nm for No. 7 rhabdomeres with the method described above (open circle in Fig. 34) as well as for other wavelengths according to Eq. (G.1) (curve intersecting open circle).

Assuming that the birefringence of rhabdomeric material is the same in each photoreceptor, we can determine V for rhabdomeres 1–6 by knowing that their diameter is twice that of rhabdomere 7 (BOSCHEK, 1971). The results are shown in Fig. 34 by the curve intersecting the oblique cross. The ratio of the effective

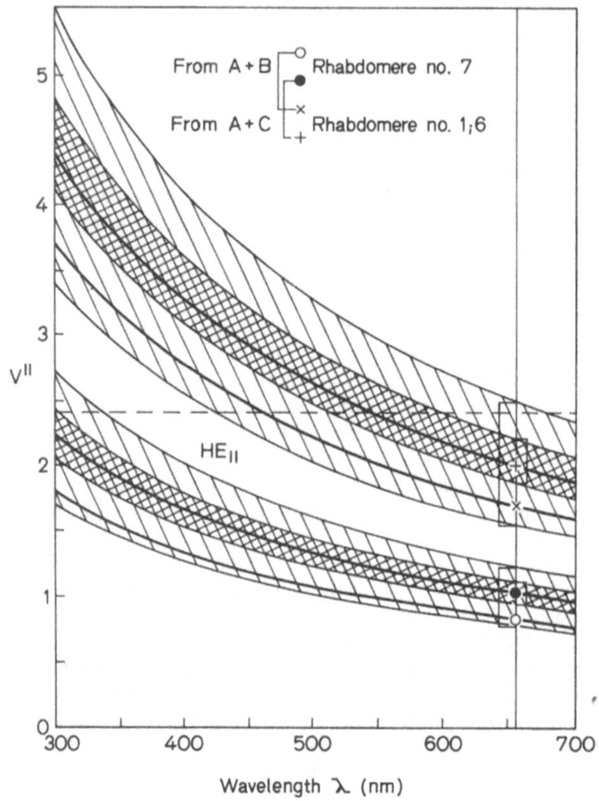

Fig. 34. The optical waveguide parameter V for fly photoreceptors as determined from the effective birefringence. For $V<2.4$, only the first mode HE_{11} may exist. The upper set of curves is for rhabdomeres 1–6 while the lower is for rhabdomere 7. The hatched areas indicate the possible deviation in measured values. Further explanation of the symbols is in the text

birefringence for Nos. 7 and 1 rhabdomeres of five ommatidia was 0.24 ± 0.07 ($\lambda = 656$ nm). From this ratio, we obtain a second determination of V for No. 7 rhabdomeres (dot in Fig. 34 and intersecting curve) and No. 1 rhabdomeres (vertical cross and intersecting curve). There is sufficient coincidence between the results determined from the two sets of data. Hatched regions represent possible deviation in values. Below the horizontal broken curve at $V=2.4$, only one mode can propagate.

Comparison of the curves shown in Fig. 34 with that for η given by Fig. 32 allows us to make some interesting observations for fly photoreceptors. First, only mode 1 can propagate on No. 7 rhabdomeres except perhaps for $\lambda < 340$ nm, consistent with direct observation of mode patterns in rhabdomere cross-sections. Waveguide effects are definitely significant for $\lambda > 350$ nm where η varies rapidly with the wavelength. On rhabdomeres 1–6, two mode families can propagate throughout most of the range of λ, also consistent with direct mode pattern observation. η for the first mode is nearly constant in rhabdomeres 1–6. The determinations of V show that the waveguide properties of No. 7 rhabdomeres are alone sufficient to explain differences in the spectral sensitivity of rhabdomeres 1–6 and 7, as theoretically predicted by SNYDER and MILLER (1972) and SNYDER and PASK (1973). Nevertheless, there is increasing evidence for the view that differences in spectral sensitivities between receptors 1–6 and 7 and/or 8 cannot be explained by waveguide effects *alone*. This is supported by ERG measurements with spectral adaptation in *Drosophila* (STARK, 1975; MINKE et al., 1975; COSENS and WRIGHT, 1975), by analysis of *Drosophila* mutants lacking rhabdomeres in receptors 1–6 or 7 (STARK et al., 1975), by evaluation of an inhibitory influence of receptors 7 and/or 8 onto receptors 1–6 by means of single rhabdomere stimulation in *Musca* (KIRSCHFELD and LUTZ, unpublished), and by microspectrophotometry on individual No. 7 rhabdomeres (KIRSCHFELD and FRANCESCHINI, unpublished) in *Musca* and *Drosophila.*

To summarize our findings, waveguide effects are pronounced when the parameter V is less than 2. Assuming that all rhabdomeric membrane is similar to that of *Musca, significant* waveguide effects occur when the rhabdom diameter is less than or equal to 1 μm. One important fact emerges from this study. Waveguide effects are present in rhabdomeres principally because of the relatively low refractive index, $n_r \cong 1.39$. If n_r were 20% larger, waveguide effects would be insignificant, i.e., $\eta \cong 1$. Since the refractive index of a photoreceptor as well as its photoabsorption depend on the density of the membrane, it is quite probable that rhabdomeric membrane is as densely packed as physically possible.

II. Influence of Waveguide Effects on the Optical Properties of the Rhabdom

If the optical properties of a rhabdom are determined by passing light on-axis through it, then waveguide effects will influence the measurement. The absorption coefficient α then becomes $\alpha\eta$, where η is the fraction of light within the photoreceptor as given by Fig. 32. But, when taking ratios such as dichroic ratios $\alpha_{\|}/\alpha_{\perp}$, the η values cancel. We have already shown that the birefringence becomes $\eta\Delta n$ in the presence of waveguide effects. STAVENGA (1974) has shown that the measurement of refractive index is also modified by waveguide effects but not simply with a multiplication by η. Instead, the measured refractive index $\langle n_r \rangle$ of the rhabdomere is

$$\langle n_r \rangle = k n_r + (1-k) n_s , \tag{G.4}$$

where $k = 1-(U/V)^2$ and U is a waveguide parameter (SNYDER, 1969). Figure 35 shows k vs. V. We emphasize that waveguide effects distort measurements only when light is passed on-axis through a narrow photoreceptor and not when it is passed side-on.

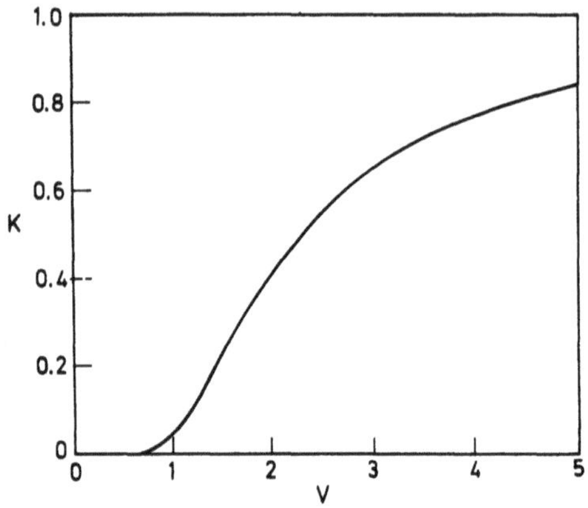

Fig. 35. The waveguide parameter k as a function of V for the first (HE_{11} mode)

Appendix A.

Influence of Waveguide Effects on Acceptance Angle $\Delta\varrho$

In Section B.I we derived an expression for the acceptance angle $\Delta\varrho$ of a retinula cell given by Eq. (3.7) as

$$(\Delta\varrho)^2 = (\Delta\varrho_l)^2 + (\Delta\varrho_r)^2\,, \tag{A.1}$$

where the symbols are defined by Fig. 4 and Table 1. Our derivation ignored the possible influence of waveguide mode effects which were described in Section G. Here, we derive $\Delta\varrho$ for a photoreceptor with pronounced waveguide mode effects. As an example, we consider retinula cell 8 in *Musca.*

In the presence of pronounced waveguide mode effects, i.e., when the parameter V of Section G is less than 2, a photon striking the center of the rhabdomere's cross-section is more likely to be accepted than if it is incident near the periphery. Furthermore, there is a finite probability that a photon will be accepted even if it misses the rhabdomere cross-section altogether (Fig. 36) (SNYDER, 1975). This is to be contrasted to the classic or top hat profile expected from simple geometric considerations. The position r_0 in Fig. 36 is $d_{Rh}/2$. Now, we next explain how to obtain the rhabdomere acceptance profile in Fig. 36.

The probability of observing a photon at a position r in the cross-section of a photoreceptor is proportional to the intensity of the electromagnetic field at that position. The intensity profile A_r is given as (SNYDER, 1969)

$$A_r = J_0^2(Ur/r_0)\,; r \leqq r_0\,, \tag{A2}$$

where J_0 is a Bessel function of order zero and U is a waveguide parameter which is approximated by (SNYDER, 1969)

$$U \cong 2.4\,e^{-1/V}\,. \tag{A3}$$

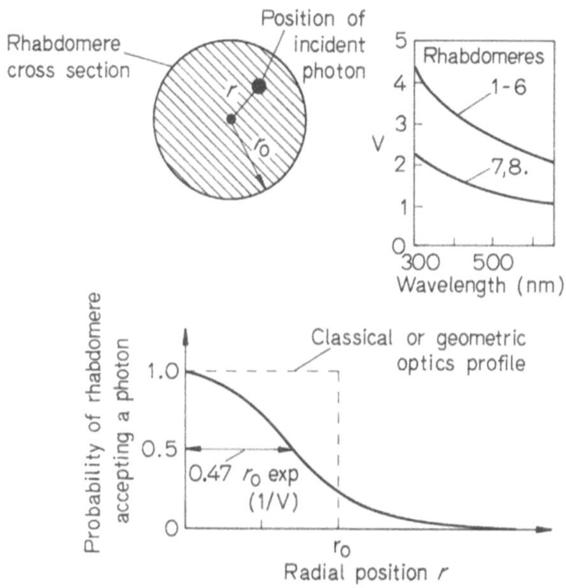

Fig. 36. Probability of rhabdomeres 7 and 8 accepting a photon at position r in its cross-section. The probability profile is proportional to the electromagnetic field intensity of a circular dielectric cylinder. The profile shown is applicable only for $V \cong 2.4$, where V is a characteristic waveguide parameter determined from direct measurements on *Musca* (KIRSCHFELD and SNYDER, 1975). The diameter of the rhabdomere $d_{Rh} = 2r_0$

Now it is a fact that the function J_0^2 is fit well by a Gaussian function of the form

$$A_r \cong e^{-2.77(r/f\Delta\varrho_r)^2}, \tag{A4}$$

where f is the focal length and $\Delta\varrho_r$ is the angular width of the acceptance profile at 50% sensitivity. Since, at 50% sensitivity, $J_0^2(x) = 0.5$, $x = Uf\Delta\varrho_r/d_{Rh} = 1.12$, we have that

$$\Delta\varrho_r \cong 1.12\,(d_{Rh}/f)/U\,. \tag{A5}$$

This expression is valid only when $V < 2$, otherwise more modes are present and $\Delta\varrho_r \rightarrow d_{Rh}/f$ the larger V. In other words, we are only considering the situation of strong waveguide effects. Thus, the problem is reduced to solving Eq. (A1) for $\Delta\varrho$. The parameter V has been measured by KIRSCHFELD and SNYDER (1976) (see Sec. G) for rhabdomeres 7 and 8 and is shown in the insert of Fig. 36. We take $\Delta\varrho_l = \lambda/D$ so that

$$\Delta\varrho = \frac{\lambda}{D}\left\{1 + 0.22\left(\frac{d_{Rh}}{\lambda F}\right)^2 e^{2/V}\right\}^{1/2}, \tag{A6}$$

where $F = f/D$. STAVENGA (1975) has found that $F \cong 1.9$ in most ommatidia of *Musca* where $d_{Rh} \cong 1\,\mu m$ in rhabdomeres 7 and 8. The most extreme waveguide effects occur in rhabdomere 8, presumed to be a blue-sensitive cell ($\lambda \cong 0.47\,\mu m$) (HARRIS et al., 1976). For $\lambda \cong 0.47\,\mu m$, $V \cong 1.4$, so that $\Delta\varrho \cong 1.61\,\lambda/D$ rather than

$1.65\,\lambda/D$ had we used the classic profile (taking $\Delta\varrho_r = d_{Rh}/f$). *Thus, the presence of pronounced waveguide effects does not significantly alter the acceptance angle $\Delta\varrho$ from that given by Eq. (B.7). However, the shape of the acceptance function should be more Gaussian shaped than cells 1–6 and the shorter cell* 7. The reason is that the larger diameter of rhabdomeres 1–6 decrease the waveguide effect as does the short length of rhabdomere 7 (SAMMUT and SNYDER, 1976).

Appendix B.

Criteria to be a Point or Extended Source

If we ignore diffraction and aberrations, then it is intuitive that, as far as a rhabdom is concerned, an extended source is one in which the angular diameter $\Delta\varrho_r$ of the rhabdomere is less than the angular diameter $\Delta\psi$ of the source, i.e., $\Delta\psi > \Delta\varrho_r$. Similarly, a point source obeys $\Delta\psi \ll \Delta\varrho_r$. We now modify these conclusions by accounting for the imperfect optics as expressed by the angular sensitivity function A of the retinula cell.

The number $n(\phi)$ of photons absorbed by a retinula cell during the integration time of the eye (due to a distant point source, inclined at angle ϕ to the axis of the ommatidium) is

$$n(\phi) = n(0)\,A(\phi) = n(0)\,e^{-2.77(\phi/\Delta\varrho)^2}\,. \tag{B1}$$

We have assumed a Gaussian acceptance function A, where $\Delta\varrho$ is the half width acceptance angle (Fig. 1). The number of photons absorbed N in one integration time, due to a uniform source of angular diameter $\Delta\psi$, is

$$N = 2\pi \int_0^{\Delta\psi/2} n(\phi)\sin\phi\, d\phi \tag{B2a}$$

$$\cong 2\pi \int_0^{\Delta\psi/2} \phi n(\phi)\, d\phi \tag{B2b}$$

$$= (\pi/2.77)(\Delta\varrho)^2 n(0)\{1 - e^{-0.69(\Delta\psi/\Delta\varrho)^2}\} \tag{B2c}$$

assuming the source is centered on the axis of the ommatidium.

Now it is intuitive that $n(0)$ is proportional to Q times the pupil area $\pi D^2/4$, where Q is the fraction of photons entering the rhabdom

$$Q = \frac{\int_0^\infty \phi A_r\, d\phi}{\int_0^\infty \phi A\, d\phi} = \left(\frac{\Delta\varrho_r}{\Delta\varrho}\right)^2. \tag{B3}$$

This result is also an excellent approximation for the case when A_r is a step function and A_l the diffraction intensity pattern.

It is possible to obtain $n(0)$ by the convolution process given in Fig. 4 when the heights of the functions are properly normalized. The A_r is correct, but A_l must be properly normalized so that the volume under A_l is proportional to the pupil area. Convolution of A_l with A_r will then give Q for $\phi = 0$.

From Eqs. (B1), (B2), and (B3) and the logic of Section I.D, we have an expression for the number of photons absorbed by a uniform source of angular diameter $\Delta\psi$

$$N = \hat{I} D^2 (\Delta\varrho_r)^2 \{1 - e^{-0.69(\Delta\psi/\Delta\varrho)^2}\} \tag{B4a}$$

$$= \bar{N} \{1 - e^{-0.69(\Delta\psi/\Delta\varrho)^2}\}, \tag{B4b}$$

where $\hat{I}$ is defined in Table 1.

When $\Delta\psi = 1.8\Delta\varrho$, to an accuracy of 90%, we can replace the bracketed portion of Eq. (B2b) by unity resulting in the expression for an infinite uniform source. When $\Delta\psi = 0.4\,\Delta\varrho$, to an accuracy of 90%, we can replace the bracketed portion of Eq. (B4) by $0.69(\Delta\psi/\Delta\varrho)^2$ so that $N = 0.69(D\Delta\psi)^2(\Delta\varrho_r/\Delta\varrho)^2$, i.e., the same dependence on D, $\Delta\varrho_r$, and $\Delta\varrho$ of a distant point source.

In conclusion, as far as the light gathering of a rhabdom is concerned, *when the angular diameter $\Delta\psi$ of the source obeys $\Delta\psi > 1.8\Delta\varrho$, the source appears infinite in extent to a rhabdom of an ommatidium centered beneath it. When $\Delta\psi < 0.4\,\Delta\varrho$, the source has the behavior of a distant point.*

Appendix C.

Influence of Waveguide Effects on Birefringence of a Rhabdomere

In Section G we stated that the birefringence $\langle\Delta n\rangle$ of a rhabdomere, when measured along its long axis, is given by

$$\langle\Delta n\rangle = \eta\Delta n\,, \tag{C1}$$

where η is the fraction of light power within the rhabdomere and Δn is the birefringence of the rhabdomeric material. Here we provide two different derivations of Eq. (C1). They both have in common the assumption that the birefringence is small, i.e., $\Delta n \ll 1$.

Birefringence is the difference in the refractive index due to light polarized in one direction and that polarized in a perpendicular direction. The refractive index of a material is the ratio of the speed of light in vacuum to its speed in the material. If mode effects are significant, some of the light travels in the medium external to the photoreceptor. Thus, the speed of light traveling along the receptor represents some average of the speed in the photoreceptor and the surrounding medium.

1. Solution by Coupled Mode Theory

The easiest way to derive Eq. (C1) is by appealing to the philosophy of coupled mode theory (SNYDER, 1972). Here we begin with a photoreceptor of refractive index $n_{||}$ and ask what happens if it changes slightly to $n_{\perp}$, i.e., a change of amount Δn. The modal propagation constant $\beta_{||}$,

$$\beta_{||} = 2\pi \langle n_{||} \rangle / \lambda; \tag{C2}$$

where $\langle n_{||} \rangle$ is the effective refractive index in the presence of mode effect, changes from $\beta_{||}$ to $\beta_{\perp}$

$$\beta_{\perp} = 2\pi \frac{\langle n_{||} \rangle}{\lambda} \left(1 + \frac{\eta \Delta n}{n_{||}}\right). \tag{C3}$$

This last result is found from Eq. (14) in SNYDER (1974), i.e., from the perturbed propagation constant. We can then calculate $\langle \Delta n \rangle$ as follows:

$$\langle \Delta n \rangle = \left(\frac{\lambda}{2\pi}\right) (\beta_{||} - \beta_{\perp}) \tag{C4a}$$

$$= \eta \Delta n \tag{C4b}$$

assuming $\langle n_{||} \rangle \cong n_{||}$, which is true because the refractive index of the rhabdomeric material is only slightly greater than the surrounding medium.

2. Solution by Effective Refractive Index

STAVENGA (1974) has shown that the measured or effective refractive index $\langle n_r \rangle$ of a rhabdomere is

$$\langle n_r \rangle = k n_r + (1-k) n_s, \tag{C5}$$

where n_r and n_s ale the refractive indices of the rhabdomere and surround respectively and

$$k = 1 - (U/V)^2, \tag{C6}$$

where U is a waveguide parameter and V is defined in Section G. Now by expressing Eq. (C5) for parallel and perpendicular components of **E** and then subtracting the two expressions we find that

$$\langle \Delta n \rangle = \Delta n + \left(\frac{U^{||}}{V^{||}}\right)^2 (m_r^{||} - n_s) - \left(\frac{U^{\perp}}{V^{\perp}}\right)^2 (n_r^{\perp} - n_s), \tag{C7}$$

where $\langle \Delta n \rangle = \left\langle n\frac{1}{\mathrm{r}}\right\rangle - \langle n_{\mathrm{s}}^{\|}\rangle$. Assuming that $\Delta n \ll 1$, we can then express all quantities involving perpendicular components in terms of parallel components by using a Taylor series expansion. For example

$$V^{\perp} \cong V^{\|}(1+\varepsilon), \tag{C8a}$$

$$\varepsilon = \Delta n/2(n_{\mathrm{r}}^{\|} - n_{\mathrm{s}}), \tag{C8b}$$

and also

$$\begin{aligned} U^{\perp} &= U(V^{\|} + \varepsilon V^{\|}) \\ &\cong U(V^{\|}) + \varepsilon V^{\|}\left(\frac{dU}{dV}\right)_{V^{\|}}. \end{aligned} \tag{C9}$$

Using the relationships

$$\frac{dU}{dV} = \frac{U}{V}\left(1 - \frac{1}{\xi}\right), \tag{C10}$$

$$\eta = \left(\frac{U}{V}\right)^2 \left\{\frac{V^2 - U^2}{U^2} + \frac{1}{\xi}\right\} \tag{C11}$$

both taken from SNYDER (1969) where ξ is a waveguide parameter, we can then show, after tedious algebra, that Eq. (C1) is correct.

Appendix D.

Two-Point Resolution

In this section, we consider two-point resolution. We know from Appendix B that the number of photons $n(\phi)$ absorbed by a rhabdom due to a distant point source is given as

$$n(\phi) = n(0)\,A(\phi), \tag{D1}$$

where $n(0)$ is proportional to $(D\Delta\varrho_{\mathrm{r}}/\Delta\varrho)^2$ (see Table 1 for definitions) and

$$A(\phi) = \mathrm{e}^{-2.77(\phi/\Delta\varrho)^2}. \tag{D2}$$

Now suppose there are two-point sources separated by angle ϕ. In order to resolve them, there must be one relatively unilluminated ommatidium between the two directly illuminated ones. Thus, the point sources are separated by an angle $\phi = 2\Delta\phi$. The rhabdom of the central relatively unilluminated ommatidium absorbs n_0 photons, where

$$n_0 = 2n(\Delta\phi) = 2n(0)\,A(\Delta\phi). \tag{D3}$$

The two directly illuminated rhabdoms each absorb n_+ photons, where

$$n_+ = n(0) + n(2\Delta\phi) \tag{D4a}$$

$$= n(0)\{1 + A(2\Delta\phi)\} \cong n(0). \tag{D4b}$$

We can now define a signal amplitude σ_{sig} in analogy with that due to the sinusoidal grating discussed in Section A.4.

$$\sigma_{\text{sig}} = (n_+ - n_0)/2 \tag{D5a}$$

$$\cong \{n(0)/2\}\{1 - 2A(\Delta\phi) + A(2\Delta\phi)\} \tag{D5b}$$

while the noise amplitude σ_{noise} is given by

$$\sigma_{\text{noise}} = \left\{\frac{n_0 + n_+}{2}\right\}^{1/2} \tag{D6a}$$

$$= \left\{\frac{n(0)}{2}\right\}^{1/2} \{1 + 2A(\Delta\phi) + A(2\Delta\phi)\}^{1/2}. \tag{D6b}$$

Thus, the signal to noise ratio SNR is

$$\text{SNR} = \left\{\frac{n(0)}{2}\right\}^{1/2} \frac{\{1 - 2A(\Delta\phi) + A(2\Delta\phi)\}}{\{1 + 2A(\Delta\phi) + A(2\Delta\phi)\}^{1/2}}. \tag{D7}$$

The eye design that optimizes the resolution of two points is found by determining the value of $\Delta\varrho_r$ that maximizes Eq. (D7). The optimum value of $\Delta\phi$ is that which gives threshold resolution (SNR = 1) at each given intensity. Unlike the acuity task of sinusoidal gratings or a white noise pattern, it is not advantageous to increase $\Delta\phi$ indefinitely as intensity decreases.

Appendix E.

Arbitrary Lattice of Ommatidia

In this section, we consider ommatidia with an arbitrary (but periodic) lattice of visual axes. The pupil is assumed to be circular and to have the same area for each arrangement. The highest angular spatial frequency ν_s that can be sampled by an array of photoreceptors is

$$\nu_s = 1/2\phi_{sd}, \tag{E1}$$

where ϕ_{sd} is the angular distance between the centers of neighboring *sampling* photoreceptors. For example, from Fig. 2, $\phi_{sd} = \Delta\phi$ for square lattice and $\phi_{sd} = \sqrt{3}\,\Delta\phi/2$ for an hexagonal lattice.

Now, using Eq. (E1), we repeat the steps of Section D leading to Eqs. (B.21) and (B.22) with $\Delta\phi$ replaced by ϕ_{sd}, so that p becomes $D\phi_{sd}$. Thus, all of the results in this paper for p of a square lattice of ommatidia apply equally to $D\phi_{sd}$ for an arbitrary lattice. The diffraction-limited case $v_s = v_{co} = D/\lambda$ from Eq. (E1) gives $D\phi_{sd} = \lambda/2$ as in the square lattice. Thus, *the intensity necessary for a diffraction-limited eye to be the optimum design is the same for all packing geometries.* However, the highest anatomic resolving power v_s results from hexagonal packing.

The *minimum* number of photons $\bar{N}$ and intensity parameter $\hat{I}$ necessary for a diffraction-limited eye to be the optimum design is found from Eqs. (B.15) and (B.16) setting $v = v_{co} = D/\lambda$ with $\Delta\varrho_r = 0.75\phi_{sd}$ (with $D\phi_{sd} = \lambda/2$) leading to

$$\log \tilde{m}^2 \bar{N} = 3.53\,, \tag{E2}$$

$$\log \tilde{m}^2 \hat{I} = 4.38 - \log \lambda^2\,, \tag{E3}$$

with λ in μm, e.g., $\log \tilde{m}^2 \hat{I} = 4.98$ when $\lambda = 0.5$ μm. *Note that the longer the wavelength, the greater the intensity must be.*

Appendix F.

Consequences of Parameter p Constant in all Regions of the Eye

In this section, we consider the consequences of having the parameter $p = D\Delta\phi$ [see Eq. (B.2)] constant in all regions of an eye. We assume the eye is designed at optimum for the light-adapted state discussed in Section II.C so that $\Delta\varrho_r = 0.75\Delta\phi$ and $\bar{N} = 0.56\hat{I}(D\Delta\phi)^2$. *Thus, when $D\Delta\phi$ is constant over the eye, every ommatidium has the same light-gathering capacity for extended uniform sources.*

The number N of photons captured by a rhabdom due to a distant point source is proportional to the facet diameter squared D^2 times a fraction Q that depends only on the ratio $\Delta\varrho_r/\Delta\varrho_l$, i.e., to within a constant $N = D^2 Q(\Delta\varrho_r/\Delta\varrho_l)$, (see Appendix B). For a light-adapted eye with a diffraction-limited lens pupil, the ratio of the two acceptance angles (see Table 1) is

$$\frac{\Delta\varrho_l}{\Delta\varrho_r} = \frac{\lambda/D}{0.75\Delta\phi} = \frac{1.33\lambda}{D\Delta\phi}\,. \tag{F1}$$

Thus, when $D\Delta\phi$ is constant over all regions of the eye, each ommatidium receives the same fraction of the total light entering the facet due to a distant point source.

Appendix G.

Polarization Sensitivity of Individual Retinula Cells of a Twisted Fused Rhabdom

In this section, we derive an expression for the polarization sensitivity of individual rhabdomeres of a fused rhabdom, twisted uniformly along its length. The analysis follows from the theory of SNYDER (1973) and SNYDER and MCINTYRE

(1975). The rhabdom is assumed to have the property that the total number of photons absorbed is independent of the direction of **E**. This condition is approximated by many fused rhabdoms, e.g., rhabdoms like that of Fig. 21 with eight rhabdomeres of equal cross-sectional area, equal dichroism, and equal absorption coefficients. Although we use this symmetric rhabdom for our following example, the results apply to any rhabdom that is *macroscopically* isotropic.

When considering the total number of photons absorbed by the entire rhabdom, we can assume that the rhabdom is isotropic with an absorption coefficient $(\alpha_{\|}+\alpha_{\perp})/2$, where $\alpha_{\|}$ and $\alpha_{\perp}$ are the absorption coefficients of rhabdomeric membrane when **E** is parallel and perpendicular to the axis of the microvilli membrane. Thus, the total number of photons $\tilde{n}(z)$, in the cross-section, at any position z along the length of the rhabdom is

$$\tilde{n}(z)=\tilde{n}(0)\,\mathrm{e}^{-(\alpha_{\|}+\alpha_{\perp})z/2} \tag{G.1}$$

independent of the direction of **E**. Now the number of photons $d\tilde{n}$ absorbed in a differential length of *rhabdomere* dz is proportional to $\tilde{n}(z)$ as well as the fractional cross-sectional area of the rhabdomere (taken here to be one-eighth because all eight rhabdomeres are assumed to have equal area) multiplied by the absorption coefficient of the rhabdomere. However, we must distinguish two cases. When **E** is parallel to the microvilli of the rhabdomere

$$(d\tilde{n})_{\|}=\frac{1}{8}\alpha_{\|}\tilde{n}(z)\,dz \tag{G2}$$

and when **E** is perpendicular to the microvilli of the rhabdomere

$$(d\tilde{n})_{\perp}=\frac{1}{8}\alpha_{\perp}\tilde{n}(z)\,dz \tag{G3}$$

so that when **E** is at angle ψ with the microvillus axis,

$$d\tilde{n}=(d\tilde{n})_{\|}\cos^2\psi+(d\tilde{n})_{\perp}\sin^2\psi \tag{G4a}$$

$$=\frac{dz}{8}\tilde{n}(0)\,\mathrm{e}^{-(\alpha_{\|}+\alpha_{\perp})z/2}\,\{\alpha_{\|}\cos^2\psi+\alpha_{\perp}\sin^2\psi\}\,. \tag{G4b}$$

When the rhabdom is twisted, angle ψ is a function of position z along the length of the rhabdom so that the number of photons absorbed N by a rhabdomere of a twisted rhabdom is found by integrating Eq. (G4b) throughout the total length L of the rhabdom.

$$N\{\psi(0)\}=\frac{n(0)}{8}\int_0^L \mathrm{e}^{-(\alpha_{\|}+\alpha_{\perp})z/2}\{\alpha_{\|}\cos^2\psi+\alpha_{\perp}\sin^2\psi\}\,, \tag{G5}$$

where ψ is given as $\psi(z)=\psi(0)-\Omega(z)$ and $\Omega(z)$ is the amount of twist at position z. Assuming a uniform twist rate gives $\Omega(z)=z\Omega(L)/L$. The integrals are easily evaluated in closed form by integration by parts. The polarization sensitivity ratio PS of a retinula cell is defined as

$$PS = \frac{N\{\psi_{max}(0)\}}{N\{\psi_{max}(0)+90°\}}, \tag{G6}$$

where $\psi_{max}(0)$ is the inclination of **E** to the microvilli axis that gives maximum N. Although only elementary algebra is necessary for solving these equations, the expressions are long and therefore only the results (Fig. 23) are presented.

References

Autrum, H., Wiedemann, I.: Versuche über den Strahlengang im Insektenauge. Z. Naturforsch. **17**b, 480—482 (1962).

Barlow, H.B.: The size of ommatidia in apposition eyes. J. Exp. Biol. **29**, 667—674 (1952).

Barlow, H.B.: Physical limits of visual discrimination. In: Photophysiology (ed. A.C. Giese), Vol. II, pp. 163—202. New York: Academic Press 1964.

Barros-Pita, J.C., Maldonado, H.: A fovea in the praying mantis eye. II. Some morphological characteristics. Z. Vergl. Physiol. **67**, 79—92 (1970).

Beersma, D.G.M., Stavenga, D.G., Kuiper, J.W.: Optical organization of the eye of the fly *Musca domestica* and behavioural consequences. J. comp. Physiol. **102**, 305—320 (1975).

Bernard, G.: Physiological optics of the fused rhabdom. In: Photoreceptor optics (eds. A.W. Snyder, R. Menzel). Berlin-Heidelberg-New York: Springer 1975.

Blakemore, C., Muncey, J.P.J., Ridley, R.M.: Stimulus specificity in the human visual system. Vision Res. **13**, 1915—1931 (1973).

Born, M., Wolf, E.: Principles of optics, p. 367. New York: Pergamon Press 1970.

Boscheck, C.B.: On the fine structure of the peripheral retina and lamina ganglionaris of the fly, *Musca domestica*. Z. Zellforsch. **118**, 369—409 (1971).

Bracewell, R.: The Fourier transform and its applications. New York: McGraw-Hill 1965.

Brown, P.K.: Rhodopsin rotates in the visual receptor membrane. Nature (Lond.) New Biol. **236**, 35—38 (1972).

Campbell, F.W., Gubisch, R.W.: Optical quality of the human eye. J. Physiol. (Lond.) **186**, 558—578 (1966).

Campbell, F.W., Robson, J.G.: Application of Fourier analysis to the visibility of gratings. J. Physiol. (Lond.) **197**, 551—566 (1968).

Carlson, A.B.: Communication systems, p. 337. New York: McGraw Hill 1968.

Collett, T.S., Land, M.F.: Visual control of flight behaviour in the hoverfly, *Syritta pipiens* L. J. comp. Physiol. **99**, 1—66 (1975).

Cone, R.A.: Rotational diffusion of rhodopsin in the visual receptor membrane. Nature (Lond.) New Biol. **236**, 39—43 (1972).

Cosens, D., Wright, R.: Light elicited isolation of the complementary visual input systems in white-eye *Drosophila*. J. Insect Physiol. **21**, 1111—1120 (1975).

Crawford, B.H.: The dependence of pupil size upon external light stimuli under static and variable conditions. Proc. roy. Soc. Lond. B**121**, 376—395 (1936).

Denton, E.J.: The contribution of the oriented photosensitive and other molecules to the absorption of whole retina. Proc. roy. Soc. Lond. B**150**, 78—89 (1959).

Døving, K.B., Miller, W.H.: Function of insect compound eyes containing crystalline tracts. J. gen. Physiol. **54**, 250—267 (1969).

Dvorak, D., Snyder, A.W.: The relationship between visual acuity and illumination in the fly, *Lucilia sericata*. Z. Naturforsch. **33**c, (1978).

Enoch, J. M.: Optical properties of the retinal receptors. J. opt. Soc. Amer. **53**, 71—85 (1963).

Fermi, G., Reichardt, W.: Optomotor reactions of the housefly *Musca domestica*. Kybernetik **2**, 15—28 (1963).

Fox, R., Lehmkuhle, S., Westendorf, D. H.: Falcon visual acuity. Science **192**, 263—265 (1976).

Franceschini, N.: Sampling of the visual environment by the compound eye of the fly: fundamentals and applications. In: Photoreceptor optics. (eds. A. W. Snyder, R. Menzel). Berlin-Heidelberg-New York: Springer 1975.

Franceschini, N., Kirschfeld, K.: Etude optique *in vivo* des éléments photorécepteurs dans l'oeil composé de *Drosophila*. Kybernetik **8**, 1—13 (1971a).

Franceschini, N., Kirschfeld, K.: Les phénomènes de pseudopupille dans l'oeil composé de *Drosophila*. Kybernetik **9**, 159—182 (1971b).

French, A. S., Snyder, A. W., Stavenga, D. G.: Image degradation by an irregular retinal mosaic. Biol. Cybernetics **27**, 229—233 (1977).

Georgeson, M. A., Sullivan, G. D.: Contrast constancy: deblurring in human vision by spatial frequency channels. J. Physiol. (Lond.) **252**, 627—656 (1975).

Goetz, K. G.: Die optischen Übertragungseigenschaften der Komplexaugen von *Drosophila*. Kybernetik **2**, 215—221 (1965).

Goldman, S.: Information theory. New York: Prentice Hall 1953.

Goldsmith, T. H.: The polarization sensitivity—dichroic absorption paradox in arthropod photoreceptors. In: Photoreceptor optics (eds. A. W. Snyder, R. Menzel), pp. 392—409. Berlin-Heidelberg-New York: Springer 1975.

Goldsmith, T. H., Bernard, G. D.: The visual system of insects. In: The physiology of insecta, Vol. II, 2nd ed. New York: Academic Press 1974.

Goodman, J. W.: Introduction to Fourier optics. New York: McGraw Hill 1968.

Gribakin, F. G., Govardovskii, V. I.: The role of the photoreceptor membrane in photoreceptor optics. In: Photoreceptor optics (eds. A. W. Snyder, R. Menzel), pp. 215—236. Berlin-Heidelberg-New York: Springer 1975.

Hamdorf, K., Rosner, G.: Adaptation und Photoregeneration im Fliegenauge. J. comp. Physiol. **86**, 281—292 (1973).

Harosi, R. E., MacNichol, E. F., Jr.: Dichroic microspectrophotometer: a computer assisted, rapid, wavelength-scanning photometer for measuring the linear dichroism of single cells. J. opt. Soc. Amer. **64**, 903—918 (1974).

Harris, W. A., Stark, W. S., Walker, J. A.: Genetic dissection of the photoreceptor system in compound eye of *Drosophila melangaster*. J. Physiol. (Lond.) **256**, 415—439 (1976).

Horridge, G. A.: Optical mechanism of clear zone eyes. In: The compound eye and vision of insects, (ed. G. A. Horridge). Oxford: Clarendon Press 1975.

Horridge, G. A.: The compound eyes of insects. Sci. Am. **237**, 108—120 (1977).

Horridge, G. A., Giddings, C., Stange, G.: The superposition eye of skipper butterflies. Proc. roy. Soc. Lond. B**182**, 457—495 (1972).

Horridge, G. A., Mimura, K., Hardie, R. C.: Fly photoreceptors. III. Angular sensitivity as a function of wave length and the limits of resolution. Proc. roy. Soc. Lond. B**194**, 151—177 (1976).

Israelachvili, J., Sammut, R., Snyder, A. W.: Birefringence and dichroism in invertebrate photoreceptors. J. Opt. Soc. Amer. **65**, 221—222 (1976).

Kirschfeld, K.: Die Projektion der optischen Umwelt auf das Raster der Rhabdomere im Komplexauge von *Musca*. Exp. Brain Res. **3**, 248—270 (1967).

Kirschfeld, K.: The absolute sensitivity of lens and compound eyes. Z. Naturforsch. **29**c, 592—596 (1974).

Kirschfeld, K.: The resolution of lens and compound eyes. In: Neural processing in visual systems (eds. F. Zettler, R. Weiler). Berlin-Heidelberg-New York: Springer 1976.

Kirschfeld, K., Franceschini, N.: Optische Eigenschaften der Ommatidien im Komplexauge von *Musca*. Kybernetik **5**, 47—52 (1968).

Kirschfeld, K., Franceschini, N.: Ein Mechanismus zur Steuerung des Lichtflusses in den Rhabdomeren des Komplexauges von *Musca*. Kybernetik **6**, 13—22 (1969).

Kirschfeld, K., Franceschini, N.: The possible role of photostable pigments within the membrane of photoreceptors. In: Biophysics of structure and mechanism. Contribution presented at EMBO Workshop October, 1976.

Kirschfeld, K., Snyder, A. W.: Waveguide mode effects, birefringence, and dichroism in fly photoreceptors. In: Photoreceptor optics (eds. A. W. Snyder, R. Menzel). Berlin-Heidelberg-New York: Springer 1975.
Kirschfeld, K., Snyder, A. W.: Measurement of a photoreceptor's characteristic waveguide parameter. Vision Res. **16**, 775—778 (1976).
Kirschfeld, K., Wenk, P.: Dorsal eye of simuliid flies. Z. Naturforsch. **31**c, 764—765 (1976).
Kuiper, J. W.: On the image formation in a single ommatidium of the compound eye in diptera. In: The functional organization of the compound eye (ed. C. G. Bernhard), pp. 35—50. Oxford: Pergamon Press 1966.
Kuiper, J. W., Leutscher-Hazelhoff, J. T.: Linear and nonlinear responses from the compound eye of *Calliphora erythrocephala*. In: Cold Spring Harbor Symp. Quant. Biol. **30**, 419—428 (1965).
Land, M. F., Collett, T. S.: Chasing behaviour of houseflies (*Fannia canicularis*). J. comp. Physiol. **89**, 331—357 (1974).
Lasansky, A.: Ultrastructural features of and extra-cellular ionic pathways in the *Limulus* ommatidium. Neurosci. Res. Program Bull. **8**, 467—469 (1970).
Laughlin, S. B.: Receptor function in the apposition eye. An electrophysiological approach. In: Photoreceptor optics (eds. A. W. Snyder, R. Menzel) pp. 479—498. Berlin-Heidelberg-New York: Springer 1975.
Laughlin, S. B., Menzel, R., Snyder, A. W.: Dichroism, absorption, and the polarisation sensitivity of photoreceptors. In: Photoreceptor optics (eds. A. W. Snyder, R. Menzel). Berlin-Heidelberg-New York: Springer 1975.
Le Grand, Y.: Light, colour, and vision, p. 84. London: Chapman and Hall, 2nd ed. 1968.
Liebman, P. A.: *In situ* microspectrophotometric studies on the pigments of single retinal rods. Biophys. J. **2**, 161—178 (1962).
Liebman, P. A.: Microspectrophotometry of photoreceptors. In: Handbook of sensory physiology (ed. H. J. A. Dartnall), Vol. VII/1. Berlin-Heidelberg-New York: Springer 1972.
Liebman, P. A.: Birefringence, dichroism, and rod outer segment structure. In: Photoreceptor optics (eds. A. W. Snyder, R. Menzel). Berlin-Heidelberg-New York: Springer 1975.
Liebman, P. A., Entine, G.: Lateral diffusion of visual pigment in photoreceptor disc membrane. Science **185**, 457—459 (1974).
Liebman, P. A., Jagger, W. S., Kaplan, M. W., Borgoot, F. G.: Membrane structure changes in rod outer segments associated with rhodopsin bleaching. Nature (Lond.) **251**, 31—36 (1974).
Lillywhite, P. G., Laughlin, S. B.: Transducer noise in a photoreceptor. Nature (Lond.) (submitted) (1977).
Mallock, A.: Proc. roy. Soc. Lond. B**55**, 85 (1894).
Mallock, A.: Divided composite eyes. Nature (Lond.) **110**, 770—771 (1922).
Mason, W. T., Fager, R. S., Abrahamson, E. W.: Structural response of vertebrate photoreceptor membranes to light. Nature (Lond.) **247**, 188—191 (1974).
Mazokhin-Porshnyakov, G. A.: Insect vision. New York: Plenum Press 1969.
McIntyre, P., Kirschfeld, K.: Chromatic aberration of a Dipteran Corneal Lens (in preparation) (1978).
Melamed, J., Trujillo-Cenóz, O.: The fine structure of the central cells in the ommatidia of dipterans. J. Ultrastruct. Res. **21**, 313—334 (1968).
Menzel, R.: Polarization sensitivity in insect eyes with fused rhabdoms. In: Photoreceptor optics (eds. A. W. Snyder, R. Menzel). Berlin-Heidelberg-New York: Springer 1975.
Menzel, R., Blakers, M.: Colour receptors in bee eye: morphology and spectral sensitivity. J. comp. Physiol. **108**, 11—33 (1976).
Menzel, R., Snyder, A. W.: Polarized light detection in the bee, *Apis mellifera*. J. comp. Physiol. **88**, 247—270 (1974).
Minke, B., Wu, C. F., Pak, W. L.: Isolation of light induced response of the central retinula cells from the electroretinogram of *Drosophila*. J. comp. Physiol. **98**, 345—355 (1975).
Moody, M. F.: Photoreceptor organelles in animals. Biol. Rev. **39**, 43—86 (1964).
Moody, M. F., Parriss, J. R.: The discrimination of polarised light by *Octopus*: a behavioural and morphological study. Z. vergl. Physiol. **44**, 268—291 (1961).
Mote, M. I.: Polarization sensitivity. A phenomenon independent of stimulus intensity or state of adaptation in the retinula cells of the crabs *Carcinus* and *Callinectes*. J. comp. Physiol. **90**, 389—403 (1974).
O'Neill, E. L.: Introduction to statistical optics. Reading, Massachusetts: Addison-Wesley 1963.

Petersen,D.P., Middleton,D.: Sampling and reconstruction of wave number limited functions in N-dimensional euclidean spaces. Information and Control. **5**, 279—293 (1962).

Pierce,J.R.: Symbols, signals, and noise. New York: Harper and Row 1961.

Poo,M., Cone,R.A.: Lateral diffusion of rhodopsin in the photoreceptor membrane. Nature (Lond.) **247**, 438—441 (1974).

Portillo,J. del: Beziehungen zwischen den Öffnungswinkeln der Ommatidien, Krümmung und Gestalt der Insektenaugen und ihrer funktionellen Aufgabe. Z. vergl. Physiol. **23**, 100—145 (1936).

Raubach,R.A., Nemes,P.P., Dratx,E.A.: Molecular organisation of rod disc membrane. Exp. Eye Res. **18**, 1—19 (1974).

Robson,J.G.: Receptive fields. In: Handbook of perception. New York: Academic Press 1976.

Rodieck,R.W.: The vertebrate retina, principles of structure and function. San Francisco: W.H.Freeman and Co. 1973.

Rose,A.: Vision, human, and electronic. New York-London: Plenum Press 1973.

Sammut,R., Snyder,A.W.: Ambiguity in determination of waveguide parameter *V* from mode cutoffs. Vision Res. **16**, 881—882 (1976).

Schmidt,W.J.: Polarisationsoptische Analyse eines Eiweiß-Lipoid-Systems, erläutert am Außenglied der Sehzellen. Kolloid-Z. **85**, 137—148 (1938).

Seitz,G.: Der Strahlengang im Appositionsauge von *Calliphora erythrocephala* (Meig.). Z. vergl. Physiol. **59**, 205—231 (1968).

Seitz,G.: Polarisationsoptische Untersuchungen am Auge von *Calliphora erythrocephala* (Meig.). Z. Zellforsch. **93**, 525—529 (1969).

Shaw,S.R.: Interreceptor coupling in ommatidia of drone honey bee and locust compound eyes. Vision Res. **9**, 999—1029 (1969a).

Shaw,S.R.: Sense cell structure and interspecies comparisons of polarized light absorption in arthropod compound eyes. Vision Res. **9**, 1031—1041 (1969b).

Sherk,T.E.: Development of the compound eyes of dragonflies (Odonata). III. Adult compound eyes. J. exp. Zool. **203**, 61—80 (1978).

Sidman,R.L.: The structure and concentration of solids in photoreceptor cells studied by refractometry and interference microscopy. J. Biophys. Cytol. **3**, 15—29 (1957).

Snyder,A.W.: Asymptotic expressions for eigenfunctions and eigenvalues of dielectric or optical waveguides. IEEE Transactions on Microwave Theory and Techniques **17**, 1130—1138 (1969).

Snyder,A.W.: Coupled mode theory for optical fibers. J. opt. Soc. Amer. **62**, 1267—1277 (1972).

Snyder,A.W.: Polarization sensitivity of individual retinula cells. J. comp. Physiol. **83**, 331—360 (1973).

Snyder,A.W.: Leaky ray theory of optical waveguides of circular cross section. Appl. Phys. **4**, 273—298 (1974).

Snyder,A.W.: Photoreceptor optics—theoretical principles. In: Photoreceptor optics (eds. A.W.Snyder, R.Menzel). Berlin-Heidelberg-New York: Springer 1975a.

Snyder,A.W.: Acuity of compound eyes: physical limitations and design. J. comp. Physiol. **116**, 161—182 (1977).

Snyder,A.W., Laughlin,S.B.: Dichroism and absorption by photoreceptors. J. comp. Physiol. **100**, 101—106 (1975).

Snyder,A.W., Laughlin,S.B., Stavenga,D.G.: Information capacity of eyes. Vision Res. **17**, 1163—1175 (1977a).

Snyder.A.W., Love,J.D.: Optical Waveguide Theory, Chapman and Halls; Wiley 1980.

Snyder,A.W., McIntyre,P.D.: Polarization sensitivity of twisted fused rhabdoms. In: Photoreceptor optics (eds. A.W.Snyder, R.Menzel). Berlin-Heidelberg-New York: Springer 1975.

Snyder,A.W., Menzel,R., Laughlin,S.B.: Structure and function of the fused rhabdom. J. comp. Physiol. **87**, 99—135 (1973).

Snyder,A.W., Miller,W.H.: Fly colour vision. Vision Res. **12**, 1389—1396 (1972).

Snyder,A.W., Miller,W.H.: Photoreceptor diameter and spacing for highest resolving power. J. opt. Soc. Amer. **67**, 696—698 (1977).

Snyder,A.W., Miller,W.H.: Telephoto lens system of falcoriform eyes, *Nature* **275**, 128, 129 (1978).

Snyder,A.W., Pask,C.: Light absorption in the bee photoreceptor. J. opt. Soc. Amer. **62**, 998—1008 (1972).

Snyder,A.W., Pask,C.: Spectral sensitivity of dipteran retinula cells. J. comp. Physiol. **84**, 59—76 (1973).

Snyder, A. W., Sammut, R.: Direction of *E* for maximum response of a retinula cell. J. comp. Physiol. **85**, 37—45 (1973).
Snyder, A. W., Sammut, R.: Contribution of unbound modes to light absorption in visual photoreceptors. J. opt. Soc. Amer. **64**, 1711—1714 (1974).
Snyder, A.W., Srinivasan, M.V.: Human psychophysics: functional interpretation for contrast sensitivity vs spatial frequency curve. *Biological Cybernetics* (in press) (1978).
Snyder, A. W., Stavenga, D. G., Laughlin, S. B.: Spatial information capacity of compound eyes. J. comp. Physiol. **116**, 183—207 (1977b).
Srinivasan, M.V., Bernard, G.D.: The effect of motion on visual acuity of the compound eye: a theoretical analysis. Vision Res. **15**, 515—525 (1975).
Stark, W. A.: Spectral selectivity of visual response alterations mediated in interconversions of native and intermediate photopigments in *Drosophila*. J. comp. Physiol. **96**, 343—356 (1975).
Stark, W. A., Harris, W., Walker, J.: Genetic isolation of three receptor spectral sensitivities with different photopigment conversions in *Drosophila*. Abstract of paper presented at Spring, 1975, meeting of the Association for Research in Vision and Ophthalmology, Sarasota, Fla. (1975).
Stavenga, D. G.: Visual receptor optics, rhodopsin, and pupil in fly retinula cells. Thesis, Rijksuniversiteit Groningen, Groningen, The Netherlands (1974).
Stavenga, D. G.: Optical qualities of the fly eye. An approach from the side of geometrical, physical, and waveguide optics. In: Photoreceptor optics (eds. A. W. Snyder, R. Menzel), pp. 126—144. Berlin-Heidelberg-New York: Springer 1975.
Stockhammer, K.: Zur Wahrnehmung der Schwingungsrichtung linear polarisierten Lichtes bei Insekten. Z. vergl. Physiol. **38**, 30—83 (1956).
Täuber, U.: Photokinetics and dichroism of visual pigments in the photoreceptors of *Eledone* (*Ozoena*) *moschata*. In: Photoreceptor optics (eds. A. W. Snyder, R. Menzel). Berlin-Heidelberg-New York: Springer 1975.
Trujillo-Cenóz, O.: The structural organization of the compound eye in insects. In: Handbook of sensory physiology (ed. M. G. F. Fuortes), Vol. VII/2, pp. 5—62. Berlin-Heidelberg-New York: Springer 1972.
Tunstall, J., Horridge, G. A.: Electrophysiological investigation of the optics of the locust retina. Z. vergl. Physiol. **55**, 167—182 (1967).
Varela, G., Wiitanen, W.: The optics of the compound eye of the honeybee (*Apis mellifera*). J. gen. Physiol. **55**, 336—358 (1970).
Vries, H. de: Physical aspects of the sense organs. Prog. Biophys. **6**, 208—264 (1956).
Wald, G.: Visual pigment and photoreceptor physiology. In: Biochemistry and physiology of visual pigments (ed. H. Langer), pp. 1—16. Berlin-Heidelberg-New York: Springer 1973.
Walcott, B.: Anatomical changes during light adaptation in insect eyes. In: The compound eyes and vision of insects (ed. G. A. Horridge). London: Oxford Press 1975.
Watanabe, A., Mori, T., Nagata, S., Hiwatashi, K.: Spatial sinewave responses of the human visual system. Vision Res. **8**, 1245—1263 (1968).
Wehner, R.: Pattern recognition. In: The compound eye and vision of insects (ed. G. A. Horridge), pp. 75—113. Oxford: Clarendon Press 1975.
Wehner, R.: Polarized light navigation by insects. Sci. Amer. **235**, No. 1, 106—115 (1976).
Wehner, R., Bernard, G. D., Geiger, E.: Twisted and non-twisted rhabdoms and their significance for polarization detection in the bee. J. comp. Physiol. **104**, 225—245 (1975).
Worthington, C. R.: Structure of photoreceptor membranes. Ann. Rev. Biophys. Bioeng. **3**, 53—109 (1974).
Wright, W. E.: Orientation of intermediates in the bleaching of sheer-oriented rhodopsin. J. gen. Physiol. **62**, 509—523 (1973).
Wyszecki, G., Stiles, W. S.: Color science: concepts and methods, quantitative data, and formula. New York: John Wiley 1967.

Chapter 6

Receptor Potentials in Invertebrate Visual Cells

M. JÄRVILEHTO, Oulu, Finland

With 19 Figures

Contents

A. Introduction

Most of our knowledge of invertebrate visual systems has come from studies on arthropods and molluscs. Relatively little is known about the function of visual structures in other invertebrates. Thus many of the principles of invertebrate visual information processing have been evolved by using the eye of the horseshoe crab *Limulus* as an experimental preparation (HARTLINE and GRAHAM,

1932; HARTLINE et al., 1952). These and subsequent excellent studies have been reviewed by numerous authors. Elsewhere in this handbook series other ways of viewing can be found concerning the receptor potential of the visual cells, theoretical considerations about the process leading to electric potential response, and the generation of action potential trains related to behavioral responses of intact animals (Vol. I, Ch. 8; Vol. VII/2, Ch. 6–8, 10).

Disregarding the study of optomotor behavior, the analysis of the function of visual cells can be divided here into three overlapping research areas, viz photoreceptor optics, photochemistry, and electrophysiology. Light impinging on receptor cells is converted into electric potentials that can be recorded using electrophysiologic techniques. The process of conversion of light energy into electric energy is the specific role of each receptor cell, and each cell is assumed to possess a constant physiologic function in the retina.

The possibility of correlating physiologic function to particular anatomic cell types in the visual system by using electrophysiologic staining techniques has begun a new era in the analysis of the receptor cell functions.

Our present view about the function of the receptor cell net is predominantly mechanistic. We assume that certain morphologic structures are the basis of certain functions, i.e., there is a morphologic and functional constancy in the visual organ.

For this reason a method of precise identification of the recorded cell is essential. The method is based on the favorable "Golgi"-staining technique, which reveals the anatomic network. This is followed by correlation of the physiologically unique group of cells with anatomic class of neurons by using selective intracellular staining. Unfortunately all cells in the network cannot be recorded simultaneously, but the basic assumption is made about the invariant function of one seemingly morphologically invariant class of cells. This gives a further basis for succesful analysis of the visual system's complicated network.

Knowledge of the anatomic wiring diagram of the visual system is of great relevance for electrophysiologists in classifying different potential responses of complex structures. On the other hand, the knowledge of classes characterized physiologically leads to a search for new morphologic connections between cells.

Probably a suitable start for this review is a short synopsis of the photoreceptor organization, although a detailed morphology of invertebrate photoreceptor cells is to be found in different volumes of this handbook.

I. Survey of Photoreceptor Organization

Although some cells that respond to light stimulation lack any apparent morphologic specialization (e.g. *Amoeba*), the surface membrane of the photoreceptor cells forming the photoreceptor organ usually has complicated cytoplasmic specializations.

The membranous structures in the cells provide an increased surface area for the photopigment. The photopigment may be spread in a single layer of molecules, each possibly of a fixed orientation for capturing photons (WALD et al., 1963).

The same structures also can be used to divide photoreceptors into two general classes, depending on whether or not they are formed in association with the basal body or cilium (EAKIN, 1965, 1972). The distinction has some interesting features. Most of the photoreceptors derived from cilia are found in vertebrates or in those phyla related to the vertebrate ancestral stock, whereas nonciliary (rhabdomeric) endings are characteristically found among annelids, molluscs, and arthropods.

In both groups, but especially in the rhabdomeric line, the visual system involves a large amount of variation. However, despite the diversity of construction they all have the same function: to inform the central nervous system about the state of light in the environment.

The modification of cilia to form sensory endings is not just limited to photoreceptors, for certain mechanoreceptors and olfactory endings have also ciliary endings.

In the distal retina of the scallop (an exception of the phylogenetic generalization), the photoreceptor cells bear a number of saclike projections each one consisting of a greatly flattened cilium (BARBER et al., 1967). In coelenterates and echinoderms, the membrane of each cilium may be thrown into fingerlike folds, whereas in the photoreceptor cells of ascidians the ciliary membrane is invaginated to form a series of parallel discs. The receptor cells in the vertebrate retina are the most highly organized ciliary photoreceptors.

The second major class of photoreceptor cells is characterized by the presence of microvilli. Generally one end of the cell has a large number of microvilli, whereas the other end has an axon. In relatively unspecialized photoreceptor cells the microvilli are located in several surfaces of the soma. In the squid photoreceptors the distal end of the cell bears microvilli on two sides, but in arthropods microvilli accumulate in fascicles on one side. Each arthropod photoreceptor cell projects the microvilli towards a center to form an axial rodlike structure (rhabdom). Each small group of photoreceptor cells (3–22) surrounded by pigment cells is known as an *ommatidium*. The compound eyes are constructed through the aggregation of hundreds or thousands of ommatidia.

At its distal end an ommatidium has a small lenslet and an accessory dioptric structure. The whole compound eye has complicated optics, which can be variant in different parts of the eye (see this volume, Ch. 3).

Although there are usually eight or nine retinular cells in an ommatidium, in some species there are only seven (some crustaceans) and in certain forms there may be over 20 (*Limulus*). One or two of the retinular cells may be reduced in size and occupy a characteristic location. Examples are the basal retinular cell of the drone honeybee and the superior and inferior central cells in the ommatidium of flies. In the compound eye of *Limulus* each ommatidium contains one, two, or no eccentric cells, which are second-order neurons located according to their function (BEHRENS and WULFF, 1965; FAHRENBACH, 1975).

The retinular cells have an axon synapsing at several levels with second-order neurons in the first or second optic ganglia. The optic lobes consist of several layers of neuropile surrounded by neuron cell bodies and connected by tracts of axons (review: STRAUSFELD, 1976).

Depending on the evolutionary level and the habitat of the animal, its visual system is capable of processing different kinds of photic information, e.g., intensity, wavelength, the *e* vector of linearly polarized light, contrast, and movement. The first sign of these processes can be recorded in the photoreceptors in the form of electric potentials.

B. Receptor Potential Analysis with Microelectrodes

I. Intracellular Recording

The technique of extracellular measurements is partly adequate for the study of single nerve cells that respond with impulse discharge; the discrete nature of the impulses provide a means of differentiating one unit from another.

The situation is different in the cells which constitute the retina distal to the medulla and which respond only with slow and graded potentials. In this case there is no decisive way to isolate single-unit activity recorded extracellularly from that of neighboring units. The impossibility of isolating unitary responses has been a barrier to analysis of retinal functions other than ERG. In order to surmount this barrier it was necessary to develop a technique of intracellular recording in small cells and a technique of marking the cell and recording site. Such a technique permits the physiological and anatomical identification of single functional units.

Although several handbooks of electrophysiology detail the techniques of intracellular recording using glass microelectrodes (BURES et al., 1960; FRANK and BECKER, 1964; LAVALLÈE et al., 1969; THOMPSON and PATTERSON, 1973), every practitioner in the field has his own methods. In this very limited essay we are able to go into only our special means for intracellular measuring techniques. For more detailed knowledge, the reader is advised to refer to the above-mentioned books.

The first requirement for a successful penetration of the cell membrane is a very finely tapered glass micropipette, with a tip diameter of less than 0.1 μm. For stable recordings and easy penetrations it is better to have a pipette with a long shank and very gradual taper. Recent studies on certain penetrations have shown that a bevelled tip form may facilitate the penetration (BARRETT and GRAUBARD, 1970). The bevelling might give a better tip form, but certainly not a smaller tip opening, and may cause leakage by recording from small axons. The same kind of favorable tip form may also be obtained with conventional electrode pullers by using an asymmetric heater element instead of an axial symmetric heating coil and inserting two fine glass capillaries into the electrode pipe. The electrode is sometimes slightly bent and its form is asymmetric. All the above mentioned electrodes are very easily filled with the desired electrolyte by means of a syringe and hypodermic needle (TASAKI et al., 1968). If the electrolyte is filtered and clean, the electrodes have a lower resistance (depending on the specific resistance of electrolyte) and greater stability.

The microelectrode is connected to the recording system via a preamplifier that ideally allows the testing of the resistance and current injection through the electrode in addition to its normal function.

For successful penetration and recording it is necessary to have an appropriate micromanipulator with motor or hydraulic drive. It is convenient to connect the microelectrode to the micromanipulator via a piezo vibrator.

Under such conditions it is possible to perform stable intracellular recordings from nerve cell axons of less than 1 μm diameter in insect eyes (AUTRUM et al., 1970).

II. Intracellular Staining Technique

Receptor cells in different visual systems are almost unexceptionally inhomogeneous cell populations. The cells can be divided into groups according to different morphologic and functional criteria. The distribution of certain cell types in the retina and their anatomic configuration is believed to be the basis for information processing at deeper levels of the visual system. In particular studies on lateral processing of visual information and identification of the information-carrying cell type requires intracellular staining techniques.

The technique of intracellular staining with Procion dyes was introduced in 1968 by several authors (KRAVITZ et al., 1968; STRETTON and KRAVITZ, 1968), and has found wide popularity.

The Procion dyes have many of the properties of an ideal marking stain. The aqueous solution (6%) of Procion Yellow is suitable as an electrolyte. The resistance of the electrode is about 90–250 MΩ. The dyes do not precipitate any common intracellular salts. The molecules are small enough to move rapidly out from the electrode, but they are still large enough to stay in the cell and its smallest branches. The fluorescence is intense enough to permit the detailed examination of the cell structure against the contrasting tissue autofluorescence and the dye withstands some forms of histologic routines. The toxicity is low enough to avoid modification of cellular electric activity.

The problems of intracellular staining have been discussed in great detail (KATER and NICHOLSON, 1973) and various alternative methods have been considered.

The dye injection method was first applied to giant neurons in the more simple nervous systems of molluscs and annelids, but now there are available data also about complicated nerve nets of both invertebrates and vertebrates.

By using two different electrodes with two different colors (Procion Yellow M4RAN and Procion Brilliant Red H3BNS) it has been possible to show morphologic and functional connexions between cells in the dogfish retina (KANEKO, 1971) and in the barnacle lateral eye (SHAW, 1972).

Intracellular staining is not always unambiguous and it is important to follow several phenomena during injection. The membrane potential, response amplitude, and form of the potential response should not change and receptive field should stay constant during the recording. If only one cell appears stained it is likely that the recording arises from that one stained cell.

In practice it is important to control the current during the injection and to inject with low current (10^{-9}–10^{-8} A) for a long time without disturbing the cell function. The sensitivity of the staining method is increased by using freeze-drying and waterless preparation (see below).

III. Identification of Recording Site

It is a well-known fact that the potential response carrying information may be changed in different parts of the receptor cell. Slow potentials are encoded to spike potentials in certain active parts of the cell or the potential form is changed by passive spread over the receptor cell axon.

In the peripheral retina most of the cells do not generate spikes but carry light information actively or passively in different forms of slow potential, which in addition may have different characteristic forms in different parts of the same cell. Therefore it is pertinent to know from which part of the cell the recording is originating.

The problem is quite easily solved if penetration of the cell is observed under the microscope, but very often it is not practicable to expose the tissue this much. In some preparations the whole eye was sliced to get the electrode into the desired part of the eye (SMITH and BAUMANN, 1969; SHAW, 1968; IOANNIDES and WALCOTT, 1971; LAUGHLIN, 1973, 1974). For conclusions about the receptive fields of certain structures this kind of preparation is not suitable because of dramatic structural distortion of the tissue.

A small hole in the eye capsule outside the cornea does not damage the optical system of the eye but on the other hand, makes it difficult to locate the electrode to the desired part of the eye.

This problem can be overcome and the electrode tip visualized precisely afterward by using freeze-fixation (ZETTLER and JÄRVILEHTO, 1970; JÄRVILEHTO and ZETTLER, 1970, 1973; JÄRVILEHTO, 1971) and subsequent freeze-drying. The method is suitable for small specimens (e.g., insect head) because of instantaneous freezing. The freezing velocity determines the size of ice crystals and the size of specimen, the freezing velocity. For freezing the specimen it is necessary to use cryogens, isopentane or Halon (TIFFE et al., 1970) to avoid the Leidenfrost's phenomenon. The frozen tissue is very hard, the electrode can be broken off without deformation of tissue, and the chitin removed to facilitate drying without thawing the tissue (−45° C). The specimen is impregnated in the freeze-drying apparatus with wax and cut into sections at room temperature. To prevent washing out the water-soluble dye and denaturating, the tissue sections have to be pressed directly without alcohol series onto a dry slide and cleared in xylene before being mounted in Eukitt or Fluoromount. As dye injection is carried out in conjunction with tip localization, the sections should be examined immediately to preserve the maximal fluorescence intensity of the injected dye. The tip of the electrode is usually found in one of the sections examined under the fluorescence microscope. Figure 1 shows the ending of a short visual fiber in the lamina of the fly, along with a piece of the recording electrode.

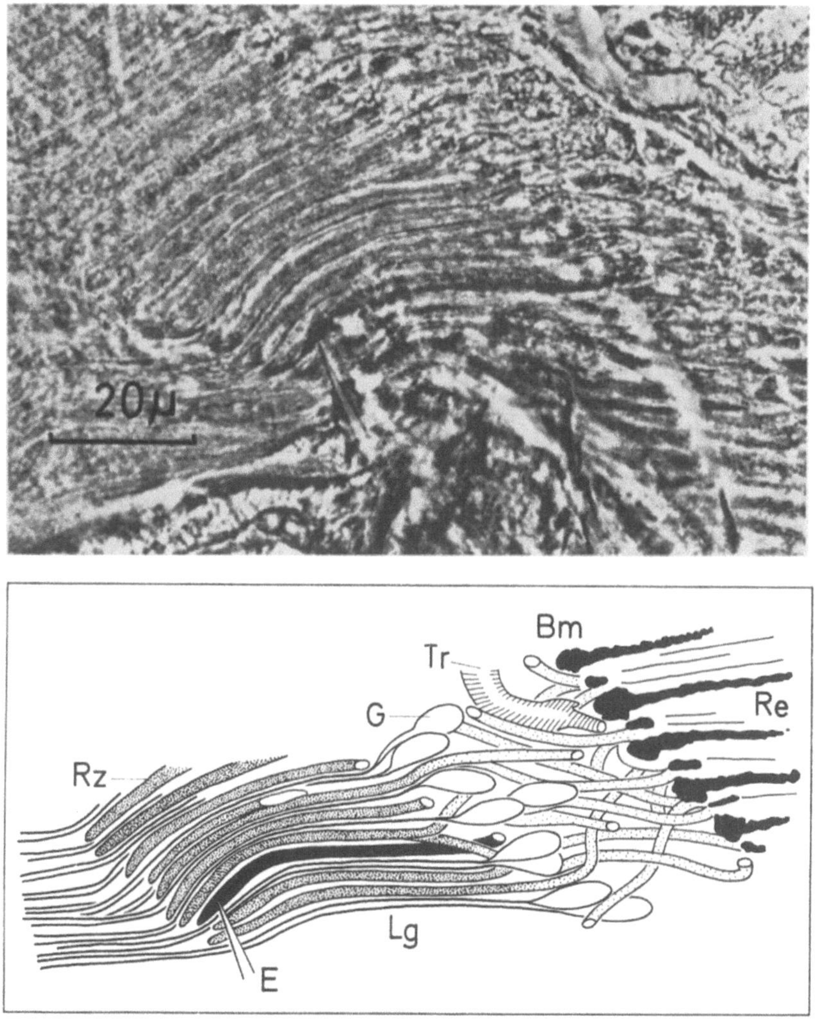

Fig. 1. Short visual fiber by intracellular staining identified with the recording electrode in the first optic ganglion of the fly, *Calliphora erythrocephala*. The histologic details are shown in the schematic drawing. (*Bm*) basement membrane, (*E*) electrode, (*G*) ganglion cell body, (*Lg*) lamina ganglionaris, (*Re*) retinular cell layer, (*Rz*) retinular cell axon, (*Tr*) trachea (JÄRVILEHTO and ZETTLER, 1970)

C. Electric Activity in Photoreceptor Cells

I. Generation of Electric Activity

Absorption of light quanta is the first step in the chain of events leading in an inadequately comprehended way, to the activation of the photoreceptor membrane and the generation of a graded receptor potential. After a light quantum has been absorbed the structure of the photopigment molecule changes and its absorption properties are different (HUBBARD et al., 1965). The illuminated photoreceptor membrane thus always contains a mixture of the original photopigment and its intermediates. Many pigment intermediates (e.g., in cephalopods, insects, and *Limulus*) are able to recycle to their starting point by absorption of a later light quantum. That means that they are photoisomerizable. The steps coupling quantum absorption to the electric processes, which can be recorded across the membrane, are not known (see HAMDORF, this volume, Ch. 4).

Functionally, the structure of rhabdomeres provides a system of electrically excitable membranes, in which the photocurrent is produced locally, close to the site of photon absorption (HAGINS et al., 1962; LASANSKY and FUORTES, 1969). The following hypothetical chain of events leads to photoactivation of the receptor.

1. Arriving photons are captured randomly by the visual pigment.
2. All photons are not absorbed, but once one is absorbed the complicated process that follows is divided into two parts on the time scale.
3. The first process happens on the order of nanoseconds and involves the initial conformational changes of chromophore from the 11-cis to the all-trans-configuration in the visual pigment. The electric manifestation is known as *early receptor potential* (ERP).
4. The second process, taking place on a time scale of milliseconds, is triggered by the conformational change and leads to a short-term conductance change of the membrane (quantum bump).
5. The number of quantum bumps is proportional to the light intensity.
6. The receptor potential consists of the accumulated bumps, each produced by absorption of one photon.

II. Early Receptor Potentials

Two types of electric activity have been demonstrated in photoreceptor cells. One type of response appears after a certain latency associated with conductance changes on the membrane and is called *late receptor potential* (LRP) or simply *receptor potential* (RP). The other type of response observed by BROWN and MURAKAMI (1964) is recorded only when very intense light flashes are used (CONE, 1964; PAK and CONE, 1964; SMITH and BROWN, 1966; BROWN et al., 1967). The potential change starts without detectable latency and was termed ERP to distinguish it from the LRP. The ERP was found later in different species of both vertebrates and invertebrates (review: CONE and PAK, 1972; FEIN and CONE, 1973; FEIN and DEVOE, 1973; HILLMAN et al., 1973; MINKE et al., 1973a).

The following characteristics of ERP distinguish it (except for the last) from those of the LRP (HILLMAN et al., 1973): (1) it has a very short or hardly detectable latency (Fig. 2), (2) waveform depends on temperature; at high temperatures it is purely negative and develops an increasingly positive phase with decreasing temperature, (3) survives chemical fixation, (4) amplitude is linearly proportional to the light intensity up to saturation, (5) is insensitive to the ionic medium, but (6) has the same action spectrum as that of the LRP (Fig. 3).

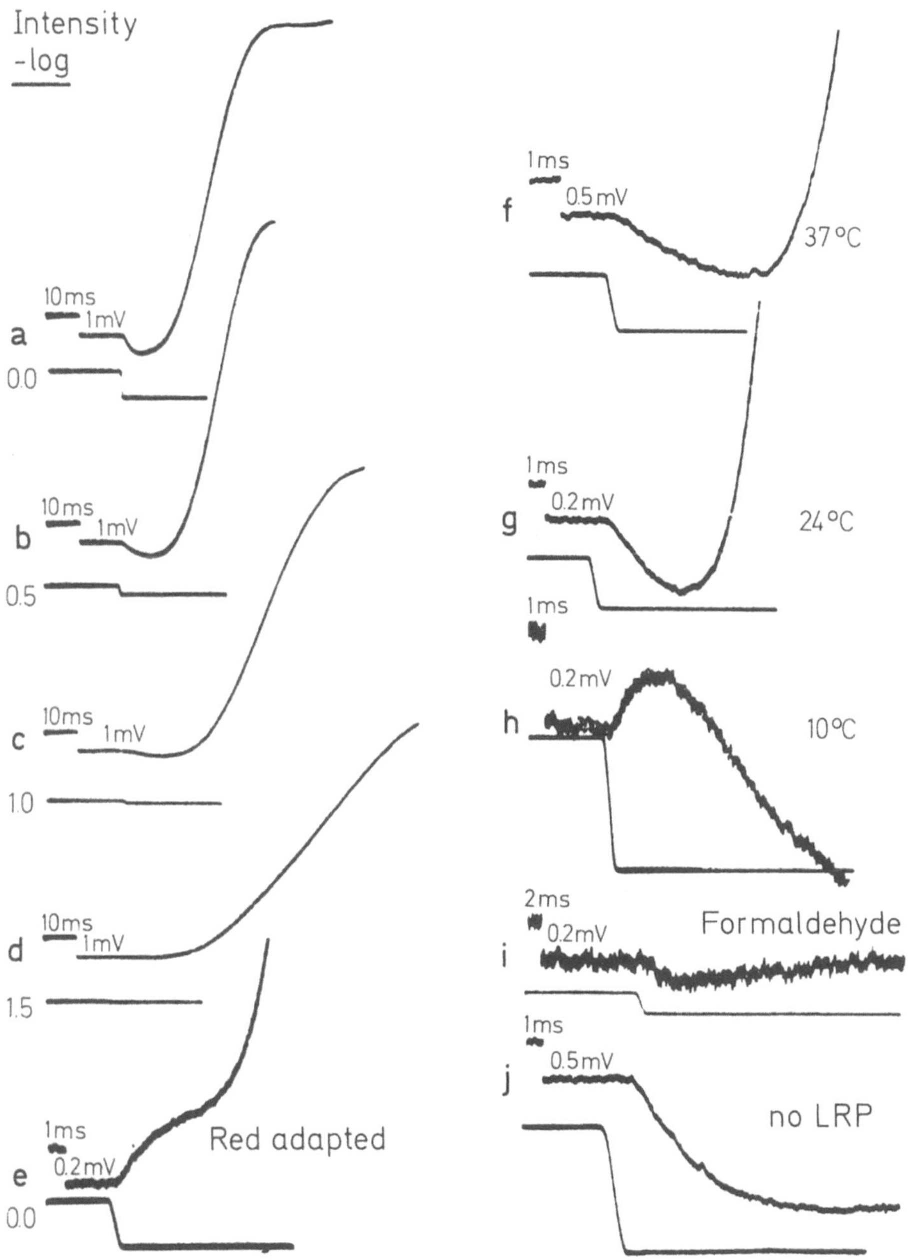

Fig. 2. The characteristics of the early receptor potential ERP in the barnacle (*Balanus*) photoreceptor. Traces A, B, C, D show the response at 24° C to white light of relative intensity log I = 0.0, −0.5, −1.0, −1.5, respectively. The small fast hyperpolarizing component, the ERP, is followed by slower positive part, the LRP. Trace E shows the response after previous adaptation to red light. Trace F, G, and H display the temperature dependence of the ERP. Trace I shows the ERP after formaldehyde fixation. In trace J the ERP is not followed by LRP (the same stimulus condition as in G). The output of a photocell is displayed under each response and the calibration before each recording (HILLMAN et al., 1973)

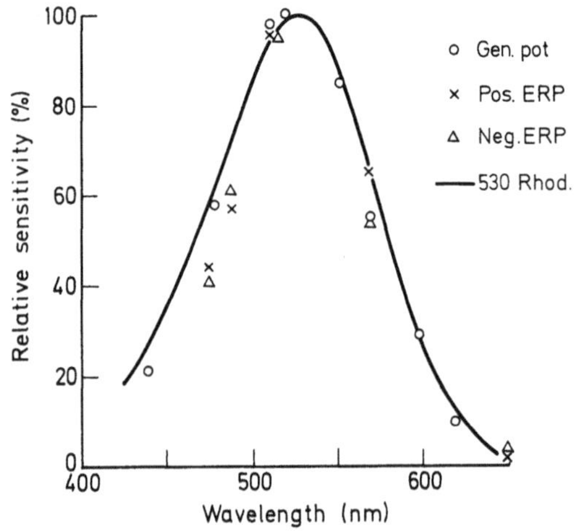

Fig. 3. The action spectrum of ERP and the physiologic response measured in *Limulus* lateral photoreceptor. *Solid line* is Dartnall's nomogram for pigment with maximum absorbancy at 530 nm. *Open circles* show the physiologic response, *squares* and *triangles* the depolarizing components of the ERP (BROWN et al., 1967)

It is known that the ERP is a complex waveform associated with first steps of breakdown and also of regeneration of rhodopsin (CONE, 1967; HAGINS and MCGAUGHY, 1967). The ERP is not generated by changes in membrane permeability (BRINDLEY and GARDNER-MELWIN, 1966) but attributed to a capacitive current resulting from charge shift on similarly oriented pigment molecules that have captured quanta. The ERP from *Balanus* median photoreceptor is about twice as fast as that recorded from the lateral photoreceptor. This difference was found to be similar to that shown for late receptor potentials (LRP) and could be attributed to different proportioning of microvillar and perikaryon membrane areas and thus to different values of capacitance in different photoreceptors (HAGINS and RÜPPEL, 1971; DAHL et al., 1977). The ERP in vertebrate rods is probably lost by disorientation of rhodopsin molecules on the membrane after heating (CONE and BROWN, 1967). The energy ratio required to produce a just detectable ERP and LRP is about $10^6:1$. Thus very high flash intensities, which already saturate the LRP, are needed to produce an ERP response. Consequently the ERP is not believed to be a direct, causal link leading to the permeability changes associated with the LRP. Nevertheless, it is a useful experimental tool for measuring the fraction of unbleached pigment in the receptor in a manner not possible by spectrophotometric methods.

III. Quantum Bumps

The existence of relatively small, discrete depolarizations of the photoreceptor cell membrane in the lateral eye of *Limulus* was discovered by YEANDLE (1958), who called them *bumps*. They occur spontaneously and in response to low intensity light stimuli. More recently they have been called *discrete waves* or *slow*

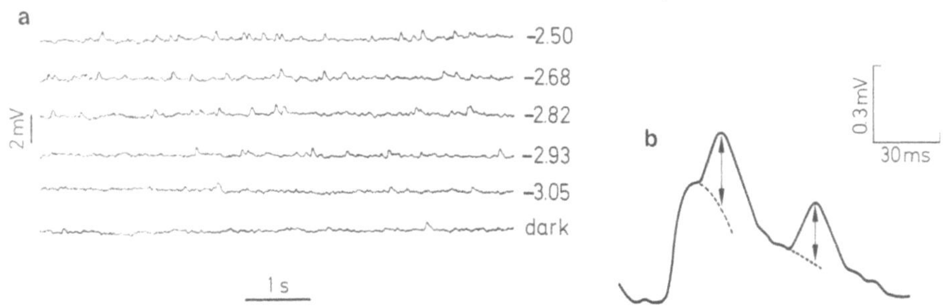

Fig. 4. (A) Intracellularly recorded bumps in photoreceptors of *Drosophila melanogaster*. Relative increasing stimulus intensity (450 nm) indicated by numbers. (B) A recording showing hypothetical bump superposition. The amplitudes (*arrows*) of the superimposing bumps were measured from extrapolated base lines (WU and PAK, 1975)

potential fluctuations and are found in a variety of different arthropods during dim illumination (review: in this handbook, FUORTES and O'BRYAN, 1972b; KIRSCHFELD, 1966; SHAW, 1968; SMOLA and GEMPERLEIN, 1973; WU and PAK, 1975; MINKE et al., 1975). The possibility that each light-evoked bump represents a single photon absorption has attracted considerable interest in explaining the origin of the quantal unit of the RP (HAGINS, 1965).

In the lateral eye of *Limulus*, bumps must exceed a threshold level for a propagated nerve impulse to be produced in the eccentric cell (YEANDLE, 1958). Thus the sum potential of bumps would be equivalent to the RP of the cell (Fig. 4). It seems probable in terms of frequency response function that the total light response is a linear summation of bumps at low intensities (FRENCH and JÄRVILEHTO, 1978). The shot-noise analysis also provided evidence (DODGE et al., 1968) that the RP is simply a summation of bumps. However, the shot-noise model provides no information about spontaneous events on the membrane. The rate of bump occurrence increases proportionally to light intensity (SCHOLES, 1964, 1965; KIRSCHFELD, 1966), but the time of occurrence appears to be random and independent (FUORTES and YEANDLE, 1964; ADOLPH, 1964). The assumption that a light-induced bump is triggered by absorption of a single photon is based on the frequency distribution of bumps after brief flashes. Recently discovered phenomenologic differences between spontaneous and light-evoked bumps suggest two different generation mechanisms (YEANDLE and SPIEGLER, 1973). Spontaneous bumps may have their origin in thermal isomerization of visual pigment molecules (SREBRO and BEHBEHANI, 1972). Whether or not quantum bumps are produced by single photons still remains unproven. But bumps are certainly electric responses of the cell with very high sensitivity to light. It seems likely, for this reason alone, that their properties are closely related to the membrane and molecular mechanisms of phototransduction.

The bumps, at least in the receptors of *Limulus*, have two characteristic properties, viz highly variable, relatively long latency and two different amplitude sizes. ADOLPH (1964) reported that bumps in *Limulus* occur small or large. The distribution in insect photoreceptors is more indistinct (SCHOLES, 1965; KIRSCHFELD, 1966; SHAW, 1968; WU and PAK, 1975). In *Limulus* (an experimental animal widely used on studies of bumps), the large bumps are generally more than 2 mV in peak amplitude. But since in different recordings the average size of bumps varies from cell to cell it is relevant to compare peak amplitude histograms, which are usually bimodal. Small bumps differ from the large ones in a slower time course (ADOLPH, 1964; BORSELLINO and FUORTES, 1968; SREBRO and BEHBEHANI, 1971). Recently two possible receptor mechanisms were also reported to arise from small or large potential fluctuations on the membrane (BARLOW and KAPLAN, 1977). Their significance was found in encoding large range (9 log units) of light intensities by single photoreceptor cell.

Bump latency distribution is very broad, from 50 to 300 ms, averaging 120 ms (SREBRO and BEHBEHANI, 1971). The Q_{10} of the latency is between 2 and 3. Extracellular calcium was found to influence the latency and variability; lowering it prolongs the latency and increases variability (MARTINEZ II and SREBRO, 1976). The correlation between latency and amplitude is contradictory in different reports. BORSELLINO and FUORTES (1968) (see also this handbook, Vol. VII/2, FUORTES and O'BRYAN, 1972b) reported that small bumps have a shorter and less dispersed latency than large ones.

On the other hand, SREBRO and YEANDLE (1970) and SREBRO and BEHBEHANI (1971) found no significant latency difference between two amplitude groups.

The relative number of large and small waves is variable from cell to cell. BORSELLINO and FUORTES (1968) found that small bumps predominate in the lateral eye photoreceptor. However, SREBRO and BEHBEHANI (1971) claim that large bumps predominate. Interestingly enough, it has been shown (YEANDLE and SPIEGLER, 1973) that in the ventral photoreceptor of *Limulus* small bumps are more likely to occur spontaneously than large ones when a light stimulus is applied. It was suggested (BEHBEHANI and SREBRO, 1974) that small bumps may result from the activation of damaged membrane areas. But the question of the unique bump size still remains open.

Large bumps have an average amplitude of approximately 4 nA in voltage-clamped photoreceptors. To support this much photocurrent it probably requires several square microns of the cell membrane (BEHBEHANI and SREBRO, 1974). Thus more than a single channel controlled by a single photopigment molecule in a single microvillus is needed to explain the individual bump: The effect of the single light quantum is spread far wider than the span of one pigment molecule.

BASS and MOORE (1970) show that if each photon activated one highly conductive channel in the membrane of the microvilli, it would create a local depolarization. Further, assuming that microvillar membrane is electrically excitable (like axonal membrane in a nerve cell) the depolarizing potential could spread over an adequate area of the membrane depolarizing the whole cell by several millivolts. The free energy required to amplify the original photochemical reaction is provided by the usual differences in electrochemical potential across neuronal membranes.

No electric excitability of the microvillar membrane would be required if numerous membrane active "particles" were produced by a single photon as suggested by BORSELLINO and FUORTES (1968).

CONE (1973) estimated on the basis of difference between "limited diffusion" and bump-induced conductance that one bump should open 100 or more sodium channels. In rods and cones, free calcium as internal transmitter could decrease sodium permeability, but for invertebrates calcium seems to play only a delayed role. In any case, it seems necessary that any process postulated to intervene between absorption of light and production of electric responses must involve some amplification mechanism. For the multiplication of the hypothetical particles the energy could be brought by a sequence of enzymatic reactions on the membrane (WALD, 1965).

IV. Graded Receptor Potentials

The slow photovoltage of the retinal photoreceptors is a continuous function of time and light intensity. This function is a membrane depolarization in many receptor cells and has been called the *generator potentials* (BERNHARD and GRANIT, 1946) or *receptor potential* (RP) (DAVIS, 1961). The term generator potential is nowadays mostly reserved for cells in which the graded potentials generate spikes, and RP is used if they are evoked by sensory stimuli.

The large number of different invertebrate visual systems gives an enormous variety of photoreceptor cells. In spite of relatively few intracellular recordings from invertebrate visual cells some common features are obvious.

In the majority of receptor cells investigated so far, the membrane presents normally resting potentials of −30 to −70 mV in the absence of light stimulation and becomes depolarized following application of stimuli. The first intracellular recordings of electric activity from visual receptors were performed by HARTLINE et al. (1952) on the visual cells of *Limulus*. Since then, depolarizing RPs have been recorded from most invertebrate visual cells.

Systematic studies on visual cells of different species show that the vertebrate visual cells in general have hyperpolarizing and invertebrates, depolarizing RPs evoked by light; there are only a few exceptions to this rule (Table 1).

Depolarizing RPs are always associated with the microvillous type of receptor cells and hyperpolarizing RPs, usually with cells of ciliary origin, but this apparent correlation between the fine structure of the photosensitive membrane of the visual cell and the polarity of the RP should be interpreted with care.

GORMAN et al. (1971) found that two species of primitive chordates with microvillar receptors hyperpolarize by illumination, as higher vertebrates do. Membrane in *Salpa* receptor cells is composed of microvilli. In *Amaroucium*, the photoreceptive membranes are modified cilia. Interestingly enough, in the microvillous-type membrane the conductance was increased during the response to light, whereas the conductance of the ciliary type membrane slightly decreased. In the eye of scallop two types of receptor cells are located at two different levels of the retina. The distal layer contains cells with ciliary structures and the proximal layer, cells with large additional microvilli (BARBER et al., 1967; MILLER, 1958).

TOYODA and SHAPLEY (1967) reported that the cells in the retina of the scallop gave hyperpolarizing potentials upon illumination. This finding was confirmed by GORMAN and MCREYNOLDS (1969) and they showed also that there is a second group of cells with depolarizing receptor potentials. MCREYNOLDS and GORMAN

Table 1.

Animal or group	Type of receptor	Type of potential response		Conductance		References
		Depol.	Hyperpol.	Increase	Decrease	
Arthropods	Microvilli	+		+		TOMITA, 1970
Molluscs						
Scallop	Ciliary		+	+		MCREYNOLDS and GORMAN, 1970
	Microvilli	+		+		MCREYNOLDS and GORMAN, 1970
Aplysia	Giant neuron		+	+		BROWN and BROWN, 1973
Squid	Microvilli	+?		+		HAGINS, 1965
Cephalochordates						
Salpa	Microvilli	+		+		GORMAN, MCREYNOLDS and BARNES, 1971
Amaroucium	Ciliary	+			+	GORMAN, MCREYNOLDS and BARNES, 1971
Vertebrates	Ciliary		+		+	TOMITA, 1970

(1970b) also showed that both responses are directly produced by photoreceptors, not, e.g., by synaptic action, and that the depolarizing responses are presumably from proximal cells (microvillous type), whereas the hyperpolarizing responses arise from distal cells (ciliary type) (Fig. 5). Furthermore, both the depolarizing and hyperpolarizing responses (Fig. 6) are associated with an increase in membrane conductance independent of the morphologic structure (cf. *Amaroucium* above) and have the same spectral sensitivity.

RP amplitude and time course vary greatly in different receptor cells according to the cell and species. Potential amplitude increases in a graded fashion with intensity of the light stimulus. At higher intensities the RP gets an initial component (also graded), then swings back through a negative trough, and finally establishes a maintained depolarizing potential response of several millivolts. At low intensities the voltage-light intensity relationship is linear. At medium intensities the relationship is quasi-linear with log-intensity and becomes highly nonlinear at intensities causing saturation of the response voltage.

The variety of different stimulation and recording methods extremely complicates the adequate comparison of visual responses in different structures. Examples of different visual cell responses can be found in this handbook, Vol. VII/2 (FUORTES and O'BRYAN, 1972a, p. 308).

RP are not generally accompanied by regenerative spike discharges. In some unidentified structures in the arthropod, retina spikes occur infrequently on the RP. More frequently a single spike can occur on the rising phase of the transient component of the RP (see below).

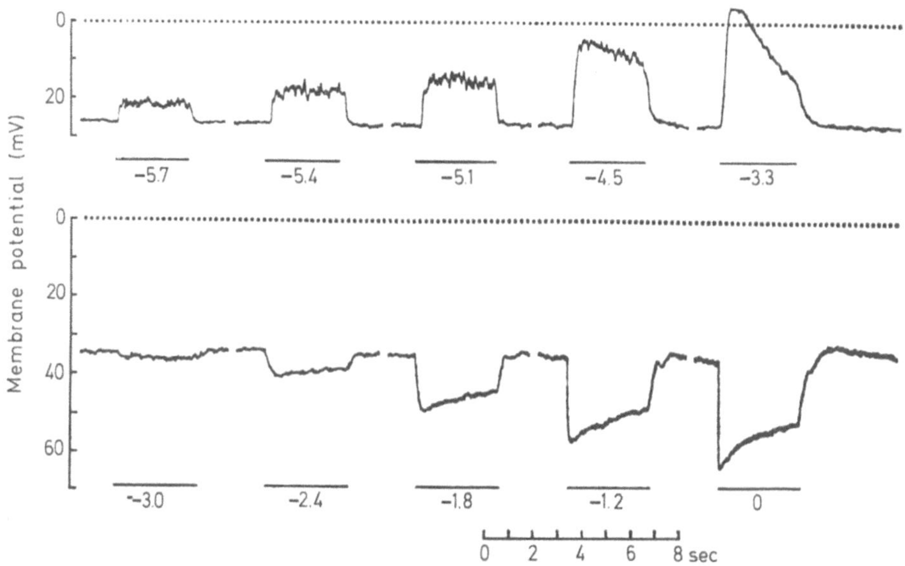

Fig. 5. Responses from a scallop (*Pecten*) eye: a microvillous type, depolarizing cell (*above*) and a ciliary type, hyperpolarizing cell (*below*) to long flashes of light. Duration and relative intensity of each flash shown below response. *Dashed line* indicates zero membrane potential level. The depolarizing cell saturates in the flash intensities above -3.3 (McReynolds and Gorman, 1970a)

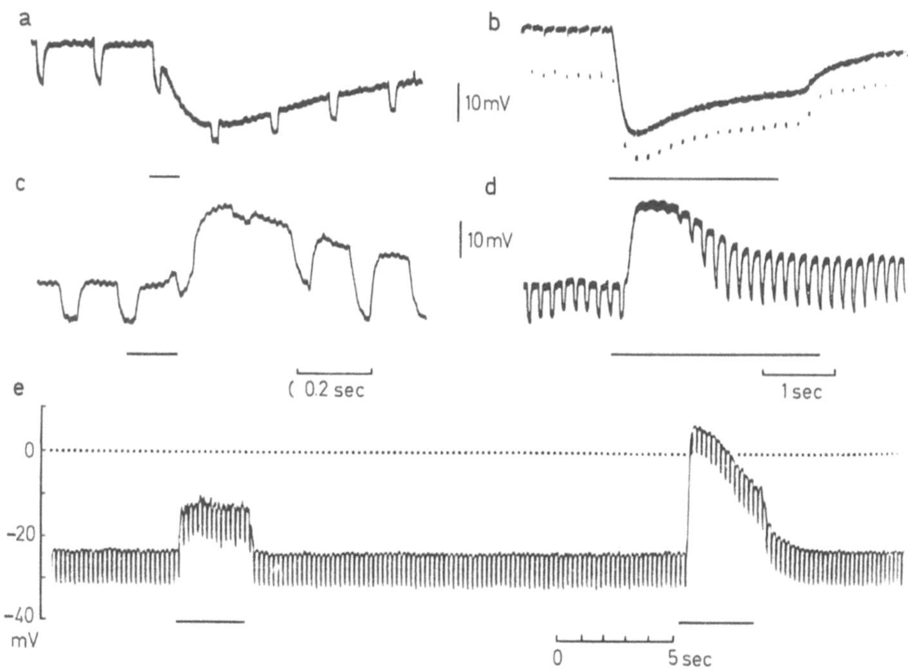

Fig. 6. Conductance changes in scallop (*Pecten*) photoreceptor during depolarizing and hyperpolarizing responses to light. The membrane responses to constant current pulses are shown before, during, and after illumination. Records are from five different cells. Duration of light stimulus is indicated by *horizontal line* under each response. Relative light intensities: $\log I = -2.4$ in A; -1.2 in B and C; 0.0 in D; -4.2 and -3.6 in E. *Dashed line* in E indicates zero membrane potential (McReynolds and Gorman, 1970b)

Two competing mechanisms have been postulated to generate the RP in the photoreceptor cell (The ion pump plays a different role in both mechanisms) (Fig. 7): (1) *Conductance Increase Mechanism* (CIM). The increase in membrane conductance caused by light generates the RP (FUORTES, 1959; BENOLKEN, 1961; HAGINS et al., 1962; KIKUCHI et al., 1962). (2) *Electrogenic Pump Mechanism* (EPM). The RP is caused by a change in the activity of an electrogenic ion pump which is controlled by light (SMITH et al., 1968a, b). An ion pump is electrogenic if, for example, it pumps sodium ions from inside to outside and the potassium ions in the opposite direction in a charge ratio which is not unity. This simplified example would also be true if other ion species were involved. Both possibilities have been discussed by several authors (BROWN et al., 1970; FUORTES and O'BRYAN, 1972a) and many experiments argue against the EPM (KOIKE et al., 1971; STIEVE, 1971a, b). Ouabain, believed to block the electrogenic pump, does not affect the depolarizing RP of the crayfish but both pump and resting potential of the cells disappear the earlier and the more intensely the cell is stimulated (STIEVE et al., 1971a). The afterpotential observed to occur following cessation of illumination (BROWN et al., 1968) in barnacle photoreceptors is abolished by application of ouabain at a concentration of 10^{-5} M. This afterhyperpolarization is believed to be due to activation of an electrogenic Na^+ pump. This is not directly triggered by illumination but by an increase of the internal Na^+ concentration due to the Na^+ influx during the depolarizing response (BROWN et al., 1970, 1971; KOIKE et al., 1971). In the absence of external sodium the depolarizing potential is drastically reduced but does not disappear entirely (FULPIUS and BAUMANN, 1969; BROWN et al., 1970). Photon absorption itself does not change the membrane permeability, but by inducing an amplification process more and more ion-selective channels are opened. The internal transmitter involved in the amplification processes for increased sodium conductance is not known, but it does not appear to be calcium, since increase in internal calcium decreases sodium conductance (Fig. 8).

External K^+ and Cl^- concentrations in barnacle photoreceptors do not affect the amplitude of the light-induced current, but only the resting potential of the dark-adapted cell. Depletion of external calcium changes the potential form during illumination and increases the amplitude of the plateau phase of the potential (MILLECCHIA and MAURO, 1969a; WULFF, 1973; LISMAN and BROWN, 1975). A delayed effect of calcium on membrane potential was also shown by iontophoretic injection of Ca^{++}, by intracellular injection of EGTA, and directly by intracellular injection of aequorin (BROWN and BLINKS, 1974). Sodium "pores" are thought to be opened by a mediator removing calcium from the membrane phase to the intracellular space. According to this it was suggested that, at least in the bee drone, an increase in intracellular sodium concentration could lead to increase in intracellular free calcium (BADER et al., 1976). Intracellular calcium would also be later responsible for photovoltage shutoff by gating sodium influx at stimulus end. At very high light intensities calcium movement gives rise to the fast spikelike potential transient.

In some light-sensitive neurons of *Aplysia* light probably acts upon a single pigment, releasing calcium, which in turn increases potassium conductance, thereby hyperpolarizing the neuronal membrane (BROWN and BROWN, 1973).

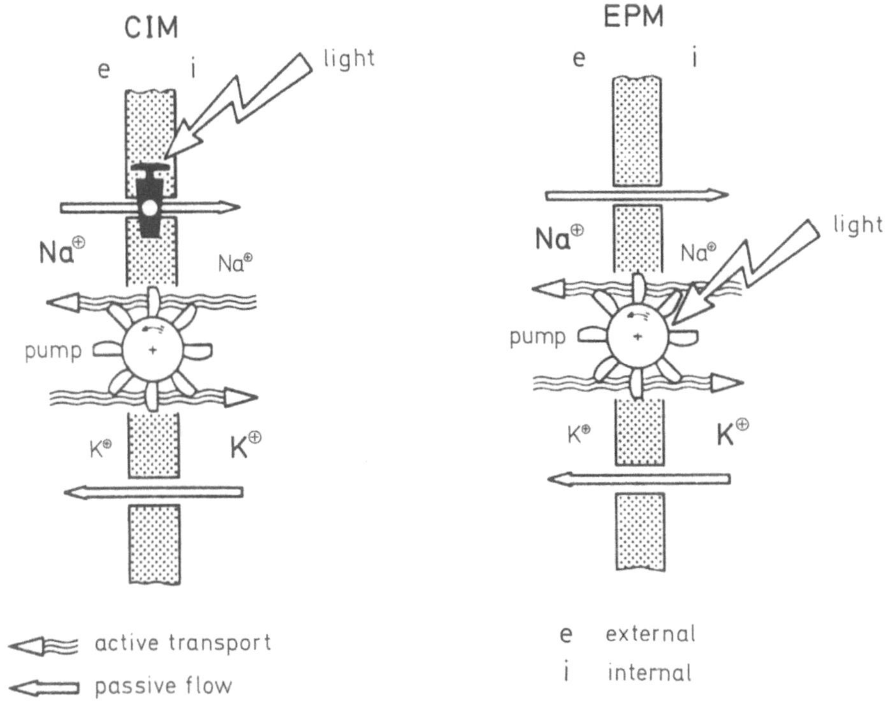

Fig. 7. Two hypotheses postulated to explain the generation of the RP in the photoreceptor cell: CIM = Conductance Increase Mechanism, EPM ≙ Electrogenic Pump Mechanism (STIEVE and WIRTH, 1971)

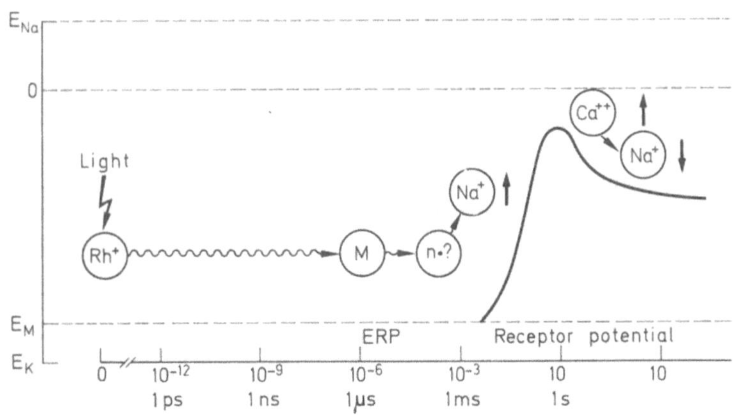

Fig. 8. A hypothetical chain of events to explain conductance changes in invertebrate photoreceptor after illumination. *Solid line* shows the RP in time and voltage scale. Process starts when light is absorbed by rhodopsin (Rh^+) and changed to metarhodopsin (M). This reaction causes ERP and induces an amplification process producing internal transmitter molecules (?), which in turn are responsible for sodium influx $(Na^+)_{in}$. Membrane potential decrease or spontaneously increased affinity to calcium in sodium pores causes closing of sodium pores and rapid decrease of sodium influx at steady level. When light is switched off, calcium excess will close all sodium pores and potential returns to its resting level. E_K is the equilibrium potential for potassium, E_{Na}, for sodium, and E_M is the resting potential of the cell

The response to a short flash of light is usually a slower wave of depolarization with a peak amplitude significantly delayed after the initial flash. In the model proposed by FUORTES and HODGKIN (1964) to explain this behavior in the eye of *Limulus*, ten identical resistor-capacitor low pass filters were cascaded together with high impedance buffer amplifiers between each stage. Experiments on the fly suggest only five stages. The time constants of the linear filters and their rate of change with temperature are similar to the electric behavior of cell membranes (FRENCH and JÄRVILEHTO, 1978). However, a series of chemical reactions with similar activation energies could also explain the observed behavior.

Membrane properties of photoreceptor cells have also been studied using voltage clamp techniques in *Limulus* (MILLECCHIA and MAURO, 1969b) and in barnacle (BROWN et al., 1968, 1969, 1970, 1971; review: FUORTES and O'BRYAN, 1972a). The light-induced current depends nonlinearly on the clamped membrane voltage, time and light intensity. The results obtained by the voltage-clamp methods seem to support the idea of a CIM on the membrane.

V. Spikes in Receptor Cells

Most photoreceptor membranes do not carry the information in the form of spike trains. In the axons of arthropod photoreceptors the potentials are conducted electrotonically to the cell terminal (JÄRVILEHTO and ZETTLER, 1970; IOANNIDES and WALCOTT, 1971; ZETTLER and JÄRVILEHTO, 1971; SHAW, 1972).

Occasionally spikes have been reported in the photoreceptor cell region of a variety of arthropod species. The photoreceptor cell axons of molluscs and annelids are thin and very long (several millimeters or centimeters), but nevertheless, an active mechanism is present. The spikes are generated on the axon membrane either spontaneously (and the light stimulus increases the activity) (*Aplysia*: JACKLET, 1969), or only by light (Scallop: HARTLINE, 1938; LAND, 1966; MCREYNOLDS and GORMAN, 1970a; File clam, *Lima scabra*: MPITSOS, 1973) in which case the proximal cells, having depolarizing RP, produce the "on" discharge, and the distal cells an "off" discharge by hyperpolarizing RP. Spontaneous discharges of spikes in cephalopod optic nerve have been reported only as a response to illumination (MACNICHOL and LOVE, 1960), but no controls were done to show that these spikes were not efferent. If they are photoreceptor axon responses the information from the eye to the optic lobe is transmitted in a spike code on the receptor cell axons.

In simple mollusc eyes like in *Hermissenda* (DENNIS, 1967; ALKON and FUORTES, 1972) or *Tritonia* (CHASE, 1974), in both of which only five cells make up the whole eye, the photoreceptor response to light is a depolarizing RP from which spikes often arise. Impulses are antidromically conducted to the cell body and superimposed on the depolarization. Some cells also showed spontaneous activity inhibited by light (DENNIS, 1967). In *Tritonia* spikes were seen in about 25% of the receptor cells. The spike amplitudes were below 12 mV and were often produced when the cell was depolarized either by injury, by illumination, or by extrinsic current.

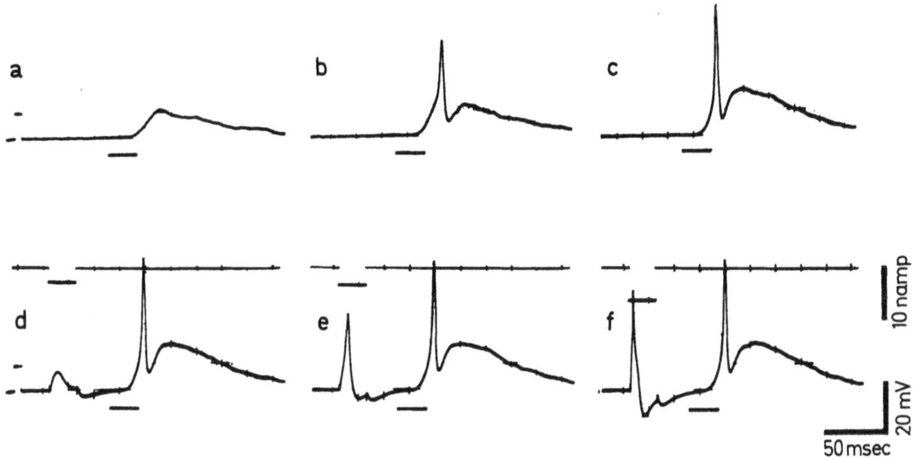

Fig. 9. Spikes triggered with light and current. In the *upper recording*, stimulation only with light of increasing intensity. In the *lower row*, the light flash used in (c) is preceded by current pulses of increasing strength (BAUMANN, 1968)

In the visual cells of the leech *(Hirudo medicinalis)* repetitive spikes are produced by illumination (LASANSKY and FUORTES, 1969; FIORAVANTI and FUORTES, 1972), but these are easily lost upon impalement of the cell with a microelectrode. Spikes may arise in the axon, at some distance from the cell body.

Photoreceptor cell axons of insects do not generally produce spike trains, but occasionally by intra- and extracellular measurements some have been recorded (Dragonfly *(Hemicordulia)*: LAUGHLIN, 1974; Locust *(Locusta)*: SCHOLES, 1965; SHAW, 1968; Beetle *(Creophilus)*: MEYER-ROCHOW, 1974a). In various insects this has been interpreted as an artifact and is to be found only under "unusual" experimental conditions. Discharges of spikes have been found in drone honeybee retinular cells (BAUMANN, 1968) towards the end of a long experiment, during impalement, when the preparation was exposed to air and allowed to dry, and at the beginning of responses to steps of strongly light-adapted receptors. The imprecise identification of the recording site may certainly also lead to misinterpretation of the origin of the intracellularly recorded potentials in different preparations.

Spikes recorded on the RP from the retinular cells of *Limulus* have their origin in the eccentric cell, which is a second-order neuron located between retinular cells. It is electrically coupled with retinular cells, presumably via its apical process (BEHRENS and WULFF, 1965; TOMITA et al., 1960; SMITH and BAUMANN, 1969).

RPs of various insects evoked by bright flashes are sometimes characterized by a large spike or spikelike peak superimposed on the rising phase of the slow potential (NAKA, 1961; NAKA and EGUCHI, 1962; AUTRUM and v. ZWEHL, 1964; SCHOLES, 1965; NAKA and KISHIDA, 1966; BAUMANN, 1968; SHAW, 1969; ZETTLER and JÄRVILEHTO, 1970; JÄRVILEHTO, 1971; CHAPPELL and DOWLING, 1972; CHAPPELL and DEVOE, 1975). The initial spikelike component is destroyed by application of tetrodotoxin to the honeybee drone compound eye (Fig. 9;

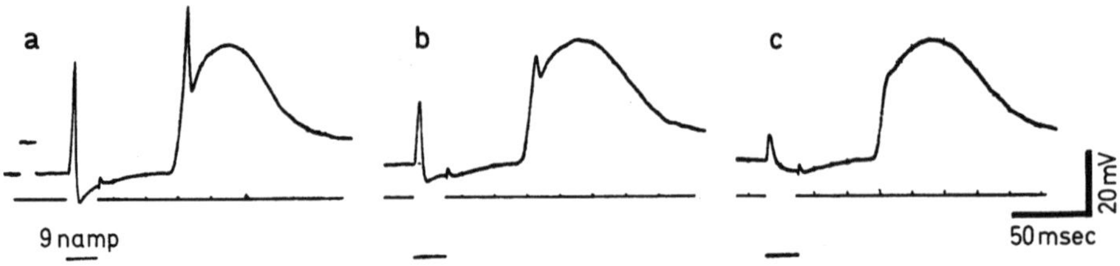

Fig. 10a–c. Action of tetrodotoxin (10^{-5} g/ml) on spikes evoked by current and light. (a) Recorded before application of the drug, a spike is triggered by a constant current pulse of 9 nA, and the second response by a bright light flash. (b) Recorded 7 min after the beginning of the perfusion with tetrodotoxin. (c) Obtained 3 min after (b) (BAUMANN, 1968)

BAUMANN, 1968) and to the dragonfly ocellus (CHAPPELL and DOWLING, 1972) whereas the slow RP is not affected. This implies that two separate processes are involved in generation of these components of the receptor response.

The initial spike may be triggered with either light or current (Fig. 10). The amplitude of the spike increases rapidly and reaches a value which is not further affected by the intensity of the flash or current. The shape of the spike in both cases is quite similar and the difference in amplitude is most probably due to summation of the spike with the slow potential.

The origin and significance of the spike (sometimes there are one or two small additional spikes) is unknown. The spike is generated actively, but whether or not it carries information of the light stimulus is an open question. Recent studies on second-order neurons of the dragonfly ocellus (CHAPPELL and DOWLING, 1972) show that this is improbable. After the spike is blocked by tetrodotoxin, the normal response to the light stimulus is still found in the postsynaptic cell.

D. Functional Organization of the Photoreceptor Cells

I. Univariance and Retinal Differentiation

The principle of univariance (NAKA and RUSHTON, 1966; RUSHTON, 1972) is essentially a specific formulation of the method of response invariance and expresses the view that the response depends only on the absorbed number of photons, not on their energy. The absorption of photons further depends on the quality and array of photopigment molecules. By changing the intensity of light with different stimulus qualities the response-/intensity-functions should run parallel. This has been demonstrated in several invertebrate visual cells (BURKHARDT, 1962; AUTRUM and v. ZWEHL, 1964; WASSERMAN, 1969; NOLTE and BROWN, 1969).

The principle of univariance predicts functional organization of receptor cells in the retina, if all parameters of light information are transmitted. Sensitivity

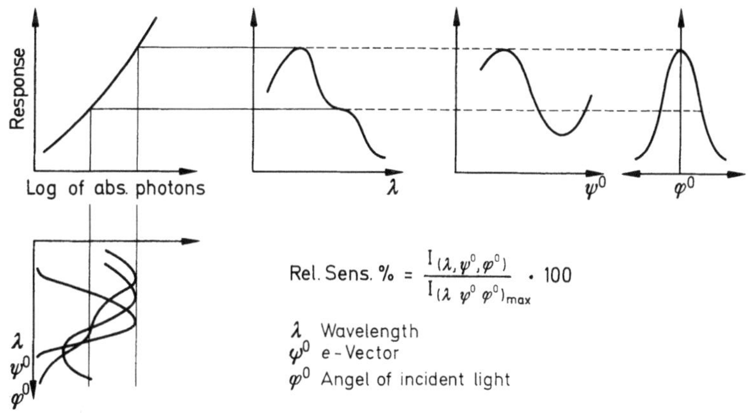

Fig. 11. Response parameter under different stimulus conditions converted through the intensity/response function of the same cell to the sensitivity function by using the principle of univariance. The point is that a cell postsynaptic to the receptor cell cannot distinguish the *type* or quality of stimulus that caused the response and can only gain knowledge of the total quantum catch of the receptor. Information about quality requires some sort of interaction between at least two differently tuned cells

function is widely used for comparisons of stimulus effectivity in different cells and preparations. This is schematically demonstrated in Fig. 11, where it is obvious that the sensitivity function of the cell for different light information parameters is related to efficiency function of the cell by voltage/log-intensity transform. The number of absorbed photons allows us to compare the sensitivity of different cells by their transducer properties. According to all mentioned before it seems apparent that one cell alone is not capable of figuring out all parameters of light information. Even just for unambiguous resolving of the *e*-vector of linearly polarized light three different receptor systems are needed. Also, the more cells with different photopigments in the retina, the better the color resolution. The limited number of different spectral and polarization sensitivities is produced by spatial resolution of the stimulus needed.

In different visual systems we can find functional units. In some species they are ommatidia or, as in the fly, they are neuro-ommatidia, so-called optic cartridges in the lamina ganglionaris. The composition of these units is very uniform throughout the retina. In the fly's retina the neural superposition has the advantage of improving the signal-to-noise ratio of the light stimulus (Kirschfeld, 1967; Smola and Gemperlein, 1973). This finding and also measurements of receptive fields from optic cartridges and long visual fibers imply two basically different functional units in the fly's visual system. The one unit (R 1–R 6) containing six cells would possess broad spectral sensitivity in the region of 460–520 nm with additional sensitivity in the UV region (350 nm), the polarized light sensitivity in individual cells would be canceled in lamina cartridges by summation of opposed responses and angle sensitivity of incident light at ~2.5°. The other unit R7, 8) would contain two cells with blue and UV receptors, with a narrow angle of incident light (1.5°) and with polarized light sensitivity (Kirschfeld, 1971). The R 1–6 system would work during dim light with high photon-capture properties and the R 7–8 system during bright light.

The above described functional differentiation is only in part proved experimentally (ECKERT, 1971; MCCANN and ARNETT, 1972) and recent experiments with electrodes filled with Procion Yellow show a rather incoherent functional anatomy (JÄRVILEHTO and MORING, 1976). According to those few experiments the same anatomic cell type (e.g., R 1) may have different spectral and polarization properties, even if in different preparations but under identical experimental conditions. Anatomically identified R 7 and R 8 also did not show any blue sensitivity as assumed for R 7/8-subsystem (HARDIE, 1977), which is inconsistent with the theory of scotopic (R 1–6) and photopic (R 7/8) vision in the fly. Recordings from second-order neurons in flies show interesting spectral properties, which are perhaps due to two different types of cartridges (MORING, 1978), thus perhaps explaining the functional incoherence in receptor cells in different ommatidia.

If a cell contains two or more photopigments producing nonuniform bumps, then the response becomes highly nonlinear, thus not necessitating a parallel run of intensity functions. Such cells assumed to contain two photopigments have been reported (NOLTE and BROWN, 1969; HORRIDGE and MIMURA, 1975). If bumps produced by any kind of photopigment are uniform the only effect of photopigment mixture in one cell is the broad-band sensitivity. However, there is no evidence for one size of bumps, but rather the opposite (ADOLPH, 1964; DOWLING, 1968), even in a single-pigment cell. Even if the response/intensity function, the widely used V-log I function, is superimposable, it alone does not prove univariance. Not only the amplitude but also the whole potential form produced by the same number of absorbed photons, although with different energies, should be the same. The identical shape of the potentials depends on the accuracy of the comparison, as indicated by behavioral experiments. They show that at least 1% or less difference in the number of absorbed photons can be perceived (FERMI and REICHARDT, 1963; LABHART, 1972).

The energy not the number of absorbed quanta is relevant for the comparison of the nonunivariant cell responses.

Partly nonunivariant cells were reported by STRECK (1972) in the fly, *Calliphora*; this finding was explained by possible ERG contamination or screening effects of pigment granules.

II. Adaptation

Adaptation in terms of sensitivity regulation in the receptor cells describes the constancy of intensity discrimination at different levels of light. Several mechanisms are coupled together for optimizing the function. Two basically different groups of mechanisms can be found: (1) The light flux is regulated through the pupil reaction (KIRSCHFELD and FRANCESCHINI, 1969) and pigment migration (HÖGLUND, 1966; review: SNYDER and MENZEL, 1975), with relatively slow time constants; and (2) the sensitivity is changed at the photoreceptor membrane by modifying quantum absorption affinity (photochemical cycle) or by neural feedback mechanisms on membrane permeability. Light adaptation is not simply a reversal of events during the dark adaptation although they are logically coupled.

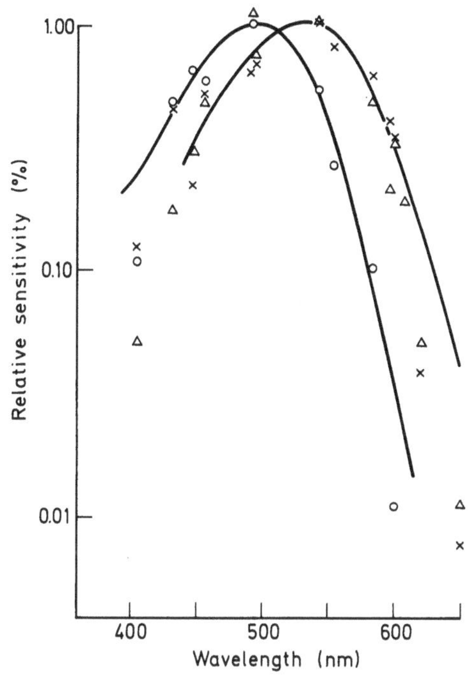

Fig. 12. The single pigment system in barnacle (*Balanus*) photoreceptor cell with two thermally stable states demonstrated by selective adaptation. Two different components of the ERP are used as criteria for the stable states of photopigment. *Circles* show the positive and *crosses*, the negative amplitude of ERP. DARTNALL's nomograms with peaks at 495 and 532 nm are given for comparison (MINKE et al., 1973)

Membrane polarization (FUORTES and HODGKIN, 1964) or other factors such as the concentration of intracellular free calcium (LISMAN and BROWN, 1975) are clearly related to the light adaptation in invertebrate visual cells (see Fig. 8).

ERG and intracellular recording experiments on insects have revealed that the long-lasting light adaptation evoked by blue light can be broken by subsequent long-wavelength stimulus (HAMDORF et al., 1971; COSENS and BRISCOE, 1972; HAMDORF and ROSNER, 1973). In other invertebrates parallel phenomena can be found (NOLTE and BROWN, 1972b; MINKE et al., 1973a, b; HOCHSTEIN et al., 1973). Photometric measurements indicate that a balance exists between photopigment and thermostable metarhodopsin (Fig. 12) possibly regulated by a short-lived photoactive intermediate. However, this is not the case in the barnacle (BROWN and CORNWALL, 1975a, b).

Immediately on the on- or offset of an adapting light, a photoreceptor starts to decrease or recover, respectively, its sensitivity. The process usually exhibits a approximately exponential time course with a time constant that can vary from a few milliseconds to about half an hour depending on the preparation and the stimulus conditions (FUORTES and HODGKIN, 1964; DODGE et al., 1968; LISMAN and BROWN, 1971; SREBRO and BEHBEHANI, 1972; AUTRUM and KOLB, 1972; DÖRRSCHEIDT-KÄFER, 1972; ROSNER, 1975). This accounts for the sodium pump activity in *Limulus* (WULFF et al., 1975).

Neuronal adaptation processes may affect the sensitivity of single receptor cells, but a system of synaptic connections between different cells may also affect larger areas of the photosensitive elements (pooling effect).

The adaptation properties of anatomically different receptor cell types are not the same and adaptation may be coupled with different functions [e.g., polarization sensitivity with the time course of dark recovery (MORING and JÄRVILEHTO, 1977]. In the fly's retina two subsystems, photopic, bright light, and scotopic, dim light, are postulated (KIRSCHFELD, 1971; ECKERT, 1971). This might explain the observed difference in dark recovery functions. Additionally, intermediate unknown adaptive processes interposed between the photochemical cycle and the electric properties of the membrane have been postulated (FEIN and DEVOE, 1973).

E. Transmission of Electric Activity in the Photoreceptor Cells

The electric activity of the photosensitive membrane must be transmitted to postsynaptic cells. The distance from the photoactive locus to the synapse of the next cell is different in different species as well as within one species in different eyes. It is shortest in *Limulus*, where the excitation of the receptor cell is transmitted to the eccentric cell via electrotonic junctions probably located in the microvilli apical endings (BEHRENS and WULFF, 1965; LASANSKY, 1967; see below "Electric coupling").

Generally the cell has an axon connecting it with the next synaptic layer. The distance to the postsynaptic cell as well as the length and diameter of the axons vary from a few tens of microns (fly) up to 10 μm *(Limulus)*. As mentioned above, the axons do not as a rule generate spikes ("Spikes in photoreceptor cells"), but slow depolarizing potentials. An exception seems to be the molluscs that transfer light information from the eye to the optic lobe, not by electrotonic spread of the RP but by active propagated spikes along the axons (*Pecten:* HARTLINE, 1938; Cephalopods: MACNICHOL and LOVE, 1960). Impulse activity has also not been detected in the receptor cell bodies of vertebrates which has lead to the supposition that the slow potential itself carries the light information.

Both extra- and intracellular recordings support the idea of passive spread of the potential on the membrane (BAUMANN, 1968; SHAW, 1968, 1972; SCHOLES, 1969; IOANNIDES and WALCOTT, 1971; JÄRVILEHTO and ZETTLER, 1970, 1971; ZETTLER and JÄRVILEHTO, 1971, 1973).

In some cells the receptor axon is short enough to minimize losses due to decremental conduction, but some visual cells (e.g., long visual fibers in the fly's, e.g., *Calliphora, Musca,* retina) have very long and relatively thin axons, thus raising problems of an insufficiently long space constant. However, there are no direct measurements of the minimum size and shape of potential necessary to transmit signals successfully.

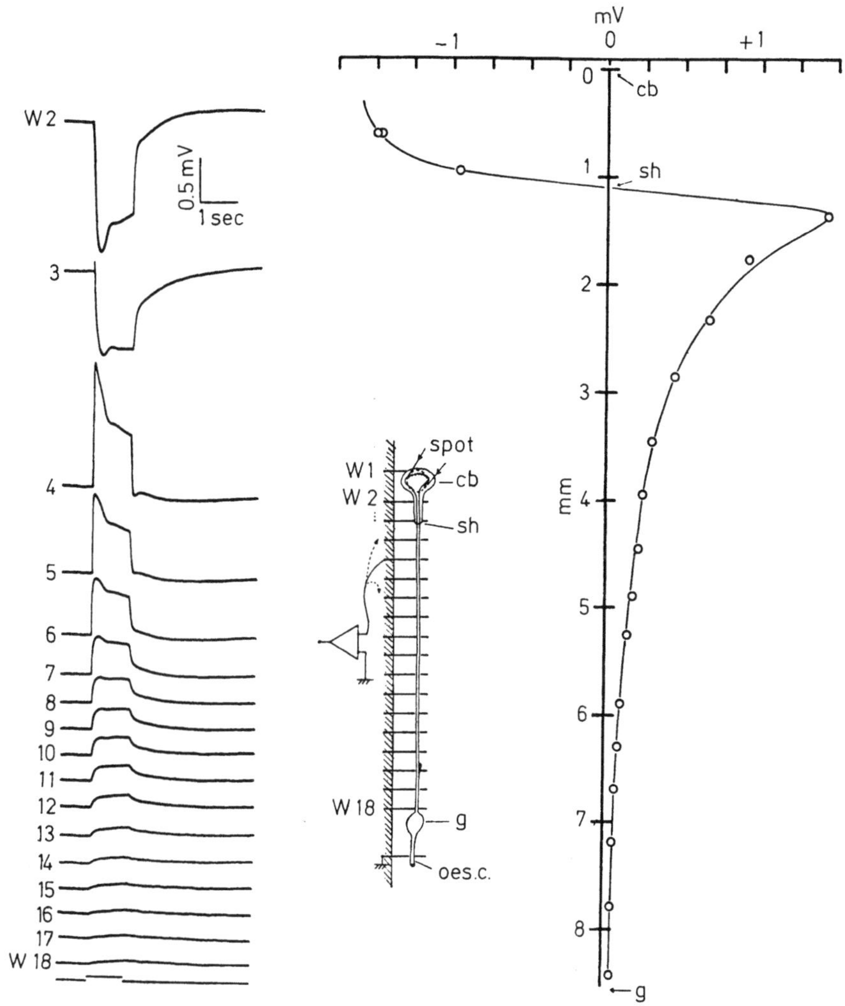

Fig. 13. The passive spread of RPs along the axons of the barnacle photoreceptors. One lateral eye and its optic nerve is laid under oil across a grid of 19 recording wires (center). The potential difference was measured in turn between each wire and the remaining grounded wires. At the *left* appears a sequence of records and at the *right*, a plot of peak response against the distance along the nerve, in which each point is the average of three to six measurements. Current enters the cell bodies during illumination and leaves along the rest of the nerve. *sh*, limit of the loose connective sheath; *g*, supraesophageal ganglion; *cb*, cell bodies; *oes.c.*, esophageal connective (SHAW, 1972)

The attenuation of the passively propagated DC signal along the membrane is inversely related to the space constant of the membrane. The space constant for DC potential components depends on the membrane resistance r_m, the specific resistance of the axoplasm r_i, extracellular specific resistance r_o and the radius of the axon R as follows:

$$\lambda=\sqrt{\frac{r_m}{r_i+r_o}\cdot\frac{R}{2}}$$

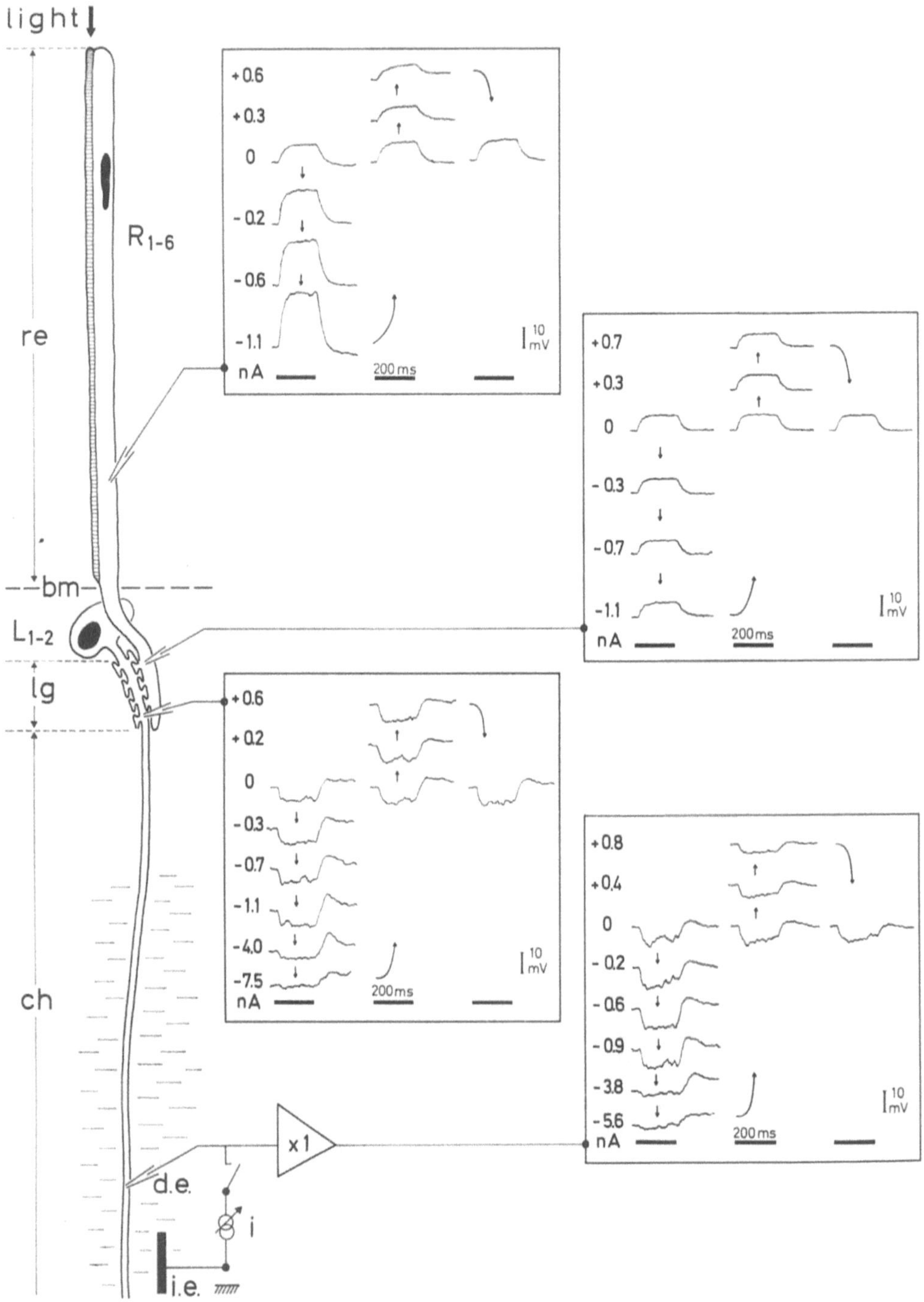

Fig. 14. Active and passive membrane properties in the visual system of the fly, *Calliphora*. Light-induced potentials under clamped current in different parts of two cell types. The recording sites are shown in the drawing. The relative intensity of light stimulus was 9.10^{-4} for *R* 1–6 and 3.10^{-5} for *L* 1–2. In each recording the numbers show the intensity and polarity of the current flowing through the electrode. *Arrows* show chronological order of recording. *re*, retinula cell soma; *R* 1–6, retinula cells 1–6; *bm*, basement membrane; *L* 1–2, second order neurons; *lg*, synaptic region in lamina ganglionaris; *ch*, axon in chiasma region; *d.e.*, different electrode; *i.e.*, indifferent electrode (ZETTLER and JÄRVILEHTO, 1973)

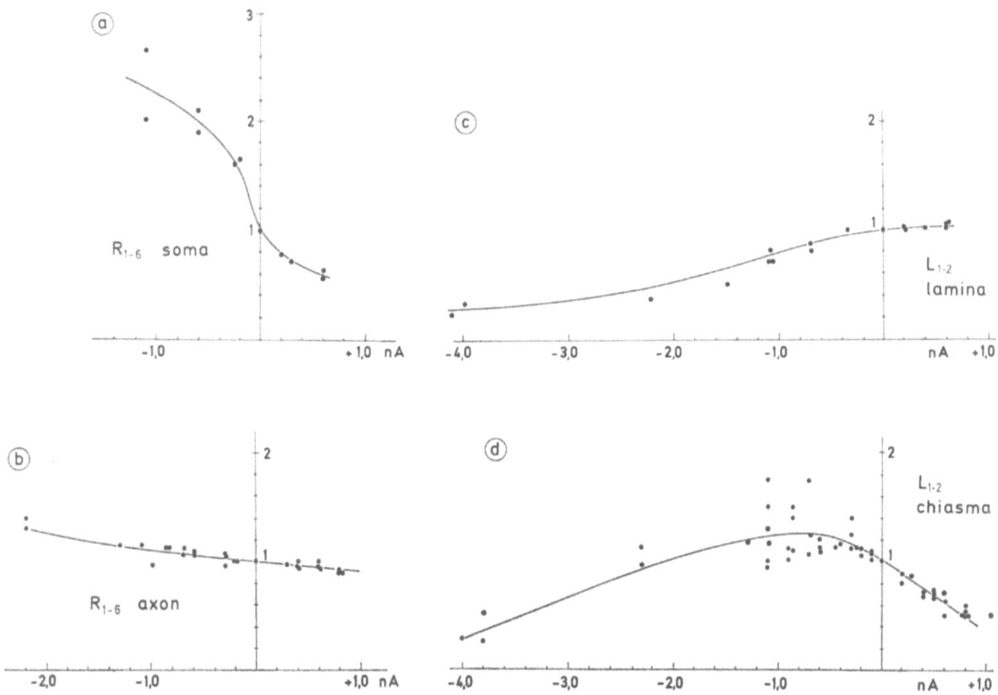

Fig. 15. Dependence of light-induced tonic potential amplitudes on clamped membrane current, averaged for several cells of types and regions indicated: (a) two cells, (b) four cells, (c) two cells, (d) six cells (ZETTLER and JÄRVILEHTO, 1973)

If extracellular resistance r_o is low, it does not influence the space constant. The high frequency components (e.g., initial phase, spikelike component) are more attenuated (JACK and REDMAN, 1971), i.e., the space constant becomes shorter because passive membrane properties can be depicted by a series of parallel R-C-links.

The attenuation for an axon of finite length, of constant diameter, and with sealed ends is

$$\frac{U_T}{U_O} = 1/\cosh(L/\lambda)$$

where U_O is the voltage at the beginning and U_T at the termination of the cable of length L and space constant λ (WEIDMANN, 1952).

Decremental conduction of the visual signal has been nicely demonstrated intra- and extracellularly by SHAW (1972) in the barnacle lateral eye (Fig. 13). The passive properties of photoreceptor axonal membrane have been demonstrated in the fly by current injection (ZETTLER and JÄRVILEHTO, 1973). Short visual fibers (50 μm) in the fly's retina are five to ten times shorter than long visual fibers, which have about the same length as second-order neurons. The second-order neurons ($L1$ and $L2$), which are postsynaptic elements, do not show any spike activity either, but slow hyperpolarizing potentials. How these potentials are conducted to the next synaptic level in the medulla is unknown. The conduction does not show measurable decrement and the membrane properties differ from those of the short type receptor cell axons, thus suggesting some sort of unknown active mechanism (Figs. 14 and 15). Therefore, there is no reason to assume that all receptor cell axons in different preparations would have the same membrane properties, but besides decremental conduction and all-or-none impulses other active membrane mechanisms may exist.

F. Receptor Cell Interaction

I. Anatomic Background of the Interaction

In many cases the anatomic arrangement of invertebrate photoreceptors gives a good basis for the assumption of certain optical, electric, and synaptic coupling of photoreceptor cells. In one ommatidium, microvilli of different receptor cells often have close contacts in different levels of the retinular cell layer. Where the receptor cells are more separated (e.g., in the open rhabdom of dipterans) or totally isolated by other cells in the retinular cell layer, the cell endings have close contacts or even synapses in the next optic ganglion. In the vertebrate retina this kind of contacts are typical and even at the first synaptic layer have formed a complex network.

Because of the large variation in visual cell morphology in invertebrates we have to limit discussion to certain obvious examples. Although there are only a few direct measurements of receptor interaction, morphologic studies are more numerous. For example, in the dragonfly *(Anax, Aeschna)* and locust *(Schistocerca)* ocelli there are some interesting anatomic clues of a complex feedback system (DOWLING and CHAPPELL, 1972; PATTERSON and GOODMAN, 1974).

In this wiring diagram it is easy to imagine many different loops between different receptors or second-order neurons depending on the efficiency of synaptic pathways. This kind of synaptic complexity between receptor cell terminals has been observed in the spider *(Lycosa)* visual system (TRUJILLO-CENÓZ, 1965) and in *Limulus* plexus (WHITEHEAD and PURPLE, 1970), where the physiologic evidence for reciprocal synaptic interaction between units has been provided (RATLIFF et al., 1963; LANGE et al., 1966). It has not yet been observed in the insect compound eye. The similarity of observations on the first synaptic region of the spider eye and the dragonfly dorsal ocellus is not very surprising since the spider visual system consists only of ocellus-type eyes. In the ocellus of the dragonfly, it appears that reciprocal, feedback synapses onto the photoreceptor axons (Fig. 16) are made both by processes from adjacent photoreceptor terminals and the dendrites of second-order terminals (TRUJILLO-CENÓZ, 1965). However, all the elements postsynaptic to the dyad synapses of spider photoreceptor terminals were described as "second-order fibers" and it was not suggested that the postsynaptic element feedback could be a process from an adjacent photoreceptor terminal.

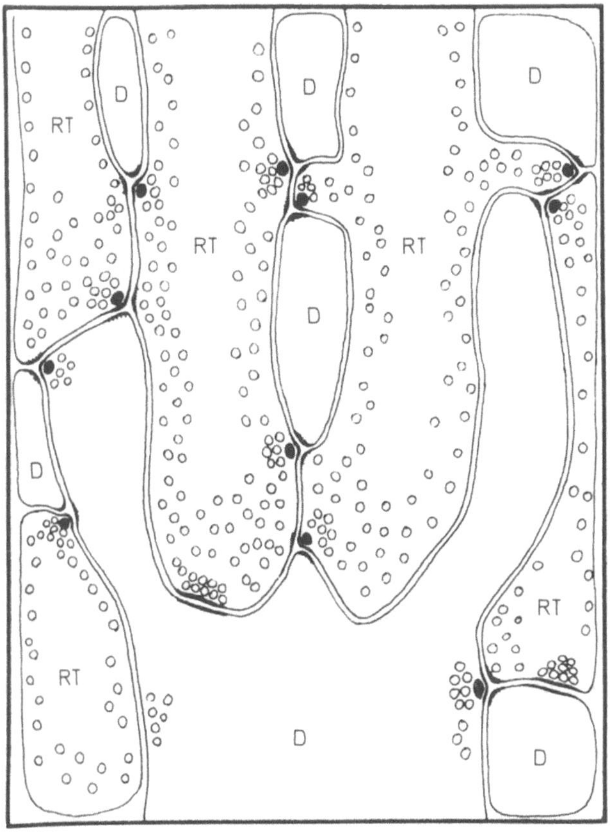

Fig. 16. A summary diagram from the neuronal plexus of the dragonfly median ocellus. Button synapses connect the receptor terminals (*RT*) and receptor axons or dendrites of ocellar nerve fibers (*D*). Several feedback loops are made by button synapses. Conventional contacts observed occasionally are not shown (DOWLING and CHAPPELL, 1972)

II. Optical Coupling

Depending on the anatomic arrangement of the microvilli, receptor cells can be more or less optically coupled. This means that several photoreceptor cells share the same dioptric system, through which light has to pass before entering the cells. In certain cases optical coupling of the cells may physically process the stimulus before it enters the receptor cell (centrally located tandem rhabdomeres of dipterans or lateral filter effects in the fused rhabdom; SNYDER et al., 1973).

If all physical constants are known, the actual stimulus to the receptor cell can be described. For the reaction of the receptor cell it is not important in what way the light stimulus is processed before eliciting the RP. Important for the cell function is only that certain physical requirements are satisfied. The significance of visual system optics is certainly in the improvement of photoreceptor cell efficiency and transmitting the image of the external world onto the receptive screen.

For further discussion of the quite extensive research of the last few years on the complicated optics of invertebrate visual systems see SNYDER, Ch. 5, this volume.

III. Electric Coupling

If the individual receptor cells in the eye do not all have the same function, it would follow that the cells work quasi-independently and therefore as electrically isolated units. By recording intracellularly from two separate cells simultaneously and stimulating one cell, it is possible to study the electric interaction between the two cells. Coupling has been found to exist (WATANABE and GRUNDFEST, 1961) and depends on a low-resistance pathway between the cells (tight electric coupling). Generally, in eyes where the receptor cells have close contacts (especially in the fused rhabdom), this is highly likely. In several other preparations this kind of electric connection is known to exist (crayfish giant motornerve synapse: FURSHPAN and POTTER, 1959; fish electromotor nuclei: BENNETT et al., 1967; horizontal cells in the dogfish retina: KANEKO, 1971).

The lateral eye of *Limulus* was the object of the first successful attempt to measure the electric interaction between visual receptors (BORSELLINO et al., 1965; SMITH et al., 1965; SMITH and BAUMANN, 1969). The equivalent experiments have been repeated later on other species of invertebrates and vertebrates. But divergent methods complicate the comparison of results.

SMITH et al. (1965) demonstrated that the cells within each ommatidium of the lateral eye of *Limulus* are electrically coupled by virtue of low-resistance junctions between neighboring retinular cells and between retinular cells and dendrite of the eccentric cell. The properties of these junctions are such that the light-evoked RP (depolarization of retinular cells) causes depolarization of the eccentric cell, which, in turn, leads to the initiation of spike potentials, which are propagated along the eccentric cell axon.

The ommatidium in the ventral eye of *Limulus* can be modeled as an electrotonic network (SMITH and BAUMANN, 1969). The connection is not dependent on the current passing through the cell membrane. Sufficiently large depolarizations of a retinular cell produce an increase in the coupling resistance of the cell membrane, and as a result the retinular cell becomes partly uncoupled from its neighbors.

The coupling resistance between retinular cell and eccentric cell is nonsymmetric, so that the current passes easier from the retinular cell to the eccentric cell than vice versa. But the connection between retinular cells is symmetric. The coupling resistance depends also on the constant light stimulus being lower at higher intensities (SMITH and BAUMANN, 1969). The electrically coupled eye of *Limulus* seems to be well adapted for receptor pooling. The receptor cells are believed to all have the same functional characteristics and are all coupled to the eccentric cell, which transmits the pooled information centrally. The function of the retinular cell axons is unknown.

The median ocellus of *Limulus* contains two spectral types of cells. One type is maximally sensitive to UV light, the other, to green light; they are called UV and

VIS cells, respectively (NOLTE and BROWN, 1972a). The retina is divided into groups where both UV cells and VIS cells can occur in the same group. UV cells in the same group are electrically coupled to one another and to a spiking arhabdomeric cell, analogous to the eccentric cell in the lateral eye ommatidia. The anatomic basis for connection was confirmed by injecting dyes of different colors into the cells.

Parallel results have been found in crayfish retina. Each ommatidium has a tiered rhabdom, with orthogonal microvilli located in alternating layers. In Crustacea polarization sensitivity is maximal if the *e* vector is parallel with the microvilli (KUWABARA and NAKA, 1959; SHAW, 1966). Two spectral types of cells were found to be polarization sensitive in orthogonal directions (MULLER, 1973).

By simultaneous double electrode recordings combined with dye injection it was found that in each ommatidium cells that are sensitive to the same orientation of the *e* vector of polarized light are electrically coupled. Cells having orthogonal polarized light sensitivity are not coupled. Furthermore, nearly all cells studied were maximally sensitive to orange light, but also a few cells were found to be sensitive to blue light. Blue-sensitive cells were not electrically coupled to orange-sensitive cells (MULLER, 1973).

The significance of this kind of coupling seems to be the possibility of increasing the absolute sensitivity of each functional group by sacrificing some spatial acuity. (Cells in the same ommatidium can be coupled with no loss of visual acuity, but possibly of other functions.)

In the more simple visual systems electric coupling between cells is obvious. Barnacle lateral eye has only three receptor cells (FAHRENBACH, 1965), all of which have the same general functions (SHAW, 1972). The receptor cells can be stimulated separately in the preparation, but the intact eye does not possess optics for images since the shell acts as an efficient diffuser. The spectral sensitivity of different receptor cells is the same.

To study the electric coupling one of the three photoreceptors in the *Balanus* lateral ocellus was penetrated under visual control with two microelectrodes for polarization and recording of the membrane potential and a second photoreceptor was penetrated with a recording microelectrode (BROWN et al., 1971). To show that the recordings were in fact intracellular and that the electrodes were in different cells, Procion Yellow was injected into the cell in some preparations (SHAW, 1972).

SHAW found that voltage change in one cell led to voltage change in the other recorded cell. The relation between voltage changes (V_2/V_1) in the cells, the coupling ratio, varied in different cell pairs in the same and different preparations (SHAW, 1972). The variation is seen in Fig. 17 and is about the same as found before in the barnacle eye by BROWN et al. (1971).

Parallel to the findings in *Limulus* the receptor cells uncouple with illumination. SHAW (1972) reported a drop in coupling ratio from 0.88 in the dark to 0.32 in steady illumination ($\log I = -2.7$). The same phenomenon was also found by BROWN et al. (1971) and explained by the nonlinear characteristics of the current-voltage relations of the membrane in the dark (Fig. 18). The dependence of the coupling on the light intensity, or possibly on the adaptation level, might be the partial explanation of pooling effects in different layers of the retina. Our knowl-

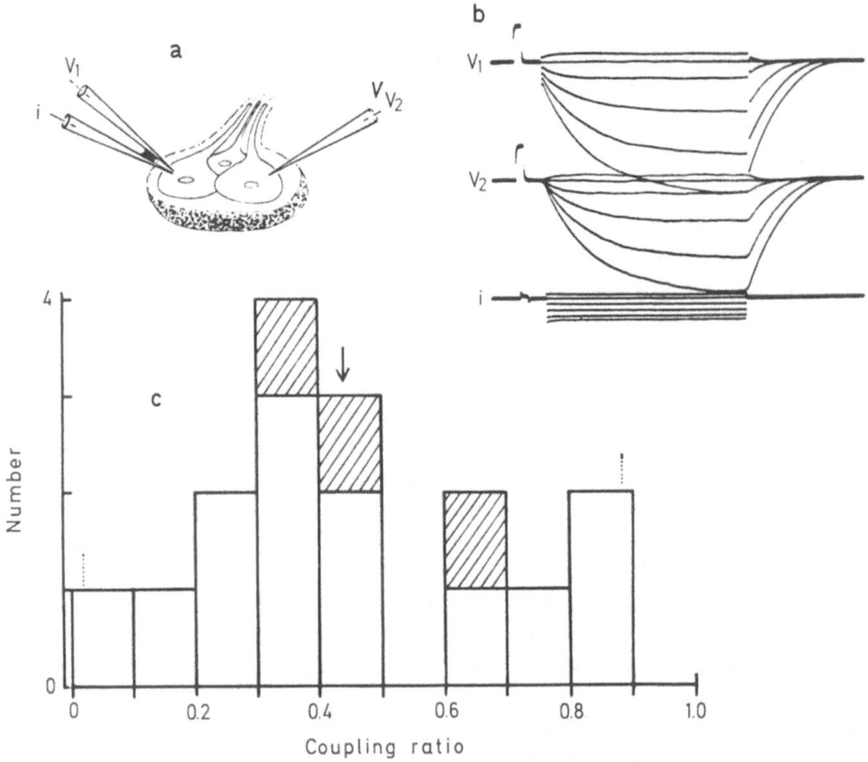

Fig. 17a–c. Electric coupling between barnacle (*Balanus*) photoreceptors. (a and b) Current *i* produces a voltage V_1 in one cell and slightly smaller voltage V_2 in another cell. Six sweeps are superimposed to show the coupling if current strength is increased. (c) Variation of the coupling ratio, V_2/V_1, in 16 different pairs of cells. In the *shaded blocks* one of the cells is believed to be the small central cell on grounds of its receptive field or dye-marking, the remaining blocks representing the two larger cells. The *arrow* gives the average of the original measurements, the *dotted lines*, the extremes of variation. Calibrations in b are 20 mV, 0.1 s and the maximum current is 12 nA (SHAW, 1972)

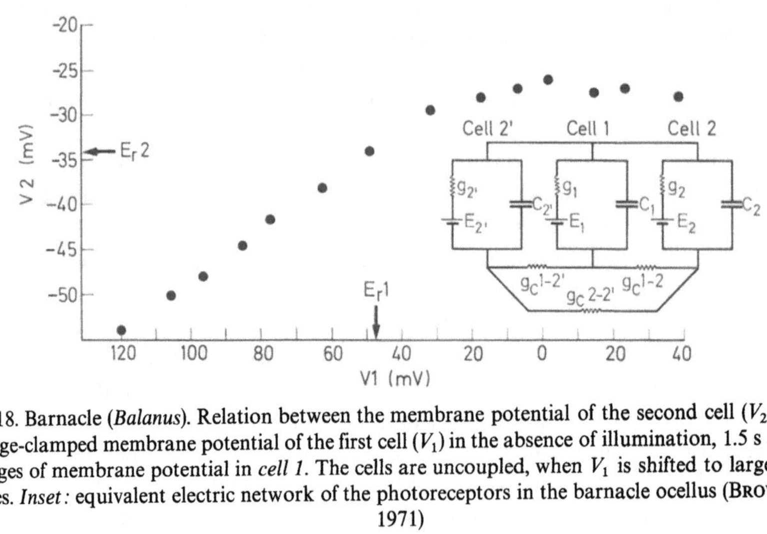

Fig. 18. Barnacle (*Balanus*). Relation between the membrane potential of the second cell (V_2) and the voltage-clamped membrane potential of the first cell (V_1) in the absence of illumination, 1.5 s after step changes of membrane potential in *cell 1*. The cells are uncoupled, when V_1 is shifted to large positive values. *Inset:* equivalent electric network of the photoreceptors in the barnacle ocellus (BROWN et al., 1971)

edge of electric coupling in insect retinular cells is limited to only a few recordings from locust *(Locusta)*, dronebee (SHAW, 1967, 1969), and fly (*Calliphora;* REHBRONN, 1972) retinular cells.

Cells in the same ommatidium of the locust *Schistocerca*, having identical receptive fields and different planes of the maximum polarization sensitivity, are slightly electrically coupled. In both locust and dronebee no histologic identification was made to show that recordings originated from adjacent cells. This was assumed on the basis of the following criteria: the cells in the same ommatidium should have identical receptive fields, the spikes on one record should always occur or fail in synchrony with those on the other record, with small differences, however, in amplitude (SHAW, 1969). The cells were strongly but not completely DC-coupled.

The experiments do not show unambiguously that cells are de facto electrically coupled or in the same ommatidium, because the cells were always, though variably, electrically coupled. Strong coupling obviously influences the mutual electric functions of the cells and produces the similarities in the recordings.

Electric coupling in highly developed retinae could provide a powerful mechanism for regulating size of receptive fields (pooling of receptor cells or their second-order cells) and sensitivity of the eye.

IV. Electric Field Interaction

Conceptually, the simplest form of cellular interaction is electric field interaction. In a tightly packed array of cells, the light-induced current passing a photoreceptor cell through its active membrane can flow along extracellular channels or can flow into the neighboring cells. If the resistance to the flow of current through the extracellular channels becomes relatively large, more current will tend to flow across the membranes of neighboring cells opposed to the active region. This flow of current will tend to depolarize these neighboring cells.

This hypothesis is consistent with the findings that the photocurrent of crayfish *(Procambarus)* retinular cells can produce large (10 mV) negative extracellular potentials when measured near the photoreceptive membranes within ommatidia (MULLER, 1973). Also by extracellular field interaction between first- and second-order cells it is possible to explain the polarity change of spectral sensitivity curves obtained intracellularly in the median ocellus of *Limulus* (NOLTE and BROWN, 1972) and in the compound eye of the fly *(Calliphora)* (JÄRVILEHTO and MORING, 1976). The intracellularly recorded potential is affected by optimally stimulating the neighboring cells. Also, in the first optic ganglion of the fly *(Calliphora)* the field effect of extracellular potentials on second-order responses have been proposed (AUTRUM et al., 1970).

The extracellular currents have been measured recently in the locust *(Schistocerca)* eye and the whole basement membrane region seems to form a barrier for extracellular current flow (SHAW, 1975). It was proposed that extracellular current flow from the maximally stimulated receptor causes an inhibitory surround thus providing an explanation for narrow receptive fields, which the second-order

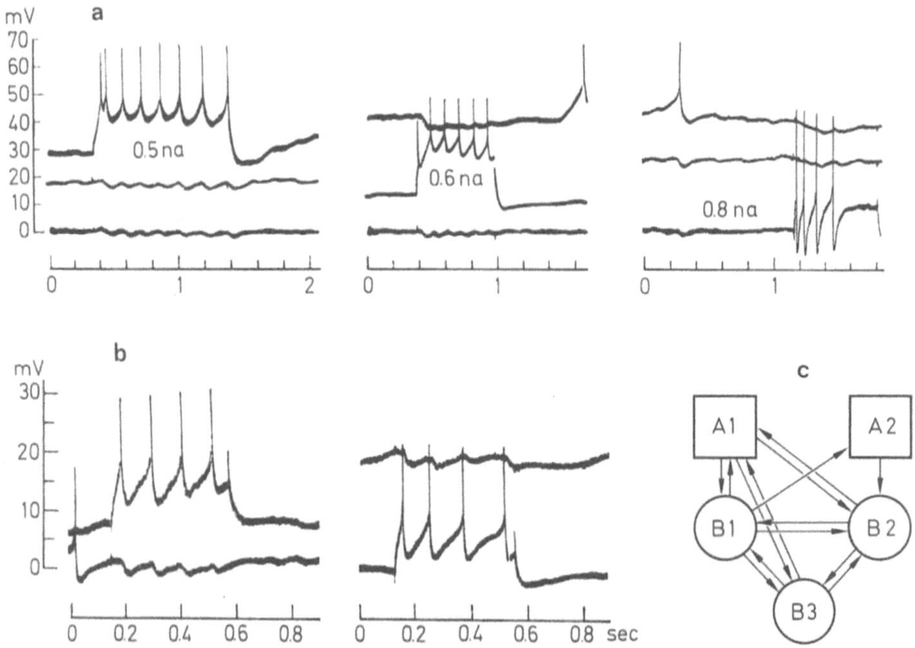

Fig. 19 A–C. Synaptic interactions between type *A* and *B* photoreceptor cells in *Hermissenda*. (A) One type *A* and two type *B* receptors were simultaneously impaled. Depolarizing current evokes spikes in one cell and hyperpolarizing waves in the other two receptor cells. (B) Same as in A but pair of type *B* cells. (C) Presumed diagram of the inhibitory interconnections (ALKON and FUORTES, 1972)

neurons show over the receptors (ZETTLER and JÄRVILEHTO, 1972a; LAUGHLIN, 1974). Such a mutual nonsynaptic interaction would also increase the sensitivity of the cell more than would be predicted from the cell's differential absorption of, e.g., polarized light and wavelength (MULLER, 1973), but the decisive experiments are still missing.

V. Synaptic Interreceptor Coupling

When cells are coupled by chemical synapses the interreceptor interaction has more plasticity than in the electrically coupled situation. The synaptic interaction can be either excitatory or inhibitory depending on the secretory output on the presynaptic membrane (ECCLES, 1964).

Interreceptor contacts are in general located on the axonal membrane and in the first synaptic layer, but little is known about the function of these synapses.

Some studies on the simple eye of the mollusc *(Hermissenda)* show that synaptic interaction of receptor cell endings in the cerebropleural ganglion is inhibitory in its nature and reciprocal (DENNIS, 1967). A spike in one cell is always followed by a hyperpolarizing synaptic potential in the other.

Figure 19 shows different forms of synaptic interaction found by ALKON and FUORTES (1972) in this simple visual system. The five receptor cells are divided into two groups, two *A* cells and three *B* cells. The basis for the division is the spike height; *A* cells produce spikes of about 45 mV and *B* cells, of about 15 mV.

Inhibitory interactions between cells in these two groups are not all equal, so that *A* cells do have a very weak inhibitory interaction and *B* cells are all mutually interconnected. The features of these interconnections are schematically shown in Fig. 19. The physiologic consequence of these inhibitory interactions is that the cell with the higher response will decrease the response of the other cells. ALKON and FUORTES (1972) suggest that the interactions on the eye of *Hermissenda* can facilitate the discrimination of light intensities, movement, and possibly, therefore, the position of stimulus. The same kind of functions are also possible in the ocelli of insects and arachnids, but in general synaptic interaction of photoreceptor cells is mediated by neurons, probably the system possessing more degrees of freedom.

G. Concluding Remarks

In spite of several extensive studies on invertebrate visual pigments, quantum absorption, and electric events in the membrane, the step (transduction process) coupling quantum absorption and the electric process is still not understood. For example, the ERP is certainly a powerful tool for explaining the photointerconversion of visual pigments, but it does not give any reasons for the long time delay (latency) between stimulus onset and RP. The permeability change for sodium and possibly also for potassium and calcium in the photoreceptor membrane seems to be responsible for RP generation. The RP is made up of quasi-linear summation of so-called quantum bumps, which are believed to be the unitary events on RP. The statistical properties of events suggest, that they are indeed caused by single photon absorptions.

Calcium in particular is increasingly regarded as important for RP dynamics: if the calcium permeability of the membrane (Ca-influx) is increased the potential amplitude of the cell rapidly decreases (Ca-inactivation).

Recently developed techniques of electrophysiologic staining provide an excellent method for studying relationships between structure and function. The functional organization is predicted by the "principle of univariance," according to which every absorbed photon acts uniformly on the photoreceptor membrane. The transduction properties of the receptor cell determine the specific function (e.g., spectral and polarization sensitivity) of the cell. Information processing in the system depends on the transmission properties of the cell membrane, on the characteristics of the electric signal, and on the types of interaction, synaptic or otherwise, between cells.

Passive interaction between individual receptor cells may be mediated by tight junctions (electric coupling) or by extracellular field potentials. This kind of interaction, if the efficiency is dynamically changed, allows an effective mechanism for regulating either receptive field size (acuity) or sensitivity, both precisely and with minimum delay.

Chemical synapses may be associated with a complex anatomic array of cells, with synaptic delay, and with a complex genetic code to provide a specific transmitter, but allow a greater degree of freedom in the system.

Which form of interaction is most important for the information processing in the invertebrate retina will be one of the central questions in the future. Simultaneous multichannel recordings from several elements will probably help to answer this question.

H. Acknowledgements

I am grateful to Professor G. A. HORRIDGE and his associates in the Department of Neurobiology, School of Biological Sciences, Australian National University, Canberra, for their hospitality and help during the preparation of this manuscript. In particular I am indepted to Drs. S. R. SHAW, D. STAVENGA, M. LEGGETT, and J. MORING for their constructive criticism of the present work, and to Mrs. ANNELI RAUTIO for her invaluable technical assistance.

References

Adolph, A.: Spontaneous slow potential fluctuations in the *Limulus* photoreceptors. J. gen. Physiol. **48**, 297—322 (1964).

Alkon, D. L., Fuortes, M. G. F.: Responses of photoreceptors in *Hermissenda*. J. gen. Physiol. **60**, 631—649 (1972).

Autrum, H., Kolb, G.: The dark adaptation in single visual cells of the compound eye of *Aeschna cyanea*. J. comp. Physiol. **79**, 213—232 (1972).

Autrum, H., Zwehl, V. v.: Die spektrale Empfindlichkeit einzelner Sehzellen des Bienen-Auges. Z. vergl. Physiol. **48**, 357—384 (1964).

Autrum, H., Zettler, F., Järvilehto, M.: Post-synaptic potentials from a single monopolar neuron of the ganglion opticum I of the blowfly *Calliphora*. Z. vergl. Physiol. **70**, 414—424 (1970).

Bader, C. R., Baumann, F., Bertrand, D.: Role of intracellular calcium and sodium in light adaptation in the retina of the honey bee drone (*Apis mellifera* L.). J. gen. Physiol. **67**, 475—491 (1976).

Barber, V. C., Evans, E. M., Land, M. I.: The fine structure of the eye of the mollusc *Pecten maximus*. Z. Zellforsch. **76**, 295—312 (1967).

Barlow, Jr., R. B., Kaplan, E.: Properties of visual cells in the lateral eye of *Limulus in situ*. J. gen. Physiol. **69**, 203—220 (1977).

Barrett, J. N., Graubard, K.: Fluorescent staining of cat motoneurons in vivo with bevelled micropipettes. Brain Res. **18**, 565—568 (1970).

Bass, L., Moore, W. J.: An electrochemical model for depolarization of a retinula cell of *Limulus* by a single photon. Biophys. J. **10**, 1—19 (1970).

Baumann, F.: Slow and spike potentials recorded from retinula cells of the honeybee drone in response to light. J. gen. Physiol. **52**, 855—875 (1968).

Behbehani, M., Srebro, R.: Discrete waves and phototransduction in voltage-clamped ventral photoreceptors. J. gen. Physiol. **64**, 186—200 (1974).

Behrens, M. E., Wulff, V. J.: Light-initiated responses of retinula and eccentric cells in *Limulus* lateral eye. J. gen. Physiol. **48**, 1081—1093 (1965).

Bennett, M. V. L., Pappas, G. D., Gimenez, M., Nakajima, Y.: Physiology and ultrastructure of electronic junctions. IV. Medullary electromotor nuclei in Gymnotid fish. J. Neurophysiol. **30**, 236—300 (1967).

Benolken, R. M.: Reversal of photoreceptor polarity recorded during graded receptor potential response to light in the eye of *Limulus*. Biophys. J. **1**, 551—564 (1961).

Bernhard, C. G., Granit, R.: Nerve as a model temperature organ. J. gen. Physiol. **29**, 257—265 (1946).

Borsellino, A., Fuortes, M. G. F.: Responses to single photons in visual cells of *Limulus*. J. Physiol. **196**, 507—539 (1968).

Borsellino, A., Fuortes, M. G. F., Smith, T. G.: Visual responses in *Limulus*. Cold Spr. Harb. Symp. quant. Biol. **30**, 429—443 (1965).

Brindley, G. S., Gardner-Melwin, A. R.: Origin of early receptor potential of the retina. J. Physiol. **182**, 185—194 (1966).

Brown, A. M., Brown, H. M.: Light response of a giant *Aplysia* neuron. J. gen. Physiol. **62**, 239—254 (1973).

Brown, H. M., Cornwall, M. C.: Spectral correlates of a quasi-stable depolarization in barnacle photoreceptors following red light. J. Physiol. **248**, 555—578 (1975a).

Brown, H. M., Cornwall, M. C.: Tonic mechanism of a quasi-stable depolarization in barnacle photoreceptors following red light. J. Physiol. **248**, 579—593 (1975b).

Brown, H. M., Hagiwara, S., Koike, H., Meech, R. M.: Membrane properties of a barnacle photoreceptor examined by the voltage-clamp technique. J. Physiol. **208**, 385—413 (1970).

Brown, H. M., Hagiwara, S., Koike, H., Meech, R. W.: Electrical characteristics of a barnacle photoreceptor. Fed. Proc. **30**, 69—78 (1971).

Brown, H. M., Meech, R. M., Sakata, H., Hagiwara, S.: Voltage-clamp of light receptor cells in the barnacle eye. Int. Cong. Physiol. Sci. **7**, 63 (1968).

Brown, H. M., Meech, R. W., Koike, H., Hagiwara, S.: Current-voltage relations during illumination: Photoreceptor membrane of a barnacle. Science **166**, 240—243 (1969).

Brown, J. E., Blinks, J. R.: Changes in intracellular free calcium concentration during illumination in invertebrate photoreceptors. Detection with aequorin. J. gen. Physiol. **64**, 643—665 (1974).

Brown, J. E., Murray, J. R., Smith, T. G.: Photoelectric potential from photoreceptor cells in ventral eye of *Limulus*. Science **158**, 665—666 (1967).

Brown, K. T., Murakami, M.: A new receptor potential of the monkey retina with no detectable latency. Nature **201**, 626—628 (1964).

Bureš, J., Petráñ, M., Zachar, J.: Electrophysiological methods in biological research. New York: Academic Press 1960.

Burkhardt, D.: Spectral sensitivity and other response characteristics of single visual cells in the arthropod eye. Symp. Soc. exp. Biol. **16**, 86—109 (1962).

Chappell, R. L., DeVoe, R. D.: Action spectra and chromatic mechanisms of cells in the median ocelli of dragonflies. J. gen. Physiol. **65**, 399—419 (1975).

Chappell, R. L., Dowling, J. E.: Neural organization of the median ocellus of the dragonfly. I. Intracellular electrical activity. J. gen. Physiol. **60**, 121—147 (1972).

Chase, R.: The initiation and conduction of action potentials in the optic nerve of *Tritonia*. J. exp. Biol. **60**, 721—734 (1974).

Cone, R. A.: Early receptor potential of vertebrate retina. Nature **204**, 736 (1964).

Cone, R. A.: Early receptor potential: photoreversible charge displacement in rhodopsin. Science **155**, 1128—1131 (1967).

Cone, R. A.: The internal transmitter model for visual excitation: some quantitative implications. In: Biochemistry and physiology of visual pigments (ed. H. Langer). Berlin-Heidelberg-New York: Springer 1973.

Cone, R. A., Brown, P. K.: Dependence of early receptor potential on orientation of rhodopsin. Science **156**, 536 (1967).

Cone, R. A., Pak, W. L.: The early receptor potential. In: Handbook of sensory physiology, Vol. I. Principles of receptor physiology (ed. W. R. Loewenstein). Berlin-Heidelberg-New York: Springer 1972.

Cosens, D., Briscoe, D.: A switch phenomenon in the compound eye of the white-eyed mutant of *Drosophila melanogaster*. J. Insect Physiol. **18**, 627—632 (1972).

Dahl, R. D., Krischer, C. C., Meissen, R., Rittinghaus, K. F.: Early receptor potentials and comparison of the electrical properties of median and lateral photoreceptors of the barnacle. Vision Res. **17**, 519—525 (1977).

Davis, H.: Some principles of sensory receptor action. Physiol. Rev. **41**, 391—416 (1961).

Dennis, M. J.: Electrophysiology of the visual system in a nudibranch mollusc. J. Neurophysiol. **30**, 1439—1465 (1967).

Dodge, F. A. Jr., Knight, B. W., Toyoda, J.: Voltage noise in *Limulus* visual cells. Science **160**, 88—90 (1968).

Dörrscheidt-Käfer, M.: Die Empfindlichkeit einzelner Photorezeptoren im Komplexauge von *Calliphora erythrocephala.* J. comp. Physiol. **81**, 309—340 (1972).

Dowling, J. E.: Discrete potentials in the dark adapted eye of *Limulus.* Nature (Lond.) **217**, 28—31 (1968).

Dowling, J. E., Chappell, R. L.: Neural organization of the median ocellus of the dragonfly. II. Synaptic structure. J. gen. Physiol. **60**, 148—165 (1972).

Eakin, R. M.: Evolution of photoreceptors. Cold Spr. Harb. Symp. quant. Biol. **30**, 363—370 (1965).

Eccles, J. C.: The physiology of synapses. New York: Springer 1964.

Eckert, H.: Die spektrale Empfindlichkeit des Komplexauges von *Musca* (Bestimmung aus Messungen der optomotorischen Reaktion). Kybernetik **9**, 145—156 (1971).

Fahrenbach, W. H.: The micromorphology of some simple photoreceptors. J. cell. comp. Physiol. **65**, 233—254 (1965).

Fahrenbach, W. H.: The visual system of the Horseshoe crab *Limulus polyphemus.* Int. Rev. Cytol. **41**, 285—349 (1975).

Fein, A., Cone, R. A.: *Limulus* rhodopsin: rapid return of transient intermediates to thermally stable state. Science **182**, 495—497 (1973).

Fein, A., DeVoe, R. D.: Adaptation in the ventral eye of *Limulus* is functionally independent of the photochemical cycle, membrane potential, and membrane resistance. J. gen. Physiol. **61**, 273—289 (1973).

Fermi, G., Reichardt, W.: Optomotorische Reaktionen der Fliege *Musca domestica.* Kybernetik **2**, 15—28 (1963).

Fioravanti, R., Fuortes, M. G. F.: Analysis of responses in visual cells of the leech. J. Physiol. **227**, 173—194 (1972).

Frank, K., Becker, M. C.: Microelectrodes for recording and stimulation. In: Physical techniques in biological research, Vol. V. Electrophysiological methods (ed. W. L. Nastuk), Part A. New York: Academic Press 1964.

French, A. S., Järvilehto, M.: The dynamic behaviour of photoreceptor cells in the fly in response to random (white noise) stimulation at a range of temperatures. J. Physiol. **274**, 311—322 (1978).

Fulpius, B., Baumann, F.: Effects of sodium, potassium and calcium ions on slow, and spike potentials in single photoreceptors. J. gen. Physiol. **53**, 541—561 (1969).

Fuortes, M. G. F.: Initiation of impulses in visual cell of *Limulus.* J. Physiol. **148**, 14—28 (1959).

Fuortes, M. G. F., Hodgkin, A. L.: Changes in time scale and sensitivity in the ommatidia of *Limulus.* J. Physiol. **172**, 239—263 (1964).

Fuortes, M. G. F., O'Bryan, P. M.: Generator potentials in invertebrate photoreceptors. In: Handbook of sensory physiology, Vol. VII/2 (ed. M. G. F. Fuortes), pp. 279—320. Berlin-Heidelberg-New York: Springer 1972a.

Fuortes, M. G. F., O'Bryan, P. M.: Responses to single photons. In: Handbook of sensory physiology, Vol. VII/2 (ed. M. G. F. Fuortes), pp. 321—338. Berlin-Heidelberg-New York: Springer 1972b.

Fuortes, M. G. F., Yeandle, S.: Probability of occurrence of discrete potential waves in the *Limulus.* J. gen. Physiol. **47**, 443—463 (1964).

Furshpan, E. J., Potter, D.: Transmission at the giant motor synapse of the crayfish. J. Physiol. **145**, 289—325 (1959).

Gorman, A. L. F., McReynolds, J. S.: Hyperpolarizing and depolarizing receptor potentials in the scallop eye. Science **165**, 309—310 (1969).

Gorman, A. L. F., McReynolds, J. S., Barnes, S. N.: Photoreceptors in primitive chordates; fine structure, hyperpolarizing receptor potentials, and evolution. Science **172**, 1052—1054 (1971).

Hagins, W. A.: Electrical signs of information flow in photoreceptors. Cold Spr. Harb. Symp. quant. Biol. **30**, 403—418 (1965).

Hagins, W. A., McGaughy, R. E.: Molecular and thermal origins of fast photoelectric effects in squid retina. Science **157**, 813—816 (1967).
Hagins, W. A., Rüppel, H.: Fast photoelectric effects and properties of vertebrate photoreceptors as electric cables. Fed. Proc. **30**, 64—68 (1971).
Hagins, W. A., Zonana, H. V., Adams, R. G.: Local membrane current in the outer segments of squid photoreceptors. Nature (Lond.) **194**, 844—846 (1962).
Hamdorf, K., Gogala, M., Schwemer, J.: Beschleunigung der „Dunkeladaptation" eines UV-Rezeptors durch sichtbare Strahlung. Z. vergl. Physiol. **75**, 189—199 (1971).
Hamdorf, K., Rosner, G.: Adaptation und Photoregeneration im Fliegenauge. J. comp. Physiol. **86**, 281—292 (1973).
Hardie, R.: Electrophysiological properties of R7 and R8 in dipteran retina. Z. Naturforsch. **32c**, 887—889 (1977).
Hartline, H. K.: The discharge of impulses in the optic nerve of *Pecten* in response to illumination of the eye. J. cell. comp. Physiol. **11**, 465 (1938).
Hartline, H. K., Graham, C. H.: Nerve impulses from single receptors in the eye. J. cell. comp. Physiol. **1**, 277—295 (1932).
Hartline, H. K., Wagner, H. G., MacNichol, E. F. Jr.: The peripheral origin of nervous activity in the visual system. Cold Spr. Harb. Symp. quant. Biol. **17**, 125—141 (1952).
Hillman, P., Dodge, I. A., Hochstein, S., Knight, B. W., Minke, B.: Rapid dark recovery of the invertebrate early receptor potential. J. gen. Physiol. **62**, 77—86 (1973).
Hochstein, S., Minke, B., Hillman, P.: Antagonistic components of the late receptor potential in the barnacle photoreceptor arising from different stages of the pigment process. J. gen. Physiol. **62**, 105—128 (1973).
Horridge, G. A., Mimura, K.: Fly photoreceptors. I. Physical separation of two visual pigments in *Calliphora* retinula cells 1—6. Proc. R. Soc. (Lond.) B **190**, 211—224 (1975).
Höglund, G.: Pigment migration, light scattering and receptor sensitivity in the compound eye of nocturnal Lepidoptera. Acta physiol. scand. **69**, Supp. 282 (1966).
Hubbard, R., Bownds, D., Yoshizawa, T.: Chemistry of visual photoreception. Cold Spr. Harb. Symp. quant. Biol. **30**, 301—315 (1965).
Ioannides, A. C., Walcott, B.: Graded illumination potentials from retinula cell axons in the bug *Lethocerus*. Z. vergl. Physiol. **71**, 315—325 (1971).
Jack, J. J. B., Redman, S. J.: The propagation of transient potentials in some linear cable structures. J. Physiol. **215**, 283 (1971).
Jacklet, J. W.: Electrophysiological organization of the eye of *Aplysia*. J. gen. Physiol. **53**, 21—42 (1969).
Järvilehto, M.: Lokalisierte intrazelluläre Ableitungen aus den Axonen der 8. Sehzelle der Fliege *Calliphora erythrocephala*. Diss. Univ. München 1971.
Järvilehto, M., Moring, J.: Spectral and polarization sensitivity of identified retinal cells of the fly. In: Neural principles in vision (eds. F. Zettler, R. Weiler). Berlin-Heidelberg-New York: Springer 1976.
Järvilehto, M., Zettler, F.: Micro-localisation of lamina-located visual cell activities in the compound eye of the blowfly *Calliphora*. Z. vergl. Physiol. **69**, 134—138 (1970).
Järvilehto, M., Zettler, F.: Localized intracellular potentials from pro- and postsynaptic components in the external plexiform layer of an insect retina. Z. vergl. Physiol. **75**, 422—440 (1971).
Järvilehto, M., Zettler, F.: Electrophysiological-histological studies on some functional properties of visual cells and second order neurons of an insect retina. Z. Zellforsch. **136**, 291—306 (1973).
Kaneko, A.: Electrical connexions between horizontal cells in the dogfish retina. J. Physiol. **213**, 95—105 (1971).
Kater, S. B., Nicholson, C.: Intracellular staining in neurobiology. New York: Springer 1973.
Kikuchi, R., Naito, K., Tanaka, T.: Effect of sodium and potassium ions on the electrical activity of single cells in the lateral eye of the horseshoe crab. J. Physiol. **161**, 319—343 (1962).
Kirschfeld, K.: Discrete and graded receptor potentials in the compound eye of the fly *(Musca)*. In: The functional organization of the compound eye (ed. C. G. Bernhard). Oxford: Pergamon Press 1966.
Kirschfeld, K.: Die Projektion der optischen Umwelt auf das Raster der Rhabdomere im Komplexauge von *Musca*. Exp. Brain Res. **3**, 248–270 (1967).
Kirschfeld, K.: Aufnahme und Verarbeitung optischer Daten im Komplexauge von Insekten. Naturwissenschaften **58**, 201—209 (1971).

Kirschfeld, K., Franceschini, N.: Ein Mechanismus zur Steuerung des Lichtflusses in den Rhabdomeren des Komplexauges von *Musca*. Kybernetik **6**, 13—21 (1969).

Koike, H., Brown, H. M., Hagiwara, S.: Hyperpolarization of a barnacle photoreceptor membrane following illumination. J. gen. Physiol. **57**, 723—737 (1971).

Kravitz, E. A., Stretton, A. D. W., Alvarez, J., Furshpan, E. J.: Determination of neuronal geometry using an intracellular dye injection technique. Fed. Proc. Fed. Am. Soc. exp. Biol. **27**, 749 (1968).

Kuwabara, M., Naka, K.: Response of single cells to polarized light. Nature (Lond.) **184**, 455—456 (1959).

Labhart, T.: The discrimination of light intensities in the honey bee. In: Information processing in the visual systems of arthropods (ed. R. Wehner). Berlin-Heidelberg-New York: Springer 1972.

Land, M. F.: Activity in the optic nerves of *Pecten maximus* in response to changes in light intensity, and to pattern and movement in the optical environment. J. exp. Biol. **45**, 83—100 (1966).

Lange, D., Hartline, H. K., Ratliff, F.: The dynamics of lateral inhibition in the compound eye of *Limulus*. In: Functional organization of the compound eye (ed. C. G. Bernhard). Oxford: Pergamon Press 1966.

Lasansky, A.: Cell junctions in ommatidia of *Limulus*. J. Cell Biol. **33**, 365—383 (1967).

Lasansky, A., Fuortes, M. G. F.: The site of origin of electrical responses in visual cells of the leech, *Hirudo medicinalis*. J. Cell Biol. **42**, 241—252 (1969).

Laughlin, S. B.: Neural integration in the first optic neuropile of dragonflies. I. Signal amplification in dark-adapted second-order neurons. J. comp. Physiol. **84**, 335—355 (1973).

Laughlin, S. B.: Neural integration in the first optic neuropile of dragonflies. II. Receptor signal interactions in the lamina. J. comp. Physiol. **92**, 357—375 (1974).

Lavallée, M., Schanne, O. I., Hébert, N. C.: Glass microelectrodes. New York: John Wiley 1969.

Lisman, J. E., Brown, J. E.: Two light-induced processes in the photoreceptor cells of *Limulus* ventral eye. J. gen. Physiol. **58**, 544—561 (1971).

Lisman, J. E., Brown, J. E.: Effects of intracellular injection of calcium buffers on light adaptation in *Limulus* ventral photoreceptors. J. gen. Physiol. **66**, 489—506 (1975).

MacNichol, E. F. Jr., Love, W. E.: Electrical responses of the retinal nerve and optic ganglion of the squid. Science **132**, 737—738 (1960).

Martinez II, J. M., Srebro, R.: Calcium and the control of discrete wave latency in the ventral photoreceptor of *Limulus*. J. Physiol. **261**, 535—562 (1976).

McCann, G. D., Arnett, D. W.: Spectral and polarization sensitivity of the dipteran visual system. J. gen. Physiol. **59**, 534—558 (1972).

McReynolds, J. S., Gorman, A. L. F.: Photoreceptor potentials of opposite polarity in the eye of the scallop, *Pecten irradians*. J. gen. Physiol. **56**, 376—391 (1970a).

McReynolds, J. S., Gorman, A. L. F.: Membrane conductances and spectral sensitivities of *Pecten* photoreceptors. J. gen. Physiol. **56**, 392—406 (1970b).

Meyer-Rochow, V. B.: Structure and function of the larval eye of the sawfly *Perga*. J. Insect Physiol. **20**, 1565—1591 (1974).

Millecchia, R., Mauro, A.: The ventral photoreceptor cells of *Limulus*. II. The basic photoresponse. J. gen. Physiol. **54**, 310—330 (1969a).

Millecchia, R., Mauro, A.: The ventral photoreceptor cells of *Limulus*. III. A voltage-clamp study. J. gen. Physiol. **54**, 331—351 (1969b).

Miller, W. H.: Derivates of cilia in the distal sense cells of the retina of *Pecten*. J. Biophys. Biochem. Cytol. **4**, 227—228 (1958).

Minke, B., Hochstein, S., Hillman, P.: Early receptor potential, evidence for the existence of two thermally stable states in the barnacle visual pigment. J. gen. Physiol. **62**, 87—104 (1973a).

Minke, B., Hochstein, S., Hillman, P.: Antagonistic process as source of visible-light suppression of afterpotential in *Limulus* UV photoreceptors. J. gen. Physiol. **62**, 787—791 (1973b).

Minke, B., Wu, C.-F., Pak, W. L.: Induction of photoreceptor voltage noise in the dark in *Drosophila* mutant. Nature (Lond.) **258**, 84—87 (1975).

Moring, J.: Spectral sensitivity of monopolar neurons in the eye of *Calliphora*. J. comp. Physiol. **123**, 335—338 (1978).

Moring, J., Järvilehto, M.: Dark recovery of polarized light sensitive and insensitive receptor cells in the retina of the fly. J. comp. Physiol. **122**, 215—226 (1977).

Mpitsos, G. J.: Physiology of vision in the mollusk *Lima scabra*. J. Neurophysiol. **36**, 371—383 (1973).

Muller, K. J.: Photoreceptors in the crayfish compound eye: electrical interactions between cells as related to polarized-light sensitivity. J. Physiol. **232**, 573—595 (1973).

Naka, K.-I.: Recording of retinal action potentials from single cells of the insect compound eye. J. gen. Physiol. **44**, 571—584 (1961).

Naka, K.-I., Eguchi, E.: Spike potentials recorded from the insect photoreceptor. J. gen. Physiol. **75**, 663—680 (1962).

Naka, K.-I., Kishida, K.: Retinal action potentials during dark and light adaptation. In: Functional organization of the compound eye (ed. C. G. Bernhard). Oxford: Pergamon Press 1966.

Naka, K.-I., Rushton, W. A. H.: An attempt to analyse colour reception by electrophysiology. J. Physiol. **185**, 556–586 (1966).

Nolte, J., Brown, J. E.: The spectral sensitivities of single cells in the median ocellus of *Limulus*. J. gen. Physiol. **54**, 636—649 (1969).

Nolte, J., Brown, J. E.: Electrophysiological properties of cells in the median ocellus of *Limulus*. J. gen. Physiol. **59**, 167—185 (1972a).

Nolte, J., Brown, J. E.: Ultraviolet-induced sensitivity to visible light in ultraviolet receptors of *Limulus*. J. gen. Physiol. **59**, 186—200 (1972b).

Pak, W. L., Cone, R. A.: Isolation and identification of the initial peak of the early receptor potential. Nature **204**, 836—838 (1964).

Patterson, J. A., Goodman, L. J.: Intracellular responses of receptor cells and second order cells in the ocelli of the desert locust *Schistocerca gregaria*. J. comp. Physiol. **95**, 251—262 (1974).

Ratliff, F., Hartline, H. K., Miller, W. H.: Spatial and temporal aspects of retinal inhibitory interaction. J. Opt. Soc. Am. **53**, 110—120 (1963).

Rehbronn, W.: Gleichzeitige intrazelluläre Doppelableitungen aus dem Komplexauge von *Calliphora erythrocephala*. Z. vergl. Physiol. **76**, 285—301 (1972).

Rosner, G.: Adaptation und Photoregeneration im Fliegenauge. J. comp. Physiol. **102**, 269—295 (1975).

Rushton, W. A. H.: Visual pigments in man. In: Handbook of sensory physiology, Vol. VII/1 (ed. H. J. A. Dartnall), pp. 364—394. Berlin-Heidelberg-New York: Springer 1972.

Scholes, J. H.: Discrete subthreshold potentials from the dimly lit insect eye. Nature (Lond.) **202**, 572—573 (1964).

Scholes, J. H.: Discontinuity of the excitation process in locust visual cells. Cold Spr. Harb. Symp. quant. Biol. **30**, 517—527 (1965).

Scholes, J. H.: The electrical responses of the retinal receptors and the lamina in the visual system of the fly *Musca*. Kybernetik **6**, 149—162 (1969).

Shaw, S. R.: Polarized light responses from crab retinula cells. Nature (Lond.) **211**, 92—93 (1966).

Shaw, S. R.: Simultaneous recording from two cells in the locust retina. Z. vergl. Physiol. **55**, 183—194 (1967).

Shaw, S. R.: Organization of the locust retina. Symp. zool. Soc. Lond. **23**, 135—163 (1968).

Shaw, S. R.: Interreceptor coupling in ommatidia of drone honeybee and locust compound eyes. Vision Res. **9**, 999—1029 (1969).

Shaw, S. R.: Decremental conduction of the visual signal in barnacle lateral eye. J. Physiol. **220**, 145—175 (1972).

Shaw, S. R.: Retinal resistance barriers and electrical lateral inhibition. Nature **255**, 480—483 (1975).

Smith, T. G., Baumann, F.: The functional organization within the ommatidium of the lateral eye of *Limulus*. Prog. Brain Res. **31**, 313—349 (1969).

Smith, T. G., Baumann, F., Fuortes, M. G. F.: Electrical connections between visual cells in the ommatidium of *Limulus*. Science **147**, 1446—1448 (1965).

Smith, T. G., Brown, J. E.: A photoelectric potential in invertebrate cells (*Limulus*). Nature **212**, 1217 (1966).

Smith, T. G., Stell, W. K., Brown, J. E.: Conductance changes associated with receptor potentials in *Limulus* photoreceptors. Science **162**, 454—456 (1968a).

Smith, T. G., Stell, W. K., Brown, J. E., Freeman, J. C., Murray, G. C.: A role for the sodium pump in photoreception. Science **162**, 456—458 (1968b).

Smola, U., Gemperlein, R.: Rezeptorrauschen und Informationskapazität der Sehzellen von *Calliphora erythrocephala* und *Periplaneta americana*. J. comp. Physiol. **87**, 393—404 (1973).

Snyder, A. W., Menzel, R. (Eds.): Photoreceptor optics. Berlin-Heidelberg-New York: Springer 1975.

Snyder, A. W., Menzel, R., Laughlin, S. B.: Structure and function of the fused rhabdom. J. comp. Physiol. **87**, 99—135 (1973).

Srebro, R., Behbehani, M.: A stochastic model for discrete waves in the *Limulus* photoreceptor. J. gen. Physiol. **58**, 267—286 (1971).

Srebro, R., Behbehani, M.: The thermal origin of spontaneous activity in the *Limulus* photoreceptor. J. Physiol. **224**, 349—361 (1972).

Srebro, R., Yeandle, S.: Stochastic properties of discrete waves of the *Limulus* photoreceptor. J. gen. Physiol. **56**, 751—767 (1970).

Stieve, H., Bollmann-Fischer, H., Braun, B.: The significance of metabolic energy and the ion pump for the receptor potential of the crayfish photoreceptor cell. Z. Naturforsch. **26**b, 1311—1321 (1971a).

Stieve, H., Wirth, C.: Über die Ionen-Abhängigkeit des Rezeptorpotentials der Retina von *Astacus leptodactylus*. Z. Naturforsch. **26**b, 457—470 (1971b).

Strausfeld, N. J.: Atlas of an insect brain. Berlin-Heidelberg-New York: Springer 1976.

Streck, P.: Der Einfluß des Schirmpigmentes auf das Sehfeld einzelner Sehzellen der Fliege *Calliphora erythrocephala* Meig. Z. vergl. Physiol. **76**, 372—402 (1972).

Stretton, A. O. W., Kravitz, E. A.: Neuronal geometry: determination with a technique of intracellular dye injection. Science **162**, 132—134 (1968).

Tasaki, K., Tsukahara, Y., Ito, S., Wayner, M. J., Yu, W. Y.: A simple, direct and rapid method for filling microelectrodes. Physiol. Behav. **3**, 1009—1010 (1968).

Thompson, R. F., Patterson, M. M. (eds.): Bioelectric recording techniques, Vol. I-A. New York-London: Academic Press 1973.

Tiffe, H. W., Zelles, L., Paucksch, H., Ernst, D. E. W.: Quick-freeze fixation with halogenated hydrocarbons. Exp. Cell Res. **63**, 207 (1970).

Tomita, T.: Electrical activity of vertebrate photoreceptors. Quart. Rev. Biophys. **3**, 179—222 (1970).

Tomita, T., Kikuchi, R., Tanaka, T.: Excitation and inhibition on lateral eye of horseshoe crab. In: Electrical activity of single cells (ed. Y. Katsuki). Tokyo: Igakushiin 1960.

Toyoda, J. I., Shapley, R. M.: The intracellularly recorded response in the scallop eye. Biol. Bull. **133**, 490 (1967).

Trujillo-Cenóz, O.: Some aspects of the structural organization of the arthropod eye. Cold Spr. Harb. Symp. quant. Biol. **30**, 371—382 (1965).

Wald, G.: Visual excitation and blood clotting. Science **150**, 1028—1030 (1965).

Wald, G., Brown, P. K., Gibbons, I. R.: The problems of visual excitation. J. Opt. Soc. Amer. **53**, 20 (1963).

Wasserman, G. S.: *Limulus* receptor action spectra. Vision Res. **9**, 611—620 (1969).

Watanabe, A., Grundfest, H.: Impulse propagation at the septal and commissural junctions of crayfish lateral giant axons. J. gen. Physiol. **45**, 267—308 (1961).

Weidmann, S.: The electrical constants of Purkinje fibres. J. Physiol. **118**, 348—360 (1952).

Whitehead, R., Purple, R. L.: Synaptic organization in the neuropile of the lateral eye of *Limulus*. Vision Res. **10**, 129—133 (1970).

Wu, C.-F., Pak, W. L.: Quantal basis of photoreceptor spectral sensitivity of *Drosophila melanogaster*. J. gen. Physiol. **66**, 149—168 (1975).

Wulff, V. J.: The effect of sodium, potassium and calcium on *Limulus* lateral eye retinular cells. Vision Res. **13**, 2309—2326 (1973).

Wulff, V. J., Stieve, H., Fahy, J. L.: Dark adaptation and sodium pump activity in *Limulus* lateral eye retinular cells. Vision Res. **15**, 759—765 (1975).

Yeandle, S.: Electrophysiology of the visual system. Discussion. Amer. J. Ophthalmol. **46**, 82—87 (1958).

Yeandle, S., Spiegler, J. B.: Light-evoked and spontaneous discrete waves in the ventral nerve photoreceptor of *Limulus*. J. gen. Physiol. **61**, 552—571 (1973).

Zettler, F., Järvilehto, M.: Histologische Lokalisation der Ableitelektrode. Belichtungspotentiale aus Retina und Lamina bei *Calliphora*. Z. vergl. Physiol. **68**, 202—210 (1970).

Zettler, F., Järvilehto, M.: Decrement-free conduction of graded potentials along the axon of a monopolar neuron. Z. vergl. Physiol. **75**, 402—421 (1971).

Zettler, F., Järvilehto, M.: Lateral inhibition in an insect eye. Z. vergl. Physiol. **76**, 233—244 (1972).

Zettler, F., Järvilehto, M.: Active and passive axonal propagation of non-spike signals in the retina of *Calliphora*. J. comp. Physiol. **85**, 89—104 (1973).

Chapter 7

Pseudopupils of Compound Eyes

DOEKELE G. STAVENGA, Groningen, The Netherlands

With 50 Figures

Contents

A. Introduction

"I come to the discussion of a remarkable optical phenomenon the unravelling of which has caused me a good deal of worry." In this way SIGMUND EXNER (1891, p. 162) begins the chapter on pseudopupils in his invaluable monograph „Die Physiologie der facettirten Augen von Krebsen und Insecten". His worry has proved worthwhile because his recognition that pseudopupils are optical revelations of the interior of compound eyes has been of longlasting and immense

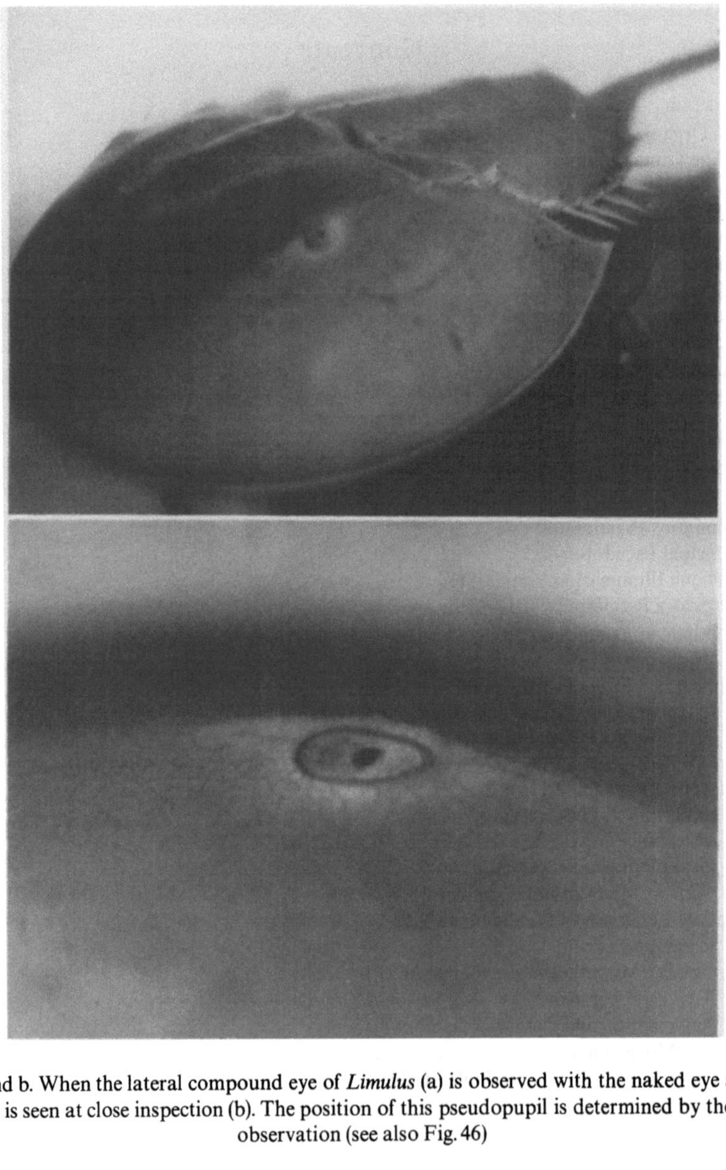

Fig. 1 a and b. When the lateral compound eye of *Limulus* (a) is observed with the naked eye a striking dark spot is seen at close inspection (b). The position of this pseudopupil is determined by the angle of observation (see also Fig. 46)

importance. The aims of this chapter are to show that pseudopupil phenomena provide a direct and deep insight, both figuratively and literally, into compound-eye structure and function, and also to indicate possible paths for future research.

A discussion of pseudopupils can do no better than start with the animal to which vision research is much indebted, the horseshoe crab *Limulus*. Anybody who has observed it must have been struck by the marked dark spot (Fig. 1) which wanders over the eye when one looks at the eye from a changing angle. LEIDIG (1855) noticed this phenomenon and termed it the pseudopupil by reason of its apparent resemblance to the pupil within the coloured iris of the human eye. Parenthetically, the word pupil (little girl, doll) originated from a name used by children when referring to the miniature mirror image of oneself seen by looking closely at another's eye. Interestingly, the extension of meaning of the word pupil into the areas of diaphragming and light control function has resulted in the word becoming common usage in compound eye studies too. However, the actual pupil mechanisms in compound eyes, i.e., the pigment movement systems, have little correspondence to the light control systems of vertebrates eyes. To avoid confusion, in this chapter a pseudopupil is always an optical phenomenon, a pupil a light control system.

EXNER (1891) classified and explained a number of different pseudopupil types. The wandering single dark spot which is conspicuous in the eyes of many arthropods, including mantids, grasshoppers, and crabs was named the principal pseudopupil (Hauptpupille). The eye shine visible in dark-adapted lepidopterans and crustaceans he termed the luminous pseudopupil (leuchtende Pseudopupille), and the term accessory pseudopupils (Nebenpupillen) was coined for the crown of dark spots surrounding the principal pseudopupil in the eyes of butterflies and dragonflies.

After EXNER'S innovative study, contributions to understanding pseudopupils remained negligible for many decades. It was only recently that YAGI and KOYAMA (1963) in their extensive study on lepidopteran eyes clarified the close relation of the elaborately patterned pseudopupils visible in the eyes of diurnal butterflies with the fine structure at the retinal level.

A new thrust in pseudopupil research was made by FRANCESCHINI and KIRSCHFELD in the course of their brilliant studies on the optics of the neural superposition eye of flies (KIRSCHFELD, 1965, 1967, 1969, 1971, 1972, 1973a; KIRSCHFELD and FRANCESCHINI, 1968, 1969; FRANCESCHINI, 1972a, b, 1975; FRANCESCHINI and KIRSCHFELD, 1971a, b, 1976). The past few years have proven unequivocally that the corneal and deep pseudopupil (terms coined by the latter authors) are most valuable tools in research into compound eyes.

A recapitulation of the now classic studies is included in the following general treatise on optics and mosaic organization of compound eyes.

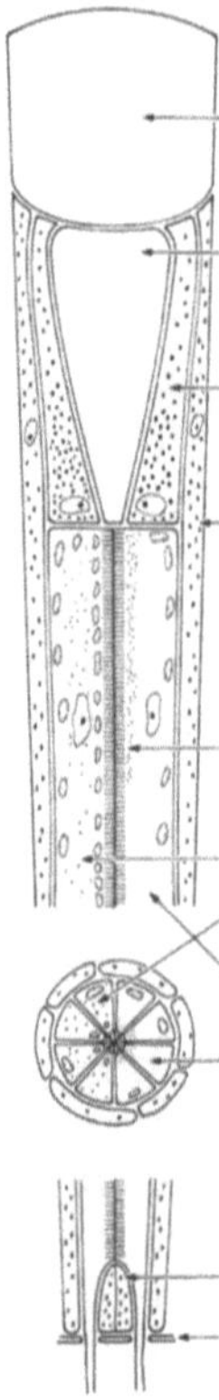

Fig. 2. The ommatidium of a hymenopteran compound eye representing a typical apposition eye. Incident light is focussed by the dioptrical apparatus (lens and crystalline cone) on the photopigment-containing rhabdom. In the dark-adapted state the rhabdom is surrounded by vacuoles which vanish upon light adaptation while retinular cell pigment granules then gather against the rhabdom (from STAVENGA and KUIPER, 1977)

B. Mosaic Organization of Compound Eyes

I. The Ommatidium

Pseudopupil phenomena are of specific value in the study of apposition eyes and therefore superposition eyes will scarcely be dealt with here. (For a treatise on the different compound eye types, see the chapter of KUNZE, Ch. 8). To provide a figure of reference for the following discussion Fig. 2 presents a schematical drawing of an ommatidium, which is the building block of the compound eye distally of the basement membrane. The dioptric apparatus consists of the facet lens and the crystalline cone. The visual sense cells together create a retinula, i.e., a small retina; the cells are also called retinular cells and photoreceptor cells. The retinula is enveloped by pigment cells; the primary pigment cells primarily surround the cone, while the secondary pigment cells are extended onto the basement membrane.

The rhabdom deserves special attention: this structure is the assembly of rhabdomeres, i.e., the strongly invaginated parts of the membrane of the retinular cells. In the ommatidium of Hymenoptera (which is closely approximated by Fig. 2) the rhabdomeres are merged. The fused rhabdom acts as one functional light guide. In flies, where the rhabdomeres are separate (in which case the rhabdom is termed "open") each individual rhabdomere is a functional light guide. Proximal to the rhabdom strongly light-absorbing basal pigment is usually found, which is located in extensions of the cone cells in bees (GRIBAKIN, 1975; RIBI, personal communication). However, in diurnal butterflies a tapetum is created proximal to the rhabdom by strongly folded trachea (DEMOLL, 1909a; MILLER and BERNARD, 1968). For general information on compound eyes valuable readings include WATERMAN (1961), BULLOCK and HORRIDGE (1965), MAZOKHIN-PORSHNYAKOV (1969), EAKIN (1972), WIGGLESWORTH (1972), and GOLDSMITH and BERNARD (1974).

II. The Facet Lens Mosaic as Reference System

Looking at an animal with compound eyes, the most striking feature apart from the protruded shape of the eyes is the mosaic of the corneal facet lenses. The facet lens mosaic is perfectly followed by the neural organization of the retina, lamina, and subsequent strata of the medulla (BRAITENBERG and STRAUSFELD, 1973; STRAUSFELD, 1976a, b). Hence, the array of facet lenses is an obvious reference system in experimental investigations of the proximally situated visual system. An unambiguous notation system will facilitate neural unit studies which employ neurophysiological, anatomical, and, of course, optical methods. A number of different description systems have come in vogue for the visual system of insects. In the following an appropriate formal scheme will be applied.

III. Coordinate Axes

The two fundamental coordinates of the plane through the mosaic of facet lenses can be directly deduced from the exquisite eyes of some species of flies which are outstanding in the phyla with compound eyes in that the facet lenses are not simply transparent light collectors, but spectral filters as well (BERNARD and MILLER, 1968; BERNARD, 1971). Long-legged flies of the family Dolichopodidae have corneas with alternating rows of yellow- and red-coloured facet lenses (Fig. 3, from TRUJILLO-CENÓZ and BERNARD, 1972). These rows are approximately parallel to a horizontal or a vertical plane. The corresponding H–V axes (Fig. 3), forming an orthogonal (curvilinear) coordinate system, introduced for the crab by WATERMAN and HORCH (1966), has been extended by STAVENGA (1975a, b) and STAVENGA and BEERSMA (1975) (see below).

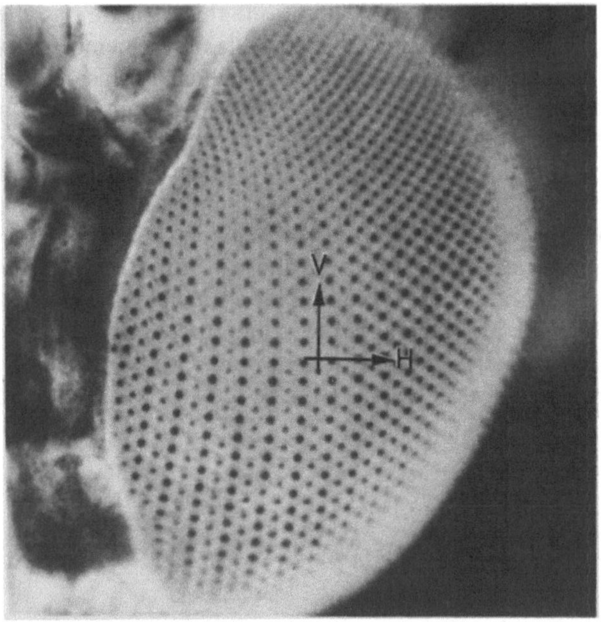

Fig. 3. Antero-lateral view of the left eye of the female long-legged fly *Sympycnus lineatus* Loew. The yellow facets appear as black spots while the red ones appear as smaller grey spots. The horizontal and vertical axes of the eye *H* and *V* are indicated by two perpendicular arrows (from TRUJILLO-CENÓZ and BERNARD, 1972)

Oblique rows of facets have been assigned axes by BRAITENBERG (1967, 1970), namely *x*- and *y*-axes; his *z*-axis is identical to the *H*-axis. Slightly different nomenclatures have been used by others; e.g., VON GAVEL (1939) and WIEDEMANN (1965) call the *H*- and *V*-axes *A* and *B*; HORRIDGE and MEINERZHAGEN (1970) use A(nterior)—P(osterior) and V(entral)—D(orsal) (cf. BRAITENBERG and STRAUSFELD, 1973, Fig. 5; BUTLER, 1971, Fig. 1); TRUJILLO-CENÓZ (1972) denotes the axes with α (or α_1) and β (or β_1); and GEMPERLEIN (1969) calls BRAITENBERG'S *x*- and *y*-axes *d*- and *v*-axes.

The *H–V* orthogonal organization of the visual system of invertebrates is recognizable at the neural and even at the fine structural level. For instance, the lamina of flies is composed of horizontal layers of cells and axons (cartridges); each layer projects into the medulla through a chiasm, in which the respective order of the axonal columns is completely reversed (STRAUSFELD, 1971, 1976a, b; see also BRAITENBERG, 1970).

Rhabdomeric microvilli oriented orthogonally have been recognized in many invertebrate species (EAKIN, 1972; SNYDER, 1973). In cephalopods and decapod Crustacea the rhabdomeric microvilli are oriented parallel to either the anteroposterior *H*-axis or the dorsoventral *V*-axis of the whole animal (see EAKIN, 1972; WATERMAN, 1975); other examples are the water strider (SCHNEIDER and LANGER, 1969), pierid butterflies (MEYER-ROCHOW, 1971a) and hymenopteran insects (WEHNER, 1976). Furthermore, orthogonality exists between the microvilli of the two central retinular cells of flies (see Sec. B.V), and even between two sets of microvilli of one and the same (eighth) retinular cell in the rock crab *Grapsus* (EGUCHI and WATERMAN, 1973).

An extensive bibliography and discussion on the function of the orthogonal microvilli in polarization vision has been published recently by BERNARD and WEHNER (1977); for a possible wiring diagram of photoreceptor cell axons onto the polarization sensitive neurons detected by LEGGETT (1976), see MEYER-ROCHOW and NÄSSEL (1977) and STOWE and LEGGETT (1978).

The orthogonal organization of the neural network of the visual system and of the microvilli of the photoreceptor cells is reflected in the result of electrophysiological and behavioural experiments. MCCANN and DILL (1969) describe extracellular recordings of cells (called class I) located in the medulla of *Musca domestica* and *Calliphora phaenicia* which serve form discrimination. These cells respond strongest at either horizontal or vertical edges. Furthermore, in the lobula and lobular plate, motion-sensitive units (class II) are found reacting strongest to either horizontally or vertically moving patterns. Simple projections of the class II units in the brain have been classified as class III (see also McCANN and FOSTER, 1971).

More recent intracellular recordings in the optic lobe of flies (DVORAK et al., 1975a, b; HAUSEN, 1976; rev. STRAUSFELD, 1976b) emphasize the dominantly horizontal or vertical movement sensitivity found in the behavioural experiments by GOETZ (1968, 1972) and GOETZ and WENKING (1973).

IV. Lattices and Indices

The assembly of facet lens centres (Fig. 3) can be considered as a two dimensional lattice which is of the centred rectangular type (Fig. 4a; STAVENGA, 1975b). The distance between the facets along the *H*- and *V*-axis respectively determine the primitive translation vectors $\mathbf{a}_h$ and $\mathbf{a}_v$. The vectors in turn directly estimate the shape of the Wigner-Seitz cell, which corresponds to a facet lens (Fig. 4c).

The facet lens shape in true compound eyes is often quite constant over the eye, but marked variations sometimes occur. Several investigators of houseflies and blowflies found that, going from anteromedial to posterolateral, the facet lens mosaic gradually changes from a hexagonal into a square facet lens pattern, and then into another hexagonal pattern (KUIPER, 1966; KIRSCHFELD, 1967; BRAITENBERG, 1967, 1970; TRUJILLO-CENÓZ, 1972; BRAITENBERG and STRAUSFELD, 1973; FRANCESCHINI, 1975). This progression can be easily understood as a gradual compression of the facet dimensions along the *H*-axis (STAVENGA, 1975b). In Fig. 5 the facet distance a_v is maintained constant for simplicity's sake; in reality it slightly increases towards the posterior (see TRUJILLO-CENÓZ, 1972, Fig. 4; BEERSMA, unpublished measurements). Concomitant with the shape of the facets varies the shape of the eye; the underlying relation is discussed in Section B.VII.

Since there is one to one correspondence of a facet lens in the cornea to a retinula in the retina, to a neuro-ommatidium in the lamina, and to a neural column in the medulla (STRAUSFELD, 1976b), the lattice description of the cornea can be straightforwardly extended to the subsequent neuron layers. Moreover, even a retinula of the fly visual system can be regarded as part of a centred rectangular lattice (Fig. 6). Coordinate axes H^* and V^* are introduced here to account for the fact that although the principal axes of the fly retinula parallel the axes in the cornea, the directions reverse in the four quadrants of the eyes, as indicated in Fig. 7 (see DIETRICH, 1909; BRAITENBERG, 1972). Left and right eyes and also upper and lower halves of each eye show mirror symmetry in respect to the fine structure of the retinulae. The imaginary horizontal symmetry line is called the equator. Indices for indicating the lattice points are introduced. Absolute indices (i, j) allocate a certain facet, with i giving the number of the column of facets and j that of the row (Fig. 7). Relative indices (i^*, j^*) define the position of an element relative to an origin (0,0) (see Fig. 6). An example of the use of the indices is given in the following formal description of the neural superposition principle. (For further details and examples, see STAVENGA and BEERSMA, 1975; and BEERSMA et al., 1975).

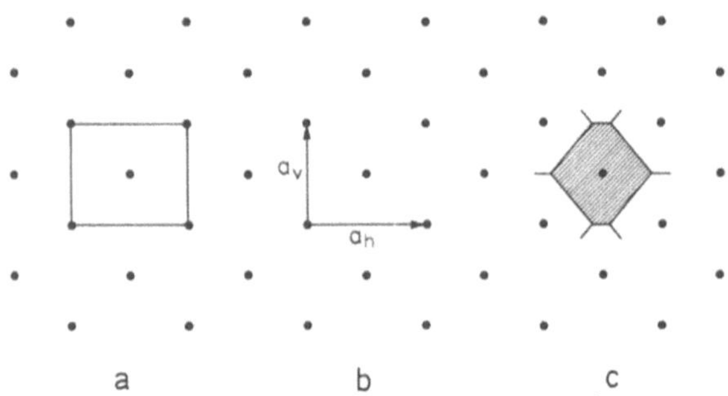

Fig. 4a–c. Part of a centred rectangular lattice. From left to right are shown: (a) The so-called unit-cell, which is a rectangle with a lattice point in its centre and a point on each corner; (b) the primitive translation vectors $\mathbf{a}_h$ and $\mathbf{a}_v$ and (c) the Wigner-Seitz or proximity cell which contains all points closer to the central lattice point than to any other lattice point; note the similarity to the cross-section of a facet lens (from STAVENGA, 1975b)

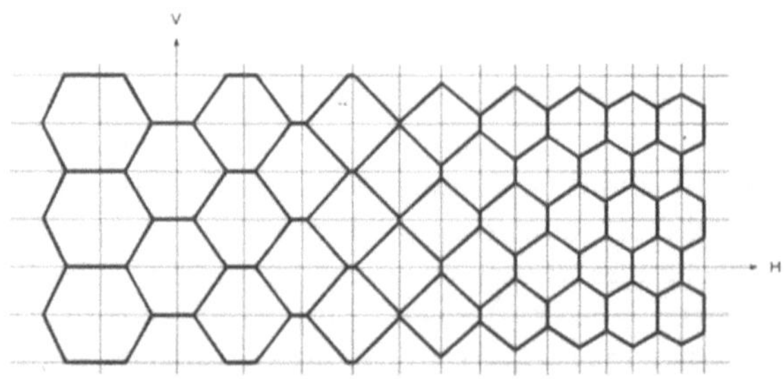

Fig. 5. Pattern of facet lenses in the cornea of the left eye of a housefly (schematized). The *H*-axis points from anterior to posterior, the *V*-axis from ventral to dorsal. The two-dimensional lattice structure of the cornea is revealed by the horizontal and vertical lines. The gradient in facet lens pattern is due to a gradient of the facet distance along the horizontal axis (from STAVENGA, 1975b)

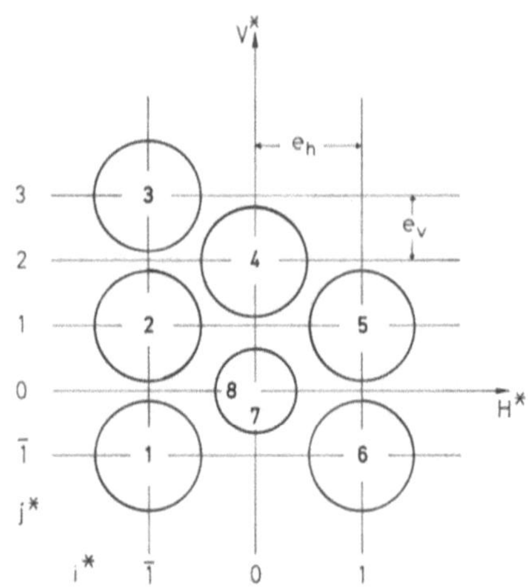

Fig. 6. Fly retinular lattice. The rhabdomere pattern of the retinular cells R 1–8 in the distal part of the retinula is given. Actually, only the rhabdomere of R 7 is present distally; the rhabdomere of R 8 is located proximally, basal to that of R 7. Horizontal and vertical lines (interdistance e_v and e_h respectively) reveal that a retinula can be considered as a part of a centred rectangular lattice with H^* and V^* as the axes. The position of the retinular cells can be characterized by relative indices i^* and j^* (from STAVENGA, 1975a)

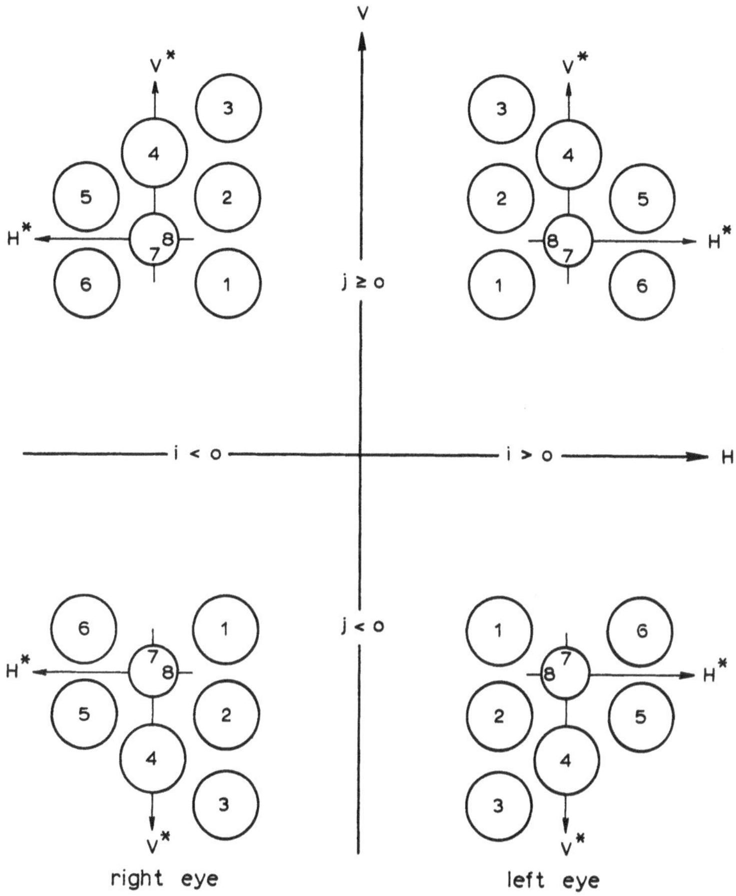

Fig. 7. Diagram of the four quadrants of the fly visual system as seen from the outside. The left and right eye are mirror-symmetrical with respect to the vertical V-axis. Upper and lower parts of the eye show mirror-symmetry with respect to the horizontal H-axis. Absolute indices i and j can be affixed to each retinula so that $i>0$ in the left eye and $i<0$ in the right eye, while $j \geqq 0$ in the upper parts of the eyes, and $j<0$ in the lower parts (from STAVENGA, 1975a)

V. Neural Superposition Theory

As a result of anatomical, optical, electrophysiological and behavioural investigations, the best understood compound eye is undoubtedly the neural superposition eye of flies. Since excellent reviews are on hand (KIRSCHFELD, 1971, 1973a; TRUJILLO-CENÓZ, 1972; BERNHARD et al., 1972; BRAITENBERG and STRAUSFELD, 1973; GOLDSMITH and BERNARD, 1974; KUNZE, this volume, Ch. 8), only the essential core of neural superposition theory will be reiterated.

The projection of retinular axons into the lamina is shown in Fig. 8. A set of neighbouring ommatidia is shown, each comprising the retinula of Fig. 6. A neuro-ommatidium, i.e., the characteristic set of neuronal elements of the lamina (also called a cartridge), receives the axons of the peripheral retinular cells R 1–6

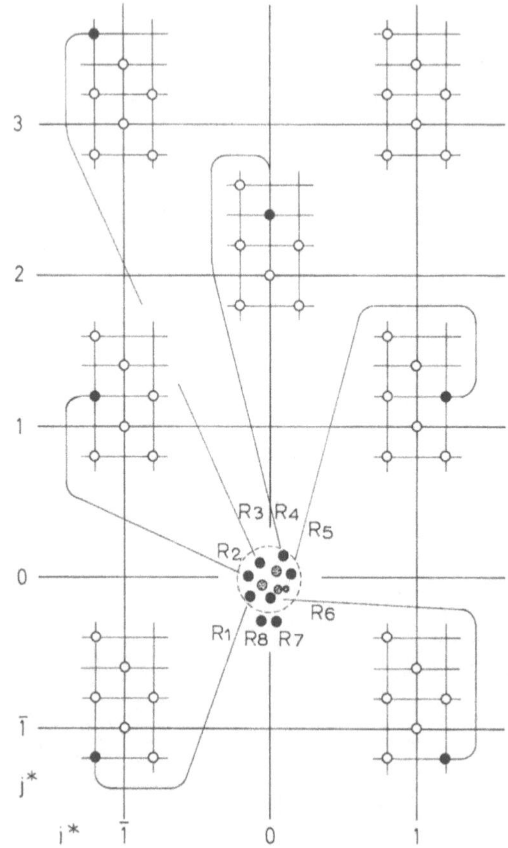

Fig. 8. Formalized projection of retinular cell axons from a conjugated set of retina-lattice elements (i.e., a set of retinulae) into the lamina lattice, as occurs in the compound eye of flies. Relative indices i^* and j^* characterize the relative positions of the retinulae. Comparison with Fig. 6 shows that a retinular cell axon with index pair (i^*, j^*) from an ommatidium having the identical index pair (i^*, j^*) projects into the neuro-ommatidium with $(i^*, j^*) = (0,0)$. The corresponding ommatidium has been omitted from the diagram (from STAVENGA and BEERSMA, 1975)

of six individual ommatidia, which are connected synaptically to second order neurons. The axons of the second order neurons run parallel with those of the two central retinular cells, R 7 and R 8 of one ommatidium, to a specific column in the medulla. Formally, a neuro-ommatidium can be described as follows: the axons of retinular cell Rk (k = 1–8), of which the relative index pairs (i^*, j^*) can be read from Fig. 6, project from ommatidia with the same index pair (i^*, j^*) into one neuro-ommatidium. Neural superposition theory states that all retinular cells which join their axons in one and the same neuro-ommatidium are looking into the same direction. Since in the lamina neurons the signals transmitted from the peripheral retinular cells R 1–6 are summed (SCHOLES, 1969), neural superposition implies a substantially improved sensitivity and signal to noise ratio (see GEMPERLEIN and SMOLA, 1972; LAUGHLIN, 1973).

A singular utilization of the neural superposition organization of flies has been revealed for the dolichopodid *Sympycnus lineatus* (TRUJILLO-CENÓZ and BERNARD, 1972). In this species the eyes contain two main populations of ommatidia which differ from each other not only in the structural colour of their corneal facets (already shown in Fig. 3), but also in the fine structure of their superior central cell. In muscoid flies the rhabdomeric microvilli of the superior (R 7) and inferior (R 8) central cells are usually oriented at 90° with respect to each other. The microvilli of R 7 are parallel to the *V*-axis and those of R 8 parallel to the *H*-axis (MELAMED and TRUJILLO-CENÓZ, 1968). In *S. lineatus* approximately half of the ommatidia (the type B or "yellow") maintain this structural organization, while in the other half (the type A or "red") the microvilli of R 7 and R 8 are both parallel to the H-axis. Consequently, two types of neuro-ommatidia occur (Fig. 9). One type (A) is characterized by the predominance of peripheral retinular cell axons R 1–6 arising from red-faceted ommatidia (5 red and 1 yellow), while the two central retinular cell axons R 7 and R 8 are yellow. The other type (B) of neuro-ommatidium shows the reverse situation (5 yellow and 1 red peripheral, together with two red central retinular cell axons). It seems that the anatomical basis for a colour contrast enhancement system in the long-legged fly is manifest (TRUJILLO-CENÓZ and BERNARD, 1972).

Another remarkable consequence of the wiring system in neural superposition eyes is encountered in the equatorial region, due to the mirror symmetry of the retinulae in the two eye halves. Some retinular cells project their axon into a neuro-ommatidium located across the equator, as has been predicted from optical studies by KIRSCHFELD (1967) and has been proven to occur histologically by HORRIDGE and MEINERZHAGEN (1970) and BOSCHEK (1971). This curious fact is represented schematically in Fig. 10. Outside the equatorial region in the lamina a neuro-ommatidium contains six (peripheral R 1–6) and two (central R 7 and R 8) retinular cell axons (Fig. 8). At the equator, a band of four rows of neuro-ommatidia having the axons of eight plus two retinular cells is sandwiched between two rows of neuro-ommatidia with seven plus two retinular cell axons. However, in the retina each ommatidium in a band totalling twelve rows of ommatidia delivers at least one axon to one of the enlarged neuro-ommatidia. Interestingly, certain horseflies (BERNARD and MILLER, 1968, Fig. 1 c) have a band of coloured facets approximately twelve rows broad in the cornea. It seems that the action of the corneal spectral filters (BERNARD, 1971) is in this case fortified by the gain inherent from the equatorial region. The value of this hypothesized combination for the horsefly is a question which has yet to be tackled. It should be added that the mirror symmetry of upper and lower eye parts has been observed directly with pseudopupil methods in the housefly *Musca*, the fruitfly *Drosophila*, and the blowfly *Calliphora* (KIRSCHFELD, 1967; FRANCESCHINI and KIRSCHFELD, 1971 b; FRANCESCHINI, 1975; BEERSMA et al., 1975; cf. Fig. 10). The fact that the single band of coloured facets of certain flies is indeed centred around the equator could be convincingly demonstrated in an unidentified Australian species which displayed upon light adaptation the yellow pupil (Sec. D.III.4) typical of most flies (unpublished observations).

The anatomical investigations quoted above have provided the secure basis of only half of the neural superposition theory, namely the conjunction of a fixed set

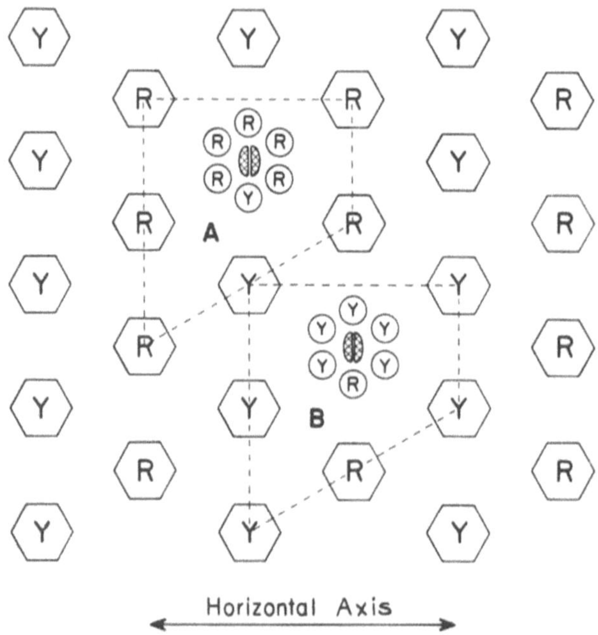

Fig. 9. In the long-legged fly *Sympycnus lineatus* two types of neuro-ommatidia occur. In type A five of six peripheral retinular cell axons descend from ommatidia with red facets, while in type B these axons descend from yellow-faceted ommatidia (from TRUJILLO-CENÓZ and BERNARD, 1972)

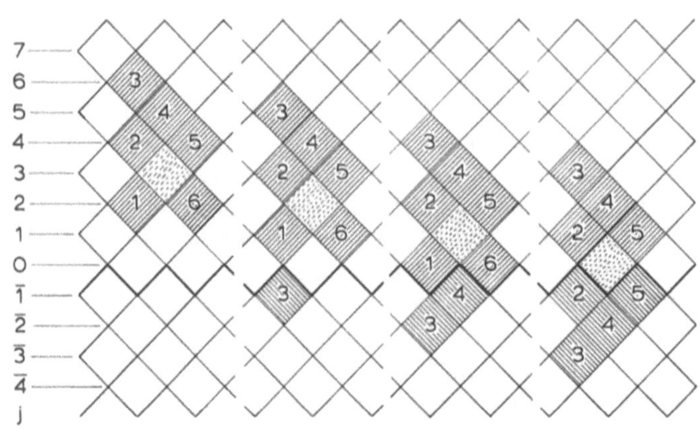

Fig. 10. Retina–lamina projection in the equatorial region of the left eye of flies. Above the (drawn) equator the retinular pattern is that of Fig. 6. Below the equator this pattern is mirrored. The ommatidial row number is indicated by the absolute index j. Going from left to right the set of distinguished squares represents the set of ommatidia which deliver axons to a neuro-ommatidium with $j=3, 2, 1$, and 0 respectively. The number in the hatched ommatidia gives the number of the retinular cell, the axon of which joins the axons of sense cells R 7 and R 8, which leave the speckled ommatidium. Neuro-ommatida having $j=2, 1, 0$ (or $-1, -2, -3$) receive axons from the opposite half of the eye. The appearance of the different stages as drawn in the figure is not unlike the reduced corneal pseudopupil phenomenon observable near the equator (see KIRSCHFELD, 1967); from STAVENGA and BEERSMA (1975)

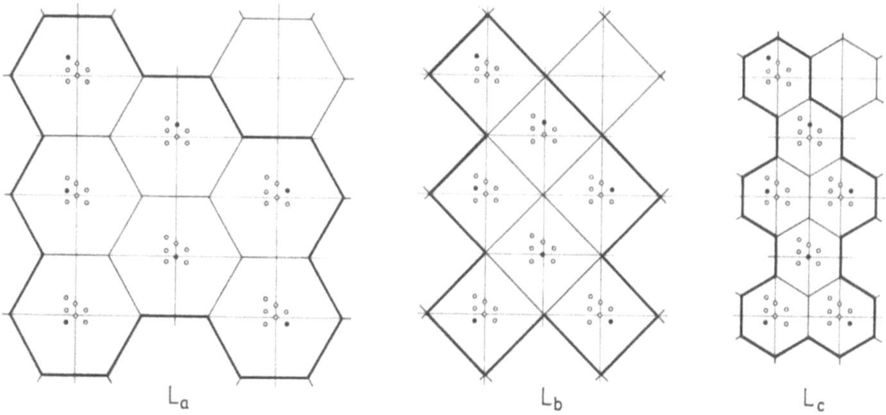

Fig. 11. The three possible regular facet lens patterns of compound eyes L_a, L_b, and L_c occur when a_v/a_h equal $1/\sqrt{3}$, 1, and $\sqrt{3}$ respectively. The fly rhabdomere pattern is drawn in to clarify the necessary distinction between pattern L_a and L_c. The black dots indicate which axons join into one neuro-ommatidium (from STAVENGA, 1975b; see also BRAITENBERG, 1967)

of retinular cell axons into one and the same neuro-ommatidium. The other half consists of the proof that the corresponding retinular cells have one and the same visual axis. This proof requires actual optical measurements for which the next two sections supply the theoretical background.

VI. Ideal Facet Lens Patterns

From the lattice approach to the cornea presented in Section B. IV it follows that there are three regular facet lens patterns (Fig. 11).

1. Pattern L_a, regular hexagons "lying" on the H-axis, occurs when $a_v/a_h = 1/\sqrt{3}$. This facet mosaic is found in butterflies, the fruitfly *Drosophila* (BRAITENBERG, 1970) and the flowerfly *Syritta pipiens* (BEERSMA et al., 1977).
2. Pattern L_b, squares with diagonals parallel to the H- and V-axis, occurring when $a_v/a_h = 1$, is encountered in crayfish (e.g., *Galathea:* EXNER, 1891, Fig. 65; *Procambarus:* WATERMAN and FERNÁNDEZ, 1970) and in the dorsal eye of the mayfly *Atalophlebia* (MEYER-ROCHOW, 1971b; HORRIDGE, 1975).
3. Pattern L_c, regular hexagons "standing" on the H-axis ($a_v/a_h = \sqrt{3}$) is found in the worker bee (BRAITENBERG, 1970) and in the spider crab (EGUCHI and WATERMAN, 1968).

Facing the variety of facet lens patterns the question emerges as to why an animal has been equipped with its specific cornea. For only one case, the square facet lens pattern of the superposition eye of crayfish, has a clear explanation been deduced from optical considerations (VOGT, 1975; see KUNZE, this volume, Ch. 8). Generally, from knowing an animal's behaviour one may be able to understand the way of sampling its optical environment — i.e., the organization of the photoreceptors. The optical constraints for the optimal construction of compound eyes have been amply treated by SNYDER (this volume, Ch. 5). The special constraint put on the eyes of flies by the neural superposition principle is that the optical alignment of those retinular cells which join their axons in one and the same neuro-ommatidium must be maintained even with the huge gradient in facet pattern over the eye from L_a via L_b into L_c as shown in Fig. 5. Essentially, there is a demand for a special arrangement of the ommatidial axes as is demonstrated below.

VII. Latticing of Ommatidial Axes

The important parameter to characterize spatial sampling by the compound eye is the interommatidial angle $\Delta\phi$ (see SNYDER, this volume, Ch. 5), or rather, the set of its components $\Delta\phi_h$ and $\Delta\phi_v$ along the horizontal and vertical coordinates respectively.

The relation between the interommatidial angles and the interfacet lens distances is presented in Fig. 12 for two cases. Figure 12a represents a facet lens pattern L_a in a spherically shaped eye, i.e., the ommatidial axes intersect at point I. Figure 12b represents a facet lens pattern L_c in an ovally shaped eye, so that axes of ommatidia located in a horizontal row or a vertical column intersect at points I_h or I_v respectively. The interommatidial angles will be generally defined by $\Delta\phi_h = D_h/R_h$ and $\Delta\phi_v = D_v/R_v$ where $D_h = a_h/2$ and $D_v = a_v/2$ (see Fig. 4).

For the spherical eye of Fig. 12a $R_h = R_v = R$. Since facet lens mosaic L_a is defined by $a_v/a_h = D_v/D_h = 1/\sqrt{3}$, it follows that $\Delta\phi_v/\Delta\phi_h = 1/\sqrt{3}$. In words, in a spherical eye with a facet lens mosaic L_a the ommatidial axes are arranged in the hexagonal pattern L_a.

In the ovally shaped eye of Fig. 12b, $R_h = R_v/3$ and $a_v/a_h = D_v/D_h = \sqrt{3}$. It thus follows that $\Delta\phi_v/\Delta\phi_h = 1/\sqrt{3}$; or, in ovally shaped eye, the curvature of which in the horizontal direction is three times that in the vertical direction, a facet lens mosaic L_c combines with a hexagonally L_a-type ommatidial axes array. In conclusion, one and the same type of spatial latticing of ommatidial axes can occur with quite different facet lens arrangements.

Actually, FRANCESCHINI (1975) utilized pseudopupil phenomena to demonstrate that in all eye regions of the blowfly the ommatidial axes are arranged in the L_a pattern, although the facet lens pattern varies greatly over the eye as described above (Fig. 5). Essentially the same result has been obtained for the housefly (BEERSMA et al., 1975). The general L_a type latticing of ommatidial axes is directly predicted by neural superposition theory (STAVENGA, 1975b), for the theory demands that the interommatidial angles equal the angles between the visual axes of retinular cells within one ommatidium defined by $\Delta\phi_h = e_h/f$ and $\Delta\phi_v = e_v/f$, where e_h and e_v are the distances of the rhabdomeres (Fig. 6) in the focal plane, and f is the focal distance. Since $e_v/e_h = 1/\sqrt{3}$ (approximately) in all retinulae, $\Delta\phi_v/\Delta\phi_h = 1/\sqrt{3}$ in all eye parts.

The interommatidial angles defined here by $\Delta\phi_h$ and $\Delta\phi_v$ have previously been termed α_h and α_v respectively (STAVENGA, 1975a, b; BEERSMA et al., 1975). The present nomenclature is more in line with the common usage to designate the interommatidial angle with $\Delta\phi$ (e.g., GOETZ, 1964; ECKERT, 1973; WEHNER, 1975). However, it should be noted that the $\Delta\phi$ from GOETZ (1964) actually corresponds to $\Delta\phi_h$ since in the optomotor experiments performed on the fruitfly *Drosophila* vertical stripe patterns moving in the horizontal directions were used. Recordings from motion sensitive neurons in the lobula

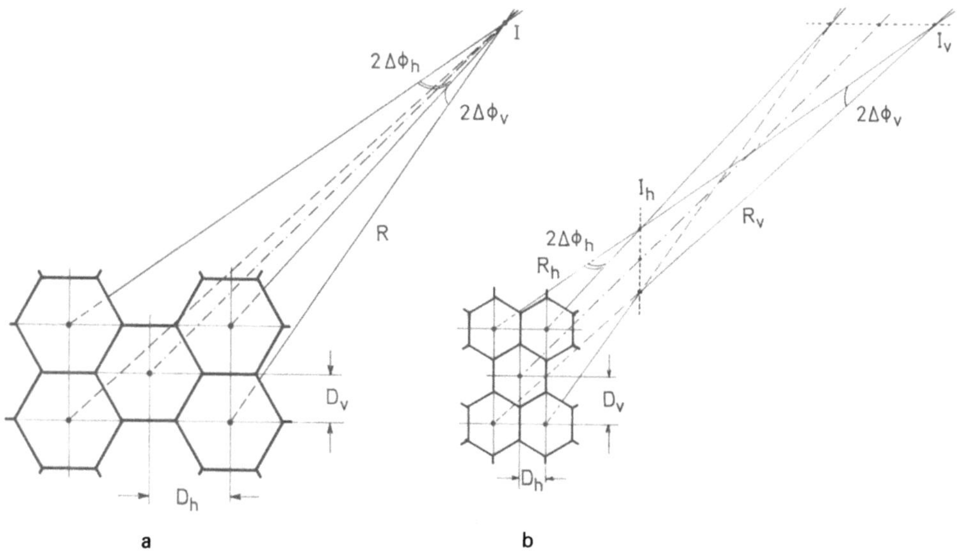

Fig. 12a and b. Facet lens pattern and interommatidial angles. In (a) facet lens pattern L_a of Fig. 11 is conjugated to a spherical eye with radius R. In (b) facet lens pattern L_c of Fig. 11 is conjugated to an oval eye with vertical radius of curvature R_v three times the horizontal radius of curvature R_h. It is discussed in the text that in this way facet lens pattern and eye shape must correlate when a certain ratio between the interommatidial angles $\Delta\phi_h$ and $\Delta\phi_v$ has to be maintained

of the blowfly *Calliphora* confirm that $\Delta\phi_h$ is the effective sampling angle of the fly visual system (ZAAGMAN et al., 1977). Confusion might be unavoidable since in KIRSCHFELD (1973b) $\Delta\phi_h$ and $\Delta\phi_v$ are twice the angles defined here.

Furthermore, in a recent report PICK (1977) has argued that the housefly *Musca* neural superposition theory holds essentially, but not exactly. The photoreceptive elements, the rhabdomeres, are slightly displaced outwardly with respect to their theoretical position. The resulting loss in acuity is overruled by a gain in photon capture capacity realized by increased rhabdomere diameters.

A corollary of the formulae derived above for the neural superposition eye is the relation between facet dimensions and eye shape given by $D_v/D_h = R_v/R_h\sqrt{3}$. It is important to notice that R_h and R_v are correct measures for eye shape only when the visual axes of the ommatidia are perpendicular to the eye surface; in real eyes this is certainly not generally true, the reason for which will be explained below.

VIII. Skew Ommatidia

In the simple conception of compound eyes given in the previous sections it was implicitly assumed that the ommatidia are circular-symmetrical (tapering) cylinders juxtaposed in such a way that the anatomical axes are perpendicular to the eye envelope. Anatomical and visual axes then coincide, but this identity no longer occurs when the anatomical axis stands skew to the eye surface. Corneal lens and the image space proximal to the lens have a refractive index different from that of the object space (at least when the animal lives in air; in aquatic animals the difference is smaller), and therefore refraction can cause substantial

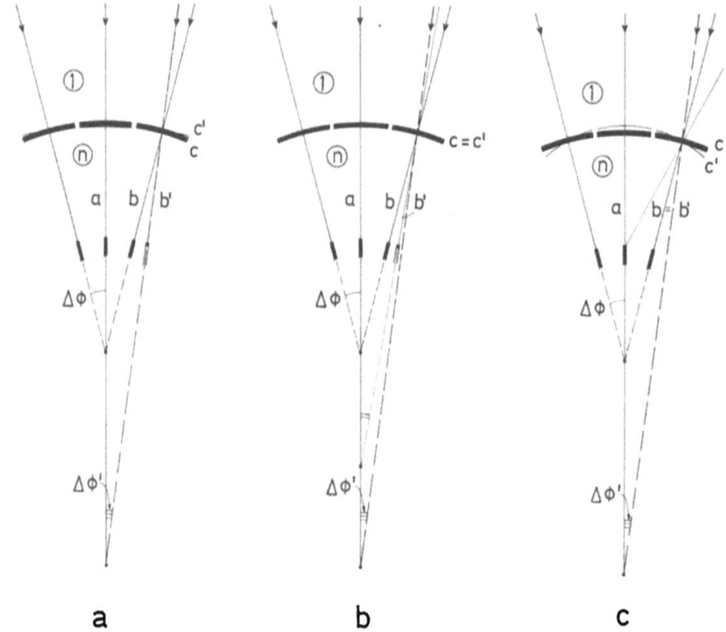

Fig. 13a–c. Design strategies for realizing a fovea, i.e., an area of the eye where the interommatidial angle $\Delta\phi'$ is smaller than the value $\Delta\phi$ elsewhere. In (a) this is achieved by flattening the corneal envelop *c* into *c'* while the anatomical axis *b* of the ommatidium is kept in a normal relation to the cornea, giving *b'*. In (b) only the anatomical axis *b* is shifted into a skew position *b'*. Due to refraction the visual axis is skewed even more. In (c) the anatomical axis is maintained, but the corneal surface protrudes. Also in this case refraction causes a decrease in the value of the interommatidial angle

deviations between the direction of the anatomical and that of the visual axis. This basic physical characteristic is advantageously applied in the construction of compound eyes (EXNER, 1891; SHERK, 1977).

A first application is illustrated in Fig. 13 which indicates how to design a fovea (MAZOKHIN-PORSHNYAKOV, 1969; HORRIDGE, 1977a, b; SHERK, 1977). Three facets are drawn with normal axes, perpendicular to the corneal envelope, of a spherically shaped compound eye, having a uniform interommatidial angle $\Delta\phi$. A local fovea, i.e., an area with smaller $\Delta\phi$, can be realized in at least three different ways. First by flattening the eye surface and maintaining the anatomical axis of the ommatidium *b* normal to that surface, resulting in axis *b'* (Fig. 13a). Secondly, keeping the cornea fixed and changing the anatomical axis *b* into a position *b'*, skew to the eye surface (Fig. 13b). Refraction adds to the oblique setting, since the new visual interommatidial angle $\Delta\phi'$ becomes substantially smaller than the new anatomical angle. A different way to obtain the same result is by bending or protruding the cornea (Fig. 13c). In both Figs. 13b and 13c the acuity is improved by a skewing ommatidial axis. However, it should be noted that with $\Delta\phi$ the facet lens diameter *D* also has to change, since eye performance depends on the eye parameter $p = D\Delta\phi$ (SNYDER, this volume, Ch. 5).

An example where skew ommatidia are designed in order to achieve a small $\Delta\phi$ together with a large facet diameter *D* is given in Fig. 14 (from DIETRICH,

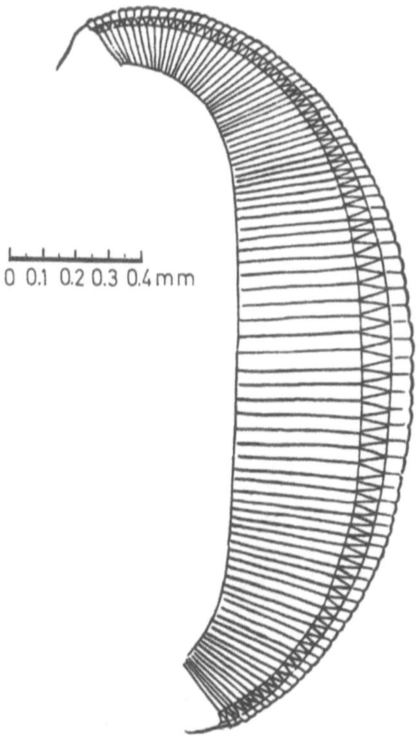

Fig. 14. Organization of the compound eye of the robberfly *Laphria flavia* (from DIETRICH, 1909). The large fovea where the interommatidial angle is small leaves little room for the peripheral ommatidia, which accordingly have a much larger interommatidial angle. Note the correlation of facet diameter and interommatidial angle

1909). The enormous fovea of the robberfly *Laphria flavia* demonstrates an important characteristic of predatory animals. Note that the fovea occupies a substantial part of the available room in the retina, while the total angle spanned is small. The extra-foveal ommatidia have to make up for this characteristic of the fovea. Their interommatidial angle becomes much larger than the foveal $\Delta\phi$, which correlates with a distinctly smaller facet diameter D (Fig. 14). It thus seems that the eye parameter p is fairly constant in robberflies, but accurate measurements have not been reported for this case. As shown in Fig. 15 the fovea of robberflies is noticeable from the outside in the facet lattice. The foveal pattern of type L_a suddenly changes into pattern type L_c, presumably at the boundary of the fovea; this is in striking contrast with the gradual changes occurring in the muscoid and calliphorid flies (see above). It might be assumed that the demand for a highly powerful fovea is satisfied to a limit where another demand can just be met, namely that of spanning the complete solid angle with the eye, i.e. obtaining an extended field of view.

Three possible strategies for realizing an extended visual field are indicated in Fig. 16. Again, three facets of a spherically shaped compound eye (see Fig. 13), are drawn with normal axes, perpendicular to the corneal envelope. A larger $\Delta\phi$

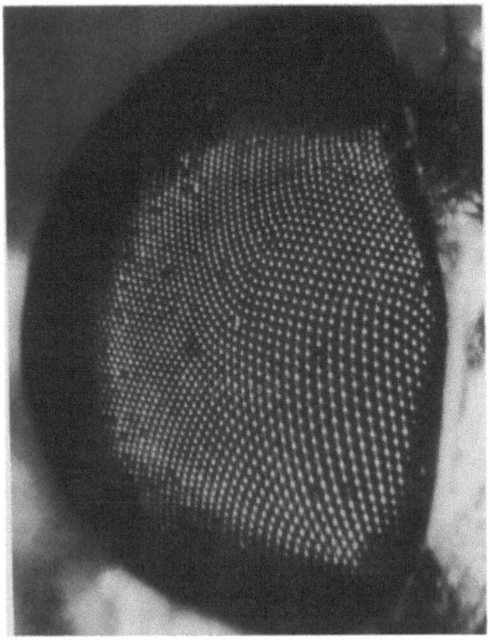

Fig. 15. Facet lattice in the fovea frontally in the right eye of a robberfly. The sudden densification of the lattice points towards the eye periphery indicates the foveal boundary (cf. Fig. 14)

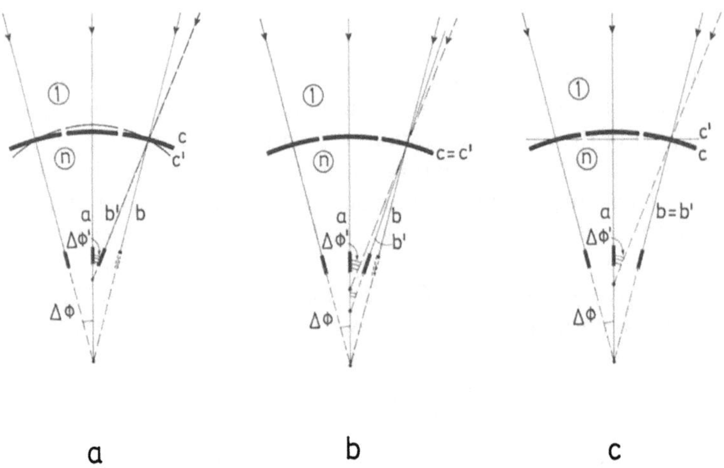

Fig. 16a–c. Design strategies for realizing an extended visual field. The interommatidial angle $\Delta\phi$ is then widened into $\Delta\phi'$. In (a) this is achieved by protruding the cornea *c* into *c'* while keeping the interommatidial axis in a normal relation to the cornea. In (b) the cornea is fixed, but the anatomical axis *b* is skewed into *b'*. Refraction gives a bonus to the skew setting. (c) demonstrates that flattening of the cornea is also an effective means of realizing a larger interommatidial angle

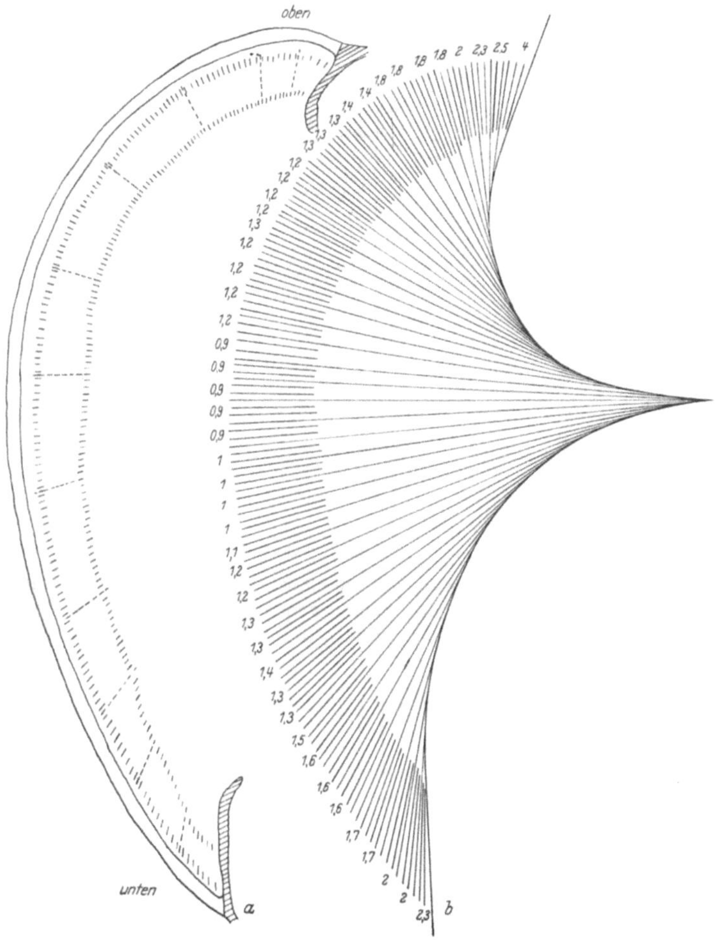

Fig. 17a and b. Diagrams of the organization of the ommatidial axes in the eye of a worker bee. In (a) the outer crown of stripes indicates the borders of the crystalline cone, the inner crown the rhabdom axes as it appears in a histological section. In (b) the resulting ommatidial axes are drawn. The local interommatidial angles are given in degrees (from BAUMGAERTNER, 1928). It is clear from the ventral part of the eye that flattening the corneal envelope together with skewing the anatomical axes is applied here to realize enlarged interommatidial angles and ultimately an extended visual field

is obtained by protrusion of the eye when the anatomical axis b is maintained normally (Fig. 16a). Keeping the cornea fixed while skewing the anatomical axis b into b' yields a bonus from refraction (Fig. 16b). Flattening of the cornea and keeping the anatomical axis fixed is a similar beneficial method (Fig. 16c). The latter method is applied ventrally in the eye of the bee as is evident from Fig. 17; an almost flat corneal envelope is combined with fairly skewed ommatidia. Skewed ommatidia at the eye periphery clearly realize the goal of an extended visual field as was acutely recognized by EXNER (1891) from his study on the eye of *Limulus*.

The effectiveness of the refraction effect by the cornea strongly depends on the actual angle β between the anatomical axis of an ommatidium and the normal to the corneal surface (Fig. 18). When α is the angle between the visual axis and the

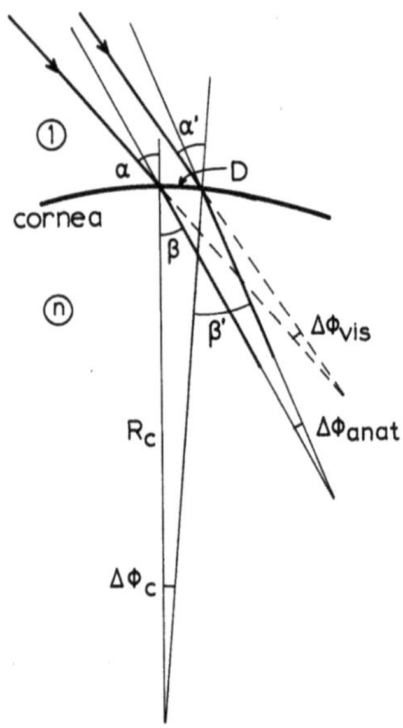

Fig. 18. Diagram of a compound eye with skew anatomical axes which cause even more skew visual axes. The visual interommatidial angle $\Delta\phi_{vis}$ can substantially differ from the anatomical interommatidial angle $\Delta\phi_{anat}$. D is the anatomical interfacet distance, R_c the local radius of curvature of the eye

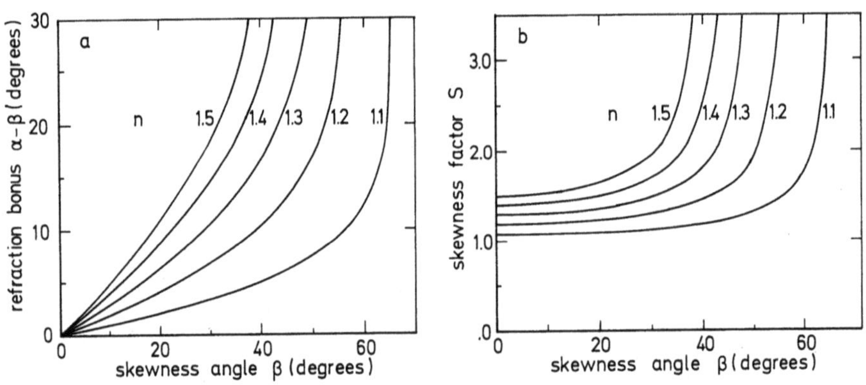

Fig. 19a and b. Refraction bonus $\alpha-\beta$, i.e., the shift in axial direction (a), and the skewness factor S (b), as a function of the skewness angle β, i.e., the angle between the anatomical ommatidial axis and the normal to the eye surface

normal to the surface, and n the refractive index of the image space, Snell's law gives $\alpha = \arcsin(n \sin\beta)$.

The change in axial direction, the refraction bonus, is $\alpha-\beta = \arcsin(n \sin\beta)-\beta$. The interommaditial angles are affected accordingly. The anatomical interommatidial angle is $\Delta\phi_{anat} = \Delta\phi_c + (\beta'-\beta)$, where $\Delta\phi_c = D/R_c$; D is the anatomical interfacet distance, and R_c the local radius of eye curvature (Fig. 18). Furthermore, the visual interommatidial angle is $\Delta\phi_{vis} = \Delta\phi_c + (\alpha'-\alpha)$.

Hence (small angles)

$$S = \frac{\Delta\phi_{vis}-\Delta\phi_c}{\Delta\phi_{anat}-\Delta\phi_c} = \frac{n\cos\beta}{\sqrt{1-n^2\sin^2\beta}} = \frac{n\cos\beta}{\cos\alpha}.$$

The dependence of this skewness factor S on the skewness of the anatomical angle β is shown in Fig. 19b for a number of refractive index values. In the extreme situation obtained with $\Delta\phi_c=0$, i.e., a flat corneal surface (see Fig. 16c), the skewness factor directly gives the factor of magnification of the visual in respect of the anatomical interommatidial angle: $\Delta\phi_{vis}=S\Delta\phi_{anat}$.

Skew ommatidia at the eye margin result in a widening of the field of vision by an amount of $\alpha-\beta$ (Fig. 19a). When the anatomical angle β approaches the critical angle of the refracting medium, the field is opened up substantially (Fig. 19a). Of course, the extreme situation of $\alpha=90°$ will never be realized, since then the effective facet diameter $D_{eff}=D\cos\alpha$ equals zero, and consequently the photon flux entering the ommatidium is annihilated. An additional constraint for constructing the ommatidia skew to the eye surface is the reflection on the lens front surface, which is normally a few percent at most, but progressively increases with larger α. Optical tricks to suppress reflection like corneal nipples (BERNHARD et al., 1965; MILLER, this volume, Ch. 3) seem to be of little interest; corneal nipples are particularly encountered in the eyes of Lepidoptera, which provide the best examples of uniform, spherical eyes.

The construction of skew ommatidia in compound eyes is a widespread occurrence. This is most fortunate for optical investigations, especially those into eyes with smooth, reflecting facet lenses; the corneal reflections from axial illuminations can then completely obscure poorly reflecting structures proximal to the facet lens. In eye regions with sufficiently skew ommatidia the corneal reflections become displaced from the ommatidia aligned with the microscope, thus facilitating studies of the proximal structures. The experimental example of the photographed pigment migration in bee retinular cells (Fig. 40) is a case in point (see Sec. D.III.4).

The foregoing has made it clear that anatomical data for interommatidial angles (e.g., BAUMGAERTNER, 1928; FRIEDERICHS, 1931, PORTILLO, 1936; BUTLER, 1971) must be corrected for skewness (EXNER, 1891; BAUMGAERTNER, 1928; FRANTSEVICH and PICHKA, 1976). However, reliable data cannot easily be obtained by a correction procedure: usually pseudopupil methods will be favoured.

C. Fundamentals of Pseudopupils

I. Focus at Different Levels

1. Spherical Eyes

The main purpose of this introductory section on pseudopupils is to show that the organization of the ommatidial axes can be revealed experimentally by using pseudopupil techniques. The best compound eyes in which to observe pseudopupils are undoubtedly the eyes of diurnal butterflies. Use of a microscope with axial illumination reveals brilliant, colourful reflections from the eyes when dark

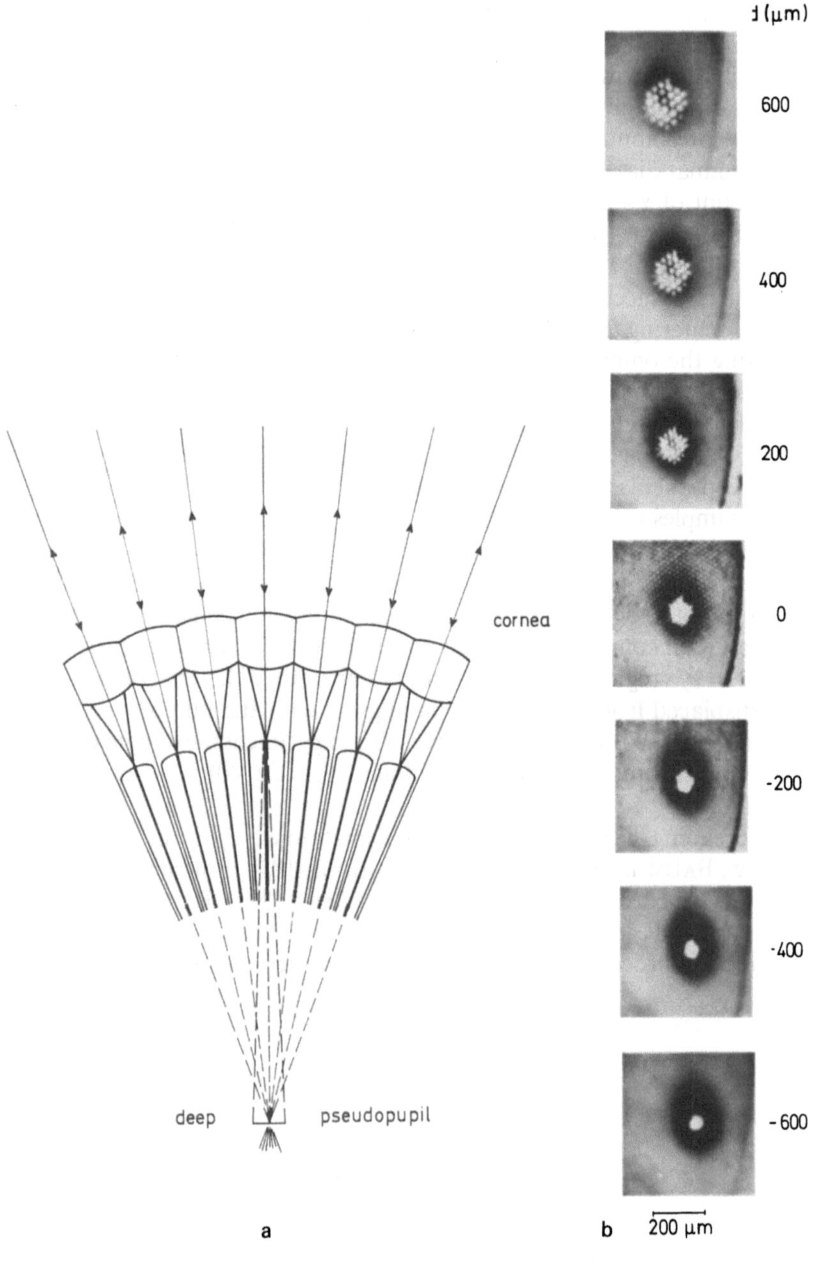

Fig. 20a and b. Eye shine and pseudopupils in the eye of the common dusky blue *Erina hyacynthina*. (a) diagrammatically shows that in butterfly eyes incident light is reflected after propagation through the fused rhabdom at the proximally situated tapetum, and subsequently partly leaves the eye via the reverse route. The diverging light beams (b) seem to emerge from a single source located at the level of the centre of curvature of the eye. This phenomenon is called the deep pseudopupil. The image seen at the level of the cornea is called the corneal pseudopupil. The distance of the level of the photographs from the cornea is given by *d*. From the distance of the light beams at different levels the value of the interommatidial angle can be deduced adapted (EXNER, 1891). The light paths in butterfly eyes giving rise to this glow or eye shine are shown diagrammatically in Fig. 20a.

In each ommatidium incident light is focussed by the facet lens and subsequently propagated in the (fused) rhabdom. Trachea are located proximal to the rhabdom and act as light reflectors (DEMOLL, 1909a, 1917) according to the principle of quarter wavelength interference (MILLER and BERNARD, 1968). Therefore, light which is not absorbed in the rhabdom can travel a double route through the rhabdom and leave the eye again. The colour of the reflected light is determined by the mirror characteristics of the tapetal trachea and the total absorbance of the visual pigment in the rhabdom (see MILLER, this volume, Ch. 3).

The light paths in a butterfly eye are demonstrated in the case of the common dusky blue *Erina hyacinthina* in Fig. 20b. As indicated, the levels from which the photographs have been taken are 200 μm apart. Light beams diverge from each facet lens when going from the level of the cornea towards the outside. The beams seem to originate from a unique light source located in the centre of the eye, i.e., where the ommatidial axes intersect (see Fig. 12a). The image seen at the level of the intersection of the ommatidial axes is of great value. FRANCESCHINI and KIRSCHFELD (1971b), the first investigators to appreciate this fact, have called this image the deep pseudopupil.

2. Deep Pseudopupil

The importance of the deep pseudopupil resides in the property of apposition eyes that the distal plane of the retinula approximately coincides with the focal plane of the dioptric apparatus of the ommatidium (proven experimentally by KIRSCHFELD and FRANCESCHINI, 1968; see also KUIPER, 1966; SEITZ, 1968; SEITZ and BURKHARDT, 1974; KUNZE, this volume, Ch. 8). Since the focal distance is small in respect of the radius of eye curvature a virtual image of (approximately) the distal plane of the retinula is created at the level of the intersection of the ommatidial axes. Additionally, since retinulae are often juxtaposed in perfect order, the images from the distal retinular planes in different ommatidia superimpose so that one intensified and enlarged image is seen (FRANCESCHINI, 1972a, b, 1975). This image, the deep pseudopupil, can be observed from the outside with a microscope; of course, the working-distance of the objective has to be longer than the distance from cornea to ommatidial axes-intersection. The deep pseudopupil thus provides an optical means for studying photoreceptor processes in living and intact animals (FRANCESCHINI and KIRSCHFELD, 1976) yielding a significant increase of signal to noise ratio when measuring from specific retinular cell types (FRANCESCHINI, 1972a).

3. Corneal Pseudopupil

As important as the deep pseudopupil is the image seen at the level of the cornea, the corneal pseudopupil (FRANCESCHINI and KIRSCHFELD, 1971b). Only a number of facets (see Fig. 20) appear bright, one reason being that the solid angle of light reflected from each ommatidium is narrow, as can be seen from the slight increase in spread in the light patches when travelling from the cornea towards levels outside the eye. However, the main limiting factor is the aperture of the microscope objective. When the microscope aperture is closed down, rays from the peripheral ommatidia are obstructed and ultimately only one bright ommatidium remains. This case is called the reduced corneal pseudopupil (FRANCESCHINI and KIRSCHFELD, 1971b); because of the special optical organization of the neural superposition eye the reduced corneal pseudopupil in flies is made up by seven facets, and even by eight or nine in the equatorial region (KIRSCHFELD, 1967). Reduction of the number of bright facets can also be achieved by reducing the aperture of the illumination beam (see Fig. 24 b).

The experiment shown in Fig. 20 b indicates that the eye of the common dusky blue (frontally) is satisfactorily described by the simple spherical model of Fig. 12 a. More important, the photographs embody quantitative data from which interommatidial angle values can be extracted. Evaluation of the data is found in Section D.II.2.

4. Oval Eyes

Many compound eyes do not approximate to the simple spherical shape (EXNER, 1891). The pseudopupil phenomena in ovally shaped eyes become slightly less clear-cut than those treated above. An experimental example is presented in Fig. 21 (see KIRSCHFELD, 1973 b).

The left eye of a honey bee is observed frontally. The illumination is applied antidromically, i.e., from the inside, in Fig. 21 a–e. In this case, light is transmitted through the rhabdom without reflection at its proximal end because honeybees lack a tapetum. The light beams radiating from the individual facets can be seen clearly separated at levels outside the eye, but more proximally they merge. The phenomena occurring at two special levels require comment (cf. FRANCESCHINI and KIRSCHFELD, 1971b; FRANCESCHINI, 1972a): in one case a bright line extends vertically (Fig. 21 c), and in another a horizontal line is seen (Fig. 21 e). In between, the beams together create a more or less circular spot (Fig. 21 d) similar to the deep pseudopupil observed in the butterfly case (Fig. 20 b). From a comparison with the diagrams of Fig. 12, it is clear that only in the case of a spherical eye can a unique deep pseudopupil be defined.

At this stage the formation of deep pseudopupils in compound eyes might be compared with image formation by lenses in geometrical optics. A point source (located at infinity) is imaged by a spherical lens into one focal point. The parallel with the spherical eye is the single deep pseudopupil. At a cylindrical lens with orthogonal axes two focal lines are distinguished, the vertical and the horizontal focal line. The two deep pseudopupils formed in the case of an oval eye can accordingly be called respectively the vertical deep pseudopupil (at the level of I_h in Fig. 12b) and the horizontal deep pseudopupil (at the level of I_v).

The pseudopupil types distinguished at different levels in the compound eye are here illustrated exclusively on fused rhabdom apposition eyes. For the more spectacular, but less simple case of the neural superposition eye of flies, the reader

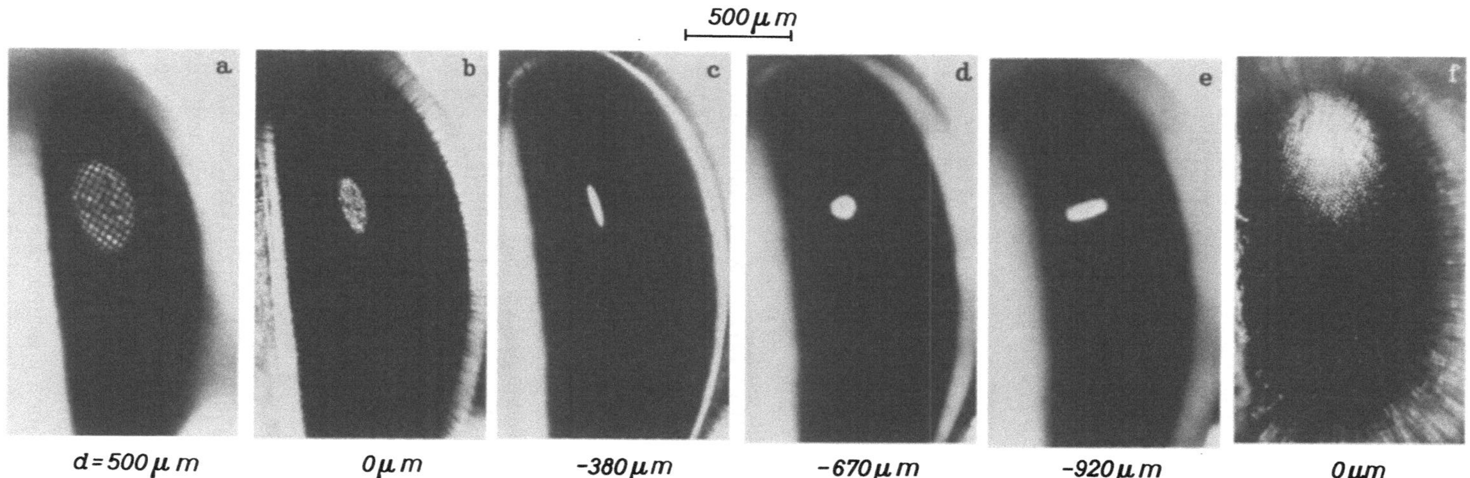

Fig. 21a–f. Pseudopupil phenomena frontally in the left eye of a honeybee photographed at a number of levels with distance *d* from the cornea. Antidromic illumination, i.e., in the back of the head, was applied in (a–e). Light which has propagated through the rhabdoms leaves the facet lenses (b). The light beams diverge towards the outside (a). At levels proximal to the cornea, a vertical and a horizontal deep pseudopupil (c, e) are observed (cf. Fig. 12b). In (f) normal, orthodromic illumination shows the facet lens pattern and the interfacetal hairs (Figure by courtesy of Prof. KIRSCHFELD)

is referred to the treatises of FRANCESCHINI and KIRSCHFELD (1971a, b, 1976) and FRANCESCHINI (1972a, b, 1975). These authors focussed their attention especially on the optical aspects of the formation of pseudopupils, in contrast to EXNER (1891) whose attention was rather for the phenomenological side; his approach will be reviewed next.

II. Luminous Pseudopupil

The eye shine (also called glow) observable in dark-adapted butterflies (Fig. 20) and moths (see KUNZE, this volume, Ch. 8) was termed a luminous pseudopupil by EXNER (1891).

Several nineteenth century accounts deal with this interesting phenomenon. A historically interesting paper with old references is that of LEYDIG (1877). The main theme is his complaint that the scientific community in its appraisal of BOLL's discovery of the purple-red colouring of the retina completely neglects (together with BOLL) his publications on the red colour perceivable in the retina of several vertebrates. Still, however modern LEYDIG's eagerness for scientific credit may be, more interesting is his reference to SPALLANZANI, who writes that "a lady of great understanding and excellent knowledge of nature, Marchese Olympia Agnelli Sessi, Signora di Nolo, revealed to him the observation of phosphorous moth eyes," and that "these discoveries probably will culminate into a new branch of optics."

The term glow (MILLER and BERNARD, 1968) is reminiscent of the association with "glowing coals" of moth eyes by the first investigators (see LEYDIG, 1864; EXNER, 1891, p. 141; MILLER et al., 1968). However, the physical origin of the eye shine is not some luminescent material. EXNER (1891, p. 153) proved that the origin was purely dioptrical as it is in the eyes of vertebrates, and that no light production occurred in the eye. From his anatomical and optical observations he was sure that no luminous bodies existed in the eye in contrast to the luminous organs of fireflies. In order to demonstrate that phosphorescence was not the cause of eye shine, he constructed an apparatus (similar to a phosphoroscope of BECQUEREL) so that a dark-adapted moth could be illuminated without adapting the observer. By rapidly rotating the gearwheel and observing the eye immediately after the bright illumination, EXNER convinced himself of the veracity of his belief from the absence of any persistent light production.

The eye shine in lepidopteran eyes is actually due to the existence of a tapetum, a proximally located reflecting layer created by the invaginations of the tracheal system (LEYDIG, 1864; DEMOLL, 1917; MILLER et al., 1968).

Striking luminous pseudopupils are also observable in the eyes of crustacea, where the tapetum is a layer of strongly reflecting pigment (EXNER, 1891). A peculiarity of several species is that the eye shine can be detected only in dark-adapted animals at night, which is due to a circadian rhythm controlling the location of both absorbing (screening) pigment and reflecting (tapetum) pigment (Sec. D.IV.2). A general characteristic of both insect and crustacean eye shine is its disappearance upon light adaptation. [Notable exceptions are the ever present eye shine of skippers (MILLER et al., 1968; HORRIDGE et al., 1972) and of the dorsal eye of the male mayfly *Cloeon* (WOLBURG-BUCHHOLZ, 1977).] The responsible pigment migrations will be described in Section D.IV.2.

On the other hand, upon light adaptation, a shine is evoked in the eye of flies; paradoxically the cause is also pigment migration, namely the migration of retinular cell pigment granules (Sec. D.III.4). Besides the related reflection phenomena observed in the eyes of, for example, *Libellula*, EXNER also classified fly eye shine as luminous pseudopupils. It seems preferable though to use the term luminous pseudopupils in reference only to those reflections which are of tapetal origin.

It must be reiterated here that the appearance of the luminous pseudopupil depends on the level of focus of the microscope (Sec. C.I.1,2,3). Thus, at the corneal level one observes the corneal luminous pseudopupil and in the centre of the eye the deep luminous pseudopupil.

When the luminous pseudopupil of the common dusky blue butterfly has disappeared by light adaptation (see Figs. 28 and 29), a dominant black spot already quite noticeable in the deep pseudopupil remains: this black spot is the principal pseudopupil.

III. Principal Pseudopupil

The principal pseudopupil in apposition eyes has its origin in the pigment enveloping the proximal part of the crystalline cone (EXNER, 1891) and not (exclusively) in the absorbing rhabdom as stated by MAZOKHIN-PORSHNYAKOV (1969). Usually the pigment situated proximally around the cone is jet-black. (For a discussion on the functional background and on the aberrant red in fly eyes, see Sec. D.IV.3.) Figure 22 illustrates the situation when a compound eye is observed with a narrow aperture microscope. When the axis of observation is approximately aligned with the ommatidial axis, the observer will perceive no or little light scattered from the black primary pigment. Off-axial observation will result in a bright facet when the underlying secondary pigment substantially scatters light. Consequently, the principal pseudopupil marks the area of the eye where the ommatidia look into the aperture of the observer's microscope. Figure 22 furthermore shows that, in the deep pseudopupil, images from structures in the set of neighbouring ommatidia are superimposed. Even with a wide aperture optical apparatus a rather crisp pseudopupil can thus be observed at the level of intersection of the ommatidial axes. (To understand why one can see a deep pseudopupil with the two naked eyes and not with one, the reader should refer to FRANCESCHINI and KIRSCHFELD'S (1971 b) analysis.)

As was stated above for the case of the luminous pseudopupil one should distinguish also for the principal pseudopupil whether observations are performed at the level of the cornea, i.e., in the corneal principal pseudopupil, or in the deep, at the level of the intersection of the ommatidial axes, i.e., in the deep principal pseudopupil.

An experimental example of the use of the principal pseudopupil is given in Fig. 23a. A damselfly was photographed from three directions (in steps of 15°

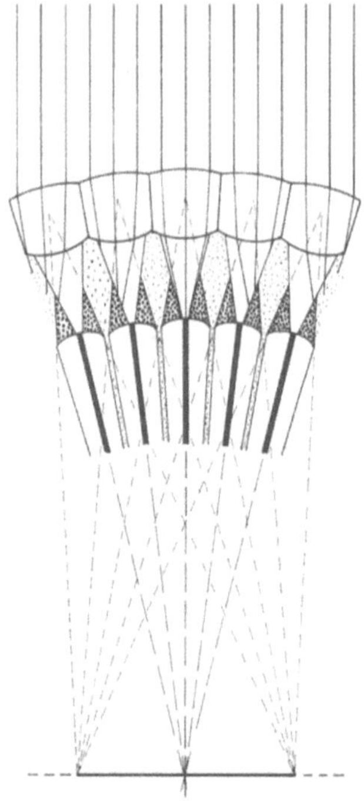

Fig. 22. Optics of the principal pseudopupil. A schematic apposition compound eye is taken with black-pigmented primary pigment cells and with lightly-pigmented secondary pigment cells. When the eye is illuminated with a distant point source the incident light is focussed on the distal end of the rhabdom in the ommatidium aligned with the source. In adjacent or more peripheral ommatidia the light will hit the light-absorbing primary pigment cells or the light-scattering secondary pigment cells. In a transition zone light entering the facet lens on the central side, i.e., that part nearest to the central ommatidium, will be projected into the primary pigment cells; the light entering more off-centre travels into the secondary pigment cells. Hence, when the eye is observed at the level of the cornea with a narrow aperture optical instrument, a patch of central facet lenses appear completely black since little light is scattered by the rhabdom and the primary cells. The peripheral facet lenses appear bright due to scattering in the secondary pigment cells. In the transition region only part of the facet lens appears black, namely a part nearest to the centre. Facet lens and cone are here assumed to be homogeneous. When gradients in refractive indices occur the optics becomes more complicated (EXNER, 1891). A pronounced black spot is observed at the level of the centre of curvature of the eye, especially when the aperture of the observation instrument is enlarged. This phenomenon of a distinct spot in the depth of the eye is due to the superposition of images in the deep pseudopupil. It is not perceived with the naked eye, only with two naked eyes (FRANCESCHINI and KIRSCHFELD, 1971b). The central black spot is called the deep principal pseudopupil in contrast with the corneal principal pseudopupil, which is observed at the level of the cornea

---

Fig. 23a–d. Principal pseudopupils and corneal reflections. In (a–c) the eyes of a damselfly were photographed after rotating the animal in steps of 15° around an axis perpendicular to the vertical plane of symmetry. Axial illumination was applied, i.e., the line of illumination and that of observation were identical. It will be appreciated from Fig. 22 that the size of the principal pseudopupil increases with an increase of the radius of curvature of the eye (or more precisely of the distance cornea—ommatidial axes intersection). In other words, the large principal pseudopupils of the middle picture (b) mark foveal regions. Principal pseudopupil and area of corneal reflections are more or less coincident here, but they are quite displaced from each other above (a) and below (c) for fovea. Since the centre of the principal pseudopupil marks the ommatidium which is aligned with the axis of observation this means that only in the case of the fovea are the ommatidial axes approximately perpendicular to the eye surface. (d) Schematic explanation of (a) ▶

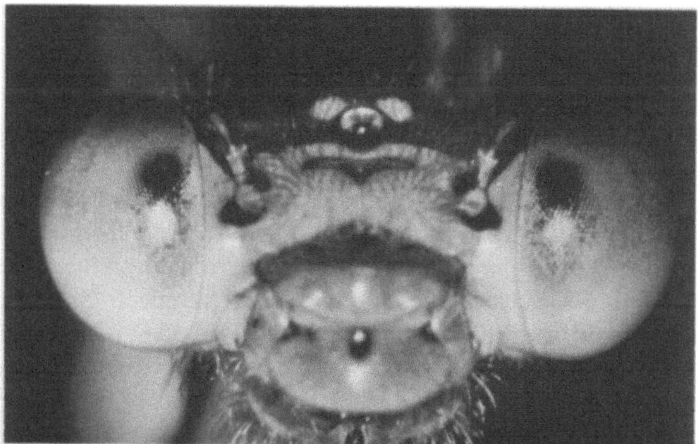

a

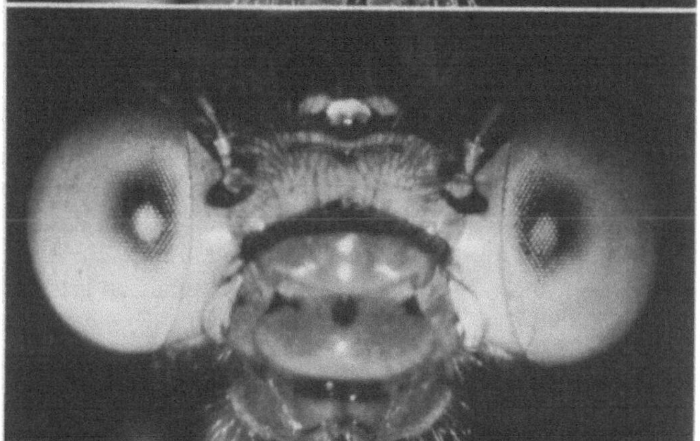

b

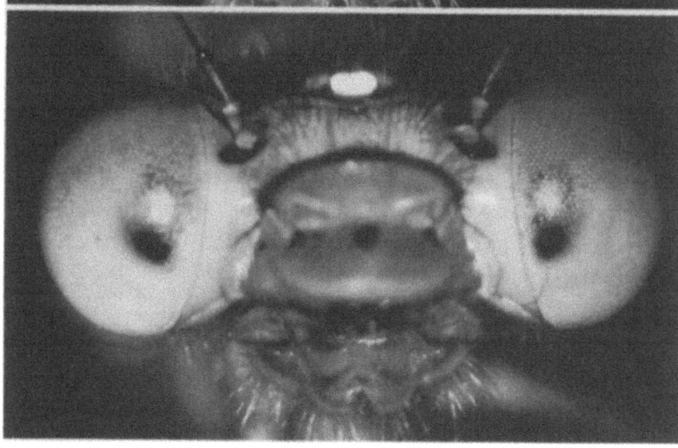

c

d

rotated around an axis perpendicular to the vertical symmetry plane) while axial illumination was applied. The middle photograph gives the approximately frontally looking fovea. The corneal reflections coincide here with a large principal pseudopupil. In the case of the more dorsally taken (upper) photograph corneal reflections and principal pseudopupil are quite displaced from each other. A similar situation occurs in the more ventrally taken (lower) picture. The diagram (Fig. 23d) shows that the relative locations of corneal reflections and pseudopupil can be interpreted as indicating that in the fovea the ommatidial axes are approximately perpendicular to the eye surface, while towards the periphery the ommatidial axes are skewed. (That the extension of the principal pseudopupil stands in direct relation with the interommatidial angle can be appreciated from Fig. 22.)

Figure 23 illustrates the fact that the corneal principal pseudopupil is a useful instrument for determining ommatidial axes in eyes which lack a clear luminous pseudopupil. Since the latter is the most direct indicator of ommatidial axes, it proves that the centre of the principal pseudopupil is a tool of similar quality. Recognition of this fact is important, for it has great potential for experimental work. The somewhat involved technique of axial illumination, necessary for observing the luminous pseudopupil, can be avoided and illuminations from oblique directions can be applied. However, with oblique illumination interesting events occur.

IV. Oblique Illumination Pseudopupil

Axial illumination of the common dusky blue butterfly, *Erina hyacinthina*, (Fig. 24) with a point source yields a luminous pseudopupil equivalent to a reduced corneal pseudopupil (Sec. C.I.3) from the aligned ommatidium (Fig. 24b). Rotation of the point source so that its direction of incidence becomes outside the aperture of the microscope extinguishes the luminous pseudopupil and thus results in a filled-in principal pseudopupil. However, additionally a black spot is evoked by the off-axis light source (Fig. 24c). This phenomenon will be called here the oblique illumination pseudopupil.

In order to explain the oblique illumination pseudopupil it has to be understood that the colour of an eye is due to the distal scattering pigments. The principal pseudopupil is apparent as a contrasting dark spot because the black pigment clustered proximally around the cones in the ommatidia aligned with the observing microscope does not scatter light into the aperture of the microscope. Equivalently, no scattered light enters the microscope from those ommatidia where the light incident from the point source hits the black pigment. The oblique illumination pseudopupil thus marks the ommatidia aligned with the light source. Movement of the illumination around the eye gives rise to a wandering spot over the eye for exactly the same reasons as the principal pseudopupil wanders over the eye when the observer turns the eye around.

An experimental example of the wandering oblique illumination pseudopupil is given in Fig. 25b–d. The extremely elongated principal pseudopupil (Fig. 25a) is a consequence of the oval shape of the eye in locusts when viewed frontally (notice

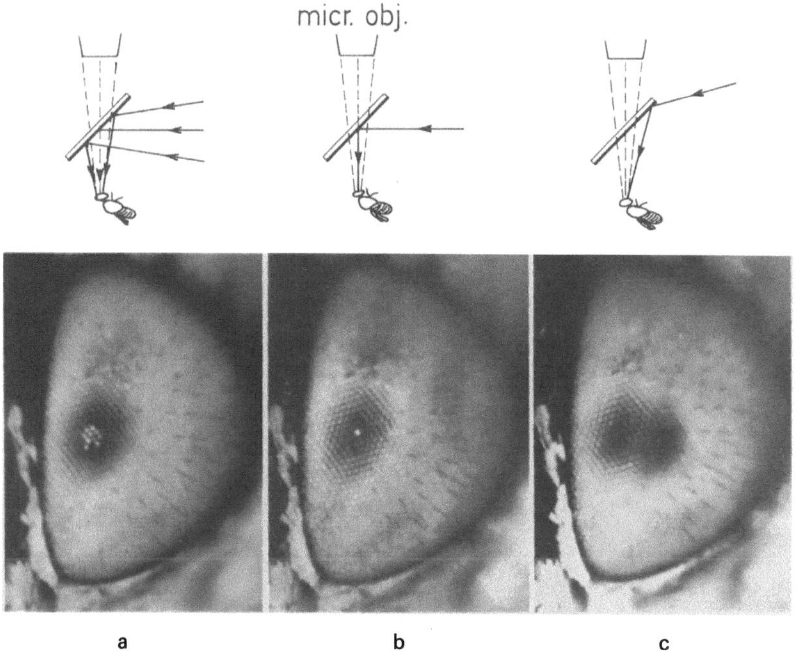

Fig. 24a–c. Pseudopupil phenomena in the compound eye of the common dusky blue *Erina hyacynthina*. When the light source completely fills the aperture of the microscope (a) a luminous pseudopupil consisting of several shiny facets is seen in the centre of the black principal pseudopupil. In the case of a point source only one facet shines, namely that of the ommatidium whose axis is aligned with the point source (b). When the point source is rotated so that light is no longer incident within the aperture of the microscope (c), the luminous pseudopupil vanishes. However, an additional dark spot then arises, which wanders over the eye when the light source is rotated. (Similarly, the principal pseudopupil wanders over the eye when the observer rotates around the eye.) The additional spot is called the oblique illumination pseudopupil

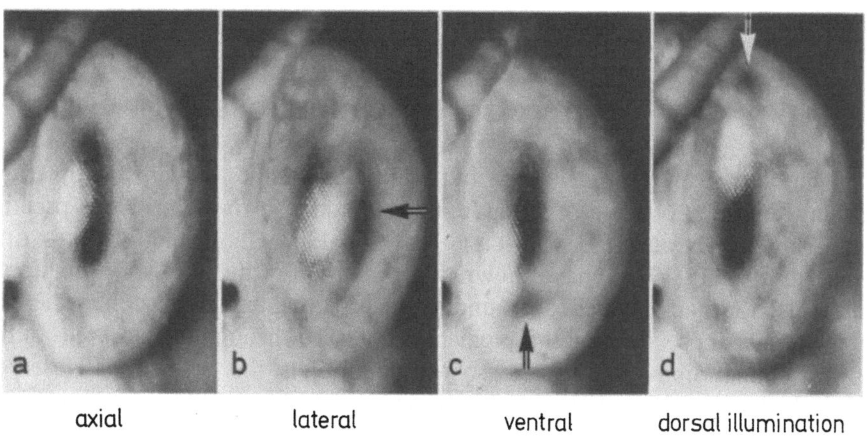

Fig. 25a–d. Principal and oblique illumination pseudopupil in the eye of some lightly coloured locust. In (a) axial illumination shows an elongated principal pseudopupil in the frontal fovea. This pseudopupil maintains its place when the eye is illuminated from oblique directions (b–d). The area of the corneal reflections and the location of the oblique illumination pseudopupil are determined by the direction of illumination

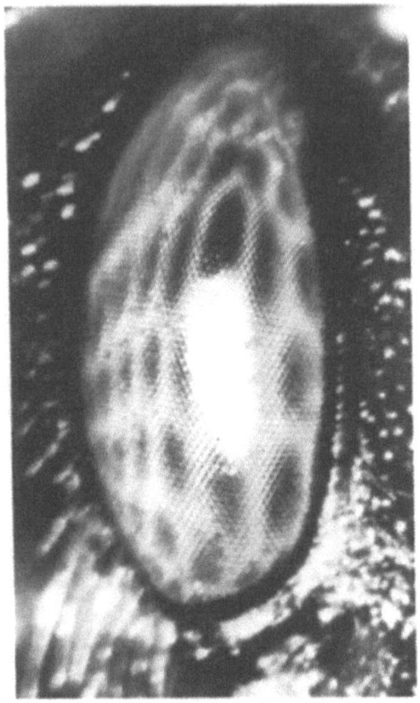

Fig. 26. Accessory pseudopupils are displayed abundantly by several hymenopteran insects which have brightly coloured eyes. The black pigment surrounding a specific cone is observed through a facet lens of adjacent ommatidia (see Fig. 27); the right eye of a hymenopteran insect is shown

the corneal reflections), but also of skewed ommatidia, especially in vertical directions. The oblique illumination pseudopupil emerging at lateral illumination (Fig. 25b) retains the elongated form, but at ventral (Fig. 25c) or dorsal (Fig. 25d) illumination the rounded form indicates a more spherical shape of the eye in those regions.

V. Accessory Pseudopupils

The eyes of many insects, e.g., butterflies, dragonflies, and hymenopterans, display a number of accessory pseudopupils in addition to the principal pseudopupil (Fig. 26). EXNER (1891), YAGI and KOYAMA (1963, Ch. 3), and FRANCESCHINI and KIRSCHFELD (1971b) have shown that the origin of the accessory pseudopupils is also the black pigment surrounding the proximal part of the cone; where as in the principal pseudopupil this pigment is observed through the facet lens of the "own" ommatidium, in the accessory pseudopupils the pigment is observed through the facet lens of the neighbouring ommatidia (Fig. 27, from YAGI and KOYAMA, 1963). EXNER (1891) classified the accessory pseudopupils in a numerical order, depending on the distance from the principal pseudopupil (see FRANCESCHINI and KIRSCHFELD, 1971b); this classification is not quite accurate, but is satisfactory as an indication. (It is easily seen that the index formalism of section B. IV enables an unambiguous classification of the accessory pseudopupils.)

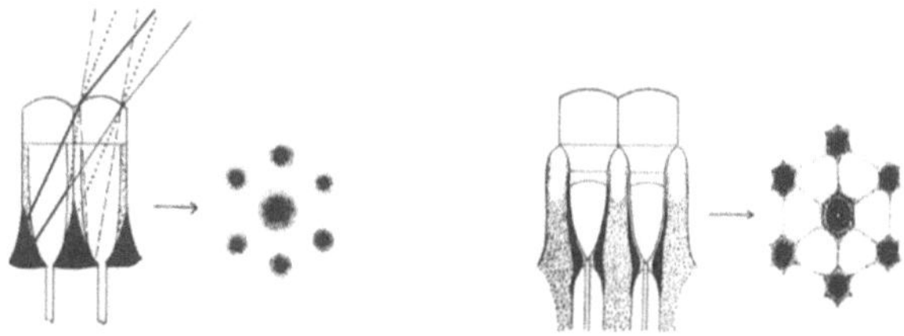

Fig. 27. Diagrams showing how eye structure and pseudopupil patterns are related (from YAGI and KOYAMA, 1963; slightly modified). The central, principal pseudopupil is due to imaging of the black pigment surrounding the cones in ommatidia approximately aligned with the axis of observation. In more off-axis ommatidia the bright scattering distal pigment is observed. Even more off-axis, the light-absorbing black pigment is seen imaged by the dioptric apparatus of the neighbouring ommatidium. This causes an accessory pseudopupil of first order (EXNER, 1891). Higher order accessory pseudopupils, caused by imaging of the pigment by not immediately neighbouring ommatidia, are perceived when the distal part of the eye is sufficiently transparent. More elaborate pseudopupil patterns arise due to gradients in the density and colour of the pigmentation in the pigment cells. The pattern of the image reveals the place of the pigment cells in the lattice structure of the compound eye (see FRANCESCHINI and KIRSCHFELD, 1971 b)

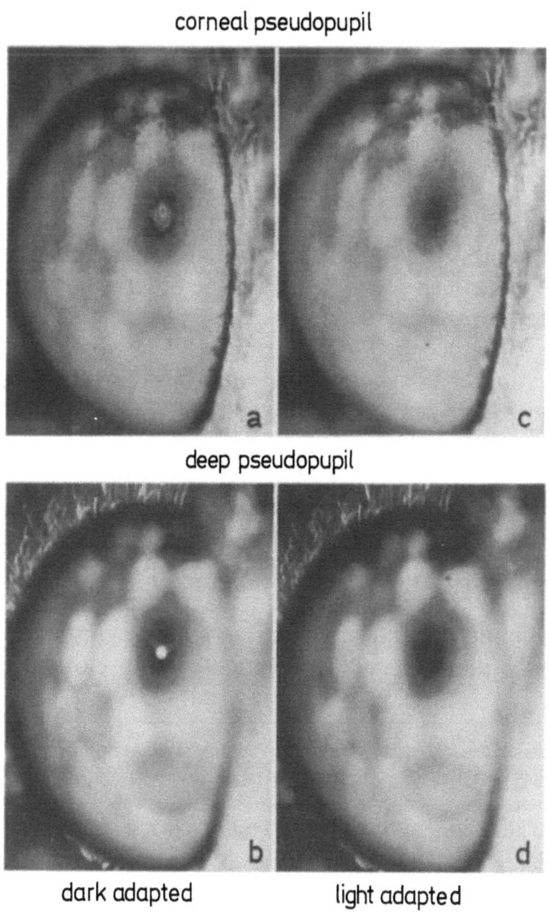

Fig. 28a–d. Corneal (a, c) and deep (b, d) pseudopupil in the eye of the butterfly *Precis* in the dark(a, b)- and light(c, d)-adapted state respectively. Luminous, principal and accessory pseudopupils can be recognised

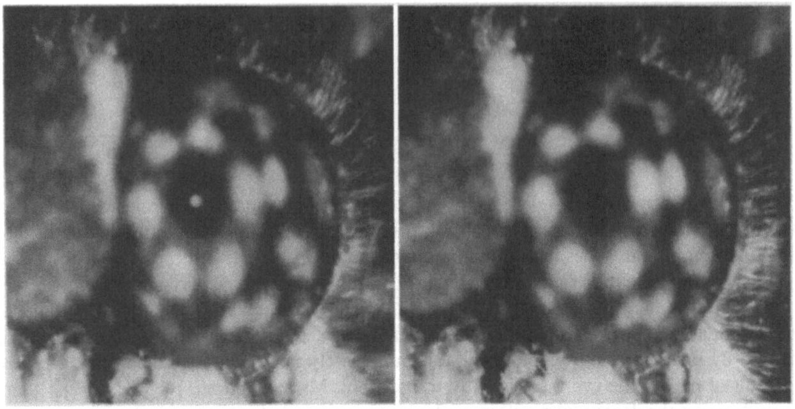

Fig. 29. The eye featured is that of the Red Admiral *Vanessa atalanta* (deep pseudopupil). The cause of the vanishing eye shine is discussed in Section D.III.4

An illuminating way of explaining the rich pseudopupil patterns of butterflies (Figs. 28, 29, 30, 43) is Franceschini and Kirschfeld's (1971 b) use of the deep pseudopupil. In the deep pseudopupil one sees the superimposed image projected by the set of facet lenses from the underlying periodic structure. Consequently, the ommatidial mosaic as it is in the focal planes of the facet lenses is seen magnified in the deep pseudopupil. It is still useful to realize that an image approximately the same as that of the deep pseudopupil can be observed at any level (e.g., at the level of the cornea), when a narrow aperture microscope is used (as was the case in for example Fig. 25).

Axial illumination of a dark-adapted butterfly eye yields a luminous pseudopupil only within the principal pseudopupil. Its appearance at the level of the cornea and in the deep pseudopupil is shown for *Precis* (Fig. 28a and b). Its extinguishment upon illumination is demonstrated by Fig. 28c and d. Fig. 29 shows the case of *Vanessa*. Accessory pseudopupils never display a luminous spot in their centre. The narrow directionality of the rhabdoms precludes such an effect (Exner, 1891; Franceschini and Kirschfeld, 1971b; Miller, this volume, Ch. 3).

Oblique illumination of an eye displaying accessory pseudopupils sometimes induces complicated phenomena. Illumination of the eye of the cabbage butterfly

---

Fig. 30a–g. The appearance of the right eye (ventrally) of the cabbage butterfly *Pieris rapae* under axial (a), (b) and oblique (c–g) illuminations. Photographs (a), (c) and (e) are taken at the level of the cornea, (b), (d), (f) and (g) represent the corresponding situations photographed at the level of intersection of ommatidial axes, i.e., the deep pseudopupil. The central black spot, the principal pseudopupil, is surrounded by only six spots (the accessory pseudopupils) at axial illumination. With oblique illumination additional accessory pseudopupils seem to appear, but in fact the oblique illumination pseudopupil arises: a set of spots homologous to the pattern observed at axial illumination. In (c) and (d), some spots of the oblique illumination pseudopupil coincide with accessory pseudopupils. Arrows towards the centre of the oblique illumination pseudopupils indicate the directions of illumination (the state was always light adapted) ►

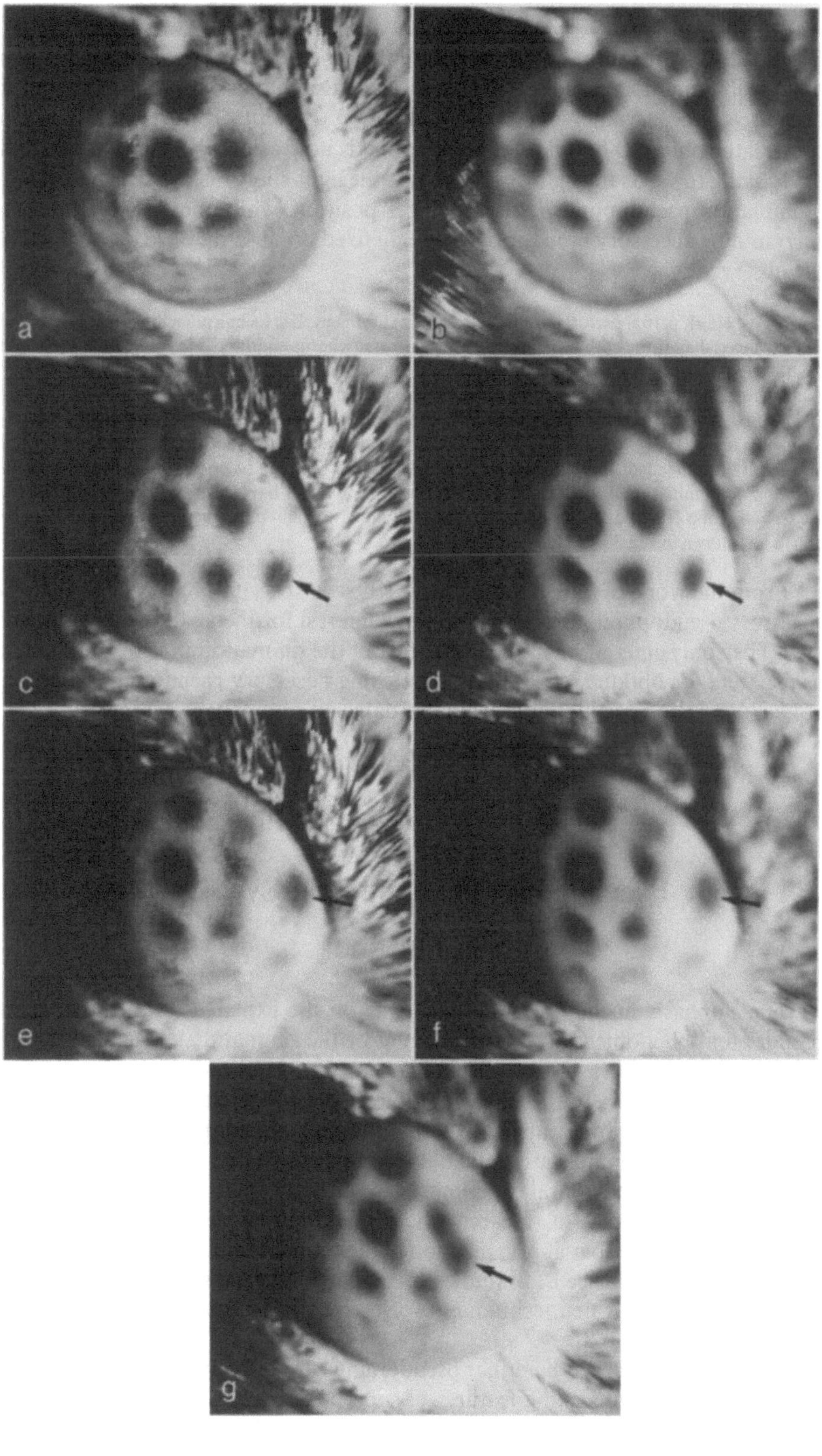
a
b
c
d
e
f
g

Pieris with a light source not aligned with the direction of observation induces an oblique illumination pseudopupil consisting of a set of dark spots homologous to that already observable with axial illumination (Fig. 30). However, the crown of spots of the oblique illumination pseudopupil is now centred around the ommatidia looking towards the light source. When the light source is rotated this oblique illumination pseudopupil pattern changes position, while the principal and the surrounding accessory pseudopupils remain fixed because they are determined by the observer.

EXNER (1891, p. 165) performed this very same experiment. His technical facilities were somewhat less advanced than those with which the present-day researcher is endowed. He studied the pseudopupils through his ophthalmoscope equipped with a lens as telescope, using a candle flame as his light source (see EXNER, 1891, p. 144). In *Pieris rapae* he noticed in addition to the principal pseudopupil and the first order of accessory pseudopupils, second and third order accessory pseudopupils. Moving the candle around the eye EXNER became confused by his observation that the first group of pseudopupils only depended on the direction of observation and that the second group depended on the direction of illumination. EXNER's confusion probably would not have occurred had he noticed that the second group of pseudopupils is absent at axial illumination and only emerges at oblique illumination; a fact shown in Fig. 30.

From the work of FRANCESCHINI and KIRSCHFELD (1971 b) it is clear that the accessory pseudopupils are of relatively low interest for the analysis of compound eyes. They only emerge when the distal part of the ommatidium is to some extent transparent for oblique rays. Still, by matching accessory pseudopupils with oblique illumination pseudopupils, valuable data can probably be gathered.

D. Pseudopupils in Compound Eye Research

I. Outline

The emphasis in the preceding part was on the explanation of the different pseudopupil phenomena from the imaging of the ommatidial structures by the facet lenses of the more or less spherical cornea. In the following an opposite goal will be pursued, namely the interpretation of the properties of the ommatidial structures from the appearance of the pseudopupils. Additionally, it will be explained that pseudopupils are potential resources in studies of higher order neurons and even in behavioural experiments.

The survey of areas of research where pseudopupils are useful will include 1) the analysis of the spatial organization of the retinal samplers and of the optical properties of retinular cells, 2) studies of adaptation processes in the retina, and 3) the highly sophisticated techniques of selective stimulation of individual receptor cells.

It has to be remarked here that pseudopupils are of main interest for apposition eyes. For the interesting features of superposition eyes, see the chapters by MILLER (this volume, Ch. 3) and KUNZE (this volume, Ch. 8).

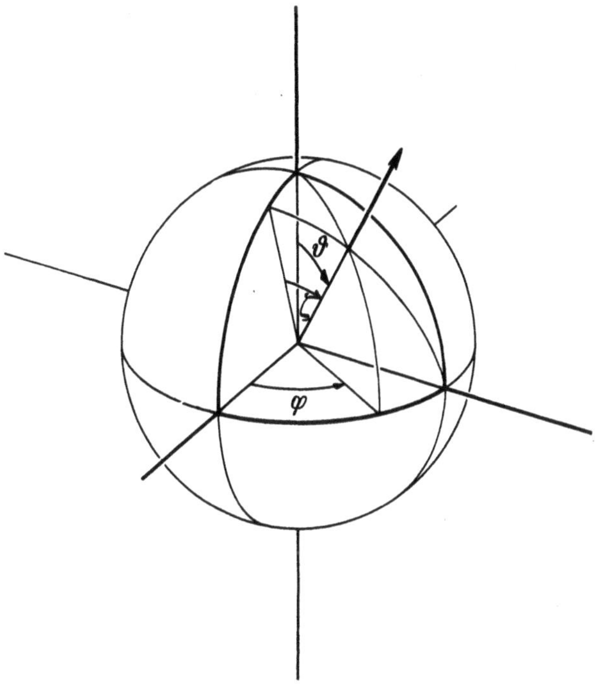

Fig. 31. Spherical coordinate system for characterizing visual axes of a compound eye. The intersection of the ommatidial axes is centred in the sphere. The plane $\phi = 0°$ is defined by the vertical symmetry plane of the animal. ζ is the angle between the visual axis of an ommatidium and that plane (from BEERSMA et al., 1977)

II. Optical Qualities of Compound Eyes

1. Extent of Visual Field and Binocular Overlap

The centre of the principal pseudopupil, being approximately equivalent to that of the luminous pseudopupil when existent (Fig. 20), marks the particular ommatidium the axis of which is aligned with the direction of observation. The extent of the visual field covered by an eye can consequently be determined by placing the animal (or eye) on a goniometer, then allowing the pseudopupil to wander around the marginal ommatidia, meanwhile reading the spherical coordinates of the goniometer (Fig. 31). When the eyes are fixed in the head, as is the case in *Limulus* and in insects, the binocular overlap is also fixed. (Of course, the latter does not hold for movable eyes like those of crab and crayfish.)

As an example of a determination of visual field and binocular overlap the case of the male blowfly *Calliphora erythrocephala* is presented (see BEERSMA et al., 1977). The visual axes in this experiment were actually determined with the aid of the pupil mechanism (Sec. D.III.4). The margins of the visual fields of the two eyes are drawn in Fig. 32. Each eye covers approximately a hemisphere with a small blind spot posteriorly and a limited binocular overlap frontally. These characteristics vary little among different fly genera (BEERSMA et al., 1977).

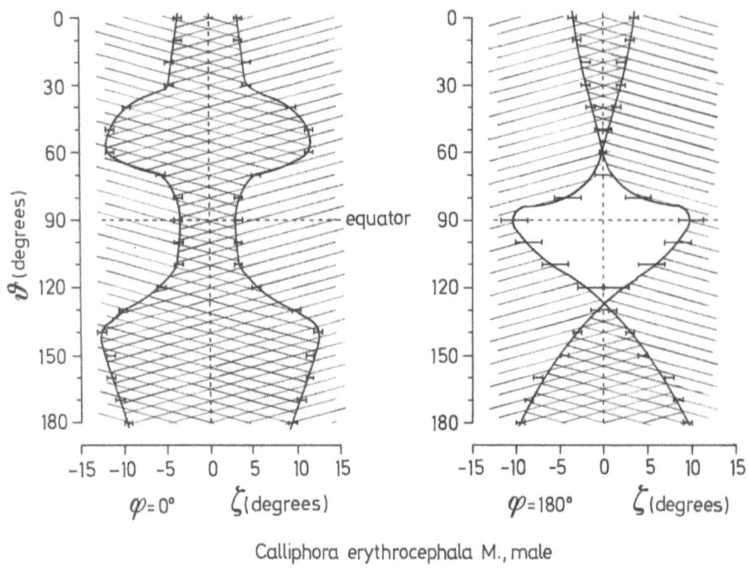

Fig. 32. Extent of the visual field of a male blowfly *Calliphora erythrocephala*. The spherical coordinates of the visual axes of the marginal ommatidia were determined. Only dorso-frontally and ventrally binocular overlap amounts to more than a few degrees. A small blind spot exists posteriorly (from BEERSMA et al., 1977)

Although the binocular overlap is approximately the same at angles about 45° upwards and downwards ($\theta = 45°$ respectively 135°), visual axis density (or visual resolution) is much higher in upward directions, as can be appreciated from the eye chart (Fig. 33), where each facet is represented by a black square (see the legend for further information). Male flies generally have an upward-looking fovea (COLLETT and LAND, 1975; KIRSCHFELD and WENK, 1976). Interestingly, STRAUSFELD (1976a, p. 149; 1976b, Fig. 14) has traced a unique giant visual neuron which invades the upper frontal columns in the lobula, corresponding to the dorsal region of overlap between the two compound eyes.

Quantitative estimates of binocular overlap with the aid of pseudopupils have been made since EXNER (1891) by DEMOLL (1909b), DEMOLL and SCHEURING (1912), BALDUS (1926), WALLACE (1959), SEITZ and BURKHARDT (1974), COLLETT and LAND (1975), FRANTSEVICH and PICHKA (1976), VIA (1977), HORRIDGE (1977a, b) and SHERK (1977, 1978a, b, c); tables have been compiled by MAZOKHIN-PORSHNYAKOV (1969) and BURKHARDT et al. (1973). Comparing the binocular overlap of a variety of insects, FRANTSEVICH and PICHKA (1976) have put forward the following general view. Predators as well as honeybee drones and male flies which perceive their potential prey or mate merely as a (moving) point in the sky have a small binocular zone and thus are unable to utilize stereopsis. Anthophiles (pollinators: bees, syrphids, butterflies, bombyliid flies, and hawk moths) have a large binocular "window" in the landing direction, or pointed towards the apex of the proboscis in hovering forms; the overlap is negligible in frontal and dorsal directions. Predators hunting amidst vegetation have a large

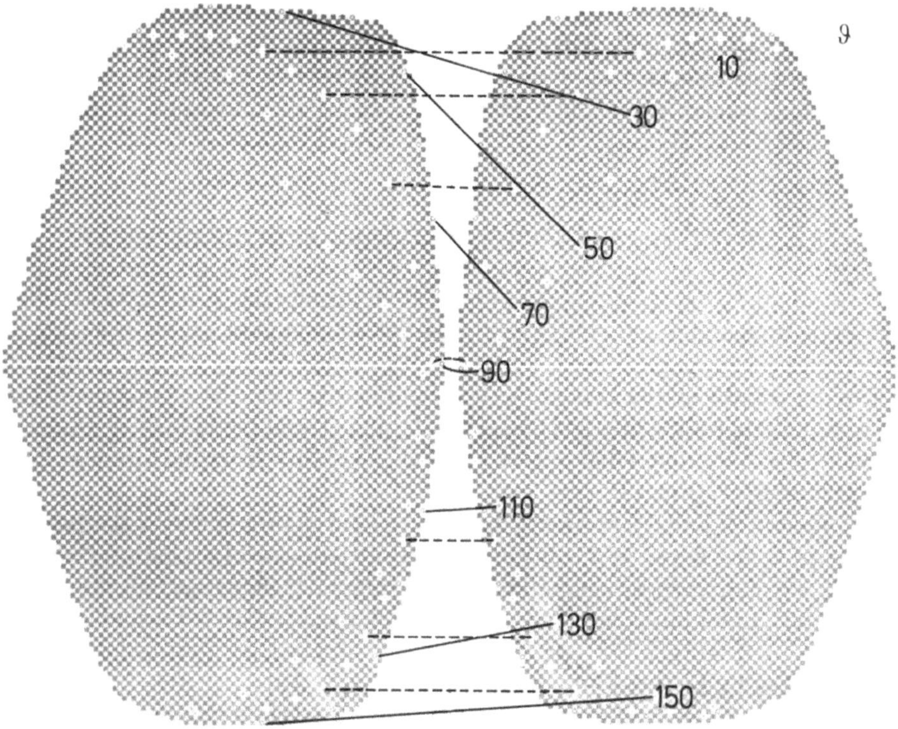

Fig. 33. Eye chart of the blowfly of Fig. 32. Each ommatidium is represented by a black square, except for 1) the ommatidia having an axis parallel to the vertical symmetry plane, which are indicated by a single dot, and 2) marginal ommatidia, indicated by a circle, of which the axis is parallel to that of a specific ommatidium in the opposite eye, namely one out of a series the ommatidia of which are indicated by a set of four dots, and which have axes diverging 10° from each other. These latter ommatidia mark the border line of the area of binocular overlap. The number of ommatidial axes in the overlap area is much higher dorsally compared to ventrally. Since binocular overlap is dorsally approximately the same as ventrally (Fig. 32), it appears that the interommatidial angles are much smaller dorsally (cf. BEERSMA et al., 1977)

binocular overlap in the direction of attack. About half or even a larger part of the eye can be involved. Binocular vision possibly provides the means to detect a prey against a complex background.

The actual mechanism of prey capture is far from being understood in detail. The possibility of stereopsis in insects must be generally discounted (BALDUS, 1926; FRIEDERICHS, 1931; WIGGLESWORTH, 1972). Intersection of ommatidial axes does not provide an accurate means of distance judgement (BURKHARDT et al., 1973). Since prey capture behaviour is triggered when certain ommatidia in the binocular region are activated by a moving object (*Aeschna* larvae: BALDUS, 1926; *Cicindela:* FRIEDERICHS, 1931; mantids: BARRÓS-PITA and MALDONADO, 1970, and LEVÍN and MALDONADO, 1970; bulldog ant: VIA, 1977), the neurons wired specifically to these special eye regions (STRAUSFELD, 1976a, b) may be assumed to relay the essential triggering signals. For a quantitative analysis of visually mediated behaviours, involving specialized eye regions, e.g., chasing (fly dorsal eye: COLLETT and LAND, 1975, KIRSCHFELD and WENK, 1976; dragonfly dorsal eye: EXNER, 1891, SHERK, 1978c), fixation (fly ventral eye: REICHARDT, 1972), *e*-vector orientation (ant dorsal eye: DUELLI, 1975), knowledge of the interommatidial angles is necessary. Methods to gain that knowledge are discussed next.

2. Visual Axes of Ommatidia; Interommatidial Angles

A number of methods can be applied to determine the visual axes of neighbouring ommatidia and consequently the interommatidial angles. Preferably, a luminous pseudopupil is used, e.g., in the way shown in Fig. 20. The visual axes then are defined by the axes of the beams radiating from the ommatidial facets. From the value of the interfacet distances D_h and D_v and the radii of curvature R_h and R_v the interommatidial angles $\Delta\varphi_h$ and $\Delta\varphi_v$ follow from the formulae given in Section B.VII (Fig. 12).

Of course, D_h and D_v are directly determined from photographs of the cornea. The values of R_h and R_v can be directly obtained also in the case of a spherical eye where $R=R_h=R_v$ and where the ommatidial axes are normal to the corneal envelope. Then R equals R_c, the radius of curvature of the corneal envelope (Fig. 18). A simple, optical method to determine R_c was described and applied to *Drosophila*, yielding $R_c=210\,\mu m$, by FRANCESCHINI and KIRSCHFELD (1971b).

A second method to determine R is to measure the distance from corneal to deep pseudopupil. In the case of *Drosophila* $R=180\,\mu m$ was obtained (FRANCESCHINI and KIRSCHFELD, 1971b). In the common dusky blue (Fig. 20) the distance corneal-deep pseudopupil is $R=600\pm100\,\mu m$. It follows from the photographs of Fig. 20 that the interfacet distances are $D_h=15.5\pm0.5\,\mu m$ and $D_v=10.2\pm0.4\,\mu m$ (cf. Fig. 12). Thus, one calculates for the interommatidial angles, $\Delta\phi_h=1.5\pm0.3°$ and $\Delta\phi_v=1.0\pm0.2°$. (The interommatidial angle $\Delta\phi=[\Delta\phi_h^2+\Delta\phi_v^2]^{\frac{1}{2}}\cong1.8°$ relates well with the applied microscope aperture of about 7°, since a row of about five ommatidia is shining.)

More accurately than from the depth of the deep pseudopupil, the interommatidial angle can be determined from the distances D_h^* and D_v^* of the beams which radiate from the facets at a level d above the cornea; then $\Delta\phi_h=(D_h^*-D_h)/d$ and $\Delta\phi_v=(D_v^*-D_v)/d$. For $d=600\,\mu m$ it results with $D_h^*=29.4\pm1.0\,\mu m$ and $D_v^*=17.2\pm0.8\,\mu m$ that $\Delta\phi_h=1.3\pm0.1°$ and $\Delta\phi_v=0.7\pm0.1°$ Fig. 20).

This method of interommatidial angle estimation has been introduced by KIRSCHFELD (1973b). By applying antidromic illumination (Fig. 21), he showed that frontally in the bee $\Delta\phi_h=1.6°$ and $\Delta\phi_v=0.9°$ (see the note in Sec. B.VII), or $\Delta\phi=[\Delta\phi_h^2+\Delta\phi_v^2]^{\frac{1}{2}}=1.8°$; EDRICH and VON HELVERSEN (1976) who applied the very same technique obtained dorsally in the bee a value $\Delta\phi=3.2°$ (Fig. 20).

The most elegant and sophisticated instrument for determining visual axes and interommatidial angles is FRANCESCHINI'S two stage pseudopupil telescope (FRANCESCHINI, 1975); the instrument and its achievements will not be described here, because the original paper should not be missed by any interested reader.

The only drawback of both KIRSCHFELD'S and FRANCESCHINI'S methods is that they are not widely applicable, since the existence of either a luminous pseudopupil (such as that of butterflies) or the possibility of antidromic illumination is necessary. However, the distinct luminance displayed by the rhabdoms of many compound eyes (FRANCESCHINI, 1977; see also the sections D. III.4, D. IV.4 and Fig. 40, 48) offers a still unexploited means to apply these methods to a large variety of animals, and especially onto completely intact eyes. The most widely applicable instrument for interommatidial angle measurement, yet, is the principal pseudopupil.

The method procedure is as follows: rotate the animal (or the photomicroscope) in fixed steps of a few degrees and take a photograph of the cornea at each step (BARRÓS-PITA and MALDONADO, 1970; HORRIDGE, 1977a, b). Since the ommatidium whose facet lens is located in the centre of the principal pseudopupil has its visual axis aligned with the axis of the microscope, one then knows the angles between the subsequent central ommatidia. Although this procedure is laborious, the method is a reliable way of mapping the visual axes lattice.

A slight uncertainty exists in the estimation of the centre of the principal pseudopupil. Due to the shape of the eye the principal pseudopupil can take a deformed elliptical or hexagonal form (EXNER, 1891; HORRIDGE, 1977b; see Fig. 25); the centre is then not accurately defined, especially in transition areas of the eye, e.g., at the margin of a fovea.

Recently, extensive interommatidial angle measurements where the principal pseudopupil was employed, were executed by SHERK (1977, 1978a, b, and c), especially in various dragonflies, and by HORRIDGE (1977a, b) in a large number of insects and crustaceans. SHERK'S studies clearly show that the distribution of visual axes and the specialization of certain eye regions is strongly coupled to the lifestyle of the animal. Dragonflies in their larval state can be distinguished according to their predatory behaviour in visual or tactile, and in active or sedentary. Larvae of all species show a concentration of visual axes in a binocular fovea and in a band along the horizontal meridian. This is more sharply defined in visual predators than in tactile predators. Sedentary species have fewer ommatidia viewing the ventral sectors of the visual field than viewing the dorsal sectors, while the most active species were approximately symmetric. In all species there is a distinct compromise between increased resolution in all directions, binocular overlap, and the size of the visual field, in order to match the prey capture and predator avoidance behaviour (SHERK, 1977).

An interesting constraint for dragonflies in the development of their eyes is that of being hemimetabolous insects. Ommatidia contributing to the fovea in an early instar are found displaced in a posterior direction at a later stage. Consequently, the benefit/cost analysis of eye construction must be extended over all larval instars including the adult stage (see SHERK, 1978).

It follows from the dragonfly measurements that in the specialization of certain regions of compound eyes the optical construction rules of Section B.VIII operate, e.g., see Fig. 34. The eye of larval dragonflies is strongly curved in the frontal fovea (see Fig. 13c).

That specialization of eye regions is widespread and always directly related to the animal's behaviour, is also forcefully argued by HORRIDGE (1977a, b), who furthermore showed that simultaneous measurement of interommatidial angles and facet diameters provides essential insight into the properties of compound eyes.

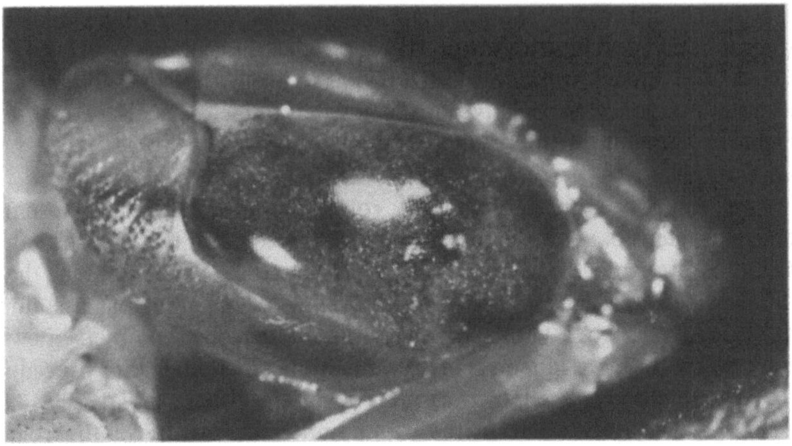

a

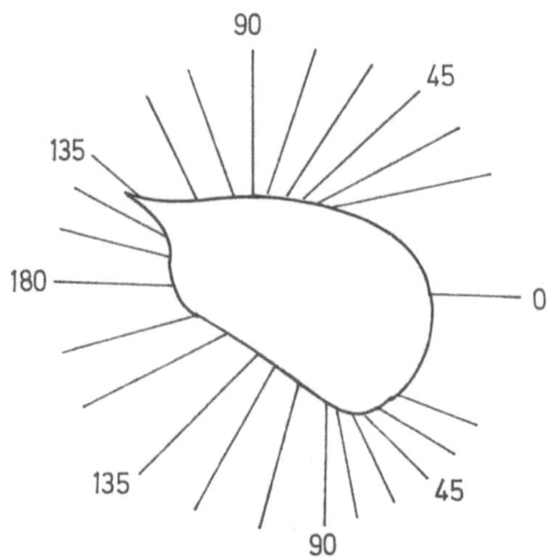

b

Fig. 34a and b. Eye shape and distribution of visual axes in larval dragonfly. A side view of the right eye of the larval *Anax junius* shows that it has a strongly curved eye frontally [(a) courtesy Dr. TRUMAN SHERK]. Pseudopupil measurements yield a visual axes organisation as drawn in (b) (modified from SHERK, 1977); the spacing of axes is 15°. It seems that a frontal fovea is constructed according to the design principle of (13c)

Another tool for measuring interommatidial angles is the dark pseudopupil observable in screening-pigmentless fly mutants (BURKHARDT et al., 1966; KIRSCHFELD, 1967). The eyes of these mutants are white coloured due to substantial scattering of light by the clear, pigmentless granules in the pigment cells. The comparatively optically homogeneous retinular cells scatter much less light, and consequently those retinular cells aligned with the observer give rise to a dark area in the eye. This marker of visual axes was used by BURKHARDT et al. (1966; see also BERNHARD et al., 1972).

More important than the latter method—pigmentless mutants are not abundantly present—is the one which employs the pupil mechanism that lies in the retinular cells. By applying axial illumination pigment granules inside the retinular cells are driven into the light path (Sec. D.III.4), and thus can be used as axis markers. BEERSMA et al. (1975) (see also STAVENGA, 1975a) measured the gradient in interommatidial axes over the housefly eye in this way.

As a last means to mark the ommatidial axes one might mention the oblique illumination pseudopupil. However, since eyes in which this type of pseudopupil can be perceived always display a much clearer principal pseudopupil, the latter marker will be generally preferred.

3. Eye Parameters, Effective Facet Diameter and Radius of Curvature of the Eye

The performance of the eye depends on: 1) the spatial resolution by the array of photoreceptor cells and 2) the light absorbed by the receptor. An important parameter in both aspects is the effective cross section of the facet lens which is often not identical with the anatomical cross section. Pigment cells can create an additional diaphragm in the dioptric apparatus, and the periphery of the facet lens is sometimes less transmittant than the central area. Another diminishing factor, particularly in lateral regions of the eye, is the skewness of the ommatidial axes to the eye envelope (Sec. B.VIII). Since a smaller cross section of the facet lens means a loss in both light capture and resolution, the effective cross section should not differ substantially from the achievable maximum.

The effective facet diameter can be estimated by photographing the cornea when a shining medium exists proximal to the lenses. That medium is provided in, for example, the case of flies, by the pupillary granules when they are assembled at the distal end of the visual sense cells as a consequence of bright illumination. Experiments on the housefly have shown an effective diameter of about 0.9 times the diameter of the inner circle of the facet in virtually all eye regions (BEERSMA, unpublished).

A similar difference to that encountered between anatomical and effective facet diameter exists between anatomical and effective curvature of the eye. The eye shape, or the anatomical curvature follows from the corneal reflections. More important for eye performance is the effective radius of curvature R given by $R = D/\Delta\phi$ (Sec. B). Yet, the crucial parameter is the eye parameter p (see SNYDER, this volume, Ch. 5), which comprises both acuity and signal to noise ratio of the photoreceptors. Although not explicitly stated in the original papers (SNYDER et al., 1977a, b), for the diameter D in the definition of $p = D\Delta\phi$ the effective diameter must be taken. However, in considering how compound eyes have to be designed for optimal functioning this is no essential difference. HORRIDGE (1977a, b) estimated p values for a large number of compound eyes. The theoretical prediction that conjugated to a certain average ambient luminance, there is a unique value for p at which eye performance is optimized was quite satisfactorily confirmed.

III. Photoreceptor Qualities

1. Definition of Photoreceptor

The following sections concern the use of pseudopupils in revealing the qualities of photoreceptors. Some definition is necessary here because an investigator's understanding of what constitutes a photoreceptor rather depends on the level where his research is concentrated. Sometimes the whole eye is considered to be the photoreceptor, but commonly an individual visual sense cell is thus thought of. Still, in the field of photoreceptor optics (SNYDER and MENZEL, 1975) the focus is upon the optics of the light-guiding structure of the visual sense cell, namely the rhabdomere (or a rhabdom, when rhabdomeres are fused together; in vertebrate vision the analogous structures are the rods and cones). Investigators working on a more basic level will regard the visual pigment molecule to be the photoreceptor, and finally the chromophore or the chain of atoms most determinant for the characteristic absorption process can be selected as being the photoreceptor.

It will be no surprise that pseudopupil techniques can be employed on virtually all levels mentioned. The first (whole eye) level has already been exemplified. The level of the visual sense cell is taken up in Section 4 after a discussion of the optical properties of rhabdom(ere)s and visual pigment processes.

2. Acceptance Angle

The spatial sensitivity of a visual sense cell is generally indicated by its acceptance angle, i.e., the half width of the angular sensitivity curve. This curve is determined from intracellular recordings by rotating a point source around the eye in small steps and then recording the response to a brief test flash at each step (e.g. WATERMAN, 1954; KIRSCHFELD and REICHARDT, 1964; WASHIZU et al., 1964; TUNSTALL and HORRIDGE, 1967). In apposition eyes, the directionality of the visual sense cells is due to both the dioptrics and the light acceptance properties of the rhabdom(ere) (see SNYDER, this volume, Ch. 5). Slight distortions of the subtle optical system during preparation and experimentation can easily result in too large values of the acceptance angle $\Delta\varrho$. Improvement of electrophysiological recording techniques often yielded narrower angular acceptance curves. For instance, TUNSTALL and HORRIDGE (1967) found in locust $\Delta\varrho = 6.6°$ and $3.4°$ for the dark and light-adapted state, respectively, while WILSON (1975) determined $\Delta\varrho$ values of 2.5° and 1.5°, respectively.

Careful recordings in bee (LAUGHLIN and HORRIDGE, 1971) have yielded $\Delta\varrho \simeq 2.5°$. This value agrees well with angular acceptance curves measured optically by EHEIM (1972) and EHEIM and WEHNER (1972) who illuminated slices of bee eyes with a rotating point source and measured the light flux leaving the cut end of a single rhabdom. However, a few critical remarks should be made. First, comparing data obtained in different experiments without precisely knowing from which area of an eye the data are taken is dangerous, because it is likely that substantial variations of acceptance angles over the eye exist; this is at least the case with the interommatidial angles (see Sec. D.II.2). Secondly, the optical system is prone to distortions, especially in eye slices. Thirdly, in the electrophysiological experiments acceptance curves are determined from the light absorbed by the visual molecules harboured within the boundary of the rhabdom, while in the optical experiments the transmitted light has been measured. Light propagation occurs partly inside and partly outside the rhabdom boundary due to waveguide effects, and the light distributions (modes) depend on the angle of incidence (SNYDER, 1975). A simple relation between absorption and transmission exists when the lowest order mode is propagated only; whether this is the case in bee rhabdoms has yet to be proven.

Combined optical and electrophysiological experiments are needed to fully understand the angular acceptance properties of visual sense cells in compound eyes. Optical acceptance angles can be measured in intact eyes in the case of butterflies using the luminous pseudopupil [1]. Figure 24 demonstrates that the eye shine from a facet lens, i.e., the amount of light propagated back and forth through its rhabdom, depends on the position of an illuminating point source. SWIHART (1974) performed some useful experiments on this phenomenon, although in general his way of reflecting on butterfly eyes must be treated with care (see STAVENGA, 1975c; STAVENGA et al., 1977; MILLER, this volume, Ch. 3).

Optical measurements of the angular dependence of the reflectivity in the superposition eye of skipper butterflies were performed by HORRIDGE et al. (1972). The close correspondence of the obtained data with angular acceptance curves determined with electrophysiological means is somewhat remarkable in view of photochemical experiments. One should expect that the light reflected from the eye and the light absorbed by the visual sense cells should have a very similar dependence on the angle of incidence only when the reflected light had traversed a substantial part of the visual sense cells. However, in skipper eyes very little photochemistry can be demonstrated by reflection experiments (unpublished observations). Much more comforting in that respect are diurnal butterflies, and certainly flies, of which the visual pigment processes will be described next.

3. Visual Pigment Processes

The properties of the visual pigments can be successfully investigated in intact compound eyes by microspectrophotometrical methods when the optics of pseudopupils are understood. This opportunity affords a valuable alternative to the standard extraction methods commonly applied in photochemical studies of visual pigments (see DARTNALL, 1972). The latter approach has its restrictions for the study of compound eyes, since the amount of visual pigment contained in one eye is relatively small, except for a few outstanding cases such as that of the neuropteran *Ascalaphus macaronius* and the sphingid moth *Deilephila elpenor* (GOGALA

[1] BURKHARDT et al. (1966; see BERNHARD et al., 1972) used the dark pseudopupil in the eye of the blowfly mutant chalky to determine the acceptance angle (called the opening angle) of an ommatidium. Since that pseudopupil is not only caused by the rhabdomeres but also due to the little scattering retinular cell bodies, those measurements are questionable. Similarly, the facet admission function determined by SHERK (1977) in dragonfly eyes is only a representation for the angles from which the black primary pigment can be seen and thus has little to do with the admission (or acceptance) of light by the photoreceptor cells.

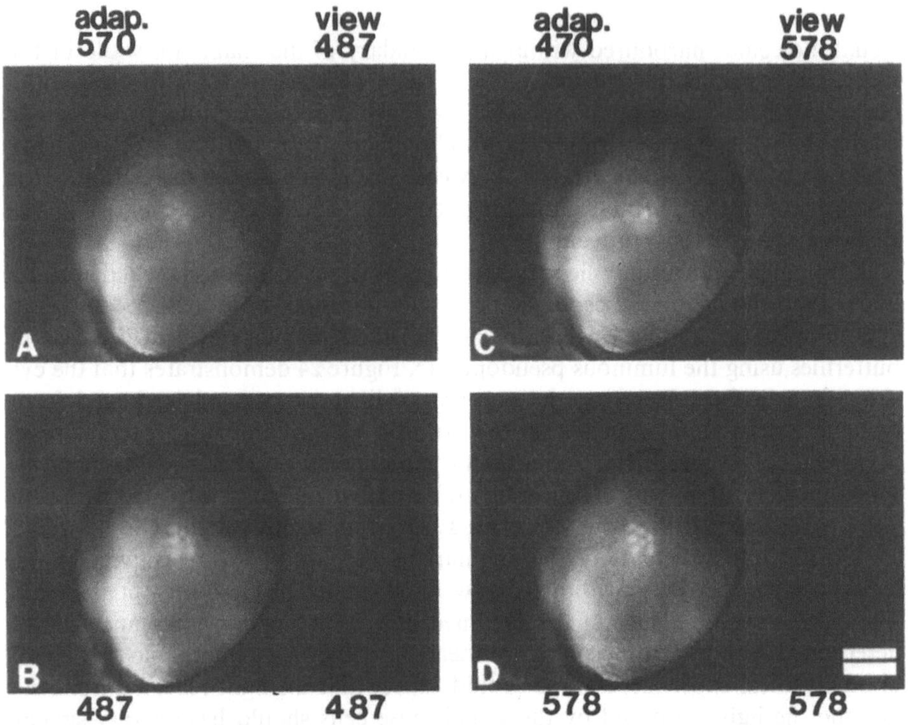

Fig. 35a–d. Deep pseudopupil of a *Drosophila* white-eyed mutant *(cnbw)* in two different stages of photoequilibrium of the visual pigments, established respectively by short wavelength (487 nm in (b), 470 nm in (c)) and by long wavelength adapting light [570 nm in (a), 578 nm in (d)]. These stages are photographed with antidromic illumination of 487 nm (a, b) and 578 nm (c, d) respectively. Since the peripheral retinular cells R 1–6 contain a visual pigment R 480–M 580 (OSTROY et al., 1974) long wavelength adapting light results in a high rhodopsin content, thus giving a low transmittance for the peripheral rhabdomeres in the blue (a) and a high transmittance in the yellow (d). Conversely, short wavelength adapting ligt yields a high metarhodopsin content, resulting in a high transmittance in the blue (b) and a lower transmittance in the yellow (c). The transmittance of the central rhabdomeres is high in the yellow and lower in the blue, probably due to the existence of pigments absorbing in the shorter wavelength range only (KIRSCHFELD and FRANCESCHINI, 1977; HARDIE et al., 1979; SMOLA and MEFFERT, 1979). Calibration mark 100 μm (figure by courtesy of Dr. WILLIAM STARK)

et al., 1970; HAMDORF et al., 1971, 1972, 1973; SCHWEMER and PAULSEN, 1973; HAMDORF, this volume, Ch. 4).

Microspectrophotometry of visual pigments in situ has proved to be a powerful addition to the standard procedures on extracts (rev. LIEBMAN, 1972). HAMDORF and LANGER (1965) were the first to explore this method in sections of eyes of the blowfly mutant chalky. The light-guiding properties of the rhabdomeres, clearly recognizable in eye slices, were revealed previously (DE VRIES, 1956; DE VRIES and KUIPER, 1958; KUIPER, 1962), and thus had strengthened the view that visual pigments are harboured in these structures. Indeed, HAMDORF and LANGER (1965) obtained evidence for photochemical changes when monitoring the light flux transmitted through the rhabdomeres. Subsequent to these early experiments LANGER and THORELL (1966) measured vertebrate-rhodopsinlike absorp-

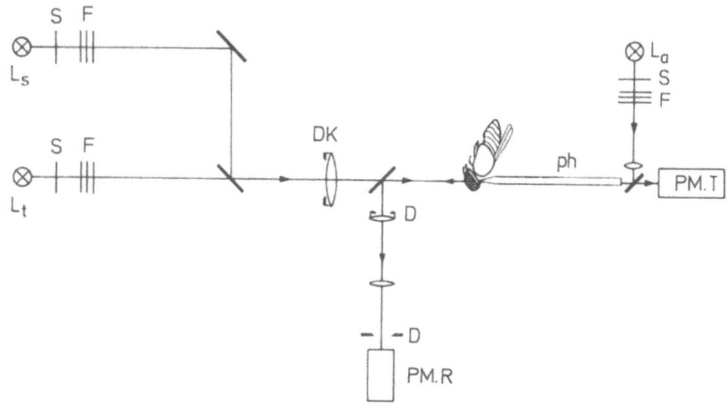

Fig. 36. Set-up for investigation of the optics of compound eyes. Orthodromic test and adapting beams are focussed on the eye by the condensor K. Light reflected and back-scattered from the eye is measured via a half-silvered mirror and a microscope by a photomultiplier (PM.R), which can be replaced by a camera. Light transmitted by the eye is guided by a quartz rod (ph), of which the naked tip is inserted in the black of the head, to a second photomultiplier (PM.T). Antidromic light can be applied through the very same quartz rod and detected with the aid of the microscope and the first photomultiplier. *D*, *S*, and *F* indicate diaphragms, shutters and filters; L_t, L_s, and L_a are lamps (from STAVENGA and KUIPER, 1977)

tion curves, which, however, appeared to represent mixtures of rhodopsin and metarhodopsin after the first microspectrophotometrical experiments on blowfly visual pigments performed *in vivo* by STAVENGA et al. (1973).

In order to exemplify the actual way of microspectrophotometry on an eye of a living fly, the series of photographs of Fig. 35 is included here. A *Drosophila* white-eyed mutant (*cn bw*) was illuminated with weak antidromic light of (view) wavelength 487 nm (Fig. 35a, b) and 578 nm (Fig. 35c, d) respectively. Simultaneously, intense, orthodromic adapting light was applied having a long (Fig. 35a, d) or a short wavelength (Fig. 35b, c). It appears that the transmittance of the various receptor types R1–6 vs R7, 8 depends on both the adapting and the viewing wavelength.

Quantitative experiments on these phenomena have been executed by STAVENGA (1976). The antidromic transmission from the rhabdomeres of different retinular cell types was measured in the deep pseudopupil by the set-up of Fig. 36. By establishing various photoequilibria with monochromatic illumination, difference spectra could be obtained from which the absorption spectra of the visual pigment states could be calculated, since the pigment is a photochrome (STAVENGA, 1975d; for related methods to calculate visual pigment spectra from electrophysiological measurements, see STARK and ZITZMANN, 1976 and TSUKAHARA and HORRIDGE, 1977). It was thus concluded that the rhodopsin of blowfly visual sense cells R1–6 absorb maximally in the blue green, $\lambda_{max} \simeq 495$ nm, and that this native visual pigment state is photointerconvertible with a yellow-absorbing metarhodopsin with $\lambda_{max} \simeq 580$ nm (STAVENGA 1974a, 1976); the visual pigment states are accordingly indicated by R495 and M580 respectively.

Essentially the same result was obtained by spectrophotometry of blowfly eye slices performed by SCHWEMER (see HAMDORF et al., 1973; HAMDORF and ROSNER,

1973; ROSNER, 1975; RAZMJOO and HAMDORF, 1976; for related experiments in *Drosophila*, see OSTROY et al., 1974, and PAK et al., 1976). In these latter experiments, difference spectra are determined with light more or less randomly passing through the retina, which is, however, certainly not a homogeneous medium. Distortions of the spectra due to wavelength dependent scattering are, then, a serious possibility. (For a discussion on inherent advantages and disadvantages of the various methods of spectrophotometry, see LIEBMAN, 1972.)

Microspectrophotometry of individual rhabdomeres was performed by KIRSCHFELD et al. (1977) in eye-cup preparations of the housefly *Musca* (white-eyed mutant). The obtained difference spectra revealed that the peripheral photoreceptors contain a visual pigment with stable states R490 and M580. In view of the inaccuracy of the difference spectra it cannot be concluded that blowfly and housefly have different visual pigments.

Still, those microspectrophotometrical experiments which rely on the light guiding-properties of fly rhabdomeres have another distorting effect to account for, namely the effect of the boundary wave. The light flux propagated by the rhabdomere is only partly conducted inside its boundary and only this fraction (which depends on the wavelength of the light), gives rise to the absorption spectrum of the visual pigment. However, it is probable that the correction factor with which to obtain a realistic visual pigment spectrum depends only slightly on wavelength (see STAVENGA and VAN BARNEVELD, 1975; STAVENGA, 1976); actual values for this correction factor depend on the rhabdomere dimensions, the refractive index of the rhabdomere (STAVENGA, 1974b, 1975e; KIRSCHFELD and SNYDER, 1975, 1976) and that of its surroundings (SEITZ, 1968; BEERSMA, 1979).

The white-eyed mutants used in the experiments mentioned above have a major advantage, namely, the lack of photostable screening pigments in the eye. Normal wild type flies possess pigment granules inside the visual sense cells. These granules can gravely influence the transmittance of the rhabdomeres. In fact, the assembly of granules acts as a pupil. Preluding the account on the pupil mechanism (Sec. D.III.4) Fig. 37 presents an experiment on a wild type blowfly where both the photoconversions and the pupil effect are apparent.

The influence of the pupil mechanism can be by-passed by working outside its intensity and/or temporal range. It was thus shown (wild type blowfly) that the rhodopsin state is regenerated from the metarhodopsin state in the dark with a half time of about 25 min at room temperature (STAVENGA et al., 1973). For a number of cases of dark regeneration of visual pigments in other invertebrates, see STAVENGA (1975f) and BRUNO et al. (1977). Furthermore, it was found that the visual pigment states of the drone fly *Eristalis* are: rhodopsin R 460 and metarhodopsin M550; both shifted towards shorter wavelengths some 30 nm in respect of the corresponding blowfly and housefly visual pigment states (STAVENGA, 1976). Flies thus generally seem to have a similar astonishingly huge bathochromic shift of the metarhodopsin with respect to the rhodopsin state as was first discovered in the UV-visual pigment of *Ascalaphus*, which has the stable states R345 and M475 (HAMDORF et al., 1973).

In this respect a quite interesting case is worth mentioning, namely that of the male hoverfly *Syrphus balteatus*. The peripheral retinular cells in the dorsal part of the eye contain a visual pigment different from that of the homologous cells in the

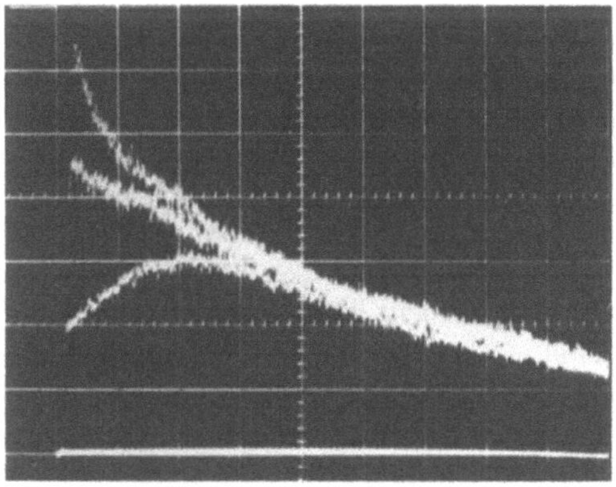

Fig. 37. Orthodromic transmission of the eye of a wild type blowfly *Calliphora erythrocephala* showing photochemical and pupil effects. Green light of wavelength 550 nm was applied after a dark adaptation time of 2 min. Preceding this, the eye was irradiated with light of wavelength 457, 550, and 603 nm yielding the lower, middle, and upper curve respectively. In the first second, photochemical effects are superimposed on those of the closing pupil. The preceding illumination of 550 nm is sufficiently intense to establish a photochemical equilibrium. Hence, when the subsequent illumination is at 550 nm no shift of the photochemical equilibrium occurs; only a pure pupillary response is induced, i.e., a drop in transmission due to the congregation of light-absorbing and scattering pigment granules against the light-guiding rhabdomeres. The same effect on the pupil occurs through the 550 nm illumination after the previous blue (457 nm) and red (603 nm) light. However, superimposed on the pupil effect is a shift of the photochemical equilibrium. The initial rise in transmission induced by the 550 nm light after the 457 nm light means a decrease in absorption at 550 nm; this is due to a conversion of a substantial fraction of metarhodopsin M 580 into rhodopsin R 495. On the other hand, after the 603 nm light virtually all the visual molecules are in the rhodopsin state. The subsequent 550 nm light establishes a moderate fraction of metarhodopsin resulting in an increase in absorption at 550 nm, i.e., an initial decrease in transmission. (The absorption of M 580 exceeds that of R 495 at 550 nm.) Note that the time constant of the pupil is in the order of 1 s

ventral part of the eye, as follows from *in vivo* difference spectra (STAVENGA, 1979). The dorsal visual pigment approximates the characteristics of the pigment generally found in the blowfly: R 495–M 580. The ventral visual pigment approximates that of *Eristalis*: R 460–M 550. In females, however, in all eye parts the same *Eristalis*-like visual pigment is found in the peripheral retinular cells (see also BERNARD and STAVENGA, 1977). It may be hypothesized that the divided eye features (differences in eye structure and in screening-pigmentation between upper and lower eye parts), which are noticeable in males of many fly genera (DIETRICH, 1909; KIRSCHFELD and WENK, 1976), are expressed in the properties of the visual pigments also.

That fly peripheral retinular cells R 1–6 in addition to the blue-green-absorbing rhodopsin contain a UV-visual pigment was proposed by HORRIDGE and MIMURA (1975) in order to explain the repeatedly observed high UV-peak in fly spectral sensitivity curves as well as anomalous polarization sensitivity measurements. The question of the high UV sensitivity recently became even more intriguing due to the discovery by KIRSCHFELD and FRANCESCHINI (1977) of photostable pigments in fly rhabdomeres; KIRSCHFELD et al. (1977) proposed that the photostable pigment molecules absorb light quanta and then transfer the energy to the visual pigment molecules (see also STARK et al., 1977, and KIRSCHFELD, 1979).

Microspectrophotometry of single rhabdomeres of the central cell R 7 *(Musca)* revealed a UV-visual pigment like that of *Ascalaphus* (KIRSCHFELD and FRANCESCHINI, 1977), in agreement with electrophysiological measurements on *Drosophila* (HARRIS et al., 1976; STARK, 1977). However, the majority of R 7 cells additionally harbour a photostable pigment, probably a carotene. This pigment can have various functions: 1) spectral sensitivities and dichroic sensitivities can be affected in both R 7 and R 8, 2) the pigment might act also in this case as a sensitizing pigment, and 3) the fluidity of the photoreceptor membrane might be modified (KIRSCHFELD and FRANCESCHINI, 1977; KIRSCHFELD, 1979). Rapid progress in revealing the characteristics of the central retinular cells of flies is recently achieved, notably by genetical methods (*Drosophila:* HARRIS et al., 1976; STARK, 1977) and by intracellular recordings (*Eristalis:* BISHOP, 1974, HORRIDGE et al., 1975; *Calliphora:* BURKHARDT, 1962, MEFFERT and SMOLA, 1976, HARDIE, 1977, HARDIE et al., 1979, SMOLA and MEFFERT, 1979). For discussions on the spectral filtering effect by the photostable pigment in the superior central retinular cell R 7, inducing the green-shifted sensitivity spectrum of the inferior cell R 8, see KIRSCHFELD et al. (1978) and HARDIE et al. (1979).

In relation to this, the most startling recent experiments on the fluorescence of fly rhabdomeres have to be taken into account. In *Musca* (white) all R 1–6 show a red fluorescence (λ_{max} between 600 and 650 nm) in all parts of the eye under blue excitation (400–500 nm) and a less saturated red fluorescence under UV excitation (300–400 nm). Some R 7–8 behave in this way too, but the great majority appear either green (70%) or black (30%) under blue excitation, whilst they are always black under UV excitation. The colours observed are interpreted to be the characteristic fluorescences of the various rhodopsins and metarhodopsins (FRANCESCHINI, 1977) or that of the photostable pigment (STARK et al., 1977).

It must be noted here that the heterogenous population of central rhabdomeres cannot properly be studied in the deep pseudopupil with a wide aperture microscope. The superposition of the images of rhabdomeres from different ommatidia will blend the different properties. Optical neutralization, i.e., nullifying the effect of the dioptric apparatus by a medium with a proper refractive index (immersion oil or nail polish) yields individually observable rhabdomeres (FRANCESCHINI and KIRSCHFELD, 1971 a, 1976; FRANCESCHINI, 1975).

For the study of visual pigment processes in intact eyes with the aid of pseudopupils, the ideal animal is perhaps the fruitfly *Drosophila.* Here the head capsule is rather transparent, and antidromic illumination can be successfully applied after insulting the animal no more than putting it "comfortably" in a well-fitting holder (FRANCESCHINI and KIRSCHFELD, 1971 a, b, 1976; FRANCESCHINI, 1975). Students of visual mutants have intelligently enriched their tool box with the optical techniques developed by the investigators mentioned, and so have been able to add substantially to the present-day understanding of the fly visual system (for details, see HARRIS et al., 1976; PAK and PINTO 1976; HARRIS and STARK, 1977).

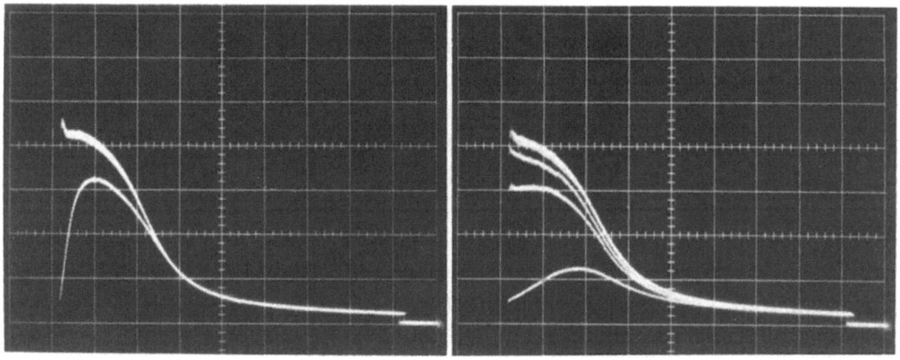

Fig. 38 a and b. Photochemical processes and pupillary response measured in reflection from the eye of a Red Admiral *Vanessa atalanta* (see Fig. 29). Blue light of 488 nm was applied orthodromically and the reflection was measured in the deep pseudopupil from the central spot, the luminous pseudopupil. Generally, light adaptation causes a drop in reflection due to the effect of the pupillary granules in the retinular cells [(a, b) upper curves]. Initially, photochemical processes are traced because the eye was pre-illuminated with UV light. Blue-absorbing metarhodopsin was thus created, being reconverted into the blue transparent, UV-absorbing rhodopsin state by the 488 nm illumination. In (a) intensity is ten times higher than in (b). Since photochemical processes are linearly dependent on intensity, conversion proceeds at a much slower rate in (b) and repeated illuminations (separated by 1 min darkness to let the pupil relax) are necessary to reach the photoequilibrium at 488 nm (sequence from lower to upper curve). The speed of the pupil hardly depends on light intensity (from STAVENGA et al., 1977)

This long excursion in the area of fly visual pigments is a reflection of the extensive achievements and the quick developments made in this specific insect order, not in the least due to the pseudopupil methods. Visual pigments of other invertebrates have been studied by microspectrophotometry on retinal tissues or isolated rhabdoms (see GOLDSMITH, 1972, HAMDORF, this volume, Ch. 4, for reviews), but there is only one other example—butterflies—where visual pigments in the eyes of completely intact living animals have been studied.

A first approach on the photochemical processes in butterflies using the luminous pseudopupil was undertaken by STAVENGA (1975c, f). Evidence for a green-absorbing rhodopsin having a blue-absorbing metarhodopsin was obtained, as was evidence for a UV rhodopsin with a blue-absorbing metarhodopsin (STAVENGA et al., 1977) (see Fig. 38). Intermediate states living on the order of several tens of milliseconds are clearly present in butterflies (STAVENGA et al., 1977), and also in hymenopterans (STAVENGA and KUIPER, 1977). Similar intermediates were not detected in the blowfly (STAVENGA, 1976; MUIJSER and STAVENGA, 1979).

A thorough investigation of butterfly visual sense cells was undertaken recently by BERNARD (1977, 1979). By measuring the reflection from the luminous pseudopupil and thus obtaining difference spectra and action spectra of the pupillary response (Sec. D.III.4) he made the startling discovery of a red-absorbing visual pigment of which the spectrum peaks at about 610 nm. This λ_{max} value is only slightly less than that of the pigment in red sensitive goldfish cones, which is the longest wavelength-absorbing visual pigment presently known (LIEBMAN, 1972; LYTHGOE, 1972). The sensitivity of the butterfly red receptor is usually enhanced by the action of a red-reflecting tapetum proximal to the rhabdom. A pronounced red sensitivity of butterflies has been known for sometime from behavioural experiments (for references see BERNHARD et al., 1972); this fact now can be understood at the level of the photoreceptors (BERNARD, 1979).

It might be useful to reiterate here that butterfly eyes have ommatidia with fused rhabdoms. Visual pigment studies therefore cannot be executed on single cells or single cell types. Flies are outstanding in this respect because they have an open rhabdom. The fusion of rhabdomeres into one light-guiding rhabdom may be a special design for enhancing colour contrast by lateral filtering (SNYDER et al., 1973). Selective filtering will occur in tiered rhabdoms such as those in the eyes of dragonflies (HORRIDGE, 1969), where rhabdomeres of different cell types are stratified. An effective design for sharpening sensitivity spectra generally will be to stack short wavelength-absorbing pigments distally and subsequently to put the visual pigments which absorb at longer wavelengths into more proximal layers. Curiously enough, EGUCHI (1971) states that in the dorsal eye of the dragonfly *Aeschna*, the green receptors are located distally and the UV receptors are situated proximally. Whether this is indeed the actual organization has to be answered by microspectrophotometry of dragonfly rhabdoms in eye cup preparations. Some care has to be taken, then, to avoid contamination with the photomechanical adaptation processes surveyed below.

4. Pigment Migration in Retinular Cells

Migration of pigment granules inside the retinular cells of compound eyes as a function of light/dark adaptation were demonstrated in a wide variety of insect orders and in several Crustacea as well (for a list of references see MAZOKHIN-PORSHNYAKOV, 1969; FRANCESCHINI and KIRSCHFELD, 1976). Classically, pigment migration is demonstrated by histological methods. The optical techniques dealt with in this chapter have provided the means of following the processes in the living eye. The latter line of research, although started almost a century ago (EXNER, 1891) only recently gained momentum with KIRSCHFELD and FRANCESCHINI'S (1969) investigation of the migration of pigment granules inside the visual sense cells of the housefly (see also KIRSCHFELD, 1969). The present state of knowledge of this pigment migration system in flies will be surveyed in this section together with related work on Hymenoptera and butterflies.

KIRSCHFELD and FRANCESCHINI (1969) demonstrated that in the peripheral retinular cells of houseflies small pigment granules are dispersed in the dark in the sense cell soma, and that during the first few seconds of light adaptation these granules move towards the rhabdomere, so inducing a decrease of the light flux transmitted through the rhabdomere. Thus, the absorption of light by visual pigment molecules located in more proximal parts of the rhabdomere is diminished. Parallel to the decrease in transmission an increase in reflection occurs, i.e., an increase in scattered light from the pigment granules which assemble in the path of illumination. Electronmicroscopy by BOSCHEK (1971) fortified the optical findings.

The name pupil (FRANCESCHINI, 1972b) is justified because the functional analogy to the pupil mechanism of the human eye is close. FRANCESCHINI (1972a, b, 1975) has provided firm evidence that the pigment migration system of flies acts as an automatic gain control mechanism. The details of the system are far from clarified, but one important point, the nature of the absorbing pigment, was studied by FRANCESCHINI and KIRSCHFELD (1976) in the fruitfly *Drosophila*. Normally the pigment granules in the retinular cells contain only ommochromes; the other class of absorbing pigments of fruitfly, that of the pterines, is absent in these granules (FUGE, 1967). Strengthened with this knowledge FRANCESCHINI and KIRSCHFELD studied mutants in which the synthesis of ommochrome pigments is selectively blocked. They observed that these mutants do not show up pseudopupil changes upon light/dark adaptation. The absence of an optical effect is probably due to an absence of the pigment granules in those mutants (FRANCESCHINI and KIRSCHFELD, 1976).

Quite revealing are FRANCESCHINI'S (1972a, 1975) findings on the *Drosophila* mutant white apricot, which completely lacks the drosopterines. The effective absorbance spectrum of the pupil which peaks in the blue–violet matches very well the absorbance spectrum of the ommochrome containing pigment of the pigment cells measured in situ. An elegant accessory outcome of these experiments is the immediate proof that the main action of the pupil granules is to absorb light from the boundary wave (FRANCESCHINI and KIRSCHFELD, 1976) and not particularly to impair the light-guiding properties by diminishing the refractive index difference between rhabdomere and surrounding medium (cf. BERNHARD et al., 1972; KOLB and AUTRUM, 1972; MILLER, this volume, Ch. 3). In the blowfly *Calliphora* the pupil absorbs maximally in the blue at about 480 nm (STAVENGA et al., 1973). This value does not relate particularly well with LANGER'S (1975) microspectrophotometry of eye sections which yielded a photoreceptor cell pigment absorbing maximally at about 420 nm. LANGER (1975) suggests from these data that the visual sense cell pigments are pteridines, but apparently further research is needed here.

The spectral absorption characteristics of the pupil are strikingly attuned to the visual pigment properties. The high absorption in the blue prevents rhodopsin from excessive conversion at bright light levels, while the steep fall in absorption towards the yellow leaves relatively unaffected those wavelengths which are most effective in photoregenerating rhodopsin from the created metarhodopsin. In other words, a closing pupil both diminishes the total light flux in the rhabdomeres and shifts the photoequilibrium considerably towards the side of the rhodopsin state (STAVENGA et al., 1973). This might be of vital importance for the functioning of the photoreceptor cell because excessive conversion of rhodopsin can result in a paralyzed state of the cell which can only be undone by long wavelength light (STAVENGA, 1974a; MINKE et al., 1975; MUIJSER et al., 1975; STAVENGA et al., 1975; see further Sec. D.IV.3 and the treatise on the prolonged depolarizing afterpotential (PDA) by HAMDORF, this volume, Ch. 4).

Interestingly, the pupil granules scatter light quite substantially in the long wavelength range (Fig. 39). (However, the relative intensity of the scattered light is probably too small to add to the photoregeneration effect.) The reflection spectrum of the green pupil of *Drosophila* peaks at 540 nm (FRANCESCHINI, 1975;

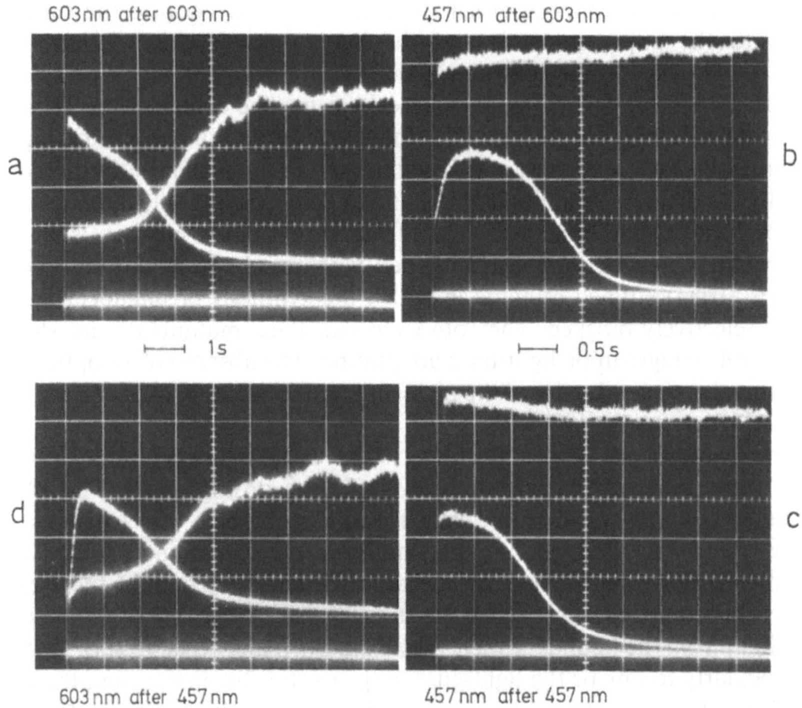

Fig. 39a–d. Simultaneous recordings of transmission and reflection of the eye of the blowfly *Calliphora erythrocephala* demonstrating the effect of the pupil and visual pigment photochemistry. Orthodromic illumination was applied during 0.5 min after a darkness time of 2 min. Light adaptation causes the migration of the retinular cell pigment granules towards the rhabdomeres where they absorb and scatter light thus yielding a decrease in the transmission, which is moderate in the red (a) and substantial in the blue (c). When the eye was previously irradiated with a different wavelength (b, d), transients in the transmission curve reveal photochemical processes in the visual pigments. These transients are noticeable in reflection (d). However, the large rise in the red reflection (a, d) is due to an increase in scattering of the pupillary pigment granules. In the blue reflection, curves are complicated because in addition to an increase in reflected light from the pigment granules, their absorbing action on light reflected from within the rhabdomeres also becomes important; cf. Fig. 41 (from STAVENGA, 1975a)

FRANCESCHINI and KIRSCHFELD, 1976). The yellow granules in *Musca* (KIRSCHFELD and FRANCESCHINI, 1969) and *Calliphora* (unpublished) yield a reflection peak near 580 nm. The relative shift of the absorbance and reflection spectra can largely be understood from the phenomenon of anomalous dispersion (FRANCESCHINI, 1972a, 1975; FRANCESCHINI and KIRSCHFELD, 1976).

Both absorption and reflection are also quite noticeable in the rhabdomeres. The absorption is, of course, due to the visual pigments. The reflection from the rhabdomeres (see FRANCESCHINI and KIRSCHFELD, 1976, Fig. 3a and d, for photographs of the deep pseudopupil of *Drosophila* and *Musca*) is possibly in part due to fluorescence (FRANCESCHINI, 1977). Visual pigment conversions are encountered in both transmission and reflection (Fig. 39). Furthermore, that reflection occurs

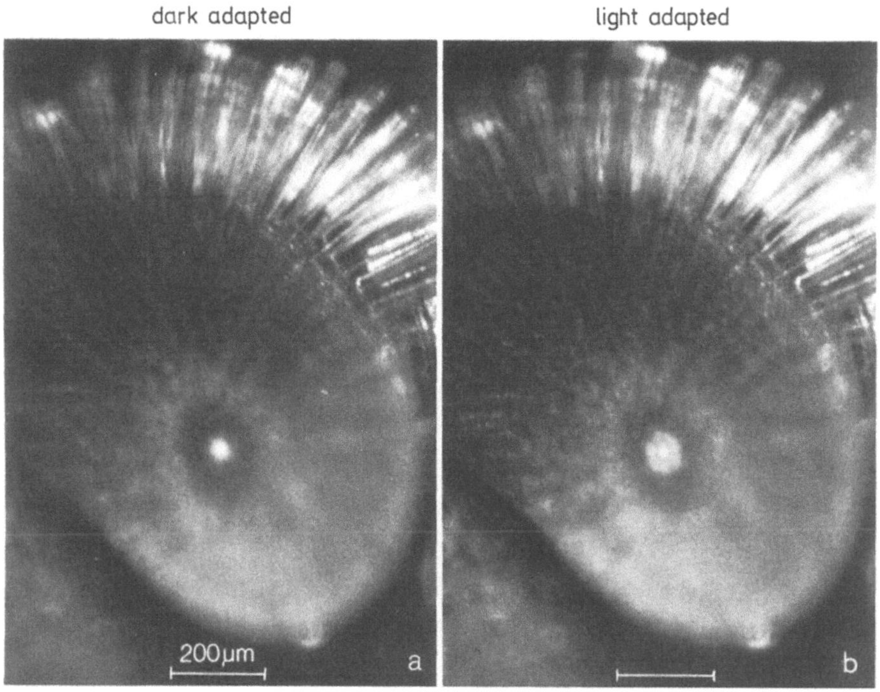

Fig. 40a and b. The deep pseudopupil in the ventral region of the eye of the worker bee *Apis* under orthodromic (axial) illumination (from FRANCESCHINI, 1975). In the dark-adapted eye purplish light is reflected from within the rhabdom (a). This reflection is substantially diminished by the action of light-absorbing and scattering pigment granules present inside the retinular cells which surround the rhabdom during light adaptation (b). In the case of butterflies, the reflection from the rhabdoms is much more intense at wavelengths where the basal tapetum is an effective reflector (compare Figs. 28, 29, and 43). The reflection from the rhabdoms in the bee eye is probably due to fluorescence and inhomogeneities existing inside and/or near the rhabdoms

within the rhabdomere is clear from the reduction in reflected blue light induced by a closing pupil (KIRSCHFELD and FRANCESCHINI, 1969; STAVENGA, 1975a).

The absorption and reflection phenomena for flies described so far are basically the same in the eyes of hymenopteran insects. The photographs of Fig. 40 will help to clarify this claim. The deep pseudopupil of the eye of a worker bee has been photographed with orthodromic (axial) illumination. The ventral region of the eye was selected because in that region the disturbing effect of corneal reflections is most easily avoided (see Sec. B.VIII). In the deep pseudopupil one observes the superimposed image of (approximately) the focal plane in adjacent ommatidia, i.e., of the plane of the rhabdom tip. In the centre of the deep pseudopupil one thus observes the light reflected back from the rhabdom. Actually, with the term reflection all kinds of observable light are lumped. The dominant contribution may be from fluorescence, but also backscattering will occur (STAVENGA and KUIPER, 1977). This reflection from the rhabdom is substantial in the dark-adapted state (Fig. 40a), but diminishes when absorbing retinular cell pigment

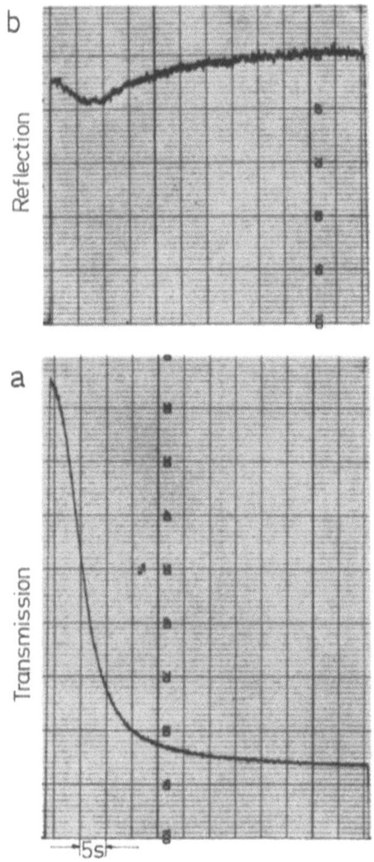

Fig. 41 a and b. Simultaneously recorded orthodromic transmission and reflection from the wasp *Paravespula germanica.* Intense white light is applied for 1 min after 1 min dark adaptation time. Transmission falls monotonically due to light loss from the rhabdom caused by retinular cell pigment granules which assemble against the rhabdom (a). A similar effect occurs with the reflection from the rhabdom ((b), initial phase). The scattering properties of the pigment granules are responsible for the rise in reflection ((b), second phase) (from STAVENGA and KUIPER, 1977)

granules cluster against the rhabdomere of their cell during light adaptation (Fig. 40 b, cf. Fig. 2). Histological evidence for this pigment migration has been documented for bee (KOLB and AUTRUM, 1972) as well as for ant (MENZEL and LANGE, 1971; MENZEL, 1972a; BRUNNERT and WEHNER, 1973).

Experimental recordings demonstrating the dynamic course of the adaptation process are presented in Fig. 41. Figure 41 a shows that light adaptation of the eye of a wasp involves a decrease in transmission (see the case of flies above). The reflection course is biphasic; the initial fall is the decrease in light emitted from the rhabdom, and the later rise is due to the increasingly scattering retinular cell pigment (STAVENGA and KUIPER, 1977). Only the latter component of the reflection curve is observable in recordings of a bumblebee (Fig. 42). Once more, the double-sided effect of the migrating pigment granules during light adaptation is

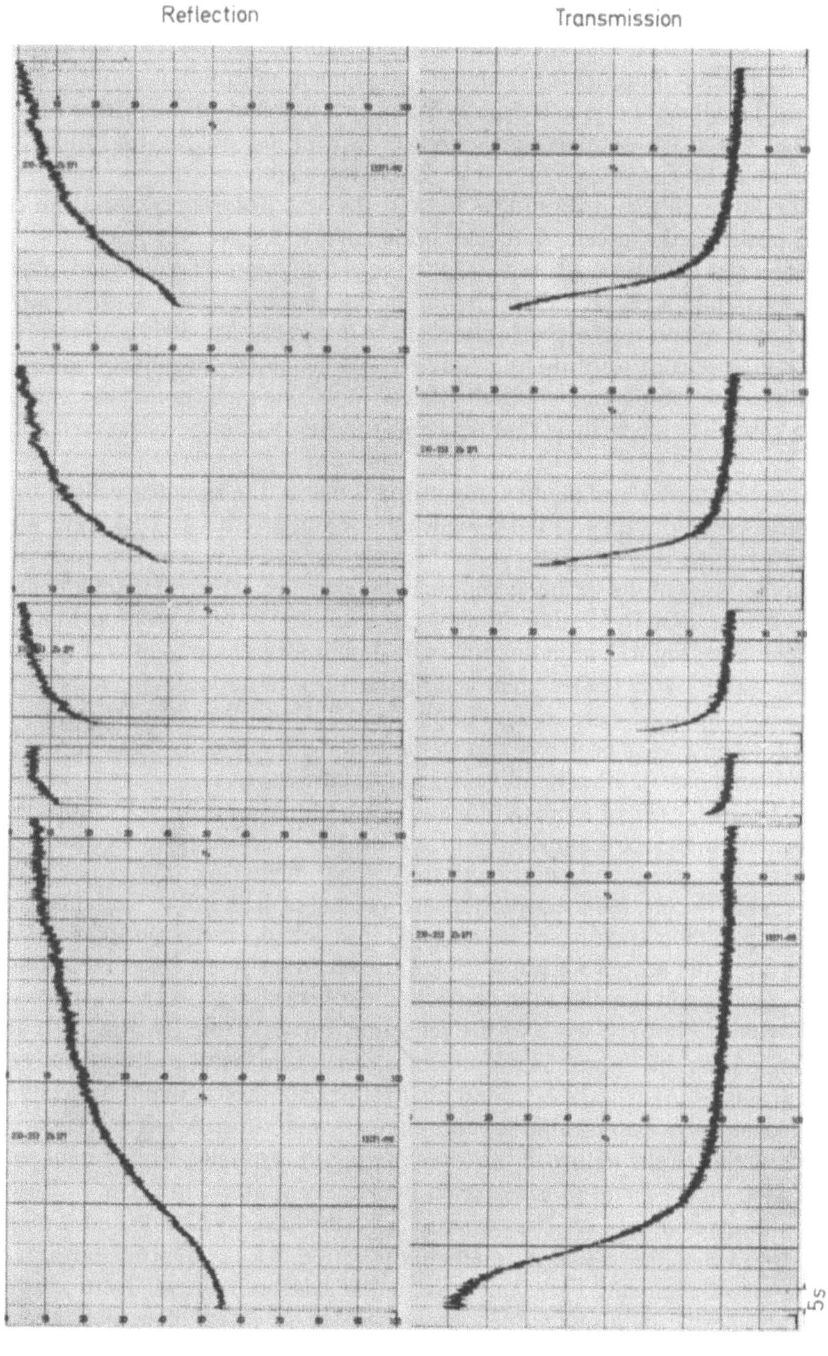

Fig. 42. Simultaneous recordings of transmission and reflection of the bumblebee *Bombus pascuorum* under white orthodromic illumination applied after 1 min dark adaptation. A drop of the signals to zero occurs during intercalated dark times of 2.5, 5, 10, and 15 s. The recordings show that the pupil in hymenopteran photoreceptor cells closes and opens again with a time constant of 10–15 s (from STAVENGA and KUIPER, 1977)

demonstrated, i.e., the decrease in transmission (by an increase in absorption) and an increase in reflection; the effects are the opposite of those occurring during dark adaptation. Because the retinular cell pigment reflection is so vivid in bumblebees, presumably due to a slightly better transparency of the primary pigment through which the retinular cell pigment is seen (cf. Fig. 2), the initial phase apparent in the wasp experiment (Fig. 42) is overruled.

In relation to this it should be noticed that the absorbance spectrum of the pupil peaks in the green. Still, the peak hardly deserves its name; the pupil spectrum lies as a broad cover over the spectral range of the visual pigments (STAVENGA and KUIPER, 1977). The rather flat shape conforms with spectra obtained with in situ microspectrophotometry (e.g., LANGER, 1975). A steep fall of the pupil spectrum towards long wavelengths would be hazardous indeed, because green visual pigments usually convert into blue metarhodopsins and therefore a specialization such as that developed in flies would be of no advantage at all.

The third example of pigment migration is that in the retinular cells of diurnal butterflies. Photographs of dark- and light-adapted butterfly eyes were already presented above (Figs. 28 and 29). In the nymphalids *Precis* and *Vanessa*, only the absorbing effect of the retinular cell pigment granules, which accumulate around the rhabdom upon light adaptation, is conspicuous. In the Question Mark (also Nymphalidae; Fig. 43) the extinction of the luminous pseudopupil is also obvious, but the scattering by the retinular cell pigment is now apparent from the annulus in the light-adapted deep pseudopupil. That such a clear annulus is generally difficult to observe is probably due to the very dense pigment sleeve surrounding the proximal part of the crystalline cone in butterfly eyes.

The pigment cells become gradually more transparent towards longer wavelengths and therefore scattered light from the retinular cell pigment granules can be measured rather easily in the infrared. Experimental recordings of the reflection course during light adaptation are presented in Fig. 44. The effect of the absorbing pupil is clear from the steep fall in reflection in the blue and yellow. In the infrared the scattering pupil causes the second, rising phase. Note that the latter curve is quite similar to the reflection curve measured in the wasp (compare Fig. 44c with Fig. 41b; see also the reflection curves of the fly, Figs. 39b and c).

Comparing the optical measurements of the three insect orders, flies, hymenopterans and butterflies, it becomes obvious to hypothesize that a basically similar mechanism is responsible for the pigment migration. A crucial question, of course, is that of the nature of the force driving the granules. The present state of discussion on the problem is reviewed by MILLER (this volume, Ch. 3). It is clear from the experiments on *Drosophila* (FRANCESCHINI, 1972a, b) and *Calliphora* (STAVENGA et al., 1975) that the pigment migration directly depends on activation of the visual pigment. The same conclusion has been drawn from chromatic adaptation experiments on hymenopterans where the extent of pigment migration in the retinular cells was measured in histological sections of the ant *Formica* (MENZEL, 1972b; MENZEL and KNAUT, 1973) and the bee *Apis* (KOLB and AUTRUM, 1974); cf. the case of the cockroach (BUTLER, 1971).

Although the study of the pigment migration processes in the retinular cells is valuable in its own right, an exciting side line must be indicated. The intracellular

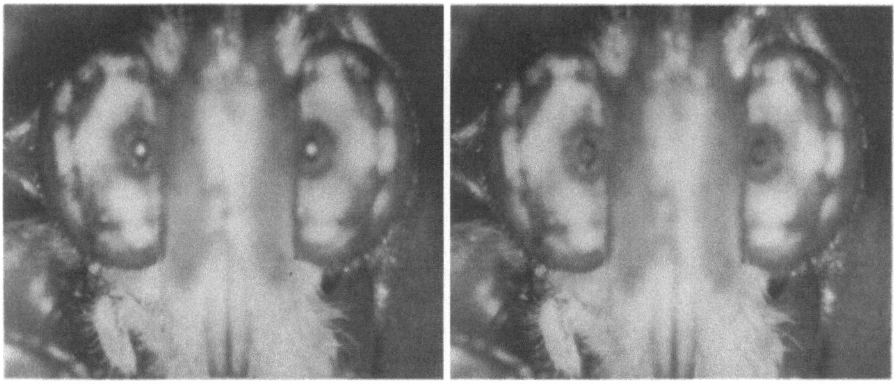

Fig. 43. Dark- and light-adapted deep pseudopupils of the Question Mark (Nymphalidae). This butterfly is a rare example where the vanishing deep luminous pseudopupil is very clearly replaced by a reflecting annulus (which seems to resolve into four spots)

pupil mechanism offers the possibility of studying visual sense cell properties in completely intact animals, and thus provides an optical probe in the sense cells at a stage after the first steps of phototransduction. This potential source of knowledge has only just started to be tapped. A point in case is the study of fly visual pigments via the pupil mechanism. The action spectrum of the pupillary response of receptor cells R1–6 determined in the fruitfly *Drosophila* (FRANCESCHINI, 1972a, b) is very similar to the action spectrum of the (late) receptor potential (e.g., MINKE et al., 1975; HARRIS et al., 1976). Furthermore, the electrophysiological sensitivity spectrum of R 1–6 is approximately equal to the absorption spectrum of the rhodopsin. This holds for instance for *Drosophila* (OSTROY et al., 1974; MINKE et al., 1975), *Calliphora* (DÖRRSCHEIDT-KÄFER, 1972; HAMDORF et al., 1973; STAVENGA, 1976), *Eristalis* (STAVENGA, 1976; TSUKAHARA and HORRIDGE, 1977) and *Musca* (GOLDSMITH, 1965; KIRSCHFELD et al., 1977). Closely corresponding sensitivity spectra were recently obtained in these latter species by BERNARD and STAVENGA (1977) by determining as a function of wavelength the intensity of monochromatic flashes necessary to induce a criterion change in pupil reflection. Hence, by measuring action spectra of the pupillary responce the characteristics of the visual pigments can be identified *in vivo* with an alternative method (FRANCESCHINI, 1972a, b).

In a comparative study on a number of fly species using pupillary reflection BERNARD and STAVENGA (1977) have found that all spectral sensitivity curves exhibit a major peak in the UV (at $\leqq 360$ nm), and a somewhat lower peak at $\geqq 450$ nm, similarly as found by receptor potential measurements (e.g., BURKHARDT, 1962, DÖRRSCHEIDT-KÄFER, 1972, TSUKAHARA and HORRIDGE, 1977). The curves fall into three groups based on the long wavelength peak: 1) 490–500 nm for a muscid fly, a calliphorid, and the dorsal region of a syrphid (see previous section D.III.3); 2) 475 nm for a chloropid and an ephydrid; 3) 450 nm for the ventral region of two different syrphids and a bombyliid.

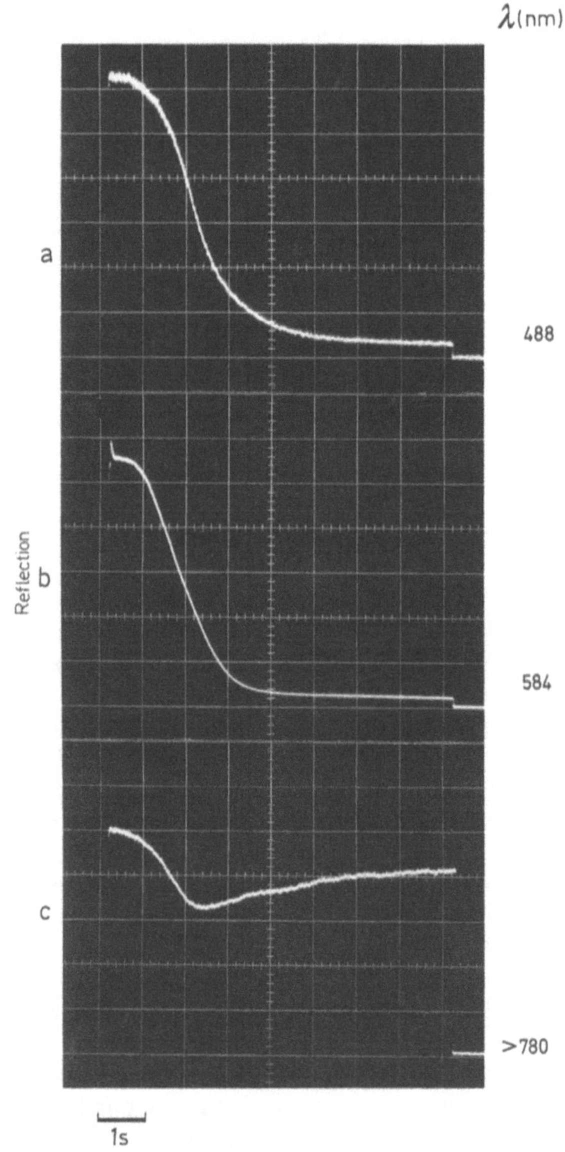

Fig. 44a–c. The course of the reflection from the compound eye of the Red Admiral *Vanessa atalanta* during light adaptation after 1 min of darkness. The eye was orthodromically illuminated for 9 s by intense white light. The reflected light was filtered before entering the photomultiplier (PM.R, Fig. 36) with respectively an interference filter 488 nm (a), an interference filter 584 nm (b), and an infrared transmitting filter RG 780 (c). The scale of the ordinate is linear (from STAVENGA et al., 1977)

Due to the strongly scattering pupil pigment of bumblebees (Fig. 42), action spectra could also be determined in this species (BERNARD and STAVENGA, 1977, 1978). In the medio-equatorial region of the eye, the three typical hymenopteran receptor types (UV, blue and green) were encountered, but dorsally only green visual sense cells could be disclosed; of course, the possibility that in the dorsal eye parts UV and blue cells occur having no pupil must be considered.

Spectral sensitivities of the pupillary response in butterflies have been measured by BERNARD (1977) using the decrease in reflectance of the luminous pseudopupil upon light adaptation. Except for the discovery of a red rhodopsin (see Sec. D.III.3), BERNARD (1977, 1979) has been able to demonstrate the existence of no less than four different receptor types, or visual pigments, in certain butterflies. RIBI (1978) has investigated the receptor properties in the ommatidium of the cabbage butterfly *Pieris rapae* with histological methods. Pigment position indicated the state of adaptation. Chromatic adaptation experiments revealed two UV- and two green receptors, containing dark-brown pigment granules, and four yellow-green receptors with red pigment granules. The spectral properties of a ninth cell could not be determined due to absence of intracellular pigment granules.

Naturally, characteristics other than the spectral sensitivity of the receptor cell can be studied with the aid of the pupil mechanism. For example, it has been found that the polarization sensitivity is the same as that of the dichroic-absorbing rhabdomeres of *Drosophila* (FRANCESCHINI, 1972a, b). The birefringence of the rhabdomeres emerges, on the other hand, when the light-adapted eye of a blowfly is observed under polarized light with a crossed analyser (unpublished).

The angular sensitivity of the visual sense cell is open to investigation via the pupillary response (cf. SWIHART, 1974). An interesting aspect here is that angular sensitivity actually depends on the state of the intracellular pupil as was first inferred from intracellular recordings and histological studies by SNYDER and HORRIDGE (1972) and BUTLER and HORRIDGE (1973) for cockroach. The acceptance angle narrows with increasing light adaptation. The same phenomenon was found in locust (TUNSTALL and HORRIDGE, 1967; WILSON, 1975) and blowfly (HARDIE, 1978). Furthermore, narrowing of the acceptance angle by retinular cell pigment migration can be assumed for honeybee from optical measurements (FRANCESCHINI, 1975; see Fig. 40).

In connection with this, another system must be mentioned which might affect the acceptance angle, namely the palisade. This is an assembly of fluid-filled vesicles surrounding the rhabdom(ere) in the dark-adapted state (HORRIDGE and BARNARD, 1965; SEITZ, 1970; MENZEL, 1972b; HORRIDGE, 1975; see BERNHARD et al., 1972, GOLDSMITH and BERNARD, 1974, and FRANCESCHINI and KIRSCHFELD, 1976 for reviews). The palisade diminishes upon light adaptation, and through the change in refractive index (SEITZ, 1970) could affect the light-guiding properties of the rhabdom(ere); this change is enhanced by accumulation of mitochondria (locust: HORRIDGE, 1966). However, a more likely function than to change the refractive index of the boundary medium seems to be to clear the surrounding medium from the absorbing pigment granules. Whether the palisade is actively involved in moving the retinular cell pigment granules has not yet been unequivocally proven (MILLER and CAWTHON, 1974; MILLER, 1975; see also FRANCESCHINI and KIRSCHFELD, 1976). One may assume from the present high interest in photoreceptor physiology and the accordingly rapid accumulation of new data that some crucial questions will soon be answered.

IV. Optical Functions of the Pigment Cells

1. Screening Pigment

In the following part, the function of pigment cells will be discussed, as will the ways in which pseudopupil methods can help us to understand this function. The pigment cells in compound eyes are often called screening pigment cells (LANGER,

1975; MENZEL, 1975) or shielding pigment cells (MILLER, 1975), indicating what is assumed to be their main function. Indeed the situation of the pigment cells around the cones and the retinulae provides these cells with an outstanding opportunity to form an optical screen for the visual sense cells.

In the following it will be attempted to put the screening function in correct perspective in the light of a number of different lines of evidence. First, displacements of the pigment in the pigment cells can affect the angular sensitivity of the photoreceptor cells. Acceptance angle changes due to pigment migration will be described in Section IV.2. Secondly, the angular sensitivity can be impaired at wavelengths where the pigment cells become leaky. The advantages and disadvantages of the partially leaky pigment screens will be discussed in Section IV.3. Another recognizable quality of pigment cells, namely that of screening off screening pigment for attentive observers, will be emphasized in Section IV.4.

2. Pigment Migration in the Pigment Cells

Migration of pigment in the pigment cells of compound eyes was investigated by EXNER (1891) in insects and crustaceans using histological and optical techniques. More recent work is reviewed by PARKER (1932), WATERMAN (1961), GOLDSMITH and BERNARD (1974), WALCOTT (1975), KUNZE (this volume, Ch. 8) and MILLER (this volume, Ch. 3). The physiological effect of the pigment migration in pigment cells has been studied mainly in superposition eyes. Substantial effects were demonstrated on absolute light sensitivity of the photoreceptor cells (1–2 log units in the moth: HÖGLUND and STRUWE, 1970) and on the acceptance angle ($\Delta\varrho = 24 \pm 8°$ dark adapted and $\Delta\varrho = 4 \pm 2°$ light adapted in crayfish: WALCOTT, 1974; for photographs of the luminous pseudopupil demonstrating the effect of the pigment migration see HÖGLUND, 1966a, b, and PAGE and LARIMER, 1975).

The same adaptation phenomenon might be expected to occur in crepuscular animals with apposition eyes, since it is advantageous in any type of eye to open up the acceptance angle when the ambient light level drops, and thus gather more light (WALCOTT, 1975; SNYDER, this volume, Ch. 5). Recently, it was shown in the apposition eye of the crab *Scylla* that the acceptance angle changes from $\Delta\varrho \approx 4°$ (light) to $\Delta\varrho \approx 10°$ (dark) due to movement of both distal and proximal pigment (LEGGETT, 1977). The position of the pigment in the light-adapted state is such that light can then only enter the retinal cells through the field stop formed by the tip of the rhabdom, since the distal pigment completely envelopes the cones. In the dark-adapted state the pigment is aggregated distally, so that the field stop is much enlarged. Light can then enter the retina alongside the rhabdoms. This extra light can be absorbed in rhabdoms of adjacent ommatidia either directly or after being scattered by the tapetum, a layer of highly reflective pigment which is exposed in the dark-adapted state, but which is retracted upon light adaptation behind a screen of dark, absorbing pigment. EXNER (1891, p. 125) extensively investigated the pigment position and the varying luminous pseudopupils in several brachyuran crabs and also in the hermit crab, but he failed to interpret the effect of pigment migration on the photoreceptors. [In fact, in a discussion of the eye of *Limulus* (p. 65) he states that migration of distal pigment would be useless in an apposition eye.]

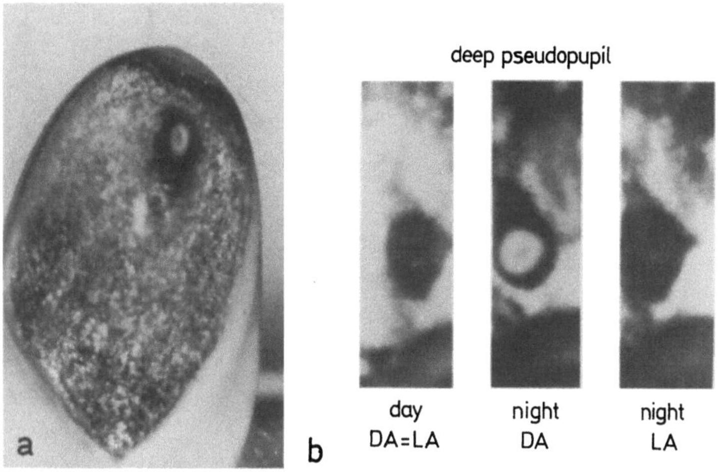

Fig. 45 a and b. A luminous pseudopupil is observed at night in the dark-adapted eye of crabs, e.g., in *Cancer borealis* (a) and *Scylla* (b, middle). Neither during day time in both dark- and light-adapted eye nor in the light-adapted night eye can this eye shine be seen, due to pigment migration induced by either illumination or by a diurnal rhythm system [(b) from LEGGETT, 1977]

The phenomenon of light scattering on the tapetum can be observed through the opened field stop in the deep luminous pseudopupil from the outside (Fig. 45; for drawings, see EXNER, 1891). The darker centre in Fig. 45a is the image of the tip of the rhabdom (LEGGETT, 1977).

The pigment migration in the pigment cells proceeds slowly (15–60 min) and therefore it is simple to fixate the pigment in intermediate states by histological techniques. In this way, WELSH (1930) and BENNITT (1932) (in macrurans) and KLEINHOLZ (1937) (in brachyurans) discovered that pigment migration in crustaceans is often under control of a circadian or diurnal rhythm system. (For related processes in insect eyes, see below.) Based on these early investigations a productive line of research has been established (reviews BROWN, 1961, 1973). However, it is irritating to notice that EXNER'S (1891) illuminating investigations have been virtually neglected by workers in this field. Also, or perhaps related to this fact, a minimal application of optical methods has occurred.

The luminous pseudopupil as a tool for investigating the circadian rhythm was employed by SMITH (1948) on grapsoid crabs and recently by PAGE and LARIMER (1975) on the crayfish *Procambarus*. In the latter study it was shown by lesions in optic tract and brains in neural recordings that circadian pigment migration is controlled via axons descending from the brain through the optic tract. In *Limulus* (lateral eyes) BARLOW et al. (1977) demonstrated diurnal changes in the electroretinogram, and hypothesized that these changes are correlated to morphological changes in (among other things) the position of screening pigment. The first indications for diurnal pigment migrations in *Limulus* were published by BEHRENS (1974). A detailed histological study undertaken by CHAMBERLAIN and BARLOW (1977) yielded impressive morphological changes, e.g., pigment migrations in the ommatidia, occurring on a day/night basis and appearing to be

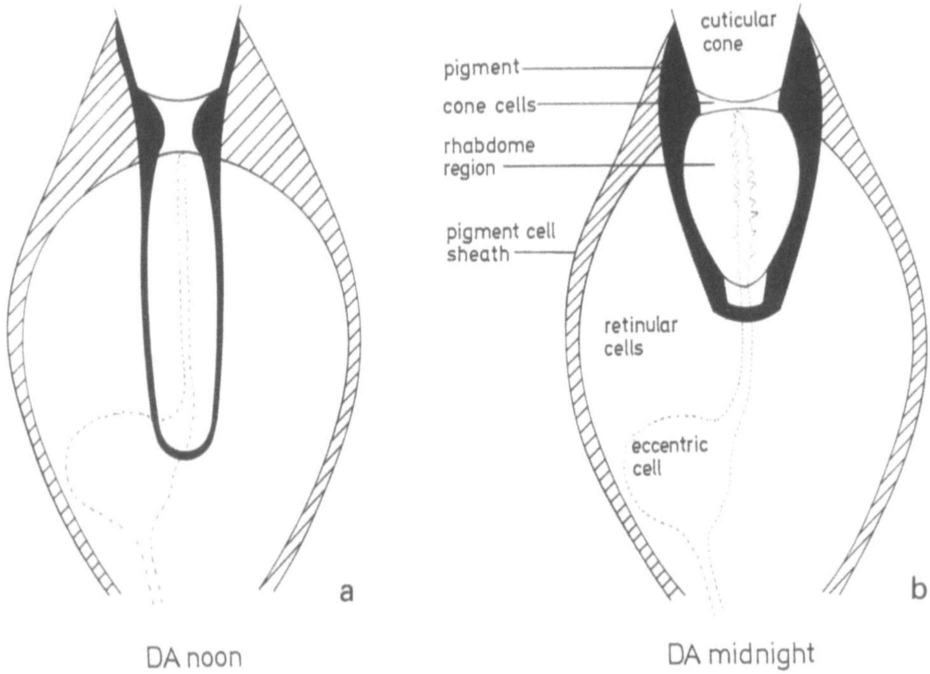

Fig. 46 a–f. Diurnal changes occurring in the compound eye of *Limulus*. Diagram (b) presents the ommatidium schematically in the dark-adapted night phase. The cone cell region separates the cone from the rhabdom region by no more than 4 μm. The aperture formed by the cone and the surrounding pigment cells is over 40 μm wide. At noon (a) cone cells separate the rhabdom of a dark-adapted ommatidium from the base of the circular cone by nearly 30 μm, forming an aperture (diameter 17 μm) with the surrounding pigment cells. The midnight state is reached while the optic nerve has an elevated activity. The noon dark-adapted state can be shifted into the midnight-adapted state by shocking the optic nerve, i.e., by stimulating the efferent input to the eye, thus confirming the hypothesis that pigment state is controlled by optic nerve activity. (c–e) present the pseudopupil phenomena accompanying the diurnal changes (observed with axial illumination). In the dark-adapted night eye a blue eye shine is seen (c), which is substantially diminished after 30 min light adaptation (d). In (e) the day eye is pictured and (g) shows that at oblique illumination the luminous pseudopupil (f) vanishes [(a, b) by courtesy of Dr. STEVEN CHAMBERLAIN]

controlled by efferent activity in the optic nerve trunk (Fig. 46). The pigment migrations are reflected in substantial changes in the pseudopupil. With axial illumination a yellow–rose ring with a dark centre is exposed in the day (both light and dark) adapted state (Fig. 46e). At night (in the dark-adapted state), the yellow ring expands, and a blue central area emerges (Fig. 46c and f). The day state can be approached during the night by way of prolonged illumination (Fig. 46d). That the pseudopupil changes follow a circadian rhythm is easily confirmed (STAVENGA, 1977). One might wonder why the marked changes in the lateral eyes of *Limulus* were not observed before, because numerous investigators have devoted attention to the visual system of this animal. The answer probably lies in the usual way of illumination. A rather dull eye appears when oblique illumination is applied (Fig. 46g; cf., e.g., RATLIFF et al., 1966, or HARTLINE et al.,

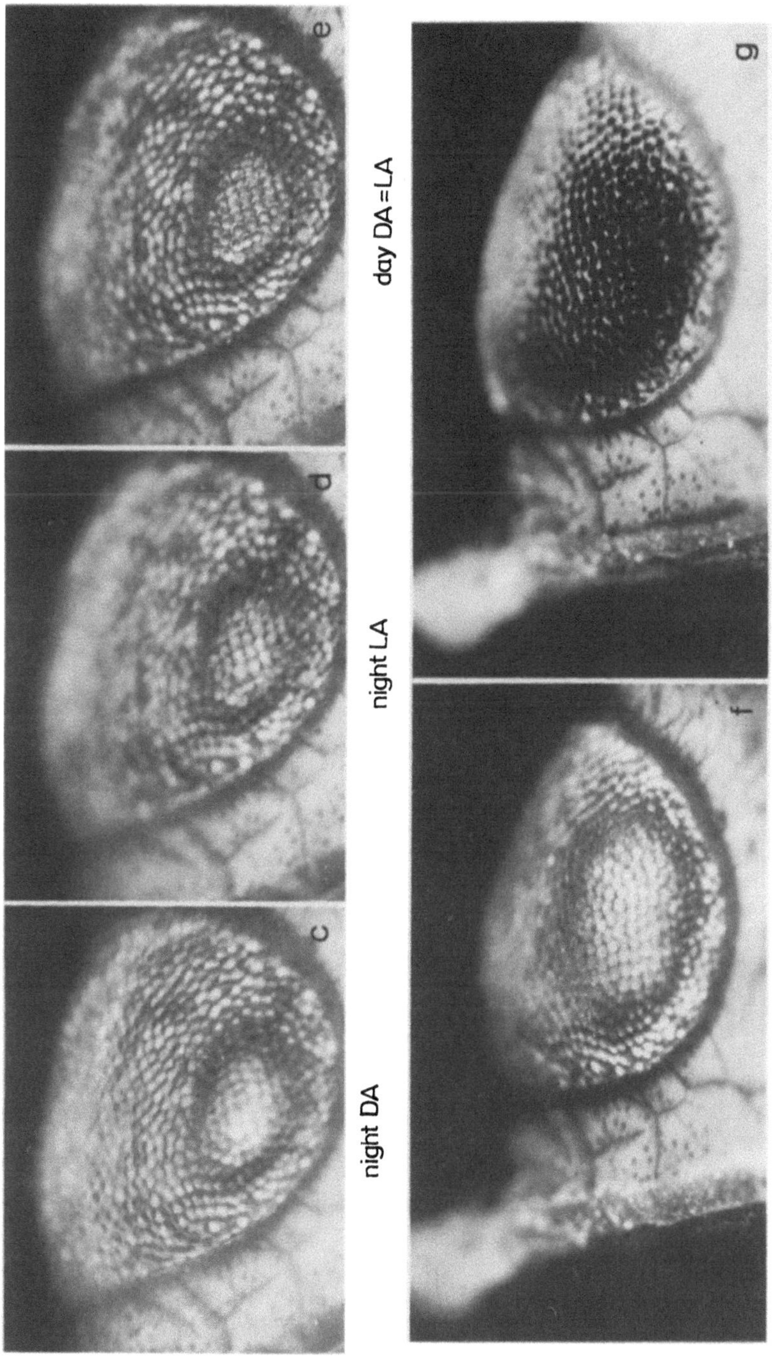
c
night DA
d
night LA
e
day DA=LA
f
axial illumination
g
oblique illumination

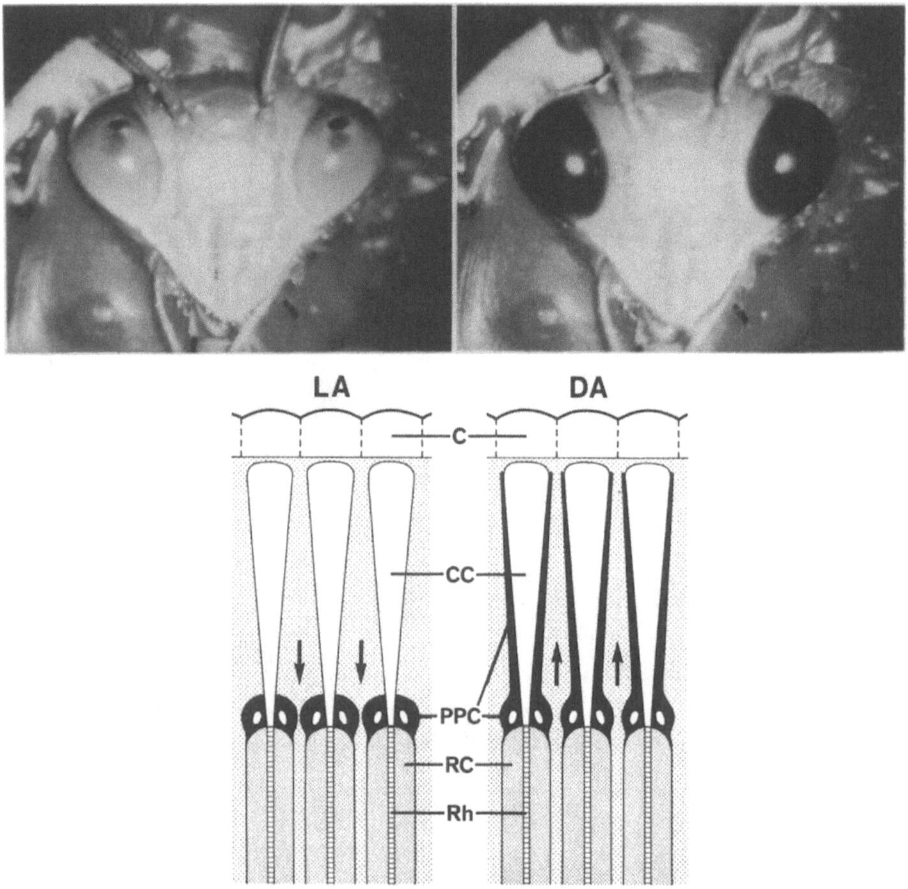

Fig. 47. Diurnally occurring pigment migrations of the primary pigment cells in the praying mantis eye have a dramatic effect on eye colour. In the day time and in the light-adapted eye at night (left) the eye colour is very similar to that of the body. In the dark-adapted eye at night (right) the eyes turn black due to a distad migration of pigment (arrows). Light adaptation causes the pigment to retract into the proximal position (arrows left). Whether the complete pigment cells move (drawn) or only the pigment granules migrate within the cell body is unclear. C indicates the corneal facet lens, CC the crystalline cone, PPC the primary pigment cells, RC the retinular cell, and Rh the rhabdom (diagram courtesy of Dr. WILLI RIBI)

1972). Neither the rose pigment cells nor the blue-scattering retinular cells are then seen.

Circadian rhythms in pigment migration in the eyes of insects have been studied less extensively, although the first observations of such processes have been reported for moths (KIESEL, 1894; DEMOLL, 1911; see PARKER, 1932). Dramatic effects can be readily recognized, though, in the eyes of the praying mantis. During the day time the eyes take the same green (or brown) colour as the body. At night the eyes turn black (FRIZA, 1929); this phenomenon can be reversed by illumination within about half an hour (Fig. 47).

A similar, but less marked change occurs in another orthopteran family, the Tettigoniidae (FRIZA, 1929). The example of a cone-headed bush cricket (Fig. 48)

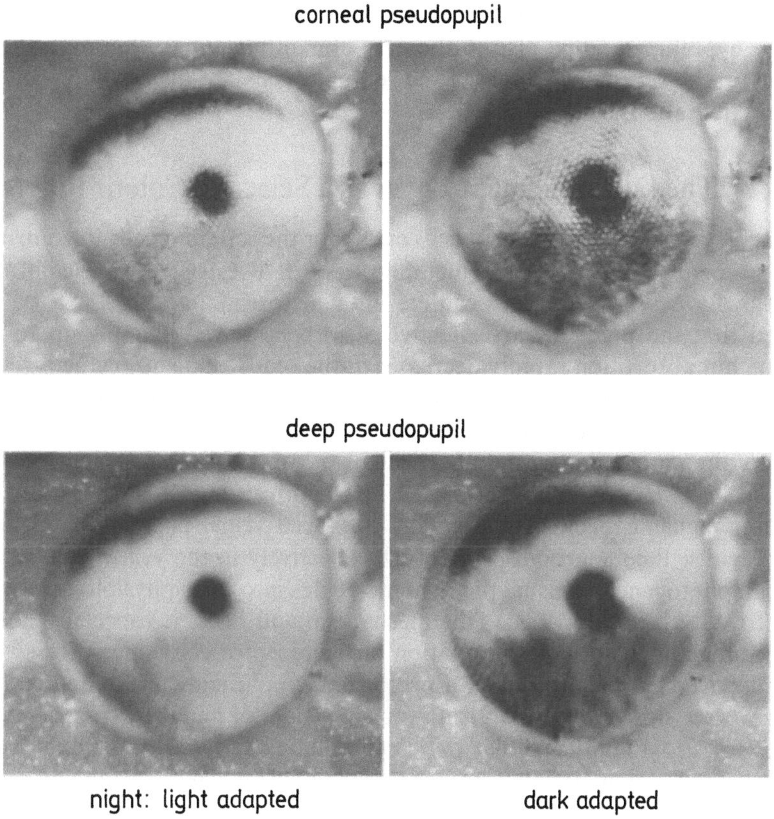

Fig. 48. Right eye of a cone-headed bush cricket (Tettigoniidae). The secondary pigment cells ventrally colour the eye green, similar to the body. A central white band is proceeded posterially at the body. Dorsally, a handsome brown stripe is noticed. The principal pseudopupil is the only optical phenomenon which moves over the eye upon rotation (FRIZA, 1929). Dominant eye colour changes do not occur upon variations of illumination in the day time. However, at night the eye darkens and the principal pseudopupil enlarges. Illumination shifts the eye back into the day state. The centre of the principal pseudopupil displays a whitish spot in the dark-adapted eye at night, presumably due to fluorescence and light scattering on inhomogeneities (palisade) near the rhabdom

is interesting in two respects. First, the dark-adapted eye displays a whitish spot in the centre of the black principal pseudopupil. This is possibly partly due to fluorescence; another cause may be backscattering of light through and around the rhabdom from a palisade, vacuoles which are known to exist in the dark-adapted retinular cells of the closely related locust (HORRIDGE and BARNARD, 1965; HORRIDGE, 1966). Secondly, the principal pseudopupil is larger in the dark-adapted state than in the light-adapted state; apparently the primary pigment cells and/or their pigments are displayed distally in the dark-adapted eye along the cones. Both primary and secondary pigment cells may move during adaptation, but histological investigations are needed to clarify the details of the actual processes.

No luminous pseudopupil was observed in the eyes of Orthoptera, probably because neither a tracheal tapetum nor a reflective pigment tapetum exists. Still, from the analogy with the processes occurring in the apposition eyes of crabs it may be hypothesized that also in the eyes of mantids and Tettigoniidae diurnal changes in the acceptance angle of the retinular cells are the consequence of the pigment migrations.

3. The Distal Pigment Layer as a Selective Colour Filter

The screen which the pigment cells create for the retinular cells to control light flux and angular sensitivity is sometimes leaky in a restricted wavelength range. The blowfly *Calliphora* is an example with historical interest. LANGER (1967) showed that the pigments are effective absorbers at wavelengths up to about 600 nm, but absorption rapidly decreases in the red. The distal pigment screen is hence leaky in the red, resulting in an artificial red peak in ERG recordings (BURKHARDT, 1962; GOLDSMITH, 1965).

After the unravelling of the photochemistry of the fly visual pigments (HAMDORF et al., 1973; STAVENGA et al., 1973), the advantage of having a red leaky screening pigment together with widely displaced visual pigment states became clear. Because the metarhodopsin absorbs effectively in the yellow-red, stray red light can photogenerate visual pigment molecules into their physiologically active state (STAVENGA et al., 1973). This mechanism will become especially advantageous at high intensities of illumination of the receptor cells since excessive rhodopsin conversion then results in a prolonged depolarizing afterpotential (PDA: see HAMDORF, this volume, Ch. 4) and thus in a state of impaired contrast transfer by the eye. Also, the intracellular pupil can remain as "glued" to its light-adapted state (STAVENGA et al., 1975). PDA and "glueing" effect can be undone by red light which reconverts metarhodopsin into the rhodopsin state (see COSENS and BRISCOE, 1972; MINKE et al., 1975; STAVENGA et al., 1975; HAMDORF and RAZMJOO, 1977; WRIGHT and COSENS, 1977).

Visual pigments working according to the photochromic flip-flop principle were described for the UV receptors of *Ascalaphus* (HAMDORF et al., 1973) and *Limulus* (NOLTE and BROWN, 1972; MINKE et al., 1973), and for the green receptor of the barnacle *Balanus* (HILLMAN et al., 1972; HOCHSTEIN et al., 1973). Hence it is obvious to assume that the phenomenon of a selective light filter as developed in flies will be encountered in other invertebrates too. Good evidence for a selective distal pigment filter was provided by LANGER (1975) for the purely UV sensitive dorsal eye of *Ascalaphus*. (A parallel but still more hypothetical example is the mayfly *Cloeon* (HORRIDGE, 1976).)

It will be clear that a pigment filter which has to favour metarhodopsin conversion over rhodopsin conversion by stray light must be either a sharp cut-off or a band filter transmitting only at wavelengths where the metarhodopsin dominantly absorbs. The construction of such a filter will occur with more ease the better the spectra of the two visual pigment states are separated. UV visual pigments and the main fly visual pigments are good cases, since the metarhodopsin is bathochromic-shifted. In such a condition the advantage of photoregeneration by stray light will overrule the disadvantage that part of the stray light activates the cell, widening angular sensitivity. Which one of the two aspects is dominant will determine the actual absorption spectrum of the pigment cells. For example, when perfect acuity is primarily called for, the pigment will be made as black as possible: it therefore may not be fortuitous that robberflies have very darkly pigmented eyes. When photoregeneration of (bathochromic-shifted) metarhodopsin is most important, a bright red eye will be realized, as occurs with most sun-loving syrphids.

From the preceding discussion it will be clear that a red (or generally long wavelength) leaky distal pigment layer is only advantageous when the longest wavelength sensitive visual pigment in the eye has a metarhodopsin which is (strongly) bathochromic-shifted in respect of the rhodopsin. Of the invertebrate visual pigments hitherto analyzed, bathochromic-shifted metarhodopsins have only been found when the rhodopsin has $\lambda_{max} \lesssim 500$ nm (HAMDORF et al., 1973; STAVENGA, 1976). Green and red receptors have hypsochromic-shifted metarhodopsins (BROWN and WHITE, 1972; HOCHSTEIN et al., 1973; SCHWEMER and PAULSEN, 1973; BERNARD, 1979). Screening pigments in eyes with the latter types of receptors are expected to have an extended absorption spectrum so that no substantial stray light can enter the layer of retinular cells.

Green receptors are very commonly found in animals with compound eyes (see GOLDSMITH, 1972). Black or dark brown distal screening pigment accordingly should be the rule in most eyes, yet many species have compound eyes with bright colours, e.g., locusts, dragonflies, and cabbage butterflies. The screening function of the light-coloured distal pigments is the topic of discussion in the following section.

4. Pigmentation and Colour of Compound Eyes

As a rule the pigment cells determine the colour of a compound eye. FRIZA (1929) pointing to this fact remarked that the stripe and colour patterning in the eyes of a variety of orthopteran insects is continuous with the patterning of the head cuticle surrounding the eyes. Furthermore, the pigment cells of insect eyes are virtually indistinguishable from the hypodermis cells of the body (FRIZA, 1929). Hence, the obvious hypothesis is that the external appearance of the compound eyes, i.e., their colouring and patterning, is an integral part of the general display system of the whole animal. Indeed, the continuity of body and eye colour is a dominant feature in many species (e.g., *Limulus*, ghost crab, locust, damselfly, and cabbage butterfly). It seems reasonable, therefore, to gain an understanding of the eye colouring with the aid of the recognized general principles of animal colouration. Camouflage is a clear goal for grasshoppers and praying mantids with their green bodies and eyes. Disruptive patterning is seen in the eyes of locusts and shore crabs. Countershading, a property seen throughout the animal kingdom, i.e., a much darker colouring dorsally than ventrally, is often apparent in the body and the eyes of grapsoid crabs.

However, in virtually all light-coloured apposition eyes a black principal pseudopupil is a dominant feature. This fact has to be regarded as inevitable by the animal (and as a generous gift from nature by the investigator). The reason why a black principal pseudopupil must exist is as follows: When the distal pigment layer is composed of pigment granules which predominantly scatter but do not absorb light the screening action becomes poor. Vision would be much impaired was an additional curtain not drawn behind the one displayed to the outside world. The layer of black pigment juxtaposed to the light-coloured pigment enacts the necessary screening function for the retinular cells (Sec. D.IV.1 and 2); the most distal layer of light-coloured pigments then probably functions as to hide the contrasting black pigment from the attentive observer, predator and/or prey.

The division of labour between the different pigment cells is usually such that the shorter primary pigment cells form a tight envelope around the cone, and thus contain black pigment granules; the layer of secondary pigment cells contain the lighter eye-colouring pigment. A noticeable exception to this rule are the pierid butterflies, where the brightly scattering primary pigment cells surround the cone distally and the black pigmented secondary pigment cells bulge proximally around the cone (see YAGI and KOYAMA, 1963).

Accessory pseudopupils are also visible in the eye of pierids, because the most distal screen is somewhat translucent. The appearance of these accessory pseudopupils is not unavoidable in contrast with the case of the principal pseudopupil. [The common dusky blue *Erina* (Fig. 24a) has a brightly coloured eye with only a principal pseudopupil.] One may therefore ask why these accessory pseudopupils have been developed. This question is of special interest for the case of butterflies because their eyes display the richest pseudopupil patterns. An answer can be given tentatively from the line of thought developed in this section: The eyes and wings of *Pieris* are white with black spots; the monarch has eyes with a black- and white-blocked pattern not dissimilar to that of the body and the wing periphery. In other words, camouflage or display is also supported by the colouration of the eyes in butterflies.

A most curious colouration is encountered in the flowerfly *Lathyrophthalmus aeneus* (Fig. 49). Flowerflies are renowned for their capacity to mimic Hymenoptera: the dronefly *Eristalis* has even gained its name from this property. Many syrphids enjoy a striking similarity to hornets and wasps. Still, the acute visual naturalist readily recognizes these flies by their red-pigmented eyes. The flowerfly *Lathyrophthalmus* on the other hand seems to form a unique example of an eye mimic. A casual, distant observer gets the impression that the eyes belong to a bee, when covered with pollen. (The body is not very bee like). This pollinated appearance is achieved by a pointillistic technique. Carmine red spots are more or less randomly distributed over a pale yellow eye, with the exception of the foveal region (Fig. 49b). The red pigment is present, though, proximal to the lighter pigment, as can be seen by both axial and oblique illumination (Fig. 49c and d). Thus, the advantage of having an eye with a red leaky pigment as in other flies seems to be overruled in this animal by another, namely that of Batesian mimicry.

Dragonflies form a group of insects whose eyes combine a number of the discussed principles of pigmentation. It may be recalled that dragonfly eyes are divided into a ventral and a dorsal eye. The facets are larger dorsally (ZIMMERMANN, 1914; see also TRUJILLO-CENÓZ, 1972), and acuity is extreme there (EXNER, 1891; HORRIDGE, 1977a, b; SHERK, 1978b). Except for the black principal pseudopupil, a huge number of accessory pseudopupils are often visible in a lightly-coloured ventral eye. The ventral eye can thus be compared with most fused rhabdom apposition eyes. The existence of UV, blue, and green receptors in the

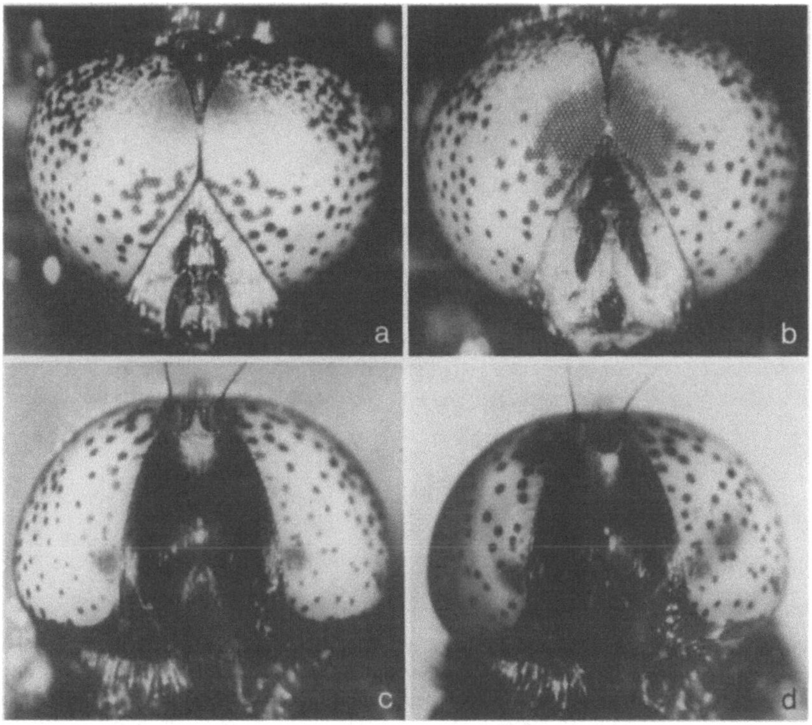

Fig. 49 a–d. The flowerfly *Lathyrophthalmus aeneus* observed with axial (a–c) and oblique illumination (d). The eye has an overall pale yellow colour but is studded with carmine spots. The red principal pseudopupil is large fronto-dorsally (b), indicating an area of small interommatidial angles, and small dorsally (a) and ventrally (c). Oblique illumination (d) ventrally creates an additional red spot, the oblique illumination pseudopupil

ventral retina of the corduliid *Hemicordulia tau* (LAUGHLIN, 1975) clearly necessitates black primary pigment cells. The same three types of receptors have been found in the dorsal eye of *Aeschna* (EGUCHI, 1971), in correspondence with the dark pigmentation one can observe there. On the other hand, dorsally in Corduliids and Libelluliids a yellow–red primary pigment layer can always be recognized below a cover of varying hue (Fig. 50). In view of the foregoing discussion, it seems likely that those dragonflies have only short wavelength receptors in their dorsal eyes. In fact, MAZOKHIN-PORSHNYAKOV (1959) reported that the dorsal eye of *Libellula* harbours violet receptors only. The biological significance of having exclusively short wavelength receptors is related to the high acuity of the dorsal eye. Thus, prey is most effectively spotted against blue sky (MAZOKHIN-PORSHNYAKOV, 1969). Related findings and discussions on this theme are presented by SHAW (1969), GOGALA et al. (1970), KIRSCHFELD and WENK (1976), SHERK (1978 b), HORRIDGE and MCLEAN (1978), and LAUGHLIN and MCGINNESS (1978).

At the end of this section on eye colouration it will be recalled that other factors than the pigmentation are sometimes most important for external appearance (GOLDSMITH and BERNARD, 1974). Interference of light in the layered corneal facet lenses causes the metallic colours in the eyes of, for example, horseflies, long-legged flies and lacewings (BERNARD and MILLER, 1968; BERNARD, 1971; see also FRIZA, 1929). Of minor importance for the appearance of the eyes are the reflections on the

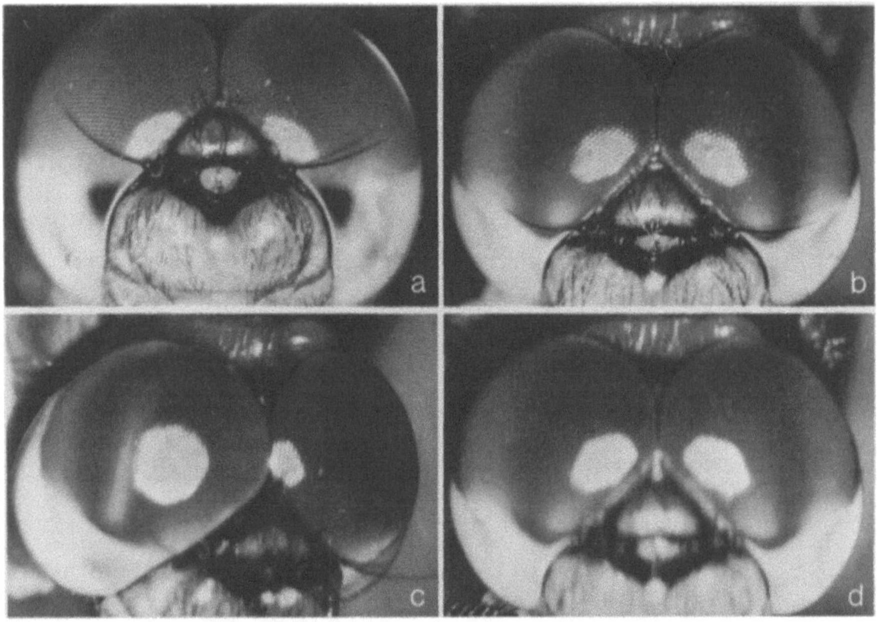

Fig. 50a–d. Pseudopupils in the compound eye of a libellulid dragonfly under axial illumination. In the ventral eye a black principal pseudopupil together with a hint of accessory pseudopupils is observed (a). In the dorsal eye (b–d) an orange principal pseudopupil appears. Frontally in the deep pseudopupil (d) an annulus is seen. The black centre is most probably the image of the rhabdom. Laterally (c) a remarkably elongated principal pseudopupil is apparent, thus revealing the curious organization of the visual axes in the dorsal eye (HORRIDGE, 1977a, b; SHERK, 1977, 1978a, b, c). Note from the corneal reflections that the eye shape is rather spherical, and also that axes are skew to the eye surface

corneal facet lenses. Corneal nipples in the facet lenses are probably installed when this glare has to be reduced (BERNHARD et al., 1965). Furthermore, the existence of interfacetal hairs can have some modifying influence on eye appearance.

Finally, it may not be superfluous to state that the pigment cells fulfil other than optical functions. Their function as supportive elements is clear from their tight structuring around the sensory cells; the pigment cells also serve as nutritive and metabolic centres for the eye (EAKIN, 1972). On the basis of the large changes in the membrane potential which occur upon illumination of the eye (e.g., BAUMANN, 1975) it may be hypothesized that the pigment cells play a role in the regulation system for the ions in the retina.

V. Retinal Movements

Retinomotor phenomena occurring upon light or dark adaptation can be studied with pseudopupil methods. Of these phenomena, pigment migrations inside retinular cells and pigment cells have been mentioned above. However, gross movements of cells are no exception, and distinct changes in rhabdom cross section and length occur (reviews MAZOKHIN-PORSHNYAKOV, 1969; and WALCOTT, 1975). Knowledge about the forces involved is negligible.

Other movements of the retina, due to mechanical forces acting inside the head capsule, also occur and are noticeable as a shift in the position of the pseudopupil. Movements of the proboscis (KIRSCHFELD and FRANCESCHINI, 1969) or the antennae induce substantial pseudopupil displacements in flies. The internal eye muscle (BURTT and PATTERSON, 1970; HENGSTENBERG, 1971; VAN BARNEVELD, 1971) effectively counteracts the unwanted distorting effect on image formation in the eye (unpublished observations by KUIPER and co-workers).

VI. Selective Stimulation of Retinular Cells

Selective stimulation of retinular cells employed in behavioural studies of flies belong to the most sophisticated applications of pseudopupil methods. (For a more classical approach in *Limulus*, see RATLIFF, 1966.) KIRSCHFELD (1972) managed to evoke optomotor responses in tethered houseflies by stimulating individual retinular cells belonging to one and the same ommatidium. KIRSCHFELD and LUTZ (1974) studied inhibitions of the motion detector units by illuminating specific retinular cells in neighbouring ommatidia. Stimulation of restricted arrays of receptors by sinusoidal gratings with specific wavelengths and orientations was performed by BUCHNER (1976) and WEHRHAHN (1976) by applying the principle of the double telescope system contrived by FRANCESCHINI (1975). The results were discussed in a wide scope by POGGIO and REICHARDT (1976) and HEISENBERG and BUCHNER (1977).

E. Acknowledgement

Enlightening discussions with Drs. MARK LEGGETT, WILLI RIBI, and MARTIN WILSON much encouraged the writing of this chapter, which started in the stimulating environment of the Departments of Neurobiology and Applied Mathematics of the Australian National University (Canberra). Important material was gathered during a (net) swinging time with Prof. GARY BERNARD at the Marine Biological Laboratory, Woods Hole. Valuable support was given by Prof. KUNO KIRSCHFELD, Dr. NICOLAS FRANCESCHINI, Dr. WILLIAM STARK and DOMIEN BEERSMA. This work was supported by the Netherlands Organization for the Advancement of Pure Research (Z.W.O.).

References

Baldus, K.,: Experimentelle Untersuchungen über die Entfernungslokalisation der Libellen *(Aeschna cyanea)*. Z. vergl. Physiol. **3**, 475—505 (1926).

Barlow, R. B., Bolanawski, S. J., Brachman, M. L.: Efferent optic nerve fibers mediate circadian rhythms in the *Limulus* eye. Science **197**, 86—89 (1977).

Barneveld, H. H. van: Statistical analysis of repetitive neural activity. Clock-spikes in *Calliphora erythrocephala*. Thesis, Groningen 1971.

Barrós-Pita, J. C., Maldonado, H.: A fovea in the praying mantis eye. II. Some morphological characteristics. Z. vergl. Physiol. **67**, 79—92 (1970).

Baumann, F.: Electrophysiological properties of the honey bee retina. In: The compound eye and vision of insects (ed. G. A. Horridge), pp. 53—74. Oxford: Clarendon Press 1975.

Baumgaertner, H.: Der Formensinn und die Sehschärfe der Bienen. Z. vergl. Physiol. **7**, 56—143 (1928).

Beersma, D. G. M.: Spatial characteristics of the visual field of flies. Thesis, Groningen 1979.

Beersma, D. G. M., Stavenga, D. G., Kuiper, J. W.: Organization of visual axes in the compound eye of the fly *Musca domestica* L. and behavioural consequences. J. comp. Physiol. **102**, 305—320 (1975).

Beersma, D. G. M., Stavenga, D. G., Kuiper, J. W.: Retinal lattice, visual field and binocularities in flies. Dependence on species and sex. J. comp. Physiol. **119**, 207—220 (1977).

Behrens, M. E.: Photomechanical changes in the ommatidia of the *Limulus* lateral eye during light and dark adaptation. J. comp. Physiol. **89**, 45—57 (1974).

Bennitt, R.: Physiological interrelationship in the eyes of decapod Crustacea. Physiol. Zool. **5**, 49—64 (1932).

Bernard, G. D.: Evidence for visual function of corneal interference filters. J. Insect Physiol. **17**, 2287—2300 (1971).

Bernard, G. D.: Discovery of red-receptors in butterfly retinas. Invest. Ophthalmol. [Suppl.] p. 6 (1977).

Bernard, G. D.: Red-absorbing visual pigment of butterflies. Science, in press (1979).

Bernard, G. D., Miller, W. H.: Interference filters in the corneas of Diptera. Invest. Ophthalmol. **7**, 416—434 (1968).

Bernard, G. D., Stavenga, D. G.: The pupillary response of flies as an optical probe for determining spectral sensitivities of retinular cells in completely intact animals. Biol. Bull. **153**, 415 (1977).

Bernard, G. D., Stavenga, D. G.: Spectral sensitivities of retinular cells measured in intact, living bumblebees by an optical method. Naturwissenschaften **65**, 442 (1978).

Bernard, G. D., Wehner, R.: Functional similarities between polarization vision and color vision. Vision Res. **17**, 1019—1028 (1977).

Bernhard, C. G., Miller, W. H., Møller, A. R.: The insect corneal nipple array. Acta physiol. scand. **63**, Suppl. **243**, 1—79 (1965).

Bernhard, C. G., Gemne, G., Seitz, A. G.: Optical properties of the compound eye. In: Handbook of sensory physiology, Vol. VII/2 (ed. M. G. F. Fuortes), pp. 357—380. Berlin-Heidelberg-New York: Springer 1972.

Bishop, L. G.: An ultraviolet receptor in a dipteran compound eye. J. comp. Physiol. **91**, 267—275 (1974).

Boschek, C. B.: On the fine structure of the peripheral retina and lamina ganglionaris of the fly, *Musca domestica*. Z. Zellforsch. **118**, 369—409 (1971).

Braitenberg, V.: Patterns of projection in the visual system of the fly. I. Retina-lamina projections. Exp. Brain Res. **3**, 271—298 (1967).

Braitenberg, V.: Ordnung und Orientierung der Elemente im Sehsystem der Fliege. Kybernetik **7**, 235—242 (1970).

Braitenberg, V.: Periodic structures and structural gradients in the visual ganglia of the fly. In: Information processing in the visual systems of arthropods (ed. R. Wehner), pp. 3—15. Berlin-Heidelberg-New York: Springer 1972.

Braitenberg, V., Strausfeld, N. J.: Principles of the mosaic organization in the visual system's neuropil of *Musca domestica*. In: Handbook of sensory physiology, Vol. VII/3A (ed. R. Jung), pp. 631—660. Berlin-Heidelberg-New York: Springer 1973.

Brown, F. A.: Physiological rhythms. In: The physiology of Crustacea, Vol. 2 (ed. T. H. Waterman), pp. 401—430. New York-London: Academic Press 1961.

Brown, F. A.: Biological rhythms. In: Comparative animal physiology (ed. C. L. Prosser), pp. 429—456. Philadelphia: Saunders 1973.

Brown, P. K., White, R. H.: Rhodopsin of the larval mosquito. J. gen. Physiol. **59**, 401—414 (1972).

Brunnert, A., Wehner, R.: Fine structure of light- and dark-adapted eyes of desert ants, *Cataglyphis bicolor* (Formicidae, Hymenoptera). J. Morph. **140**, 15—30 (1973).

Bruno, M. S., Barnes, S. N., Goldsmith, T. H.: The visual pigment and visual cycle of the lobster, *Homarus*. J. comp. Physiol. **120**, 123—142 (1977).

Buchner, E.: Elementary movement detectors in an insect visual system. Biol. Cybern. **24**, 85—101 (1976).

Bullock, T. H., Horridge, G. A.: Structure and function in the nervous systems of invertebrates. San Francisco-London: Freeman 1965.

Burkhardt, D.: Spectral sensitivity and other response characteristics of single visual cells in the arthropod eye. Symp. Soc. exp. Biol. **16**, 86—109 (1962).

Burkhardt, D., Darnhofer-Demar, B., Fischer, K.: Zum binokularen Entfernungssehen der Insekten. 1. Die Struktur des Sehraums von Synsekten. J. comp. Physiol. **87**, 165—188 (1973).

Burkhardt, D., Motte, I. de la, Seitz, G.: Physiological optics of the compound eye of the blowfly. In: The functional organization of the compound eye (ed. C. G. Bernhard), pp. 51—62. Oxford: Pergamon Press (1966).

Burtt, E. T., Patterson, J. A.: Internal muscle in the eye of an insect. Nature **228**, 183—184 (1970).

Butler, R.: The identification and mapping of spectral cell types in the retina of *Periplaneta americana*. Z. vergl. Physiol. **72**, 67—80 (1971).

Butler, R., Horridge, G. A.: The electrophysiology of the retina of *Periplaneta americana* L. 1. Changes in receptor acuity upon light/dark adaptation. J. comp. Physiol. **83**, 263—278 (1973).

Chamberlain, S. C., Barlow, R. B.: Morphological correlates of efferent circadian activity and light adaptation in the *Limulus* lateral eye. Biol. Bull. **153**, 418—419 (1977).

Collet, T. S., Land, M. F.: Visual control of flight behaviour in the hoverfly, *Syritta pipiens* L. J. comp. Physiol. **99**, 1—66 (1975).

Cosens, D., Briscoe, D.: A switch phenomenon in the compound eye of the white eyed mutant of *Drosophila melanogaster*. J. Insect Physiol. **18**, 627—632 (1972).

Dartnall, H. J. A.: Photosensitivity. In: Handbook of sensory physiology, Vol. VII/1 (ed. H. J. A. Dartnall), pp. 122—145. Berlin-Heidelberg-New York: Springer 1972.

Demoll, R.: Über eine lichtzersetzliche Substanz im Facettenauge, sowie über eine Pigmentwanderung im Appositionsauge. Arch. ges. Physiol. **129**, 461—475 (1909a).

Demoll, R.: Über die Beziehungen zwischen der Ausdehnung des binokularen Sehraumes und dem Nahrungserwerb bei einigen Insekten. Zool. Jb. (System. Geogr. Biol. Tiere) **28**, 523—530 (1909b).

Demoll, R.: Über die Wanderung des Irispigments im Facettenauge. Zool. Jb. (Allgem. Zool. Physiol.) **30**, 169—180 (1911).

Demoll, R.: Die Sinnesorgane der Arthropoden, ihr Bau und ihre Funktion. Braunschweig: Vieweg 1917.

Demoll, R., Scheuring, L.: Die Bedeutung der Ocellen von Insekten. Zool. Jb. (Allg. Zool. Physiol.) **31**, 519—628 (1912).

Dietrich, W.: Die Facettenaugen der Dipteren. Z. wiss. Zool. **92**, 465—539 (1909).

Dörrscheidt-Käfer, M. J.: Die Empfindlichkeit einzelner Photoreceptoren im Komplexauge von *Calliphora erythrocephala*. J. comp. Physiol. **81**, 309—340 (1972).

Duelli, P.: A fovea for e-vector orientation in the eye of *Cataglyphis bicolor* (Formicidae, Hymenoptera). J. comp. Physiol. **102**, 43—56 (1975).

Dvorak, D. R., Bishop, L. G., Eckert, H. E.: Intracellular recording and staining of directionally selective motion detecting neurons in fly optic lobe. Vision Res. **15**, 451—453 (1975a).

Dvorak, D. R., Bishop, L. G., Eckert, H. E.: On the identification of movement detectors in the fly optic lobe. J. comp. Physiol. **100**, 5—23 (1975b).

Eakin, R. M.: Structure of invertebrate photoreceptors. In: Handbook of sensory physiology. Vol. VII/1 (ed. H. J. A. Dartnall), pp. 625—684. Berlin-Heidelberg-New York: Springer 1972.

Eckert, H. E.: Optomotorische Untersuchungen am visuellen System der Stubenfliege *Musca domestica* L. Kybernetik **14**, 1—23 (1973).

Edrich, W., Helversen, O. von: Polarized light orientation of the honey bee: the minimum visual angle. J. comp. Physiol. **109**, 309—314 (1976).

Eguchi, E.: Fine structure and spectral sensitivities of retinular cells in the dorsal sector of compound eyes in the dragonfly *Aeschna*. Z. vergl. Physiol. **71**, 201—218 (1971).

Eguchi, E., Waterman, T. H.: Cellular basis for polarized light perception in the spider crab, *Libinia*. Z. Zellforsch. **84**, 87—101 (1968).

Eguchi, E., Waterman, T. H.: Orthogonal microvillus pattern in the eighth rhabdomere of the rock crab *Grapsus*. Z. Zellforsch. **137**, 145—157 (1973).

Eheim, W. P.: Directional intensity distribution in single ommatidia of the honeybee, *Apis mellifera*. In: Information processing in the visual system of arthropods (ed. R. Wehner), pp. 83—87. Berlin-Heidelberg-New York: Springer 1972.

Eheim, W. P., Wehner, R.: Die Sehfelder der zentralen Ommatidien in den Appositionsaugen von *Apis mellifica* und *Cataglyphis bicolor* (Apidae, Formicidae; Hymenoptera). Kybernetik **10**, 168—179 (1972).

Exner, S.: Die Physiologie der facettirten Augen von Krebsen und Insecten. Leipzig-Wien: Deuticke 1891.

Franceschini, N.: Sur le traitement optique de l'information visuelle dans l'oeil à facettes de la drosophile. Thesis, Grenoble 1972a.

Franceschini, N.: Pupil and pseudopupil in the compound eye of *Drosophila*. In: Information processing in the visual system of arthropods (ed. R. Wehner), pp. 75—82. Berlin-Heidelberg-New York: Springer 1972b.

Franceschini, N.: Sampling of the visual environment by the compound eye of the fly: Fundamentals and applications. In: Photoreceptor optics (eds. A. W. Snyder, R. Menzel), pp. 98—125. Berlin-Heidelberg-New York: Springer 1975.

Franceschini, N.: *In vivo* fluorescence of the rhabdomeres in an insect eye. Proc. Int. Union Physiol. Sci. **13**, p. 237. 27th Int. Congr. (Paris) 1977.

Franceschini, N., Kirschfeld, K.: Etude optique *in vivo* des éléments photorécepteurs dans l'oeil composé de *Drosophila*. Kybernetik **8**, 1—13 (1971a).

Franceschini, N., Kirschfeld, K.: Les phénomènes des pseudopupille dans l'oeil de *Drosophila*. Kybernetik **9**, 159—182 (1971b).

Franceschini, N., Kirschfeld, K.: Le contrôle automatique du flux lumineux dans l'oeil composé des Diptères. Propriétés spectrales, statiques et dynamiques du mécanisme. Biol. Cybern. **21**, 181—203 (1976).

Frantsevich, L. I., Pichka, V. E.: The size of binocular zone of the visual field in insects. J. Evol. Biochem. Physiol. (USSR) **12**, 461—465 (1976).

Friederichs, H. F.: Beiträge zur Morphologie und Physiologie der Sehorgane der Cicindelinen (Col.). Z. Morphol. Ökol. Tiere **21**, 1—171 (1931).

Friza, F.: Zur Frage der Färbung und Zeichnung des facettierten Insektenauges. Z. vergl. Physiol. **8**, 289—336 (1929).

Fuge, H.: Die Pigmentbildung im Auge von *Drosophila melanogaster* und ihre Beeinflussung durch den white-locus. Z. Zellforsch. **83**, 468—507 (1967).

Gavel, L. von: Die „kritische Streifenbreite" als Maß der Sehschärfe bei *Drosophila melanogaster*. Z. vergl. Physiol. **27**, 80—135 (1939).

Gemperlein, R.: Grundlagen zur genauen Beschreibung von Komplexaugen. Z. vergl. Physiol. **65**, 428—444 (1969).

Gemperlein, R., Smola, U.: Übertragungseigenschaften der Sehzelle der Schmeißfliege *Calliphora erythrocephala*. 3. Verbesserung des Signal-Störungs-Verhältnisses durch präsynaptische Summation in der Lamina ganglionaris. J. comp. Physiol. **79**, 393—409 (1972).

Goetz, K. G.: Optomotorische Untersuchung des visuellen Systems einiger Augenmutanten der Fruchtfliege *Drosophila*. Kybernetik **2**, 77—92 (1964).

Goetz, K. G.: Flight control in *Drosophila* by visual perception of motion. Kybernetik **4**, 199—208 (1968).

Goetz, K. G.: Processing of clues from the moving environment in the *Drosophila* navigation system. In: Information processing in the visual system of arthropods (ed. R. Wehner), pp. 255—263. Berlin-Heidelberg-New York: Springer 1972.

Goetz, K. G., Wenking, H.: Visual control of locomotion in the walking fruitfly *Drosophila*. J. comp. Physiol. **85**, 235—266 (1973).

Gogala, M., Hamdorf, K., Schwemer, J.: UV-Sehfarbstoff bei Insekten. Z. vergl. Physiol. **70**, 410—413 (1970).

Goldsmith, T. H.: Do flies have a red receptor? J. gen. Physiol. **49**, 265—287 (1965).

Goldsmith, T. H.: The natural history of invertebrate visual pigments. In: Handbook of sensory physiology, Vol. VII/1 (ed. H. J. A. Dartnall), pp. 685—719. Berlin-Heidelberg-New York: Springer 1972.

Goldsmith, T. H., Bernard, G. D.: The visual system of insects. In: The physiology of Insecta, Vol. II (ed. M. Rockstein), pp. 165—272. San Francisco: Academic Press 1974.

Gribakin, F. G.: Functional morphology of the compound eye of the bee. In: The compound eye and vision of insects (ed. G. A. Horridge), pp. 154—176. Oxford: Clarendon Press 1975.

Hamdorf, K.: The physiology of invertebrate visual pigments. In: Handbook of sensory physiology Vol. VII/6A (H. Autrum, ed.), this volume. Berlin-Heidelberg-New York: Springer 1979

Hamdorf, K., Langer, H.: Veränderungen der Lichtabsorption im Fazettenauge bei Belichtung. Z. vergl. Physiol. **51**, 172—184 (1965).

Hamdorf, K., Razmjoo, S.: The prolonged depolarizing afterpotential and its contribution to the understanding of photoreceptor function. Biophys. Struct. Mech. **3**, 163—170 (1977).

Hamdorf, K., Rosner, G.: Adaptation und Photoregeneration im Fliegenauge. J. comp. Physiol. **86**, 281—292 (1973).

Hamdorf, K., Schwemer, J., Gogala, M.: Insect visual pigment sensitive to ultraviolet light. Nature **231**, 458—459 (1971).

Hamdorf, K., Paulsen, R., Schwemer, J., Täuber, U.: Photoreconversion of invertebrate visual pigments. In: Information processing in the visual systems of arthropods (ed. R. Wehner), pp. 97—108. Berlin-Heidelberg-New York: Springer 1972.

Hamdorf, K., Paulsen, R., Schwemer, J.: Photoregeneration and sensitivity control of photoreceptors of invertebrates. In: Biochemistry and physiology of visual pigments (ed. H. Langer), pp. 155—166. Berlin-Heidelberg-New York: Springer 1973.

Hardie, R. C.: Electrophysiological properties of R7 and R8 in dipteran retina. Z. Naturforsch. **32**c, 887—889 (1977).

Hardie, R. C.: Peripheral visual function in the fly. Thesis. Canberra 1978.

Hardie, R. C., Franceschini, N., McIntyre, P. D.: Electrophysiological analysis of the fly retina. II. Spectral mechanisms in R 7 and R 8. J. comp. Physiol., in press (1979).

Harris, W. A., Stark, W. S.: Hereditary retinal degradation in *Drosophila melanogaster*. A mutant defect associated with the phototransduction process. J. gen. Physiol. **69**, 261—291 (1977).

Harris, W. A., Stark, W. S., Walker, J. A.: Genetic dissection of the photoreceptor system in the compound eye of *Drosophila melanogaster*. J. Physiol. **256**, 415—439 (1976).

Hartline, H. K., Ratliff, F.: Inhibitory interaction in the retina of *Limulus*. In: Handbook of sensory physiology, Vol. VII/2 (ed. M. G. F. Fuortes), pp. 381—447. Berlin-Heidelberg-New York: Springer 1972.

Hausen, K.: Functional characterization and anatomical identification of motion sensitive neurons in the lobula plate of the blowfly *Calliphora erythrocephala*. Z. Naturforsch. **31**c, 629—633 (1976).

Heisenberg, M., Buchner, E.: The rôle of retinula cell types in visual behavior of *Drosophila melanogaster*. J. comp. Physiol. **117**, 127—162 (1977).

Hengstenberg, R.: Das Augenmuskelsystem der Stubenfliege *Musca domestica*. I. Analyse der "clock spikes" und ihrer Quellen. Kybernetik **9**, 56—77 (1971).

Hillman, P., Hochstein, S., Minke, B.: A visual pigment with two physiologically active stable states. Science **175**, 1486—1488 (1972).

Hochstein, S., Minke, B., Hillman, P.: Antagonistic components of the late receptor potential in the barnacle photoreceptor arising from different stages of the pigment process. J. gen. Physiol. **62**, 105—128 (1973).

Höglund, G.: Pigment migration and retinular sensitivity. In: The functional organization of the compound eye (ed. C. G. Bernhard), pp. 77—88. Oxford: Pergamon Press 1966a.

Höglund, G.: Pigment migration, light screening and receptor sensitivity in the compound eye of nocturnal Lepidoptera. Acta Physiol. Scand. **69** [Suppl. 282] 1—56 (1966b).

Höglund, G., Struwe, G.: Pigment migration and spectral sensitivity in the compound eye of moths. Z. vergl. Physiol. **67**, 229—237 (1970).

Horridge, G. A.: The retina of the locust. In: The functional organization of the compound eye (ed. C. G. Bernhard), pp. 513—541. Oxford: Pergamon Press 1966.

Horridge, G. A.: Unit studies on the retina of dragonflies. Z. vergl. Physiol. **62**, 1—37 (1969).

Horridge, G. A.: Arthropod receptor optics. In: Photoreceptor optics (eds. A. W. Snyder, R. Menzel), pp. 459—478. Berlin-Heidelberg-New York: Springer 1975.

Horridge, G. A.: The ommatidium of the dorsal eye of *Cloeon* as a specialization for photoreisomerization. Proc. R. Soc. Lond. B. **193**, 17—29 (1976).

Horridge, G. A.: The compound eye of insects. Sci. Am. **237**, 108—121 (1977a).

Horridge, G. A.: Insects which turn and look. Endeavour **1**, 7—17 (1977b).

Horridge, G. A., Barnard, P. B. T.: Movement of palisade in locust retinula cells when illuminated. Quart. J. Micr. Sci. **106**, 131—135 (1965).

Horridge, G. A., Giddings, C., Stange, G.: The superposition eye of skipper butterflies. Proc. R. Soc. Lond. B. **182**, 457–495 (1972).

Horridge, G. A., McLean, M.: The dorsal eye of the mayfly *Atalophlebia* (Ephemeroptera). Proc. R. Soc. Lond. B. **200**, 137—150 (1978).

Horridge, G. A., Meinerzhagen, I. A.: The accuracy of the patterns of connections of the first and second-order neurons of the visual system of *Calliphora*. Proc. R. Soc. Lond. B. **175**, 69—82 (1970).
Horridge, G. A., Mimura, K.: Fly photoreceptors. I. Physical separation of two visual pigments in *Calliphora* retinula cells 1—6. Proc. R. Soc. Lond. B. **190**, 211—224 (1975).
Horridge, G. A., Mimura, K., Tsukahara, Y.: Fly receptors. II. Spectral and polarized light sensitivity in the dronefly *Eristalis*. Proc. R. Soc. Lond. B. **190**, 225—237 (1975).
Kiesel, A.: Untersuchungen zur Physiologie des facettirten Auges. Sitz. Ber. Akad. Wiss. Wien. (Math. Nat.) **103** (3), 97—139 (1894).
Kirschfeld, K.: Das anatomische und das physiologische Sehfeld der Ommatidien im Komplexauge von *Musca*. Kybernetik **2**, 249—257 (1965).
Kirschfeld, K.: Die Projektion der optischen Umwelt auf das Raster der Rhabdomere im Komplexauge von *Musca*. Exp. Brain Res. **3**, 248—270 (1967).
Kirschfeld, K.: Optics of the compound eye. In: Processing of optical data by organisms and by machines (ed. W. Reichardt), pp. 116—136. New York: Academic Press 1969.
Kirschfeld, K.: Aufnahme und Verarbeitung optischer Daten im Komplexauge von Insekten. Naturwissenschaften **58**, 201—209 (1971).
Kirschfeld, K.: The visual system of *Musca*: studies on optics, structure and function. In: Information processing in the visual systems of arthropods (ed. R. Wehner), pp. 61—74. Berlin-Heidelberg-New York: Springer 1972.
Kirschfeld, K.: Das neurale Superpositionsauge. Fortschr. Zool. **21**, 229—257 (1973a).
Kirschfeld, K.: Optomotorische Reaktionen der Biene auf bewegte „Polarisations-Muster". Z. Naturforsch. **28**c, 329—338 (1973b).
Kirschfeld, K.: The function of photostable pigments in fly photoreceptors. Biophys. Struct. Mech., in press (1979).
Kirschfeld, K., Feiler, R., Franceschini, N.: A photostable pigment within the rhabdomere of fly photoreceptors no. 7. J. comp. Physiol. **125**, 275—284 (1978).
Kirschfeld, K., Franceschini, N.: Optische Eigenschaften der Ommatidien im Komplexauge von *Musca*. Kybernetik **5**, 47—52 (1968).
Kirschfeld, K., Franceschini, N.: Ein Mechanismus zur Steuerung des Lichtflusses in den Rhabdomeren des Komplexauges von *Musca*. Kybernetik **6**, 13—22 (1969).
Kirschfeld, K., Franceschini, N.: Photostable pigments within the menbrane of photoreceptors and their possible role. Biophys. Struct. Mech. **3**, 191—194 (1977).
Kirschfeld, K., Franceschini, N., Minke, B.: Evidence for a sensitising pigment in fly photoreceptors. Nature **269**, 386—390 (1977).
Kirschfeld, K., Lutz, B.: Lateral inhibition in the compound eye of the fly *Musca*. Z. Naturforsch. **29**c, 95—97 (1974).
Kirschfeld, K., Reichardt, W.: Die Verarbeitung stationärer optischer Nachrichten im Komplexauge von *Limulus* (Ommatidien-Sehfeld und räumliche Verteilung der Inhibition). Kybernetik **2**, 43—61 (1964).
Kirschfeld, K., Snyder, A. W.: Waveguide mode effects, birefringence and dichroism in fly photoreceptors. In: Photoreceptor optics (eds. A. W. Snyder, R. Menzel), pp. 56—77. Berlin-Heidelberg-New York: Springer 1975.
Kirschfeld, K., Snyder, A. W.: Measurements of a photoreceptor's characteristic waveguide parameter. Vision Res. **16**, 775—778 (1976).
Kirschfeld, K., Wenk, P.: The dorsal compound eye of simuliid flies: an eye specialized for the detection of small rapidly moving objects. Z. Naturforsch. **31**c, 764—765 (1976).
Kleinholz, L. H.: Studies in the pigmentary system of Crustacea. II. Diurnal movements of the retinal pigments of Bermudan decapods. Biol. Bull. **72**, 176—189 (1937).
Kolb, G., Autrum, H.: Die Feinstruktur im Auge der Biene bei Hell- und Dunkeladaptation. J. comp. Physiol. **77**, 113—125 (1972).
Kolb, G., Autrum, H.: Selektive Adaptation und Pigmentwanderung in den Sehzellen des Bienenauges. J. comp. Physiol. **94**, 1—6 (1974).
Kuiper, J. W.: The optics of the compound eye. In: Biological receptor mechanisms (ed. J. W. L. Beament), Symp. Soc. Exp. Biol., pp. 58—71. London: Cambridge University Press 1962.
Kuiper, J. W.: On the image formation in a single ommatidium of the compound eye in Diptera. In: The functional organization of the compound eye (ed. C. G. Bernhard), pp. 35—50. Oxford: Pergamon Press 1966.

Kunze, P.: Apposition and superposition eyes. In: Handbook of sensory physiology Vol. VII/6A (ed. H. Autrum). Berlin-Heidelberg-New York: Springer 1979.

Langer, H.: Über die Pigmentgranula im Facettenauge von *Calliphora erythrocephala*. Z. vergl. Physiol. **55**, 354—377 (1967).

Langer, H.: Properties and functions of screening pigments in insect eyes. In: Photoreceptor optics (eds. A. W. Snyder, R. Menzel), pp. 429—455. Berlin-Heidelberg-New York: Springer 1975.

Langer, H., Thorell, B.: Microspectrophotometry of single rhabdomeres in the insect eye. Exp. Cell Res. **41**, 673—677 (1966).

Laughlin, S. B.: Neural integration in the first optic neuropile of dragonflies. I. Signal amplification in dark-adapted second-order neurons. J. comp. Physiol. **84**, 335—355 (1973).

Laughlin, S. B.: Receptor function in the apposition eye—an electrophysiological approach. In: Photoreceptor optics (eds. A. W. Snyder, R. Menzel), pp. 479—498. Berlin-Heidelberg-New York: Springer 1975.

Laughlin, S. B., Horridge, G. A.: Angular sensitivity of the retinula cells of dark-adapted worker bee. Z. vergl. Physiol. **74**, 329—335 (1971).

Laughlin, S. B., McGinness, S.: The structures of dorsal and ventral region of a dragonfly retina. Cell Tissue Res. **188**, 427—448 (1978).

Leggett, L. M. W.: Polarised light-sensitive interneurons in a swimming crab. Nature **262**, 709—711 (1976).

Leggett, L. M. W.: Some visual specialisations of a crustacean eye. Thesis, Canberra 1977

Levín, L., Maldonado, H.: A fovea in the praying mantis eye. III. The centring of the prey. Z. vergl. Physiol. **67**, 93—101 (1970).

Leydig, F.: Zum feineren Bau der Arthropoden. Müller's Arch. Anat. Physiol. 406—444 (1855).

Leydig, F.: Das Auge der Gliedertiere. Tübingen: Laupp und Siebeck 1864.

Leydig, F.: Die Farbe der Retina und das Leuchten der Augen. Arch. Naturg. **43**, 1—6 (1877).

Liebman, P. A.: Microspectrophotometry of photoreceptors. In: Handbook of sensory physiology, Vol. VII/1 (ed. H. J. A. Dartnall), pp. 481—528. Berlin-Heidelberg-New York: Springer 1972.

Lythgoe, J. N.: List of vertebrate visual pigments. In: Handbook of sensory physiology, Vol. VII/1 (ed. H. J. A. Dartnall), pp. 604—624. Berlin-Heidelberg-New York: Springer 1972.

Mazokhin-Porshnyakov, G. A.: Colorimetric study of vision in dragonflies. Biophysica **4**, 427—436 (1959).

Mazokhin-Porshnyakow, G. A.: Insect vision. New York: Plenum Press 1969.

McCann, G. D., Dill, J. C.: Fundamental properties of intensity, form and motion perception in the visual nervous system of *Calliphora phaenicia* and *Musca domestica*. J. gen. Physiol. **53**, 385—413 (1969).

McCann, G. D., Foster, S. F.: Binocular interactions of motion detection fibers in the optic lobes of flies. Kybernetik **8**, 193—203 (1971).

Meffert, P., Smola, U.: Electrophysiological measurements of spectral sensitivity of central visual cells in eye of blowfly. Nature **260**, 342—344 (1976).

Melamed, J., Trujillo-Cenóz, O.: The fine structure of the central cells in the ommatidia of dipterans. J. Ultrastruct. Res. **21**, 313—334 (1968).

Menzel, R.: Feinstruktur des Komplexauges der roten Waldameise, *Formica polyctena* (Hymenoptera, Formicidae). Z. Zellforsch. **127**, 356—373 (1972a).

Menzel, R.: The fine structure of the compound eye of *Formica polyctena* – Functional morphology of a hymenopterean eye. In: Information processing in the visual systems of arthropods (ed. R. Wehner), pp. 37—47. Berlin-Heidelberg-New York: Springer 1972b.

Menzel, R.: Colour receptors in insects. In: The compound eye and vision of insects (ed. G. A. Horridge), pp. 121—153. Oxford: Clarendon Press 1975.

Menzel, R., Knaut, R.: Pigment movement during light and chromatic adaptation in the retinula cells of *Formica polyctena* (Hymenoptera, Formicidae). J. comp. Physiol. **86**, 125—138 (1973).

Menzel, R., Lange, G.: Änderungen der Feinstruktur im Komplexauge von *Formica polyctena* bei der Helladaptation. Z. Naturforsch. **26**b, 357—359 (1971).

Meyer-Rochow, V. B.: A crustacean-like organization of insect-rhabdoms. Cytobiologie **4**, 241—249 (1971a).

Meyer-Rochow, V. B.: Fixierung von Insektenorganen mit Hilfe eines Netzmittels. Das Dorsalauge der Eintagsfliege *Atalophlebia costalis*. Mikrokosmos **60**, 348—351 (1971b).

Meyer-Rochow, V. B., Nässel, D. R.: Crustacean eyes and polarization sensitivity. Vision Res. **17**, 1239—1240 (1977).

Miller, W. H.: Mechanisms of photomechanical movement. In: Photoreceptor optics (eds. A. W. Snyder, R. Menzel), pp. 415—428. Berlin-Heidelberg-New York: Springer 1975.
Miller, W. H.: Ocular optical filtering. In: Handbook of sensory physiology, Vol. VII/6A (ed. H. Autrum). Berlin-Heidelberg-New York: Springer 1979.
Miller, W. H., Bernard, G. D.: Butterfly glow. J. Ultrastruct. Res. **24**, 286—294 (1968).
Miller, W. H., Cawthon, D. F.: Pigment granule movement in *Limulus* photoreceptors. Invest. Ophthalmol. **13**, 401—405 (1974).
Miller, W. H., Bernard, G. D., Allen, J. L.: The optics of insect compound eyes. Science **162**, 760—767 (1968).
Minke, B., Hochstein, S., Hillman, P.: Antagonistic process as source of visible-light suppression of afterpotential in *Limulus* UV photoreceptors. J. Gen. Physiol. **62**, 787—791 (1973).
Minke, B., Wu, C.-F., Pak, W. L.: Isolation of light induced response of the central retinula cells from the electroretinogram of *Drosophila*. J. comp. Physiol. **98**, 345—355 (1975).
Muijser, H., Stavenga, D. G.: Rapid photopigment conversions in blowfly visual sense cells. Consequences for receptor potential and pupillary response. Biophys. Struct. Mech., in press (1979).
Muijser, H., Leutscher-Hazelhoff, J. T., Stavenga, D. G., Kuiper, J. W.: Photopigment conversions expressed in receptor potential and membrane resistance of blowfly visual sense cells. Nature **254**, 520—522 (1975).
Nolte, J., Brown, J. E.: Ultraviolet-induced sensitivity to visible light in ultraviolet receptors of *Limulus*. J. Gen. Physiol. **59**, 186—200 (1972).
Ostroy, S. E., Wilson, M., Pak, W. L.: *Drosophila* rhodopsin: photochemistry, extraction and differences in the *norp* Ap^{12} phototransduction mutant. Biochem. Biophys. Res. Commun. **59**, 960—966 (1974).
Page, T. L., Larimer, J. L.: Neural control of circadian rhythmicity in the crayfish. II. The ERG amplitude rhythm. J. comp. Physiol. **97**, 81—96 (1975).
Pak, W. L., Pinto, L. H.: Genetic approach to the study of the nervous system. Ann. Rev. Biophys. Bioeng. **5**, 397—448 (1976).
Pak, W. L., Ostroy, S. E., Deland, M. C., Wu, C.-F.: Photoreceptor mutant of *Drosophila*: Is protein involved in intermediate steps of phototransduction? Science **194**, 956—959 (1976).
Parker, G. H.: The movement of the retinal pigment. Ergeb. Biol. **9**, 239—291 (1932).
Pick, B.: Specific misalignments of rhabdomere visual axes in the neural superposition eye of dipteran flies. Biol. Cybern. **26**, 215—224 (1977).
Poggio, T., Reichhardt, W.: Visual control of orientation behaviour in the fly. Part. II. Towards the underlying neural interactions. Q. Rev. Biophys. **9**, 377—438 (1976).
Portillo, J. del: Beziehungen zwischen den Öffnungswinkeln der Ommatidien, Krümmung und Gestalt der Insektenaugen und ihrer funktionellen Aufgabe. Z. vergl. Physiol. **23**, 100—145 (1936).
Ratliff, F.: Selective adaptation of local regions of the rhabdom in an ommatidium of the compound eye of *Limulus*. In: The functional organization of the compound eye (ed. C. G. Bernhard), pp. 187—191. Oxford: Pergamon Press 1966.
Ratliff, F., Hartline, H. K., Lange, D.: The dynamics of lateral inhibition in the compound eye of *Limulus*. I. In: The functional organization of the compound eye (ed. C. G. Bernhard), pp. 399—424. Oxford: Pergamon Press 1966.
Razmjoo, S., Hamdorf, K.: Visual sensitivity and the variation of total pigment content in the blowfly photoreceptor membrane. J. comp. Physiol. **105**, 279—286 (1976).
Reichardt, W.: First steps in a behavioral analysis of pattern discrimination in Diptera. In: Information processing in the visual systems of arthropods (ed. R. Wehner), pp. 213—215. Berlin-Heidelberg-New York: Springer 1972.
Ribi, W. A.: Ultrastructure and migration of screening pigments in the retina of *Pieris rapae* L. (Lepidoptera, Pieridae). Cell Tiss. Res. **191**, 57—73 (1978).
Rosner, G.: Adaptation und Photoregeneration im Fliegenauge. J. comp. Physiol. **102**, 269—295 (1975).
Schneider, L., Langer, H.: Die Struktur des Rhabdoms im „Doppelauge" des Wasserläufers *Gerris lacustris*. Z. Zellforsch. **99**, 538—559 (1969).
Scholes, J.: The electrical responses of the retinal receptors and the lamina in the visual system of the fly *Musca*. Kybernetik **6**, 149—162 (1969).
Schwemer, J., Paulsen, R.: Three visual pigments in *Deilephila elpenor* (Lepidoptera, Sphingidae). J. comp. Physiol. **86**, 215—229 (1973).

Seitz, G.: Der Strahlengang im Appositionsauge von *Calliphora erythrocephala* (Meig.). Z. vergl. Physiol. **59**, 205—231 (1968).

Seitz, G.: Nachweis einer Pupillenreaktion im Auge der Schmeißfliege. Z. vergl. Physiol. **69**, 169—185 (1970).

Seitz, G., Burkhardt, D.: Bau und optische Leistungen des Komplexauges der Stielaugenfliege *Cyrtodiopsis dalmanni* Wiedemann. J. comp. Physiol. **95**, 49—59 (1974).

Shaw, S. R.: Interreceptor coupling in ommatidia of drone honeybee and locust compound eyes. Vision Res. **9**, 999—1029 (1969).

Sherk, T. E.: Development of the compound eye of dragonflies (Odonata). I. Larval compound eyes. J. Exp. Zool. **201**, 391—416 (1977).

Sherk, T. E.: Development of the compound eyes of dragonflies (Odonata). II. Development of the larval compound eyes. J. Exp. Zool. **203**, 47—60 (1978a).

Sherk, T. E.: Development of the compound eyes of dragonflies (Odonata). III. Adult compound eyes. J. Exp. Zool. **203**, 61—80 (1978b).

Sherk, T. E.: Development of the compound eyes of dragonflies (Odonata). IV. Development of the adult compound eyes. J. Exp. Zool. **203**, 183—200 (1978c).

Smith, R. I.: The role of the sinus glands in retinal pigment migration in grapsoid crabs. Biol. Bull. **95**, 169—185 (1948).

Smola, U., Meffert, P.: The spectral sensitivity of the visual cells in the eye of the blowfly *Calliphora erythrocephala*. J. comp. Physiol., in press (1979).

Snyder, A. W.: Polarization sensitivity of individual retinula cells. J. comp. Physiol. **83**, 231—360 (1973).

Snyder, A. W.: Optical properties of invertebrate photoreceptors. In: The compound eye and vision of insects (ed. G. A. Horridge), pp. 179—235. Oxford: Clarendon Press 1975.

Snyder, A. W.: The physics of vision in compound eyes. In: Handbook of sensory physiology Vol. VII/6A (H. Autrum, ed.), this volume. Berlin-Heidelberg-New York: Springer 1979.

Snyder, A. W., Horridge, G. A.: The optical function of changes in the medium surrounding the cockroach rhabdom. J. comp. Physiol. **81**, 1—8 (1972).

Snyder, A. W., Menzel, R. (eds.): Photoreceptor optics. Berlin-Heidelberg-New York: Springer 1975.

Snyder, A. W., Menzel, R., Laughlin, S. B.: Structure and function of the fused rhabdom. J. comp. Physiol. **87**, 99—135 (1973).

Snyder, A. W., Laughlin, S. B., Stavenga, D. G.: Information capacity of eyes. Vision Res. **17**, 1163—1175 (1977a).

Snyder, A. W., Stavenga, D. G., Laughlin, S. B.: Spatial information capacity of compound eyes. J. comp. Physiol. **116**, 183—207 (1977b).

Stark, W. S.: Sensitivity and adaptation in R7, an ultraviolet photoreceptor, in the *Drosophila* retina. J. comp. Physiol. **115**, 47—59 (1977).

Stark, W. S., Zitzmann, W. G.: Isolation of adaptation mechanisms and photopigment spectra by vitamin A deprivation in *Drosophila*. J. comp. Physiol. **105**, 15—27 (1976).

Stark, W. S., Ivanyshyn, A. M., Greenberg, R. M.: Sensitivity and photopigments of R1-6, a two-peaked photoreceptor, in *Drosophila*, *Calliphora* and *Musca*. J. comp. Physiol. **121**, 289—305 (1977).

Stavenga, D. G.: Visual receptor optics, rhodopsin and pupil in fly retinula cells. Thesis, Groningen 1974a.

Stavenga, D. G.: Refractive index of fly rhabdomeres. J. comp. Physiol. **91**, 417—426 (1974b).

Stavenga, D. G.: Optical qualities of the fly eye – An approach from the side of geometrical, physical, and waveguide optics. In: Photoreceptor optics (eds. A. W. Snyder, R. Menzel), pp. 126—144. Berlin-Heidelberg-New York: Springer 1975a.

Stavenga, D. G.: The neural superposition eye and its optical demands. J. comp. Physiol. **102**, 297—304 (1975b).

Stavenga, D. G.: Visual adaptation in butterflies. Nature **254**, 435—437 (1975c).

Stavenga, D. G.: Derivation of photochrome absorption spectra from absorbance difference measurements. Photochem. Photobiol. **21**, 105—110 (1975d).

Stavenga, D. G.: Waveguide modes and refractive index in photoreceptors of invertebrates. Vision Res. **15**, 323—330 (1975e).

Stavenga, D. G.: Dark regeneration of invertebrate visual pigments. In: Photoreceptor optics (eds. A. W. Snyder, R. Menzel), pp. 290—295. Berlin-Heidelberg-New York: Springer 1975f.

Stavenga, D. G.: Fly visual pigments. Difference in visual pigments of blowfly and dronefly peripheral retinula cells. J. comp. Physiol. **111**, 137—152 (1976).

Stavenga, D. G.: Optics of compound eyes and circadian pigment movements studied by pseudopupil observations *in vivo*. Biol. Bull. **153**, 446 (1977).
Stavenga, D. G.: Visual pigment processes and prolonged pupillary responses in insect photoreceptor cells. Biophys. Struct. Mech., in press (1979).
Stavenga, D. G., Barneveld, H. H. van: On dispersion in visual photoreceptors. Vision Res. **15**, 1091—1095 (1975).
Stavenga, D. G., Beersma, D. G. M.: Formalism for the neural network of visual systems. Biol. Cybern. **19**, 75—81 (1975).
Stavenga, D. G., Kuiper, J. W.: Insect pupil mechanisms. I. On the pigment migration in the retinula cells of Hymenoptera (suborder Apocrita). J. comp. Physiol. **113**, 55—72 (1977).
Stavenga, D. G., Zantema, A., Kuiper, J. W.: Rhodopsin processes and the function of the pupil mechanism in flies. In: Biochemistry and physiology of visual pigments (ed. H. Langer), pp. 175—180. Berlin-Heidelberg-New York: Springer 1973.
Stavenga, D. G., Flokstra, J. H., Kuiper, J. W.: Photopigment conversions expressed in pupil mechanisms of blowfly visual sense cells. Nature **253**, 740—742 (1975).
Stavenga, D. G., Numan, J. A. J., Tinbergen, J., Kuiper, J. W.: Insect pupil mechanisms. II. Pigment migration in retinula cells of butterflies. J. comp. Physiol. **113**, 73—93 (1977).
Stowe, S., Leggett, M.: Retina-lamina connectivity and polarization sensitivity in Crustacea. Vision Res. **18**, 1087 (1978).
Strausfeld, N. J.: The organization of the insect visual system (light microscopy). II. The projection of fibers across the first optic chiasma. Z. Zellforsch. **121**, 442—454 (1971).
Strausfeld, N. J.: Atlas of an insect brain. Berlin-Heidelberg-New York: Springer 1976a.
Strausfeld, N. J.: Mosaic organizations, layers, and visual pathways in the insect brain. In: Neural principles in vision (eds. F. Zettler, R. Weiler), pp. 245—279. Berlin-Heidelberg-New York: Springer 1976b.
Swihart, S. L.: Acceptance angles of butterfly ommatidia. J. Insect Physiol. **20**, 1027—1036 (1974).
Trujillo-Cenóz, O.: The structural organization of the compound eye in insects. In: Handbook of sensory physiology, Vol. VII/2 (ed. M. G. F. Fuortes), pp. 5—61. Berlin-Heidelberg-New York: Springer 1972.
Trujillo-Cenóz, O., Bernard, G. D.: Some aspects of the retinal organization of *Sympycnus lineatus* Loew (Diptera, Dolichopodidae). J. Ultrastruct. Res. **38**, 149—160 (1972).
Tsukahara, Y., Horridge, G. A.: Visual pigment spectra from sensitivity measurements after chromatic adaptation of single fly retinula cells. J. comp. Physiol. **114**, 233—251 (1977).
Tunstall, J., Horridge, G. A.: Electrophysiological investigation of the optics of the locust retina. Z. vergl. Physiol. **55**, 167—182 (1967).
Via, E.: Visually mediated snapping in the bulldog ant: a perceptual ambiguity between size and distance. J. comp. Physiol. **121**, 33—51 (1977).
Vogt, K.: Optics of the crayfish eye. Z. Naturforsch. **30c**, 691 (1975).
Vries, Hl. de: Physical aspects of the sense organs. Prog. Biophys. **6**, 207—264 (1956).
Vries, Hl. de, Kuiper, J. W.: Optics of the insect eye. Ann. NY Acad. Sci. **74**, 196—203 (1958).
Walcott, B.: Unit studies on light-adaptation in the retina of the crayfish, *Cherax destructor*. J. comp. Physiol. **94**, 207—218 (1974).
Walcott, B.: Anatomical changes during light-adaptation in insect compound eyes. In: The compound eye and vision of insects (ed. G. A. Horridge), pp. 20—33. Oxford: Clarendon Press 1975.
Wallace, G. K.: Visual scanning in the desert locust *Schistocerca gregaria* Forskål. J. exp. Biol. **36**, 512—525 (1959).
Washizu, Y., Burkhardt, D., Streck, P.: Visual field of single retinula cells and interommatidial inclination in the compound eye of the blowfly *Calliphora erythrocephala*. Z. vergl. Physiol. **48**, 413—428 (1964).
Waterman, T. H.: Directional sensitivity of single ommatidia in the compound eye of *Limulus*. Proc. Nat. Acad. Sci. (Wash.) **40**, 252—257 (1954).
Waterman, T. H.: Light sensitivity and vision. In: The physiology of Crustacea, Vol. II (ed. T. H. Waterman), pp. 1—64. New York: Academic Press 1961.
Waterman, T. H.: The optics of polarization sensitivity. In: Photoreceptor optics (eds. A. W. Snyder, R. Menzel), pp. 339—371. Berlin-Heidelberg-New York: Springer 1975.
Waterman, T. H., Fernández, H. R.: *E*-vector and wavelength discrimination by retinular cells of the crayfish *Procambarus*. Z. vergl. Physiol. **68**, 154—174 (1970).

Waterman, T. H., Horch, K. W.: Mechanism of polarized light perception. Science **154**, 467—475 (1966).
Wehner, R.: Pattern recognition. In: The compound eye and vision of insects (ed. G. A. Horridge), pp. 75—113. Oxford: Clarendon Press 1975.
Wehner, R.: Structure and function of the peripheral visual pathway in hymenopterans. In: Neural principles in vision (eds. F. Zettler, R. Weiler), pp. 280—333. Berlin-Heidelberg-New York: Springer 1976.
Wehrhahn, C.: Evidence for the role of retinal receptors R7/8 in the orientation behaviour of the fly. Biol. Cybern. **21**, 213—220 (1976).
Welsh, J. H.: Diurnal rhythm of the distal pigment cells in the eyes of certain crustaceans. Proc. Natl. Acad. Sci. (USA) **16**, 386—395 (1930).
Wiedemann, I.: Versuche über den Strahlengang im Insektenauge (Appositionsauge). Z. vergl. Physiol. **49**, 526—542 (1965).
Wigglesworth, V. B.: The principles of insect physiology. London: Chapman and Hall 1972.
Wilson, M.: Angular sensitivity of light and dark adapted locust retinula cells. J. comp. Physiol. **97**, 323—328 (1975).
Wolburg-Buchholz, K.: The superposition eye of *Cloeon dipterum*: The organization of the lamina ganglionaris. Cell Tissue Res.'77, 9—28 (1977).
Wright, R., Cosens, D.: Blue-adaptation and orange-adaptation in white-eyed *Drosophila:* evidence that the prolonged afterpotential is correlated with the amount of M 580 in R_{1-6}. J. comp. Physiol. **113**, 105—128 (1977).
Yagi, N., Koyama, N.: The compound eye of Lepidoptera. Tokyo: Shinkyo Press 1963.
Zaagman, W. H., Mastebroek, H. A. K., Buyse, T., Kuiper, J. W.: Receptive field characteristics of a directionally selective movement detector in the visual system of the blowfly. J. comp. Physiol. **116**, 39—50 (1977).
Zimmermann, K.: Über die Facettenaugen der Libelluliden, Phasmiden und Mantiden. Zool. Jb. (Anat. Ontog.) **37**, 1—36 (1914).

Chapter 8

Apposition and Superposition Eyes

PETER KUNZE, Stuttgart, Germany

With 46 Figures

Contents

A. Introduction

The basic concept of the physiology of compound eyes was laid down 150 years ago by J. MÜLLER (1826). From their typical facetted structure, compound eyes could be expected to function in a way quite different from vertebrate eyes. Of course, it is well established that in facetted eyes specialized optical systems convey the light to visual sense cells, that screening pigments are used to restrict ray propagation, and that basically the same visual pigments absorb the light. But all these means are used in very distinct and sometimes unique ways.

This article concerns physiologic optics of compound eyes, but almost entirely within the limits set by the application of geometric optics. The main objective will be to consider the way in which the visual environment is represented as an image in compound eyes. This cannot be done without taking up the lines of investigation laid out by nineteenth century physiologists, notably S. EXNER, whose impact on contemporary research in arthropod vision is still obvious.

After an account of MÜLLER's concept, I will restate EXNER's view on the physiologic optics of compound eyes from which the title "apposition and superposition eyes" was derived. Following this a number of recent investigations with new results on the optical properties of insect and crustacean compound eyes will be presented and EXNER's conclusions on arthropod vision reconsidered.

B. Compound Eye and Compound Vision
(J. MÜLLER)

In his chapter on arthropod vision, MÜLLER (1826) describes two possible ways in which visual organs select directional rays. (The following are more or less freely translated excerpts.)

"The one is based on light collection by refracting media, i.e., lenses. Thus, light leaving an object point in different directions and meeting the collecting system will be united again at the image point. The focal points of light, which is identical with respect to its origin, will therefore appear on the retina in the same order as in the object."

"Regarding the second way, it is comprehensible that an image can be formed on a surface if only light perpendicular to that surface is admitted. All other light from object points not meeting the sensitive surface perpendicularly is somehow intercepted. An indistinct but true image can be formed on a retina in this way. Apparently, such a retina should be spherical in order to admit light from various directions. The amount of light coming in perpendicular to the surface from all directions will be small but comparable to that admitted by the most restricted pupil of a simple eye. The distinctness of the image will increase as the non-perpendicular light rays are excluded from the retina. It all depends on an organ in front of the spherical retina, which can perform that separation. The existence of this kind of an eye seems compelling by the nature of the laws governing the propagation of light."

MÜLLER also seems to be the first who gave a correct assessment of the anatomy of compound eyes. He describes them as composed of identical units each consisting of a distal transparent cornea followed by a transparent cone and a visual fiber ("Sehfaser"), the light-sensitive structure in close contact with the cone. All three components have the same axis and, as a unit, are surrounded by dark pigment. Proximally, the visual fibers continue into the optic lobe. In the following, MÜLLER is bringing together his anatomic observations and his theory on the vision of compound eyes.

"Vision of insects is based neither on dioptric nor on catoptric systems. Instead, light coming from object points and illuminating the retina from all directions is selected and restricted to single points corresponding to the object. Thus, an image is formed. In the compound eyes of insects and Crustacea, this is achieved by way of the transparent cones that are situated between cornea and visual fibers, and which are coated with opaque pigment. Each of these cones only admits as much light to the visual fibers as penetrates near the axis. All other light from the same object point meeting the cornea at oblique angles will not reach the proximal end of the cones and therefore will not be perceived by the other visual fibers but will be absorbed by the pigment-coated sides of the cones."

"Figure 1 shows a radial section through a compound eye and the light paths into it. Light from the point *d* will completely illuminate cone *h*, the axis of which contains *d*. All other cones will not be illuminated by *d* down to their tip. The farther they are from the line *md* the less light will enter. Therefore, light from the point *d* will only be perceived by the fiber *m*. Similarly, light from *b* will—via the cones *f*—only be perceived by the fibers *k*."

"In this way an image corresponding to the object is formed. Each visual fiber conveys its specific perception to the optic nerve; and as the separate fibers unite in a continuous nerve mass, the single impressions of the single fibers merge into a continuous image."

"The more cones that make up a sector of the eye sphere, the more distinct the image in the eye will be; the distinctness will also increase with the length of the cones.

The convex cornea of the single facets will refract the axoparallel light towards the axis and thus concentrate it in the cone tip. The refraction of the convex surface of the cornea is, however, not sufficient to form separate small images for each facet. This would actually hamper a distinct image perception because each small image would be inverted and, therefore, in a wrong position with respect to the next small image."

Of MÜLLER's views, the following basic statements are still valid:

Each subunit of a compound eye (ommatidium) receives light only from a restricted spatial angle.
The retinal image is composed of the light contributions of all subunits.
The retinal image is upright.

From this general concept of compound vision, it sufficed to consider only the rectilinear propagation of light in a homogeneous medium. However, MÜLLER was

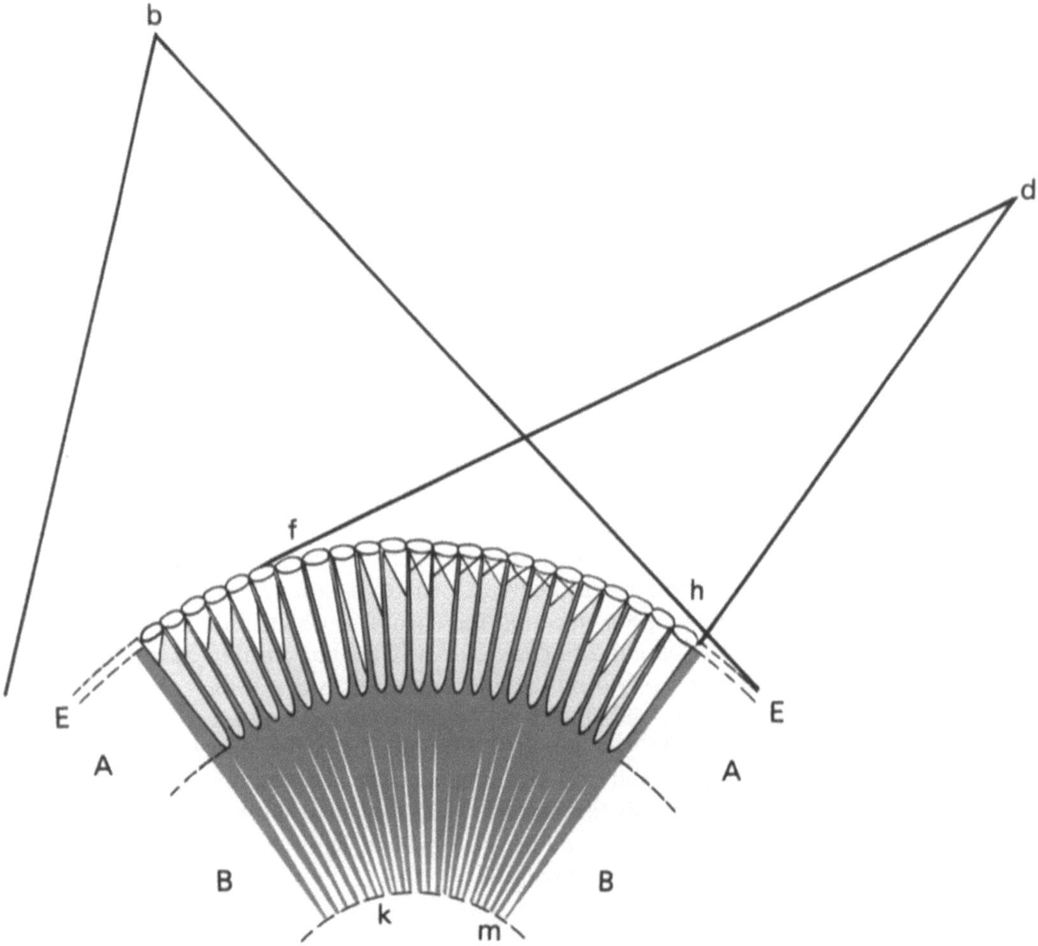

Fig. 1. Radial section through a constructed compound eye. *EE*, eye periphery with the facets of the cornea; *AA*, the transparent cones surrounded by opaque pigment; *BB*, the visual fibers in contact with the cones and surrounded by opaque pigment. Light from point *b* will only illuminate the cones *f* down to their proximal ends, likewise point *d* to the cone *h*. (after MÜLLER, 1826)

mistaken in believing that refraction and reflexion of light played no considerable role in compound vision. It was EXNER (1891) who firmly established their importance for compound eyes. GRENACHER (1879) had already pointed out that the existence of inverted images at the distal retinula ends was not contradictory to Müller's theory. (For half a century the observation of such images had been taken as proof against Müller's theory.) His argument was that in an ommatidium of most arthropod compound eyes one fused rhabdom is found which acts as a receptive unit. Therefore, only the amount of light being projected onto the rhabdom will be perceived in each ommatidium. Still, the problem remained unsolved for such eyes in which the retinula contains an unfused rhabdom. Only since 1962 has research of the dipteran eye closed this gap in our knowledge of the physiology of facetted eyes. Although the proven existence and reception of an inverted image by several retinula cells in each ommatidium is contradictory to some of MÜLLER's statements, the neural representation of an upright image (in the lamina) is in full agreement with his concepts.

C. The Optics of Compound Eyes

(S. EXNER)

Research of the light-transmitting structures of compound eyes with respect to their optical properties was, up to the end of the eighteenth century, more or less restricted to the observation of corneal images. These real images, already reported by LEEUWENHOOK and others, are produced by the positive refracting power of each facet of an isolated cornea. In intact eyes, however, light is transmitted not only by the cornea but also by the cone (MÜLLER). Furthermore, according to GRENACHER's findings it seemed very probable that the actual light-sensitive structure is the rhabdom, which is formed by the retinular cells. Starting from this basis, EXNER explored how light was transmitted to the rhabdom by cornea and cone. He began to measure the optical properties of these structures (e.g., the curvature of refracting surfaces, and refractive indexes), which are necessary for constructing ray paths. In so doing, he was immediately confronted with the problem that the cones, contrary to the cornea, were difficult to handle because of their soft consistency. Also, in dissected eyes, they did not stay in place, thus hindering microscopic observations. Suitable compound eyes apparently were those in which cornea and cones formed a unit, as, for example, in *Limulus* and certain beetles. He derived his basic results from the investigation of these eyes, and on the basis of comparative anatomy arrived at a general anatomic and functional classification of compound eyes. "In my experience, the compound eyes can be divided into three types according to their optical function; all of them form an upright retinal image but in different ways. Two types work dioptrically, one mainly catoptrically."

I. The Lens Cylinder

The refractive index of a lens can be determined from measurements of the radii of the curved surfaces and of the focal distance. EXNER did this for the corneal lens of a water beetle *(Hydrophilus)* and arrived at a value of $n = 1.8$ which seemed unbelievably high. He soon found, with the help of a self-constructed microrefractometer (EXNER, 1885), that the refractive index of the investigated cornea was not homogeneous (a prerequisite for the above determination of n), but decreased from

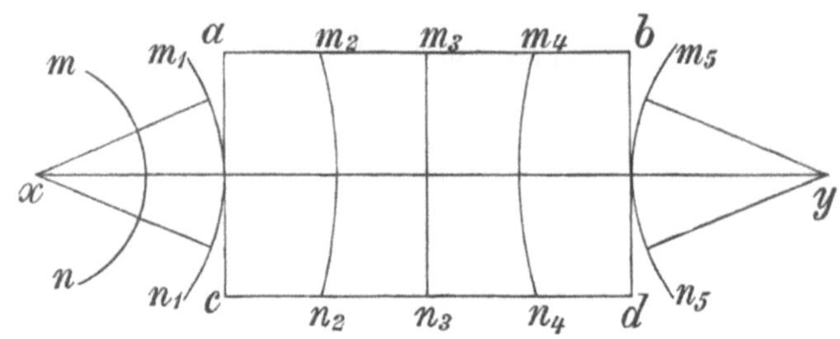

Fig. 2. Light propagation in a lens cylinder. A wave front *mn* of light from the axial object point *x* is successively deformed when passing the lens cylinder *abcd*, and focused in image point *y*. (EXNER, 1891)

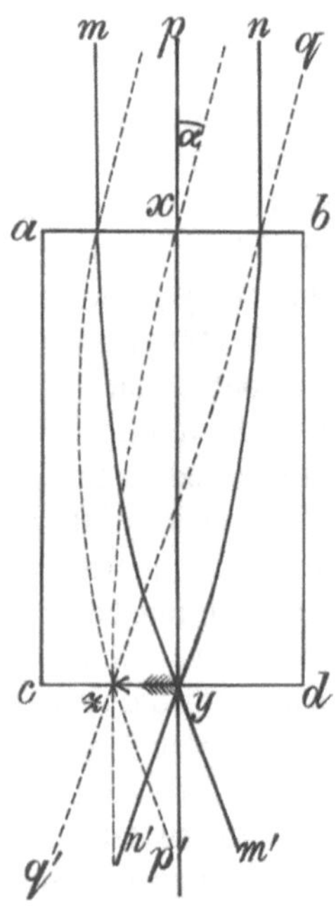

Fig. 3

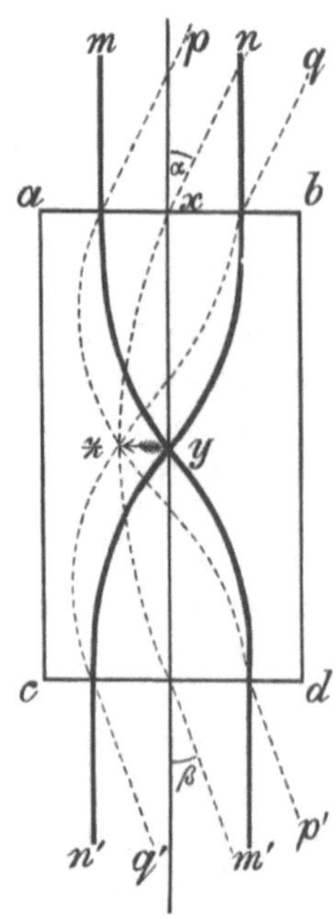

Fig. 4

Fig. 3. Ray path in a lens cylinder *abcd* as long as its focal length. *xy*, axis; *mn*, axoparallel rays focused in *y*; *pq*, parallel rays with incident angle α focused in *z*. (EXNER, 1891)

Fig. 4. Ray path in a lens cylinder *abcd* twice as long as its focal length. *mn*, axoparallel incident rays; *pq*, parallel rays with incident angle α. *n'm'*, axoparallel emerging rays; *q'p'*, parallel rays with emerging angle $\beta = \alpha$. (EXNER, 1891)

the axis to the periphery. In addition, he found that the cornea still produced inverted real images, even after the curved surfaces had been removed. He concluded that the observed inhomogeneity of the refractive index acts like a lens.

The dioptric action of a cylindrical body in which the refractive index decreases radially from the axis—EXNER called it a "lens cylinder"—is explained in Fig. 2. Light from an axial object point *x* meets a lens cylinder in the form of a spherical wave front m_1n_1. Since the wave front is travelling at different velocities in the cylinder due to the differences in refractive index (slower in the axis than in the periphery), it will become deformed. In the given case this leads to an emerging wave front which is concave. In other words, rays from the object point x are united at the image point y. If this is so for all rays from x, the radial decay of the refractive index must follow a specific function. This function was determined by KARL EXNER (1886), the brother of SIGMUND EXNER, (applying FERMAT'S principle to a thin lens) to be of the form of a square parabola.

From Fig. 2 it is apparent that rays from one object point can leave the lens cylinder either divergent, parallel, or convergent, depending on the length of the cylinder. EXNER thought two specific lengths of a lens cylinder to be the most relevant. If the length is such that parallel rays, after the passage through the cylinder, are united in the plane of its basis, an inverted image will be formed there (Fig. 3). The actual length of the cylinder is then equal to its focal length. In the second case, parallel rays are united halfway inside the cylinder. An inverted image is formed there, and the rays from each image point leave the cylinder again parallel (Fig. 4). The resultant ray path is erect. This lens cylinder is twice its focal length.

EXNER regards the two model cases of lens cylinders as the simplest possible, and in this form probably not existent in compound eyes. Lens cylinders seem to occur generally in combination with spherical lenses. Additionally, in the second case, the two halves of the lens cylinder may not be identical; usually the intermediate image seems to be closer to the basis, and the exit angle β larger than the entrance angle α, as in a magnifying telescope.

II. Apposition Images

The lateral eyes of *Limulus* are facetted. Cornea and cones form a unit (Fig. 5). Next to each cone tip lie a number of unpigmented short cells proximally followed by the retinula. The retinula is composed of a varying number of sense cells which are typically arranged around the ommatidial axis and which form a star-shaped rhabdom. The proximal part of the cone and the retinula are surrounded by densely pigmented cells.

If an eye is excised and the inner soft parts gently removed, the dioptric systems—cornea and cones—can be tested optically under the microscope. EXNER mounted the preparation on a hole in an object slide, filled the concave side with a glycerole solution of refractive index $n = 1.346$, and covered it with a cover glass. Light leaving the proximal ends of the cones was observed.

With this technique, EXNER observed behind each cone the inverted images of objects which had been placed in front of the eye preparation. The images are not very sharp due to the fact that the image plane is curved. The images were found at, or shortly behind, the cone tip. The images could also be seen behind preparations in which the distal and proximal portions of the cones had been removed. The remainder of each cone functioned as a collecting lens. Apparently, the cornea-cone system of the *Limulus* eye functions predominantly like a lens cylinder of a length approximately equal to its focal length (Fig. 6).

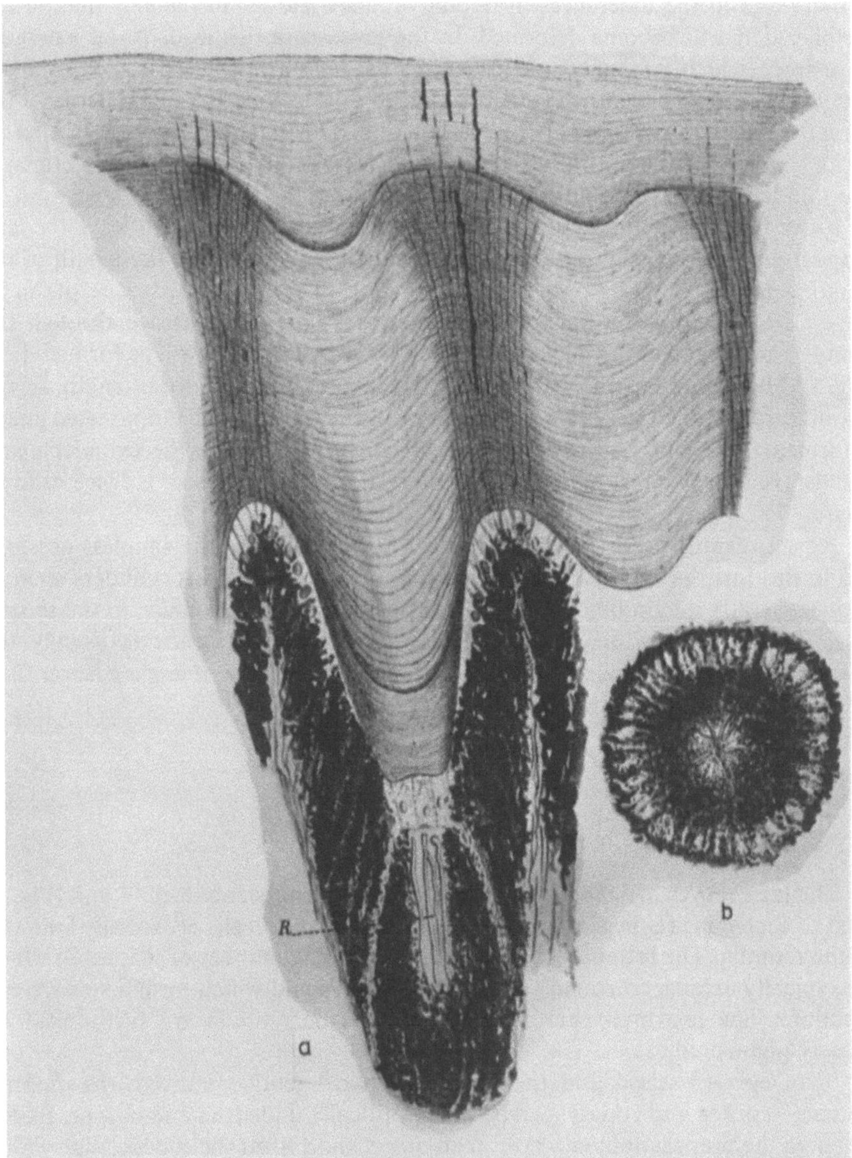

Fig. 5. *a* Axial section through an ommatidium of the lateral eye of *Limulus polyphemus*; *R*, rhabdom. *b* Cross section. (EXNER, 1891)

In an optical determination the visual field of each dioptric system was found to be about 8°, and the visual fields of adjacent systems overlapping. It must be kept in mind, however, that only the small section of each single image, that falls on the distal rhabdome end, can act as a stimulus.

The perceived retinal image is an upright apposition image consisting of the single contributions of all the dioptric systems. It can be directly observed in the eye

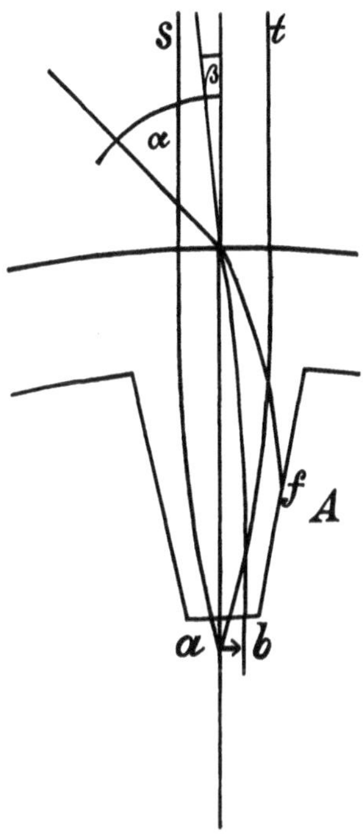

Fig. 6. Ray path in a cornea-cone unit of *Limulus*. *st*, axoparallel rays focused in *a*, *b* focus for rays with incident angle *β*. Rays with incident angle *α* are eliminated into the screening pigment at *f*. (EXNER, 1891)

preparation. Its somewhat fuzzy appearance is apparently due to the lack of screening pigments. Nevertheless, a dark object in front of a bright background can be discerned behind the cones; when the object is moved they are darkened in the direction of the object's movement.

In the compound eye of *Limulus* the retina is in close contact with the dioptric apparatus. EXNER concludes that this is characteristic for eyes with apposition images. He was able to observe apposition images also in preparations (with the soft parts in situ) of such eyes that correspond in the above sense to that of *Limulus*, but in which cornea and cones are separate structures. They included compound eyes of different (diurnal) insects and Crustacea. It is of minor importance whether their rhabdomes are found to be longer than those in the *Limulus* eye; decisive is the direct connection of the rhabdome ends with the dioptric apparatus.

"As far as the functioning of the rhabdoms is concerned, I have to call to mind the optical properties of the rods in the vertebrate retina as they were pointed out by E. BRÜCKE. Because of the high refractive index of the rhabdom with respect to the surround, a light ray which enters the rhabdom at an acute angle will be conducted down its length by total reflection. Whether the rhabdom is digging into pigment (as in *Limulus*), or not, it is sufficient that the light can enter it at the distal end."

III. Superposition Images

"Having described the optical composition of one kind of eye, I will proceed to a second type of compound eye. The erect retinal image of the *Limulus* eye came into existence because the light portions, which enter each facet, meet the retina adjacent to each other. I tried to indicate this by the term apposition image. In the eye described in the following, the light from each facet is to a great extent superimposed on the retina. Therefore, I call this retinal image a superposition image."

As an example of this type of eye, EXNER uses the eye of the firefly, *Lampyris*. Here, cornea and cones form a unit too, facilitating the investigation. In other respects, this insect eye differs considerably from that of the Xiphosuran, *Limulus*. A radial section (containing an eye radius) through a peripheral eye portion is shown in Fig. 7. Each facet has a spherical surface. The cones, which adhere strongly to the cornea, are surrounded by dense pigment and only their tips are free of it. The cones reach into a deep zone of unpigmented elongate cells. This zone is bordered by the retina, which basally contains a dense pigment and which ends at a membrane. From there, nerves extend to the optic ganglia. Unlike apposition eyes, a wide separation exists between cones and rhabdoms.

For optical investigation a *Lampyris* eye is excised, and—as with the *Limulus* eye—its inner parts are removed with a soft brush. The eye cup is transferred to a drop of dilute glycerole ($n = 1.346$; the refractive index of beetle hemolymph) on a cover glass in such a way that the concave side is filled with the glycerole and the

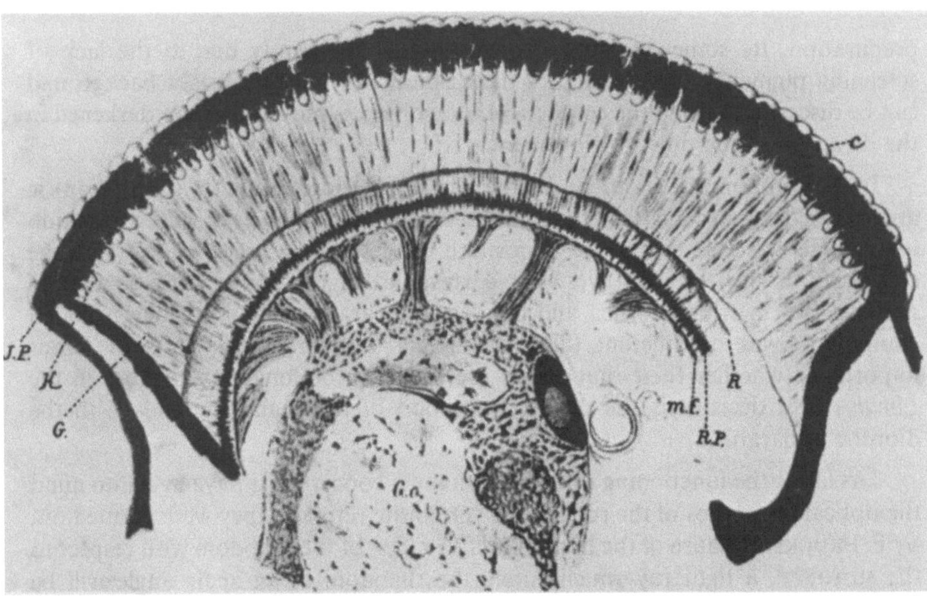

Fig. 7. Radial section of Celloidin embedded eye of *Lampyris splendidula*, eye periphery. Distance $KR \approx 640$ μm. *IP*, iris pigment; *K*, cone tip; *G*, clear zone; *R*, retina; *RP*, retina pigment; *mf*, membrana fenestrata; *Go*, ganglion opticum. (EXNER, 1891)

convex side is in air. This preparation is then placed on a hole in an object slide with the eye cup hanging in the hole. Under a microscope, light passing from the convex side through the preparation can be observed.

EXNER's microscope was without a condenser, and when he used a mirror it was always a plane mirror. With the microscope focus in a high position, above the cover glass, he observed an erect image of the objects in front of the eye preparation (the inverting action of the microscope and the effect of the mirror having been taken into account). He was able to recognize in this image the landscape in front of the window of his room: the brown wooden walls of a barn, its red tiled roof, and twigs of the little plum trees in front of blue sky. To him, a surprisingly distinct image.

EXNER determined the approximate position of the image by focusing on the cone tips and on the image. From the readings of the microscope fine adjustment, the distance could be calculated. He found the image to lie 0.23 mm behind the cones. Compared with the eye dimensions in Fig. 7 this position is at a considerable distance from the retina. This discrepancy might have two causes: (1) the eye dimensions in the periphery (from where the section was taken) are somewhat smaller than in the center where the cones produced the observed image, (2) strong tissue shrinkage by the histological procedure placed the retina closer to the cones. In principle, EXNER assumes that the distances cone–retina and cone–image coincide approximately, in other words, that the observed image resembles the retinal image.

The observed image is an erect one. Additional simple experiments with point light sources show that each cornea-cone unit is more or less an afocal optical system which forms an erect ray path. EXNER points out that this effect can be obtained by a pair of convex lenses in air which are arranged on one axis at a distance equal to the sum of their focal lengths. Rays from a distant object point will emerge from the system again parallel and to the same side of the axis from which they entered. Their exit angle β increases with the entrance angle α according to $tg\beta/tg\alpha = c$. "Now, the situation in the cones is not as simple as in such a lens combination. Initially I believed that the convex cornea surface forms the first lens of such a combination, and the curvature of the cone tip the second. This is not the case, however, as I learned by intensive studies. Rather, the action of the cone is caused, at least predominantly, by its layering which makes it a lens cylinder. Of course the curved surfaces will support this action."

In Fig. 8 a cone attached to its cornea has been drawn to scale. From a distant object *ab*—with point *a* being on axis—parallel light bundles meet the cornea at different angles: *mn* axoparallel and *pq* at an acute angle with the axis. If a homogeneous refractive index of 1.55 is assumed, the spherical surface of the cornea would bring parallel rays to a focus far behind the cone, as indicated by the dotted lines m_2 and n_2. If in addition to this first weak refraction the action of a lens cylinder is assumed, the bundle *mn* will be united in the image point a'. Similarly, the rays *p* and *q* coming from the object point *b* will form image point b'. This real image $a'b'$ is not directly observable, being inside the cone. However, when the cone tips of celloidin embedded material are cut away, a small inverted image appears near the section plane of a cone. Since the refractive index of the embedding medium celloidin ($n = 1.56$) compensates the refractive power of the cornea surface, the

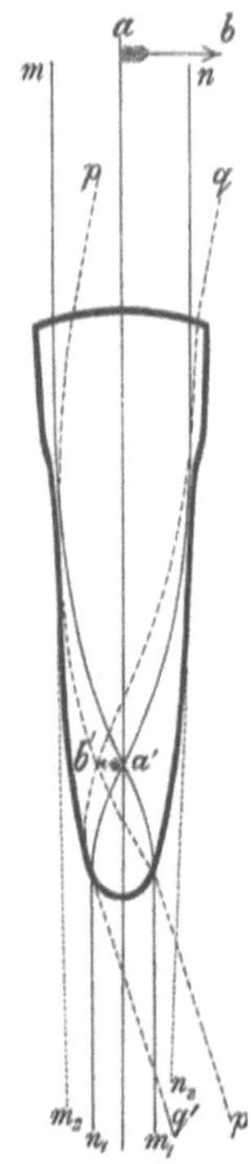

Fig. 8. Ray path in a cornea-cone unit of *Lampyris* drawn to scale. *ab*, object at infinity; *b'a'*, its image inside the cone. *mn*, incident axoparallel rays from *a*; *pq*, incident parallel rays from *b*; n_1m_1, and *q'p'*, respective emerging rays. The emerging rays m_2n_2 would result if the rays *mn* are refracted only by the surface of the cornea. (EXNER, 1891)

observed little image must be due to refraction by the inhomogeneous medium of cornea and cone. The position of the intermediate image *a'b'* is in the rear half of the cone.

Apparently, each cornea-cone unit acts like an astronomic telescope adjusted for infinity in which the refractive power of curved surfaces and of the lens cylinder are combined. In Fig. 9 identical cornea-cone units *kk* are radially arranged with their axes *a–h* intersecting in o. Parallel light meets each facet at a different angle. It is deflected by each unit toward that axis which is parallel to the direction of incidence. The deflected rays intersect approximately at *B*. Likewise, rays from a far object point on the axis *b* will intersect approximately at *B'*. In this way an image *BB'* is formed. In the eye of *Lampyris* about 30 dioptric systems combine for one image point.

The rays from one object point do not intersect in a stigmatic image point. The geometric construction of Fig. 9 shows them to form a caustic surface. In addition, the distinctness of the superposition image will depend on the diameter of the ray bundles which are superimposed.

It should be pointed out that the dioptric systems need not be afocal to form a superposition image. When a lens cylinder is longer than its focal length or an additional convex surface is present, the chief rays will be inclined in that way

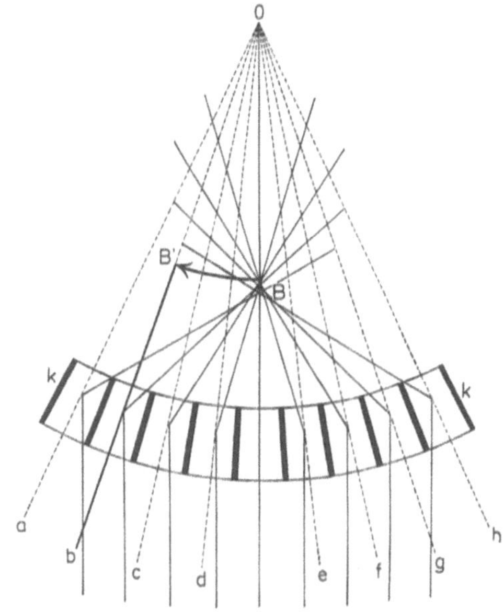

Fig. 9. Superposition ray path. *kk*, radially arranged dioptric systems with axes *a–h* which intersect in *o*. *BB'*, superposition image. (EXNER, 1891)

which is necessary for a superposition image. The rays from one object point might then even diverge slightly when they leave the dioptric system. In this case, the retinal image would be less sharp than in the ideal superposition image. This defect would be corrected at least in the light-adapted eye by the interposing iris pigment which will be discussed presently. This or similar modifications of the superposition image are conceivable.

The formation of a superposition image requires a wide transparent space between dioptric systems and the retina, as was shown in the histological section of Fig. 7. There, the distal pigment, contained in pigment cells, is packed around the cones. The pigment position shown is typical for a dark-adapted eye. In the light-adapted eye, this pigment is positioned in the zone proximal to the cones. However, cone tips and the space around the ommatidial axis remain free. Thus, a sleeve of pigment moves backward.

The functional significance of this pigment movement becomes apparent when the interaction of the distal pigment with the superposition ray path is considered. In Fig. 10, a schematic eye is illuminated by light parallel to the axis *ab*. Cornea and cones deflect it toward point *b* in the retina *nn*. In the lower portion of the eye, where the pigment is in the position for dark adaptation, light from all cones can reach point *b*. In the upper half, however, the proximally displaced pigment intercepts the light rays converging on *b*, with the exception of rays near the axis. It is apparent that during the transition of the pigment from the dark-adapted to the light-adapted state, more and more light is intercepted, starting at cones far from the axis. Accordingly, the relative brightness at point *b* will decrease. This process can be compared to the iris in the vertebrate eye which also reduces the diameter of

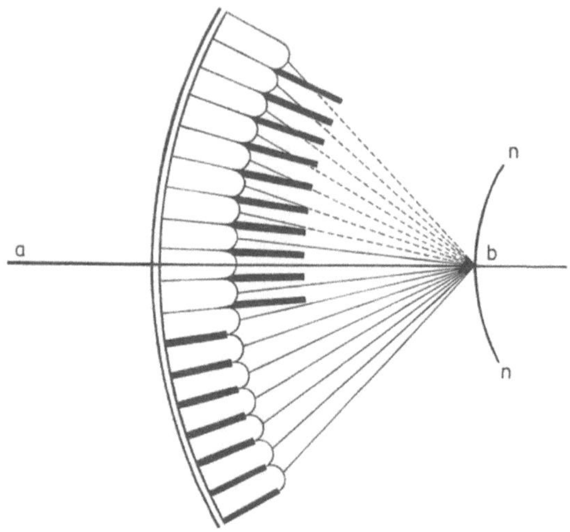

Fig. 10. Superposition ray path and pigment migration. *ab*, direction of incident light; *nn*, retina. *Lower portion:* dark-adapted state with pigment in distal position. *Upper portion:* light-adapted state with pigment in proximal position intercepting strongly inclined rays *(dotted lines)*. (EXNER, 1891)

the light cone illuminating one point in the retina. Therefore, EXNER called the distal screening pigment "iris pigment".

This effect of the moving iris pigment is only possible in eyes which form a superposition image. Interestingly, this pigment movement is found only in nocturnal arthropods, i.e., in insects and Crustacea which have to use their eyes by night and also under daylight conditions.

Besides *Lampyris*, EXNER investigated a number of insects for the existence of superposition images. He was able to observe superposition images in other beetles in whose eyes cornea and cones are united *(Agrypnus, Luciola, Cantharis, Rhagonycha)*. More difficult is the optical investigation of compound eyes in which cornea and crystalline cones are separate. Preparations of these eyes had to be investigated with the soft tissue in situ or else the crystalline cones would have been disarranged. Therefore, EXNER prepared eye cups either by freeze sectioning or by sectioning of alcohol-fixated eyes. In this way he investigated *Hydrophilus* and observed superposition images in some nocturnal Lepidoptera *(Sphinx, Ocneria, Leucoma, Cnophria)*. However, he also succeeded in slicing off eye cups from the fresh eyes of some beetles *(Cetonia, Tropinota)*, and in finding superposition images.

In crustacean eyes, EXNER failed to obtain superposition images due to the soft, delicate consistency of the crystalline cones which disintegrated in the preparation even after freeze sectioning. He proposed, however, the existence of superposition images for these decapod eyes that have a wide transparent zone in front of the retina, that show the typical light-induced movement of the iris pigment, and in which the "eye glow" can be observed.

The eye glow of arthopods was first reported for Lepidoptera (KLEEMANN, 1792). It can most frequently be observed in compound eyes in which the rhabdoms are embedded in a reflecting tapetum. In insects, the tapetum consists of tracheoles (LEYDIG, 1864); in Crustacea, of a reflecting pigment. The most perfect glow appears in eyes which form superposition images. (EXNER describes eye glow in eyes which form apposition images too, and discusses it in connection with their optical properties.) As a rule, the glow can be seen when the dark-adapted compound eye, the eye of the observer, and a light source are approximately arranged in one line. This is best achieved with an ophthalmoscope. Incident and reflected light rays are therefore more or less parallel. In both cases a sizable luminous disc is seen irrespective of the direction from which the animals eye is viewed.

The explanation of the arthropod eye glow is analogous to that of the vertebrate eye. The tapetum behind the retinular cells reflects light diffusely and enables light to pass through the rhabdoms twice. Thus, if we consider point *B* in Fig. 9 as a luminous object, rays going out from it are deflected by the cones and cornea in the same way as the incoming rays were deflected. The result is again parallel emerging light. If the conditions were exactly as just described, the eye glow would only be visible with the help of an ophthalmoscope. For the compound eye this is even less true than for the vertebrate eye. The directions of light incidence and of observation may deviate a good deal depending on the observed species and the pigment position in the eye. Also, with parallel incident light the reflected light forms a cone when emerging from the eye. The reason is that even for a point light source the retinal image is not an ideal point, particularly as it extends into the depth of the retina, which is of considerable thickness. In addition, diffusion of light in the rhabdoms and the tracheoles must be taken into account.

The eye glow of the dark-adapted eye changes when the eye is continuously illuminated with a strong light. The glowing disc constricts more and more, becomes less bright, and finally disappears. Apparently, this is caused by the moving pigment, as explained in Fig. 10. The light reflected by the tapetum is affected by the pigment in the same way as the incoming light. Beginning with rays most inclined to the eye radius through the point of reflection, an increasing number of reflected rays are hindered from reaching the cones.

"The two forms of compound eyes described above—of *Limulus* and *Lampyris*—constitute types to which almost all arthropod compound eyes studied by myself can be traced back, despite their enormous diversity. In the mixed forms discussed below, which perhaps could be viewed as special types, I see affirmation of the concept that those two are basic types.

The essential and anatomically easily recognizable difference between both forms of eyes is the functionally determined position of the retina, at least in so far as (1) a superposition image is only possible in eyes with a thick layer of transparent tissue between the dioptric apparatus and the sensitive layer of the retina so that a retinal element can be met by rays emanating from many cone tips and (2) an eye with retinal elements that are situated in close vicinity to the cone tips can form an apposition image."

Some observations suggest that eyes exist that in the dark receive a superposition image but by day an apposition image. Actually, the retinal image in the *Lampyris* eye is in bright sunlight an apposition image, if one assumes that the pigment sleeve is dense and narrow enough to channel light from one facet to only the corresponding rhabdom. However, numerous compound eyes exist that are intermediates between the two described forms. In these eyes the rhabdoms consist of two parts: a thick, fully developed proximal one and a very thin distal one. The layer of the thick proximal part, by form and position, can be compared to the *Lampyris* retina. With the iris pigment in light position only an apposition image will be formed. With the pigment in dark position a superposition image can be formed in the region of the thick rhabdom portions. Although the thin rhabdom portions seem to be rudimentary, they might still be light sensitive. In this case the visual acuity of the dark-adapted eye will be decreased because oblique rays from one object point can stimulate a number of rhabdoms in their thin distal parts.

IV. The Catoptric Eye

EXNER explained the formation of retinal images in the two types of eyes described so far exclusively by the dioptric action of lens cylinders and spherical surfaces. This kind of explanation appeared to him insufficient for the eyes of *Phronima* (Crustacea). *Phronima*, like some Crustacea and insects, has a lateral and

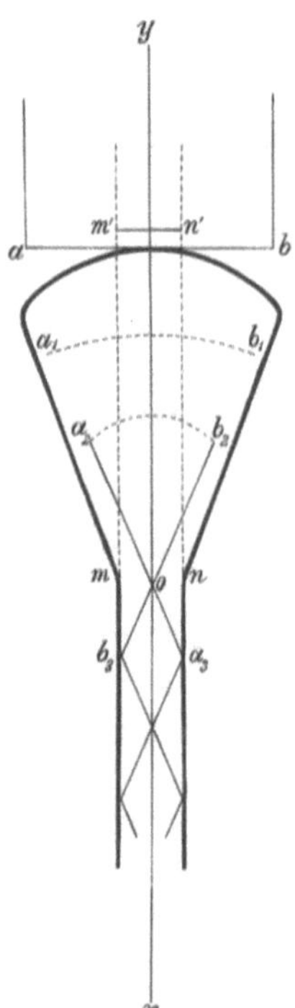

Fig. 11. Scheme of light propagation in the crystalline cone of *Phronima*. *yx*, axis; *ab*, bordering rays of axoparallel incident light bundle and their successive states; *mn*, distal end of light-guide. (EXNER, 1891)

a dorsal compound eye on each side. The crystalline cones in these eyes are remarkably elongated. They consist, in the dorsal eye, of a little distal bead tapering into a long thread. EXNER assumed that the distal beads act like collecting lens cylinders and that the threads conduct the light to the rhabdoms by way of total reflection. His explanation of the light-collecting and guiding system is given in Fig. 11. Incident light parallel to the axis *yx* of the crystalline cone is collected by the curved surface and the lens cylinder. The light bundle *ab* comes to a focus in o. If no refraction takes place, only a light bundle of the diameter *m'n'* can enter the thin thread at *mn*. The refraction of light leads to a considerable increase of the light flux in the thread. Rays slightly inclined to the axis are also collected in the thread, all others eliminated. EXNER tested his hypothesis in thick sections of a Celloidin-embedded lateral eye. He found that illumination of the distal end of a crystalline cone makes the cut end appear bright "as in a glass rod".

V. Discussion

EXNER'S papers and especially his book on the physiology of facetted eyes are still, after eighty years, interesting and rewarding (but, not always easy) to read. I have tried to represent his work as concisely as possible, although in a very abridged form. "There is much more in his book, which is a record of by far the most complete research yet made on the physiology of the compound eye..." (ELTRINGHAM, 1919). This citation seems to indicate the impact of EXNER'S ideas on the physiologists of his time and the following decades. The terms apposition eye and superposition eye[1] soon became part of the standard vocabulary of comparative physiology.

In the following seven decades only comparatively few studies in the field of arthropod physiologic optics were reported. There still remained a number of problems. Most important, detailed quantitative measurements of the refractive indices in the dioptric systems of arthropod eyes did not exist. Although conclusively deduced by EXNER on the basis of the optical performance of these elements in *Lampyris* and *Limulus*, direct evidence for the proposed gradient of the refractive index was incomplete. This made the general concept of lens cylinders for all arthropod eyes questionable. (EXNER himself found to his surprise that the pseudocone of the dipteran eye had apparently no dioptric properties.) Furthermore, in eucone insect superposition eyes only poor superposition images had been observed by EXNER and in those of Crustacea none at all. (Only PARKER, 1897, reported a superposition image in *Astacus.*)

The concept of the apposition eye remained relatively undisputed. Points of interest were the relevance of the small image behind the cone in each ommatidium and the overlap of visual fields of adjacent ommatidia.

By contrast, new research of superposition eyes (DE BRUIN and CRISP, 1957; NUNNEMACHER, 1959, 1961; KUIPER, 1962), with results discordant to the super-

[1] I was unable to find out who coined or defined these terms. EXNER himself never used them in his papers. The earliest mention of "Superpositionsauge" (superposition eye) seems to be in DEMOLL (1910). General agreement exists for the use of these terms instead of EXNER'S diction "eyes which form apposition images" or "superposition images".

position theory, started to throw doubt on it. The main points of criticism were the lack of observation of superposition images in crustacean superposition eyes and the failure to observe refractive index gradients in crystalline cones despite the use of the new method of phase-contrast microscopy. The issue is well covered in reviews by GOLDSMITH (1964) and GOLDSMITH and BERNARD (1974). As an alternative for eyes, anatomically of the superposition type, the new theory was proposed that structures of high refractive index (crystalline tracts), extending between cone and rhabdom of one ommatidium, act as light-guides (essentially in the same way as proposed by EXNER for the eye of *Phronima*). This and other alternatives are discussed by HORRIDGE (1971, 1975).

The issue became controversial when KUNZE (1969) reported results which were in accordance with EXNER'S views: the existence of erect images behind each cone and a superposition image in situ as well as eye glow with partial illumination of the eye. These observations could not be explained on the basis of light conduction in crystalline tracts, but were in accordance with the superposition theory. However, KUNZE (1969) also stated that the apparent ray path of the superposition theory found in the moth eye does not disprove the possibility of an optical function of the crystalline tracts. Both light conduction and image superposition can take place simultaneously. Research by several workers followed on a variety of Coleoptera, Lepidoptera and decapod Crustacea which confirmed the superposition principle of EXNER. There is no doubt that the superposition principle is realized in insect eyes in its original form.

D. Superposition Eyes

Superposition eyes are characterized by—at least in the dark-adapted state—a wide transparent space (clear zone) distal to the rhabdom layer allowing the formation of an erect image by superposition of light entering the eye through different facets.

I. The Eye of *Ephestia*

1. Anatomy

The compound eye of *Ephestia* has almost the form of a half sphere with a radius of approximately 340 μm; in the rear portion the eye radius is somewhat shorter. The eye contains 2000–2500 ommatidia hexagonally arranged with frequent irregularities. From anatomic measurements the divergence angle between neighbouring ommatidia is $\approx 3°$.

An average ommatidium (Fig. 12) consists of a hexagonal cornea, four crystalline cone cells (Semper cells) which contain the crystalline cone, two

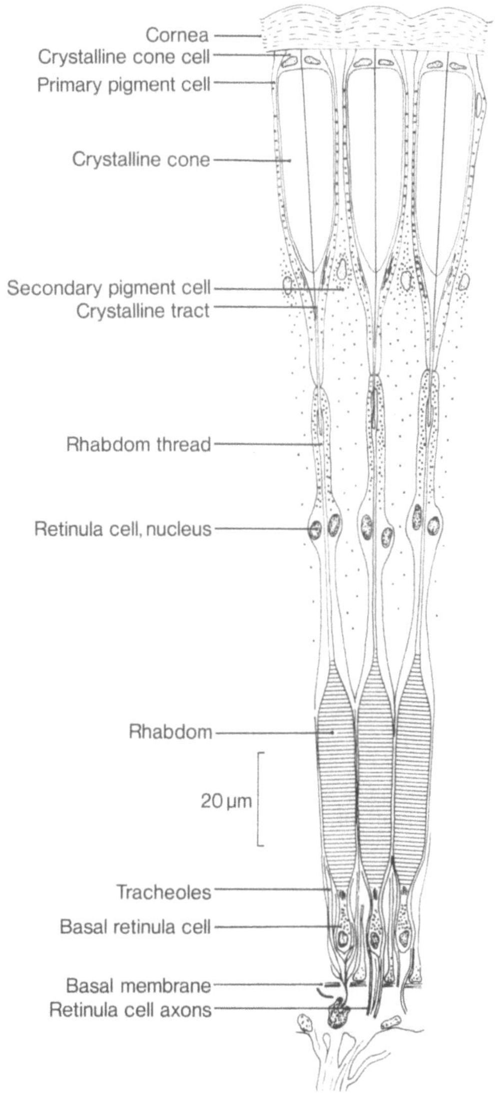

Fig. 12. Axial section through ommatidia of *Ephestia*. (FISCHER and HORSTMANN, 1971, slightly changed)

primary pigment cells surrounding the cone cells, ten retinula cells (including a basal one), two secondary pigment cells, and a number of tracheoles (FISCHER and HORSTMANN, 1971).

Occasionally, ommatidia with a differing number of cone cells (3–5) are found. The number of retinula cells varies consistently between 9 and 12. One cell in each retinula is a basal cell next to the basal membrane with microvilli parallel to the axis of the ommatidium. The microvilli of the retinula cells form a fused rhabdom which, in a radial section, is cigar shaped. At its biggest diameter it is only

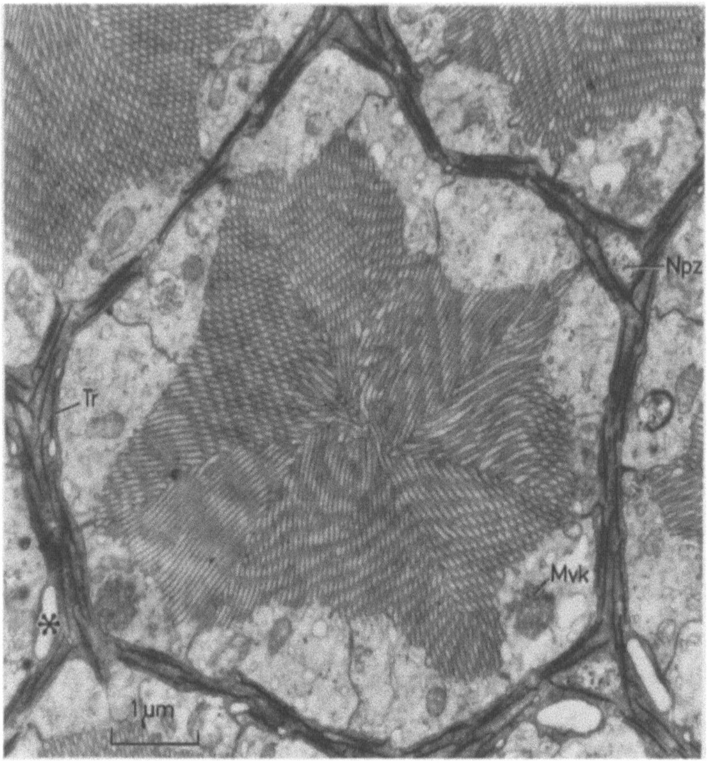

Fig. 13. Cross section of ommatidium in the region of distal tracheole ends. *Tr*, closed tracheole; *, open tracheole; *Npz*, secondary pigment cell; *Mvk*, multivesicular body. (FISCHER and HORSTMANN, 1971)

surrounded by a narrow cytoplasmic seam (Fig. 13). The tracheoles that form the tapetum also extend to this distance from the basal membrane. The distal conical portion of the rhabdom narrows into a thin thread (the "Achsenfaden" of UMBACH, 1934) of 0.5 µm diameter that extends over the distal length of the retinula cells (Fig. 14). Each retinula cell continues beyond the basal membrane as a postretinal fiber (axon).

The retinula cells contain pigment granules in the region distal to the nuclei. Sharing the secondary pigment cells with the neighbor ommatidia, each retinula is surrounded by six secondary pigment cells, a number that also seems to vary between 6 and 10 (HORRIDGE and GIDDINGS, 1971). Basally, the retinula cells are embedded in a basket of tracheoles which surround it densely and form a reflecting tapetum.

In Fig. 12 the ommatidia are presented in the light-adapted state. The pigment granules in the secondary pigment cells are distributed from the region of the cone tips to the rhabdoms. The nuclei of the retinula cells are situated about halfway between distal cell end and the widest rhabdom diameter.

On dark adaptation, drastic changes occur (Fig. 15). The pigment granules of the secondary pigment cells move completely into the cell portions situated between the cones leaving a wide transparent space between cones and rhabdom

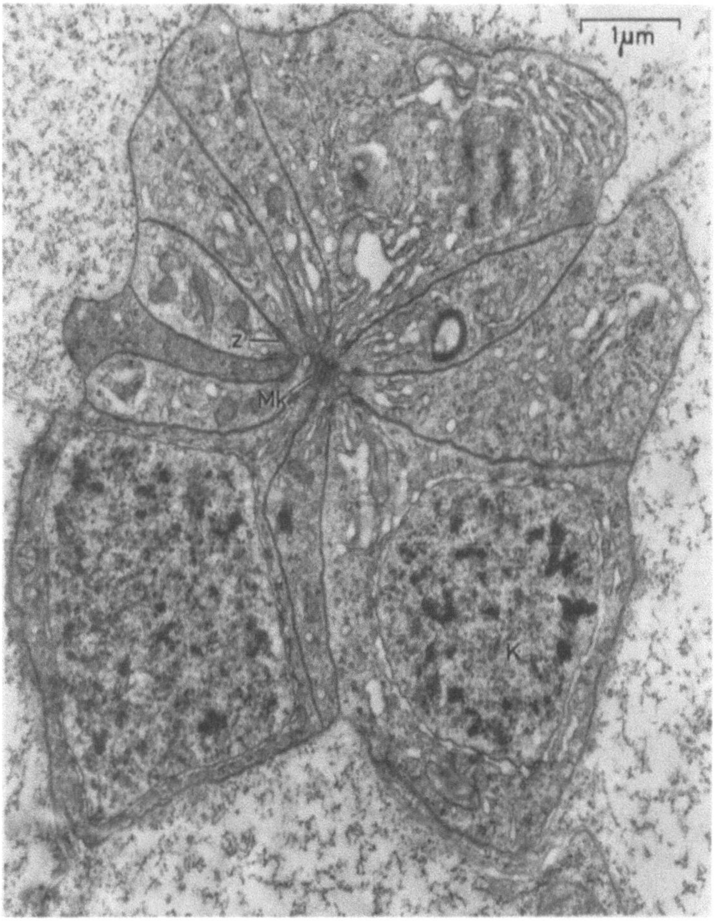

Fig. 14. Cross section of retinula in the nuclear zone. *K*, nucleus; *Mk*, microvilli; *Z*, zonula adhaerens. (FISCHER and HORSTMANN, 1971)

layer. The retinula cells extend towards the cones and the nuclei move into the distal cell end. Complete light adaptation requires ~12 min, dark adaptation 20 min, as determined by histologic control (DAY, 1941).

2. Optical Parameters

The optical parameters—refractive indexes and geometry—of the light-transmitting components in the *Ephestia* eye are not easily available. The minuteness of structures and the necessity of taking measurements from histological preparations or excised components, which may have changed their properties, are considerable handicaps.

However, reliable data for cornea and cone of the *Ephestia* ommatidium do exist now which allow the construction of ray paths in the dioptric system.

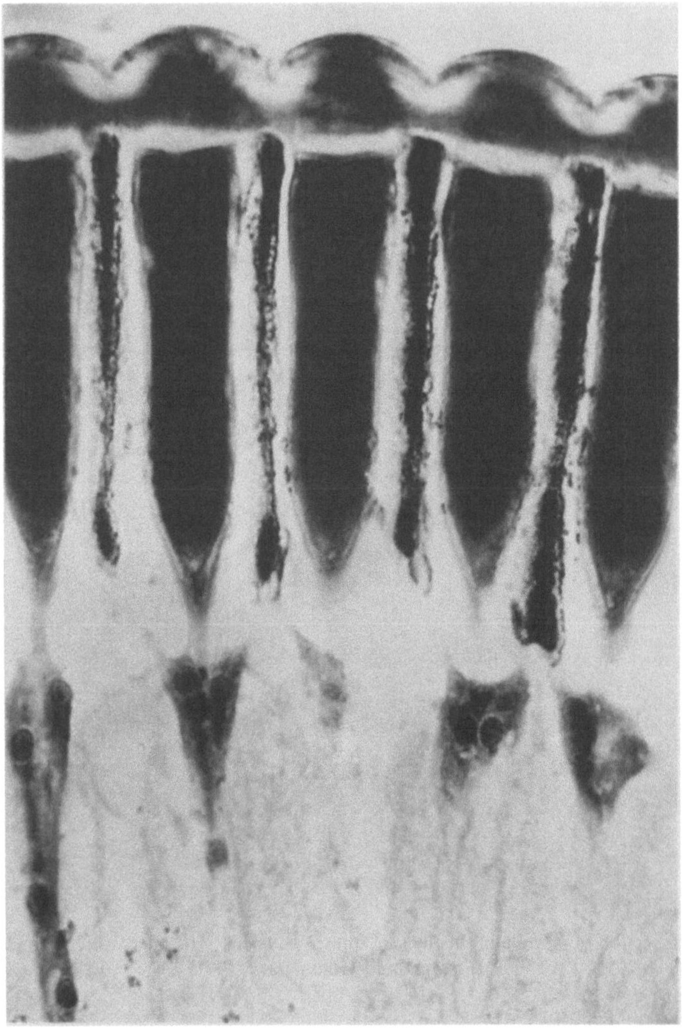

Fig. 15. Radial section through periphery of dark-adapted eye of *Ephestia*. Araldite, 5 μm, toluidin-blue

Refractive index measurements of other eye structures (e.g., crystalline tract, distal retinula cell column, and pigment cell contents) are not available.

a) Cornea

A thorough investigation of the *Ephestia* cornea was performed by VOGT (1974). The smallest diameter of the hexagonal cornea facet is 20 μm and has the shape of a plano-convex lens. The spherical surface which forms the boundary to air has a radius of 14 μm. The axial height is 10 μm.

Measurements of the refractive index were taken in the interference microscope after Jamin–Lebedeff. Observation of the facet in axial direction shows the refractive index distribution to be radially symmetrical. Thus, from an axoparallel

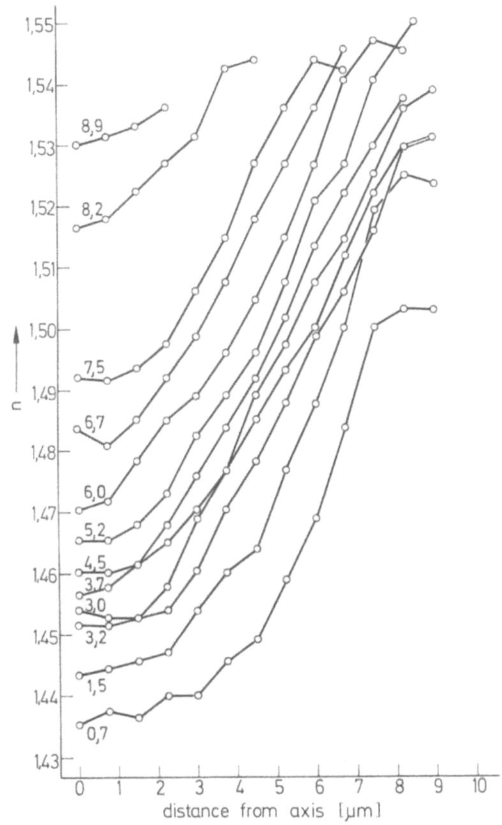

Fig. 16. Refractive index measurements in one half of an axial section, 3-μm thick, through the cornea of *Ephestia*. *Ordinate:* refractive index n; *abscissa:* distance of measuring points from cornea axis. Parameter: distance from inner cornea surface in micrometers. (VOGT, 1974)

section containing the axis all information can be obtained about the refractive index distribution in the entire cornea. An exact method of obtaining such sections is shown in VOGT (1974).

The cornea facet shows a radial as well as an axial variation of refractive index (Figs, 16, 17). The refractive index increases from the axis to the periphery and from proximal to distal. Lowest values of $n = 1.43$–1.44 are found in the center of the facet basis, maximal values of $n = 1.54$–1.55 at the cornea surface. The integral over the axial values measured in cornea sections ($n = 1.47$) coincides with the axial refractive index measured on intact cornea facets.

Optical studies of the isolated cornea show that each facet produces a sharp real inverted image at a focal distance of 37.5 ± 3.5 μm from the basis, with water in the image space. With the formula for thick homogeneous lenses (since the radius of curvature and thickness are also known) an effective refractive index (total index) of $n = 1.397$ can be calculated. This value is significantly lower than any measured refractive index in the cornea. Thus, the refractive index distribution has the effect of a divergent lens. This can be tested by compensating the refractive

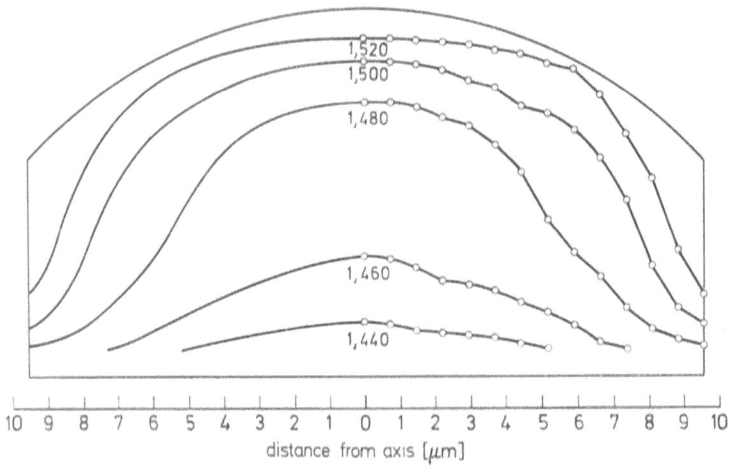

Fig. 17. Refractive index distribution constructed from measurements as in Fig. 16. (VOGT, 1974)

power of the spherical cornea surface with a medium of $n=1.41$. Although it is much smaller than the surface refractive index of 1.54 only virtual images can be observed. Thus, direct observations on the unsectioned lens confirm the interferometric measurements of the refractive index distribution.

b) Crystalline Cone

HAUSEN (1973) has studied in detail the optical properties of the crystalline cone.

The cone has a length of almost 42 μm. The distal end is slightly curved. 5–6 μm from it the biggest diameter of 14 μm is found. From there the cone tapers gradually to the proximal tip. The cone is enclosed in the cone cells, the nucleus containing parts of which separate it from the cornea.

To obtain cross sections of fresh cones, an eye was covered with a transparent varnish and allowed to dry. It was opened with a vibrating razor blade and filled with a warm gelatine solution. After cooling the gelatine was solid enough for cutting sections of the eye 10–15 μm thick. Circular cross sections of cones were used for refractive index determinations in the interference microscope after Jamin–Lebedeff.

In all cross sections a radial variation of the refractive index was obvious (Fig. 21a). Measurements (one example is given in Fig. 18) showed that in first approximation the refractive index variation can be described by

$$n = n_A - a x^2 \tag{1}$$

with n = refractive index, n_A = axial refractive index, x = distance from axis, and a being a constant. In all preparations an edge refractive index of $n_E = 1.383$ was found, irrespective of the diameter of the cross section. Therefore this index can be assumed to be the edge refractive index for the entire cone. Highest axial values found were around $n = 1.50$.

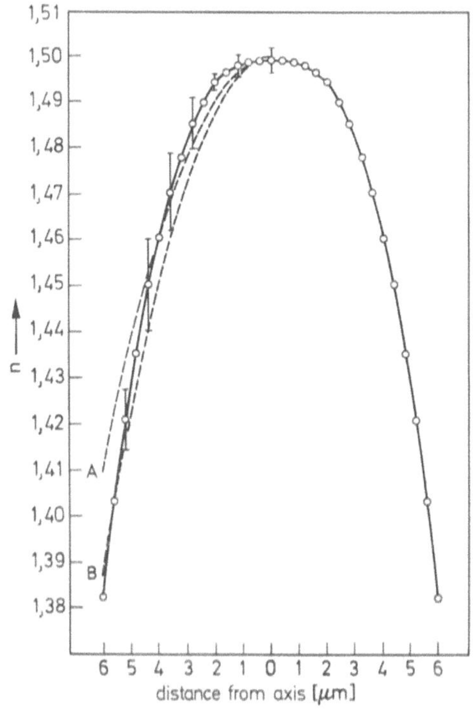

Fig. 18. Refractive index measurements in cross sections from the distal half of the crystalline cone of *Ephestia*. Mean values from four preparations. *A*, *B*, calculated curves of the form $n = n_A - ax^2$. (HAUSEN, 1973)

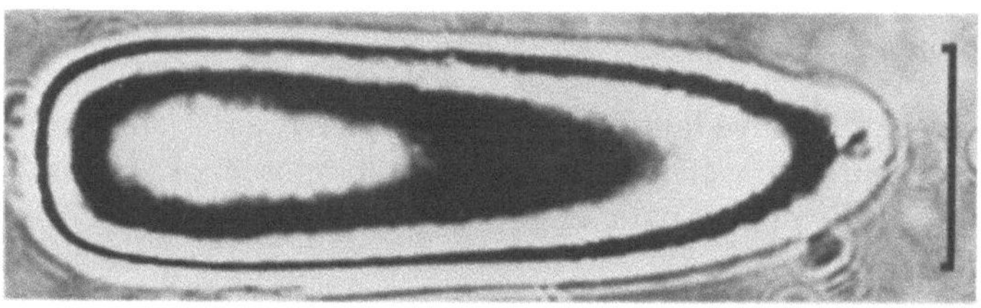

Fig. 19. Appearance of an isolated complete cone under the interference microscope. *Marker* 10 μm. (HAUSEN, 1973)

The sections were much to thick to determine the index distribution in the cone with acceptable accuracy. For this purpose an indirect method was used which was based on the assumption that function (1) is valid for each cross section. Whole fresh cones can be freed mechanically from the cone cells in an isotonic solution. When transferred to gelatine for preservation they showed no changes for 1–2 days. Their appearance in the interference microscope (Fig. 19) is due to different optical path lengths which depend on changes in geometry and refractive index. Refractive index determinations in such a complete cone are only possible in the median line (with the exception of the very distal and proximal ends). For each point on this line the geometric path length can be regarded as being approximate-

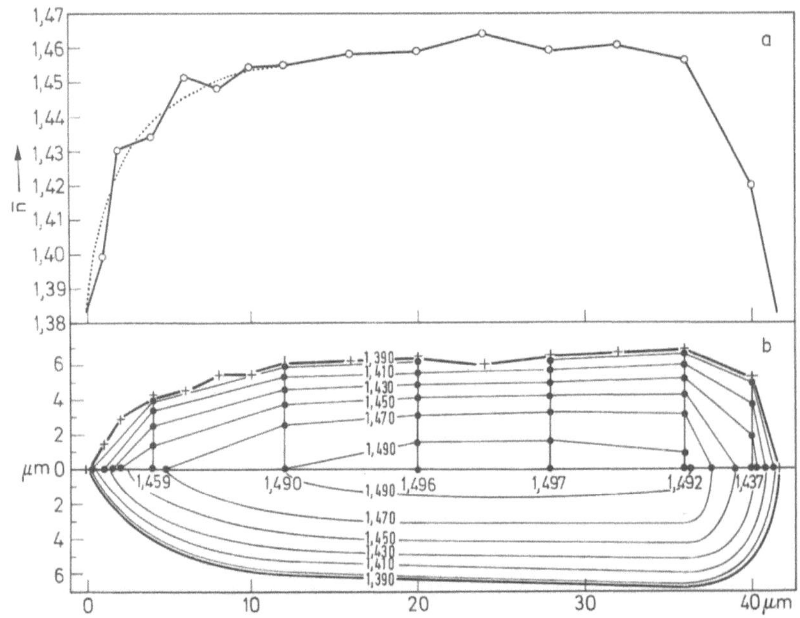

Fig. 20. *a* Average refractive index $\bar{n}$ in median line of complete cone. *Abscissa*: distance on cone axis, o = cone tip. *b* Construction of refractive index distribution in the cone. Same abscissa as in *a*. (HAUSEN, 1973)

ly equal to the cone diameter. Thus, the optical path length for a measuring ray in a median point C is

$$W = 2\bar{n}_c r_c \tag{2}$$

with $\bar{n}_c$ = average refractive index for C, and r_c = cone radius containing C. For a ray in the median plane of a cone which is composed of infinitesimally thin concentric layers the optical path length is

$$W = 2 \int_{x=0}^{x=r_c} n dx \tag{3}$$

From (1), (2), and (3) the following equation is obtained:

$$\bar{n}_c = n_A - \tfrac{1}{3} a r_c^2 . \tag{4}$$

For point C in the cone edge ($n = n_E$) follows from (1)

$$n_E = n_A - a r_c^2 . \tag{5}$$

From (4) and (5) follows

$$n_A = \tfrac{1}{2}(3\bar{n}_c - n_E) \tag{6}$$

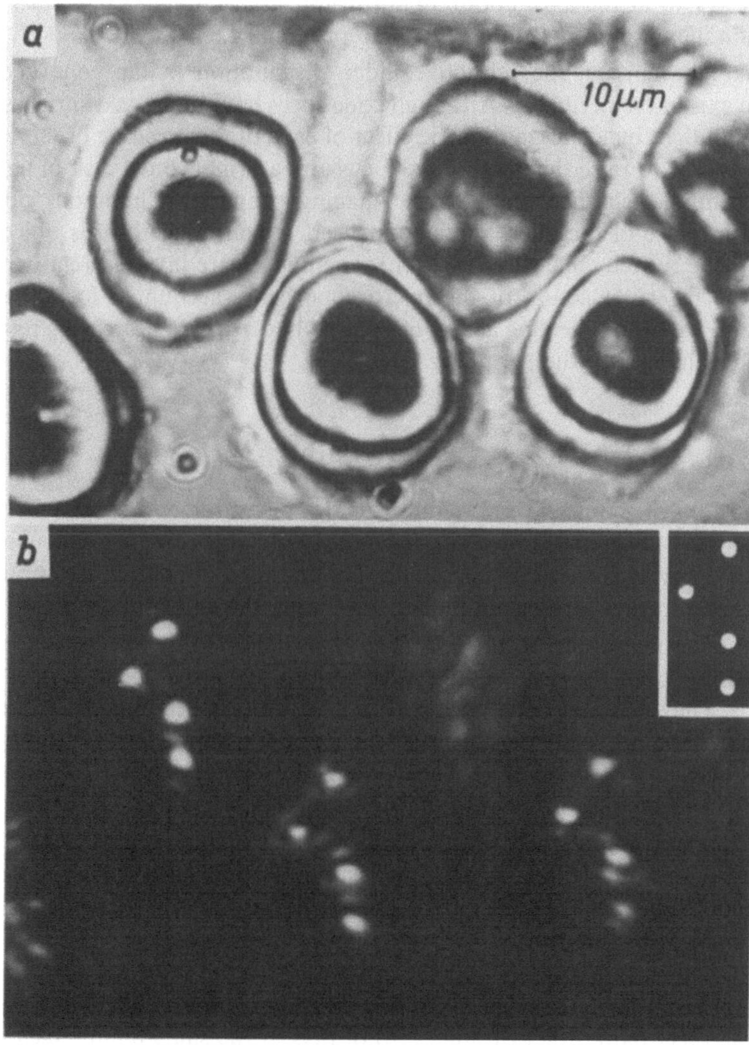

Fig. 21. *a* Cross sections (10–15-µm thick) of crystalline cones as observed through the interference microscope. The interference rings in the cone slices are the expression of radial variation of refractive index. *b* Real images of an object which is situated in front of the same preparation that was used in *a*. Same scale as in *a*. The inset shows the object; maximal distance of light points is 15 mm. (KUNZE and HAUSEN, 1971)

and

$$a=\frac{n_A-n_R}{r_c^2}. \tag{7}$$

The measurements of whole cones yield the values $\bar{n}_c$ and r_c (Fig. 20a). Since n_E is known, the refractive index distribution of the cone can be constructed. This has been done for a cone in Fig. 20b calculating six profiles of the refractive index distribution (indicated by *vertical lines*). The cone has a core of highest refractive index of ≈1.50 in the axis, about 14 μm from the distal end. The values of the constant a increase from distal $2.0 \cdot 10^{-3}$ to proximal $4.6 \cdot 10^{-3}$.

Unlike the cornea, the refractive index distribution in the cone decreases from the axis to the periphery. Although cone slices have no curved surfaces in the direction of light incidence, a dioptric effect of each of them can be observed. The simplest way is to raise the microscope focus and a bright, small disk of light will appear behind each slice. To test the imaging property, a luminous object is installed in place of the microscope condenser. A real image of this object now appears behind each cone slice (Fig. 21b). The images are inverted relative to the object (the inverting action of the microscope being accounted for); this is in accordance with the collecting action of a lens cylinder with a decreasing refractive index from the axis to the periphery.

3. Ray Path

The results of geometric and refractive index measurements by HAUSEN (1973) and VOGT (1974) were used by CLEARY et al. (1977) for a computer calculation of ray paths applying Snell's law for infinitesimally thin sections. (The computation pertains only to rays incident in planes containing the axis of the system.)

Light entering parallel to the axis (Fig. 22) is collected by the corneal lens. In the crystalline cone, the rays follow curved trajectories and come to a focus in the cone axis. They intersect between $x = 31$ and 35 μm (smallest bundle diameter at $x = 34$ μm). From there the rays diverge and are bent towards the axis so that they leave the cone almost axoparallel. The maximal divergence of a ray from the axis is 1.1° in the exit bundle. Two rays suffer total reflection at the cone boundary.

Parallel light bundles with incident angles other than zero also form a focus each in the crystalline cone (Fig. 23). Smallest bundle diameters are found between $x = 34$ and 36 μm. Their distance from the axis increases in good approximation proportional to the tangent of the input angle.

Smallest bundle diameters do not exceed 0.6 μm in any instance. The rays leave the cone on the same side from which they entered the cornea. Thus, each ray follows an erect path. The exit angles (β) of rays in the same bundle have—with one exception—a maximal divergence of not more than 5°.

The exit angle of a ray increases with the input angle (Fig. 24). The regression line has a slope of 1.32 (standard error 0.04, correlation coefficient 0.98). Thus the investigated optical system can be said to have an angular magnification of 1.32.

With input angles more than 12° increasingly less rays are processed by the entire system due to the fact that rays miss the crystalline cone. This results in smaller exit bundles. From input angles of 23° no computed ray passes through the entire cone. Thus, between 22° and 23° the limiting angle for ray processing of the system is reached. In the model, rays were neglected which suffered total reflection at the boundary cone surround. In the material system, total reflection at this boundary does not necessarily occur and has not been observed. Alternatively, it seems possible that these rays are absorbed by pigment granules at this boundary. For lack of evidence, the rays in question were not processed further in the model.

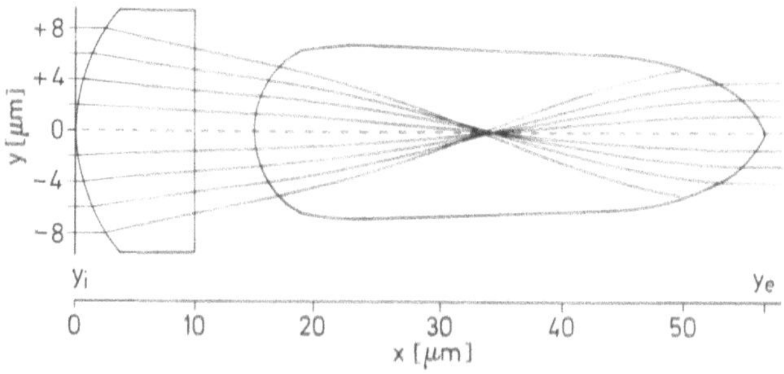

Fig. 22. Ray path in dioptric system of *Ephestia* for input angle $\alpha = 0°$. Starting y are defined for $x = 0$. (CLEARY et al., 1977)

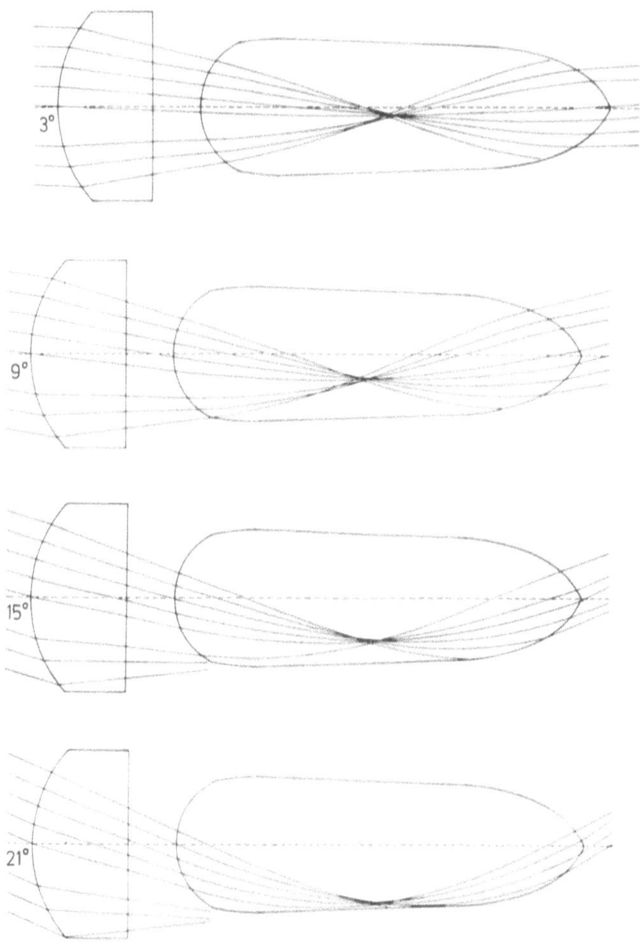

Fig. 23. Ray paths in dioptric system of *Ephestia* for input angles from 3–21°. Same dimensions as in Fig. 22. (CLEARY et al., 1977)

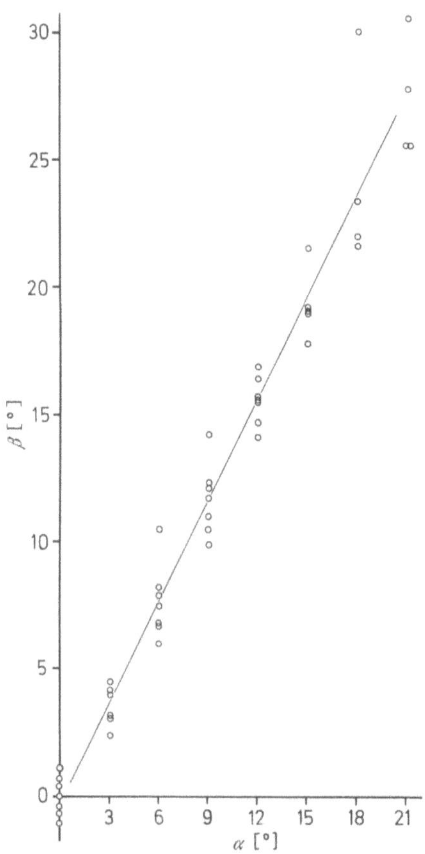

Fig. 24. Exit angle β versus input angle α for dioptric system of *Ephestia*. (CLEARY et al., 1977)

Even if totally reflected, they do not seem to be of determining influence on the performance of the dioptric system.

A result of the ray path construction is that the investigated optical system can, at first approximation, be regarded as a telescopic system with properties analogous to a Kepler telescope adjusted for infinity. Parallel entering light is focused in the plane of the intermediate image and leaves the system parallel. The angular magnification is the same for all input angles. Unlike a Kepler telescope, the ray path is not entirely produced by the refracting power of spherical lenses. A combination of refraction by curved surfaces (predominantly the cornea surface) with refraction by material of inhomogeneous refractive index (predominantly the crystalline cone) determines the ray path. From the model calculation exit light bundles can be selected so as to simulate parallel light falling on to the eye surface. In Fig. 25 they are drawn leaving dioptric systems which are arranged at 3° intervals on a circle of 340 μm radius. They are rectilinearly propagated assuming a homogeneous eye medium. The exit light bundles superimpose; the envelope forming a caustic curve (compare Fig. 9).

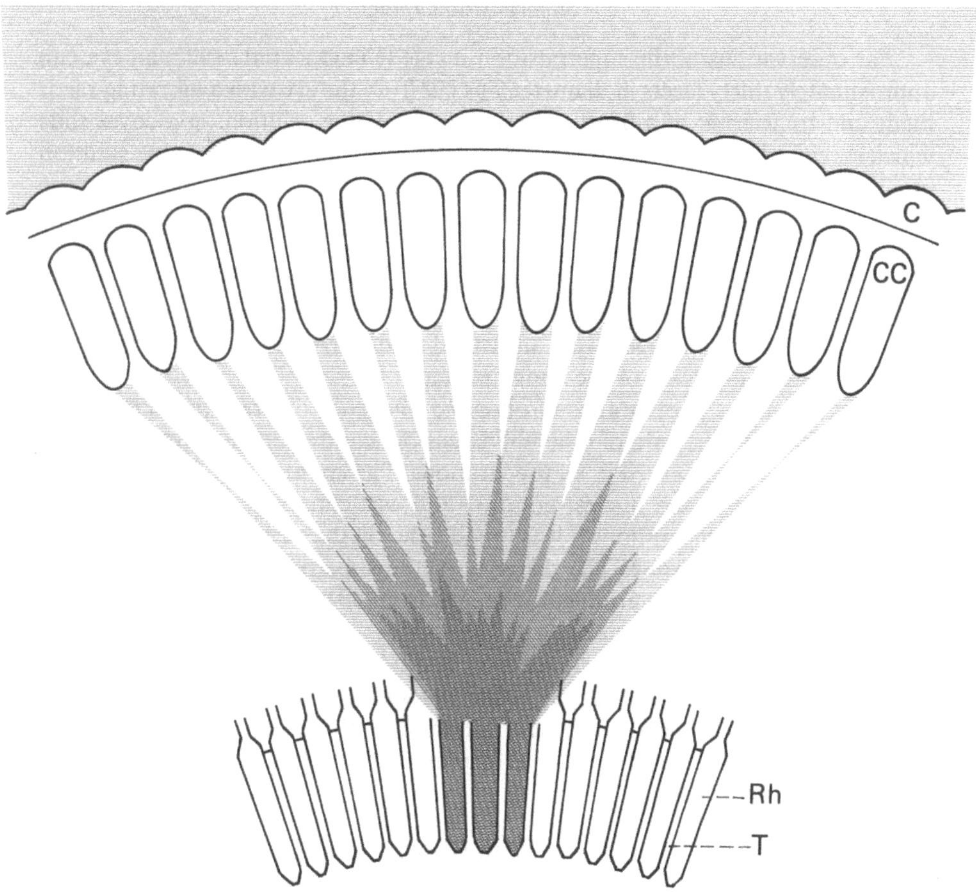

Fig. 25. Superposition of parallel incident light in the eye of *Ephestia*. *C*, cornea; *CC*, crystalline cone; *Rh*, rhabdom: *T*, tapetum. (CLEARY et al., 1977)

The smallest diameter formed by the intersecting bundles is the constructed superposition image of the incident parallel rays. It is situated at a distance of 105 μm from the axial cone tip and is ~25 μm wide. On comparison with the eye anatomy it is found in the region of widest rhabdom diameter with an extension of about three rhabdom diameters.

4. Experimental Tests

Various methods have been used to test the properties of the *Ephestia* eye. They were performed with dissected as well as with intact eyes. Advantageous for such tests is the availability of eye mutants of *Ephestia*, which do not contain screening pigment. With such mutants prolonged observations are possible without the long dark-adaptation times that for the wild type are necessary for obtaining the superposition state.

a) Superposition Image

For observations of the superposition image (CLEARY et al., 1977) an eye cup approximately thirty facets in diameter was cut from the eye with a vibrating razor blade and transferred to a drop of fresh warm gelatine solution on a cover slide which stabilized the attached eye cup. The refractive index of the gelatine solution ($n = 1.339$ at room temperature) was assumed to be close to that of the clear zone in the intact eye. The inverted slide formed the lid of a shallow moist chamber; the lower face of the chamber was also a cover slide. The device was placed under an incident light microscope. Four miniature light bulbs in front of the preparation served as objects for light propagation by the dioptric systems of the eye cup. The position of observed images were measured and corrected accounting for the different refractive indexes of the image media.

Figure 26 shows photographs of observed images at different positions in the same preparation. They are of different extension, contrast, and distinctness. The same variation was observed in all preparations. The image in Fig. 26b appears to be of least extension and highest contrast and was chosen as the superposition image. In different preparations it was found either in the cover glass (as the one in Fig. 26b) or in the gelatine. The distance of the superposition image from the crystalline cone tip was determined. It ranged from 90 µm to 165 µm with a mean of 125 ± 15 µm (standard deviation).

By comparison with the eye anatomy (Fig. 12) 80% of all superposition images were found to be within the region of the rhabdom, which is 85–140 µm distant from the cone tip. The mean distance of 125 µm from the cone tip also corresponds well with the position of the smallest common diameter of superimposing ray bundles from ray tracing.

b) Eye Glow

Another line of evidence is available from observations of the intact eye. Although of an indirect nature, it is obtained from eyes not disturbed by histological and surgical procedures.

Eye glow is readily observed in the *Ephestia* eye, which has a well-developed tracheal tapetum. If the dark-adapted eye is illuminated and observed with an incident light microscope (HOMANN, 1925) using an objective of small numerical aperture (0.15), a circular disc of about 16–18 facets in diameter appears brightly illuminated. The diameter of this disc is essentially equal between *Ephestia* eyes of the dark-adapted wild type and mutants devoid of dark screening pigment. Thus the screening pigment does not seem to interfere with light propagation in the dark-adapted wild-type eye.

The light beam leaving the eye in the eye glow is not exactly parallel as could be assumed from reversing the superposition ray path. HORRIDGE et al. (1972) observed eye glow in *Ephestia* and found the glowing patch to disappear when the angle between the incident parallel light and the direction of observation was 10°–20°. Apart from diffraction effects which can account for about 2° divergence it must be remembered that incident parallel light is superimposed in the rhabdom layer with a diameter of approximately three rhabdoms. This must therefore result in a divergence of light in the eye glow.

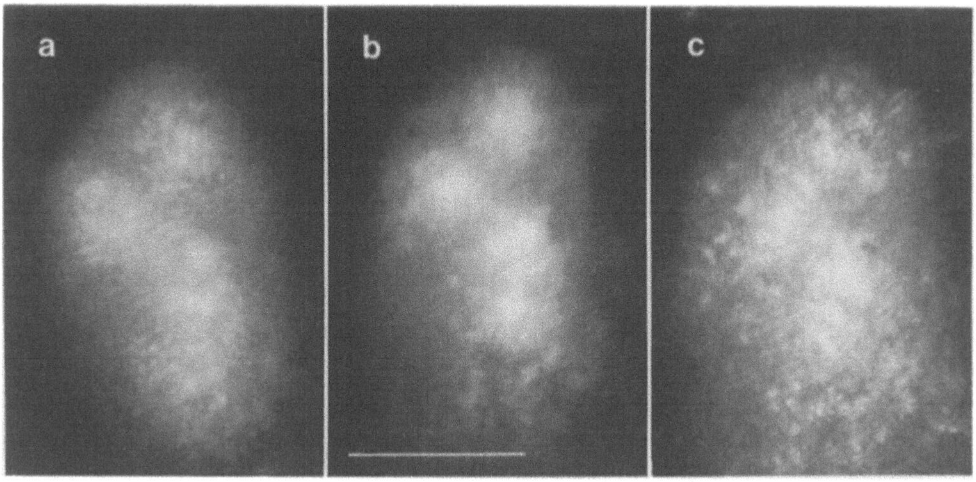

Fig. 26. Superposition images behind an eye cup preparation of *Ephestia*. They were photographed at *a* 110 µm, *b* 130 µm, and *c* 150 µm proximal to the central crystalline cone tip with equal exposure times. Same object as in Fig. 21, 33 mm in front of preparation. (CLEARY et al., 1977)

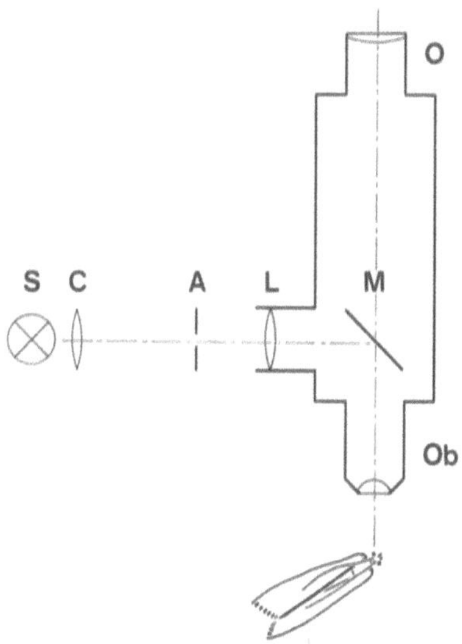

Fig. 27. Setup for observation of eye glow. *O*, ocular; *S*, light source; *C*, collector; *A*, circular aperture which is projected by lens, *L*, and objective, *Ob*, onto the eye of the animal; *M*, semitransparent mirror. (KUNZE, 1972a)

When a small pinhole is inserted in the illuminating light path and projected onto the eye (Fig. 27) it is possible to illuminate only one facet. According to the superposition ray path, light shining into the eye through only one facet should also be reflected out in the entire glow area (Fig. 28). The area within which the glow appears should stay the same when light of constant direction of incidence enters the eye through different facets. When the light pencil is displaced peripherally, the outermost facet will be reached, which takes part in the

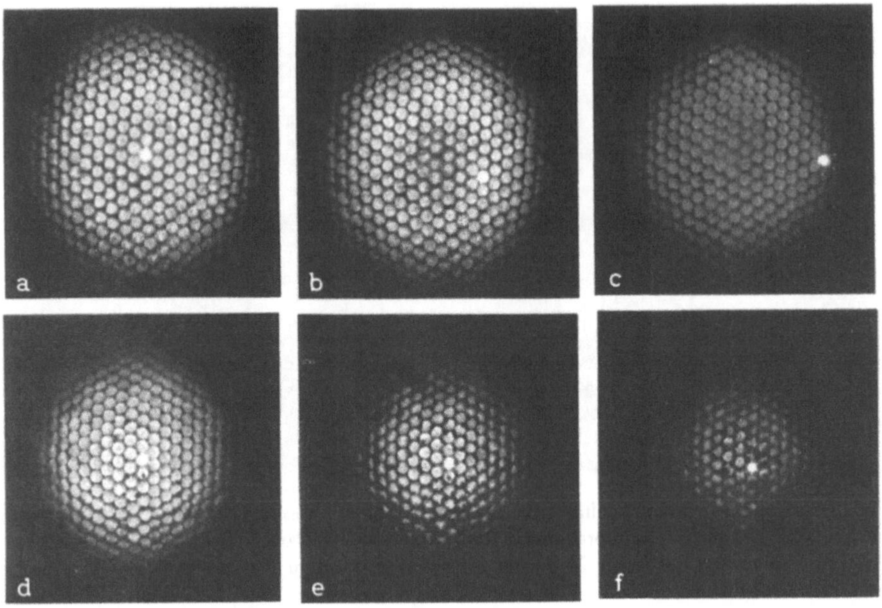

Fig. 28. Eye glow in *Ephestia*. Illumination through only one facet (note cornea reflex). *a–c* mutant transparent (without screening pigment). *d–f* wild type. *d* dark-adapted, *e* 3 min, *f* 5 min after beginning of illumination. (KUNZE, 1970)

superposition (Fig. 28c). Illuminating the eye through this facet results in an eye glow of much reduced intensity, since only a fraction of light incident under such an angle with the ommatidial axis will be transmitted by the dioptric system. This observation, as well as the extension of the disc within which the eye glow can be elicited, check well with the calculated ray path (Fig. 23).

In the dark-adapted wild-type eye prolonged bright illumination will cause the screening pigment to move. Under the microscope focused on the cornea it can be observed that each glowing facet is gradually constricted. Simultaneously the glow intensity and the radial extent of the glow area decrease (Fig. 28d–f). The latter two effects have already been explained by EXNER (see Fig. 10).

The constriction of each single facet is apparently caused by the pigment migration. As has been shown, the screening pigment, which in the dark-adapted eye is concentrated between the crystalline cones, migrates with illumination into the area proximal to the cones. It must be concluded that not a material object but an image of an object was observed, which, lying proximal to the dioptric system, was projected into the plane of the cornea. The explanation can be based on the model of the dioptric system in superposition eyes that acts like a Kepler telescope. With no stops in the ray path, the boundary of the object glass will serve as the entrance pupil and the image of it formed by the eye piece as the exit pupil. In this case the entire object glass is used for the passage of light. If, however, the ray path is limited in the plane of the exit pupil, the image of this pupil will serve as the entrance pupil. Assuming that with the microscope the entrance pupil or a plane close to it was focused, and that with the pigment movement the exit pupil or a plane close to it was constricted, the image of this constricting pigment was observed. This explanation can be confirmed experimentally by introducing an object into the eye close to the cone tip and observing its inverted image in the plane of the cornea (KUNZE, 1972).

c) Optomotor Experiment

The observations on the performance of the *Ephestia* eye reported up to now were obtained entirely through optical methods. They involved only the light propagating and restricting parts of the eye. The following experiment (KUNZE, 1970) was designed to test if light transmitted in the superposition ray path is of receptive significance.

According to the superposition theory a single facet projects an extended part of the surround (in *Ephestia* $\approx 42°$) onto the rhabdom layer. If this part of the surround consists of a moving pattern, a single facet projects a moving pattern on many rhabdoms and the animal should be able to react to it.

Ephestia, like other insects, is able to react with head movements to visual movement stimuli. To measure this reaction and at the same time be able to project the movement stimulus through only one facet, the moth was affixed to a holder and the head connected to a torque compensator (KUNZE, 1961). Thus, the head was fixed in space and its reaction could be measured as a torque in two directions (right and left).

For the presentation of the movement stimulus the experimental setup for the projection of a pinhole (Fig. 27) was extended in so far as a moving pattern was projected through the pinhole into the principal plane of the microscope objective (in this case with numerical aperture = 0.6). Thus, an eye in the object plane of the microscope can see the image of the moving object. (The image extends over an angle which is determined by the aperture of the microscope objective.) In this setup the eye receives light only through a small circular opening (the image of the pinhole) of 15 μm diameter. This opening is focused onto one facet.

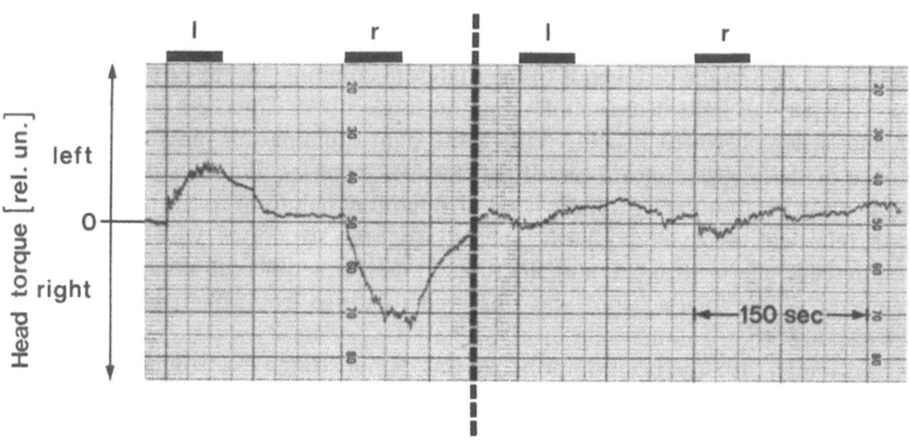

Fig. 29. Optomotor reaction of *Ephestia* to moving pattern projected onto one facet. Duration of movement stimulus indicated by *horizontal bars*. Right of *dashed line* test to stray light. (KUNZE, 1970)

In Fig. 29 the reactions of an *Ephestia* mutant devoid of screening pigment is presented. *Ephestia* reacts accordingly to movement stimuli from the right and left. Although stimulus conditions were constant for the left and right stimulus there is a slight asymmetry in the reactions due to asymmetric attachement of the head to the torque compensator.

A possible transfer of movement information by stray light through facets other than the intended one had to be controlled. For this purpose a drop of opaque varnish covering about three facets was placed on the eye next to the illuminated facet. When the illuminating light bundle was projected onto this varnish, movement stimuli did not elicit reactions.

The reactions are well reproducible. The wild-type *Ephestia* reacts only with dark-adapted eyes; the reaction ceases rapidly with pigment migration into the light position.

The experiment shows that image information of functional significance to the central nervous system of the moth was transferred via a superposition ray path.

II. Other Insect Superposition Eyes

1. The Eye of Skipper Butterflies

Through extensive research HORRIDGE et al. (1972) found that the eyes of a number of Hesperioidea function according to the superposition principle of EXNER. The anatomy of these eyes (Fig. 30) is the typical one for eyes in which a superposition image can be formed. It is remarkable that the screening pigment in the secondary (accessory) pigment cells remains, under all light conditions, in the position typical of a dark-adapted eye, i.e., in the space between the crystalline cones. Observations on these eyes therefore are undisturbed by pigment migration. Of the various methods of research two will be presented here showing new experimental approaches to the superposition eye.

Firstly, in skipper eyes it was possible to remove parts of cornea and cones, thereby exposing the receptor layer. If the eye is illuminated with parallel light, this light is focused in the distal ends of the rhabdom columns. In Fig. 31 this is shown for two parallel light beams subtending an angle of 8°. Thus, the superposition image is demonstrated in situ. With white light illumination the superposition image in these eyes appears bluish (FERNANDEZ-MORÁN, 1958), as does the eye glow, showing that the light has passed through the rhabdoms to the tapetum and was reflected there by regularly spaced tracheal structures which produced interference effects (MILLER and BERNARD, 1968). For different skipper eyes the diameter of the focused spot was observed to range from 40 to 50 μm which corresponds to 5–7 ommatidia or 8.5°–13.3°.

Secondly, the superposition principle was demonstrated electrophysiologically using intracellular recording of receptor potentials in *Trapezites symmomus* and *T. peroni*. After penetration of an unit a point light source was aligned for maximum response. Then the eye was screened safe for a slit which allowed illumination of a row of facets three facets wide. This slit was movable at right angles to the light

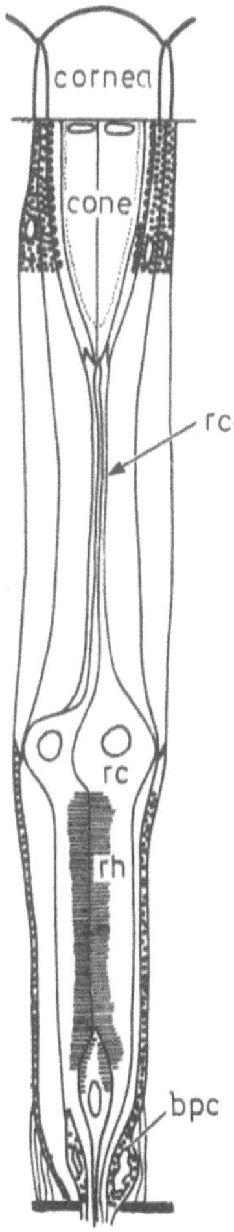

Fig. 30

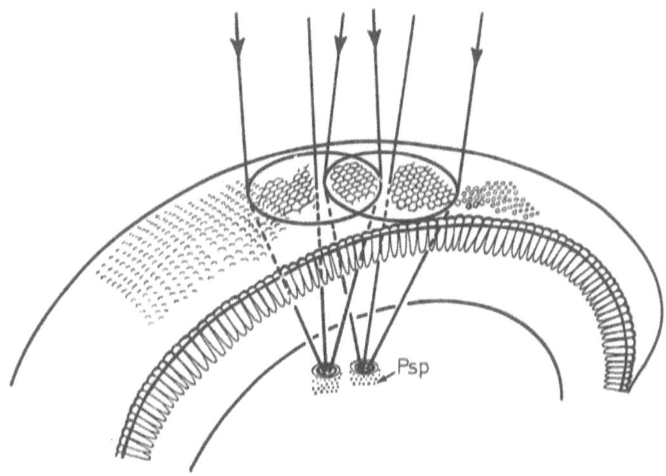

Fig. 31

Fig. 30. Diagram of an ommatidium, as found in skipper butterflies by electron microscopy. *rcc*, retinula cell column; *rc*, retinula cell; *rh*, rhabdom; *bpc*, basal pigment cell. (HORRIDGE et al., 1972)

Fig. 31. View onto the rhabdom layer through a cut in the top of the eye of *Trapezites symmomus*. Two beams of light subtending an angle of 8° form two separable spots of light on the distal ends of the rhabdom columns. *Psp*, rhabdom pattern. (HORRIDGE et al., 1972)

direction. In this way it was possible to illuminate the eye in different positions with flashes of parallel light of constant direction. The result was that parallel light which entered over a circular area about 20 facets in diameter elicited a response in the same retinula cell. The response was maximal for the center position of the beam and fell symmetrically when the beam was moved peripherally in this area. The experiment directly proves that parallel light which entered the eye through different facets was directed to one retinula cell, a necessary condition for the superposition principle.

2. The Eye of *Rhagonycha* (Cantharidae)

As already mentioned, EXNER found the superposition image in cornea preparations of *Lampyris* at a distance which would have placed it far behind the rhabdom layer if measured histologically. Probably, his use of a glycerin solution as an artificial eye medium changed the optical properties (see Discussion).

In our laboratory the eye of another Cantharoid beetle, *Rhagonycha fulva*, was investigated by E. KANDZIORA. This eye (Fig. 32) closely resembles that of *Lampyris*. The method was exactly that used for *Ephestia* by CLEARY et al. (1977): Eye cups were carefully sectioned from dark-adapted fresh eyes and stabilized with gelatine. They were observed in a moist chamber from the sectioned side with a water-immersion objective (n.a. 0.65). Superposition images (Fig. 33) were recorded in 59 preparations of male eyes and their position behind the cone tip determined. The mean distance was found to be 79 ± 5 µm. The rhabdom was found from histology to extend from 69 to 125 µm from the cone tip. Therefore the superposition image can be assumed to be formed in the distal part of the rhabdom layer. The average number of ray bundles which converge from different cones to form the superposition image was 56 for all preparations. At least this number of dioptric systems combines in this eye for superposition. With an interommatidial angle of $\approx 5°$ they subtend a maximal angle of 42°.

III. The Catoptric Superposition Eye

Recent research by VOGT (1975) on various crayfish, and, independently, LAND (1976) on the shrimp *Oplophorus spinosus*, showed that in superposition eyes of decapod Crustacea the superposition of light is the result of radially arranged reflectors. This is in contrast to EXNER's concept of radially arranged lens cylinders.

Actually, EXNER mentions the possibility of reflectors. In connection with his observations of an erect ray path in the *Lampyris* eye (Fig. 9) he writes (EXNER, 1889, p. 28): "Thereafter one should believe that it was a reflection in the crystalline cone. This is not the case. The described effect is based on refraction." He was right for the crystalline cone of the eye of *Lampyris* and other insects, but he was not right in extending the lens cylinder concept to the cones in the eyes of macruran decapods.

In this group of Crustacea eyes are found that are characterized as superposition eyes by a clear zone distal to the rhabdom layer. Typical for these eyes is the flat surface of the cornea facets and the shape of the crystalline cone cells (see survey of WATERMAN, 1961). The four cells form a square base pyramid with a very acute tip angle. Therefore the sides of the pyramid include almost right angles (Fig. 34). The cone has a distal position in the cone cells which taper into a proximal tail, the crystalline cone stalk. A new observation of VOGT (1975) and LAND (1976) was that the sides of the cone over part of their length are paralleled by highly reflecting layers.

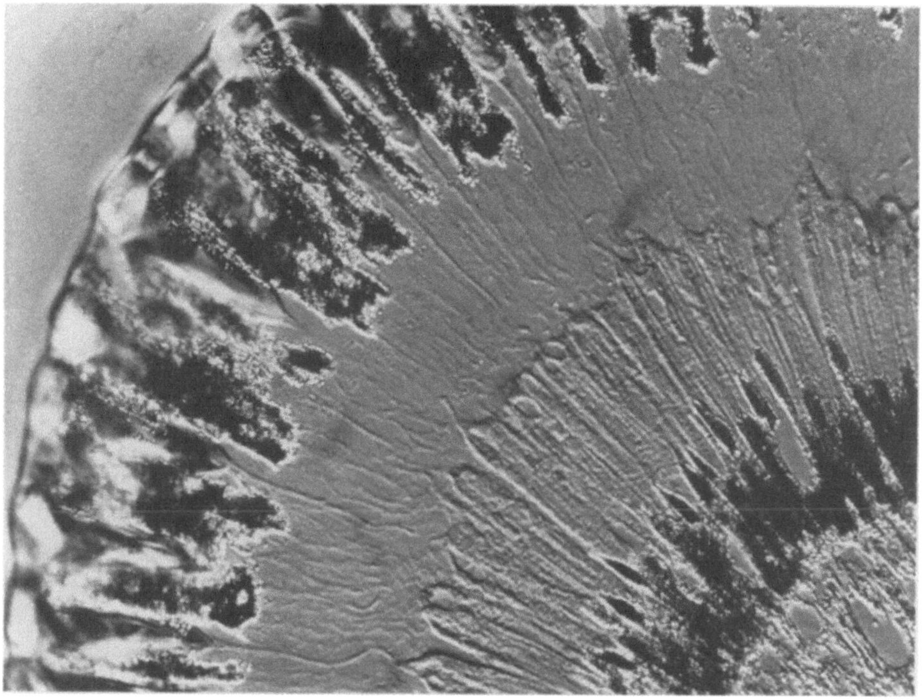

Fig. 32. Radial section through eye of *Rhagonycha fulva*. Interference contrast. (Courtesy E. KANDZIORA)

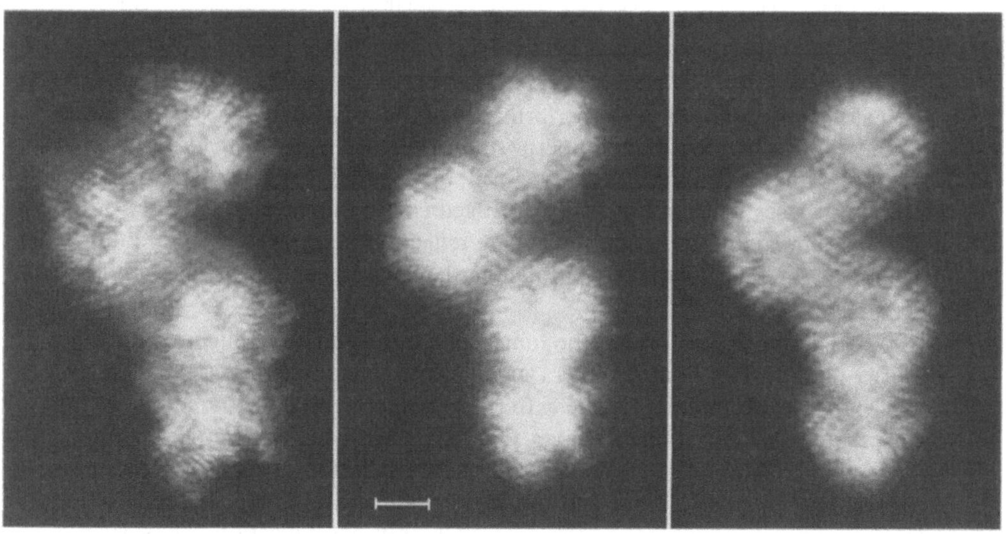

Fig. 33. Superposition images in *Rhagonycha* eye photographed at 10-μm depth intervals. The image in the middle was regarded the best focused one. *Marker* 10 μm. (Courtesy E. KANDZIORA)

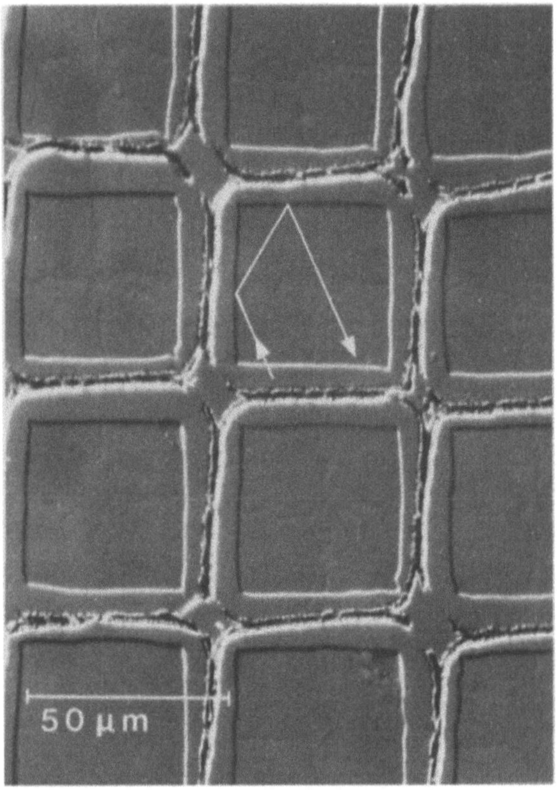

Fig. 34. Cross section through cones in crayfish eye. Cone cells are lined with reflecting layer. Path of a ray, twice totally reflected at the cone sides, is projected into the cross-sectional plane. (VOGT, 1977)

The following account of the performance of the crayfish eye is based on VOGT (1975, 1977). He showed that a superposition ray path can be three-dimensional when the radially arranged mirrors form orthogonal systems. Two types of reflecting surfaces which are radially arranged exist in the eye. 1. the sides of the highly refracting cone and cone stalk; here reflection takes place as total internal reflection. 2. the reflecting layers paralleling the cone sides; they act as multilayer reflectors.

Both reflector types form orthogonal compound mirrors which produce an erect ray path. In Fig. 35 a ray enters such an orthogonal mirror system from the side of the observer under an angle φ to the axis and an angle χ to one side of the system. After two reflections it leaves the system with the same angles φ and χ again in the direction of the observer. In Fig. 34 a reflected ray is projected into a cross-sectional plane of a cone. Naturally, rays entering in a plane parallel to one side of the cone will suffer the same fate after only one reflection. Thus, such a system is afocal, produces an erect ray path, and has an angular magnification of 1/1. This makes it suitable in radial arrangement for the formation of a superposition image at about 1/2 of the radius of the entire arrangement.

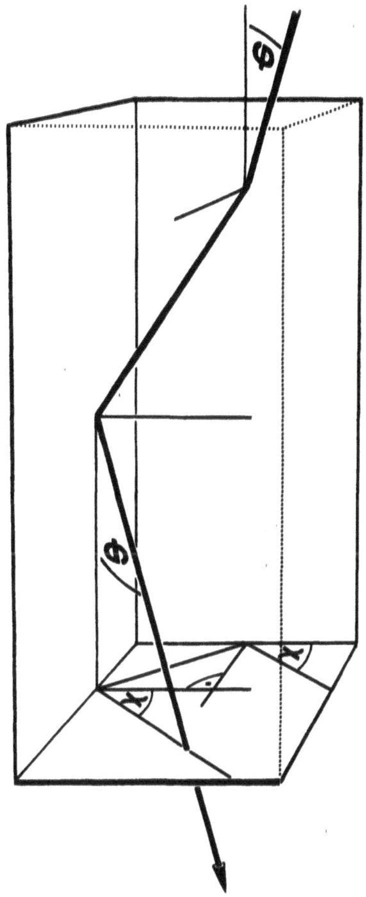

Fig. 35. Schematic drawing of a ray path in a crystalline cone of the crayfish eye. The ray is twice reflected at two sides of the cone. φ is the angle between incident (or exit) ray and direction of the cone axis. Ray and perpendiculars in the reflection points are projected into a cross-sectional plane at right angles to the cone axis. χ is the angle between projected incident ray and the hind edge of the cross-sectional plane and also between the projections of exit ray and perpendicular in the second reflection point. (VOGT, 1977)

Necessary conditions for imaging in such an arrangement are that rays are only twice reflected for the case $\chi \neq 0$ and only once for the case $\chi = 0$ (incidence in a median plane).

In Fig. 36 the ray path is constructed in a schematic eye with the described catoptric systems. For simplicity, only two selected cases are shown with ray incidence in a median plane and therefore only one reflection per ray. Light incidence is parallel to the axis of the central ommatidium *o*.

Bundle *t* is supposed to enter the cone under an angle to its axis which is smaller than the critical angle for total internal reflection at the sides of the cone cells. After one reflection all rays are directed towards ommatidium *o*. Apparently it is of no consequence whether in the cone the reflection occurs at the sides of the cone or at the reflecting layer; the described effect is the same. It is also evident

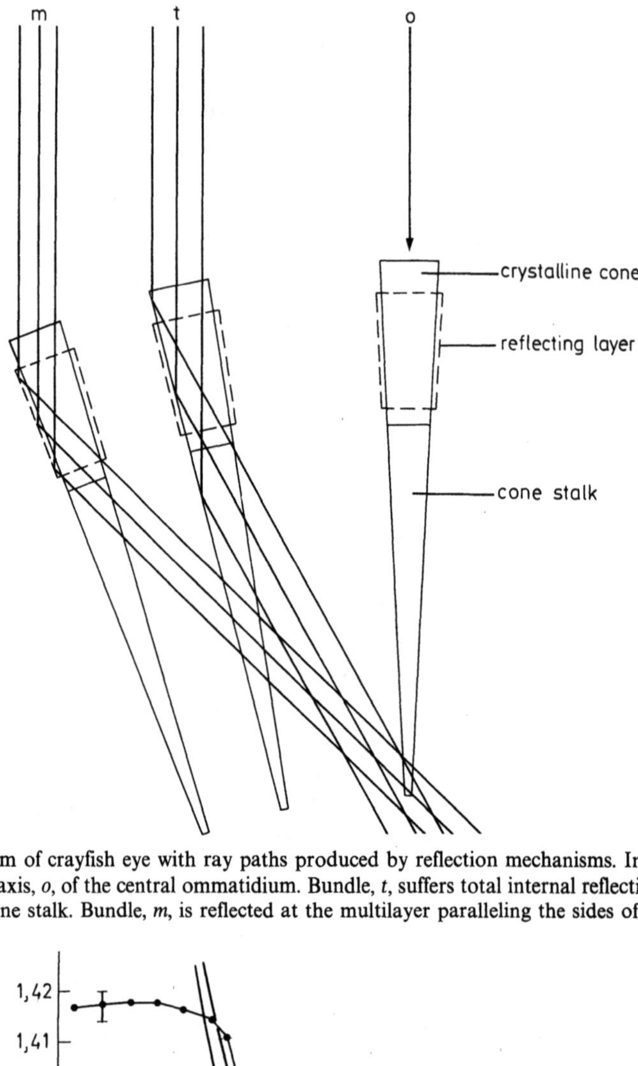

Fig. 36. Diagram of crayfish eye with ray paths produced by reflection mechanisms. Incident light is parallel to the axis, *o*, of the central ommatidium. Bundle, *t*, suffers total internal reflection at one side of cone and cone stalk. Bundle, *m*, is reflected at the multilayer paralleling the sides of the cone cells

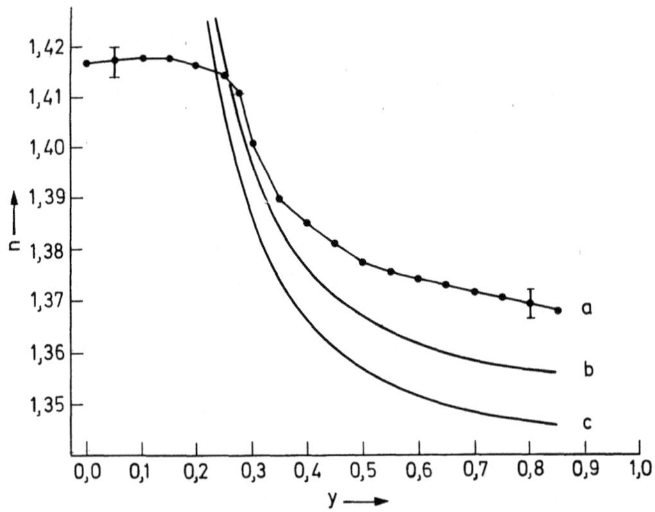

Fig. 37. Refractive index as a function of axial distance in cone and cone stalk of crayfish. ***Ordinate:*** refractive index *n*. *Abscissa:* Relative distance on axis of ommatidium. *O*, distal cone end, from 0.29 to 0.89 cone stalk. *a*, measured refractive index (mean of 6 preparations with typical standard deviations). *b* and *c*, refractive indices calculated from Eq. (8) with $n_M = 1.35$ and 1.34. (Vogt, 1977)

from the drawing that a ray which is not reflected at the cone sides has to be reflected at the side of the cone stalk in order to fulfill the imaging condition. If we suppose that conditions for total reflection are the same over the entire length of the cone stalk, rays could be reflected again at the opposite side. This can be prevented when the conditions for total internal reflection vary axially in the cone stalk in such a way that the limiting angle for a ray with the reflecting plane decreases proximally.

According to VOGT (1977) most rays will fulfill the imaging conditions when the refractive index in the cone stalk decreases axially from the cone following the function:

$$n_y = n_M \left[1 + \left(\frac{\sin\delta}{y+\cos\delta-1}\right)^2\right]^{\frac{1}{2}} \tag{8}$$

with y the distance from the distal cone end divided by the length of the pyramid, n_y the refractive index at y, n_M the refractive index of the surrounding medium, and δ the angle between opposite sides of the pyramid. (Naturally, an equivalent function can be assumed for constant n of the cone stalk medium and varying n in the surround medium.)

Measurements of the refractive indexes in cone and cone stalk (Fig. 37) show that the refractive index is constant in the cone. (This is in accordance with the findings of CARRICABURU, 1968.) However in the cone stalk, the refractive index varies over the entire length, between ≈ 1.37 proximally and 1.41 distally; the form of the variation resembling the above function.

A continuation of the refractive index variation in the cone would lead to unphysiologically high values of n. Here, reflection is secured by the reflecting layers outside the cone. This is important for rays entering the cone at high values of φ (ray bundle m, Fig. 36). The refractive index proximal to the cone will always be small enough for rays to pass through the opposite side of the pyramid.

VOGT (1977) shows with electron microscopy that the distal reflecting layer is a multilayer consisting of crystals. Isolated together with the cones they show interference colors also observed by LAND (1976). Although only three to four layers are found, reflectance is high at great angles of light incidence on the multilayer.

In the catoptric superposition eye an imaging ray path is formed by a restricted number of reflections at orthogonal reflectors in or close to the sides of the cone cells.

IV. Discussion

1. Methods

As it turns out, the reinvestigation of superposition eyes was a question of methods rather than of theoretical concepts. A few comments on methods might therefore be helpful for future work.

a) Determination of Refractive Index

For quantitative refractive index measurements in arthropod optical systems the use of the phase contrast microscope as a "null detector" seems to be inadequate (HAUSEN, 1973). With interference microscopy SEITZ (1968) was the first to demonstrate refractive index variations in an arthropod dioptric system. This method is at present the most reliable. But its application to arthropod optical systems involves difficulties which may lead to serious systematic errors.

The method allows the measurement of the optical path length which is the product of refractive index and geometric path length. For refractive index determination, therefore, the condition is that the measuring rays pass parallel to the microscope axis through the specimen and that the specimen thickness is known.

For unsectioned cornea-cone systems this is the case for a measurement through the axis of the system. An approximation is possible in cones with none or small curvatures in axial direction for measurements in the median line (HAUSEN, 1973). However, curved surfaces or equivalent refractive index distributions will inevitably lead to trajectories of the measuring ray, which are longer than those determined by thickness measurements; and the n-determinations will be too high.

Thickness determination by microtome setting is problematic and has to be controlled (MEYER-ROCHOW, 1973) but a high degree of uncertainty remains. A reliable method of higher sensitivity is to embed the object in a medium of known refractive index, section both simultaneously, and determine the thickness of the section in the interference microscope and also the planeness of the section (HAUSEN, 1973; VOGT, 1974).

A source of possible errors is the mounting medium which must not change the refractive index to be measured. By the method of multiple embedding VOGT (1975) was able to show that distilled water as well as NaCl solutions and solutions of bovine serum albumin did not change the refractive index in the cornea. This is not the case with glycerin solutions. In the cornea, glycerin raises the refractive index generally and decreases refractive index differences (down to 24% of the original value). Similar effects have to be expected for other materials also containing water which will be replaced by glycerin.

This is significant since many authors (including EXNER) used such solutions as medium in connection with the determination of optic parameters in preparation of compound eyes. In an eye-cup preparation a flattening of refractive index variations can result in decreased refractive power of the dioptric systems and consequently in an increased distance of the superposition image.

The use of a gelatine solution (5%–10%) as a mounting medium for eye cup preparations (KUIPER, 1962) repeatedly proved to be advantageous also for refractive index measurements. Isolated crystalline cones embedded in gelatine retained their optical properties for 1–2 days, as observed with interference microscopy (HAUSEN, 1973).

b) Ray Tracing

Ray tracing is the necessary consequence of having measured refractive indexes in a dioptric system. Although simple for homogeneous systems, the task becomes

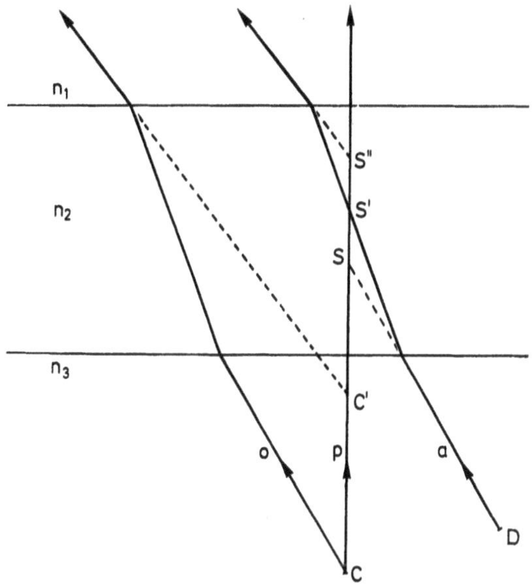

Fig. 38. Refraction of rays *o*, *p*, *a* from cone tips *C* and *D* when passing through media of different refractive indices ($n_2 > n_3 > n_1$). Rays, *p*, and *a*, are exit rays for parallel incident light; they form a superposition image at *S*. (CLEARY et al., 1977)

difficult for systems with a continually changing refractive index. For approximation, such a system can be subdivided into a finite number of layers of constant refractive index and Snell's law applied at each phase boundary. Systematic errors, however, are inevitable with this method. (e.g., rays parallel to the axis of a cone with axially constant and radially varying refractive index would not be refracted at all). Consequently, conclusions drawn from such ray tracings are merely preliminary.

The method of choice is to apply Snell's law for the continuous case as first done by MEGGITT and MEYER-ROCHOW (1975) for the dioptric system of an arthropod eye using a computer implementation. Although satisfactory for geometric ray tracing, the extension to Fresnel's reflection formulas is highly desirable.

c) Eye Cup Preparations

The preparation of an eye cup has become a standard method for investigation of arthropod eyes. The small and delicate eyes are difficult to section without disarranging the optical systems. Besides patience, a well-mounted vibrating razor blade (KIRSCHFELD, 1967) is a very useful tool for this preparation.

In such a preparation not only the optical systems of the eye cup influence the direction of light. The media outside the eye have also to be accounted for (CLEARY et al., 1977) since at phase boundaries rays not normal to them change direction according to Snell's law. In Fig. 38 ray bundles from cones *C* and *D* are supposed to intersect in situ at *S* (as in the formation of a superposition image). With the eye cup on a cover glass ($n_2 > n_3$) the rays are refracted and intersect at *S'*. This point is optically shifted to S'' by the phase transition n_2 to n_1 and will appear there when viewed from the direction of n_1. Similarly C is observed at C'. The wanted distance CS is changed to the observed distance $C'S''$. It is easy to deduce that for small angles to the miscroscope axis the wanted distance $CS = C'S''n_3/n_1$.

Measurements of distances and angles in eye cup preparations on cover glasses with oil immersion or dry objectives can yield considerable errors (more than 50% in depth determinations) if not corrected. The use of water-immersion objectives reduces the error to an acceptable amount when the refractive indexes of mounting medium and eye cup content are close to that of water.

The above correction has never been reported in past literature and large errors in depth and angle determinations from eye cup preparations have to be assumed.

2. Optical Mechanisms in Superposition Eyes

Two dioptric mechanisms with imaging properties have been shown to exist in insect superposition eyes, spherical lenses and inhomogeneous lenslike media. Ray tracing proved that in the eye of *Ephestia* both mechanisms combine for the production of an erect telescopic ray path. The same can be expected for other arthropod eyes with dioptric surfaces and lens cylinders. As EXNER has already pointed out, each mechanism by itself could produce the same result (see also WINTHROP and WORTHINGTON, 1966; HORRIDGE, 1971).

In certain beetle superposition eyes with flat cornea facets refraction is almost entirely due to inhomogeneous media. Ray tracing in such eyes (MEYER-ROCHOW, 1973; MEYER-ROCHOW and HORRIDGE, 1975) shows that the dioptric systems produce erect ray paths but with highly diverging exit angles (up to 14°) on parallel ray incidence. Thus, the superposition focus is poor. The result, however, is not conclusive since the use of glycerine makes the refractive index determinations questionable. The eyes are likely to focus much better.

A Kepler-telescope-like system using only lenses has not yet proven to be functional in superposition eyes as little as the use of a pinhole instead of the second lens, as suggested by HORRIDGE (1971).

The radially arranged mirror systems producing a superposition image as discovered by VOGT (1975) and LAND (1976) in crustacean eyes are the only new optical mechanisms found since EXNER (1891) to be functional in superposition eyes. Thus, we know at present two types of superposition eyes: one, only described for insects with dioptric systems, and the other only described for crustaceans with catoptric systems.

Other optical mechanisms cannot be excluded to play a role in visual perception in superposition eyes, as already stated. The possibility that different light-transmitting systems (e.g., superposition and light-guide action) can be at work simultaneously was pointed out by KUNZE (1969) and HORRIDGE (1975). However, evidence is scarce for light-transmitting systems other than superposition. How effective light transmitted via light-guides could be is still an open question. Positive proof that light transferred in this way is of functional significance for a receptor cell in a superposition eye is lacking.

Accessory mechanisms accompanying changes of light flux are very diverse in superposition eyes. Insect eyes exist, e.g., skipper eyes, where the screening pigment does not interfere under any light conditions with the superposition ray path (HORRIDGE et al., 1972). In others, the movement of screening pigment in secondary pigment cells and retinula cells is pronounced and its effect on the superposition ray path has already been described. In particular the eyes of nocturnal Lepidoptera have been investigated with respect to light-dependent movement of screening pigment (TUURALA, 1954; YAGI and KOYAMA, 1963; HÖGLUND, 1966). It is certain that movement of screening pigment into dark position increases the admittance of light to the receptor cells. Other anatomic changes during light–dark adaptation occur with retinula and cone cells (see survey of WALCOTT, 1975). A controlling effect on light flux in the eye has not yet been shown by experiment.

3. Superposition Focus

How good a superposition focus can be expected to exist? Since a theoretical treatment of this question is not available at present only preliminary considerations can be presented.

Let us first discuss a single dioptric system in *Ephestia.* The performance of such a system can be described as being limited by diffraction, the determining aperture being the exit pupil (see WESTHEIMER, 1972, Vol. VII/2 of this handbook; KIRSCHFELD, 1976). The smallest common area of all possible exit rays is chosen for the exit pupil. The diameter d of this area can be derived from the result of the computer ray tracing (CLEARY et al., 1977) and is 10 μm. The position of this area is ≈3 μm distal to the cone tip (i.e., the area lies only partially outside the cone). If we assume, deviating from the ray tracing result, that the dioptric system is focused in the plane of the constructed superposition focus (Fig. 25), the radius of Airy's disk can be calculated in radial measure to be $1.22\,\lambda/nd$. With a wavelength in vacuum $\lambda = 0.55$ μm, refractive index $n = 1.36$, and $d = 10$ μm the radius of Airy's disk is 2.85°. At a distance of 108 μm from the exit pupil the diameter of Airy's disk is 10.8 μm. This is a little more than the distance of 8–10 μm between the centers of neighbouring rhabdoms in this plane as measured by FISCHER and HORSTMANN (1971). This is the smallest focus behind the dioptric systems we can reasonably expect in a superposition eye with the dimensions used. In bigger eyes Airy's disk would only be smaller when the diameter of the exit pupil is increased more than proportional to the distance exit pupil—rhabdom.

Limitations to the superposition focus can also be expected from geometric optical considerations. EXNER (1889, 1891) has shown that dioptric systems with constant angular magnification arranged on a spherical surface superimpose parallel incoming rays on a caustic surface (Fig. 9). Such a system inevitably has no stigmatic focus and the smallest diameter of the caustic bundle has to be regarded as the superposition focus. A construction of only center rays using the geometric data of *Ephestia* gives a caustic bundle with a smallest diameter of 6–7 μm. This caustic bundle diameter depends on the "inner aperture angle" of a superposition eye, defined as the angle between the axis of the central ommatidium and the most inclined axis of a dioptric system admitting light incident parallel to the central axis (KIRSCHFELD, 1974). This aperture angle can be reduced by reducing the diameter of the eye surface admitting parallel light, the "effective aperture," or with constant effective aperture, by increasing the distance from cone tip to rhabdom layer, the "effective focal length." For constant rhabdom diameter, the latter only seems possible with an increased eye radius. Thus, for constant effective aperture bigger superposition eyes can be expected to have lower magnifying dioptric systems. In this case the caustic diameter will be reduced (which should be observable as a decreased divergence angle of the eye glow).

The relation of eye size and angular magnification of the dioptric system in a superposition eye approximating to small angles of incidence was formulated by EXNER (1889, 1891) to be

$$G_2 = \frac{r \cdot \varphi_1}{\varphi_1 + \varphi_2}$$

with G_2 the distance from eye center to the superposition focus, r the eye radius, and φ_1 and φ_2 the two focal distances in the telescopic dioptric system numbered in the direction of light incidence.

From the above relation follows that $G_2 = r/2$ when $\varphi_1 = \varphi_2$. Like spherical mirrors, this is always the case in the catoptric superposition eyes excluding refraction distal of the cones. These eyes, by necessity, have an angular magnification of unity irrespective of their size.

E. Apposition Eyes

Apposition eyes are characterized by an almost complete optical isolation of the individual ommatidia. A retinula cell in such a compound eye primarily receives—contrary to the superposition eye—light which entered through the facet of its own ommatidium, independent of light conditions. With respect to anatomic structure, apposition eyes come closest to MÜLLER's concept of a compound eye (Fig. 1). However, recent research of apposition eyes with unfused rhabdomeres are not in accordance with the concept that all retinula cells of one ommatidium should receive light only from the same, more or less extended, point of the optical surround. Although the existence of compound eyes with unfused rhabdomeres has been known since GRENACHER (1879), the functional significance of separate receptors within one ommatidium has only now become of interest. Interest and work concentrated primarily on the dipteran eye. The eye of *Musca* will subsequently be used to show the anatomic and functional properties of an apposition eye as far as the scope of this article is concerned.

I. The Eye of *Musca*

1. Anatomy of the Ommatidium

A *Musca* eye contains 3000–3200 ommatidia in an hexagonal array. The facet diameters vary between 36–38 μm in the dorsoanterior region and 19–21 μm in the ventroposterior region of the eye (BRAITENBERG, 1967, 1970).

A diagram of the distal portion of an ommatidium is shown in Fig. 39. The spherical surface of the cornea is covered with an antireflex coating (BERNHARD et al., 1965). In direct contact with the cornea proximally follows the cone which is a pseudocone (GRENACHER, 1879) of very soft consistency. Its height varies between 30 and 50 μm in different regions of the eye (BRAITENBERG, 1967). The cone shows no structural differentiations in the light microscope and has a homogeneous refractive index (EXNER, 1891; KIRSCHFELD, 1967) that seems close to that of water. Proximal to the cone four cone cells (Semper cells) are interposed between cone and retinula.

The retinula is a long column formed by eight cells. The cell bodies of six retinula cells (numbered 1–6 by DIETRICH, 1909) extend all the way from the Semper cells to the basal membrane and penetrate it with their axons. The cross-sectional arrangement of cells 1–6 is the same throughout the length of the retinula. The body of the 7th retinula cell is only found in the distal region of the retinula between cells 1 and 6; it extends proximally as an axon. Retinula cell 8 is confined to the proximal region of the retinula and is situated between cells 1 and 2. It also extends an axon through the basal membrane.

The retinula cells form a ring around a central extracellular cavity. Rhabdomeres protrude into this space from each retinula cell. As is typical for Diptera, the rhabdomeres do not fuse. The arrangement of the rhabdomeres is typical and similar in all ommatidia. The arrangement in Fig. 40 is the one found in the dorsal half of a right eye, viewed from the outside. The arrangement shows

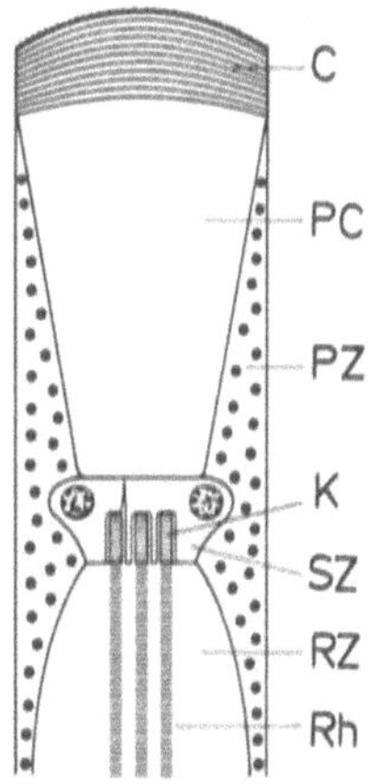

Fig. 39. Distal portion of a *Musca* ommatidium. *C*, cornea; *PC*, pseudocone; *PZ*, pigment cell; *K*, rhabdomere cap; *SZ*, Semper cell; *RZ*, retinula cell; *Rh*, rhabdomere. (KIRSCHFELD, 1967)

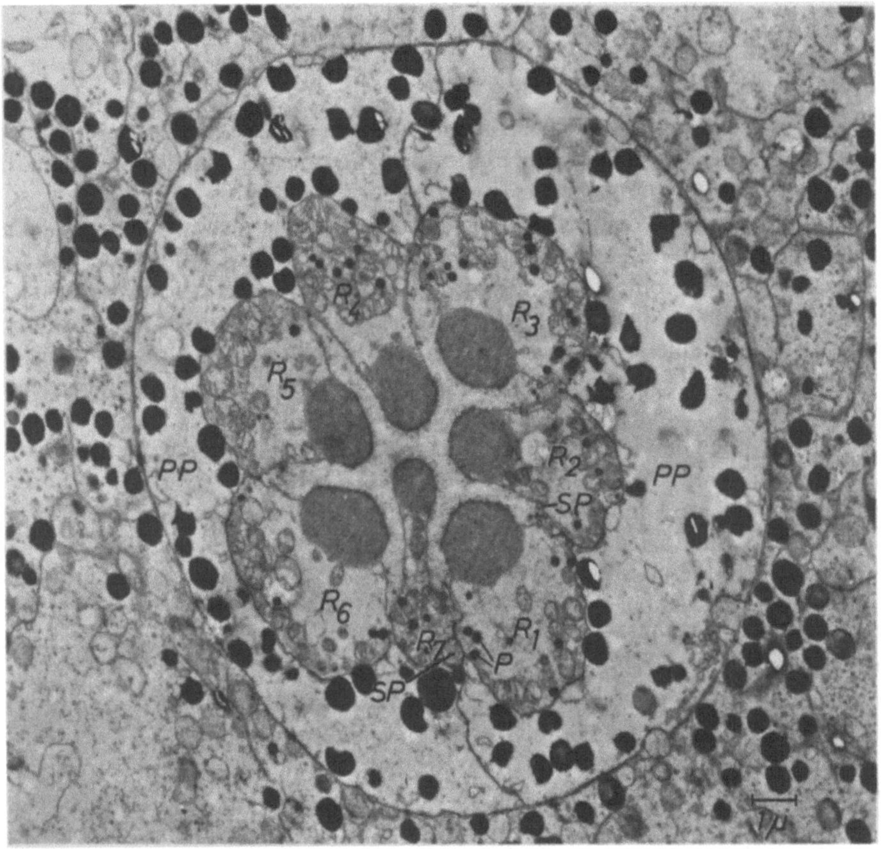

Fig. 40. Cross section through distal portion of the retinula. The retinula cells (*R* 1–7) are surrounded by two primary pigment cells (*PP*). Also the retinula cells contain pigment granules (*P*). *SP*, Semper cell processes which extend to the basal end of the retinula. The whole is surrounded by secondary pigment cells. (BOSCHEK, 1971)

mirror image symmetry in the dorsal and ventral eye halfs with the eye equator as the mirror axis. The rhabdomeres of cells 7 and 8 form a single rhabdomere strand in the center of the retinula with rhabdomere 7 in the distal portion being replaced by rhabdomere 8 in the proximal portion.

Distal rhabdomere diameters are $\approx 2\,\mu m$ for 1–6 and $1\,\mu m$ for 7 and 8 (BOSCHEK, 1971). They are distally capped by optical dense material (shown by SCHNEIDER and LANGER, 1966, to be extracellular) intruding into excavations of the Semper cells (TRUJILLO-CENÓZ and MELAMED, 1966). Probably in this way a constant and exact position of the rhabdomere ends is secured, an essential condition for the geometric optics of this eye.

Two primary pigment cells surround the distal part of the ommatidium and 12 secondary pigment cells, shared also by the neighbor ommatidia, envelope the whole. For further details see TRUJILLO-CENÓZ (1972), Vol. VII/2 of this Handbook.

2. Optics

As already mentioned in Section II, the existence of inverted images behind cornea and cone at least in the dipteran eye had posed a problem to nineteenth century physiologists (cf. GRENACHER, 1879, p. 152). It did not fit fully into MÜLLER's theory of compound vision. In an ommatidium where the unfused rhabdomeres possess light-guide properties and are placed behind a dioptric system, the optical axes of the retinula cells must necessarily diverge. This had already been recognized by VIGIER (1909) and was experimentally shown to exist in dipteran compound eyes by DE VRIES and KUIPER (1958) and by AUTRUM and WIEDEMANN (1962) using eye cup preparations. They noted that individual rhabdomeres could be alternately illuminated dependent on the position of a small light source in front of the preparation. Quantitative measurements with this method showed a divergence of rhabdomere axes within one ommatidum of *Musca* of 2°–4° (KIRSCHFELD, 1967). Apparently, each ommatidium is a little lens eye with seven discrete receptors.

The question now was whether the distal ends of receptors were located in the focal plane of the dioptric system. It can be answered without having to dissect the eye and test the single dioptric components, cornea and cone, by making use of the light-guide properties of the rhabdomeres. With "antidromic" illumination of an eye (KIRSCHFELD and FRANCESCHINI, 1968), i.e., by shining a strong light through a small hole into the head capsule, light will enter the rhabdomeres at the basal end, and be guided in the rhabdomeres to the distal end, where it is projected out of the eye by the dioptric system. Thus, the distal rhabdomere endings serve as a test object, the image of which will be accessible for observation in a microscope. (The unsymmetric rhabdomere arrangement in the *Musca* eye has the advantage of revealing image rotations.) In order to observe exclusively the rhabdomere images of one ommatidium, a diaphragm was placed on the eye which left only one facet free. Depending of the focus position of the observing microscope (with an objective of long working distance and relatively large aperture) images, as in Fig. 41, can be obtained. With the exception of Fig. 41c the images consist of seven bright discs that show precisely the rhabdomere arrangement as observed in histological and in eyecup preparations. Apparently, each luminous disc is the image of a rhabdomere ending (rhabdomere cap).

When the cornea is focused (Fig. 41c), only one bright spot of facet diameter is visible. With the focus deep in the eye at $-1000\,\mu m$, a virtual image is observed. It has the same size as the corresponding inverted real image at $+1000\,\mu m$. Both images are of comparable acuity. From these findings it is easy to deduce that the rhabdomere endings must be close to the focal plane of an inverting dioptric system.

The image of the center rhabdomere 7 appears smaller than those of 1–6. This is in accordance with the histological findings of different rhabdomere diameters (Fig. 40). The dioptric system apparently is of a quality to preserve these details. The recorded images of the rhabdomeres are an indication of the far field of the receptors. Apparently, the acceptance angles of rhabdomeres 1–6 are bigger than the acceptance angle of rhabdomere 7. This was tested and confirmed by ECKERT (1973) in an optomotor study.

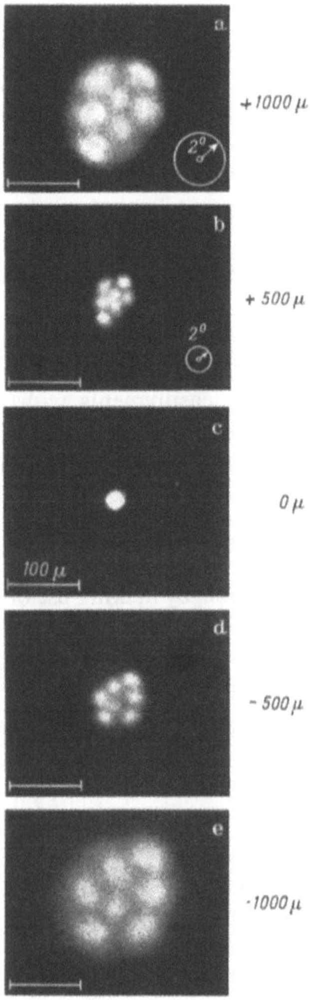

Fig. 41

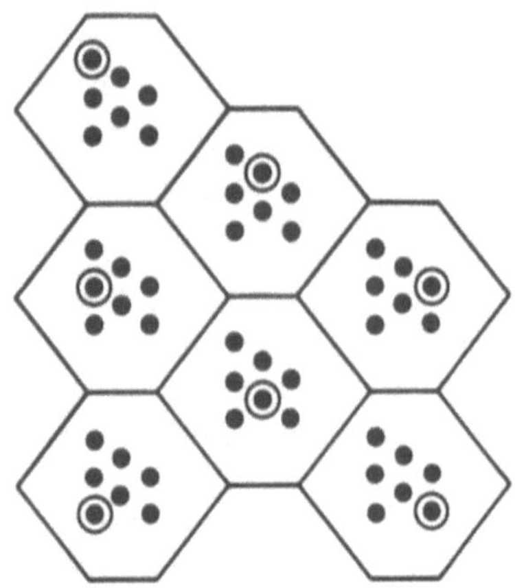

Fig. 42

Fig. 41. Optical properties of single ommatidium as observed in "antidromic" illumination. A diaphragm of about 30 μm diameter in front of the eye left only one facet uncovered. Microscope focus as indicated: 0 = cornea surface, + distal, − proximal of cornea. Living *Musca* ♀, dorsoanterior region of right eye. (KIRSCHFELD and FRANCESCHINI, 1968)

Fig. 42. Diagram of seven facets in dorsoanterior region of left eye of *Musca* viewed from the outside with underlying rhabdomere patterns. The encircled rhabdomeres receive light from one and the same point of the environment. (KIRSCHFELD, 1967)

From this result it might be assumed that the *Musca* eye is able to resolve the visual surround into 7 times more "points" than the eye has ommatidia. This could be possible provided that no coincidences between rhabdomere axes exist. The test for possible coincidence can be carried out using the eye preparation as before. To observe different facets, the blind in front of the eye has to be removed. With a small observation aperture only light parallel to the microscope axis will be observed. If under those conditions the cornea is focused, only seven facets appear brightly illuminated. They are arranged in the typical pattern with an orientation as in Fig. 41 when the same area of the eye is observed. Evidently the optical axes of seven rhabdomeres, each in a different ommatidium, are parallel to each other. The positions of the rhabdomeres with coinciding axes are shown in Fig. 42. The distant optical environment is resolved by the *Musca* eye into as many areas as it contains ommatidia, and each area is viewed by seven receptors from seven ommatidia. In the eye equator more than seven receptors look at one distant point. Due to the mirror image symmetry of rhabdomere patterns in adjacent ommatidia up to nine optical axes of rhabdomeres coincide (KIRSCHFELD, 1967).

The alignment of rhabdomere axes can also be tested with incident (orthodromic) illumination with the same result as above (KIRSCHFELD, 1965, 1967). In this case the light reflected by the rhabdomere ends is observed. Light patterns observable from outside the eye which are caused by structural and optical properties of the compound eye are known as "pseudopupils". They have proved to be of great use for the in vivo studies of optical and other functional properties of compound eyes (FRANCESCHINI and KIRSCHFELD, 1971). Visual performance and pseudopupils of compound eyes are separately treated by STAVENGA (1979) in this volume (Ch. 7) presenting further information on the optics of dipteran eyes.

3. Neural Superposition

As in other compound eyes, the lamina ganglionaris (the distal most neuropile of the optic lobe) of the *Musca* eye contains cell columns, "cartridges", consisting of second order neurons and the endings of photoreceptor axons. Photoreceptor axons converge on each cartridge from different ommatidia and make synaptic contacts with the second order neurons (CAJAL and SÁNCHEZ, 1915; TRUJILLO-CENÓZ, 1965).

In *Musca*, one cartridge receives the axons of those retinula cells R 1–6 from different ommatidia which, as shown by KIRSCHFELD (1967), have parallel optical axes. This projection of retinula axons to the lamina was revealed for *Musca* with light microscope techniques by BRAITENBERG (1967) and for *Lucilia*, also a dipteran, from serial electron microscope sections by TRUJILLO-CENÓZ and MELAMED (1966). A cross section of a *Musca* cartridge (Fig. 43) shows the six identified photoreceptor axons surrounding the axons of second order neurons with which they establish synapses (BOSCHEK, 1971; STRAUSFELD, 1971). The axons of retinula cells R 7 and R 8 bypass the cartridge without making synaptic contacts and proceed to the medulla where they end (CAMPOS-ORTEGA and STRAUSFELD, 1972).

In Fig. 44 the organization of the *Musca* eye is illustrated in a schematic radial section. Parallel light is focused in three different ommatidia on three rhabdomeres with coinciding optical axes. The corresponding retinula cells, of type 1–6, send axons to the same cartridge. From there signals are carried further by second order neurons.

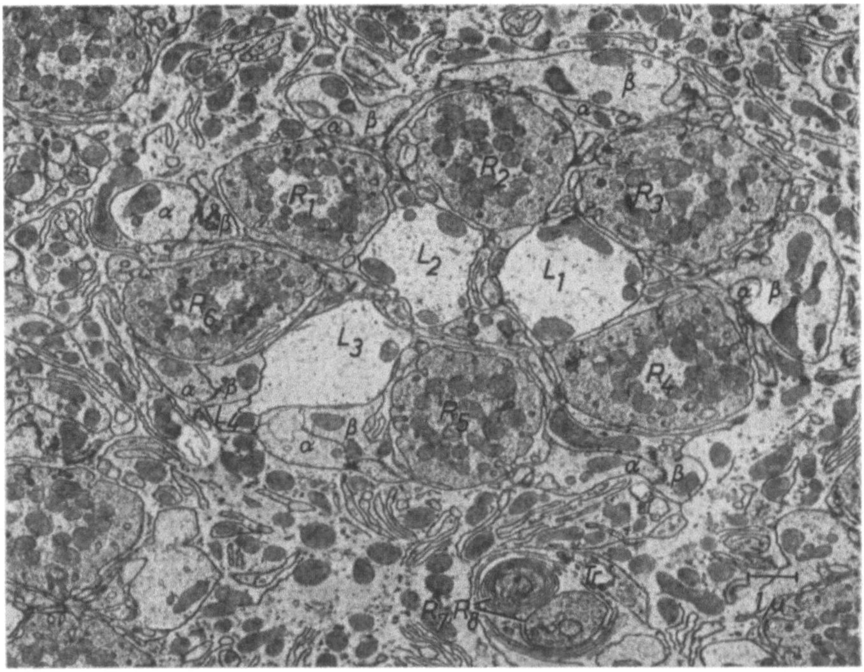

Fig. 43. Cross section through the distal portion of a cartridge in *Musca* showing the photoreceptor axons (*R* 1–6) and the second order neurons (*L* 1–4), with (*L* 3) and (*L* 4) withdrawing from the center to the periphery of the cartridge. α, β centrifugal fibers. The long visual fibers (*R* 7, 8), closely apposed by a tracheole *Tr*, bypass the cartridge. (BOSCHEK, 1971)

The functional interpretation of the *Musca* eye proposed by KIRSCHFELD (1967) is based on the assumption that light-induced signals are superimposed in the cartridges. Since the superposition is selective for signals originated in receptor cells with identical far fields, an increased absolute sensitivity of the eye is the consequence. Experimental support of signal superposition in the lamina of the fly was presented by SCHOLES (1969).

Analogous to the superposition eye as defined by EXNER, in which the superposition of light leads to an increased sensitivity, the type of eye found in *Musca* was called "neural superposition eye" (KIRSCHFELD, 1967).

4. Single Receptor Stimulation

The possibility of single receptor stimulation follows from the finding that seven separate rhabdomere endings are situated in the focal plane of a collecting dioptric system and, with the specific projection of visual fibers, single cartridges can be activated. KIRSCHFELD (1972) has used this possibility to elicit optomotor reactions of *Musca*. The movement stimulus was the alternate illumination of pairs of receptors in a single ommatidium. The optomotor reaction was measured using

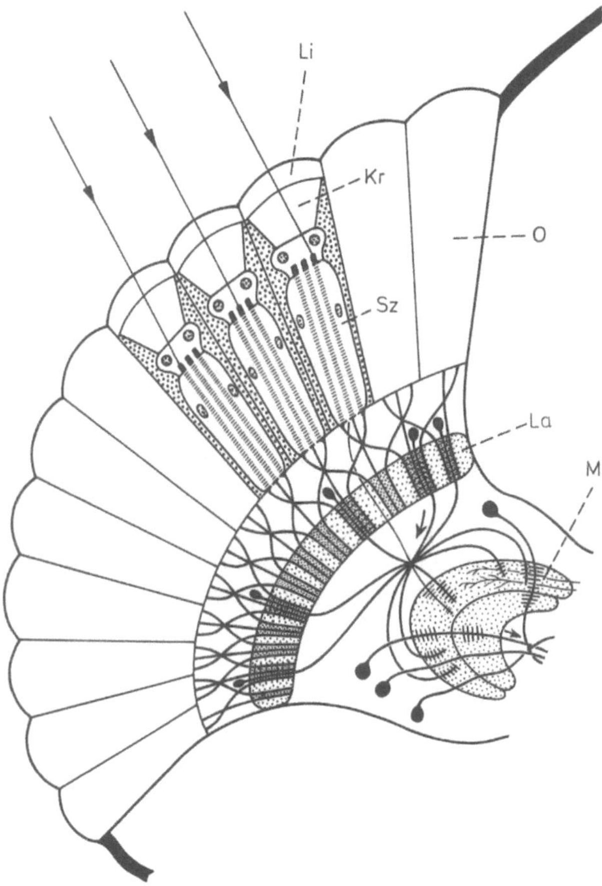

Fig. 44. Schematic drawing of a horizontal section through a fly's compound eye with parts of the optic lobe. *Li*, corneal lens; *Kr*, pseudocone; *Sz*, photoreceptor cell; *O*, ommatidium; *La*, lamina; *Me*, medulla. (KIRSCHFELD, 1971)

HASSENSTEIN'S (1950) method of the *y*-maze globe. The fly was placed in a combined observation and stimulation light path with the head fixed in space. The light path allowed observation of the deep pseudopupil. Thus it was possible to project the images of diaphragms, serving as light sources, onto the images of the rhabdomeres. With illumination of only one facet and suitable diaphragm extension, pairs of single receptors could be illuminated. Electromechanical shutters in front of the two diaphragms could be activated so as to produce an apparent movement of a luminous object.

Musca shows an optomotor reaction when only two receptors, the minimum for movement perception, receive the movement stimulus. The reaction was also used to test the cross talk between receptor cells within one ommatidium which is due to stray light. The observed reactions were corrected for the errors caused by optical cross talk.

The result of the study is that in one ommatidium three pairs of receptors combine to produce optomotor reactions: 1 and 3, 1 and 6, 2 and 5 (cf. Fig. 40). A further result is (KIRSCHFELD and LUTZ, 1974) that the additional illumination of rhabdomere 7/8 in certain neighboring ommatidia can inhibit this reaction.

These results confirm the concept that one receptor cell is a functional unit. They are further in agreement with the concept of neural superposition which states that one ommatidium should be enough to enable movement perception and that one cartridge can be activated by one of the receptor axons entering it.

II. Discussion

The concept of the neural superposition eye seems to be applicable to higher Diptera in general. A number of optical studies and histological studies (e.g., KUIPER, 1962; SEITZ, 1968; FRANCESCHINI, 1975; STAVENGA, 1974; STRAUSFELD, 1970) support this view. At present, nothing can be said about whether this concept also applies to other compound eyes with open rhabdoms which, besides Diptera, exist in the insect orders Hemiptera, Coleoptera, and Dermaptera and also in Crustacea.

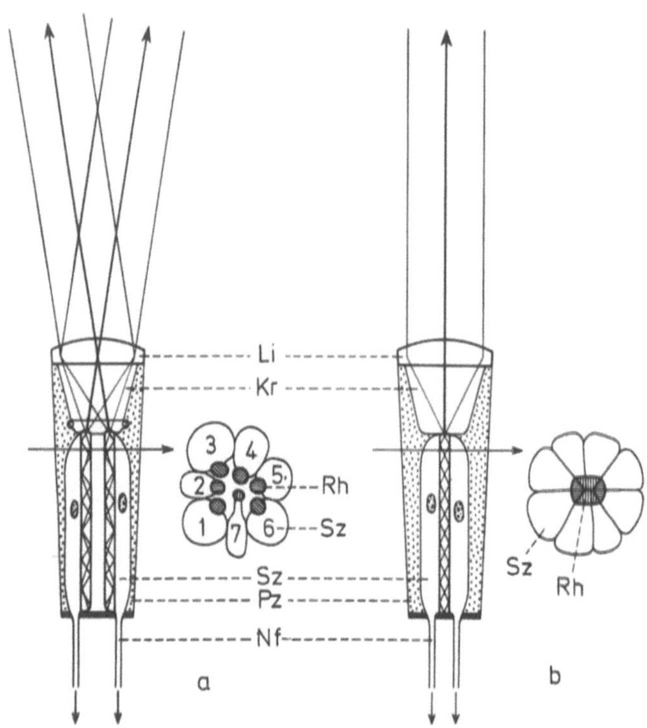

Fig. 45. Diagram of *a* ommatidium with open (unfused) rhabdom and *b* ommatidium with fused rhabdom. *Arrows* indicate the corresponding cross sections. *Li*, cornea lens: *Kr*, cone; *Sz*, photoreceptor cell; *Pz*, pigment cell; *Nf*, nerve fiber. (KIRSCHFELD, 1971)

Besides optical study of the intact eye as described for *Musca* and other Diptera direct measurements of parameters of dioptric systems also exist in some insect eyes. Geometry and refractive indexes have been determined for the dioptric system of the pseudocone eye of *Calliphora* (SEITZ, 1968) and for the eucone eye of *Apis* (VARELA and WIITANEN, 1970). In these eyes the cornea refractive index shows an axial variation with highest values in the distal region up to $n = 1.473$ and lowest values in the proximal region down to $n = 1.47$. In both animals the cone is homogeneous. In the acone eye of *Lethocerus* (Hemiptera) the cornea contains a central core of highest refractive index of $n = 1.61$ which decreases distally to $n = 1.58$, radially to 1.52, and proximally to $n = 1.47$ (IOANNIDES and HORRIDGE, 1975). Ray tracing in all three dioptric systems show them to be collecting systems with focal planes in, or close to, the plane of the distal rhabdomere endings. In *Calliphora* and *Apis* this is mainly due to the curved surface of the corneal facet. In *Lethocerus*, with a flat cornea, a "lens cylinder" is the equivalent dioptric system. This is the first eye in which the existence of a lens cylinder with collecting properties (cf. Fig. 3) has been shown to exist with methods of refractometry. (For all three dioptric systems the reported refractive indexes may have to be corrected due to measurements in glycerine embedded or frozen material.)

The performance of apposition eyes with respect to geometric optics may be summarized in two models of ommatidia (Fig. 45). They are similar in that the dioptric systems act like a collecting lens and that the rhabdomere endings sit in the focal plane of this lens. They are dissimilar in the type of rhabdom—open or fused—in that unfused rhabdomeres have differing optical axes and the fused rhabdom has one optical axis.

F. Conclusions

I. Classification of Compound Eyes

The approach to arthropod compound eyes chosen in this article was somewhat conservative in its restriction to anatomy and geometric optics. This was necessary for a critical assessment of the historical background of this field of science. We can see that EXNER's contribution to arthropod vision has stood the test of time. Rigorous research with refined methods has confirmed his findings, and his classification of compound eyes is still valid. It has proved to be well founded both on the basis of anatomy and physiology and new types of compound eyes found since EXNER fit very well into his classification. The neural superposition eye is an apposition eye according to EXNER's definition. It constitutes a type of eye in its own right on the basis of receptor optics and neural data processing (KIRSCHFELD, 1973). Superposition eyes, as defined by EXNER, may now be classed on the basis of anatomy and geometric optics as dioptric and catoptric superposition eyes.

II. Sensitivity of Compound Eyes

We are left wondering whether differences in anatomy and optical performance of the types of compound eye discussed affect the extraction of information from the visual surround. A theoretical approach, as far as photometry is concerned, is offered by KIRSCHFELD (1974) as follows.

Compound eyes are poor devices for the recognition of distant *pointlike* objects. Their small facet diameters limit the admittance of light. Animals with lens eyes are much better off in this case (BARLOW, 1952; KIRSCHFELD, 1966; SNYDER et al., 1973). The area of the bee's ommatidial facet (diameter 25 μm) compares to the pupil of a human eye (diameter 7 mm), $\approx 1/10^5$. A bee's ommatidium gathers that much less light from a distant point light source. Assuming a similar absolute sensitivity of the receptors in man and bee, a bee should not be able to see even one of the brightest stars, Venus.

Compound eyes appear in a better light when exposed to distant *extended* luminous objects. This situation seems more natural since extended objects are of primary interest to animals, especially insects and Crustacea.

For the following argumentation it is assumed that the optical environment has a Lambertian radiation pattern, that lenses are aplanatic, and that image media have the same refractive index.

In the case of extended objects, the mean number of quanta reaching a receptor per unit time will be

$$q \sim B\frac{A^2}{f^2}\delta^2$$

where B is the mean brightness, A the diameter of the entrance pupil and f the inner focal distance of the dioptric system, and δ the diameter of the receptor (RODIECK, 1973; KIRSCHFELD, 1976). In this case the number of quanta per unit time and receptor does not depend only on A but on A/f. More precisely, it depends on the angle which the entrance pupil subtends to the focal point, presumably the center of the distal receptor end. This angle is the acceptance angle $2u$. A receptor of the diameter δ will receive the same number of quanta when its acceptance angle is equal, no matter if it sits in a lens or a compound eye. The small facet diameter in an ommatidium is compensated for by a short focal distance, or a large area in the optical surround from which the receptor receives light. If we take A, f and δ as found in the eyes of man and bee (references in KIRSCHFELD, 1974), the number of quanta reaching a receptor per unit time from an extended object will be about the same in both eyes. The fly's eye will receive 7–9 times more quanta, due to a correspondingly higher number of receptors collecting light from the same area of the optical surround.

In a superposition eye with screening pigment in the distal position, a rhabdom receives light not only through the corresponding facet but through an area of facets, the diameter of which is determined by the limiting angle of the optical systems. In *Ephestia* this area consists of about 180 facets. The diameter of this area is the effective aperture $\bar{A}$ and correspondingly an effective focal length $\bar{f}$ and acceptance angle $\bar{u}$ can be defined (Fig. 46). Admittance of light does not quite increase by this factor of 180 because the peripheral facets contribute somewhat

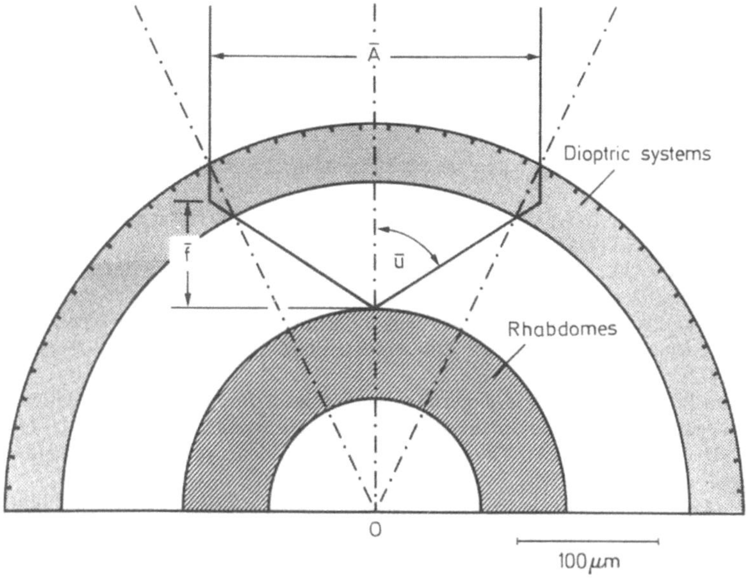

Fig. 46. Diagram of medial section through a superposition eye. $\bar{A}$, effective apertur; $\bar{u}$, aperture angle; $\bar{f}$, effective focal length. (KIRSCHFELD, 1974)

less; a factor of 100 may reasonably be assumed. Thus, a moth is about 100 times better off than a bee for pointlike objects but still 1000 times worse than man. However, the advantage of the superposition eye is prominent for the case of extended objects: the number of quanta per unit time available to the receptor of a moth is higher by a factor of about 100 compared to man and 1000 to the bee. In the catoptric superposition eye of Crustacea the effective aperture can be expected to be still bigger than in lepidopteran superposition eyes. This makes Crustacea likely to excell even moths in sensitivity.

G. Acknowledgement

I wish to thank Mrs. E. LUDWIG, Ms. M. GONNERT, and Ms. S. NERLICH for help in preparing manuscript and figures, and also Mr. K. VOGT for his critical reading of the manuscript.

References

Autrum, H., Wiedemann, I.: Versuche über den Strahlengang im Insektenauge (Appositionsauge). Z. Naturforsch. [B] **17b**, 480—482 (1962).

Barlow, H. B.: The size of ommatidia in apposition eyes. J. exp. Biol. **29**, 667—674 (1952).

Bernhard, C. G., Miller, W. H., Møller, A. R.: The insect corneal nipple array. Acta physiol. scand. **63**, Suppl. 243, 1—79 (1965).
Boschek, B. C.: On the fine structure of the peripheral retina and lamina ganglionaris of the fly, *Musca domestica*. Z. Zellforsch. **118**, 369—409 (1971).
Braitenberg, V.: Patterns of projection in the visual system of the fly. I. Retina-Lamina projections. Exp. Brain Res. **3**, 271—298 (1967).
Braitenberg, V.: Ordnung und Orientierung der Elemente im Sehsystem der Fliege. Kybernetik **7**, 235—242 (1970).
De Bruin, G. H. P., Crisp, D. T.: The influence of pigment migration on vision of higher Crustacea. J. exp. Biol. **34**, 447—463 (1957).
Cajal, S. R., Sánchez, D.: Contribución al conocimiento de los centros nerviosos de los insectos. I, Retina y centros opticos. Trab. Lab. Invest. Biol. Univ. Madrid **13**, 1—168 (1915).
Campos-Ortega, J. A., Strausfeld, N. J.: The columnar organization of the second synaptic region of the visual system of *Musca domestica* L. I. Receptor terminals in the medulla. Z. Zellforsch. **124**, 561—585 (1972).
Carricaburu, P.: Contribution à la dioptrique oculaire des Arthropodes: Détermination des indices des milieux transparents de l'ommatidie. Mém. Soc. Histoire nat. Afr. Nord **9**, 1—146 (1968).
Cleary, P., Deichsel, G., Kunze, P.: The superposition image in the eye of *Ephestia kühniella*. J. Comp. Physiol. **119**, 73—84 (1977).
Day, M. F.: Pigment migration in the eyes of the moth, *Ephestia kühniella* Zeller. Biol. Bull. **80**, 275—291 (1941).
Demoll, R.: Die Physiologie des Facettenauges. Ergebn. Fortschr. Zool. **2**, 431—516 (1910).
Dietrich, W.: Die Facettenaugen der Dipteren. Z. wiss. Zool. **92**, 465—539 (1909).
Eckert, H. E.: Optomotorische Untersuchungen am visuellen System der Stubenfliege *Musca domestica* L. Kybernetik **14**, 1—23 (1973).
Eltringham, H.: Butterfly vision. Trans. Entom. Soc. Lond. **79**, 1—49 (1919).
Exner, K.: Zur Linsenformel. Linsenwirkung nicht homogener Körper. Ann. Phys. Chem. **28**, 111—116 (1886).
Exner, S.: Ein Mikrorefraktometer. Arch. mikr. Anat. **25**, 97—112 (1885).
Exner, S.: Das Netzhautbild des Insektenauges. S.-B. Akad. Wiss. Wien **98**, 13—65 (1889).
Exner, S.: Die Physiologie der facettirten Augen von Krebsen und Insecten. Leipzig-Wien: Franz Deuticke 1891.
Fernández-Morán, H.: Fine structure of the light receptors in the compound eyes of insects. Exp. Cell. Res. Suppl. **5**, 586—644 (1958).
Fischer, A., Horstmann, G.: Der Feinbau des Auges der Mehlmotte, *Ephestia kuehniella* Zeller (Lepidoptera, Pyralididae). Z. Zellforsch. **116**, 275—304 (1971).
Franceschini, N.: Sampling of the visual environment by the compound eye of the fly: Fundamentals and applications. In: Photoreceptor Optics, eds. A. W. Snyder, R. Menzel. Berlin-Heidelberg-New York: Springer 1975.
Franceschini, N., Kirschfeld, K.: Les phénomènes de pseudopupille dans l'oeil composé de *Drosophila*. Kybernetik **9**, 159—182 (1971).
Goldsmith, T. H.: The visual system of insects. In: The Physiology of Insecta, ed. M. Rockstein, Vol. **1**, pp. 397—462. New York: Academic Press 1964.
Goldsmith, T. H., Bernard, G. D.: The visual system of Insects. In: Physiology of Insecta, ed. M. Rockstein, Vol. II. New York-London: Academic Press 1974.
Grenacher, H.: Untersuchungen über das Sehorgan der Arthropoden. Göttingen: Vandenhoek und Ruprecht 1879.
Hassenstein, B.: Wandernde geometrische Interferenzfiguren im Insektenauge. Naturwissenschaften **37**, 45—46 (1950).
Hausen, K.: Die Brechungsindices im Kristallkegel der Mehlmotte *Ephestia kühniella*. J. Comp. Physiol. **82**, 365—378 (1973).
Höglund, G.: Pigment migration, light screening and receptor sensitivity in the compound eye of nocturnal Lepidoptera. Acta. physiol. scand. **69**, Suppl. 282, 1—56 (1966).
Homann, E.: Der Vertikalilluminator als Augenspiegel bei kleinen Augen. Biol. Zbl. **44**, 582—591 (1925).
Horridge, G. A.: Alternatives to superposition images in clear-zone compound eyes. Proc. roy. Soc. B **179**, 97—124 (1971).

Horridge, G. A.: Optical mechanisms of clear-zone eyes. In: The Compound Eye and Vision of Insects, ed. G. A. Horridge. Oxford: Clarendon Press 1975.

Horridge, G. A., Giddings, C.: The retina of *Ephestia* (Lepidoptera). Proc. roy. Soc. B **179**, 87—95 (1971).

Horridge, G. A., Giddings, L., Stange, G.: The superposition eye of skipper butterflies. Proc. roy. Soc. B **182**, 457—495 (1972).

Ioannides, A. C., Horridge, G. A.: The organization of visual fields in the hemipteran acone eye. Proc. roy. Soc. B **190**, 373—391 (1975).

Kirschfeld, K.: Das anatomische und das physiologische Sehfeld der Ommatidien im Komplexauge von *Musca*. Kybernetik **2**, 249—257 (1965).

Kirschfeld, K.: Discrete and graded receptor potentials in the compound eye of the fly (*Musca*). In: The Functional Organisation of the Compound Eye, ed. C. G. Bernhard. Proc. Int. Symp. Stockholm 1965. Oxford-New York: Pergamon Press 1966.

Kirschfeld, K.: Die Projektion der optischen Umwelt auf das Raster der Rhabdomere im Komplexauge von *Musca*. Exp. Brain Res. **3**, 248—270 (1967).

Kirschfeld, K.: Aufnahme und Verarbeitung optischer Daten im Komplexauge von Insekten. Naturwissenschaften **58**, 201—209 (1971).

Kirschfeld, K.: The visual system of *Musca*: Studies on optics, structure and function. In: Information Processing in the Visual System of Arthropods, ed. R. Wehner. Berlin-Heidelberg-New York: Springer 1972.

Kirschfeld, K.: Das neurale Superpositionsauge. Fortschr. Zool. **21**, 229—257 (1973).

Kirschfeld, K.: The absolute sensitivity of lens and compound eyes. Z. Naturforsch. **29c**, 592—596 (1974).

Kirschfeld, K.: The resolution of lens and compound eyes. In: Neural Principles in Vision, eds. F. Zettler, R. Weiler. Berlin-Heidelberg-New York: Springer 1976.

Kirschfeld, K., Franceschini, N.: Optische Eigenschaften der Ommatidien im Komplexauge von *Musca*. Kybernetik **5**, 47—52 (1968).

Kirschfeld, K., Lutz, B.: Lateral inhibition in the compound eye of the fly, *Musca*. Z. Naturforsch. **29c**, 95—97 (1974).

Kleemann, C. F. C.: Beyträge zur Natur- und Insekten-Geschichte, Erster Teil. Nürnberg: Raspische Buchhandlung 1792.

Kuiper, J. W.: The optics of the compound eye. Symp. Soc. Exp. Biol. **16**, 58—71 (1962).

Kunze, P.: Untersuchung des Bewegungssehens fixiert fliegender Bienen. Z. vergl. Physiol. **44**, 656—684 (1961).

Kunze, P.: Eye glow in the moth and superposition theory. Nature (Lond.) **223**, 1172—1174 (1969).

Kunze, P.: Verhaltensphysiologische und optische Experimente zur Superpositionstheorie der Bildentstehung in Komplexaugen. Verh. dtsch. Zool. Ges. **64**, 234—238. Stuttgart: Fischer 1970.

Kunze, P.: Comparative studies of arthropod superposition eyes. Z. vergl. Physiol. **76**, 347—357 (1972a).

Kunze, P.: Pigment migration and the pupil of the dioptric apparatus in superposition eyes. In: Information Processing in the Visual Systems of Arthropods, ed. R. Wehner. Berlin-Heidelberg-New York: Springer 1972b.

Kunze, P., Hausen, K.: Inhomogeneous refractive index in the crystalline cone of a moth eye. Nature (Lond.) **231**, 392—393 (1971).

Land, M. F.: Superposition images are formed by reflection in the eyes of some oceanic decapod crustacea. Nature (Lond.) **263**, 764—765 (1976).

Leydig, F.: Das Auge der Gliederthiere. Tübingen: Laupp 1864.

Meggitt, S., Meyer-Rochow, V. B.: Two calculations on optically non-homogeneous lenses. In: The Compound Eye and Vision of Insects, ed. G. A. Horridge. Oxford: Clarendon Press 1975.

Meyer-Rochow, V. B.: The dioptric system of the eye of *Cybister* (Dytiscidae: Coleoptera). Proc. roy. Soc. Lond. B **183**, 159—178 (1973).

Meyer-Rochow, V. B., Horridge, G. A.: The eye of *Anoplognathus* (Coleoptera, Scarabaeidae). Proc. roy. Soc. Lond. B **188**, 1—30 (1975).

Miller, W. H., Bernard, G. D.: Butterfly Glow. J. Ultrastr. Res. **24**, 286—294 (1968).

Müller, J.: Zur vergleichenden Physiologie des Gesichtssinnes des Menschen und der Thiere nebst einem Versuch über die Bewegungen der Augen und über den menschlichen Blick. Leipzig: C. Cnobloch 1826.

Nunnemacher, R. F.: The retinal image of arthropod eyes. Anat. Rec. **134**, 618—619 (1959).

Nunnemacher, R. F.: The structure and function of arthropod eyes. In: Progress in Photobiology, eds. B. Chr. Christensen, B. Buchmann. Amsterdam: Elsevier Publishing Comp. 1961.

Parker, G. H.: The retina and optic ganglia in decapods, especially in *Astacus*. Mitt. Zool. Stat. Neapel **12**, 1—73 (1897).

Rodieck, R. W.: The Vertebrate Retina. Principles of Structure and Function. San Francisco: W. H. Freemann 1973.

Schneider, L., Langer, H.: Die Feinstruktur des Überganges zwischen Kristallkegel und Rhabdomeren im Facettenauge von *Calliphora*. Z. Naturforsch. **21b**, 196—197 (1966).

Scholes, J.: The electrical responses of the retinal receptors and the lamina in the visual system of the fly *Musca*. Kybernetik **6**, 149—162 (1969).

Seitz, G.: Der Strahlengang im Appositionsauge von *Calliphora erythrocephala* (Meig.). Z. vergl. Physiol. **59**, 205—231 (1968).

Seitz, G.: Untersuchungen am dioptrischen Apparat des Leuchtkäferauges. Z. vergl. Physiol. **62**, 61—74 (1969).

Snyder, A. W., Menzel, R., Laughlin, S. B.: Structure and function of the fused rhabdome. J. comp. Physiol. **87**, 99—135 (1973).

Stavenga, D.: Visual Receptor Optics, Rhodopsin and Pupil in Fly Retinula Cells. Dissertation Rijksuniversiteit Groningen 1974.

Strausfeld, N.: Golgistudies on insects. II. The optic lobes of Diptera. Phil. Trans. roy. Soc. Lond. [Biol. Sci.] **258**, 135—223 (1970).

Strausfeld, N.: The organization of the insect visual system (Light microscopy). Z. Zellforsch. **121**, 377—441 (1971).

Trujillo-Cenóz, O.: Some Aspects of the Structural Organization of the Intermediate Retina of Dipterans. J. Ultrastruct. Res. **13**, 1—33 (1965).

Trujillo-Cenóz, O.: The structural organization of the compound eye in insects. In: Handbook of Sensory Physiology, Vol. VII/2. ed. M. G. F. Fuortes, Berlin-Heidelberg-New York: Springer 1972.

Trujillo-Cenóz, O., Melamed, J.: Electron Microscope Observations on the Peripheral and Intermediate Retinas of Dipterans. In: The Functional Organization of the Compound Eye, ed. C. G. Bernhard. Proc. Int. Symp. Stockholm 1965. Oxford: Pergmon Press 1966.

Tuurala, O.: Histologische und physiologische Untersuchungen über die photomechanischen Erscheinungen in den Augen der Lepidopteren. Ann. Acad. Sci. Fenn. **24**, 1—69 (1954).

Umbach, W.: Entwicklung und Bau des Komplexauges der Mehlmotte *Ephestia kühniella* Zeller nebst einigen Bemerkungen über die Entstehung der optischen Ganglien. Z. Morph. Ökol. Tiere **28**, 561—594 (1934).

Varela, F. G., Wiitanen, W.: The optics of the compound eye of the honeybee *(Apis mellifera)*. J. gen. Physiol. **55**, 336—358 (1970).

Vigier, M. P.: Mécanisme de la synthèse des impressions lumineuses recueillies par les yeux composés des Diptères. C. R. Acad. Sci. [D] (Paris) **64**, 1221—1223 (1909).

Vogt, K.: Optische Untersuchungen an der Cornea der Mehlmotte *Ephestia kühniella*. J. Comp. Physiol. **88**, 201—216 (1974).

Vogt, K.: Zur Optik des Flußkrebsauges. Z. Naturforsch. **30c**, 691 (1975)

Vogt, K.: Ray path and reflection mechanisms in crayfish eyes. Z. Naturforsch. **32c**, 466—468 (1977).

De Vries, H., Kuiper, J. W.: Optics of the insect eye. Ann. N. Y. Acad. Sci. **74**, 196—203 (1958).

Walcott, B.: Anatomical changes during light adaptation in insect compound eyes. In: The Compound Eye and Vision of Insects, ed. G. A. Horridge. Oxford: Clarendon Press 1975.

Waterman, T. H.: Light sensitivity and vision. In: The physiology of Crustacea, ed. T. H. Waterman. New York-London: Academic Press 1961.

Westheimer, G.: Optical properties of vertebrate eyes. In: Handbook of Sensory Physiology, Vol. VII/2. ed. M. G. F. Fuortes, Berlin-Heidelberg-New York: Springer 1972.

Winthrop, J. T., Worthington, C. R.: Superposition image formation in insect eyes. Biophys. J. **6** (Proc. Bioph. Soc.), 124 (1966).

Yagi, N., Koyama, N.: The Compound Eye of Lepidoptera. Tokyo: Shinkyo-Press 1963.

Chapter 9

Spectral Sensitivity and Color Vision in Invertebrates

RANDOLF MENZEL, Berlin, Germany

With 14 Figures

Contents

A. Introduction

The visual world offers a variety of information that enables animals to orient themselves to their surroundings. One specific feature of the visual information is the wavelength difference of light *emitted* from light sources or *reflected* from objects. Under natural conditions, a general difference in the spectral composition of light is based upon the fact that light coming directly from light sources, e.g., the sun, sky, and moon, is characterized by its relatively high content of short wavelength (<450 nm), whereas light reflected from natural objects such as leaves and soil lacks UV light and is dominated by green and yellow light (>450 nm) (Figs. 1). The physical basis of these differences is the preferential absorption by pigment molecules, which—in contrast to atoms and small molecules—have resonances above 300 nm and shift the bulk of the reflected light into the green and orange region of the spectrum. In water, the wavelength range is compressed in the long-wavelength part of the spectrum due to water's high absorption of red and infrared light (Fig. 1). But in water, too, direct light is relatively richer in short-wave light. Neglecting the finer differences in the spectral content of light, this general distinction between direct and reflected light could be the cue that allows an animal to differentiate such basically different habitats as open spaces, shadowed areas, spaces rich in food, hiding places, etc. It is easy to imagine that even a primitive visual system can take advantage of the spectral differences between direct and reflected light if the photoreceptive organ differentiates between short and long wavelengths. This is based on the possibility that simple sensory-neural mechanisms selectively control general behavior patterns, such as escape, foraging, and hiding.

Like other biological processes, color vision as a neural mechanism must have evolved from primitive systems which themselves were superior to those neural mechanisms without a chromaticity-coding capacity. As primitive receptive organs had very low spatial resolution, only large fields of view differing in chromatic content could have been used as information sources. Therefore, the question arises whether it was the difference between the short- and long-wavelength content of direct and reflected light that influenced the origin of chromaticity-coding visual systems. If that was the case, early chromaticity-coding systems should have been antagonistically organized between short- and long-wavelength sensitivity. In this review we examine this central question by presenting data on the spectral properties of the various photoreceptive structures in a phylogenetic order. Invertebrates offer specially suitable material for such an approach because (1) invertebrates contain many independent phylogenetic lines, (2) photoreceptors and photoreceptor organs ranging from light sensitive dermal cells to highly complex lens and compound eyes can be found, and (3) invertebrates inhabit all existing ecological habitats.

Another question arising from the view presented above is that chromaticity-coding visual systems have evolved because they specifically elicit appropriate behavior patterns. For example, if short-wave light causes a sessile water-living animal to retract and green light causes it to bend towards the light source, one must conclude that the neural pattern generators for the different motor patterns

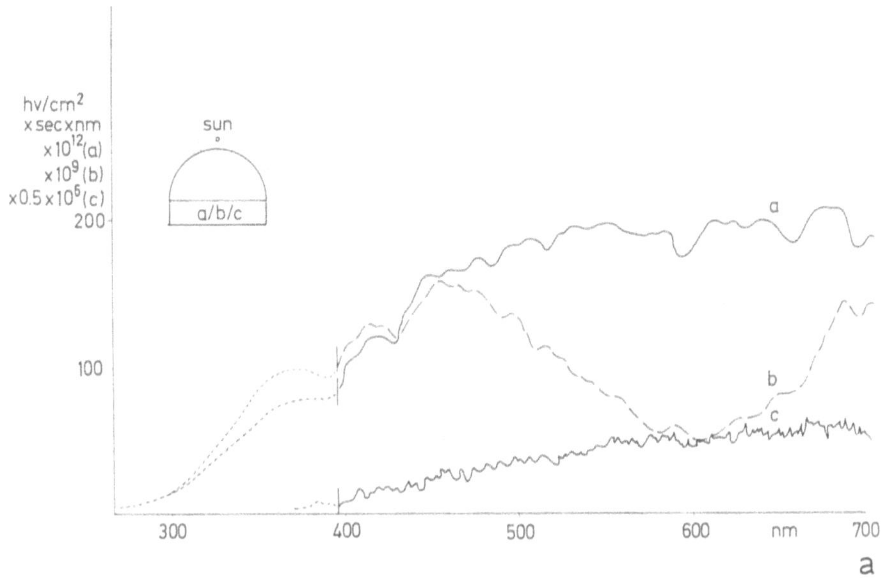

Fig. 1a–c. Examples of the spectral distribution of natural light (McFarland and Munz, 1976). All irradiance and radiance values are photons/$cm^2 \cdot s \cdot nm$. Measurements were carried out between 400 and 700 nm. For rough orientation below 400 nm the curves are extended by the author (R.M.) in the light of data reported by Hess (1939) cited in Rozenberg (1966), and Condit and Grum (1964). (*A*) Spectral irradiance during midday (*a*), evening twilight (*b*), and full moonlight (*c*), at earth's surface (Eniwetok Atoll, Marshall Islands, summer 1970). A high, thin cirrus cover and low cumulus clouds (~30% cover, *see inset*) occurred during midday. Sun unobscured when recording, low cumulus were also present at twilight and moonlight. Total irradiance between 400 and 700 nm is 53000×10^{12} (midday), 32×10^{12} (twilight) and 30×10^9 photons/$cm^2 \cdot s$ (moonlight). (*B*) Spectral radiance (*d*, *e*) and spectral irradiance (*f*) during midday of a clear sky (30th January 1974, Ithaca, New York). Sun altitude 30°, sky cloudless. Spectral radiance (*d*) within a cone of 15° plane angle (0.05° steradians) centred at 60° (90° opposite to the sun along the sun's azimuth, *see inset*) spectral radiance (*e*) with the same aperture pointing to the sun (30°) spectral irradiance (*f*) (total irradiance in the visible spectrum 400–700 nm: 47710×10^{12} photons/$cm^2 \cdot s$. (*C*) Relative numbers of photons in downwelling irradiance in a clear sea near Eniwetok Atoll (*g*) and an estuary (St. George Sound) (*h*). Depth of detection 1m. Both spectra recorded for a clear sky. Freshwater lakes have maximal transmission usually at shorter wavelength with about the same bandwidth as shown here for the St. George Sound. The λ_{max} are: very clear water: $\lambda_{max} = 470$ nm, oligotrophic to eutrophic lakes 500–530 nm, eutrophic lakes 550 nm, lakes highly stained with dissolved organic matter: practically no light below 600 nm deeper than 1 m. Spectral distribution (*i*) of visible light transmitted through a maple leaf. The maple leaf was placed over the unrestricted cosine detector

are specifically connected to the appropriate photoreceptor inputs. I shall use the term "wavelength-specific behavior pattern" to describe such situations. Many examples of wavelength-specific behavior are given in Sec. D of this review. Most interestingly, the wavelength regions <450 nm and >450 nm clearly control different sets of behavior patterns. These findings provoke the speculation that wavelength-specific behavior patterns are indicators of a neuronal strategy in the early evolutionary history of color vision. In lower invertebrates such as coelenterates, pseudocoelomates, and plathelminthes, different motor programs are often located in different neuronal networks (see Bullock and Horridge, 1965). If these

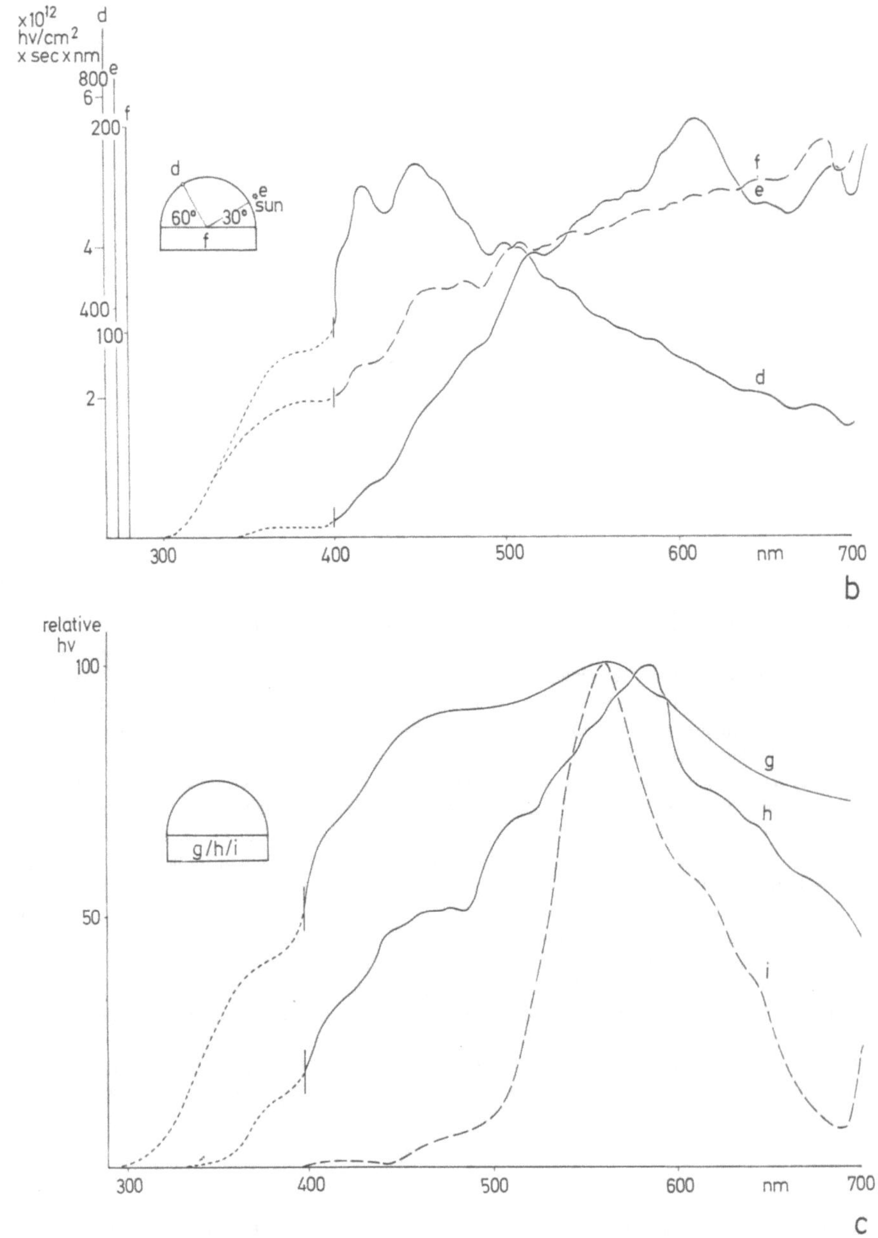

neural networks have their own visual input systems and if these input systems differ in wavelength sensitivity, differential behaviors can be elicited even though the whole system does not code "color" and the animal has no color vision. These speculations are especially interesting because the wavelength-specific behavior patterns may give us clues about the evolution of the sensory and interneuron mechanisms of color vision in invertebrates.

Many invertebrate species, especially within the arthropods, possess a highly developed color vision system; it is valuable to ask how it evolved. Unfortunately, the available data are not nearly sufficient to develop any conclusive explanation of the basic evolutionary strategies. It is especially frustrating that those groups of species that are most interesting for phylogenetic comparisons, such as the coelenterates, polychaetes, and primitive arthropods, are scarcely studied, if at all. This review intends to stimulate experimental work with species which are of special interest from a phylogenetic point of view.

I. Definition of Important Terms

There are various possibilities for expressing the wavelength dependence of the response of photoreceptors, of visual interneurons, or of the whole animal (see e.g., RODIECK, 1973, p. 266ff, DIEHN, this volume). In all cases it is essential to take into consideration the two parameters—light intensity (number of quanta) and wavelength—when determining the response. From response-intensity functions (Fig. 2a) one can define two spectral dependency functions (1) the *spectral efficiency function* (Fig. 2b); the wavelength dependence of the response height for wavelengths of equal intensity (number of quanta), and (2) the *spectral sensitivity function* [$S(\lambda)$] or action spectrum, reciprocal of the number of quanta of each wavelength which elicits a constant response (Fig. 2c). If one tries to compare spectral dependencies of receptors, interneurons, or the whole animal with the spectral absorption properties of photopigments, $S(\lambda)$ is the only appropriate function (BURKHARDT, 1962, 1964; RODIECK, 1973).

The response-intensity function can be a complex function with changing gain factors for different intensity levels (or response levels) and different wavelengths (Fig. 1a). In such a case the number of quanta absorbed is not the only factor determining the response, so the principle of univariance does not hold. Consequently the $S(\lambda)$ function differs at different response levels (Fig. 2c). It is therefore important that in any spectral experiment the response-intensity function is tested for various

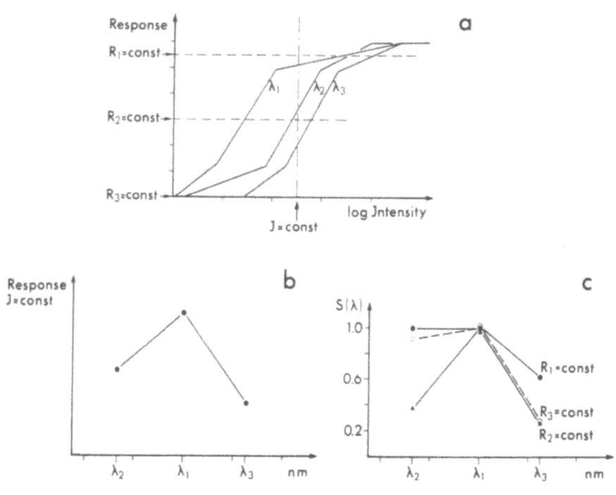

Fig. 2. Procedure involved in calculating the spectral efficiency function (*b*) and the $S(\lambda)$ (*c*) from response-intensity functions (*a*). Figure 2*a* shows that in this case univariance holds only for the middle part of the response-intensity function. So, only the $S(\lambda)$ function for R_2=constant can be directly compared with the quantum capture efficiency of the photopigment (see text). Additional factors come into play near threshold and saturation. (e.g., screening pigment absorption, scattered light, selective chromatic feedback through interneurons to the receptors, etc.), and change $S(\lambda)$ remarkably

wavelengths at different intensity levels. (More about the univariance of spectral responses, see Rodieck, 1973; and Ch. 6, this Vol.).

The term *wavelength dependence* is used when the experimental procedure, or the data reported, do not allow one to regard a response-wavelength function as a sensitivity or an efficiency function.

Color vision is defined as the ability of an animal to distinguish spectral colors on the basis of wavelength differences, but not brightness differences. Color vision cannot be proven by demonstrating that an animal has a spectral preference within one behavior pattern and another spectral preference with a different behavior pattern. Neither is it proof of color vision, if the univariance principle does not hold for a given behavior pattern. Furthermore, the finding of photoreceptor types with different spectral properties can only be considered as the premise for color vision. The most reliable method of demonstrating color vision is the training procedure to monochromatic light, in which light intensity and wavelengths are varied in the test.

If an animal displays markedly different $S(\lambda)$ for different behavior patterns, the individual behavior pattern will be called *wavelength specific*. Such wavelength-specific responses are especially interesting in species with less highly evolved visual systems.

B. Photoreceptors: Spectral Sensitivity

Sight begins with the absorption of light by a membrane-bound photopigment in a photoreceptor cell. The pigment undergoes a series of molecular changes, and an electric effect results; an ion current flows, and the voltage across the cell membrane changes. This voltage change is transmitted through a simple or complex nerve network, and neuronal processing elicits or controls the animal's behavior.

The pigment molecule causing light sensitivity of photoreceptors absorbs the naturally occurring electromagnetic waves over a relatively broad frequency band and is minimally affected by the thermal noise. Rhodopsin (retinal plus protein) was apparently selected early in the evolution of metazoans as the molecule best suited for this purpose and is therefore found to be almost unique as the light-sensing molecule both in invertebrates and vertebrates. The reasons for the evolutionary success of this molecule seem to be (1) its extremely high molar extinction, (2) the possibility of dense package in the cell's external membrane, (3) the relatively broad absorption band (80–100 nm half bandwidth), and (4) the variability of the absorption maximum within the wavelength region 330–700 nm by changes only in the protein moiety. Color vision and behavior patterns specific to various wavelengths of light require photoreceptors with different wavelength sensitivities. In general the various $S(\lambda)$ of photoreceptors result from their containing rhodopsins with varying spectral absorption properties (see Hamdorf, this volume). Absorption maxima vary as a result of interaction with the protein component. The peaks of maximal wavelength sensitivity (λ_{max}), therefore, best describe the wavelength-dependent visual input systems. More than one type of photoreceptors containing photopigments with varying λ_{max} provide the visual interneurons with the potential of extracting wavelength information independent of intensity.

The existence of rhodopsins with different absorption maxima is one but not the only prerequisite for the determination of differences in the wavelength distribution of natural light. The various rhodopsins have to be located in separate receptor input systems and these have to be wired specifically to interneuron and motocontrol systems.

In this chapter we collect the experimental data on the spectral properties of the photoreceptors and present these data in an order that follows the systematic grouping of the invertebrates. This method of presentation also enables us to compare the spectral properties of the photoreceptors of the various species from an evolutionary point of view.

I. Lower Invertebrates

(Protozoa, Porifera, Coelenterata, Plathelminthes, Pseudocoelomates, Annelida).

Procaryote organisms do not belong to the groups to be dealt with here; however, we wish to draw attention to some recent papers which are interesting from a comparative point of view. Several species respond phototactically, e.g., the motile bacterium *Rhodospirillum* and the Cyanophycean algae *Oscillatoria*. The spectral dependence of their responses resembles that of photosynthesis, suggesting that phototaxis in bacteria may be directly induced by the supply of photosynthetically derived energy to the flagella (Link's hypothesis, see CLAYTON, 1964, 1971) and thus may be unrelated to any photosensing system. However in the case of *Halobacterium halobium* this is not a sufficient explanation (see Ch. 2, this volume). HILDEBRAND and DENCHER (1974, 1975; see HILDEBRAND, 1977) have proved that two light-induced reactions occur, one of which—the inverse photophobic response (motor response to light-off)—is maximal at 565 nm and controlled by the photosystem 565 nm (undissociated retinal protein complex, bacteriorhodopsin; OESTERHELT and STOECKENIUS, 1971; OESTERHELT et al., 1973); while the other, which is the more sensitive reaction—the direct photophobic response (motor response to light-on)—has maxima at 280 and 370 nm and is controlled by another retinal protein complex (DENCHER, unpublished observation). Since illumination of phospholipid membranes containing bacteriorhodopsin produces an electric potential across the membrane as the result of an H^+ transport out of the cell (DRACHEV et al., 1974), the photoreactions of *Halobacterium* possibly represent a sensory function of the pigments in addition to their energy production. The short latency of the photophobic reactions also suggests that it is not controlled by ATP (HILDEBRAND and DENCHER, 1974). However, other explanations are possible, such as viewing the response of the flagella as the results of the proton shift preceeding the ATP production. In any case it is interesting that *Halobacterium* displays behavioural responses to UV-light and longer wavelength light, which seem to be controlled by two different photopigments. So we find wavelength-specific behavior already in a procaryotic cell.

Flagellates (see DIEHN, this Handbook, Vol. VII/6A) have been studied particularly well on their light responses (review CLAYTON, 1964; HAND et al., 1967; DIEHN, 1969a, b; HAND, 1970; FEINLEIB and CURRY, 1971). In *Euglena* it was shown that the reaction to light is controlled by flavinic chromophores, a photopigment in the paraflagellar body (CHECCUCCI et al., 1976). The spectral dependence of phototaxis and phototrapping peaks at 380 nm, and between 425 and 480 nm; wavelengths above 530 nm produce no response. An earlier reported response to red light results from increased photosynthesis and the attractiveness of oxygen. In mutants without the eye spot there is a single maximum at 410 nm and a generally greater sensitivity up to 530 nm, but streptomycin-treated white mutants contain the paraflagellar body as wild type *Euglena* and display the same spectral dependence of phototaxis as wild types (CHECCUCCI, 1976; CHECCUCCI et al., 1976). In addition, the paraflagellar body exhibits flavinlike fluorescence (CHECCUCCI, 1976). All these data support the conclusion that a flavoprotein is the light-sensing molecule in *Euglena*. DIEHN (1976), however, has shown that only the inverse photophobic response of *Euglena* is mediated by a flavin pigment, while the direct response most likely involves a different receptor pigment. Other flagellates have other light-sensing molecules, e.g., carotene protein and phycoerythrin (see DIEHN, 1976), but in all cases light sensing and photosynthesis are uncoupled. Flagellates are, therefore, the most primitive organisms that can unequivocally "see"; that is, perceive light and respond to it behaviorally using a specialized photoreceptive structure which is independent, at least functionally, of the energy-supplying photosynthetic system. The photopigment involved, however, seems to be different in different species, and also different from the visual pigment of metazoans.

Although many *ciliates* and *amoebas* also react to light with motion, a microscopically identifiable photoreceptor organelle has not yet been found, and their directional responses to light stimulation are different from the phototaxis of flagellates (JENNINGS, 1915; MAST, 1911; FOLGER, 1925; FRAENKEL and GUNN, 1961; MÜLLER, 1965). However, the findings of many older studies, particularly those dealing with ciliates, have proved impossible to verify (cf. GRABOWSKI, 1939; DEMBROWSKI, 1950; MACHEMER, 1967; HILDEBRAND, 1969, 1970). Reliably, it can only be concluded that amoebas react to illumination by slowing down the motions of their pseudopodia (wavelengths between 450 and 500 nm are the most effective, HITCHCOCK, 1961); while the ciliates, *Paramecium*, *Opalina* and *Euplotes*, respond with a phobic reaction, although only to UV light (λ_{max} = 280 nm) (HILDEBRAND, 1969, 1970) and not to visible light as has previously been supposed. In *Paramecium* and *Opalina*, the action spectrum of the phobic reaction is the same as that found for the decrease in membrane potential and membrane resistance upon illumination. Therefore reaction to UV light is thought to depend on a photochemical change in the membrane proteins, chromophoric molecules do not appear to be involved. Furthermore, it is not known what mediates the light sensitivity of amoebas and ciliates with symbiontic algae (see also DIEHN, 1976).

Porifera respond to illumination by contracting whole or part of their bodies (photophobic growth) but neither the light-sensing cells nor the photopigment involved have been studied (cf. BULLOCK and HORRIDGE, 1965 and this volume, Ch. 10).

Coelenterata are the first phylum in the evolutionary tree of which at least some members are equipped with complex light-sensing organs and show complicated light-induced behavior (cf. BULLOCK and HORRIDGE, 1965). Unfortunately, the few papers on light-induced behavior in coelenterates deal almost entirely with polyps, in which no specialized photoreceptive structures have been detected.

Hydra spirardi is especially sensitive to blue light (350–500 nm) (PASSANO and McCULLOUGH, 1962; SINGER et al., 1963), and the reaction adapts to prolonged illumination (TARDENT and FREI, 1969). Various authors assume that carotenoid pigments in the endodermal cells may participate in photoreception. NORTH (1957) showed that the sea anemone *Metridium* is able to respond to less than 10 quanta

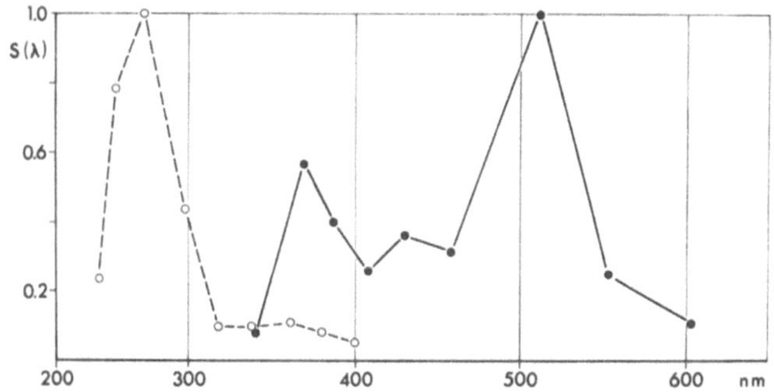

Fig. 3. Action spectrum of light-induced behavior of the sea anemone *Anthopleura xanthogramma* (CLARK and KIMELDORF, 1970). —●—: Action spectrum of the tentacle flexion: --○--: Action spectrum of the tentacle retraction. The measurement of action spectra is based on the comparison of the efficiency of the test wavelength to that of a reference wavelength. The relative efficiency given on the ordinate is calculated from the formula: $\frac{ED_R}{ED_T \cdot \frac{\lambda_T}{\lambda_R}}$ where ED_R and ED_T are the minimum effective dose (threshold) of the reference wavelength (ED_R) and the test wavelength (ED_T); λ_R and λ_T are the reference and the test wavelength respectively

of absorbed light; and NORTH and PANTIN (1958) found a single peaked $S(\lambda)$ ($\lambda_{max}=500$ nm) for the light-induced contraction response which is not inconsistent with a rhodopsinlike pigment. The spectral threshold curve for light-induced bending of the tentacles of the sea anemone, *Anthopleura xanthogramma*, shows a main peak at 500 nm and a secondary maximum at about 360 nm and is also similar to a rhodopsin resonance curve (CLARKE and KIMELDORF, 1970). However, the retraction of the tentacles, also light induced, is triggered only by wavelengths between 240 and 400 nm. The maximum at 280 nm suggests that this may be a photochemical effect of proteins minus the chromophoric group (Fig. 3). The photoreceptive cells (Fig. 3) are probably spread over the whole animal, but the region of the circumoral nervous plexus is especially sensitive to light.

Plathelminth worms are equipped with photoreceptors diffusely scattered over the whole body as well as with complex organs of sight in the form of ocelli with inverse retinas (WOLKEN, 1958; BULLOCK and HORRIDGE, 1965; RÖHLICH and TÖRÖK, 1961). Extracellular recordings from planarian ocelli are consistent with a rhodopsinlike photopigment with $\lambda_{max}=508$ nm (BROWN et al., 1968). Spectral phototaxis experiments put the sensitivity maximum of the ocelli at about 475 (VIAUD, 1948) or 530 nm (MARRIOTT, 1958). This difference might be caused by a varying contribution from the dermal photoreceptors or by the effects of screening pigments (BROWN et al., 1968).

Among the *Pseudocoelomate worms* the planktonic rotatoria in particular frequently demonstrate light-induced behavior and are equipped with one central eye, or with several eyes composed of one or more cells (BULLOCK and HORRIDGE, 1965; EAKIN and WESTFALL, 1965; HERTEL, 1972). These eyes have not yet been successfully examined physiologically. VIAUD (1940, 1943, 1948) observed photo-

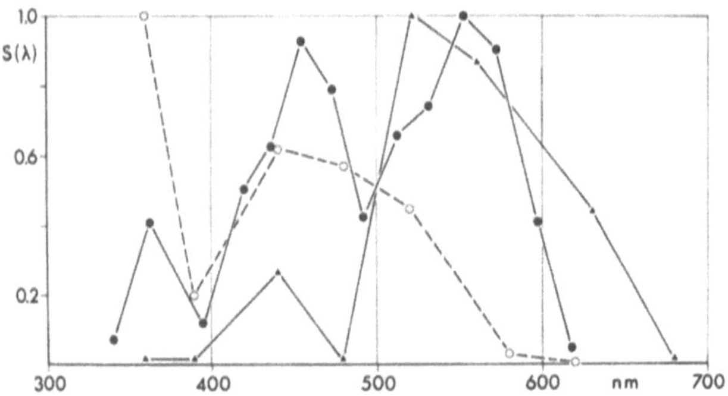

Fig. 4. $S(\lambda)$ of the phototactic response of three freshwater plankton rotifera species; —●—: *Asplanchna priodonta* (after HERTEL, 1972); --○--: *Filinia longiseta*, and —▲—: *Polyarthra remata* (MENZEL and ROTH, 1972). The rotifera were swimming in a Y maze, in which one arm was illuminated with monochromatic light through the end window, the other arm was dark. Response-intensity functions were measured for each wavelength, and the $S(\lambda)$ was calculated for a constant response of 70% choice of the illuminated arm

tactic orientation in the rotatoria species *Brachionus pala* and *Asplanchna girodi* and found the wavelengths around 520 and 360 nm most effective.

MENZEL and ROTH (1972) and HERTEL (1972) determined the $S(\lambda)$ functions of three species of rotatoria and found multipeak functions in *Asplanchna priodonta* and *Polyarthra remata* (Fig. 4). By contrast the species *Filinia longiseta* had only one sensitivity maximum in the visible light range at 440 nm. The difference of the $S(\lambda)$ function of *Filinia* to those of the other two species correlates well with the chromatic light conditions in which these species live (MENZEL and ROTH, 1972). Spectral dependencies of phototactic responses comparable with that of *Filinia* were found by ARORA (1967) for *Pedalia intermedia* and *Pedalia angularis* (λ_{max} = 465 nm) and others comparable with *Polyarthra* for *Brachionus calyciflorus* (λ_{max} = 530 nm, 650 nm) (cf. also POURRIOT and CLEMENT, 1973; CLEMENT, 1976).

VIAUD (1948) hypothesized that the UV sensitivity results from a "general" (dermal and/or neuronal) light sensitivity and predominantly stimulates the movement of the cilia, thus controlling the motor activity. By contrast the long-wavelength sensitivity is caused by the eyes and controls the swimming direction. HERTEL (1972) has carried out careful studies with *Asplanchna priodonta* and *Asplanchna breitwellii* to test VIAUD'S hypothesis. He found no evidence for different actions of UV or longwave light. The phototactic swimming track depended only on the intensity of the spectral light. The $S(\lambda)$ functions calculated for different parameters of the swimming track (straightness, velocity, interruptions, tendency to switch to negative phototaxis) were not significantly different from the spectral threshold function of the positive phototactic response. He also found no difference between the two *Asplanchna* species, an interesting observation since *A. priodonta* has one central eye and two lateral eyes whereas *A. breitwellii* has only one central eye. The conclusion from these observations is that the lateral eyes and the central eye in *A. priodonta* contain the same mixture of photopigment, or that the information from both eye types is pooled in visual interneurons which control the phototactic response. CLEMENT (1976) has tested VIAUD'S hypothesis, too, but

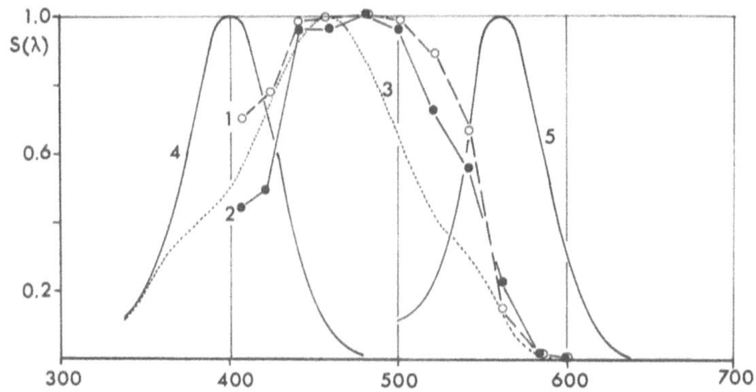

Fig. 5. Electrophysiologically determined $S(\lambda)$ of annelid eyes. *1* and *2*. $S(\lambda)$ of dark-adapted posterior (*1*) and anterior (*2*) eyes of the polychaet *Nereis mediator* in the atoke phase. The $S(\lambda)$ was measured by determining the reciprocal of the number of quanta required to elicit a constant electroretinogram of 350 μV (YINGST et al., 1972). *3*. $S(\lambda)$ of the eyes of *Vanadis* (WALD and RAYPORT, 1977) measured in the same way as in *Nereis*. *4*. and *5*. $S(\lambda)$ of the main retina (*4*) and accessory retina (*5*) of *Torrea candida* (WALD and RAYPORT, 1977)

using different methods. He also found no selective motion-stimulating effect of UV light.

Free-living *Nematodes* have a pair of sense organs, *amphids*, at the front end, which appear to have a photoreceptive function in addition to their chemoreceptive one (BURR and BURR, 1975). *Chromadorina viridis* is negatively phototactic to UV illumination (λ_{max} = 366 nm) but positively phototactic to wavelengths >400 nm (λ_{max} = 465 and 570 nm) (CROLL, 1966). Both reactions are induced via the ocelli in the amphids, but it is not clear what type of photopigment is involved.

Annelid worms possess light-sensitive structures, consisting of one or several cells, scattered all over the body as well as complex eyes at the front or on the tentacles (cf. BULLOCK and HORRIDGE, 1965). It is a pity that this rich variety of photoreceptive organs has been so neglected since it is among the annelids that the phylogenetic roots of the highly developed eyes of spiders, crustaceans and insects are to be found.

In only three species (*Nereis mediator*, *Torrea candida*, *Vanadis formosa*) has the $S(\lambda)$ been measured electrophysiologically and only then with ERG recordings (Fig. 5). YINGST et al. (1972) found in *Nereis* $S(\lambda)$ curves (λ_{max} = 450–500 nm) which are broader than one would expect from a single rhodopsin photopigment. The same is true for *Vanadis* (WALD and RAYPORT, 1977, λ_{max} = 460 or 480 nm). The authors' presumption of a bi- or multipigment system is supported by their finding that different areas of the retina differ in their $S(\lambda)$ functions. In *Torrea candida* (Fig. 5) the $S(\lambda)$ of the main retina peaks at 400 nm and is narrower than a Dartnall function. Two small patches of photoreceptors on each side of the lens (the accessory or secondary retina), however, have a $S(\lambda)$ function with λ_{max} at 560 nm. WALD and RAYPORT interpret the secondary retina as a specialized photoreceptor system to detect light reflected from the main retina. It should be extremely interesting to find out the significance of the secondary retina and the wavelength shift in its sensitivity. (Is it a mechanism for an optical determination of depth in the sea, as WALD and RAYPORT speculate?)

UNTEUTSCH (1937) and HOWELL (1939) infer from the varying reactions of the earthworm to increasing and decreasing illumination that there are different receptive systems for the detection of shadow and light, which might be characterized by different spectral dependencies. HAGIWARA and MORITA (in BULLOCK and HORRIDGE, 1965, p. 748) and LAVERACK (1969) found light-induced action potentials in the ventral cord ganglion of the leech, *Hirudo medicinalis*, the action spectrum, however, has not yet been determined. These large interneurons receive input from photosensory cells located in seven bilateral pairs in both the ventral and dorsal surface of each of the body segments. These photoreceptors contain the same single photopigment as the five pairs of eyes on the head of the leech (λ_{max} = 450 nm; KRETZ et al., 1976).

The sipunculid worm *Golfingia* responds to illumination of the cerebral ganglion with retraction of the proboscis. The action spectrum of this response peaks at 550 nm (STRATTON, 1965). Light-induced neuronal activity was recorded from the ventral side of the ganglion (O'BENAR and MATSUMOTO, 1976), but the photoreceptors were not found.

II. Arthropoda

1. Crustacea

The spectral properties of visual pigments in Crustacea have been reviewed by WALD (1968), GOLDSMITH (1961, 1972), GOLDSMITH and FERNANDEZ (1968) (see also HAMDORF, this volume) and will not be repeated here. Retinal was found to be the chromophoric group and the λ_{max} values range from 460 to 562 nm. Open-sea water species tend to have λ_{max} values at shorter wavelengths (460–480 nm), freshwater species at longer wavelengths (520–562 nm). This can be interpreted as an adaptation to the spectral distribution of light in the natural environment: open-sea water transmits best at wavelengths around 470–480 nm, fresh water around 510–530 nm. Intertidal crabs have a variety of λ_{max} values.

There are often discrepancies between absorption maxima of extracted pigments and $S(\lambda)$ determined electrophysiologically. In some cases this can be traced back to the chromatic absorption properties of the screening pigments. Other explanations are (1) methodical difficulties during extraction of the photopigment, (2) self-screening of photopigment confined into a long rhabdom, and (3) color filter effect of photostable metarhodopsin (cf. MENZEL, 1975; see also p. 523 and HAMDORF, this volume).

Open-sea water crustaceans appear to have only one sort of photopigment; however, extractions may be misleading, since small quantities of another pigment are hard to detect. One photopigment (λ = 508 nm) was found by GOLDSMITH (1968), BRUNO and GOLDSMITH (1974) and SCOTT and MOTE (1974) in *Callinectes sapidus*, the swimming blue crab, similarly in *Carcinus*, the green crab (BRUNO et al., 1973) and in *Pleuroncodes planipes*, a galatheid crab (FERNANDEZ, 1973). In ERG recordings from the small complex eyes of the amphipods *Pontoporeia affinis* and *Pontoporeia femorata* DONNER (1971) also found only a single peaked $S(\lambda)$ ($\lambda_{max}=550$ nm). Intertidal crab species may have only one visual pigment (as in *Uca pugilator, Uca pugnax, Sesarma reticulata*; SCOTT and MOTE, 1974) or two photopigments ($\lambda_{max}=435$, 565 nm), as demonstrated by WALD (1968) in *Carcinides maenas*. Fresh water crustacea contain more than one photopigment in their compound eyes. WALD (1968) used ERG recordings combined with spectral adaptation to demonstrate two pigments in *Orconectes viridis* and *Procambarus clarkii* ($\lambda_{max}=435$, 565 nm). Intracellular recordings in the crayfish *Orconectes* and *Procambarus* (NOSAKI, 1969; WATERMAN and FERNÁNDEZ, 1970) definitely established the presence of two different receptor types with different pigments ($\lambda_{max}=565$, 430 nm). It was found that two of the seven retinula cells in each ommatidium are pure blue receptors, two others yellow receptors, the remaining three may be blue or yellow. It is interesting that NOSAKI (1969) in *Procambarus* found only yellow receptors in winter, but yellow and blue receptors in summer.

Among the *Entomostrac Crustacea* the barnacle eye (*Balanus*, Cirripedia) has been studied extensively (see MILLECCHIA et al., 1972; BROWN and CORNWALL, 1975, for references). Both early receptor potential measurements (MINKE et al., 1973a; HILLMAN et al., 1972, 1973) and intracellular recordings of the late receptor potential (SHAW, 1972; HOCHSTEIN et al., 1973; BROWN and CORNWALL, 1975) reveal a dominant green photopigment ($\lambda_{max}=530$–540 nm) and a secondary UV pigment ($\lambda_{max}=370$–380 nm). Both pigments contribute to the depolarizing intracellular light response of single visual cells, so it is possible that they are mixed together in the same photoreceptive membrane. Evidence that a photochemically independent UV pigment does actually exist is taken from the fact that long-wavelength adaptation produces less sensitivity reduction in UV than in green light (BROWN and CORNWALL, 1975).

The intracellular recordings revealed a wavelength-specific response which characterizes the light-induced turn-over of the photopigments and which may be of relevance also for the wavelength-specific response pattern of the barnacle; strong illumination with red light (>580 nm) causes prolonged and stable depolarization of the cell (the "latch-up" response); blue-green light produces repolarization (λ_{max} of the repolarizing effect 495–520 nm; HOCHSTEIN et al., 1973, BROWN and CORNWALL, 1975). I will not discuss the photochemical implications of this reaction here; they are treated in detail by HAMDORF (this volume). In our context it is interesting to note that such a pigment system may enable the barnacle to respond specifically to color even though all the photoreceptors in the eye are obviously of the same kind and contain the same photopigments.

Artemia salina (Anostraca), which lives in brine, also seems to have only one photopigment ($\lambda_{max}=410$ nm) in its compound eyes (HERTEL, 1977, using ERG recordings). The $S(\lambda)$ curve of the *Artemia* eye is wider on either side of the maximum than a rhodopsin resonance curve. This could be caused by self-screening of the photopigments in the 100-µm-long rhabdoms or by reduced absorption of screening pigments in the UV and red regions. Selective adaptation experiments gave no indication of an additional photopigment.

Crustacean larvae have ocelli, but little is known about their functioning and photopigments. EATON and BROWN (1970) carried out ERG recordings in the nauplius eye of *Pandalus borealis*, a decapod crustacean. They found maximum sensitivity at 475–500 nm.

Spectral phototaxis experiments with crustacean larvae demonstrate single photopigment systems in most species, although the experimental procedures were not carefully enough controlled in the older papers (cf. GROSS, 1913; HECHT, 1921). VISSCHER and LUCE (1928) and SIDDLE (1968) found one photopigment (λ_{max} = 530–545 nm) controlling the positive phototactic response in barnacle larval eye (UV was not tested). Adult Phyllopoda (water fleas) species, display $S(\lambda)$ in their positive or negative phototactic responses which indicate the existence of more than one photopigment in their compound eyes. VON FRISCH and KUPELWIESER (1913) found color-specific differences of the phototactic response in *Daphnia*; *Daphnia* responds to a reduction of the intensity of non-colored light by a positive phototactic response, and to an increase in intensity by a negative response. In blue light, however, a dimming causes a negative response, and in yellow light an increase in intensity a positive response. Since these early studies the spectral phototaxis, and the swimming behavior, "color dances," of *Daphnia* were studied repeatedly and suggest that at least two (blue, yellow) possibly three (red) photoreceptor systems exist (HEBERDEY, 1936; KOEHLER, 1924; SMITH and BAYLOR, 1953; DAAN and RINGELBERG, 1969; STEARNS, 1975). The copepod *Diaptomus nevadensis* was found to display color-specific motion patterns comparable to *Daphnia*: high motion activity and negative phototaxis is found in blue light, and lower activity and positive phototaxis, in yellow and red (HAIRSTON, 1976), (see Sec. D).

2. Chelicerata

It is not surprising that the first measurements of $S(\lambda)$ were made in the lateral eye of *Limulus polyphemus*, since the first recordings from any single visual elements were carried out in 1932 with the *Limulus* lateral eye (HARTLINE and GRAHAM, 1932) and established a new epoch of visual physiology. Extracellular recordings of spikes from the eccentric cell axons, ERG, and single retinula cell recordings demonstrated that the lateral eye consists entirely of green receptors containing a rhodopsin with λ_{max} = 535 nm (see Table 1b). The same is true of the primitive "internal" (or ventral) photoreceptor cells (λ_{max} = 530 nm). By contrast the median ocellus contains UV as well as green receptors (λ_{max} = 360 nm; λ_{max} = 520 nm) which are partly electrically coupled through the extracellular space (NOLTE and BROWN, 1970, 1972a) indicated by the observation that UV receptors hyperpolarize in response to strong green illumination, and green receptors to strong UV light. Interestingly the UV receptors react to strong UV illumination in the same manner as was described above for the barnacle. The cell remains depolarized even after the light is switched off and repolarization only follows long-wavelength illumination (NOLTE and BROWN, 1972b). In this case, the repolarization is most effective with 480 nm light. In agreement with the finding of UV receptors only in the median ocellus, LALL and CHAPMAN (1973) demonstrated, through phototaxis experiments with *Limulus* under natural light conditions, that UV sensitivity exists only in the median ocellus.

Until now *Scorpion* eyes have been studied only with ERG recordings (see Table 1b). Their lateral eyes are more UV sensitive, while the median eyes react to green light only. FLEISSNER (1968) described an interesting change in the $S(\lambda)$ function of the main and secondary eyes of the scorpion *Buthus occitanus* during adaptation: in the dark-adapted state the $S(\lambda)$ is much flatter than in the light-adapted state; the peak sensitivity (λ_{max} = 490 nm), however, does not shift. It seems possible that the high absorption and self-screening of the photopigment in the dark-adapted eye causes this flattening.

Arachnid Chelicerata. Intracellular recording has so far detected only one receptor type (λ_{max} = 510 or 530 nm) in the secondary eyes of spiders, but various types in the main eyes: UV (λ_{max} = 360–370 nm) and green (λ_{max} = 510–532 nm) in *Phidippus regius, Lycosa baltimoriana, L. miami, L. lenta*; and UV (λ_{max} = 360 nm), blue-green (480–500 nm) green (520–540 nm) and orange (580 nm) in *Menemerus confusus* (see Table 1b). In addition, DEVOE (1972), but not YAMASHITA and TATEDA (1976), reported double peaked $S(\lambda)$ in single receptors with a highly variable ratio of UV to green sensitivity. It remains to be shown whether these double peaked curves represent an actual population of receptors or whether they result from natural or artificial coupling (via the recording electrode) between UV and green receptors.

LAND (1969a) has put forward the fascinating hypothesis that the stratification of the retinas in the main eyes of jumping spiders (Salticidae, Dendryphantinae) into four layers could be an adaptation to the strong chromatic aberration of the lens. If so the different rhabdom layers should correspond to the focal planes of different parts of the spectrum. Thus, red light would be most sharply focused in layer 1 (most proximal); blue-green light, in layer 2; and UV light, in layer 3; and according to LAND, layer 4 (most distal), being too close to the lens, would not receive a focused image. If the incident light is indeed focused in this way then a suitable arrangement of the color receptors would enable the apparent deficiency of the lens to be made positively advantageous. Further, as in the insect-fused rhabdom (see later), the upper layers would act as color filters, thus sharpening the $S(\lambda)$ of the system. DEVOE (1975) found no evidence from his intracellular recordings to support LAND's hypothesis, but YAMASHITA and TATEDA (1976), on the other hand, used ERG recordings and found greater long-wavelength sensitivity proximally, and greater short-wavelength sensitivity distally. They interpret their results as support for LAND's hypothesis.

Spectral phototaxis experiments with female two-spotted spider mites (*Tetranychus urticae*) in the dispersal phase have demonstrated two major peaks of the action spectrum of positive phototactic runs, one in the UV at 375 nm, another in the green at 530 nm (a minor peak was found in the red at 670 nm, which is most probably caused by the transmission of the brownish shielding pigment) (SUSKI and NAEGELE, 1963; NAEGELE et al., 1966; MCENROE and DRONKA, 1966a, b; MILLS, 1973). The phototactic runs are aligned higher in UV light than in green light. From this and other results (see also Sec. D.5, p. 558) the authors conclude that the anterior eyes of the spider mite contain UV and green receptors, but the posterior eyes, UV receptors only. Most interestingly, the attraction of UV and green light changes with the physiologic states of the mites (MCENROE, 1969; MILLS, 1974; see also p. 558).

3. Insecta

Several recent papers have reviewed the $S(\lambda)$ of insect compound eyes (WASSERMAN, 1973; GOLDSMITH and BERNARD, 1974; MENZEL, 1975). The enormous literature on this subject will, therefore, not be exhaustively discussed here. The photochemistry of the photoreisomerising rhodopsin pigments in insect eyes is dealt with by HAMDORF (this volume).

The various color receptor types can be divided into three large groups: UV, blue, and green receptors (Table 2, only the most recent data for each species are given). Other color receptor types, e.g., yellow or red, have seldom been reported and the evidence for their existence is in any case controversial. The mean maximum sensitivities of the UV, blue, and green receptors are around 350, 440, and 510 nm respectively. Measured $S(\lambda)$ of single retinula cells have been found to differ from the corresponding rhodopsin resonance curves in two respects; one or more secondary peaks may appear; and/or the width of the main peak may be greater or less than the rhodopsin resonance function. A variety of explanations has been proposed to account for these observations, any or all, or perhaps none, of which may be responsible in a given case.

In the following we will discuss first (a) possible mechanisms underlying the broadening of $S(\lambda)$ and the appearance of secondary peaks (sections: 1. Self-screening of photopigments, 2. Positive electric coupling between retinula cells, 3. More than one photopigment in one visual cell, 4. Selective transmission of the screening pigments). We will then discuss (b) possible mechanisms causing narrower $S(\lambda)$ functions than rhodopsin resonance functions (Sections: 1. Spectral transmission of the dioptric apparatus, 2. Selective reflection of the tapetum, respectively, 3. Selective filtering by screening pigments and Metarhodopsins, 4. Negative electrical coupling between retinula cells). Five additional chapters are concerned with (c) $S(\lambda)$ of photoreceptors of selected insect species, (d) the distribution and arrangement of color receptors in insect compound eyes, (e) the spectral phototaxis of insects, (f) the spectral properties of insect ocelli, and (g) the spectral properties of the extraretinal photoreception in insects.

a) Secondary Peaks and Broadening of $S(\lambda)$

1. Self-Screening of the Photopigments.

Suppose that 0.8–1.0% of incident light is absorbed per micron as it passes down the rhabdom (GOLDSTEIN and WILLIAMS, 1966; SNYDER, 1975; HAMDORF and SCHWEMER, 1975; STAVENGA, 1975; LIEBMAN, 1975). Rhabdoms may be up to 1000 μm long. The spectral distribution of the light reaching the deeper layers of the rhabdom is altered by the absorption of the layers above. The longer the rhabdom, the more light that is absorbed, and the broader the spectral absorption function becomes the position of the maximum also shifts a little towards longer wavelengths. This self-screening may be counteracted by serial or parallel filter effects in the rhabdom (SNYDER et al., 1973). Rhabdoms built up by laterally fused or tiered rhabdomeres in a single light guiding structure (fused rhabdom) may contain different photopigments, e.g., in the bee (GRIBAKIN 1969, 1972; KOLB and AUTRUM, 1974; MENZEL and BLAKERS, 1976), cockroach (BUTLER, 1971), ant (MENZEL and KNAUT, 1973; MENZEL and BLAKERS, 1975), dragonfly (LAUGHLIN, 1976b), and butterfly (HAMDORF et al., 1972). The differently absorbing rhabdomeres, arranged parallel or serially, act as color filters for one another thus allowing each to retain its high sensitivity, due to its high absorption probability, and resulting in $S(\lambda)$ functions for each cell which closely resemble the spectral absorption of a thin layer of their photopigments (see A. SNYDER, Ch. 5, this volume).

An accumulation of metarhodopsin should also cause a selective filtering effect shifting the maximal sensitivity according to its spectral absorption properties and concentration. The deep trough (in most cases around 400 nm) and the sharp fall in sensitivity above 590 nm led to the speculation that blue wavelengths and long wavelengths absorbing metarhodopsin may cause this deviation from broad $S(\lambda)$ functions of rhabdoms with high total absorption. However, there are no conclusive results yet to support this.

Fused and tiered rhabdoms seem specially designed to counteract self-screening (s.a.), and in those cases where broad or multipeaked $S(\lambda)$ functions were found in single photoreceptors (bee, ant, dragonflies, locust, butterflies; see Table 2b) they may result from positive electric coupling between retinula cells with different photopigments.

2. Positive Electric Coupling Between Retinula Cells

Electric coupling between retinula cells of the same ommatidium has been found in the drone bee and the locust (SHAW, 1967; 1969a). Since neighboring retinula cells may contain different photopigments (see above), electric coupling would cause broadening of $S(\lambda)$ functions and secondary peaks in single cells. However, it is not yet clear to what extent electric coupling is artificially produced by the recording electrode damaging the cells.

Several authors pointed out that long-lasting stable intracelluar recordings make it very unlikely that the visual cell is damaged by the electrode. However, artificial connection of neighboring retinula cells without an opening to the extracellular space may be possible. In such a case the resting potential of both cells may stay low and the cells may share their light-induced current sources. Intracellular marking should reveal such a recording situation by staining two cells. Unfortunately there are only few papers where the recording site was identified by intracellular marking, e.g., cockroach (MOTE and GOLDSMITH, 1971), fly (MCCANN and ARNETT, 1972; MEFFERT and SMOLA, 1976), and bee (MENZEL and BLAKERS, 1976). In the bee, MENZEL and BLAKERS (1976) marked single retinula cells with high and low secondary $S(\lambda)$. Individually marked green receptors had secondary UV sensitivity of 20–60% of the main green peak, individually marked UV cells had 5–30% secondary green sensitivity. In UV receptors the height of the secondary green sensitivity was correlated with the number of spines in the first optic neuropil (the lamina). It is concluded from these findings that some unknown mechanism of positive electric interaction between

Table 1a. *Limulus*

Order/Family	Species	Eye type	Major λ_{max} [nm]	Secondary λ_{max} [nm]	Method	Further observation and interpretation	References
Xiphosura	*Limulus polyphemus*	Median ocellus	360	530	ERG	Two separate chromatic inputs	WALD and KRAININ, 1963
	Limulus polyphemus	Median ocellus	360 550	550 360	ERG, selective adaptation	Two separate chromatic inputs	CHAPMAN and LALL, 1967
	Limulus polyphemus	Median ocellus	360 depolarization (UV cells)	530 hyperpolarization	Intracellular recording	Separate UV and VIS cells but each may contain 2 pigments (UV and green pigment)	NOLTE, BROWN and SMITH, 1968
	Limulus polyphemus	Median ocellus	530 depolarization (VIS cells)	360 hyperpolarization	Intracellular recording	Separate UV and VIS cells but each may contain 2 pigments (UV and green pigment)	NOLTE and BROWN, 1969
	Limulus polyphemus	Median ocellus	360 525	— 360	Intracellular recording (depolarizing responses only)	S(λ) of both UV and VIS cells independent of screening pigment; S(λ) of VIS cells broader than that of white lateral eye	NOLTE and BROWN, 1970
	Limulus polyphemus	Median ocellus	360 525	— —	Intracellular recording	Both UV and green receptors give a fast hyperpolarizing response to wavelengths to which the other type is most sensitive: Electrically coupled to extracellular space (local ERG-effect), direct electric coupling between green cells of the same type and between UV cells and a rhabdomic cell	NOLTE and BROWN, 1970

Limulus polyphemus	Median ocellus	360	—	Intracellular recording	After strong UV stimulation prolonged depolarization. Repolarized with longer wavelength ($\lambda_{max}=480$ nm)	NOLTE and BROWN, 1972b
Limulus polyphemus	Lateral eye	Green	?	Action potential from optic nerve		GRAHAM and HARTLINE, 1935
Limulus polyphemus	Lateral eye	520	?	Extracted photopigment	One rhodopsinlike photopigment with relatively stable metarhodopsin	HUBBARD and WALD, 1960
Limulus polyphemus	Lateral eye	520	370	ERG		WALD and KRAININ, 1963
Limulus polyphemus	Lateral eye	510	370	ERG selective adaptation	One type of photoreceptors	CHAPMAN and LALL, 1967
Limulus polyphemus	Lateral eye	520–540	UV sensitivity higher at lower stimulus intensities	Intracellular SPF (slow potential fluctuations)	UV effect: cornea fluorescence or second photopigment	ADOLPH, 1968
Limulus polyphemus	Lateral eye	525	370	Intracellular slow potentials	Two receptor classes: UV and green receptors; UV receptors do not adapt	WASSERMAN, 1969
		520–525	370	Intracellular recording	$S(\lambda)$ dependent on sreening pigment: relative higher sensitivity at longer wavelengths in pigmented eyes	NOLTE and BROWN, 1970
Limulus polyphemus	"Ventral" eye photoreceptors	520–525	370	Intracellular recording	Same $S(\lambda)$ function as in lateral eye receptors	NOLTE and BROWN, 1970
Limulus polyphemus	"Ventral" eye photoreceptors	535–545	—	Intracellular recording	One photoreceptor type	MILLECCHIA et al., 1966
Limulus polyphemus	"Ventral" eye photoreceptors	530	—	MSP difference spectrum	One green photopigment	MURRAY, 1966

Table 1b. Chelicerata

Order/Family	Species	Eye type	Major λ_{max}	Secondary λ_{max}	Method	Further observation and interpretation	References
Scorpions	*Centruroides sulpturus* *Vejovis spinigerus* *Opistha-canthus validus*	Lateral eye	375 (or shorter)	510	ERG	Adaptation produces a strong decrease in UV sensitivity independent of adaptation wavelength. Absorption of screening pigment is not selective	MACHAN, 1968
		Median eye	510	—		no change in S(λ) with adaptation	
	Heterometrus fulvipes; H. gravimanus	Neural photo-receptor in the metasoma	568	440	Extracellular spike recording	Function and photopigment involved unknown (cf. Ch. B)	GEETHABALI and PAMPAPATHI RAO, 1973
	Heterometrus fulvipes; H. gravimanus	Lateral eye	380 (or shorter)	510	ERG		
	Heterometrus fulvipes; H. gravimanus	Median eye	510	—	ERG		
	Buthus occitanus	Principal eye	490	—	ERG	Dark-adapted eyes with much broader S(λ) functions. No selective chromatic adaptation effect	FLEISSNER, 1968
Araneae Lycosidae	*Lycosa baltimoriana, L. miami, L. carolinensis*	Secondary eyes (anterior lateral posterior median, lateral) Compound eye	505	360	ERG selective adaptation	No selective adaptation effects; only one photopigment	DEVOE et al., 1969
Araneae Salticidae	*Lycosa baltimoriana* *L. miami*	Secondary eye (anterior lateral)	510	380	Intracellular recording	Only one photopigment	DEVOE, 1972
	L. lenta	Principal eye (anterior median)	510, 360–370 (variable relative heights)	510, 360–370	Intracellular recording	Each cell contains variable amounts of two pigments	DEVOE, 1972

	Phidippus regius	Principal eye	370 (UV cells)	—	Intracellular recording	Large range of UV: green sensitivity ratios for the UV green cells	DeVoe, 1975
			532 (green cells)	—			
			370 (UV green cells)	520			
	Phidippus regius	Secondary eye (anterior lateral)	530	—	ERG	Only green receptor	DeVoe, 1975
	Menemerus confusus	Secondary eyes (anterior lateral, posterior lateral)	535	—	ERG and intracellular recording	Only one photopigment, one photoreceptor type	Yamashita and Tateda, 1976
	Menemerus confusus	Principal eye (anterior median)	360, 480–500	—		Four different receptor types	
			520–540	—			
			580	—			

Table 2a. Insecta, UV-Receptors

Order/Family	Species	Eye type	Major λ_{max}	Secondary λ_{max}	Method	Further observation and interpretation	References
Odonata	*Anax junius* *Aeschna tuberculifera* *Libellula pulchella*	Median ocellus	360	500	Intracellular recording	Imago, selective reduction of green peak with yellow light. Two pigments in one cell or two cells electrically coupled	Chappell and DeVoe, 1975
	Aeschna cyanea	Compound eye	≦336 and 336	— 445	Intracellular recording	Larva	Autrum and Kolb, 1968
	Aeschna cyanea	Compound eye	356	—	Intracellular recording	Imago, dorsal part of the eye	Eguchi, 1971
	Hemicordula tau	Compound eye	350	—	Intracellular recording	Imago, "single pigment cells"	Laughlin, 1975
		Compound eye	350	450 or 500	Intracellular recording	"linked pigment cells"	Laughlin, 1975
Orthoptera	*Periplaneta americana*	Compound eye	365	—	Intracellular recording	Imago, single photopigment	Mote and Goldsmith, 1970

Table 2a (continued)

Order/Family	Species	Eye type	Major λ_{max}	Secondary λ_{max}	Method	Further observation and interpretation	References
Hemiptera	*Notonecta glauca*	Compound eye	356	—	Intracellular recording	Imago, posterodorsal part of the eye, single photopigment	BRUCKMOSER, 1968
	Notonecta glauca	Compound eye	350	?	ERG	Imago, ventral part, selective chromatic adaptation	BENNETT and RUCK, 1970
Neuroptera	*Ascalaphus macaronius*	Compound eye	332	—	ERG	Imago, dorsal eye part, only UV receptors, single photopigment	GOGALA, 1967
	Ascalaphus macaronius	Compound eye	380	505	ERG	Selective adaptation effects; two different photopigments	GOGALA, 1967
Lepidoptera	*Heliconius numata*	Compound eye	390	500	Intracellular recording	Imago, "linked photopigment system"	STRUWE, 1972
	Macroglossum stellatorum	Compound eye	348	430 and 500	ERG	Imago, dominant UV peak at low stimulus intensities	HASSELMANN, 1962
Coleoptera	*Carabus auratus*	Compound eye	348	430 and 500	ERG	Imago, equal height of UV and blue peak at low stimulus intensities	HASSELMANN, 1962
Diptera	*Drosophila melanogaster*	Compound eye	370	—	ERG	Imago, isolation of retinula cell R7 by mutant selection and blue adaptation	STARK, 1975 HARRIS et al., 1976
	Calliphora erythrocephala	Compound eye	344	—	Intracellular recording	Imago, retinula cells R7 and for R8, no marking	MEFFERT and SMOLA, 1976
	Calliphora erythrocephala	Compound eye	344	443	Intracellular recording	Imago, retinula cell R7 (intracellular marking)	MEFFER and SMOLA, 1976
	Eristalis tenax	Compound eye	340	420	Intracellular recording	Imago, single UV photopigment (?)	BISHOP, 1974
	Eristalis tenax	Compound eye	350	450 or 520	Intracellular recording	Two photopigments in one cell	HORRIDGE et al., 1975
	Eristalis tenax	Compound eye	350	—	Intracellular recording	One photopigment, rarely found	HORRIDGE et al., 1975

Hymenoptera	*Apis mellifera*	Compound eye	≦318 or 341	—	Intracellular recording	Drone bee	Autrum and Zwehl, 1964
	Apis mellifera	Compound eye	341–358	435 and 450	Intracellular recording	Worker bee; cells with double peak from damaged (?) cells	Autrum and Zwehl, 1964
	Apis mellifera	Compound eye	340	460 and 560	Intracellular recording	Worker bee; double peaks from natural electric coupling (?)	Menzel and Blakers, 1976
	Apis mellifera	Compound eye	340	—	Intracellular recording	Worker bee; retinula cell No. 9	Menzel and Blakers, 1976

Table 2b. Insecta, Blue-receptors

Order/Family	Species	Eye type	Major λ_{max}	Secondary λ_{max}	Method	Further observation and interpretation	References
Odonata	*Aeschna cyanea*	Compound eye	412–432	—	Intracellular recording	Larva	Autrum and Kolb, 1968
	Aeschna cyanea	Compound eye	445–475	519	Intracellular recording	Imago; dorsal eye part	Eguchi, 1971
	Libellula spec.	Compound eye	420	350	Intracellular recording	Larva and imago; broad $S(\lambda)$ functions caused by self-screening of photopigment	Horridge, 1969
	Hemicordula tau	Compound eye	440 440	— 350	Intracellular recording	Imago; "single pigment cells" "linked pigment cells"	Laughlin, 1975
Orthoptera	*Locusta migratoria*	Compound eye	430	—	Intracellular recording	Imago; varying amount of two photopigments in one cell	Bennett et al., 1967
Hemiptera	*Notonecta glauca*	Compound eye	420 464	Partially 500	Intracellular recording	Imago; posterodorsal eye; double peaks from damaged cells	Bruckmoser, 1968

Table 2b (continued)

Order/Family	Species	Eye type	Major λ_{max}	Secondary λ_{max}	Method	Further observation and interpretation	References
Lepidoptera	*Heliconius numata*	Compound eye	450–470	Shoulder at 380 and 560	Intracellular recording	"linked" photopigment system	STRUWE, 1972
Coleoptera	*Carabus auratus*	Compound eye	430	348 and 500	ERG	Imago; UV and blue peak with equal height at low stimulus intensities	HASSELMANN, 1962
Diptera	*Drosophila melanogaster*	Compound eye	~440	?	ERG	Imago; isolation of retinula cell R8 by mutant selection	HARRIS et al., 1976
	Musca	Compound eye	465	—	Intracellular recording	Imago; retinula cells R7 and/or R8	MCCANN and ARNETT, 1972
	Calliphora	Compound eye	443	—	Intracellular recording	Imago; retinula cell R8 (no marking)	MEFFERT and SMOLA, 1976
	Eristalis tenax	Compound eye	465	350	Intracellular recording	Imago	BISHOP, 1974
	Eristalis tenax	Compound eye	450	—	Intracellular recording	Imago; rarely found single peaked cells; cf. double peaked UV cells	HORRIDGE et al., 1975
Hymenoptera	*Apis mellifera*	Compound eye	410 or 460	partially 340, 540	Intracellular recording	Worker bee; double peak S(λ) functions from damaged cells	AUTRUM and VON ZWEHL, 1964
	Apis mellifera	Compound eye	440	—	Intracellular recording	Worker bee; no blue receptors with secondary peaks	MENZEL and BLAKERS, 1976
	Apis mellifera	Compound eye	450	358	Intracellular recording	Drone bee	AUTRUM and VON ZWEHL, 1964

Table 2c. Insecta, Green-receptors

Order/Family	Species	Eye type	Major λ_{max}	Secondary λ_{max}	Method	Further observation and interpretation	References
Odonata	*Anax junius* *Aeschna tuberculifera* *Libellula pulchella*	Median ocellus	500	360	Intracellular recording	Imago; varying ratios of UV to green sensitivities, electric coupling or two photopigments in one cell	CHAPPELL and DEVOE, 1975
	Aeschna cyanea	Compound eye	494	350	Intracellular recording	Imago; ventral part of the eye	AUTRUM and KOLB, 1968
	Aeschna cyanea	Compound eye	519	—	Intracellular recording	Imago;	AUTRUM and KOLB, 1968
	Aeschna cyanea	Compound eye	494	350	Intracellular recording	Imago; dorsal part of the eye, very broad S(λ) functions	EGUCHI, 1971
	Libellula needhami	Compound eye	538	350	Intracellular recording	Imago; ventral part of the eye; broad S(λ) functions caused by self-screening	HORRIDGE, 1969
	Anax junius	Compound eye	500	350	Intracellular recording	Imago; ventral part of the eye; broad S(λ) functions caused by self-screening	HORRIDGE, 1969
	Hemicordula tau	Compound eye	510	—	Intracellular recording	Imago; ventral part of the eye "single pigment cells"	LAUGHLIN, 1975
	Hemicordula tau	Compound eye	490–510	350–360	Intracellular recording	"Linked pigment cells" with varying heights of UV and green peaks	
Orthoptera	*Periplaneta americana*	Compound eye	510	—	Intracellular recording	Imago; only one photo-pigment with typical rhodopsin function	MOTE and GOLDSMITH, 1970
	Locusta migratoria	Compound eye	460–525	350	Intracellular recording	Imago; very broad S(λ) functions mixture of two photopigments in one cell	BENNETT et al., 1967
Hemiptera	*Notonecta glauca*	Compound eye	510–530	—	Intracellular recording	Imago; one photopigment posterodorsal part of the eye	BRUCKMOSER, 1968
	Notonecta glauca	Compound eye	530	?	ERG	Imago; anterior and dorsal part of the eye	BENNETT and RUCK, 1970

Table 2c (continued)

Order/Family	Species	Eye type	Major λ_{max}	Secondary λ_{max}	Method	Further observation and interpretation	References
Lepidoptera	*Heliconius numata*	Compound eye	540	370	Intracellular recording	Imago; "linked photopigment system"	STRUWE, 1972
	Macroglossum stellatorum	Compound eye	500	348, shoulder at 430	ERG	Imago; dominant green peak at higher stimulus intensities	HASSELMANN, 1962
	Phalera bucephala	Compound eye	500	348, shoulder at 430	ERG	Imago; dominant green peak at higher stimulus intensities	HASSELMANN, 1962
Diptera	*Drosophila melanogaster*	Compound eye	470	350	ERG	Imago; isolation of retinula cells R 1–6 by mutant selection equal height of UV and green peak	HARRIS et al., 1976
	Drosophila melanogaster	Compound eye	485	≦380	ERG and Intracellular recording	Imago; functionally isolated retinula cells R 1–6	MINKE et al., 1975; ALAWI et al., 1972; STARK and HARRIS, 1975
	Calliphora erythrocephala	Compound eye	490	350	Intracellular recording	Imago; UV and green peak of equal height; receptors R 1–6	BURKHARDT, 1972
	Musca domestica, *Phaenicia sericata*	Compound eye	480	350	Intracellular recording	Imago; UV and green peak of equal height; receptors R 1–6 identified	MCCANN and ARNETT, 1972
	Cyrtodiopsis dalmanni	Compound eye	490	360 and 450	ERG	Imago;	BURKHARDT, 1972
	Bibio marci	Compound eye	520	350	ERG	Imago;	BURKHARDT and DE LA MOTTE, 1972
	Panorpa cognata	Compound eye	540	360	ERG	Imago;	BURKHARDT and DE LA MOTTE, 1972
Hymenoptera	*Apis mellifera*	Compound eye	540	Partially 350 and shoulder at 440	Intracellular recording	Worker bee; drone bee artificial or natural electric coupling between retinula cells; varying heights of secondary sensitivity	AUTRUM and VON ZWEHL, 1964; MENZEL and BLAKERS, 1976

retinula cells with different $S(\lambda)$ causes variable heights of secondary sensitivity and broader $S(\lambda)$ functions.

Natural electric coupling might occur (1) via direct contacts between retinula cells [as proposed for the *Limulus* lateral eye (LASANSKY, 1967; SMITH and BAUMANN, 1969), the *Limulus* ocellus (NOLTE and BROWN, 1972a), and the locust and dronebee compound eye (SHAW, 1969a)], (2) via junctions to accessory (pigment) cells (as found in the dronebee eye by BERTRAND et al., 1972; BAUMANN, 1975; and in the *Limulus* ocellus by NOLTE and BROWN, 1972a), (3) via receptor-receptor synapses in the first optic neuropil (as proposed for the dragonfly ocellus: DOWLING and CHAPPELL, 1972; the fly lamina: BOSCHEK, 1970; CHI and CARLSON, 1976; and the bee lamina: VARELA, 1970), and (4) through the extracellular space in the first optic neuropil (as proposed by SHAW, 1975; MENZEL and BLAKERS, 1976).

There were several attempts to test the possibility of electric coupling between retinula cells as a dynamic process which changes with the average level of illumination (SHAW, 1969a, b; SNYDER et al., 1973). However, the advantage of such a dynamic coupling is obscure and the experimental evidence for it is inconclusive.

3. More Than One Photopigment in One Visual Cell

Theoretically double peaked or multiple peaked $S(\lambda)$ functions might result from a mixture of two or more photopigments in one cell. The advantage would be an increase in sensitivity due to the broadening of the spectral absorption function. Double peaked $S(\lambda)$ functions in the compound eye of locust (BENNETT et al., 1967), dragonfly (LAUGHLIN, 1976b), flies (*Calliphora*: HORRIDGE and MIMURA, 1975; MEFFERT and SMOLA, 1976; *Eristalis*: HORRIDGE et al., 1975), and in the median ocelli of the wolf spider (DEVOE, 1972, see p. 536) are interpreted to result from photopigment mixture in one cell. It is interesting to note in all these cases and many others, where "linked photopigment systems" cannot be attributed to electric coupling (see Table 2b), one peak of sensitivity occurs in the UV, the other, in the blue or green light.

Besides the results from flies the evidence for two or more photopigments in one visual cell is very weak. In the fly, *Calliphora*, HORRIDGE and MIMURA (1975) found different directions of sensitivities to the E vector of light in the UV and in visible light. On the prepositions that (1) electric coupling is excluded, (2) both sensitivity peaks result from a rhodopsin photopigment, and (3) the rhodopsin molecules of the two kinds have the same orientation relative to the microvilli direction, the conclusion can only be that the photopigments are located in different parts of the rhabdomere and that the rhabdomere twists around its long axis. KIRSCHFELD and FRANCESCHINI (1977), however, have found a photostable, blue-absorbing pigment in 70% of the fly's R7 cells, whose direction of maximal dichroic absorption is perpendicular to the UV-absorbing rhodopsin coexisting in the same rhabdom. Furthermore, KIRSCHFELD and FRANCESCHINI have indirect evidence that the UV sensitivity of the fly's retinula cells R1–6 is based on a photostable pigment. They conclude that both photostable pigments are not rhodopsins, but a carotene or a flavoprotein pigment in the case of the blue-absorbing pigment and an unidentified pigment for the enhanced UV sensitivity.

Absolute measurements of the bleaching of rhodopsin in these R7 rhabdomeres make it likely that the photostable pigment acts not only as an internal color and dichroic filter but also as an "antenna" pigment, which transfers the energy to the effector (rhodopsin) pigment. This view is supported by intracellular $S(\lambda)$ measurements (SMOLA and MEFFERT, 1976).

It would be extremely interesting to know if the existence of such non-rhodopsin accessory pigments (antenna pigments) is limited to the fly. Their photostability and their appearance in only a fraction of visual cells might help to explain a variety of unexplained phenomena in arthropod visual cells (e.g., the difficulty of selectively adapting the UV sensitivity in the visual cells of many species and the high sensitivity of insect eyes to UV).

4. Selective Transmission of the Screening Pigments

Increased sensitivity in the orange region in flies and butterflies is attributed to the lower long-wavelength absorption of the ommochromes and pteridines which make up the screening pigments (cf. LANGER, 1975).

b) Narrower $S(\lambda)$ Functions

$S(\lambda)$ functions are sometimes narrower than corresponding rhodopsin resonance functions. This may be due to several factors.

1. Spectral Transmission of the Dioptric Apparatus

Cornea Reflection. Light strikes the corneal surface before reaching the photosensory structures of the retina. The reflection at this surface is diminished in many species, especially in those which orient behaviorally at low light intensities, e.g., nocturnal Lepidoptera, by a regular array of protrusions (BERNHARD, 1967; BERNHARD and MILLER, 1968; BERNHARD et al., 1963, 1965; BERNHARD et al., 1968; BERNHARD et al., 1970). These protrusions or cornea nipples project 50–250 nm, i.e., less than a wavelength of light, and function as a broadbanded anti-reflection surface maximally effective for wavelengths corresponding to about twice the nipple height. By contrast smooth corneas are found in the diurnal insects *Aeschna, Apis, Bombus* and *Coccinella* (BERNHARD et al., 1970) except in the case of the butterfly *Vanessa* and the mosquito, both of which possess a nipple array, while other flies have flat protrusions (MILLER et al., 1966).

Cornea Transmission. The nipple structure of the cornea increases the light reaching the dioptric apparatus without significantly changing its spectral content (BERNHARD et al., 1963, 1965). However, obvious chromatic selective effects are seen from the bright orange, red, yellow, green, and blue reflection patterns on the compound eyes of many Tabanidae and 23 other dipteran species (BERNHARD and MILLER, 1968; STEYSKAL, 1957). These are thought to be produced by the alternate dense and light bands seen in the cornea under the electron microscope. They have a periodicity of approximately a quarter of a wavelength of the reflected light and are thought to act as interference filters. The half-bandwidth of the reflected light has been measured as 60–100 nm (BERNARD, 1971; BERNARD and MILLER, 1968) and the position of the maxima varies between species (for example the range of the blue-green facets is 55 nm and that of the red facets is 80 nm). Although the positions of the reflection maxima are variable in closely related species, the maximum for one of the facet colors always has a wavelength at which the other has its first reflection minimum. This suggests that the interference filters may be used for color vision by the consecutive comparison of contrast between adjacent facets. TRUJILLO-CENÓZ and BERNARD (1972) have found evidence for this hypothesis by analyzing the connections of the six short retinula axons in the lamina. The normal pattern of connections (BRAITENBERG, 1967; KIRSCHFELD, 1968; TRUJILLO-CENÓZ, 1966) is changed so that only those retinula cells beneath corneal lenses with the same color converge upon the same cartridge.

Eyes with a metallic sheen of different colors have been found in some Neuroptera (Chrysomelidae and Myrmeleontidae: BERNARD, 1971) and Coleoptera (FRIEDRICHS, 1931). Neither the corneal spectral transmission nor the $S(\lambda)$ of single retinula cells have so far been studied in these species with corneal interference filters.

All those insects in which color discrimination has been studied also have a layered cornea, but these have no specific color and the layers have a wider separation (e.g., 800–1200 nm for drone bee, PERRELET, 1970). The color of light reflected from the eye in these insects depends only on the color of the screening pigments, for instance, the different eye colors of *Drosophila* mutants. Measurements of corneal absorption in dragonfly (KOLB et al., 1969), show that the absorption decreases slowly from long to short wavelengths and increases rapidly below 330 nm. The same result is found in the moths *Smerinthus ocellata* and *Endromis versicolora* (BERNHARD et al., 1965), *Manduca sexta* (CARLSON and PHILIPSON, 1972), the hymenopterans *Apis mellifera* and *Xylocopa violacea* (CARRICABURU and CHARDENOT, 1967), the housefly *Musca domestica* (GOLDSMITH and FERNANDEZ, 1968), the dungbeetle *Euoniticellus* (MEYER-ROCHOW, 1975) and the fruitfly *Drosophila melanogaster* (STROTHER and SUPERDOCK, 1972). In both dipterans the transmission was measured down to 240 nm and showed a minimum at 280 nm. Therefore, with the exception of the dipterans (many Tabanidae), neuropterans, and coleopterans, which have corneal lenses with interference filters, it can be presumed for all other insects that light reaches the rhabdom through the dioptric apparatus without any chromatic change. The transmission in the far UV (280–320 nm) is more strongly attenuated and may influence the $S(\lambda)$ of UV receptors. It is important that msp transmission measurements are carried out with preparations of fresh cornea because the transmission changes considerably during dehydration by alcohol or drying (KOLB et al., 1969).

The spectral transmission of the crystalline cone is known only in the dragonfly *Aeschna cyanea* (KOLB et al., 1969) and the lepidopteran *Manduca sexta* (CARLSON and PHILIPSON, 1972). There is no large difference in absorption compared with the cornea.

2. Selective Reflection of the Tapetum

Some butterflies with apposition eyes and long fused rhabdoms possess a tracheal tapetum between the proximal end of the rhabdom and the basement membrane which gives rise to the glow of dark-adapted eyes. MILLER and BERNARD (1968) found that this glow has different colors in different parts of the eye in some butterflies (e.g., *Euptychia*, *Anartia*, *Danaus*). Typically, the dorsal facets have a blue glow, anterodorsal a green-yellow glow, and anteroventral an orange-red glow. This is because there are four tracheoles at the proximal end of each rhabdom, with a large number of plates arranged very regularly, perpendicular to the long axis. In orange facets the plates are 106 nm thick and are separated by 148 nm of air, whereas in blue-green facets they are 123 nm thick with a separation of 88 nm. As the refractive index of the cytoplasm within the plates is ~ 1.4, these parallel plates function as interference filters which reflect maximally in the first example at 590 nm (bandwidth 125 nm) and in the second example at 490 nm (bandwidth 104 nm). The nonreflected light is absorbed by the underlying screening pigment. As the reflected light passes through the rhabdom twice the sensitivity of the receptor cells is selectively increased in this part of the spectrum. The consequences have not yet been carefully studied for their influence on the $S(\lambda)$ function of individual receptors.

3. Selective Filtering by the Screening Pigments and by the Thermostable Metarhodopsins

Screening pigment granules contain a very high concentration of ommochrome molecules (see LANGER, 1975, for review). The high density of the chromophore molecules results in a metallic reflection of less-absorbed wavelengths (FRANCESCHINI, 1975). In the fly, *Musca domestica* for example, the transmitted light is reddish, and the reflected light greenish. It is likely, but unknown, that the $S(\lambda)$ of receptors is influenced by the chromatic reflection of the screening pigment. One would expect an increase in green sensitivity. The selective transmission of screening pigment causes a relative sensitivity increase above 600 nm (BURKHARDT, 1964).

As discussed above, metarhodopsin should act as a color filter. There is, however, no conclusive experimental evidence to this effect.

4. Negative Electric Coupling Between Retinula Cells

The dense packing of the photoreceptive membrane in a fused rhabdom could lead to competition among the visual cells of an ommatidium for the limited Na^+ ions in the common extracellular space (HAMDORF and SCHWEMER, 1975). Stimulation of one color receptor type with its most effective wavelength would then reduce the sensitivity of another color receptor type to this wavelength. Some evidence for such an effect was found in the honey bee (MENZEL and BLAKERS, 1976), but whether it is significant under physiologic light regimes is unknown. The actual position of maximum sensitivity of a receptor can also be influenced by factors in addition to those described above. These are the optical properties of the lens (dioptric apparatus) and the waveguide nature of the rhabdom or rhabdomeres (SNYDER, 1975; and Ch. 5, this volume).

c) Spectral Sensitivity of Photoreceptors of Selected Insect Species

All carefully examined insect species have either two (*Periplaneta, Ascalaphus, Formica, Drosophila*) or three (*Aeschna, Libellula, Hemicordula, Notonecta, Heliconius, Deilephila, Musca, Calliphora, Eristalis, Apis, Paravespula*) photoreceptor types (see Table 2b).

In flies (*Musca, Calliphora, Drosophila*) the situation is somewhat more complicated, but recent analyses of visual mutants of *Drosophila*, single cell recordings combined with intracellular staining and microspectrophotometry, have established a fairly concise picture. Retinula cells R1–6 are homogeneous and have a $S(\lambda)$ function with two maxima of about equal height at 470–490 nm and 350 nm. The two peaks most likely result from two different pigments, one a rhodopsin (R490, M560), the other a photostable molecule, which probably transfers the absorbed energy without radiation to the rhodopsin system (see above). The subsystem R7/8 (centrally tiered rhabdomeres of retinula cells R7 and R8) may be differently organized in *Drosophila* and *Calliphora*. In *Drosophila* R7 is a UV receptor (rhodopsin 350, metarhodopsin 470), R8 has a $S(\lambda)$ function very similar to subsystem R1–6, but differs insofar as it does not adapt. The blue sensitivity may therefore be based on a rhodopsin-metarhodopsin system with similar spectral absorption (rhodopsin 470, metarhodopsin 470), the photostable UV sensitivity may be based on a photostable pigment as in *Calliphora* R1–6 (see above). In *Calliphora* both retinula cell types R7 and R8 are most likely divided into those with single peaks and those with double peaks. R7 cells have single peaks in the UV (344 nm) and double peaks in the blue and UV (344, 440 nm). R8 cells have single peaks in the blue (440 nm) and single peaks in the yellow-green (550 nm) (MEFFERT and SMOLA, 1976; SMOLA, pers. communication).

d) Distribution and Arrangement of Color Receptors

The distribution and arrangement of the color receptors can differ appreciably in different regions of the compound eye. In the dorsal eye region, UV and blue cells are generally more frequent, sometimes the only receptor type present (*Ascalaphus*, drone bee, *Bibio*, *Aeschna*, *Anax*). Green receptors are more frequent in ventral (e.g., *Aeschna*) and frontal sections, in the drone bee, for example, appearing only in the most ventral part of the eye (review MENZEL, 1975). HAMDORF and GOGALA (1973) saw the advantage of the dense packing of UV receptors in the dorsal eye region for an increased contrast sensitivity to small dark objects in front of a predominantly UV-blue-emitting background (sky).

Where different color receptor types occur together in the eye region they are not isolated in separate ommatidia but mixed in various proportions in each ommatidium (see also page 522 on the lateral and tiered filter effect in the fused rhabdom). An ommatidium in the median frontal eye of the worker bee, for example, contains two UV receptors, whose long thin axons penetrate the first optic neuropile and project into the second; two green receptors with thick axons projecting into the proximal layer of the first neuropile; two more green receptors with shorter, thinner axons; two blue receptors with short axons and a variable number of spines; and another UV cell, the nineth proximal receptor (Fig. 6) (GRIBAKIN, 1972; KOLB and AUTRUM, 1974; MENZEL and BLAKERS, 1976).

e) Spectral Phototaxis

Spectral phototaxis experiments have been carried out with many insect species (reviews GROSS, 1913; WEISS, 1946; AUTRUM and THOMAS, 1973) and have given some indication of the color receptors present in the various species. UV

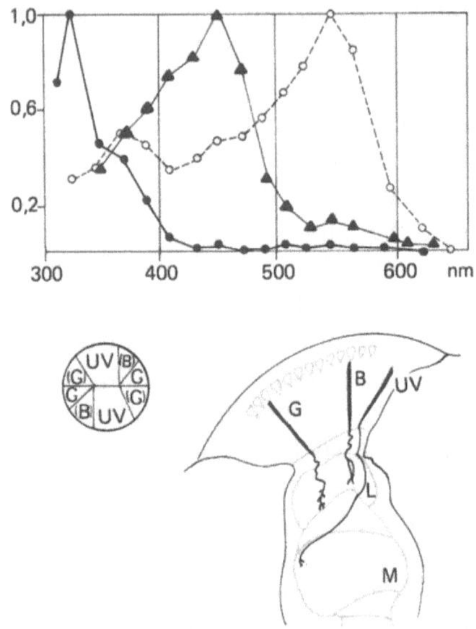

Fig. 6. $S(\lambda)$ of three individual color receptors of the honey bee compound eye. The light responses of the receptors were recorded intracellularly, and the receptors were marked afterward by injections of the dye Procion yellow. The lower part of the figure gives a reconstruction of the axons of these three receptors. (UV=UV receptor, B=blue receptor, G=green receptor, L=lamina, M=medulla). The circle shows schematically the relative position of the color receptor types within one ommatidium. Blue (B) and green (G) receptors in parentheses are morphologically indiscriminable; therefore their position may be interchanged (B with G). – ● – : $S(\lambda)$ function of an UV receptor with λ_{max} at 330 nm (the average λ_{max} of many UV receptors is at 340 nm). The long axon of this UV receptor penetrates the first optic ganglion (L), crosses the chiasma between the first and second optic ganglion and projects into the peripheral layer of the second optic ganglion (M). The axon carries short spines in the most proximal lamina layer and at its terminals in the medulla. Other UV receptors have different $S(\lambda)$ functions and different spine patterns in the lamina and medulla. —▲—: Function of a blue receptor with λ_{max} at 450 nm. The thin axons of the blue receptor (B) bifurcate in the distal lamina layer, and the two very thin branches project into the proximal middle layer of the lamina. Both branches carry no spines but few short arborizations distributed over their entire length. --○--: Function of a green receptor with λ_{max} at 543 nm (average of many green receptors: λ_{max} = 540 nm). The thick axon of the green receptor projects into the most proximal layer of the medulla, where it divides into three furcae, which carry many long spines and short blubs

light of wavelengths 330–370 nm was found to be most attractive for most species, so UV receptors have already been postulated by the early authors. Long wavelengths of light are visible for most insects up to about 650 nm, so the existence of red receptors was thought to be very unlikely. Furthermore, phototaxis experiments showed that the attraction of monochromatic light is not constant for a given species, but varies with developmental stages: sexual cycle, stages of foraging behavior, hiding or flight behavior, etc. This hinders closer comparison of the huge amount of data collected on spectral phototaxis in many species, e.g., on several hundred species by WEISS (1943a, b, 1944, 1946), WEISS et al. (1941, 1942) and MIKKOLA (1972). In addition, the physical parameters of the stimulating light (spectral transmission of filters and light intensity) and the adaptation stage of the animals were not carefully enough controlled. Nevertheless, the data by WEISS shows that in beetles about half of the Chrysomelidae species tested preferred green light to UV light whereas all other species (Lampyridae, Coccinellidae, Cerambycidae, Scolytidae, Scarabaeidae) reacted most strongly to UV light. In Orthoptera, the Mantidae chose green at least as often as UV, and in Hymenoptera all species tested preferred UV. Light trapping experiments under natural conditions (at dawn and at night) (MIKKOLA, 1972) are partially in disagreement to WEISS' observation. *Lampyris noctiluca* (Coleoptera), *Hepialus fusconebulosus* (Lepidoptera), and Aleurodidae (Homoptera) were found to be more attracted to green light than to UV light. But these are exceptions. Generally, UV light is most attractive to insects under natural conditions, as is known by everybody who uses a UV-emitting "black lamp" to collect insects at night.

Honey bees were frequently tested on their spectral phototaxis (BERTHOLF, 1931a, b; SANDER, 1933; HEINTZ, 1959; HARAGSIM and POSPISIL, 1969; LABHART, 1974). As bees can be trained to react to spectral colors, the spectral threshold function of learned colors was also determined (DAUMER, 1956; THOMAS and AUTRUM, 1965; MENZEL, 1967; VON HELVERSEN, 1972). The most reliable data in both experimental sets gave a dominant UV peak (Fig. 12) which proves that dark-adapted or weakly light-adapted bees are 5–10 times more sensitive to UV (350 nm) light than to green (530 nm) light. The spectral threshold function reported by VON HELVERSEN (1972) and THOMAS and AUTRUM (1965) (trained bees) and by WENDLAND (1976) and KAISER et al., 1977 (spontaneous positive phototactic runs) shows an additional third peak in the blue (440 nm). These three peaks coincide very well with the three color receptor types in the bee eye (see above), but from the point of view of the retinal organization (number, size, and arrangements of receptors) there exists no simple explanation for the high UV sensitivity. LAUGHLIN (1976b) has put forward the idea that the higher sensitivity of UV receptors is a mechanism to compensate for the lower intensity of UV (300–390 nm) light under natural light conditions. In fact, UV light reaches only about 15–20% of intensity (on a quantal basis) than other wavelength regions between 400 and 650 nm (see Sec. D).

In *Drosophila* the analysis of the spectral phototaxis is successfully used to describe the spectral properties of the photoreceptors. At low light intensities the phototaxis $S(\lambda)$ function has two peaks of nearly equal height at 360 and 500 nm; at higher intensities the maxima shift to 350 and 480 nm and the UV sensitivity is about 45 times the green sensitivity (SCHÜMPERLI, 1973). These changes can be understood on the basis of the higher sensitivity of the retinula subsystem R1–6, the lower sensitivity of the subsystem R7 and R8 (KIRSCHFELD, 1973a), and the fact that retinula cell R7 is an UV receptor (HARRIS et al., 1976; STARK, 1977) (or UV and blue-green receptor; MEFFERT and SMOLA, 1976; SMOLA and MEFFERT, 1975) and R8 a blue (or blue and green) receptor (see p. 533). The direct proof comes from phototaxis experiments with mutants (see Fig. 15, in HAMDORF, this volume). Mutants without rhabdomeres R7 lack the high UV sensitivity (STARK et al., 1976; HU and STARK, 1977). Those with R1–6 and R7 deficits have a much reduced sensitivity and a double peaked $S(\lambda)$ function with the longer wavelength peak shifted to about 440 nm (HARRIS et al., 1976). The change in $S(\lambda)$ as found in flies at different illumination levels could be interpreted as a kind of Purkinje shift, well known in human psychophysics. Indeed, in flies this change is based on the different absolute sensitivity of broadband receptors (subsystem R1–6) and short-wavelength-absorbing receptors (subsystem R7 and 8). In addition, the fly's retina displays a duality of the receptor morphology, and early behavioral experiments (FINGERMAN and BROWN, 1952) favor the idea that flies lack color vision at low light intensities (see also MAZOKHIN-PORSHNYAKOV et al., 1975, on Tabanidae). It is an interesting, but as yet unanswered question, whether other insect species also have a kind of scotopic or photopic duality. There is experimental evidence for changing $S(\lambda)$ with light intensity in other insect species. WEISS (1942) reports that most of the species he tested responded to UV light more strongly at lower intensities (cf. HEINTZ, 1959; for the bee); HASSELMANN (1962) found ERG recordings from beetle and moth eyes to have comparable results, the $S(\lambda)$ function giving rise to UV and blue sensitivities with decreasing stimulus strength. Optomotor reactions in color-striped drums (e.g., LÜDTKE, 1954; RESCH, 1954; OHNESORG, 1955, with *Notonecta*) and color training experiments with bees (THOMAS and AUTRUM, 1965) demonstrate varying chromatic sensitivities with light adaptation. Since there is no specific evidence yet for the existence of color vision or disappearance of color vision at low light intensities in any insect species, it is doubtful if these spectral shifts are comparable to the Purkinje phenomenon in humans. From the point of view of the large variety of retinal structures in insects, it seems highly probable that there are diverse mechanisms for coping with the large scale of intensity variation of natural light, e.g., specialized high-sensitive and low-sensitive receptors, selective growth of the total rhabdomere volume at low light intensity, dynamic electric coupling between visual cells with the same point of view, and neural pooling of the signals from receptors with the same point of view. Changes of $S(\lambda)$ functions may indicate a shift from a photopic to a scotopic system; the disappearance of color vision, however, has to be demonstrated separately.

f) Insect Ocelli

Most adult insect species have two or three small dorsal-lens eyes (ocelli), whose function is not known with certainty (Review: GOODMAN, 1970). Behavioral studies indicate that the ocelli might control the overall activity of the animal and the muscular tonus (criticism of this concept in BULLOCK and HORRIDGE, 1965). More quantitative data have shown that the sensitivity to light mediated through the compound eyes is controlled by the ocelli; thus the sensitivity decreases with reduced illumination of the ocelli (MÉDIONI, 1959; JANDER and BARRY, 1968; SCHRICKER, 1965; GOULD, 1975). The spectral properties have been studied in a few insect species only (RUCK, 1957, 1958, 1965; WALTHER and DODT, 1959; SUGA and KATSUKI, 1962; SONNTAG, 1971; CHAPPELL and DE VOE, 1975; EATON, 1976; KIRSCHFELD and LUTZ, 1977). Besides the fly *Calliphora* all insect ocelli so far studied show two sensitivity peaks; one in the green (500–530 nm) and one in the UV around 360 nm. In *Calliphora* the longer-wavelength peak is moved to 450 nm (KIRSCHFELD and LUTZ, 1977). As in the cockroach, spectral adaptation does not select different spectral mechanisms, whereas two or even three different receptor types exist in dragonfly ocelli (CHAPPEL and DE VOE, 1975). Intracellular recordings from dragonfly median ocelli showed double peaked $S(\lambda)$ functions (360,

500 nm) in single visual cells. No UV only or green only cells were found. The long wavelength sensitivity was selectively adaptable; and white light adaptation increased the UV sensitivity. The authors conclude that electric coupling or the existence of two pigments in one cell could account for these results; which explanation is correct is unknown. From the ERG recordings by RUCK (1965) and EATON (1976) it seems more probable that the UV and the green sensitivity result from two separate receptor types, because they found (1) qualitatively different patterns of the ERG potential to UV and green light flashes, (2) different slopes of the amplitude-response function at short and long wavelengths, and (3) selective adaptation of the green sensitivity.

g) Extraretinal Photoreception

Photoperiodic control of development and circadian rhythm of motor activity are controlled exclusively by extraretinal light-sensitive structures or, as in more primitive insect groups, in combination with retinal photoreception (review, BRADY, 1974; SAUNDERS, 1975; TRUMAN, 1976). Elegant transplant experiments and local light stimulation experiments proved that the dorsal protocerebrum is the region of extraretinal light sensitivity. In most species blue light (450–475 nm) is most effective, and red light has no effect. An exception is *Chaoborus americanus*, to which green light (550 nm) is most effective in a photoperiodic developmental response (BRADSHAW, 1972). PITTENDRIGH et al. (1970) report an interesting difference in the effect of red light on *Pectinophora gossypiella*. Circadian responses such as initiation of the hatching rhythm, adult eclosion, and oviposition are red insensitive. The photoperiodic control of diapause, however, is sensitive to red light. It seems likely, although it has not been studied in detail, that both types of responses are controlled by extraretinal receptors.

The photopigment involved has not yet been determined. If *Drosophila* is reared on a carotenoid-free medium, the sensitivity of the compound eyes drops by a 1000-fold, but the sensitivity of the eclosion response is not reduced (ZIMMERMANN and GOLDSMITH, 1971). This experiment, however, does not exclude rhodopsin as the photopigment since, as the authors point out, the small amount of carotene carried through the egg may have been selectively used for the extraretinal system. Interestingly enough the estimated absolute sensitivity of the extraretinal light sensitive systems is not much different from that of the compound eye as determined in ERG measurements (threshold $\leqq 3 \times 10^{-2}$ J/m^2; see TRUMAN, 1976).

III. Mollusca

A mollusc eye containing more than one photopigment has yet to be found (see Table 3). However, with two exceptions (the brain-eye of the opisthobranch gastropod, *Hermissenda crassicornis*, which contains only five receptors, and the small lens eyes along the mantle edge of the bivalv, *Pecten irradians*) no intracellular recordings have been published. Other studies have used ERG measurements, which like pigment extraction methods detect only the dominant

Table 3. Mollusca

Class, Order/Family	Species	Eye type	Major λ_{max}	Secondary λ_{max}	Method	Further observation and interpretation	References
Gastropoda Prosobranchia	*Strombus* spec.	Lens eye on telescope	485	—	ERG and spikes from optic nerve	Selective adaptation: no effect. A single photopigment	GILLARY, 1974
Gastropoda Opisthobranchia (Anaspidea)	*Aplysia californica*	Simple ocellus	500–550	?	Mass-spike activity from optic nerve and partially ERG	Recalculation gives evidence for a rhodopsin-like photopigment	WASER, 1968
Gastropoda Opisthobranchia Nudibranchia	*Hermissenda crassicornis*	Lens eye in the brain (five photoreceptor cells)	490	—	Intracellular recordings	Excitation to illumination, inhibition between the five photoreceptors, one photopigment	DENNIS, 1967 (s. ALKON and FUORTES, 1972)
Pulmonata Stylommatophora	*Helix pomatia*	Lens eye	475	—	ERG	One photopigment	VON BERG and SCHNEIDER, 1972
Pulmonata Stylommatophora	*Otala lactea*	Lens eye	490	—	ERG Chromatic adaptation	One photopigment	GILLARY and WOLBARSHT, 1967
Gastropoda Bivalvia	*Pecten irradians*	Lens eye	500	—	Intracellular recordings	Depolarizing and hyperpolarizing photoreceptors in different retinal layers, contain the same single photopigments	GORMAN and MCREYNOLDS, 1969 MCREYNOLDS and GORMAN, 1970a, b
Gastropoda Bivalvia	*Lima scabra*	Lens eye	470	—	Single neuron (spikes) recording from optic nerve	On and off responses, same single pigment involved (see also Ch. B)	MPITSOS, 1973

Table 3 (continued)

Class, Order/Family	Species	Eye type	Major λ_{max}	Secondary λ_{max}	Method	Further observation and interpretation	References
Cephalopoda (Decapoda)	*Loligo* *Sepia* *Sepiella*	Lens eye	493 492 500		Spectrophotometry of extracted pigments	Rhodopsin with thermostable metarhodopsin	See Rev. GOLDSMITH, 1972; HARA and HARA, 1972; ABRAHAMSON and FAGER, 1973
(Octopoda)	*Octopus* *Eledone*	Lens eye	475 470	— —	Spectrophotometry of extracted pigments	Rhodopsin with thermostable metarhodopsin	See Rev. GOLDSMITH, 1972; HARA and HARA, 1972; ABRAHAMSON and FAGER, 1973
	Octopus *Eledone*	Lens eye	475 470	 —	Spectrophotometry of extracted pigments	Rhodopsin with thermostable metarhodopsin	See Rev. GOLDSMITH, 1972; HARA and HARA, 1972; ABRAHAMSON and FAGER, 1973
	Octopus	Lens eye	480		ERG chromatic adaptation	Only one photopigment S(λ) narrower after light adaptation (reduced self-screening with high photopigment concentration)	HAMASAKI, 1968
	Eledone	Lens eye	470	—	ERG	One photopigment	HAMDORF et al., 1968

photopigment with certainty. Also, most of the ERG studies did not cover the wavelength region 300–400 nm; so the question of whether UV receptors occur in molluscs (for example in the land snail *Helix pomatia*) remains open.

The maximum $S(\lambda)$ lies between 470 and 500 nm. Neither biotopic differences among the species, nor eye type (simple ocellus, primitive lens eye, complex lens eye) are correlated with the position of λ_{max}. It is highly probable (in Gastropoda, Bivalvia) or certain (in Cephalopoda) that the photopigment is a rhodopsin. Other conclusions (e.g., *Aplysia*, WASER, 1968) are based on errors such as false determination of the $S(\lambda)$ (a corrected calculation of WASER's data produces a typical rhodopsin resonance function).

Eyes of Bivalvia and Gastropods contain two, or sometimes up to four, structurally different types of photoreceptors (cf. BRANDENBURGER, 1975), which are intermingled or separated in layers. It is interesting to note that, although the two retinal layers in the eye of *Pecten* contain the same photopigment, they respond with opposite polarity to illumination: the distal, ciliary photoreceptors hyperpolarize and the proximal, rhabdomeric receptors depolarize (McREYNOLDS and GORMAN, 1970a, b). The same seems to be true for the eyes of *Lima scabra* (MPITSOS, 1973). In line with these findings the three conductance changes found in photoreceptors of *Hermissenda*—(1) initial Na^+ permeability increase causing a depolarization, (2) the following K^+ outflow causing a re- and hyperpolarization, and (3) the delayed K^+ permeability increase producing a low depolarization—show the same spectral dependence and are therefore most likely triggered by one type of photopigment molecule (DETWILER, 1976).

There are many kinds of photoreceptors in *Aplysia*. In addition to those in the eye, photoreceptors exist in the skin of the mantle (LUKOWIAK and JACKLET, 1972; NEWBY, 1973), the abdominal ganglion (ARVANITAKI and CHALAZONITIS, 1961), the cerebral ganglion, and the oral tentacles (BLOCK and SMITH, 1973). These photoreceptors all seem to be sensitive to a broad spectrum of light ranging from UV to red, but accurate measurements have not yet been performed (LICHEY et al., 1976).

Phototaxis experiments with the nudibranch *Onchidoris fusca* demonstrate a single photopigment ($\lambda_{max} = 500$ nm) in the paired eyes, which are embedded into a thick pigmented skin (HUGHES, 1970). The $S(\lambda)$ function is narrower than a corresponding "Dartnall-curve" especially at the shorter wavelength side. This seems to result from the filtering effect of the pigmented skin. The spectral dependence of the light-induced closing response of *Pecten maximus* has two peaks (at 540 and 475 nm; CRONY-DILLON, 1966). As both receptor layers in the mantle eyes contain the same photopigment (see above and Table 3) the peak at longer wavelengths may result from the dermal photoreceptors, as in *Lima scabra*.

IV. Echinodermata

Echinodermata have not evolved highly organized light-sensing structures. In many species even the most primitive type of photoreceptor organs (single photoreceptor cells, pigment cup ocelli, eyespots) were not found, and in those, where such structures were described by earlier histologists, no physiologic evidence for their photosensitivity could be found (review, BOOLOOTIAN, 1967; YOSHIDA, 1966; YOSHIDA, Ch. 10, this volume; MILLOTT, 1968). The only direct evidence of physiologic activities of eyespots in response to light has been obtained from *Asterias* by HARTLINE et al. (1952) and from the sea urchin *Diadema setosum* by TAKAHASHI (1964). Since many species that are clearly photosensitive have no demonstrable eyes, it is assumed that dermatoptic (dermal, neural, or/and myoneural) photosensitivity plays a major role in their photic responses (see YOSHIDA, Ch. 10, this volume).

Concerning the spectral aspects of photic responses, HESS showed in 1915 that blue-green light is most effective in inducing the shadow response (spine movement) in *Centrostephanus longispinus* and the tube feet and tentacle response in *Holotheria poli*. The same was found for the tube feet response of *Psammechinus miliaris* (MILLOTT, 1968). In *Asterias amurensis*, YOSHIDA and OHTSUKI (1966) succeeded in separating the inputs of the dermatoptic light sensitivity and photoreceptors in a special structure (the optic cushion) controlling the photic response. They found a wavelength dependence with $\lambda_{max} = 485$ nm when both the optic cushion and the dermatoptic light sense are working, and a $\lambda_{max} = 505$ nm when only the dermatoptic light sense controls the behavior. Wavelength dependence was established by YOSHIDA (1957) for the response of single chromatophore cells (pigment distribution to illumination by a tiny spot of light). Wavelengths between 450–500 nm ($\lambda_{max} = 460$ nm) are most effective. It is concluded that the light-sensitive pigment is different from the screening pigment. YOSHIDA and MILLOTT (1960) speculate about the echinochrom A as the visual pigment involved.

So far it is not known if the Echinoderm species possess more than one photopigment, which would allow color-specific responses or even color vision (see Sec. D.I). RODENHOUSE and GUBERLET (1946) report an interesting aspect of the phototactic behavior of the cushion star fish *Pteraster tesselatus*; these animals are positively phototactic to white light. When monochromatic light is used, they respond positively to green light, but negatively to blue and yellow light. Unfortunately, the intensity dependence was not studied carefully enough, so the possibility cannot be ruled out that this is an intensity effect rather than a chromatic effect. Another interesting case is reported by BOLTT and EWER (1963). They found the retractor muscle of the lantern muscular system sensitive to light (myoneural light sensitivity). If two flashes of light from a tungsten filament lamp were given, the muscle contracted to both. If the second flash came from a mercury lamp, the muscle relaxed rather than contracted to the second flash. Here too, it is unknown if the effect is caused by an intensity difference. Probably the higher UV content of the flash from the mercury lamp caused the different response.

As the dermatoptic light sense and specialized photoreceptors may act as combined or possibly separated inputs to control the behavior, wavelength-selective behavior patterns could result if these two systems differ in their $S(\lambda)$. YOSHIDA and OHTSUKI (1966) describe an example for the starfish *Asterias amurensis*. Separated inputs from these two visual systems, however, have not yet been reported.

V. Other Invertebrates

The invertebrate groups not mentioned so far (Nemertini, Sipunculida, Echiurida, Onychophora, Tentaculata, Pogonophora, Chaetognatha) were scarcely studied if at all for the spectral properties of their visual systems or for the spectral properties of visually guided behavior.

C. Central Processing of Color Information

The large variety of complex nervous systems in invertebrates strongly suggests that the neuronal mechanisms involved with color coding differ to a great degree within the various systematic groups. The phylogenetic connections between these various groups, however, provide ideal material for evaluating how a complex neuronal integration mechanism like color coding has been developed during evolution and how different nervous structures solve the basically similar problems of coding wavelength independent of intensity. Unfortunately, at present there is very little experimental data available on the neuronal mechanisms of processing color information, and most of them comes from highly evolved arthropods with very complex visual neuropiles.

We have speculated in the general introduction that color vision may have evolved out of a combination of initially independent wavelength-selective receptors with interneuron pathways controlling different behavior patterns. If this is in any way correct we should expect more primitive species to have separate interneural pathways dominated by different wavelength regions and controlling different behaviors. Furthermore, as the presence or absence of UV light is the most important indicator for different visual cues, highly evolved neuronal systems may have as evolutionary relicts different neuronal strategies in processing information coming from UV receptors and long-wavelength receptors. There are indeed data to support this view, which will be presented in this chapter, but much more information is needed to draw a precise picture.

This chapter discusses two basically different types of interneurons involved in processing visual information; first, interneurons with their own endogenous light sensitivity, and, secondly, true visual interneurons, which receive their input from ocular photoreceptors and serve the central processing of visual information.

I. Neurons with Endogenous Light Sensitivity

Photosensitive neurons found in a number of invertebrates may function independently of or in cooperation with an external visual sense organ in controlling the animal's behavior. If different behaviors are controlled by an ocular and nonocular light sense or by different pathways of the nonocular light sense this might be termed a first stage of neuronal integration with the object of specific (selective) control of behavior. If these various systems of visual input differ in wavelength dependence then independent wavelength-specific behavior control should be possible.

Although the photosensitivity of interneurons (e.g., in the visceral ganglion of the marine snail *Aplysia* and in the abdominal spinal cord of crayfish) has long been known from electrophysiologic recordings (PROSSER, 1934; ARVANITAKI and CHALAZONITIS, 1949) and has stimulated a number of investigations on other molluscs and crustaceans (review, KENNEDY, 1964; MILLOTT, 1968), its role in regulating behavior is not clear. In transparent invertebrates, such as coelenterates, pelagic molluscs, nemathelminths, and crustacea, light-sensitive interneurons may be in some species the only input system, in other species an additional input system for behavior control: Furthermore the nonocular light sensitivity is not restricted to light sensitive neurons, but may be the property of epithelial cells, too. Such dermal light sensitivity is known in many invertebrates (rev. MILLOTT, 1968) and appears to function exclusively in the general perception of luminosity, wavelength differences being of little or no importance. Light-sensitive interneurons or the dermal light sense may differ in their spectral properties from that of eyes, and could therefore give rise to a chromatically independent input system (see YOSHIDA, Ch. 10, this volume).

In *Aplysia* the photosensitive neurons of the visceral ganglion respond to illumination with either increased excitation (on-response) or with inhibition and phasic excitation to light-off (off-response) or with on- and off-responses. Excitation is connected with depolarization; inhibition, with hyperpolarization of the membrane potential. Measurements of $S(\lambda)$ suggest that a carotenoid pigment is responsible for inhibition and a hemoprotein for excitation (ARVANITAKI and CHALAZONITIS, 1961). HISANO et al. (1972a, b) arrived at a similar conclusion after investigating the on- and off-responses of the giant photosensitive neurons in the sub- and supraoesophageal ganglia of the marine pulmonate mollusc *Onchidium verriculatum*. The on- and off-responses of the photosensitive elements in the pallial nerve of the shellfish *Spisula* are also maximally sensitive to different wavelengths (KENNEDY, 1960); the excitatory process to red, the inhibitory one to green.

In the shellfish *Lima scabra*, inputs from the numerous eyes arranged along the edge of the mantle are distinguishable from those originating in the diffusely scattered dermal photoreceptors. The eyes react to light flashes with an on- and off-response mediated by a single pigment ($\lambda_{max} = 470$ nm); while the dermal photoreceptors (photosensitive neurons?) give only off-responses with maximum sensitivity at 520 nm (MPITSOS, 1973). It has been shown for *Pecten* also that the on- and off-responses of the eyes, generated in various photoreceptors, originate from the same photopigment (LAND, 1966; MCREYNOLDS and GORMAN, 1970b) but here no dermal photoreceptors have been found.

Shellfish react to a sudden decrease in light intensity (over-shadowing) with a withdrawal response, drawing in the mantle and partially or completely closing the valves or operculum. In *Lima* it seems probable that this reflex is triggered by an off-response from both photoreceptive systems, eyes and dermal photoreceptors (MPITSOS, 1973). The physiologic differences in adaptational behavior induced by the two systems suggest that, in faint illumination, the eyes ($\lambda_{max}=470$ nm) and, in strong illumination, the dermal photoreceptors ($\lambda_{max}=520$ nm) predominate in controlling the reaction. Whether there is actually a change in $S(\lambda)$ of the reflex with increasing intensity has not yet been investigated.

With the exception of the photoreceptive elements in the pallial nerve of *Spisula* and in the mantle fringe of *Lima*, it is not clear whether photoreceptive neurons inside the body play any role in photoreception. Examples for light-sensitive neurons whose functions are unknown are those found in the radial nerve of the echinoid *Diadema* (YOSHIDA and MILLOTT, 1959, see also page 541), in the pleuroparietal ganglion and the abdominal nerve of the pulmonate molluscs *Onchidium verriculatum* (HISANO et al., 1972a, b) and *Aplysia* (see above), in the ventral nerve cord of the crayfish (KENNEDY, 1960, 1964), and in the metasomatic ganglion of the scorpion *Heterometrus* (GEETHABALI and PAMPAPATHI RAO, 1973). Such neurons may have a function in regulating the motor activity of the animals, but a specific behavioral role is unknown in most cases. In the crayfish, WELSH (1934) found that illumination of the caudal photoreceptor stimulates movement of the legs, whereas illumination of the compound eye causes a reduction of leg movement.

In the paired neurons of the crayfish's sixth abdominal ganglion only one pigment system is involved ($\lambda_{max}=500$ nm) (WELSH, 1934; LARIMER et al., 1966; BRUNO and KENNEDY, 1962); in other neurons there may be two systems of photopigments at work, e.g., in molluscs (see above) and the photoreceptive neurons in the sixth ganglion of *Heterometrus* which are maximally sensitive at $\lambda_{max}=420$ and 580 nm (GEETHABALI and PAMPAPATHI RAO, 1973). However, it is not clear whether the various λ_{max} play a role in wavelength-specific effects.

II. Visual Interneurons

Very little is known about the function of interneurons in the processing of information which comes from invertebrate eye. Studies have been made only in crustacea and insects, and usually these experiments have not attempted to analyze color integration. In the following chapter I will first discuss the relevance of data from crustacean visual interneurons to color coding and then those from insects.

In both crustacea and insects the information coming from the compound eyes is processed in a complex system of several layers of neuropiles (see for example STRAUSFELD, 1975). Besides those neurones which code motion of large objects (optomotor interneurons) the visual interneurons are in general very fine and are difficult to record from. Furthermore, some of these neurons produce only graded de- or hypopolarizations and no spikes and are therefore not seen individually in extracellular recordings (see LAUGHLIN, this volume Part B). Thus the results so far collected are limited, mainly by the technical problems of getting stable recordings.

In the crayfish, *Procambarus*, TREVINO and LARIMER (1970) and WOODCOCK and GOLDSMITH (1970) have measured the action spectrum of sustaining units.

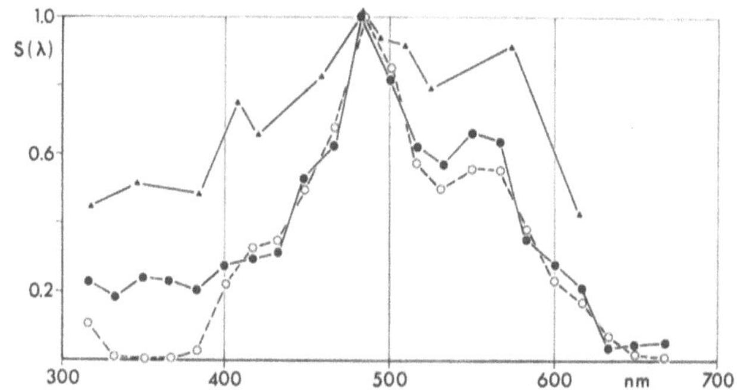

Fig. 7. $S(\lambda)$ of lamina monopolar cells in dragonflies and bees. —●—: $S(\lambda)$ of a monopolar cell in the bee; $S(\lambda)$ function closely resembles that of the average $S(\lambda)$ functions. - - ○ - -: $S(\lambda)$ of a monopolar cell in the bee; its short-wavelength sensitivity is much lower than that of most of the bee monopolar cells measured (MENZEL, 1974). —▲—: Average $S(\lambda)$ of monopolar cells in the dragonfly *Hemicordula tau* (LAUGHLIN, 1976)

These fibers give a sustaining response to stationary flashes of light (WIERSMA and YAMAGUCHI, 1966, 1967). Their function and where they fit into the information-processing hierarchy is unknown. It is not even firmly established that they are sensory interneurons. Their action spectrum indicates that most probably both receptor types in the compound eye (blue $\lambda_{max}=430$–460 nm and green $\lambda_{max}=570$ nm) provide positive input. An interesting observation that has unfortunately not been further investigated is that the spikes triggered by blue illumination, following adaptation to red light, have a considerably longer latency (120 ms) than those following green illumination (30 ms).

All other studies of color processing have been done on insects. Visual information is processed successively in the visual ganglia—lamina, medulla, lobula, lobula plate and finally protocerebrum. There are many unanswered questions concerning the color coding properties of the first interneurons in the lamina—the monopolar cells. In the fly, the large monopolar cells L1 and L2 receive input from the retinula cells 1–6 only. Those R1–6 cells which look at the same point in the environment produce excitatory input to L1 and L2, while neighboring R1–6 cells provide inhibitory inputs (lateral inhibition, ZETTLER and AUTRUM, 1975). The chromatic properties of the monopolar cells seem to be the same as those of the retinula cells R1–6 (see pp. 533), although this has not yet been accurately measured. The lateral inhibition, at least does not involve chromatic changes (ZETTLER and AUTRUM, 1975).

Large monopolar cells of the bee lamina (MENZEL, 1974) and the dragonfly lamina (LAUGHLIN, 1976a) have broad $S(\lambda)$ functions. In the dragonfly the "linked pigment" cells (see pp. 529) serve as inputs very similar to those found in the fly. In the bee either all three photoreceptor types or only the blue and green receptors are positive inputs. No spectral opponency is found in the dark-adapted large monopolar cells. In two recordings from bee monopolar cells, the UV sensitivity was reduced, possibly an indication for a negative input from UV receptors (Fig. 7).

ARNETT (1972), also MCCANN and ARNETT (1972) have reported extracellular recordings from fibers in the first optic chiasm (between lamina and medulla) of the fly. These fibers have $S(\lambda)$ which indicate that only the retinula cells R1–6 serve as input for the large monopolar cells L1 and L2. Neuroanatomic findings also suggest that the lamina should play a minor role in color coding. The lamina receives broad spectral inputs from the retinula cells R1–6 while the receptors with different $S(\lambda)$ R7+8 project to the medulla without any synaptic interactions in the lamina (BOSCHEK, 1970; STRAUSFELD, 1975).

However, a few recordings indicate that some color processing may still occur in the lamina via a centrifugal projection from the medulla to the lamina. MIMURA (1974) found efferent fibers in the lamina which are large field and are UV sensitive only. This fiber type may have excitatory or inhibitory effects on the lamina interneurons. MIMURA (1974) has also recorded extracellularly from unknown neurons in the lamina of the fleshfly *Boettcherisca peregrina*, which showed varying wavelength sensitivity and in a few cases wavelength-dependent response patterns for wavelengths between 440 and 620 nm. No conclusions can be drawn from the latter about the inputs, since no $S(\lambda)$ were determined. MIMURA also describes a neuron whose on-excitatory receptive field center has two off-excitatory neighboring fields. The on-excitation in the center is triggered by wavelengths between 440 and 520 nm; at 560 nm in the center, an off-excitation is triggered. This observation is the only one so far of a lamina neuron involved with true color contrast.

Although the neural strategy of the first steps of color coding in the insect brain are unknown, there appears to be one general rule. The shortest-wavelength-coding receptors are different from long-wavelength- or spectrally broadband-coding receptors in their projection pattern. The result is that specific information on short wavelengths is not processed in the first visual ganglion. It is tempting to interpret this as an indicator for different wiring strategies of short and long wavelength coding systems in phylogenetic ancestors of these highly evolved insect species. (The same is true for the polarization sensitive channels, which are connected to short-wavelength, narrow-band receptors.)

The neuronal mechanisms of color coding in more central parts of the brain have been studied only in the butterfly species *Papilio troilus*, *Heliconius erato*, *Epargyreus clarus*, and *Morpho amathonte centralis* (SWIHART, 1967, 1968, 1969, 1970, 1971, 1972a, b) and the honey bee *Apis mellifera* (KIEN and MENZEL, 1977a, b; ERBER and MENZEL, 1977). The complex nature of these second- or higher-order neurons is illustrated by the fact that the time-dependent parameters of their response (on, sustained, and off activity) frequently differ in $S(\lambda)$, indicating that the excitatory and inhibitory input pathways must have differing time courses. Furthermore, the amplitude and dominance of the various response parameters is dependent on the adaptation state of the animal as well as on the size and configuration of the stimulus relative to the interneurons' receptive field. In view of all these difficulties and the small number of stable recordings from color coding interneurons we cannot be very specific about the neuronal strategy of color coding in the insect brain. A few generalizations are possible, though. The following types of neurons can be tentatively defined according to $S(\lambda)$:

a) Luminosity neurons

Subclass 1: Receiving input from all three color receptor types

Subclass 2: Receiving input from two color receptor types

b) Chromaticity neurons
Subclass 1: Opponent neurons
Subclass 2: Monochromatic neurons with
λ_{max} corresponds to a color receptor type
λ_{max} does not correspond to a color receptor type

c) Polychromatic neurons
Subclass 1: different $S(\lambda)$ of the various response patterns
Subclass 2: other complexities

Luminosity neurons respond with a first stronger (on response) then lower (sustained response) increase or decrease in spike activity and can be recorded in the medulla, lobula, or protocerebrum in both bees and butterflies (*Morpho*, *Epargyreus*). Subclass 1 receives inputs of the same kind (positive or negative) from all receptor types and is, therefore, sensitive throughout the whole visual region (300–650 nm for the bee, 400–750 nm for butterflies, UV not tested) with maximal sensitivity in the bluish green to green region (490–560 nm) and a secondary maximum in the long UV–violet (370–410 nm bee, *Morpho*, see Fig. 8a). Single unit marking and spot marking in the bee (KIEN and MENZEL, 1977a) and latency measurements in *Morpho* (SWIHART, 1972a) suggest that these neurons belong to a luminosity coding system which is represented at all neural levels of the visual system. The large monopolar cells are most probably the first link in this chain. These neurons could act as a luminosity reference for color coding. The simple organization of their large receptive fields (60° in bees, no quantitative data for butterflies) support this conclusion.

Luminosity neurons of subclass 2 receive input from only two of the three receptor types (bee, KIEN and MENZEL, 1977a). Other neurons which may be subclass 1 or 2 possess properties unusual in luminosity fibers. Some examples are OFF responses only, different patterns of the sustained response for different wavelengths, long latencies (*Morpho*, SWIHART, 1972a), and different slopes of the intensity function for different wavelengths (bee, KIEN and MENZEL, 1977a).

Chromaticity fibers have been recorded in the medulla and lobula of both bees and butterflies.

Subclass 1: Opponent neurons respond to stimulation with a phasic-tonic increase or decrease (depending on the wavelength) in their relatively high, dark discharge frequency (10–20 Hz). In the bee medulla UV-excited, blue- and green-inhibited (UV^+/B^-G^-) neurons are recorded much more frequently than UV^-/B^+G^+ neurons (Fig. 8b, KIEN and MENZEL, 1977b). In the protocerebrum of *Heliconius*, *Papilio*, and *Epargyreus*, SWIHART (1969, 1970, 1972a, b) recorded B^+/G^-, G^+/B^-R^-, R^+/G^-O^-, B^+/G^-R^- types (UV not tested) (B for wavelengths 430–480 nm, G for 490–560 nm, O for 560–620 nm, R > 620 nm).

The inhibitory input can be quantified in two-color experiments. In the bee, for example, the most frequently found color opponent neurons show the greatest sensitivity changes in violet (at about 400 nm approximately 0.5 relative sensitivity change per 5 nm, see Fig. 8b). This becomes especially interesting when related to the fact that bees distinguish violet colored foodmarkers with the greatest precision (VON HELVERSEN, 1972, see Sec. C) and learn such markers quickly (MENZEL, 1967). In dark-adapted bees the color opponent property is always

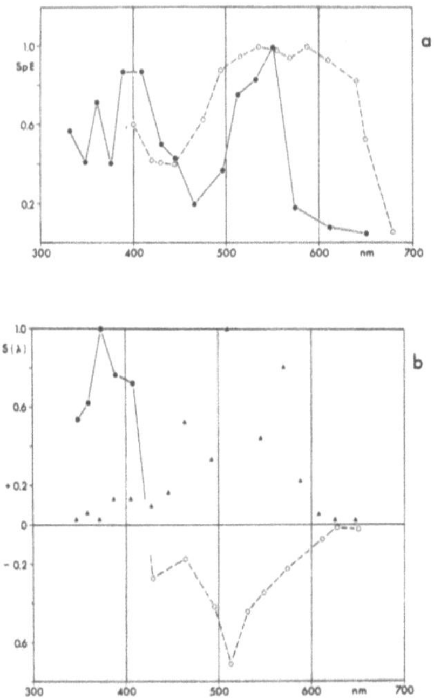

Fig. 8. (*a*) $S(\lambda)$ of broadband visual interneurons (luminosity type neurons) in the bee and butterfly brain. —●—: $S(\lambda)$ of a lobula neuron in the bee brain, which produced an excitatory sustaining response to stationary, monochromatic test flashes. Sensitivity is calculated for half the maximal response (KIEN and MENZEL, 1976a). - - ○ - -: Spectral efficiency of a neuron recorded somewhere in the brain of the butterfly *Morpho amathonte centralis.* The response is a sustained excitation with a short latency of 8–10 ms (SWIHART, 1972b). (*b*) $S(\lambda)$ of a color opponent neuron recorded in the proximal medulla of the bee. —●—: The sustained excitatory response to UV; - - ○ - -: The sustained inhibitory response in blue and green (type $UV^+ B^- G^-$). ▲: The $S(\lambda)$ of the ON response of the same neuron (KIEN and MENZEL, 1976b)

expressed in the sustained activity of the neurons. The ON-response of the same neuron displays generally a broadband sensitivity with maximum in the bluish green to green region (see Fig. 8).

Subclass 2: Monochromatic neurons. Most monochromatic neurons recorded in butterflies and bees have maximum sensitivity at a wavelength at which one of the photoreceptors is also maximally sensitive. *Heliconius* (SWIHART, 1972a) is likely to have blue (440 nm), bluish green (490 nm), and orange (600 nm) receptors; corresponding maxima of the monochromatic neurons in the protocerebrum are 430, 480, and 620 nm. In *Morpho* (SWIHART, 1972a) violet (404 nm) and green (550 nm) neurons were found which correspond to the violet (428 nm) and green receptors (564 nm). In the bee medulla and lobula, UV-, blue- or green-dominated neurons were found, which correspond to UV, blue, and green receptors.

The monochromatic neurons in both bees and the butterflies *Morpho*, *Heliconius*, and *Papilio* have little or no spontaneous spike discharge in the dark, so the possibility of inhibition at certain wavelengths is masked in a one-color experiment carried out in the dark-adapted state. In fact, there is evidence for inhibitory inputs in most of these monochromatic neurons, e.g., the fact that the $S(\lambda)$ calculated for such neurons in the bee medulla and lobula are narrower than the corresponding $S(\lambda)$ of the receptors.

In addition to these monochromatic neurons corresponding to color receptors, SWIHART (1970, 1972a, b), and KIEN and MENZEL (1977b) found neurons which show narrow band reaction to wavelengths where the $S(\lambda)$ of two receptors overlap. Examples of this type of subclass 2 neurons are (1) a violet ($\lambda_{max} = 410$ nm) neuron, (2) the "deep blue narrow-band" neuron in the protocerebrum of *Papilio*, and (3) two neurons in the bee lobula recorded simultaneously but clearly differentiable by spike size of which one was maximally sensitive in violet (400 nm) and the other one in bluish green (495 nm).

Polychromatic neurons. A variety of neurons with broadband sensitivity but all kinds of complicated responses to spectral light have been recorded in butterflies and bees. For example, SWIHART (1972a) found "variable latency color fibers" with very different latencies for various wavelengths, but no tests were made to find out whether this was an intensity effect. Frequently KIEN and MENZEL (1977b), ERBER and MENZEL found neurons in the bee lobula and protocerebrum which responded to monochromatic flashes with complex response/log intensity functions differing for different wavelengths and for the various response characteristics (ON, sustained, OFF).

Subclass 1 is characterized by its different $S(\lambda)$ of the various response patterns. For example, neurons have been recorded that have a sustained response preferential to UV but with ON and OFF responses dominated by long-wavelength light. Neurons in subclass 2 display various complexities. Regarding their spatial sensitivity, there are complicated patterns of distributions of wavelength sensitivities. Regarding their intensity dependence, two types of polychromatic neurons have been recorded in the bee protocerebrum which appear to be interesting for neuronal color coding. They have been named I- and C-band neurons (ERBER and MENZEL, 1977). The I-band neuron responds only to a specific band of intensities, the R/log I curve sloping steeply at lower and higher intensities. The region of optimal response is at different intensities for different wavelengths but the calculated $S(\lambda)$ is mostly broadband (maximum in green). The C-band neurons have very flat R/log I functions. Different wavelengths tend to produce a certain level of activity. Such neurons seem to code wavelength independently of intensity. Those I- and C-band neurons reacting to illumination of the ipsilateral eye have multimodal properties, responding also to stimulation by smell, to presentation of sugar water to antennae or proboscis, and to mechanical stimulation.

Spectral Response of Motion-Detecting Neurons. As recordings from neurons sensitive to motion are relatively easy to make in the region between the insect's lobula and brain, such neurons were examined for their role in color coding. Such units react preferentially to patterns of stripes which are moved in front of the ipsilateral or contralateral eye. They are excited when the pattern is moved in the preferred direction and inhibited when it is moved in the opposite or zero direction. They also respond to high intensity stationary flashes of light with a short discharge. BISHOP (1969, 1970) found that in the fly these neurons have a broad $S(\lambda)$ with a maximum at 500 nm, uniform intensity functions at different

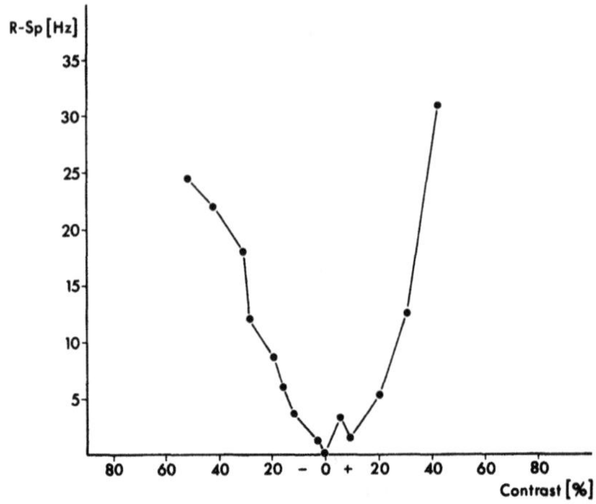

Fig. 9. Response of a directionally motion-sensitive neuron in the bee lobula to a moving two color pattern of varying contrast (abscissa). Wavelength combination 356 nm/539 nm. The ordinate gives spike response during stimulation minus spike response in the dark. Contrast is calculated by the formula $(a-b)/(a+b)$, where (a) is the intensity of one wavelength and (b) of the other. Contrast = 0 is arbitrarily moved to the point of lowest response. The lowest response equals the spontaneous response level indicating that no moving color contrast but only intensity contrast stimulates the cell (KAISER, 1975)

wavelengths, and no wavelength-related differences at various adaptation stages. He concluded, therefore, that they have no capacity for color coding. ARNETT (1972), MCCANN and ARNETT (1972) stimulated these fibers more adequately with moving stripes monochromatically illuminated. Neurons that reacted to long-wavelength spatial patterns showed the same spectral properties as BISHOP had measured. The $S(\lambda)$ was that of the retinula cells 1–6 (two sensitivity peaks at 350 and 480 nm) and $S(\lambda)$ did not change with the intensity level of the monochromatic test light, nor with selective adaptation to green light. The same intensity ratio was always found to produce a balance for forward and reverse motion at various wavelengths. On the other hand the spectral properties of those neurons responding only to short-wavelength spatial patterns had a λ_{max} at 465 nm and a UV sensitivity of 40%. At strong green adaptation and medium spatial wavelength pattern an $S(\lambda)$ maximum between 480 and 465 nm was found. This indicates that two different types of class-II motion-detecting units exist, which receive input from both the receptor subsystems R1–6 or R7 and R8—the exact type of input remaining unknown.

In bees (BISHOP, 1970; KAISER, 1972; MENZEL, 1973) and butterflies (SWIHART, 1972a, b) the $S(\lambda)$ of the movement-sensitive neurons is dominated by the green receptors. Proof that the neurons in bees respond only to brightness contrast, not to chromatic contrasts, was furnished by KAISER (1972) who used a method developed by VON BUDDENBROCK (1929). A colored stripe pattern was moved before a colored background. As Fig. 9 shows, an intensity ratio was found for each pair of colors (color of background versus color of moving stripe pattern) at which the reaction was almost zero. This result together with the fact that the neurons show identical dependence on intensity and velocity at different wavelengths proves that they can have no color coding function. Furthermore, it is probable that a reaction controlled via this visual interneuronal system will have no color properties (see Sec. D).

To summarize this section, the recordings of color coding interneurons now at our disposal—few in number considering the complex neuronal mechanisms in the insect brain—do not allow us to devise a model of neuronal strategy in the processing of color information. Generally speaking we can state that neuron types exist in the insect brain such as are also known from the vertebrate retina, lateral geniculate nucleus, tectum opticum and cortex (see review ABRAMOV, 1972; GOURAS, 1972; DEVALOIS, 1972; JUNG, 1973). The vertebrate principle of dual channels for luminosity and chromaticity coding seems to operate in the insect brain, too. Color opponency is similarly the basic principle in the chromaticity neurons. Color-specific neurons such as found by GOURAS (1972) in the vertebrate cortex (area 17) which react to specific wavelengths, but not to white light, have not yet been detected in insects. The most important difference in color-processing strategy between vertebrates and insects is in the relationship between spatial and spectral sensitivity. In dark-adapted bee neurons, by contrast to the vertebrate horizontal cells, retinal ganglion cells and LGN units, color opponency is not combined with spatial opponency. The receptive fields of most neurons are fairly large with substructures, which, however, are not organized in a center-surround antagonism. Generally the same was found by SCHÜMPERLI (1975) in various butterflies. Further experimental research is needed in order to determine whether the data obtained from dark-adapted insects will also hold good for light-adapted states in general.

D. Wavelength-Specific Behavior and Color Vision

The evolution of color vision is barely understood because of the lack of comparative data and conceptual ideas about possible phylogenetic mechanisms. Without question, the development of the complex retinal and neuronal mechanisms underlying color vision involves many steps of adaptive changes, which must have provided the animal with an advantage at every step. The advantage of using the differences of wavelength distributions in the visual world are obvious; areas with higher probability of food can be chosen, those with a higher probability of predators or unsafe ecological factors can be avoided, open space can be differentiated from enclosed areas, conspecific animals can be recognized more precisely, etc. We have put forward the idea (see Introduction) that color vision in invertebrates may have evolved from wavelength selectivity in the visual control of certain behavior patterns. This concept implies that various motor patterns underlying different behaviors are triggered and/or controlled specifically by sets of photoreceptors, which differ in their $S(\lambda)$. Originally, such different photoreceptors may have even been located in separate eyes and their neuronal machineries may also have been completely separated. Later in evolution the coexisting but independent color receptor types and neuronal machineries may have been combined to allow color vision. By contrast to wavelength specificity true color vision is then characterized by wavelength discrimination based solely on chromatic differences and independence of intensity differences. There are examples for wavelength specificity, and we will present them here. However, the small number of well analyzed cases, make our view rather speculative, especially since in highly evolved species it is questionable whether wavelength specificity is a relict or a relatively new specialization.

I. Wavelength-Specific Behavior

1. UV–VIS Antagonism

The naturally occurring wavelength distribution (Fig. 1) demonstrates a basic difference between short wavelengths (<400 nm, UV) and longer wavelengths (>400 nm, VIS). It is, therefore, not surprising that basically different behavioral responses are controlled by UV and VIS light. The saltwater bacterium *Halobacterium halobium* responds with a direct photophobic reaction to UV ($\lambda_{max}=280$, 370 nm) and with an inverse photophobic reaction to VIS ($\lambda_{max}=565$ nm, see Sec. B). The sea anemone *Anthopleura xanthogramma* retracts its tentacles to illumination with UV ($\lambda_{max}=280$ nm) and bends the tentacles in response to VIS illumination ($\lambda_{max}=500$ nm, secondary maximum at 360 nm, see Fig. 3). The nematod *Chromadorina viridis* avoids UV light ($\lambda_{max}=366$ nm) and responds positively phototactic to VIS light (400 nm) (CROLL, 1966). These three examples may stand for other less well analyzed examples of water-inhabiting invertebrates which demonstrate phobic reactions to UV light. As UV light does not penetrate turbid and densely populated water, UV illumination may signal shallow water, or a dry biotope and so cause avoidance responses. There are many other examples of the antagonistic action of UV and VIS light. These are discussed in the following paragraphs.

2. Phototaxis

Movement directed toward (positive phototaxis) or away from (negative phototaxis) a light source can be elicited in most mobile invertebrate species and has, therefore, been used intensively to characterize spectral threshold or $S(\lambda)$ (see Sec. B). Apart from one example (*Drosophila*, see below) it is unknown if the animals discriminate wavelengths in their phototactic response or if they are color-blind. Where the $S(\lambda)$ has only one narrow peak, it is quite obvious that only one receptor type controls phototaxis and color discrimination is absent, e.g., the gastropod *Onchidoris fusca* (λ_{max} at 500 nm, HUGHES, 1970; see Sec. B). $S(\lambda)$ with more than one peak are frequently reported (see for example Fig. 4 on Rotifera species) and suggest that a multiple photoreceptor system exists even in low invertebrate species. However, such functions give no information about wave-

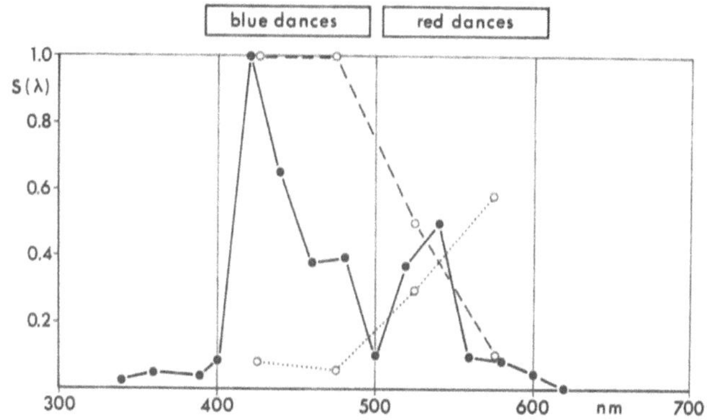

Fig. 10. Spectral properties of light-induced behavior patterns in Phyllopoda. —●—: Spectral phototaxis of *Polyphemus pediculus* (SCHAMP, 1973). The animals were swimming in a Ymaze, and the increase of choice for the illuminated arm against the nonilluminated arm was used to calculate the $S(\lambda)$. --○--: Spectral threshold for the induction of eye movement in *Daphnia magna* when the light falls on the eye through the side of the head. ----○----: Spectral threshold for the same eye response when the light falls through the top of the animal's head (see text). *Daphnia magna* displays blue dances to illumination from short wavelengths (<500 nm) and green dances to illumination from long wavelengths (>500 nm) (SMITH and BAYLOR, 1953)

length discrimination or wavelength-selective effects. In Rotifera such wavelength-selective effects were suspected for UV and green light (VIAUD, 1948) but could not be confirmed (see p. 512). In *Daphnia*, however, wavelengths shorter than 500 nm cause selectively different behavior patterns from wavelengths longer than 500 nm (Fig. 10). In blue light *Daphnia* is highly active, leans forward, and moves steadily over long distances. The direction of movement is predominantly at right angles to the propagation of light. In longwave light *Daphnia* keeps an upright position, moves slowly and predominantly in a direction parallel to the line of propagation of light (SMITH and BAYLOR, 1953; DAAN and RINGELBERG, 1969; STEARNS, 1975). Changes from blue to red, and dimming of blue or red light cause changes in the movement pattern which can be understood on the basis of these "color-specific dances". VON FRISCH and KUPELWIESER (1913) and HEBERDEY (1948) described these changes of movement patterns as a switch to negative phototaxis if blue light is reduced and a switch to positive phototaxis if yellow light or non-colored light is reduced. Similar wavelength-selective effects were found in the copepod *Diaptomus nevadensis* (HAIRSTON, 1976).

On a step further are findings by ROBERT et al. (1958) and YOUNG (1974) that the spectral threshold functions for inducing eye movement in *Daphnia* depends on the orientation of illumination with respect to the animal. If the light falls straight on the compound eye the threshold function peaks in the yellow-green, if it falls on the eye through the top of the animal's head it peaks in the violet-blue (Fig. 10). The naupliar eyes or a dermal light sense were excluded in YOUNG'S (1974) experiments as possible inputs. All these findings fit very well if one assumes a dual blue-red receptor system in the *Daphnia* compound eye, where the blue and red photoreceptors control specific behavior patterns, body position, speed of movement, positive or negative phototaxis.

The ecological background of this behavior is an accumulation of *Daphnia* in phytoplankton rich layers. The chlorophyl in the plankton absorbs blue light, and absence of blue light reduces speed of movement. The phytoplankton-poor layers with more blue light may induce foraging behavior. An alternative interpretation has been offered by HAIRSTON (1976) who found that copepods pigmented with carotenoids astaxanthin and astaxanthin ester have a much longer survival time under blue light than those animals without the carotenoids. Plankton crustacea may, therefore, be sensitive to short wavelength light and the faster swimming response under blue light illumination may be an avoidance response to damaging short-wavelength light. Interestingly, the behavior is not as stereotyped as it sounds. At low temperatures (10° C) the blue light effect disappears, and *Daphnia* responds to all wavelengths with a red dance (SMITH and BAYLOR, 1953). There is supporting evidence that in addition *Daphnia* has UV receptors in the naupliar eyes, since *Daphnia* with their compound eyes extirpated still avoid UV (SCHULTZ, 1928). The naupliar eye most likely controls the vertical migration. Its action seems to depend much more on environmental factors (pH, CO_2, redoxpotential, pressure) than that of the compound eyes (BAYLOR, 1957).

By contrast to animals living in water, arthropods living in air are, generally speaking, strongly attracted to UV light. Honey bees, flies, and many other insects are 2–100 times more sensitive to UV than to green light (Lit. see p. 535, C. Fig. 11). WEISS et al. (1942) and WEISS (1943b) tested many insect species and found that at low intensities most were strongly attracted by UV light. In competition experiments, where equal energy UV and green light were given as alternatives, all species choose the UV light. At higher intensities UV light tends to reduce positive phototactic responses so that VIS light becomes relatively more attractive. It is unknown whether these observations indicate color discrimination in the phototactic response of insects. In flies a reverse effect was described. At low intensities they are more sensitive to green light; at high intensities they are more sensitive to UV light (BERTHOLF, 1933, SCHÜMPERLI, 1973). Analyses using receptor mutants of *Drosophila* showed that all photoreceptor cells contribute to phototaxis, but the UV light (BERTHOLF, 1933; SCHÜMPERLI, 1973). Analyses using receptor mutants of (HARRIS et al., 1976; STARK et al., 1976; HU and STARK, 1977). In addition, HEISENBERG and BUCHNER (1977) demonstrated a wavelength-specific inversion of the phototactic response at higher intensities. Flies in a two-compartment chamber choose that compartment which is illuminated, e.g., with green light. If an equal UV illumination is superimposed on both compartments, the flies leave the compartment illuminated with UV and green light and choose that one illuminated with UV only. This phototactic inversion effect is produced even at a UV/green intensity ratio of 0.001, is specific for UV light, and cannot be explained by a negative phototactic response to the brighter illumination with two lights. It seems likely, therefore, that flies discriminate wavelengths (at least UV–VIS) in their phototactic response.

The phototactic response may change in sign or may alter qualitatively during different behavior patterns or in different behavioral contexts. Bees are positively phototactic when flying out of the hive and negatively phototactic when returning to the hive (VON FRISCH, 1965). The $S(\lambda)$ in positive or negative phototactic runs, however, do not differ significantly (WENDLAND, 1977). The female spider mite, *Tetranychus urticae*, in the dispersal phase is most strongly attracted by UV, and only moderately attracted by green light, whereas the spider mite in the sedentary

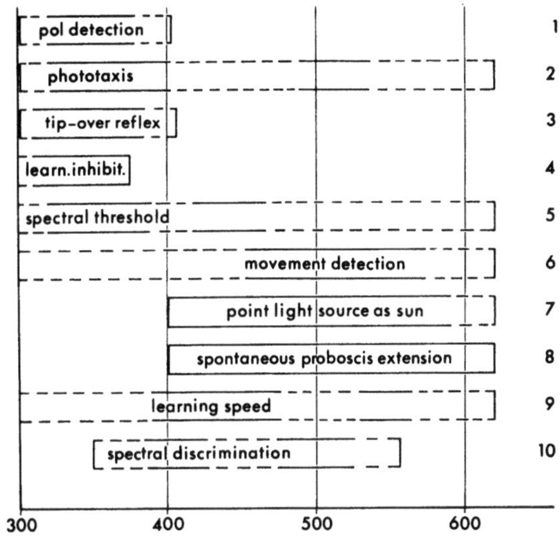

Fig. 11. Spectral range of wavelength-specific behavior in the bee (Nos. 1–8) and two examples of wavelength dependence in behavior patterns of the bee which involve color vision (Nos. *9*, *10*). *1*: Detection of polarized light is limited to the UV. The action spectrum is that of an UV receptor (VON HELVERSEN and EDRICH, 1974). *2*: Phototaxis is dominated by UV receptors but seems to involve also the blue and green receptors but with reduced weight (KAISER et al., 1977), *3*: Strong UV light from below induces a tip-over reflex. The bees end up "flying" on the ground glass dorsal side downwards (MENZEL, 1967; VON HELVERSEN, 1972). Equally, UV light at the hive entrance inhibits the departure of bees leaving the hive. The effective wavelength range seems to be the same as for the tip-over reflex (MENZEL, unpubl. observations). *4*: Learning curves become flatter at high UV light intensity (MENZEL, 1967). *5*: Spectral threshold as determined by VON HELVERSEN (1972). *6*: Movement detection in an optomotor experiment is dominated by green receptors (KAISER and LISKE, 1974). *7*: Bees dancing on a horizontal comb interpret a monochromatic point light source as the sun, if the wavelengths are >400 nm (EDRICH, 1977). The action spectrum suggests input from both blue and green receptors. *8*: DAUMER (1958) observed a spontaneous proboscis extension when walking bees crossed a border line between an area with and without UV reflection. The action spectrum has not been measured. *9*: Learning curves are steepest for violet light and flattest in the bluish green (MENZEL, 1967). *10*: Spectral discrimination is best around 400 and 490 nm (see Fig. 12; VON HELVERSEN, 1972)

phase is repelled by green light and moderately attracted by UV. The transition from the one behavior to the other depends on the humidity in the environment and is controlled by special humidity receptors (MCENROE, 1969; MILLS, 1974).

Generalizing from a few well-analyzed cases (e.g., bee and fly) one can deduce that the high effectiveness of UV in the phototactic response is not accounted for by the relative proportions of color receptors. Bees, for example, most probably have only two long and one short UV receptors, but four green and two blue receptors per ommatidium (MENZEL and BLAKERS, 1976). UV receptors may therefore be connected to the neural network which controls phototaxis in such a way that their excitation is more heavily weighted than that of the other receptors. MAZOKHIN-PORSHNYAKOW (1969) offered a plausible hypothesis, which is in line with the general view of wavelength specificity presented here. He argues that UV light signifies "open space", since most objects absorb rather than reflect UV (see Fig. 1), and sky light is the only extensive natural source of UV. So UV signifies room for free flight. Most likely the phototactic response is an escape response and so UV is the most reliable signal.

3. Optomotor Response and Wavelength Specificity

The movements of a whole animal or the response of the eyes to movement in its surroundings has frequently been used to test the animal's color discrimination capacity (see KAISER, 1974a). Older studies, mainly on insects (e.g., SCHLIEPER, 1928; SCHLEGTENDAL, 1934; MOLLER-RACKE, 1952) used a rotating drum with strips of a colored paper alternating with gray strips of various brightnesses. Optokinetic responses were found for all kinds of combinations of the gray tones and the colored paper, and it was concluded that the optomotor system of the animal is capable of color discrimination. Careful elimination of border lines and adjustment of the intensity ratios of test wavelengths and background light prove that both the fly *Phormia regina* (KAISER, 1968) and the bee (KAISER and LISKE, 1974) are insensitive to color contrast in the optokinetic response. In the fly the optomotor system receives input from the retinula cells R1–6 (see Sec. B). At high intensity and with narrow strips an unknown interaction between the R1–6 and the R7–8 input system takes place and causes a reduction of the UV sensitivity and a shift of the green maxium to shorter wavelengths (ECKERT, 1971). However, it is most unlikely that this interaction allows color contrast detection in the optomotor system even at high light intensities (HEISENBERG and BUCHNER, 1977). For the bee it has been clearly shown that the optomotor system is dominated by the input of the green receptors and lacks wavelength discrimination even at high intensities.

It would be most interesting to know if the lack of wavelength discrimination in the optomotor system of arthropods is a general strategy of neural wiring. HORRIDGE (1967) concluded from his experiments with the crab *Carcinus* that blue and green receptors in adjacent ommatidia are first separately connected for elementary, horizontal or vertical movement detection. The color-specific and movement-sensitive channels are then summed up with respect to sensitivity to movement direction, so that wavelength contrast is eliminated by the movement perception mechanism. The wiring diagram proposed by HORRIDGE (1967), however, is partially in disagreement with some of his own findings and those of others (e.g., the narrow and single peaked $S(\lambda)$ of the optomotor response, the $S(\lambda)$ of individual motion-sensitive fibers, (see Sec. B.II), the number of retinula cells per ommatidium, etc.). Furthermore, OLIVIO et al. (1975) have found some indication for at least a weak-wavelength contrast effect in crayfish. Therefore, it remains an unanswered, but highly interesting question whether there are any general rules governing the relationship of wavelength discrimination to large-field movement perception in arthropods.

4. Polarization Sensitivity and Wavelength Specificity

Polarized light detection is a common capacity of arthropods. The few cases where the wavelength aspect of polarized light detection was studied, demonstrated wavelength specificity but no wavelength discrimination. In flies the various experimental findings are most adequately interpreted by the assumption that only the receptor cells R7–8 are input to a polarized light-detecting mechanism, not the receptor R1–6 (KIRSCHFELD and REICHARDT, 1970; HEISEN-

BERG, 1972; McCANN and ARNETT, 1972; KIRSCHFELD, 1973a; KIRSCHFELD and LUTZ, 1974). Since R7–8 are UV and blue receptors, the polarization detection should be limited to the short wavelengths. However, optomotor turning tendencies can also be elicited in flies moving pol-filters illuminated with long-wavelength light (KIRSCHFELD, 1973b; HEISENBERG and BUCHNER, 1977). This effect is not yet understood. The bee has a specialized UV receptor serving as a polarization detector (MENZEL and SNYDER, 1974). Polarization orientation is, therefore, limited to short wavelengths (VON FRISCH, 1965; KIRSCHFELD, 1973c) and has an action spectrum of an UV receptor (VON HELVERSEN and EDRICH, 1974). A similar UV-limited polarization orientation was found for the fast-running desert ant, *Cataglyphis bicolor* (DUELLI and WEHNER, 1973).

Bees use polarized light or a point light source as a substitute for celestial navigation cues for their dances on a horizontal comb (VON FRISCH, 1965). EDRICH (1977) has recently shown that a point light source is interpreted as the sun if the light is unpolarized and contains predominantly longer wavelengths (>400 nm). Unpolarized light with predominance of short wavelengths (<400 nm) is interpreted as a sky area with an azimuth 180° to the sun (Fig. 11). This finding may contain the solution for the additional parameter needed to calculate the sun's position unambiguously from the *E* vector orientation in one patch of blue sky (cf. KIRSCHFELD et al., 1975). In our context these findings tell us that in bees orientation mechanisms using celestial cues are probably based on wavelength-specific integration circuits. These would reduce the complexity by giving up color discrimination, but keep the essential wavelength information by selective connections to one class of color receptors.

In crayfish and crab polarization sensitivity is equally present in all classes of color receptors (HORRIDGE, 1967; WATERMAN and FERNÁNDEZ, 1970). It is unknown how the incoming information on wavelength and *E* vector orientation is integrated in the visual ganglia. It seems most likely that detection mechanisms for the *E* vector neglects the wavelength information by summing the appropriately aligned receptors independently of their $S(\lambda)$ (HORRIDGE, 1967). In such a case polarization sensitivity would work through a broad spectral range which is indeed of obvious advantage for animals living in water.

5. Specialized Eye Regions and Wavelength-Specific Behavior

There is growing evidence that in invertebrates separated photosensitive structures, eyes, or eye regions are specialized in controlling different behavior patterns. In many cases such separated "eyes" differ in their set of color receptors and are therefore wavelength-specific in their behavior control. Earthworms have different action spectra for the shadow and light response (see p. 514), and there is indirect evidence that topologically different photosensitive structures are involved for each response pattern. "Dermal" and specialized photoreceptors in mussels may have different action spectra (e.g., in *Lima*, see p. 540) and are probably differently involved in light-induced responses (MPITSOS, 1973). Illumination of the light-sensitive neuron in the ventral cord of decapode crustacea activates leg movement, whereas illumination of the compound eyes reduces leg movement (WELSH, 1934). In *Limulus* the lateral eye is dominated by green receptors, whereas the median ocellus contains both UV and green receptors. We do not know exactly which behaviors in *Limulus* are controlled by each eye, but a few observations indicate that lateral and median eyes play roles of differing importance for straight movement and phototaxis (COLE, 1922, 1923; NORTHROP and LOEB, 1922).

A few scorpion and spider species have "color-specific eyes" (see Table 1b). In scorpions the lateral eyes may be only UV sensitive and the median eyes only green sensitive. In other species, even if the eyes contain the same photopigment, they differ in other respects. MACHAN (1966, cited by FLEISSNER, 1975) found that lateral eyes are more sensitive than median eyes, and FLEISSNER (1975) regards the lateral eyes as the major receptor organ for Zeitgeber stimuli. In spiders secondary eyes (postero-lateral and postero-median) contain only green receptors, and main eyes (antero-median-principle) at least UV and green receptors (YAMASHITA and TATEDA, 1976, found four color receptor types in jumping spiders). Jumping spiders use the lateral eyes to detect movement of an object which results in an alignment of the body, so that the object is faced by the principle eyes, in a defensive response they keep the moving object directly behind them (LAND, 1969b, 1971). Elucidation and identification of pattern and color of the object is performed by the principle eyes after the spider has made its turn (HOMANN, 1928; CRANE, 1949; KAESTNER, 1950; DREES, 1952). Movement perception is, therefore, monochromatic and specific for green light (see above 3.). Goal-oriented behavior such as prey catching or mating involves color vision and highly developed form recognition. For this purpose the principle eye is equipped with a four-layered scanning retina (see p. 517, LAND, 1969a, b). Three of these four retina layers are wavelength specific, the most proximal is orange sensitive, the next is blue-green, the next, blue sensitive. They are selectively stimulated by the strong chromatic aberration of the lens. The most distal layer could serve for polarization detection in the UV, but in contrast to the other three layers, this has not yet been studied electrophysiologically.

The spider mite, *Tetranychus urticae*, seems to have only UV receptors in the posterior eyes, but UV and green receptors in the anterior eyes. This distribution may be the reason for a higher alignment of phototactic runs in UV (MCENROE and DRONKA, 1966a).

Insect eyes frequently have an uneven distribution of color receptors across the eye. This distribution is correlated to ecological specializations to which the various species are adapted (for details see MENZEL, 1975). Dorsal eye parts or whole dorsal eyes looking upward are equipped preferentially with UV receptors; green receptors dominate the ventral eye regions. Eyes of species that do not rely on food collection from plants or prey catching (like the drone bee) seem to contain green receptors less frequently. Where small objects have to be detected against a background of bright sky highly sensitive long rhabdoms with narrow fields of view and short-wavelength-sensitive receptors are used. Long-wavelength-sensitive eye parts for looking down and sideways are commonly equipped with broad fields of view. There are many specializations of the dioptric apparatus which differ among the eyes of the different insects orders (HORRIDGE, 1977) and it would be interesting to know how these specializations are combined with the spectral properties of the underlying photoreceptors.

II. Color Vision

1. Overview

By contrast to wavelength specificity true color vision is defined as the capacity for wavelength discrimination which is based solely on the perception of the difference in the relative spectral distribution of stimuli, independent of their intensities (or absolute spectral distributions). The first question one may ask is whether an animal is capable of wavelength discrimination, and we will see that the answer to this question is affirmative for many invertebrate species. Secondly, on what type of system is the ability for wavelength discrimination based? This question ranges from the number and types of color receptors, through the strategies of the neuronal processes, to the psychophysical conditions of color vision. Sections B and C above have collected the information on the color receptor types and the neuronal strategies; here we concentrate on the psychophysical data.

Color vision has been proven only in a few invertebrate species though indirect evidence exists for many species. This indirect evidence is based on the existence of more than one color receptor type in the eye, or on more or less anecdotal reports of color preference during flower visiting, host selection, mating behavior, or hiding behavior. Several recent articles have reviewed the extensive literature on color vision in invertebrates, especially of insects, (AUTRUM, 1968; AUTRUM and THOMAS, 1973; MAZOKHIN-PORSHNYAKOW, 1969; GOLDSMITH and BERNARD, 1974; MENZEL, 1975), so I shall provide only a short summary of the main findings and will then concentrate on the bee, the best-studied invertebrate.

A Short Summary. Mya and *Pecten* respond maximally at two wavelengths with a closure of their shells during a sudden decrease of light intensity (HECHT, 1921; CRONY-DILLON, 1966), but this is in no way an indication of color discrimination and belongs in our opinion to the category of wavelength-specific behavior. KÜHN (1933, 1950) found a more colorful pigmentation of *Sepia*'s (Cephalopod) body on a green, yellow, and red background than on an uncolored background. In addition, the striking coloration of *Sepia* during courtship seems biologically meaningful only if *Sepia* has color vision. More careful studies, however, have failed to demonstrate color vision in cephalopods (e.g., *Octopus*, MESSENGER, 1973) and further studies are urgently needed.

Various authors have demonstrated wavelength-specific responses in *Daphnia* (Phyllopoda, Entomostraca) (see p. 553), but again none of these experiments prove color discrimination. Color vision in decapod crustacea has been studied mainly with moving colored-stripes (for literature see AUTRUM and THOMAS, 1973). The experiments had the drawback that contrast borders and hue differences between colored strips and gray stripes may not have been completely excluded (see above). Other observations are not conclusive, e.g., color discrimination in the fiddler crab (HYATT, 1973). From a comparative point of view, the question of color discrimination in crustacea is interesting because the various species live in biotopes with great differences in the chromatic light climate (sea water, fresh water, intertidal, terrestrial) and the evolutionary development of color discrimination could have differed greatly among the species in these biotopes. There is some indication that freshwater species are more likely to have color vision than marine species (see Sec. A), but this is not yet supported by any careful behavioral analyses.

In summary, none of the invertebrate species other than a few insect species and jumping spiders (KAESTNER, 1950) have been studied carefully enough to draw the conclusion that true color vision exists, although the probability of color vision in many other species is high. The major problem is the difficulty in using training techniques in most invertebrates. The problem of color vision in flies is an instructive example. For a long time it was unclear whether flies discriminate color because they showed no response to color contrast in optomotor experiments (see above). Also, spontaneous response patterns such as phototaxis and preferential landing on colored papers (ILSE, 1949) were not used in a way which could clearly prove the existence of color vision. Only recently were flies successfully trained to react to spectral lights and were found able to discriminate between blue and yellow (FUKUSHI, 1976; MENNE and SPATZ, 1977). Since color discrimination is not present in all aspects of visual orientation, the failure or success of demonstrating optomotor responses to wavelength contrast, or selective color mixture effects in phototaxis, does not simultaneously exclude or prove wavelength discrimination in goal-oriented behavior such as prey catching, food collection, courtship behavior, and host selection. In my view the most valuable and important information needed is not only the demonstration of color vision in an animal, but also the determination of both the behavioral contexts in which color vision exists and the informative content of the various wavelength regions.

2. Trichromatic Color Vision of the Honey Bee

VON FRISCH'S (1914) famous training experiments firmly established that bees discriminate colors as food signals. Meanwhile, the trichromaticity of the bee's color vision is supported by electrophysiological studies of the color receptor types in the retina (see p. 533), interneuron recordings (see p. 546) and psychophysical experiments. The psychophysical results are extremely valuable because bees can be presented with very specific situations and they will respond to them by choosing the signal most similar to these to which it was initially trained.

a) Spectral Threshold Function

As well as showing a high UV sensitivity during phototaxis, the bee has a high UV sensitivity in a training situation (KÜHN, 1927; DAUMER, 1956; MENZEL, 1967). VON HELVERSEN (1972) measured the spectral threshold function (see Fig. 12) by training bees to feed at a nonilluminated ground disc. Then the bees were presented with the choice between a weakly illuminated and a nonilluminated disc. By measuring the lowest quantum density at each wavelength where the bee did not choose the illuminated disc a threshold function is obtained. The threshold function has three maxima (340, 440, and 540 nm) and the relative sensitivity values at these maxima are about 16:2.7:1. It has not yet been determined if bees *discriminate* color close to visual threshold.

b) Spectral Discrimination

As one would expect from the $S(\lambda)$ of the UV, blue, and green receptors (p. 533) color discrimination is best in the violet (400 nm) and bluish green (490 nm) region (Fig. 12). Weakly light-adapted bees are able to discriminate wavelengths separated by 7 nm (~400 nm) and 4 nm (~490 nm). Wavelengths below 350 nm and above 550 nm are not discriminable (VON HELVERSEN, 1972).

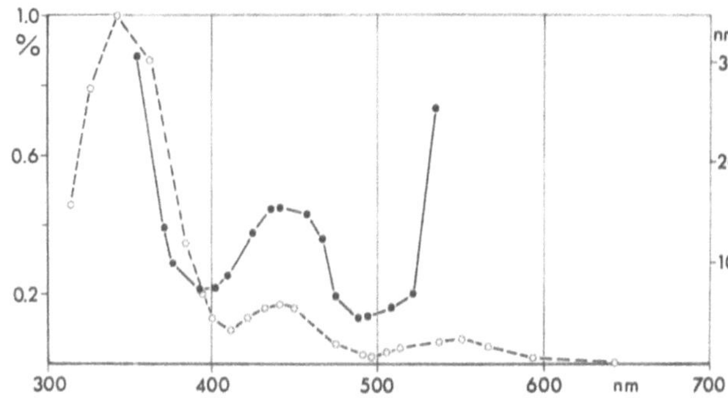

Fig. 12. Spectral threshold function (--○--, left ordinate) and spectral discrimination function (—●—, right ordinate) for the bee (VON HELVERSEN, 1972). The spectral threshold function was measured with one bee at low background light intensity. The spectral discrimination function gives the increment of wavelengths difference for 70% correct responses to the trained color

DAUMER (1956) has also found a wavelength discrimination between 530, 588, and 616 nm. It cannot be ruled out that individual bees may differ in their sets of color receptor types since there are also reports on an additional blue receptor (λ_{max} = 460 nm, AUTRUM and VON ZWEHL, 1964). However, it seems more likely that DAUMER's long wavelengths were discriminated on the basis of differential UV transmission of the interference filters used.

c) *Color Matching Experiments*

DAUMER (1956, 1963) used an apparatus which allowed wavelength mixing with various intensity relations of the wavelength used. The major results are: (cited after VON HELVERSEN, 1967).

A mixture of:	Is matched with:
0.15 (λ360) + 0.85 (white without UV)	= white light (xenon arc)
0.15 (λ360) + 0.85 (λ490)	= white light
0.35 (λ440) + 0.65 (λ588)	= white without UV
0.30 (λ440) + 0.55 (λ588) + 0.15 (λ360)	= white light
0.55 (λ474) + 0.45 (λ530)	= 490
0.50 (λ474) + 0.50 (λ588)	= 490
0.55 (λ474) + 0.45 (λ616)	= 490
0.40 (λ440) + 0.60 (λ530)	= 490
0.35 (λ440) + 0.65 (λ588)	= 490
0.50 (λ616) + 0.50 (λ530)	= 588
0.40 (λ490) + 0.60 (λ616)	= 530
0.50 (λ440) + 0.50 (λ490)	= 474
0.80 (λ360) + 0.20 (λ440)	= 375

A mixture of 0.1 to 0.4 (λ360) and 0.9 to 0.6 (λ588) produces a new quality of light, bee purple, which is discriminated from λ588, λ490, λ474, λ440, λ360. Therefore, complementary colors are: λ490 – λ360, λ360 – (0.65 × λ588 + 0.35 × λ440),

$\lambda440-(0.79\times\lambda588+0.21\times\lambda360)$, $\lambda588-(0.67\times\lambda440-0.33\lambda\,\text{UV})$. The general conclusion is that the trichromatic color vision system of the bee differs from that of man only in a shift of the visible spectrum by about 100 nm to short wavelengths.

d) Simultaneous Color Contrast

Simultaneous color contrast was demonstrated for the first time in an animal by KÜHN (1927) who showed that bees respond to a gray ring on a yellow background as if the ring had a blue color. Recent experiments by NEUMEYER and VON HELVERSEN (1976) quantified this effect and proved that color contrast plays an important role in color detection in the bee. They trained bees to choose a medium blue (or medium yellow) pigment color and tested, which blue (or yellow) the bees chose from a series of blue (or yellow) colors with increasing saturation when the background was blue or yellow. On a yellow background the blue appears more blue to the bee (more saturated) than on a gray background and on a blue background more yellow than on a gray background. Or to put it in another way, blue appears less blue on a blue background and yellow, less yellow on a yellow background. Using monochromatic light they found the sensitivity to be reduced to the wavelength region of the background. The effect of color contrast increases with the width of the background area and decreases with increasing separation of colored background and test field.

These results are best understood if one assumes color-opponent lateral interactions. Recordings in the optic lobes have shown color-opponent interneurons but, firstly, color opponency was not combined with spatial opponency, and secondly, color opponency was found mainly between UV and long wavelengths (see Sec. B.II) whereas NEUMEYER'S and VON HELVERSEN'S experiments refer to a wavelength antagonism of blue and green. Most probably the neuronal circuitry underlying the observed color contrast is located in the lamina and distal medulla, where recordings are difficult to obtain. Preliminary recordings from more distal regions of the medulla have recently demonstrated two additional results of importance in this context: (1) color opponency also exists between blue and green and (2) the receptive field size and structure critically depends on adaptation state (HERTEL, personal comment).

e) An Attempt to Formulate a Color Vision Model for the Bee

Theoretical evaluations of psychophysical data have a long and famous tradition in human color vision (cf. DEVALOIS and ABRAMOV, 1966; RUSHTON, 1972). A graphical approach, the color triangle, was found to be very useful for combining the quantum catch properties of the three receptor types with the intensity-independent location of colors. DAUMER (1956) had already collected enough information to calculate a color triangle (Fig. 13) which gives the three primary colors ($\lambda360$, $\lambda440$, $\lambda588$) and the complementary colors to the primaries. The additional information we have on the $S(\lambda)$ of the color receptor types ($\lambda_{max}=340$, 440, and 540 nm), the spectral threshold function, and the spectral discrimination function allows a more complete description of the bee's color space (Fig. 14) but these models still include uncertainties which can be solved only in new experiments.

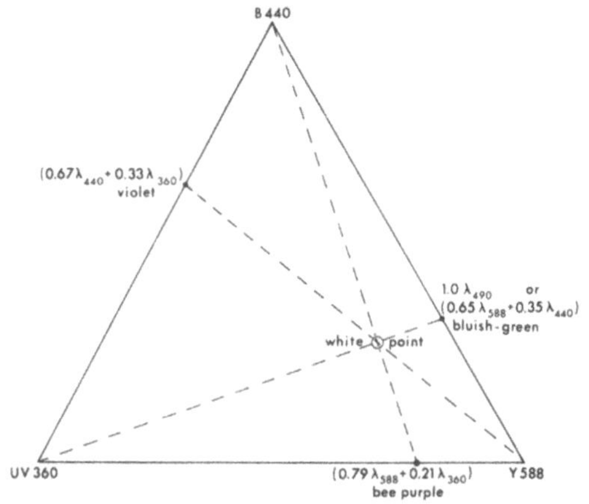

Fig. 13. Color receptor triangle calculated by DAUMER (1956) from his color matching experiments. The white point is the intersection of the line connecting the complementary colors

The uncertainties relate to (1) the existence of secondary sensitivities of the individual color receptors, and (2) the scale relationships of the three primaries. Secondary sensitivities with varying heights have been found in single receptor recordings but it is unknown if, and to what extent, they contribute to the output at the axon endings of the receptors (see MENZEL and BLAKERS, 1976, for discussion). VON HELVERSEN (1970) has modeled the spectral discrimination function using DARTNALL curves as spectral outputs of the receptors and found a divergency from the experimentally determined function, which disappears if one allows secondary sensitivity of the green and/or blue receptor in UV. The uncertainty about the scale relationship of the three color receptor outputs results from our ignorance of their absolute or even relative quantum catch properties. Figure 14 presents just one model, which I think is the most likely one, but many other models are possible with different scale relationships of the three primaries. For example, the deviation of the calculated spectral discrimination function from the measured function could be reduced if one assumes a smaller input signal from the blue receptors.

The model shown here assumes secondary $S(\lambda)$ of the individual color receptor types, because that is what was found in single receptor recordings (Sec. A, p. 533). Certainly, secondary sensitivities are a disadvantage if the receptors are inputs to a color discrimination mechanism only. But in the bee, the three photoreceptor types have to function also as "rods", since, as far as we know, there are no specialized dim-light receptors. Broader $S(\lambda)$ functions and secondary sensitivities increase the absolute sensitivity of receptors. However, it is still inclear whether the secondary sensitivities change in height with the average background light intensity and so adjust to an optimal compromise between absolute sensitivity and spectral discrimination. Furthermore, it is unknown if different eye regions differ in $S(\lambda)$ of their receptors.

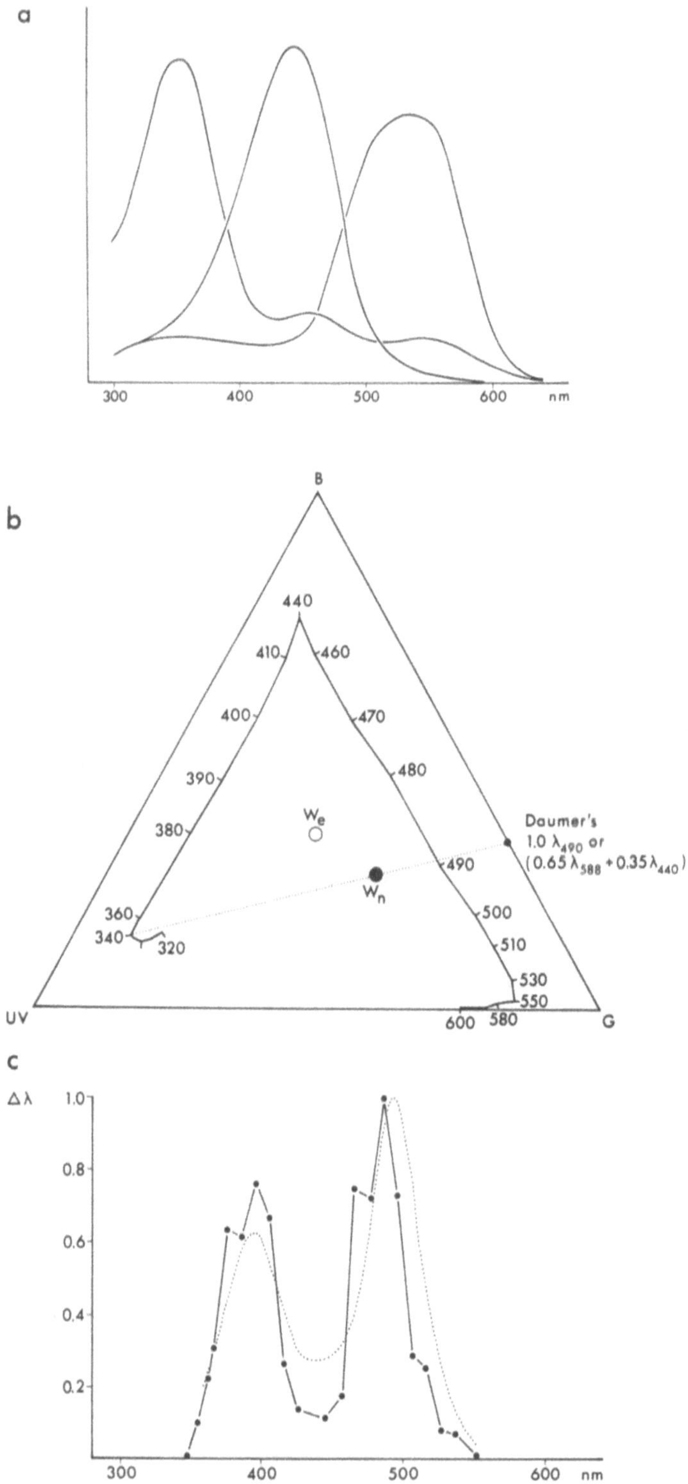
a
300
400
500
600
nm
b
B
440
410
460
400
470
390
480
We
380
Daumer's
1.0 λ_{490} or
(0.65 λ_{588} + 0.35 λ_{440})
490
Wn
360
500
340
320
510
530
550
UV
600
580
G
c
Δλ
1.0
0.8
0.6
0.4
0.2
300
400
500
600
nm

f) Ecological Aspects of Color Vision in the Bee

The bee discriminates colors in the behavioral context of collecting nectar and pollen from flowers and during orientation at the hive entrance, but is color-blind during other behaviors. It is a well-established observation that flowers visited by Hymenoptera and Lepidoptera have predominantly violet, blue, and purple (UV yellow) colors (MÜLLER, 1883; DAUMER, 1958). Violet and blue are also the colors which are learned fastest and remembered longest by the bee (MENZEL, 1967, 1968, see Fig. 11). Many flowers have contrasting patterns in the UV. Spots of nectar secretion are characterized by their UV absorption (DAUMER, 1958) and in many flowers nectar fluoresces a brilliant blue or yellow under UV light (THORP et al., 1975). The combination of UV absorption and blue (or yellow) fluorescence is most likely an effective visual signal for the presence of floral nectar. DAUMER (1958) observed a spontaneous proboscis extension when walking bees crossed from a UV reflection area to a UV-absorbing area (see Fig. 11). THORP et al. (1976) have supporting evidence that bees use the UV-absorbing and longwave fluorescing property of nectar to detect visually the presence of nectar. Coevolution of flowers and their insect pollinators has resulted in high absolute sensitivities and high-wavelength discrimination sensitivity in the violet-blue region. The excellent wavelength discrimination in the bluish green may be an adaptation to the second important food source for bees, the honey dew, produced by aphids (VON HELVERSEN, 1972). However, one should keep in mind that high discriminability in the bluish green may be more important for precise orientation at the hive entrance. The ancestors of our domesticated bee lived in hollow trees, where the hidden entrance was surrounded by greenish and brown color marks.

◀ Fig. 14*a–c*. A model of the color vision system of the bee. (a) gives the $S(\lambda)$ of the UV, blue, and green receptors (*UV, B, G*). The signal from each receptor is scaled so that the area between the curve and the baseline is the same for each. The relative secondary sensitivity for each receptor type is about the same as in Fig. 6. (*b*) Color receptor triangle (RUSHTON, 1972) using the signals from receptors *UV, B, G* in *a* to construct a three dimensional plot. The thick line shows the positions of pure spectral lights of wavelengths indicated. "W_n" is the white point for day light quanta (relationship UV:blue:green = 1:4:8), "W_e" is the white point for light with equal quanta throughout the spectrum. The dotted line connects the two complementary colors, UV and bluish green. (*c*) Spectral discrimination function calculated from the $S(\lambda)$ (*UV, B, G*) given in a (—●—) and experimentally determined spectral discrimination (------) (VON HELVERSEN, 1972, see Fig. 12). Both functions are expressed as relative spectral discrimination sensitivity and normalized to their maximum. Spectral discrimination is calculated from the transform of each $S(\lambda)$. Transform criterium for each wavelength:

$$uv_0 = \frac{UV}{UV+B+G}, \quad b_0 = \frac{B}{UV+B+G}, \quad g_0 = \frac{G}{UV+B+G},$$

where $UV+B+G=1$. By summation of all changes ($|\Delta uv_0|, |\Delta b_0|, |\Delta g_0|$) at intervals of 10 nm: $\Delta\lambda = \Sigma|\Delta uv_0| + |\Delta b_0| + |\Delta g_0|$. (For details see RUSHTON, 1972, the same symbols are used here as by RUSHTON)

E. Conclusion

Many invertebrate species have developed sets of photoreceptors with differing $S(\lambda)$, which serve two different visual orientation mechanisms: wavelength specificity, and color discrimination. Wavelength-specific wiring of receptor inputs enhances the visual contrast and selects wavelength regions for the specific control and release of certain behavioral patterns. On the contrary, color discrimination allows orientation to wavelength contrast independent of intensity contrast. The data collected in this review stress the necessity of studying multichromaticity not only for color vision, but also for wavelength specificity. It is speculated that wavelength specificity could be an early step in the evolution of mechanisms enabling color vision. However, many careful studies especially in lower invertebrates are needed to dismiss or support this view.

F. Acknowledgement

I am most grateful to many friends and collegues who willingly discussed aspects or the general idea of this article with me. It is a honor to name specifically: Drs. J. ERBER, S. LAUGHLIN, O. VON HELVERSEN, J. KIEN, and H. HERTEL.

I am obliged to Ms. M. BLAKERS, Mrs. B. CWIENCZEK, Drs. S. LAUGHLIN and J. KIEN for their critical reading of the manuscript and help with the English, also Mrs. B. CWIENCZEK for her careful help in finishing the manuscript.

References

Abrahamson, E. W., Fager, R. S.: The chemistry of vertebrate and invertebrate visual photoreceptors. Topics Bioenergetics **5**, 125—196 (1973).

Abramov, I.: Retinal mechanisms of colour vision. In: Handbook of Sensory Physiology, Vol. VII/2. Physiology of Photoreceptor Organs, ed. M.G.F. Fuortes, pp. 567—608. Berlin-Heidelberg-New York: Springer 1972.

Adolph, A. R.: Thermal and spectral sensitivities of discrete slow potentials in *Limulus* eye. J. gen. Physiol. **52**, 584—599 (1968).

Alawi, A. A., Jennings, V., Grossfield, J., Pak, W. L.: Phototransduction mutants of *Drosophila melanogaster*. In: The Visual System; Neurophysiology, biophysics and their clinic application, ed. G.B. Arden, pp. 1—21. New York: Plenum Press 1972.

Alkon, D. L., Fuortes, M. G. F.: Responses of photoreceptors in *Hermissenda*. J. gen. Physiol. **60**, 631—649 (1972).

Arnett, D. W.: Spatial and temporal integration properties of units in first optic ganglion of Dipterans. J. Neurophysiol. **35**, 429—444 (1972).

Arora, H. C.: Photodynamic action of light on rotifera. J. Zool. Soc. India **17**, 37—47 (1967).

Arvanitaki, A., Chalazonitis, N.: Réactions bioelectrique neuroniques à la photoactivation spécifique d'une hâme-proteine et d'une caroténe-proteine. Arch. Sci. Physiol. **3**, 27—44 (1949).

Arvanitaki, A., Chalazonitis, N.: Excitatory and inhibitory processes initiated by light and infrared radiations in single identifiable nerve cells (giant ganglion cells in *Aplysia*). In: Nervous Inhibition, ed. E. Florey, pp. 194—231. New York: Pergamon Press 1961.

Autrum, H.: Colour vision in Man and Animals. Naturwissenschaften **55**, 10—18 (1968).

Autrum, H., Kolb, G.: Spektrale Empfindlichkeit einzelner Sehzellen der Aeschniden. Z. vergl. Physiol. **60**, 450—477 (1968).

Autrum, H., Thomas, J.: Comparative physiology of colour vision in animals. In: Handbook of Sensory Physiology, ed. R. Jung, Vol. VII/3A, Central Processing of Visual Information, pp. 661—692. Berlin-Heidelberg-New York: Springer 1973.

Autrum, H., v. Zwehl, V.: Spektrale Empfindlichkeit einzelner Sehzellen des Bienenauges. Z. vergl. Physiol. **48**, 357—384 (1964).

Baumann, F.: Electrophysiological properties of the honey bee retina. In: The Compound Eye and Vision of Insects, ed. G. A. Horridge, pp. 53—74. Oxford: Clarendon Press 1975.

Baylor, E. R.: Diurnal migration of plancton crustaceans. In: Recent Advances in Invertebrate Physiology, ed. B. T. Scheer, p. 21—35. University of Oregon: Eugene Publications 1957.

Bennett, R., Ruck, P.: Spectral sensitivities of dark- and light adapted *Notonecta* compound eyes. J. Insect Physiol. **16**, 83—88 (1970).

Bennett, R., Tunstall, J., Horridge, G. A.: Spectral sensitivity of single retinula cells of the Locust. Z. vergl. Physiol. **55**, 195—206 (1967).

v. Berg, E., Schneider, G.: The spectral sensitivity of the dark adapted eye of *Helix pomatia* L. Vision Res. **12**, 2151—2152 (1972).

Bernard, G. D.: Evidence for visual function of corneal interference filters. J. Insect Physiol. **17**, 2287—2309 (1971).

Bernard, G. D., Miller, W. H.: Interference filters in the cornea of Diptera. Invest. Ophthalmol. **7**, 416—434 (1968).

Bernhard, C. G.: Light transmission and its regulation in the compound eye. Med. Biol. **17**, 100—107 (1967).

Bernhard, C. G., Gemne, G., Møller, A. R.: Modification of specular reflexion and light transmission by biological surface structures. Quart. Rev. Biophys. **1**, 89—105 (1968).

Bernhard, C. G., Gemne, G., Sällström, J.: Comparative ultrastructure of corneal surface topography in insects with aspects on phylogenesis and function. Z. vergl. Physiol. **67**, 1—25 (1970).

Bernhard, C. G., Miller, W. H.: Interference filters in the corneas of Diptera. Invest. Ophthalm. **7**, 416—434 (1968).

Bernhard, C. G., Miller, W. H., Møller, A. R.: Function of the corneal nipples in the compound eyes of insects. Acta. physiol. scand. **58**, 381—382 (1963).

Bernhard, C. G., Miller, W. H., Møller, A. R.: The insect corneal nipple array—a biological broad-band impedance transformer that acts as an antireflection coating. Acta. physiol. scand. **63**, Suppl. 243, 1—79 (1965).

Bertholf, L. M.: Reactions of the honey bee to light. J. Agr. Res. **42**, 379—419 (1931a).

Bertholf, L. M.: The distribution of stimulative efficiency in the ultraviolet spectrum for the honey bee. J. Agr. Res. **43**, 703—713 (1931b).

Bertholf, L. M.: The extend of the spectrum for *Drosophila* and the distribution of stimulative efficiency in it. Z. vergl. Physiol. **18**, 32—64 (1932).

Bertrand, D., Perrelet, A., Baumann, F.: Propriétés physiologiques des cellules pigmentaires de l'œil du faux bourdon. J. Physiol. (Paris) **65**, 102 (1972).

Bishop, L. G.: Spectral response of single neurons recorded in the optic lobes of the housefly and blowfly. Nature (Lond.) **219**, 1372—1373 (1968).

Bishop, L. G.: A search for colour encoding in the responses of a class of fly interneurons. Z. vergl. Physiol. **64**, 355—371 (1969).

Bishop, L. G.: The spectral sensitivity of motion detector units recorded in the optic lobe of the honey bee. Z. vergl. Physiol. **70**, 374—381 (1970).

Bishop, L. G.: A note on the preservation of chromatic information in the lamina of the worker honey bee *(Apis mellifera)*. J. comp. Physiol. **77**, 233—283 (1972).

Bishop, L. G.: An ultraviolet photoreceptor in a dipteran compound eye. J. comp. Physiol. **91**, 267—275 (1974).

Block, G. D., Smith, J. T.: Cerebral photoreceptors in *Aplysia*. Comp. Biochem. Physiol. **46A**, 115—121 (1973).

Boltt, R. E., Ewer, D. W.: Studies on myoneural physiology of Echinodermata: IV The lantern retractor muscle of *Parechinus*, responses to stimulation by light. J. exp. Biol. **40**, 713—726 (1963).

Boolootian, R. A.: Physiology of Echinodermata, 822 p. New York-London: J. Wiley 1967.

Boschek, C. B.: On the structure and synaptic organization of the first optic ganglion in the fly. Z. Naturforsch. **25**, 5 (1970).

Bradshaw, W. E.: Action spectra of photoperiodic response in a diapausing mosquito. Science **175**, 1361—1362 (1972).

Brady, J.: The physiology of insect circadian rhythms. Advanc. Insect Physiol. **10**, 1—115 (1974).

Braitenberg, V. B.: Patterns of projection in the visual system of the fly. I. Retina-lamina projections. Exp. Brain Res. **3**, 271—298 (1967).

Brandenburger, J. L.: Two new kinds of retinal cells in the eye of a snail *Helix aspersa*. J. Ultrastruct. Res. **50**, 216—230 (1975).

Brown, H. M., Ito, H., Ogden, T. E.: Spectral sensitivity of the planarian ocellus. J. gen. Physiol. **51**, 255—260 (1968).

Brown, H. M., Cornwall, M. C.: Spectral correlates of a quasi-stable depolarization in barnacle photoreceptors following red light. J. Physiol. (Lond.) **248**, 555—578 (1975).

Bruckmoser, P.: Die spektrale Empfindlichkeit einzelner Sehzellen des Rückenschwimmers *Notonecta glauca* L. (Heteroptera) Z. vergl. Physiol. **59**, 187—204 (1968).

Bruno, M. S., Goldsmith, T. H.: Rhodopsin of the blue crab *Callinectes*: Evidence for absorption differences in vitro and in vivo. Vision Res. **14**, 653—ff. (1974).

Bruno, M. S., Kennedy, D.: Spectral sensitivity of photoreceptor neurons in the sixth ganglion of the crayfish. Comp. Biochem. Physiol. **6**, 41—46 (1962).

Bruno, M. S., Mote, M. J., Goldsmith, T. H.: Spectral absorption and sensitivity measurements in single ommatidia of the green crab, *Carcinus*. J. comp. Physiol. **82**, 151—163 (1973).

v. Buddenbrock, W.: Einige Bemerkungen zum augenblicklichen Stand der Frage nach dem Farbensinn der Tiere. Zool. Anz. **84**, 189—201 (1929).

Bullock, Th. H., Horridge, G. A.: Structure and function in the nervous system of invertebrates. San Francisco-London: W. H. Freeman and Co. 1965.

Burkhardt, D.: Spectral sensitivity and other response characteristics of single visual cells in the arthropod eye. Symp. Soc. exp. Biol. **16**, 86—108 (1962).

Burkhardt, D.: Colour discrimination in insects. Advanc. Insect Physiol. **2**, 131—203 (1964).

Burkhardt, D.: Electrophysiological studies on the compound eye of a stalked-eye fly, *Cyrtodiopsis dalmanni* (Diopsidae, Diptera). J. comp. Physiol. **81**, 203—214 (1972).

Burkhardt, D., de la Motte, J.: Electrophysiological studies on the eyes of Diptera, Mecoptera and Hymenoptera. In: Information Processing in the Visual Systems of Arthropods, ed. R. Wehner, pp. 148—153. Berlin-Heidelberg-New York: Springer 1972.

Burr, A. H., Burr, C.: The amphid of the Nematode *Oncholaimus vesicarius*: Ultrastructural evidence for a dual function as chemoreceptor and photoreceptor. J. Ultrastructure Res. **51**, 1—15 (1975).

Butler, R.: The identification and mapping of spectral cell types in the retina of *Periplaneta americana*. Z. vergl. Physiol. **72**, 67—80 (1971).

Carricaburu, P., Chardenot, P.: Spectres d'absorption de la cornée de quelques arthropodes. Vision Res. **7**, 43—50 (1967).

Carlson, S. D., Philipson, B.: Microspectrophotometry of the dioptric apparatus and compound rhabdom of the moth *Manduca sexta* eye. J. Insect Physiol. **18**, 1721—1731 (1972).

Chapman, R. M., Lall, A. B.: Electroretinogram characteristics and the spectral mechanisms of the median ocellus and the lateral eye in *Limulus polyphemus*. J. gen. Physiol. **50**, 2267—2287 (1967).

Chappell, R. L., De Voe, R. D.: Action spectra and chromatic mechanisms of cells in the median ocelli of dragonflies. J. gen. Physiol. **65**, 399—419 (1975).

Checcucci, A.: Molecular sensory physiology of *Euglena*. Naturwissenschaften **63**, 412—417 (1976).

Checcucci, A., Colombetti, R., Ferrara, R., Lenci, F.: Action spectra for photoaccumulation of green and colourless *Euglena*: Evidence for identification of receptor pigments. Photochem. Photobiol. **23**, 51—54 (1976).

Chi, C., Carlson, S. D.: Close apposition of photoreceptor cell axons in the house fly. J. Insect Physiol. **22**, 1153—1157 (1976).

Clark, E. D., Kimeldorf, D. J.: Tentacle response of the sea anemone *Anthopleura xanthogramma* to UV and visible radiations. Nature (Lond.) **227**, 856—857 (1970).

Clayton, R. K.: Phototaxis in microorganisms. In: Photophysiology, ed. A. C. Giese, Vol. II, pp. 50—77. New York: Academic Press 1964.

Clayton, R. K.: Light and living matter: A guide to the study of photobiology. Vol. 2; The biological part. New York: McGraw Hill, 1971.

Clement, P.: Les influences de la lumière sur les Rotifères. Thèse Doctorat Etat, Université Lyon, France (1976).

Cole, W. H.: Circus movement of *Limulus* and the tropism theory. J. gen. Physiol. **5**, 417—426 (1922).

Cole, W. H.: The effect of laboratory age upon the phototrophic reaction of *Limulus*. J. gen. Physiol. **6**, 295—297 (1923).

Condit, H. R., Grum, F.: Spectral energy distribution of daylight. J. Opt. Soc. Amer. **54**, 937—944 (1964).

Crane, J.: Comparative biology of salticid spiders at Rancho Grande, Venezuela Part IV. An analysis of display. Zoologica **34**, 159—214 (1949).

Croll, N. A.: The phototactic response and the spectral sensitivity of *Chromadorina viridis* (Nematoda, Chromadorina) with a note on the nature of the paired pigment spots. Nematologica (Leiden), **12**, 610—614 (1966).

Crony-Dillon, J. P.: Spectral sensitivity of the scallop *Pecten maximus*. Science **151**, 345—346 (1966).

Daan, N., Ringelberg, J.: Further studies on the positive and negative phototactic reaction of *Daphnia magna* Straus. Netherland. J. Zool. **19**, 525—540 (1969).

Daumer, K.: Reizmetrische Untersuchungen des Farbensehens der Bienen. Z. vergl. Physiol. **38**, 413—478 (1956).

Daumer, K.: Blumenfarben, wie sie die Bienen sehen. Z. vergl. Physiol. **41**, 49—110 (1958).

Daumer, K.: Kontrastempfindlichkeit der Bienen für „weiß" verschiedenen UV-Gehalts. Z. vergl. Physiol. **46**, 321—335 (1963).

Dembrowski, J.: On conditioned reactions of *Paramecium caudatum* towards light. Acta biol. exp. **15**, 5—17 (1950).

Dennis, M. J.: Electrophysiology of the visual system in a nudibranch mollusc. J. Neurophysiol. **30**, 1439—1465 (1967).

Detwiler, P. B.: Multiple light-evoked conductance changes in the photoreceptors of *Hermissenda crassicornis*. J. Physiol. (Lond.) **256**, 691—708 (1976).

DeValois, R. L.: Central mechanisms of colour vision. In: Handbook of Sensory Physiology, ed. R. Jung, pp. 209—254, Vol. VII/3 A) Central processing of visual information. Berlin-Heidelberg-New York: Springer 1973.

DeValois, R. L., Abramov, J.: Colour Vision. Ann. Rev. Psychol. **17**, 337—362 (1966).

DeVoe, R. D.: Dual sensitivity of cells in the wolf spider eyes at ultraviolet and visible wavelength of light. J. gen. Physiol. **59**, 247—269 (1972).

DeVoe, R. D.: Ultraviolet and green receptors in principal eyes of jumping spiders. J. gen. Physiol. **66**, 193—ff. (1975).

DeVoe, R. D., Small, R. J. W., Zvargulis, J. E.: Spectral sensitivities of wolf spider eyes. J. gen. Physiol. **54**, 1—32 (1969).

Diehn, B.: Two perpendicularly oriented pigment systems involved in phototaxis of *Euglena*. Nature (Lond.) **221**, 366—367 (1969a).

Diehn, B.: Action spectra of the phototactic responses in *Euglena*. Biochim. biophys. Acta (Amst.) **177**, 136—143 (1969b).

Diehn, B.: Photomovement of microorganisms. Photochem. Photobiol. **23**, 455—456 (1976).

Dörrscheidt-Käfer, M.: Die Empfindlichkeit einzelner Photorezeptoren im Komplexauge von *Calliphora erythrocephala*. J. comp. Physiol. **81**, 309—340 (1972).

Donner, K. O.: On vision in *Pontoporeia affinis* and *P. femorata* (Crustacea, Amphipoda). Soc. sci. fenn. Comment. biol. **32**, 1—17 (1971).

Dowling, J. E., Chappell, R. L.: Neural organization of the median ocellus of the dragonfly II. Synaptic structure. J. gen. Physiol. **60**, 148 (1972).

Drachev, L. A., Kaulen, A. D., Ostroumov, S. A., Skulachev, V. P.: Electrogenesis by bacteriorhodopsin incorporated in a planar phospolipid membrane. FEBS Letters **39**, 43—45 (1974).

Drees,O.: Untersuchungen über die angeborenen Verhaltensweisen bei Springspinnen (Salticidae). Z. Tierpsychol. **9**, 169—207 (1952).

Duelli,P., Wehner,R.: The spectral sensitivity of polarized light orientation in *Cataglyphis bicolor* (Formicidae, Hymenoptera). J. comp. Physiol. **86**, 37—53 (1973).

Eakin,R.M., Westfall,J.A.: Ultrastructure of the eye of the Rotifer *Asplanchna brightwelli*. J. Ultrastruct. Res. **12**, 46—62 (1965).

Eaton,J.L.: Spectral sensitivity of the ocelli of the adult cabbage looper moth, *Trichoplusia ni*. J. comp. Physiol. **109**, 17—24 (1976).

Eaton,P.B., Brown,C.M.: Photoreception in the nauplius eye of *Pandalus borealis* Kröger: Decapoda, Crustacea. Canad. J. Zool. **48**, 119—121 (1970).

Eckert,M.: Die spektrale Empfindlichkeit des Komplexauges von *Musca* (Bestimmungen aus Messungen der optomotorischen Reaktion). Kybernetik **9**, 145—156 (1971).

Edrich,W.: Die Rolle einzelner Farbrezeptortypen bei den verschiedenen Lichtreaktionen der Biene. Verh. dtsch. Zool. Ges. **70**, 236 (1977).

Eguchi,E.: Fine structure and spectral sensitivities of retinular cells in the dorsal sector of the compound eyes in the dragonfly. Z. vergl. Physiol. **71**, 201—218 (1971).

Eguchi,E., Waterman,T.H., Akiyma,J.: Localization of the violet and yellow receptor cells in the crayfish retinula. J. gen. Physiol. **62**, 355—375 (1973).

Erber,J., Menzel,R.: Visual interneurons in the mediam photocerebrum in the bee. J. comp. Physiol. **121**, 65—77 (1977)

Feinleib,M.E., Curry,G.M.: The nature of the photoreceptor in phototaxis. In: Handbook of Sensory Physiology, ed. W.R.Loewenstein, Vol. I, pp. 366—395. Principles in Receptor Physiology. Berlin-Heidelberg-New York: Springer 1971.

Fernandez,H.R.: Spectral sensitivity and visual pigment of the compound eye of the galatheid crab *Pleuroncodes planipes*. Marine Biol. **20**, 148—153 (1973).

Fingerman,M., Brown,F.A.: A "Purkinje-shift" in insect vision. Science **116**, 171—172 (1952).

Fleißner,G.: Untersuchungen zur Sehphysiologie der Skorpione. Verh. dtsch. Zool. Ges. Innsbruck, 375—380 (1968).

Fleißner,G.: A new biological function of the scorpion's lateral eyes as receptors of Zeitgeber stimuli. Proc. 6th int. Arachn. Congr. 1974 (1975).

Folger,H.T.: A quantitative study of reactions to light in *Amoeba*. J. exp. Zool. **41**, 261—291 (1925).

Fraenkel,G.S., Gunn,D.L.: The orientation of animals. New York: Cover Publ. Inc. 1961.

Franceschini,N.: Sampling of the visual environment by the compound eye of the fly: Fundamentals and applications. In: Photoreceptor Optics, eds. A.Snyder, R.Menzel, pp. 98—125. Berlin-Heidelberg-New York: Springer 1975.

Friederichs,H.F.: Beiträge zur Morphologie und Physiologie der Sehorgane der Cicindelliden (Col.). Z. Morph. Ökol. Tiere **21**, 1—172 (1931).

Frisch,K.von: Der Farbensinn und Formensinn der Biene. Zool. J. Physiol. **37**, 1—238 (1914).

Frisch,K.von: Tanzsprache und Orientierung der Bienen. Berlin-Heidelberg-New York: Springer 1965.

Frisch,K.von, Kupelwieser,E.M.: Über den Einfluß der Lichtfarbe auf die phototaktische Reaktion niederer Krebse. Biol. Zentralbl. **33**, 517—552 (1913).

Friza,F.: Zur Frage der Färbung und Zeichnung des facettierten Insektenauges. Z. vergl. Physiol. **8**, 289—336 (1929).

Fukushi,T.: Classical conditioning to visual stimuli in the housefly, *Musca domestica*. J. Insect Physiol. **22**, 361—364 (1976).

Geethabali,S., Pampapathi Rao,K.: A metasomatic neural photoreceptor in the scorpion. J. exp. Biol. **58**, 189—196 (1973).

Gillary,H.L.: Light-evoked electrical potentials from the eye and optic nerve of *Strombus*: Response waveform and spectral sensitivity. J. exp. Biol. **60**, 383—396 (1974).

Gillary,H.L., Wolbarsht,L.M.: Electrical responses from the eye of a snail. Rev. canad. Biol. **26**, 125—134 (1967).

Gogala,M.: Die spektrale Empfindlichkeit der Doppelaugen von *Ascalaphus macaronius* Scop. (Neuroptera, Ascalaphidae). Z. vergl. Physiol. **57**, 232—243 (1967).

Goldsmith,T.H.: The colour vision of insects. In: Light and Life, eds. W.D.McElroy, B.Glass, pp. 771—794. Baltimore: Johns Hopkins Press, 1961.

Goldsmith, T. H.: The natural history of invertebrate visual pigments. In: Handbook of Sensory Physiology, ed. H. J. A. Dartnall, Vol. VII/1, pp. 685—719, Photochemistry of Vision. Berlin-Heidelberg-New York: Springer 1972.

Goldsmith, T. H., Bernard, G. D.: The visual system of insects (Chapter 5) In: Physiology of Insecta, ed. M. Rockstein, pp. 175—271, New York-London: Academic Press 1974.

Goldsmith, T. H., Fernandez, H. R.: Comparative studies of crustacean spectral sensitivity. Z. vergl. Physiol. **60**, 156—175 (1968).

Goldstein, E. B., Williams, T. P.: Calculated effects of "screening pigments". Vision Res. **6**, 39—50 (1966).

Goodman, L. J.: The structure and function of the insect dorsal ocellus. Advanc. Insect Physiol. **9**, 97—195 (1970).

Gorman, A. L. F., McReynolds, J. S.: Hyperpolarizing and depolarizing receptor potentials in the scallop eye. Science **165**, 309—310 (1969).

Gould, J. L.: Honey bee recruitment: The dance-language controverse. Science **189**, 685—693 (1975).

Gouras, P.: *S*-Potentials. In: Handbook of Sensory Physiology, ed. M. G. F. Fuortes, Vol. VII/2. Physiology of photoreceptor organs, pp. 513—530. Berlin-Heidelberg-New York: Springer 1972.

Grabowski, U.: Experimentelle Untersuchungen über das angebliche Lernvermögen von *Paramecium*. Z. Tierpsychol. **2**, 265—282 (1939).

Graham, C. H., Hartline, H. K.: The response of single visual sense cells to lights of different wavelengths. J. gen. Physiol. **18**, 917—931 (1935).

Gribakin, F. G.: Cellular basis of colour vision in the honey bee. Nature (Lond.) **233**, 639—641 (1969).

Gribakin, F. G.: The distribution of the long wave photoreceptors in the compound eye of the honey bee as revealed by selective osmic staining. Vision Res. **12**, 1225—1230 (1972).

Gross, A.: The reactions of arthropods to monochromatic lights of equal intensities. J. exp. Zool. **14**, 467—513 (1913).

Hagiwara, S., Morita, H.: Electrotonic transmission between two nerve cells in the leech ganglion. J. Neurophysiol. **25**, 721—731 (1962).

Hairston, N. G.: Photoprotection by carotenoid pigments in copepod *Diaptomus nevadensis*. Proc. nat. Acad. Sci. USA **73**, 971 (1976).

Hamasaki, D. J.: The ERG-determined spectral sensitivity of the Octopus. Vision Res. **8**, 1013—1021 (1968).

Hamdorf, K., Gogala, M.: Photoregeneration und Bereichseinstellung der Empfindlichkeit beim UV-Rezeptor. J. comp. Physiol. **86**, 231—245 (1973).

Hamdorf, K., Höglund, G., Langer, H.: Mikrophotometrische Untersuchungen an der Retinula des Nachtschmetterlings *Deilephila elpenor*. Verh. dtsch. Zool. Ges. **65**, 276—281 (1972).

Hamdorf, K., Schwemer, J.: Photoregeneration and adaptation process in insect photoreceptors. In: Photoreceptor Optics, eds. A. W. Snyder, R. Menzel, pp. 263—289. Berlin-Heidelberg-New York: Springer 1975.

Hamdorf, K., Schwemer, J., Täuber, K.: Der Sehfarbstoff, die Absorption der Rezeptoren und die spektrale Empfindlichkeit der Retina von *Eledone moschata*. Z. vergl. Physiol. **60**, 375—415 (1968).

Hand, W. G.: Photostatic orientation by the marine dinoflagellate *Gyrodinium dorsum*. I. A mechanism model. J. exp. Zool. **174**, 33—38 (1970).

Hand, W. G., Forward, R., Davenport, D.: Short term photic regulation of a receptor mechanism in a dinoflagellate. Biol. Bull. **133**, 150—165 (1967).

Hara, T., Hara, R.: Cephalopod retinochrome. In: Handbook of Sensory Physiology, ed. H. J. A. Dartnall, Vol. VII/1, pp. 720—746. Berlin-Heidelberg-New York: Springer 1972.

Haragsim, O., Pospisil, J.: Photopositive response of field-bees and drones to monochromatic light. Acta ent. bohemoslov. **66**, 1—8 (1969).

Harris, W. A., Stark, W., Walker, J. A.: Genetic dissection of the photoreceptor system in the compound eye of *Drosophila melanogaster*. J. Physiol. (Lond.) **256**, 415—439 (1976).

Hartline, H. K., Graham, C. H.: Nerve impulses from single receptors in the eye. J. cell. comp. Physiol. **1**, 277—295 (1932).

Hartline, H. K., Wagner, G., MacNichol, E. F.: The peripheral origin of nervous activity in the visual system. Cold Spr. Harb. Symp. quant. Biol. **17**, 125—141 (1952).

Hasselmann, E. M.: Über die relative spektrale Empfindlichkeit von Käfer- und Schmetterlingsaugen bei verschiedenen Helligkeiten. Zool. J. Physiol. **69**, 537—576 (1962).

Heberdey, R. F.: Der Farbensinn helladaptierter Daphnien. Biol. Zentralbl. **56**, 207—319 (1936).

Heberdey, R. F.: Das Unterscheidungsvermögen von *Daphnia* für Helligkeiten farbiger Lichter. Z. vergl. Physiol. **31**, (1), 89—111 (1948).

Hecht, S.: The relation between the wavelength of light and its effect on the photosensory process. J. gen. Physiol. **3**, 375—390 (1921).

Heintz, E.: La question de la sensibilité des abeilles à l'ultraviolet. Insect Soc. **6**, 223—229 (1959).

Heisenberg, M.: Comparative behavioural studies on two visual mutants of *Drosophila*. J. comp. Physiol. **80**, 119—136 (1972).

Heisenberg, M., Buchner, E.: The role of retinula cell types in visual behaviour of *Drosophila melanogaster*. J. comp. Physiol. **117**, 127—162 (1977).

v. Helversen, O.: Theoretische Untersuchungen zum Farbensehen der Insekten. Diploma-Thesis, Universität Freiburg (1967).

v. Helversen, O.: Untersuchungen zum Farbensehen der Honigbiene. Dissertation, Universität Freiburg (1970).

v. Helversen, O.: Zur spektralen Unterschiedsempfindlichkeit der Honigbiene. J. comp. Physiol. **80**, 439—472 (1972).

v. Helversen, O., Edrich, W.: Der Polarisationsempfänger im Bienenauge: Ein Ultraviolettrezeptor. J. comp. Physiol. **94**, 33—47 (1974).

Hertel, H.: Analyse der spektralen Empfindlichkeit von *Asplanchna priodonta* mit Hilfe der phototaktischen Reaktion. Diploma Thesis, Techn. Hochschule Darmstadt (1972).

Hertel, H.: Aspekte zur Photorezeption von *Artemia salina*. Verh. dtsch. Ges. Zool. Erlangen (1977).

Hess, C.: Untersuchungen über den Lichtsinn bei Echinodermen. Pflüg. Arch. ges. Physiol. **160**, 1—26 (1915).

Hildebrand, E.: Untersuchungen zur Wirkung ultravioletten Lichts auf die elektrischen Eigenschaften der Zellmembran einiger Protozoen. Dissertation Techn. Hochschule Aachen (1969).

Hildebrand, E.: Elektrophysiologische Untersuchungen der Rezeptoreigenschaften von Protozoen. Verh. dtsch. Zool. Ges. (64. Tag) 182—186 (1970).

Hildebrand, E.: What does *Halobacterium* tell us about photoreception? Biophys. Struct. Mech. **3**, 69—77 (1977).

Hildebrand, E., Dencher, N.: Photophobische Reaktionen von *Halobacterium halobium* und ihre Beziehung zum Retinal-Proteinkomplex der Zellmembran. Ber. dtsch. Botan. Ges. **87**, 93—99 (1974).

Hildebrand, E., Dencher, N.: Two photosystems controlling behavioural responses of *Halobacterium halobium*. Nature (Lond.) **257**, 46—48 (1975).

Hillman, P., Hochstein, S., Minke, B.: A visual pigment with two physiologically active stable states. Science **175**, 1486—1488 (1972).

Hillman, P., Hochstein, S., Dodge, T. A., Knight, B. W., Minke, B.: Rapid dark recovery of the invertebrate early receptor potential. J. gen. Physiol. **62**, 77—86 (1973).

Hisano, N., Tateda, H., Kuwabara, M.: Photosensitive neurons in the marine pulmonate mollusc *Onchidium verruculatum*. J. exp. Biol. **57**, 651—660 (1972a).

Hisano, N., Tateda, H., Kuwabara, M.: An electrophysiological study of the photoexcitative neurons of *Onchidium verruculatum* in situ. J. exp. Biol. **57**, 661—672 (1972b).

Hitchcock, L. jr.: Colour sensitivity of the *Amoeba* revisited. J. Protozool. **8**, 322—324 (1961).

Hochstein, S., Minke, B., Hillman, P.: Antagonistic components of the late receptor potential in the barnacle photoreceptor arising from different stages of the pigment process. J. gen. Physiol. **62**, 105—128 (1973).

Homann, H.: Beitrag zur Physiologie der Spinnenaugen, I und II. Z. vergl. Physiol. **7**, 201—268 (1928).

Horridge, G. A.: Perception of polarization plane, colour and movement in two dimension by the crab *Carcinus*. Z. vergl. Physiol. **55**, 207—224 (1967).

Horridge, G. A.: Unit studies on the retina of dragonflies. Z. vergl. Physiol. **62**, 1—37 (1969).

Horridge, G. A.: The compound eye of insects. Sci. Amer. **237**, 108—121 (1977).

Horridge, G. A., Mimura, K.: Fly photoreceptors I. Physical separation of two visual pigments in *Calliphora* retinula cells 1—6. Proc. roy. Soc. B **190**, 211—224 (1975).

Horridge, G. A., Mimura, K., Tsukahara, Y.: Fly photoreceptors. Spectral and polarized light sensitivity in the drone fly *Eristalis tenax*. Proc. roy. Soc. B **190**, 225—237 (1975).

Howell, C. D.: The response to light in the earthworm *Pheretima agrestis*, Goto and Hatai, with special reference to the function of the nervous system. J. exp. Zool. **81**, 231—259 (1939).

Hu, K. G., Stark, W. S.: Specific receptor input into spectral preference in *Drosophila*. J. comp. Physiol. **121**, 241—252 (1977).
Hubbard, R., Wald, G.: Visual pigment of horseshoe crab *Limulus polyphemus*. Nature (Lond.) **186**, 212—ff. (1960).
Hughes, H. P. J.: The spectral sensitivity and absolute threshold of *Onchidorus fusca*. J. exp. Biol. **52**, 609—618 (1970).
Hyatt, G. W.: Physiological and behavioural evidence for colour discrimination by fiddler crabs. In: Physiological Ecology of Estuarine Organisms, ed. J. Vernberg. University of South Carolina Press, 1973.
Ilse, D.: Colour discrimination in the dronefly *Eristalis tenax*. Nature (Lond.) **163**, 255—256 (1949).
Jander, R., Barry, C. K.: Die phototaktische Gegenkopplung von Stirnocellen und Facettenaugen in der Phototaxis der Heuschrecken und Grillen. Z. vergl. Physiol. **57**, 432—458 (1968).
Jennings, H. S.: Behaviour of the lower organisms. New York: Columbia University Press 1915.
Jung, R.: Visual perception and neurophysiology. In: Handbook of Sensory Physiology, ed. R. Jung, Vol. VII/3, pp. 1—152. Berlin-Heidelberg-New York: Springer 1973.
Kaestner, A.: Reaktion der Hüpfspinnen (Salticidae) auf unbewegte farblose und farbige Gesichtsreize. Zool. Beitr. **1**, 13—50 (1950).
Kaiser, W.: Zur Frage des Unterscheidungsvermögens für Spektralfarben: Eine Untersuchung der Optomotorik der königlichen Glanzfliege *Phormia regina* M. Z. vergl. Physiol. **61**, 71—102 (1968).
Kaiser, W.: A preliminary report on the analysis of the optomotor system of the honey bee. In: Information processing in the visual system of arthropods, ed. R. Wehner, pp. 163—170. Berlin-Heidelberg-New York: Springer 1972.
Kaiser, W.: The spectral sensitivity of the honey bee's optomotor walking response. J. comp. Physiol. **90**, 405—408 (1974b).
Kaiser, W.: The relationship between visual movement detection and colour vision in insects. In: The compound eye and vision of insects, ed. G. A. Horridge. Oxford: Clarendon Press 1975.
Kaiser, W., Bishop, L. G.: Directionally selective motion detecting units in the optic lobe of the honey bee. Z. vergl. Physiol. **67**, 403—413 (1970).
Kaiser, W., Liske, E.: Optomotor reactions of stationary flying bees during stimulation with spectral light. J. comp. Physiol. **89**, 391—408 (1974).
Kaiser, W., Seidl, R., Vollmar, J.: Spectral sensitivities of behavioural patterns in honey bees. J. Comp. Physiol. **122**, 27—44 (1977).
Kennedy, D.: Neural photoreceptors in a lamellibranch mollusc. J. gen. Physiol. **44**, 551—572 (1960).
Kennedy, D.: The photoreceptor process in lower animals. In: Photophysiology, ed. A. C. Giese, Vol. II, pp. 79—121. New York-London: Academic Press 1964.
Kien, J., Menzel, R.: Chromatic properties of interneurons in the optic lobes of the bee. I. J. comp. Physiol. **113**, 17—34 (1977a).
Kien, J., Menzel, R.: Chromatic properties of interneurons in the optic lobes of the bee. II. J. comp. Physiol. **113**, 35—53 (1977b).
Kirschfeld, K.: The *Musca* compound eye. Symp. Zool. Soc. Lond. **23**, 165—166 (1968).
Kirschfeld, K.: Das neurale Superpositionsauge. Fortschr. d. Zool. **21**, 229—257 (1973a).
Kirschfeld, K.: Vision of polarized light. Intern. Biophysic Congress, USSR Pushchino (1973b).
Kirschfeld, K.: Optomotor responses of the bee to moving "polarization patterns". Z. Naturforsch. **28c**, 329—338 (1973c).
Kirschfeld, K., Franceschini, N.: Photostable pigments within the membrane of photoreceptors. Biophys. Struct. Mech. **3**, 191—194 (1977).
Kirschfeld, K., Lindauer, M., Martin, H.: Problems of menotactic orientation according to the polarized light of the sky. Z. Naturforsch. **30c**, 88—90 (1975).
Kirschfeld, K., Lutz, B.: Lateral inhibition in the compound eye of the fly *Musca*. Z. Naturforsch. **29c**, 95—97 (1974).
Kirschfeld, K., Lutz, B.: The spectral sensitivity of the ocelli of *Calliphora* (Diptera). Z. Naturforsch. **32c**, 439—441 (1977).
Kirschfeld, K., Reichardt, W.: Optomotorische Versuche an *Musca* mit linear polarisiertem Licht. Z. Naturforsch. **25b**, 228 (1970).
Koehler, O.: Sinnesphysiologische Untersuchungen an Libellen-Larven. Verh. dtsch. Zool. Ges., Göttingen **29**, 83—91 (1924).

Kolb, G., Autrum, H.: Selektive Adaptation und Pigmentwanderung in den Sehzellen des Bienenauges. J. comp. Physiol. **94**, 1—6 (1974).

Kolb, G., Autrum, H., Eguchi, E.: Die spektrale Transmission des dioptrischen Apparates von *Aeschna cyanea* Müll. Z. vergl. Physiol. **63**, 434—439 (1969).

Kretz, J. R., Stent, G. S., Kristan, W. B.: Photosensory input pathways in the medicinal leech. J. comp. Physiol. **106**, 1—37 (1976).

Kühn, A.: Über den Farbensinn der Bienen. Z. vergl. Physiol. **5**, 762—800 (1927).

Kühn, A.: Über Farbensinn und Anpassung der Körperfarbe an die Umgebung bei Tintenfischen. Nachr. Ges. Wiss. Göttingen Math. phys. Kl. Biol. **1** (1933).

Kühn, A.: Über Farbwechsel und Farbensinn von Cephalopoden. Z. vergl. Physiol. **32**, 572—598 (1950).

Labhart, T.: Behavioural analysis of light intensity discrimination and spectral sensitivity in honey bee *Apis mellifera*. J. comp. Physiol. **95**, 203—216 (1974).

Lall, A. B., Chapman, R. M.: Phototaxis in *Limulus* under natural conditions: Evidence for reception of near-ultraviolet light in the median dorsal ocellus. J. exp. Biol. **58**, 213—224 (1973).

Land, M. F.: Activity in the optic nerve of *Pecten maximus* in response to changes in light intensity, and to pattern and movement in the optical environment. J. exp. Biol. **45**, 83—99 (1966).

Land, M. F.: Structure of the retina of the principle eyes of jumping spiders (Salticidae, Dendryphantinae) in relation to visual optics. J. exp. Biol. **51**, 443—470 (1969a).

Land, M. F.: Movement of the retina of jumping spiders (Salticidae, Dendryphantinae) in response to visual stimuli. J. exp. Biol. **51**, 471—493 (1969b).

Land, M. F.: Orientation by jumping spiders in the absence of visual feedback. J. exp. Biol. **54**, 119—139 (1971).

Langer, H.: Properties and functions of screening pigments in the insect eye. In: Photoreceptor Optics, eds. A. W. Snyder, R. Menzel, pp. 429—458. Berlin-Heidelberg-New York: Springer 1975.

Larimer, J. L., Trevino, D. L., Ashby, E. A.: A comparison of spectral sensitivities of caudal photoreceptors of epigeal and cavernicolous crayfish. Comp. Biochem. Physiol. **19**, 404—415 (1966).

Lasansky, A.: Cell junctions in ommatidia of *Limulus*. J. cell. Biol. **33**, 365—383 (1967).

Laughlin, S. B.: Receptor function in the apposition eye: An electrophysiological approach. In: Photoreceptor Optics, eds. A. W. Snyder, R. Menzel, pp. 479—498. Berlin-Heidelberg-New York: Springer Verlag 1975.

Laughlin, S. B.: Neural integration in the first optic neuropile of the dragonfly. IV. Interneuron spectral sensitivity and contrast coding. J. comp. Physiol. **112**, 199—212 (1976a).

Laughlin, S. B.: The sensitivities of dragonfly photoreceptors and the voltage gain of transduction. J. comp. Physiol. **111**, 221—247 (1976b).

Laverack, M. S.: Mechanoreceptors, photoreceptors and rapid conduction pathways in the leech *Hirudo medicinalis*. J. exp. Biol. **50**, 129—140 (1969).

Lichey, M. E., Block, G. D., Hudson, D. J., Smith, J. T.: Circadian oscillators and photoreceptors in the gastropod *Aplysia*. Photochem. Photobiol. **23**, 253—273 (1976).

Liebman, P. A.: Birefringence, dichroism and rod outer segment structure. In: Photoreceptor Optics, eds. A. W. Snyder, R. Menzel, pp. 199—214. Berlin-Heidelberg-New York: Springer 1975.

Lüdtke, H.: Dunkeladaptation und Verschiebung der Helligkeitswerte im Auge von *Notonecta glauca*. Z. Naturforsch. **9**, 159—163 (1954).

Lukowiak, K., Jacklet, J. W.: Habituation: A peripheral and central nervous system process in *Aplysia*. Fed. Proc. **31**, 405—412 (1972).

Machan, L.: Studies on structure, electroretinogram, and spectral sensitivity of the lateral and median eyes of the scorpion. Doct. Diss. Univ. Wisconsin, Madison 1966.

Machan, L.: Spectral sensitivity of the scorpion eyes and the possible role of shielding pigment effect. J. exp. Biol. **49**, 95—105 (1968).

Machemer, H.: Versuche zur Frage der Dressierbarkeit hypotricher Ciliaten. Z. Tierpsychol. **23**, 641—654 (1967).

Marriott, F. H. C.: The absolute light sensitivity and spectral threshold curve of aquatic flatworm *Dendrocoelum lacteum*. J. Physiol. (Lond.) **143**, 369—379 (1958).

Mast, S. O.: Reaction in *Amoeba* to light. J. exp. Zool. **9**, 265—278 (1910).

Mast, S. O.: Light and behaviour of organisms. New York (1911).

Mazokhin-Porshnyakov, G. A.: Insect Vision, p. 306. New York: Plenum Press 1969.

Mazokhin-Porshnyakov, G. A., Cherkasov, A. D., Burakova, O. V., Vishnevskaya, T. M.: On the colour vision of Tabanidae (Diptera). Zool. Zh. **54** (10), 1574—1576 (1975).

McCann, G. D., Arnett, D. W.: Spectral and polarization sensitivity of the dipteran visual system. J. gen. Physiol. **54**, 534—558 (1972).
McEnroe, W. D.: The photoresponse to skyshine of the two-spotted spider mite *Tetranychus urticae*. Acarologia **11**, 281—283 (1969).
McEnroe, W. D., Dronka, K.: Eyes of the female two-spotted spider mite *Tetranychus urticae*. II. Behavioural analysis of the photoreceptor. Ann. Entom. Soc. Amer. **62**, 466—469 (1966a).
McEnroe, W. D., Dronka, K.: Colour vision in the adult femal two-spotted spider mite. Science **154**, 782—784 (1966b).
McFarland, W. N., Munz, F. W.: The visible spectrum during twilight and its implication to vision. In: Light as an Ecological Factor, eds. G. C. Evans, R. Bainbridge, O. Rackham, Vol. II, pp. 249—270. Oxford-London-Edinburgh-Melbourne: Blackwell Sci. Publ. 1974.
McReynolds, J. S., Gorman, A. L. F.: Membrane conductances and spectral sensitivities of *Pecten* photoreceptors. J. gen. Physiol. **56**, 392—406 (1970a).
McReynolds, J. S., Gorman, A. L. F.: Photoreceptor potentials of opposite polarity in the eye of the scallop *Pecten irradians*. J. gen. Physiol. **56**, 376—391 (1970b).
Médioni, J.: Les organes de stimulation photique chez *Drosophila melanogaster*. Nonspécifité de cette stimulation relativement aux réaction à la lumière. C. R. Soc. Biol. **153**, 1845—1854 (1959).
Meffert, P., Smola, U.: Electrophysiological measurements of spectral sensitivity of central visual cells in the eye of blowfly. Nature (Lond.) **260**, 342—344 (1976).
Menne, D., Spatz, H. C.: Colour vision in *Drosophila melanogaster*. J. comp. Physiol. **114**, 301—312 (1977).
Menzel, R.: Untersuchungen zum Erlernen von Spektralfarben durch die Honigbiene, *Apis mellifica*. Z. vergl. Physiol. **56**, 22—62 (1967).
Menzel, R.: Das Gedächtnis der Honigbiene für Spektralfarben. I. Kurzzeitiges und langzeitiges Behalten. Z. vergl. Physiol. **60**, 82—102 (1968).
Menzel, R.: Spectral responses of moving detecting and "sustained" fibres in the optic lobe of the bee. J. comp. Physiol. **82**, 135—150 (1973).
Menzel, R.: Spectral sensitivity of monopolar cells in the bee lamina. J. comp. Physiol. **93**, 337—346 (1974).
Menzel, R.: Colour receptors in insects. In: The compound eye and vision in insects, ed. G. A. Horridge, pp. 121—153. Oxford: Clarendon Press 1975.
Menzel, R., Blakers, M.: Functional organization of an insect ommatidium with fused rhabdom. Cytobiology **11**, 279—298 (1975).
Menzel, R., Blakers, M.: Colour receptors in the bee eye—morphology and spectral sensitivity. J. comp. Physiol. **108**, 11—33 (1976).
Menzel, R., Knaut, R.: Pigment movement during light and chromatic adaptation in the retinula cells of *Formica polyctena*. J. comp. Physiol. **86**, 125—138 (1973).
Menzel, R., Roth, F.: Spektrale Phototaxis von Planktonrotatorien. Experientia (Basel) **28**, 356—357 (1972).
Menzel, R., Snyder, A. W.: Polarized light detection in the bee, *Apis mellifera*. J. comp. Physiol. **88**, 247—270 (1974).
Messenger, J. B.: Some evidence for colour blindness in *Octopus*. J. exp. Biol. **59**, 68—77 (1973).
Meyer-Rochow, V. B.: Axonal wiring and polarization sensitivity in eye of the rock lobster. Nature (Lond.) **254**, 522—523 (1975).
Mikkola, K.: Behavioural and electrophysiological responses of night-flying insects, especially Lepidoptera, to near-ultraviolet and visible light. Ann. Zool. Fennici **9**, 225—254 (1972).
Millecchia, R., Bradbury, J., Mauro, A.: Simple photoreceptors in *Limulus polyphemus*. Science **154**, 1199—1201 (1966).
Millecchia, R., Gwilliam, G. F.: Photoreception in a barnacle: Electrophysiology of the shadow reflex pathway in *Balanus cariosis*. Science **177**, 438—440 (1972).
Miller, W. H., Bernard, G. D.: Butterfly glow, J. Ultrastruct. Res. **24**, 286—294 (1964).
Miller, W. H., Bernard, G. D.: Skipper glow. MIT Res. Lab. Electron. qu. Progr. Rep. **88**, 114—117 (1968).
Miller, W. H., Møller, A. R., Bernhard, C. G.: The corneal nipple array. In: The Functional Organization of the Compound Eye, ed. C. G. Bernhard, pp. 21—33. London: Pergamon Press 1966.
Millott, N.: The dermal light sense, Invertebrate Receptors. Symp. Zool. Soc. London **23**, 1—6 (1968).

Mills, L. R.: On the detail morphology of the visual system, dorsal setae and glands of the two-spotted spider mite, *Tetranychus urticae*, PhD-Thesis, Stanford University (1973).

Mills, L. R.: Structure of the visual system of the two-spotted spider-mite, *Tetranychus urticae*. J. Insect Physiol. **20**, 795—808 (1974).

Mimura, K.: Analysis of visual information in lamina neurones of the fly. J. comp. Physiol. **88**, 335—347 (1974).

Minke, B., Hochstein, S., Hillman, P.: Early receptor potentials evidence for the existence of two thermally stable states in the barnacle visual pigment. J. gen. Physiol. **62**, 87—104 (1973a).

Minke, B., Wu, C. F., Pak, W. L.: Isolation of light induced response of central retinula cells from electroretinogram in *Drosophila*. J. comp. Physiol. **98**, 345—355 (1975).

Moller-Racke, J.: Farbensinn und Farbenblindheit bei Insekten. Zool. J. Physiol. (Lond.) **63**, 153—324 (1952).

Mote, M. I., Goldsmith, T. H.: Spectral sensitivity of colour receptors in the compound eye of the cockroach *Periplaneta*. J. exp. Zool. **173**, 137—145 (1970).

Mote, M. J., Goldsmith, T. H.: Compound eyes: localization of two receptors in the same ommatidium. Science **171**, 1254—1255 (1971).

Mpitsos, G. J.: Physiology of vision in the mollusc *Lima scabra*. J. Neurophysiol. **36**, 371—383 (1973).

Müller, H.: Versuche über die Farbenliebhaberei der Honigbiene. Kosmos **12**, 273—299 (1883).

Müller, H.: Zur Phototaxis von *Amoeba proteus*. Exp. Cell Res. **39**, 225—232 (1965).

Murray, G.: Intracellular absorption difference spectrum of *Limulus* extra-ocular photolabile pigment. Science **154**, 1182—1183 (1966).

Naegele, J. A., McEnroe, W. D., Soans, A. B.: Spectral sensitivity and orientation response of the two-spotted spider mite, *Tetranychus urticae* from 350 nm—700 nm. J. Insect Physiol. **12**, 1187—1195 (1966).

Neumeyer, Ch., v. Helversen, O.: Simultaner Farbkontrast bei der Honigbiene. Verh. dtsch. Zool. Ges. **69**, 260 (1976).

Newby, N. A.: Habituation to light and spontaneous activity in the isolated siphon of *Aplysia*. Pharmacological observations. Comp. gen. Pharmacol. **4**, 91—100 (1973).

Nolte, J., Brown, J. E.: The spectral sensitivities of single cells in the median ocellus of *Limulus*. J. gen. Physiol. **54**, 636—649 (1969).

Nolte, J., Brown, J. E.: The spectral sensitivity of single receptor cells in the lateral, median and ventral eye of normal and whiteeyed *Limulus*. J. gen. Physiol. **55**, 787—801 (1970).

Nolte, J., Brown, J. E.: Electrophysiological properties of cells in the median ocellus of *Limulus*. J. gen. Physiol. **59**, 167—185 (1972a).

Nolte, J., Brown, J. E.: Ultraviolet-induced sensibility to visible light in ultraviolet receptors of *Limulus*. J. gen. Physiol. **59**, 186—200 (1972b).

Nolte, J., Brown, J. E., Smith, T. G.: A hyperpolarizing component of the receptor potential in the median ocellus of *Limulus*. Science **162**, 677—679 (1968).

North, W. J.: Sensitivity to light in the sea anemone *Metridium senile*. II. Studies of reaction time variability and the effects of changes in the light intensity and temperature. J. gen. Physiol. **40**, 715—733 (1957).

North, W. J., Pantin, C. F. A.: Sensitivity to light in the sea anemone *Metridium senile*: Adaptation and action spectrum. Proc. roy. Soc. B. **148**, 385—396 (1958).

Northrop, J. H., Loeb, J.: The photochemical basis of animal heliotropism. J. gen. Physiol. **5**, 581—595 (1922).

Nosaki, H.: Electrophysiological study of colour encoding in the compound eye of crayfish *Procambarus clarkii*. Z. vergl. Physiol. **64**, 318—328 (1969).

O'Benar, J. D., Matsumoto, Y.: Light-induced neural activity and muscle contraction in the marine worm *Golfingia goulchi*. Comp. Biochem. Physiol. **55A**, 77—81 (1976).

Oesterhelt, D., Stoeckenius, W.: Rhodopsin-like protein from purple membrane of *Halobacterium halobium*. Nature (Lond.) New Biol. **233**, 149—152 (1971).

Oesterhelt, D., Meenthen, M., Schumann, L.: Reversible dissociation of purple complex in bacteriorhodopsin and identification of 13-cis and all-trans retinal as chromophores. Europ. J. Biochem. **40**, 453—463 (1973).

Ohnesorg, E.: Untersuchungen über die Entwicklung des Sehvermögens beim Rückenschwimmer *Notonecta glauca*. Z. vergl. Physiol. **37**, 424—438 (1955).

Olivio, R.F., Hiraga, Y., Koumjian, K.: Colour discrimination capability in the crayfish optokinetic system. Neuroscience **1**, 902 (1975).

Passano, L.M., McCullough, C.B.: The light response and the rhythmic potentials of *Hydra*. Proc. nat. Acad. Sci. (Wash.) **48**, 1376—1382 (1962).

Perrelet, A.: The fine structure of the retina of the honey bee drone. An electron microscopical study. Z. Zellforsch. **108**, 530—562 (1970).

Pittendrigh, C.S., Eichhorn, J.M., Minis, D.M., Bruce, V.G.: Circadian systems, VI. Photoperiodic time measurement in *Pectinophora gossypiella*. Proc. nat. Acad. Sci. (Wash.) **66**, 758—764 (1970).

Pourriot, R., Clement, P.: Photopériodisme et cycle hétérogonique chez *Notommata copens*. II. Influence de la qualité de la luminière, spectres d'action. Arch. Zool. exp. gén. **114**, 277—300 (1973).

Prosser, C.L.: Action potentials in the nervous system of the crayfish. I. Spontaneous impulses, II. Response to illumination of the eye and caudal ganglion. J. Cell. Comp. Physiol. **4**, 185—209, 363—367 (1934).

Razmjoo, S., Hamdorf, K.: Visual sensitivity and variation of total photopigment content in blowfly photoreceptor membrane. J. comp. Physiol. **105**, 279—286 (1976).

Resch, B.: Untersuchungen über das Farbensehen von *Notonecta glauca* L. Z. vergl. Physiol. **36**, 27—40 (1954).

Ribi, W.A.: The first optic ganglion of the bee. I. Correlation between visual cell types and their terminals in the lamina and medulla. Cell Tiss. Res. **165**, 103—112 (1975).

Robert, P., Scheffer, D., Médioni, J.: Réactions oculo-motrices de la Daphnie en résponse à des lumières monochromatiques d'égale énergie. Sensibilité visuelle et sensibilité dermatoptique. C. R. Seanc. Soc. Bio. **152**, 1000—1003 (1958).

Rodenhouse, J.Z., Guberlet, J.E.: The morphology and behaviour of the cushion star *Pteraster tesselatus*. Univ. Wash. Publ. Biol. **12**, 21—47 (1946).

Rodieck, R.W.: The vertebrate retina. San Francisco: Freeman & Co. 1973.

Röhlich, P., Török, L.J.: Elektronenmikroskopische Untersuchungen des Auges von Planarien. Z. Zellforsch. Mikr. Anat. **54**, 362—381 (1961).

Roth, H., Menzel, R.: ERG of *Formica polyctena* and selective adaptation. In: Information processing in the visual system of Arthropods, ed. R. Wehner, pp. 177—182. Berlin-Heidelberg-New York: Springer 1972.

Rozenberg, G.V.: Twilight—a study in atmospheric optics. New York: Plenum Press 1966.

Ruck, P.: The electrical responses of dorsal ocelli in cockroaches and grashoppers. J. Insect Physiol. **1**, 109—123 (1957).

Ruck, P.: A comparison of the electrical response of compound eyes and dorsal ocelli in four insect species. J. Insect Physiol. **2**, 261—274 (1958).

Ruck, P.: The components of the visual system of a dragonfly. J. gen. Physiol. **49**, 289—307 (1965).

Rushton, W.A.H.: Pigments and signals in colour vision. J. Physiol. (Lond.) **220**, 1—31 (1972).

Sander, W.: Phototaktische Reaktionen der Bienen auf Lichter verschiedener Wellenlängen. Z. vergl. Physiol. **20**, 267—286 (1933).

Saunders, D.S.: Spectral sensitivity and intensity threshold in *Nasonia* photoperiodic clock. Nature (Lond.) **253**, 732—733 (1975).

Schamp, R.: Untersuchungen zur spektralen Phototaxis von *Daphnia magna* und *Polyphemus pediculus*. Staatsexamensarbeit Techn. Hochsch. Darmstadt (1973).

Schlegtendal, A.: Beitrag zum Farbensinn der Arthropoden. Z. vergl. Physiol. **20**, 545—581 (1934).

Schlieper, C.: Über die Helligkeitsverteilung im Spektrum bei verschiedenen Insekten. Z. vergl. Physiol. **8**, 281—288 (1928).

Schricker, B.: Die Orientierung der Honigbiene in der Dämmerung. Z. vergl. Physiol. **44**, 420—458 (1965).

Schümperli, R.A.: Evidence for colour vision in *Drosophila melanogaster* through spontaneous phototactic choice behaviour. J. comp. Physiol. **86**, 77—94 (1973).

Schümperli, R.A.: Monocular and binocular visual fields of butterfly interneurons in response to white and coloured light stimulation. J. comp. Physiol. **103**, 273—289 (1975).

Schultz, H.: Über die Bedeutung des Lichtes im Leben niederer Krebse. Z. vergl. Physiol. **7**, 488—552 (1928).

Scott, S., Mote, M.: Spectral sensitivity in some marine crustacea. Vision Res. **14**, 659—664 (1974).

Shaw, S.R.: Simultaneous recording from two cells in the locust retina. Z. vergl. Physiol. **55**, 183—194 (1967).

Shaw, S. R.: Interreceptor coupling in ommatidia of drone honey bee and locust compound eye. Vision Res. **9**, 999—1029 (1969a).
Shaw, S. R.: Optics of arthropod compound eye. Science **164**, 88—90 (1969b).
Shaw, S. R.: Decremental conduction of the visual signal in barnacle lateral eye. J. Physiol. (Lond.) **220**, 145—175 (1972).
Shaw, S. R.: Retinal resistance barriers and electrical lateral inhibition. Nature (Lond.) **255**, 480—483 (1975).
Siddle, N.: The absolute threshold and spectral sensitivity curve of barnacle nauplii. M. Sc. Thesis Univ. of Wales (1968).
Singer, R. H., Rushforth, N. S., Burnett, A. L.: The photodynamic action of light on *Hydra*. J. exp. Zool. **154**, 169—174 (1963).
Smith, F. E., Baylor, E. R.: Colour responses in the Cladocera and their ecological significance. Amer. Naturalist. **87**, 49—55 (1953).
Smith, T. G., Baumann, F.: The functional organization within the ommatidium of the lateral eye of *Limulus*. Prog. Brain Res. **31**, 313—327 (1969).
Smola, U., Meffert, P.: A single-peaked UV receptor in the eye of *Calliphora erythrocephala*. J. comp. Physiol. **103**, 353—357 (1975).
Smola, U., Meffert, P.: Spektrale Empfindlichkeit der zentralen Sehzellen im Auge der Schmeißfliege *Calliphora erythrocephala*. Verh. dtsch. Zool. Ges. p. 282 (1976).
Snyder, A. W.: Optical Properties of Invertebrate Photoreceptors. In: The compound eye and vision in insects, ed. G. A. Horridge. Oxford: Clarendon Press 1975.
Snyder, A. W., Menzel, R., Laughlin, S. B.: Structure and function of the fused rhabdom. J. comp. Physiol. **87**, 99—135 (1973).
Sonntag, C.: Spectral sensitivity studies on the visual system of the praying mantis *Tenodera sinensis*. J. gen. Physiol. **57**, 93—112 (1971).
Stark, W. S.: Spectral sensitivity of visual response alterations mediated by interconversions of native and intermediate photopigments in *Drosophila*. J. comp. Physiol. **96**, 343—356 (1975).
Stark, W. S.: Sensitivity and adaptation in R7, an ultraviolet photoreceptor, in the *Drosophila* retina. J. comp. Physiol. **115**, 47—59 (1977).
Stark, W. S., Ivanyshyn, A. M., Hu, H. G.: Spectral sensitivities and photopigments in adaptation of fly visual receptors. Naturwissenschaften **63**, 513—518 (1976).
Stavenga, D. G.: Dark regeneration of invertebrate visual pigments. In: Photoreceptor Optics, eds. A. W. Snyder, R. Menzel, pp. 290—295. Berlin-Heidelberg-New York: Springer 1975.
Stearns, S. C.: Light responses of *Daphnia pulex*. Limnol. Oceanogr. **20**, 564—570 (1975).
Steyskal, G. C.: Notes on colour pattern of eye in Diptera. Bull. Brooklyn Entomolog. Soc. **52**, 89 (1957).
Stratton, W. P.: Light induced contraction in *Golfingia* proboscis retractor muscles. Biol. Bull. **129**, 425—432 (1965).
Strausfeld, N.: Atlas of an insect brain. Berlin-Heidelberg-New York: Springer 1975.
Strother, G. K., Superdock, D. A.: In situ absorption of *Drosophila melanogaster* visual screening pigments. Vision Res. **12**, 1545—1547 (1972).
Struwe, G.: Spectral sensitivity of single photoreceptors in the compound eye of a tropical butterfly. J. comp. Physiol. **79**, 197—201 (1972).
Suga, N., Katsuki, Y.: Vision in insects in terms of the electrical activities of the descending nerve fibres. Nature (Lond.) **194**, 658—660 (1962).
Suski, Z. W., Naegele, J. A.: Light response in the two spottet spider mite. Analysis of behavioural response. In: Advances in Acarology I, ed. J. A. Naegele, pp. 435—444. 1963.
Swihart, Chr.: Colour discrimination by the butterfly *Heliconius charitonius*. Anim. Behav. **19**, 156—169 (1971).
Swihart, S. L.: Neural adaptations in the visual pathway of certain Heliconine butterflies and related forms to variations in wing coloration. Zoologica (N.Y.) **52**, 1—14 (1967).
Swihart, S. L.: Single unit activity in the visual pathway of the butterfly *Heliconius erato*. J. Insect Physiol. **14**, 1589—1601 (1968).
Swihart, S. L.: Colour vision and the physiology of the superposition eye of a butterfly (Hesperiidae). J. Insect Physiol. **15**, 1347—1365 (1969).
Swihart, S. L.: The neural basis of colour vision in the butterfly *Papilio troilus*. J. Insect Physiol. **16**, 1623—1636 (1970).

Swihart, S. L.: Modelling the butterfly visual pathway. J. Insect Physiol. **18**, 1915—1928 (1972a).

Swihart, S. L.: The neural basis of colour vision in the butterfly *Heliconius erato*. J. Insect Physiol. **18**, 1015—1025 (1972b).

Takahashi, K.: Electrical responses to light stimuli in the isolated radial nerve of the sea urchin *Diadema setosum*. Nature (Lond.) **201**, 1343—1344 (1964).

Tardent, P., Frei, E.: Reaction patterns of dark- and light-adapted *Hydra* to leight stimuli. Experientia (Basel) **25**, 265—267 (1969).

Thomas, J., Autrum, H.: Die Empfindlichkeit der dunkel- und helladaptierten Biene *Apis mellifica* für spektrale Farben: Zum Purkinje-Phänomen der Insekten. Z. vergl. Physiol. **51**, 204—218 (1965).

Thorp, R. W., Briggs, D. L., Estes, E. H., Erickson, E. H.: Nectar fluorescence under ultraviolet irradiation. Science **189**, 476—477 (1975).

Thorp, R. W., Briggs, D. L., Estes, J. R., Erickson, E. H.: Fluorescent nectar. Science **194**, 342 (1976).

Trevino, D. L., Larimer, J. L.: The response of one class of neurons in the optic tracts of crayfish to monochromatic light. Z. vergl. Physiol. **69**, 139—151 (1970).

Trujillo-Cenóz, O.: Some aspects of the structural organization of the arthropod eye. Cold Spr. Harbor Symp. quant. Biol. **30**, 371—382 (1966).

Trujillo-Cenóz, O., Bernard, G. D.: Some aspects of the retinal organization of *Sympycnus lineatus* Loew (Diptera, Dolichopodidae). J. Ultrastruct. Res. **38**, 149—160 (1972).

Truman, J. W.: Extra retinal photoreception in insects. Photochem. Photobiol. **23**, 215—225 (1976).

Unteutsch, W.: Über den Licht- und Schattenreflex des Regenwurms. Zool. Jahrb. (allg. Zool.) **58**, 9—12 (1937).

Varela, F. G.: Fine structure of the visual system of the honey bee *Apis mellifera* II. Lamina. J. Ultrast. **31**, 178—194 (1970).

Viaud, G.: Recherches expérimentales sur la phototropisme de rotifères. Bull. Biol. **74**, 68—93 (1940); Bull. Biol. **77**, 224—242 (1943).

Viaud, G.: Le phototropisme et les deux modes de la photoréception. Experientia (Basel) **4**, 81—88 (1948).

Visscher, J. P., Luce, R. H.: Reactions of the cyprid larvae of barnacles to light with special reference to spectral colours. Biol. Bull. **54**, 336—350 (1928).

Wald, G.: Single and multiple systems in arthropods. J. gen. Physiol. **51**, 125—156 (1968).

Wald, G., Krainin, J. M.: The median eye of *Limulus*, an ultraviolet receptor. Proc. nat. Acad. Sci. (Wash.) **50**, 1011—1017 (1963).

Wald, G., Rayport, S.: Vision in annelid worms. Science **196**, 1434—1438 (1977).

Walther, J. B., Dodt, E.: Spektralsensitivität von Insekten-Komplexaugen im UV bis 290 nm. Z. Naturforsch. **14b**, 273—278 (1959).

Waser, P. M.: The spectral sensitivity of the eye of *Aplysia californica*. Comp. Biochem. Physiol. **27**, 339—347 (1968).

Wasserman, G. S.: *Limulus* receptor action spectra. Vision Res. **9**, 611—619 (1969).

Wasserman, G. S.: Invertebrate colour vision and the tuned receptor paradigm. Science **180**, 268—275 (1973).

Waterman, T. H., Fernández, H. R.: *E*-vector and wavelength discrimination by retinular cells of the crayfish *Procambarus*. Z. vergl. Physiol. **68**, 154—174 (1970).

Weiss, H. B.: Colour perception in insects. J. Ecol. Entomol. **36**, 1—17 (1943a).

Weiss, H. B.: The group behaviour of 14,000 insects to colours. Entomol. News **54**, 152—156 (1943b).

Weiss, H. B.: Insect responses to colours. J.N.Y. Entomol. Soc. **52**, 267—271 (1944).

Weiss, H. B.: Insects and the spectrum. J.N.Y. Entomol. Soc. **54**, 17—30 (1946).

Weiss, H. B., Soraci, F. A., McCoy, jr. E. E.: Notes on the reaction of certain insects to different wavelengths of light. J.N.Y. Entomol. Soc. **49**, 1—20 (1941).

Weiss, H. B., Soraci, F. A., McCoy, jr. E. E.: The behaviour of certain insects to various wavelengths of light. J.N.Y. Entomol. Soc. **50**, 1—35 (1942).

Welsh, J. H.: The caudal photoreceptor and response of the crayfish to light. J. Cell. Comp. Physiol. **4**, 379—388 (1934).

Wendland, S.: Spektrale Phototaxis weißäugiger Bienen. Diploma Thesis, Freie Universität Berlin (1977).

Wiersma, C. A. G., Yamaguchi, T.: The neuronal components of the optic nerve of the crayfish as studied by single unit analysis. J. comp. Neurol. **128**, 333—358 (1966).

Wiersma, C. A. G., Yamaguchi, T.: Integration of visual stimuli by the crayfish central nervous system. J. exp. Biol. **47**, 409—431 (1967).

Wolken, J. J.: Studies of photoreceptor structure. Ann. N.Y. Acad. Sci. **74**, 164—181 (1958).

Woodcock, A. E. R., Goldsmith, T. H.: Spectral responses of sustaining fibres in the optic tracts of crayfish *Procambarus*. Z. vergl. Physiol. **69**, 117—133 (1970).

Woodcock, A. E. R., Goldsmith, T. H.: Differential wavelength sensitivity in the receptive fields of sustaining fibres in the optic tract of the crayfish *Procambarus*. J. comp. Physiol. **87**, 247—257 (1973).

Yamashita, S., Tateda, H.: Spectral sensitivities of jumping spider eyes. J. comp. Physiol. **105**, 29—41 (1976).

Yingst, D. R., Fernandez, H. R., Bishop, L. G.: The spectral sensitivity of a littoral annelid: *Nereis mediator*. J. comp. Physiol. **77**, 225—232 (1972).

Yoshida, M.: Spectral sensitivity of chromatophores in *Diadema setosum*. J. exp. Biol. **34**, 222—225 (1957).

Yoshida, M.: Photosensitivity. In: Physiology of Echinodermata, ed. R. A. Boolootian, pp. 435—465. New York-London-Sydney: Interscience Publ. 1966.

Yoshida, M., Millott, N.: Light sensitive nerves in an echinoid. Experientia (Basel) **15**, 13—14 (1959).

Yoshida, M., Millott, N.: The shadow reaction of *Diadema antillarum*. Reexamination of spectral sensitivity. J. exp. Biol. **37**, 390—397 (1960).

Yoshida, M., Ohtsuki, H.: Compound ocellus of a starfish: Its function. Science **153**, 197—198 (1966).

Young, St.: Directional differences in the colour sensitivity of *Daphnia magna*. J. exp. Biol. **61**, 261—267 (1974).

Zettler, F., Autrum, H.: Chromatic properties of lateral inhibition in the eye of a fly. J. comp. Physiol. **97**, 181—188 (1975).

Zimmermann, W. F., Goldsmith, T. M.: Photosensitivity of the circadian rhythm and of visual receptors in carotenoid-depleted *Drosophila*. Science **171**, 1167—1969 (1971).

Chapter 10

Extraocular Photoreception

M. YOSHIDA, Tamano, Okayama, Japan 706[1]

With 39 Figures

Contents

[1] From August, 1979, Ushimado, Okayama (Japan 701-43)

A. Introduction

Light affects animal life in diverse ways. Photic information is perceived through eyes in some animals, but in others photosensitivity is mediated through an extraocular surface. Such sensitivity, called the dermal or diffuse light sense, is known to exist in most of the major phyla, and a remarkable variety of biologically significant photoreactions ensue, escape from predators, attack on prey, search for a more suitable habitat, initiation of spawning, maintenance of circadian rhythm, control of endocrine activities, etc. We will be concerned here only with the short-term effects of light, and with some of the principal types of responses.

MILLOTT (1968) defined the dermal light sense "as a widespread photic sense that is not mediated by eyes or eyespots and in which light does not act directly on an effector." Recent advances in ultrastructure research have revealed cells scattered over the body surface which could be photosensitive. Those cells may be found either in isolation, e.g., those in the skin of the earthworm (HESSE, 1896; RÖHLICH et al., 1970), or in clusters forming a so-called ocellus, such as dorsal ocellus of *Onchidium* (YANASE and SAKAMOTO, 1965), shell eyes in *Chiton* (BOYLE, 1969b), etc. According to Millott's definition, photosensitivity found in more deeply situated nerve tissues may fit the same category. Here again, progress in microelectrode techniques enables demonstration of photoexcitive neurons among many other nonphotosensitive ones, e.g., in crayfish caudal ganglion (PROSSER, 1934a; WILKENS and LARIMER, 1972), *Aplysia* ganglion (ARVANITAKI and CHALAZONITIS, 1960), and *Onchidium* pleuroparietal ganglion (HISANO et al., 1972a, b).

In this article, the terms will be used in a somewhat more restricted sense for the sake of convenience, though this has no physiologic justification. Thus, dermal photosensitivity will refer to the phenomenon in which no cells that are morphologically recognizable as a possible photoreceptive site have ever been demonstrated. If there are cells which, regardless of being either solitary or aggregated, are widespread over the body surface and can be structurally related to known types of photosensory cells, the term "diffuse photosensitivity" is more appropriate. Neural photosensitivity will be referred to as that which can be demonstrated by direct stimulation of the nervous tissue, and neuronal photosensitivity as the cases where the photoexcitable neurons have been demonstrated by intracellular recording techniques.

B. Direct Responses

Cells capable of responding to external stimuli with no intervening cellular elements are effectors as well as receptors. Protists, nematocysts, and some of the chromatophores are included in this category. Strictly speaking, any primary receptor cell may be referred to as an effector as well since its activities that excite secondary cells may be taken to be an expression of effectors (LOEWENSTEIN, 1971). Here, however, we will use the term in its more usual sense.

Apart from protists, photic responses of which are reviewed in this volume by DIEHN (Ch. 2), some chromatophores are typical photoexcitable effectors, and classic accounts have been reviewed by STEVEN (1963) with due critical consideration of the experimental methods.

KLEINHOLZ (1938) demonstrated color changes in sea urchins from pale to dark when dark-adapted animals were brought to light. MILLOTT (1952) showed later in *Diadema antillarum* that the response was independent of the CNS, and further that urchins confined to darkness for 23 hr were darker at 10 a.m. than those at 8 p.m. after 3-hrs dark treatment. These results suggest the presence of an inherent diurnal rhythm oscillating between pale (evening) and dark (morning) phases. Using *Centrostephanus longispinus*, on the other hand, DAMBACH and JOCHUM (1968) proposed the presence of a dual mechanism, fast and slow, controlling the rate of color changes at different parts of the body surface. The light-dependent color changes in Diadematids bear a special importance in connection with the dermal light sense, which is distributed in the epidermis below chromatophores. Chromatophores here are thus a sort of iris diaphragm in the lens eye. The inherent rhythm and the differential response rate should therefore be involved in regulating the sensitivity level over the body surface, but the intimate mechanism for them has not yet been worked out.

As some of the irides are known to respond directly to light, this is also the case for sea urchin chromatophores. Evidence for direct responsiveness was obtained with a Japanese species, *D. setosum* (YOSHIDA, 1956). If a pale region of a dark-adapted piece of epidermal surface was illuminated by a small spot 3 μm in diameter, pigment granules moved to cover the illuminated spot while granules in other parts of the cell remained concentrated (Fig. 1). Though the results appeared as if the hyaloplasm had been photosensitive, the spectral sensitivity measurements (YOSHIDA, 1957) located the maximal sensitivity at 470 nm, suggesting involvement of a colored substance. The fact that the peak is not greatly shifted from that of the *off*-response in spine movements (YOSHIDA and MILLOTT, 1960) indicates that the same type of a pigment can be involved in the two different kinds of response in the same animal.

The cells are rather unusual in that those which are heavily distorted by centrifugal forces are still photoresponsive (YOSHIDA, 1960). Based on light-microscope observations, MILLOTT (1964) tried to explain the unusual behavior of the chromatophores by a concept of chromatoglyphs. Later, WEBER and DAMBACH (1972) observed that the darkening was attained by stretching the chromatophore branches into intercellular spaces of the surrounding tissue (Fig. 2). This suggests an amoeboid nature. The mobility appears to be based on filamentous elements within the cell, as demonstrated with the electron microscope. The inhibition of dispersion and concentration of pigment granules by administration of cytochalasin B that is known to interfere with the cytoplasmic contractile mechanisms may be taken as an additional proof for involvement of microfilaments in pigment and/or amoeboid movements (DAMBACH and WEBER, 1975). Further, they succeeded in isolating single cells from the epidermal tissue (WEBER and DAMBACH,

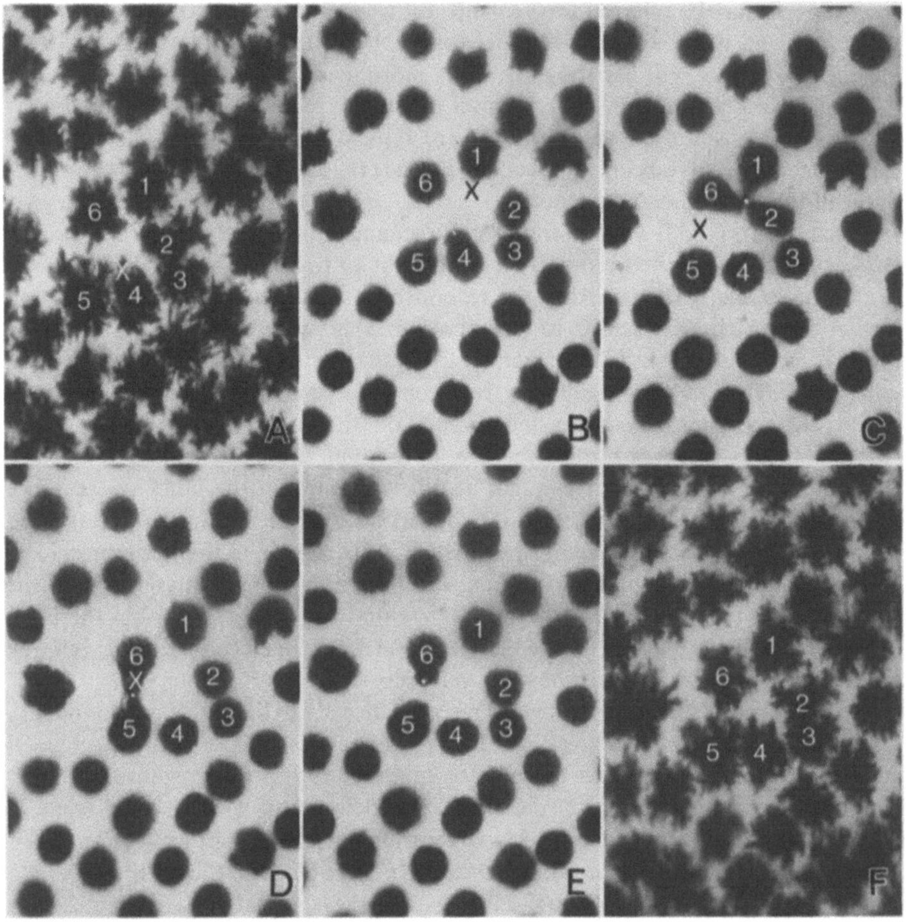

Fig. 1. Experiments in which a spot of light was projected onto various parts of dermal chromatophores of *Diadema setosum*. A: light-adapted chromatophores (initial condition); the whole preparation placed in complete darkness except for the spot of light. Duration of darkness: A to B, 1.5 hr, and B to C, C to D and D to E, 1 hr. Crosses indicate places where spot was projected at beginning of dark periods. Circles indicate places where spot had been projected until photograph was taken. Between E and F spot was removed, whole preparation was placed in darkness for 20 min, then under light for 20 min. (From YOSHIDA, 1956)

1974). When placed in darkness for 30 min, the isolated cells took an ovoid form, which started to transform rapidly into a stellate shape within a few seconds upon illumination. This is a beautiful demonstration of the amoeboid nature of the diadematid chromatophores and requires a modification of YOSHIDA's interpretation of the direct sensitivity which was assumed at that time, based on the common concept of chromatophore branches being pre-fixed. Though the direct responsiveness can stand as it was, the behavior observed by YOSHIDA was due to responses of a branch of the sensitive amoeboid chromatophore extending toward the scattered light.

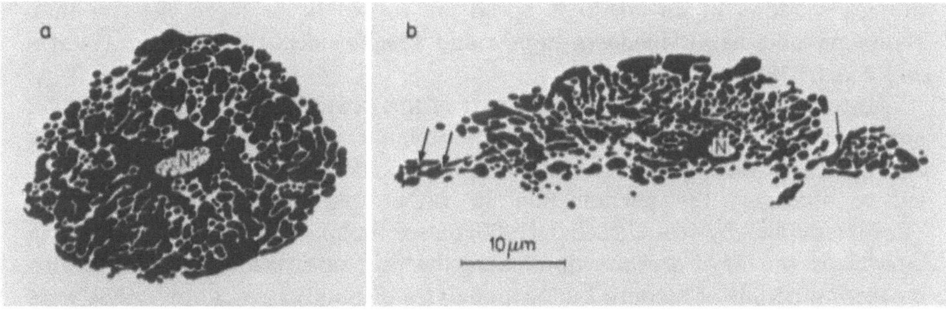

Fig. 2. Sketches of chromatophores of *Centrostephanus longispinus*, based on electron microscope observations. Pigmented areas are black. a: ball-shaped chromatophore with nucleus (N) and processes, b: 15 min after light reaction, showing dispersion with simultaneous stretching of processes (*arrows*). (From WEBER and DAMBACH, 1974)

Another type of direct response which may deserve attention is the phototropic growth reactions in a bryozoan species, *Bugula avicularia* (KAISSLING, 1963). The growth appears to be dependent on the light-directed movements of spheric cells along the capsula of the growing bud. Since the spheric cells have a refractive index higher than that of the surrounding media, the intracellular gradient in light intensity will ensue and it was suggested that such a gradient would be translated by the intracellular receptive elements into physiologic ones.

C. Dermal Photoreception

I. Types of Reactions

Categorically, dermal photosensitivity as defined above may be divided into two types: 1) reflex responses of a part of the body, for which the presence of nervous elements is needed, and 2) kinetic, tactic, or other responses, for which a more central mechanism is needed to achieve coordinated movements of the accessory organs involved.

1. Reflex Responses

There are many types of reflex responses mediated through the dermal light sensitivity, either alone or in combination with other types of photosensitive systems. The effective stimulus can be light-*on*, -*off* or both, either instantaneous or changing with time. Among those, the most prevailing are the so-called shadow reactions (VON BUDDENBROCK, 1930; STEVEN, 1963; MILLOTT, 1968), which may be related to protecting the animal from potential predators. Thus, in some cases, moving shadows at an adequate speed are shown to be more effective than stationary ones as in *Anodonta cygnea* and *Pseudanodonta complomata* (BRAUN and FAUST, 1954).

Historically, workers in the early part of this century tried to find sensitivity gradients over the body surface and photolabile pigments in the dermis, as exemplified by the work of CROZIER (1914, 1916, 1917) on holothurians, or were led to interpret the phenomenon in rather philosophic terms, such as "Reflexrepublik" by VON UEXKÜLL (1900) on sea urchins. HECHT's work on *Ciona intestinalis* and *Mya arenaria* dominated the field until the 1950s. During this pioneering period of modern biology, analyses of photochemical processes were tried in terms of changes in the reaction times under various lighting or adapting conditions. The basis for this approach was HECHT's formal scheme of photochemistry (1934), and work done along this line has been reviewed by STEVEN (1963). Below, I shall describe 1) dermatoptic responses that are governed by a multimodal mechanism, 2) electric correlates of the dermal light sense and 3) photic responses affected by the presence of symbionts. Analysis of the neural mechanism involved will be dealt with in Sec. C.III, and instances whereby dermal light sense coexists with other receptive systems, in Sec. F.II.

Several species of *Hydra* respond to light by contraction and by tactic movements (HAUG, 1933). RUSHFORTH et al. (1963) showed in *Hydra pirardi* an inverse relationship between the latencies of contraction responses and the intensity of light, the response being unaffected by removal of tentacles. The response pattern appears to be deeply influenced by adaptive conditions, so that in *H. attenuata* continuous exposure to light or darkness for 2 weeks alters contraction or extension responses of the body and tentacles to identical light stimuli in a reverse way (FREI, 1973). FELDMAN and LENHOFF (1960) claimed that photoreceptors in *H. littoralis* were mainly distributed in the hypostomes and tentacles, because sectioning below the level of tentacles left the phototactic behavior undisturbed only in the apical half. The question as to the location of the photoreceptive site appears still unsettled. Dr. K. NODA of Tokyo Metropolitan Institute of Gerontology has informed me that any fragment of chopped pieces of *H. magnipapillata* gathers in a lighted area (personal communication).

On the other hand, the rate of photoinduced contraction is inhibited by reduced glutathione, and such an inhibitory effect is markedly depressed by removal of tentacles (RUSHFORTH et al., 1964). These findings suggest that mechano- and photoreceptive mechanisms are interrelated in eliciting the contraction responses, the former being associated with the feeding activities and residing with the tentacles. According to PASSANO and MCCULLOUGH (1962), rhythmic potentials (RPs) recordable from the body surface lose their regular rhythm upon illumination and then reoccur in the dark at a more rapid rate than before, and feeding activities do not interrupt the RPs. Thus, the RPs show some similarity with the photoinduced contractions in that both are affected by light and independent of tentacular activities. Analysis of the relationship between the electric events and contraction pattern will be dealt with later (Sec. C.III).

Electric activities originating from dermal receptors have been recorded in bivalve molluscs, *Lima scabra* and *L. scabra tenera*, by MPITSOS (1973). After extirpation of eyes, which produce large spikes at the cessation of illumination

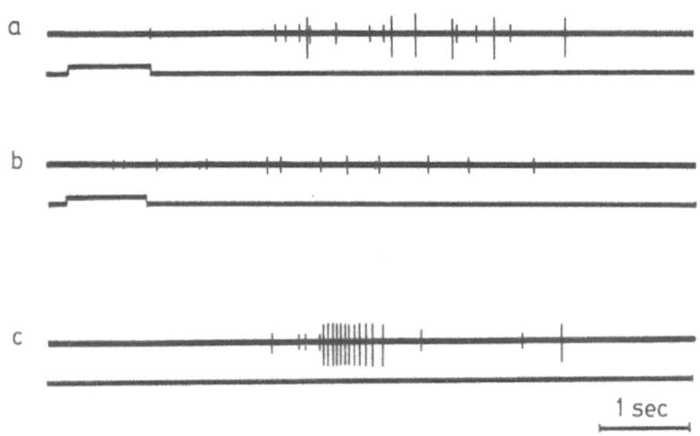

Fig. 3. Identification of dermal photoreceptor in *Lima scabra*. A: *off*-responses of one large and several small units when illumination of a large area of the mantle was withdrawn. B: small unit responses after extirpation of all eyes. C: responses of small and then large units when a small blackened flag was swept across illuminated area. (From MPITSOS, 1973)

(Fig. 3A), small spikes which in all probability are activities of the small nerve fibers originated from the dermal photoreceptors become evident in Fig. 3B. It is clear from the record C, which was obtained by sweeping a small black flag across the illuminated area of the mantle, that the presumably dermal unit has shorter latency than the large spikes which originate from the ocular system.

Some of the sea anemones respond to light without detectable photoreceptors (NORTH and PANTIN, 1958). The parietal muscle of *Metridium senile* contracts within 2–10 s after illumination by blue light at an intensity of 8.6×10^8 quanta/cm^2·s (NORTH, 1957). MARKS (1976) studied electrophysiologically the photic responses in the burrowing sea anemone, *Calamactis praelongus*, and succeeded in distinguishing two types in the responsive mechanism. The one is local and non-nervous contractions of circular and parietal muscles, leading to the column bending. The other is extensive and neurally co-ordinated responses, including contraction of retractor muscles which themselves are insensitive to light.

With these responses of non-symbiotic animals in mind, we may now consider the response pattern in sea anemones which harbour endo-symbiotic zooxanthellae.

It is well known that sea anemones which harbor endosymbiotic zooxanthellae expand under illumination. According to PEARSE (1974b), most of *Anthopleura elegantissima* respond by expansion within 20 min after illumination, regardless of whether they normally harbor the symbiotic algae or not (Fig. 4A). In darkness, those without zooxanthellae remain expanded, whereas not only those with zooxanthellae, but also those from which originally harbored zooxanthellae have been artificially removed contract in the same time course (Fig. 4B). This suggests that a crucial factor is not the actual presence of zooxanthellae—which infers a secondary effect due to photosynthesis, such as accumulation of evolved oxygen, metabolites, etc.—but that the presence of zooxanthellae conditions the photoreceptive system to be functional. The importance of the presence of zooxanthellae for reflex activities in sea anemones has been reinforced recently by GLADFELTER (1975). He observed concomitant contraction and expansion in normal and

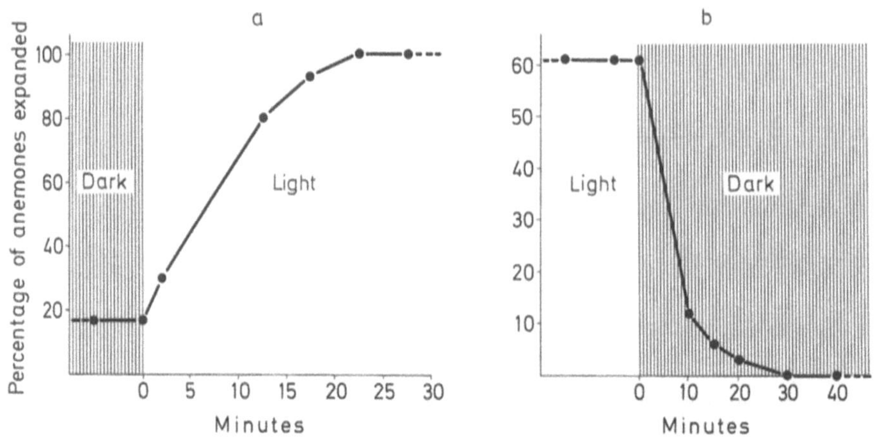

Fig. 4. Timecourse of expansion (A) and contraction (B) in running sea water by sea anemone, *Anthopleura elegantissima* with zooxanthellae. Ordinates: Percentage of anemones expanded. (From PEARSE, 1974b)

pseudotentacles of a Caribbean sea anemone, *Lebrunea coralligens*. Here, normal small tentacles start to contract when the intensity of light is increased above 4 lux, while large pseudotentacles that harbor about 20–30 times more zooxanthellae than the normal tentacles begin to extend when the intensity of light becomes higher than 1 lux, and reach their full expansion at about 500 lux. With little information about the nervous architecture and the photoreceptive site, the elucidation of sensory and neuromuscular mechanisms involved in eliciting seesaw-like swings in length of the two qualitatively different tentacles await future work.

2. Coordinated Movements

Reflex movements often proceed in a highly coordinated fashion, as in spine movements of a sea urchin, *D. antillarum* (MILLOTT and TAKAHASHI, 1963), in swimming arousal of a crinoid, *Antedon bifida* (DIMELOW, 1958), etc. Here, however, I shall restrict the subject to behavioral responses which require for their performance a mechanism of central processing of information fed in from the widespread area of the body surface.

In the case of kinetic movements, regardless of whether those are of the type of klinokinesis or orthokinesis, the topography of the photosensitive regions over the body surface affects the resultant behavioral pattern to only a small extent. On the other hand, the mechanism in inducing taxes that depend on the direction of the incident light requires a little more consideration. If the sensitive region is restricted to a localized part of the body, the animal can locate the light source by klinotactic behavior. This is not always the case, however. In a photonegative bivalve, *Lasaea rubra*, the tip of the extensible foot is sensitive. MORTON (1960) observed that there was no strictly regular alternation of the shell's swing to the right and left, which may be expected if the animal located the light source by an

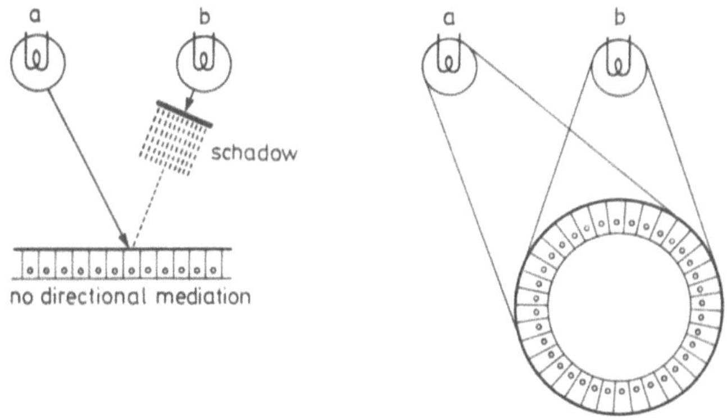

Fig. 5. Detection of direction of incident light. In flat animals, source of light cannot be located by a simple tegument, but in cylindrical organisms source of light can be located because the body itself casts a shadow that can be identified by dermal sensitivity. (From MILNE and MILNE, 1956)

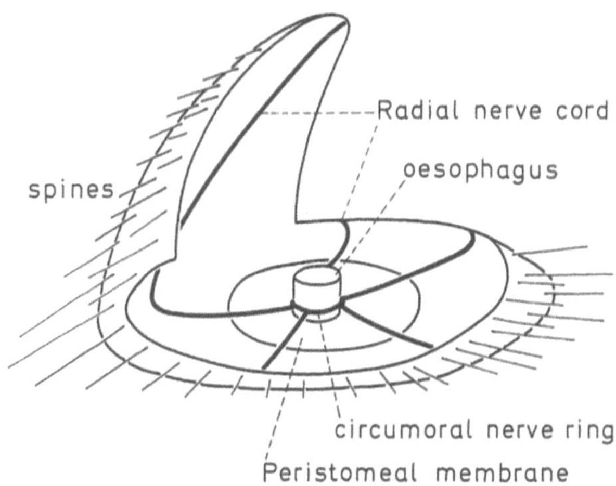

Fig. 6. The type of preparation, used for analysis of tactic movements in the sea urchin, *Temnopleurus toreumaticus*. (Unpublished)

alternate shading of the anteriormost photosensitive region, and we have to introduce some additional mechanisms.

If the entire body surface is sensitive, the light source may be located by the principle as elegantly schematized by MILNE and MILNE (1956). As shown in Fig. 5, light shone from the side results in an uneven intensity pattern due to opacity of the body. Such unbalanced information, after being processed in the CNS, will lead the animal to move toward or away from light by coordinated movements of accessory locomotory organs. A good example may be found in sea urchins.

A sea urchin, *Temnopleurus toreumaticus*, is positively phototactic (YOSHIDA, KABAYASHI and MORI, unpublished; see YOSHIDA, 1966). If aboral halves of the urchins are cut off a little below the ambitus together with viscera, leaving one or two ambulacra intact (Fig. 6), the majority of the operated animals move with the

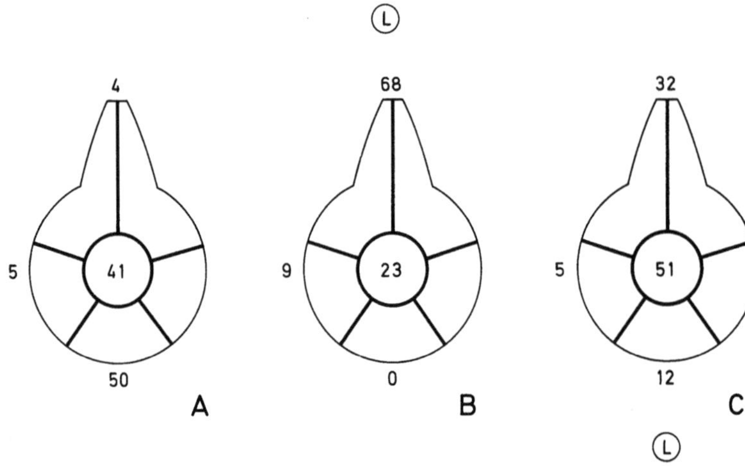

Fig. 7. Schematic representation of results from operated urchins as in Fig. 5. Circumoral nerve ring and radial nerve cords indicated by thick lines. Intact area was illuminated by a parallel beam (L) either from outside (B) or from inside (C). A: control experiments in darkness. Numbers around each figure show percentages of urchins which moved in the corresponding direction; those at center are percentages of urchins that did not move. (From YOSHIDA, 1966)

intact ambulacra on their back in darkness (control; Fig. 7A). If the light is shone from outside onto the intact ambulacral zone, 68% of the animals move towards the light, but none away from it (Fig. 7B). Similar phenomena have been observed with a European species, *Psammechinus microtuberculosis* (YOSHIDA, 1957). If the light is shone from inside, the percentage of animals which move away from the light becomes significantly high (Fig. 7C). The direction of movements in darkness can thus be reversed by illuminating the ambulacrum left intact, regardless of the direction of incident light. If urchins with one radial nerve transected near the oral nerve ring are placed in between two light sources facing each other, they never move toward the source on their operated side irrespective of its intensity. This suggests that the information should be fed into the nerve ring where the motor centers are located at each junction with the radial nerve cord (SMITH, 1950). The role of the motor centers for determining the locomotory direction has later been studied by YOSHIDA and OHTSUKI (1968) on the ocelli-bearing starfish, *Asterias amurensis*, and it is now safe to say that the information of the environmental light conditions as perceived through their dermal photosensitivity must be processed in motor centers of the nerve ring in order to achieve the coordinated tactic movements.

The foregoing accounts may appear self-evident, but some of the coordinated movements in sea urchins do not require the presence of the nerve ring. A number of regular as well as irregular sea urchins take up opaque objects such as eel grass, shell gravel, and so forth (for review, see YOSHIDA, 1966). This is called the covering reaction (MILLOTT, 1955). Using *Lytechinus variegatus*, MILLOTT (1956) observed a beautifully coordinated action in which tube feet and spines carried covering objects over various routes in response to movements of light source (Fig. 8). However, removal of the lantern, together with the oral nerve ring did not

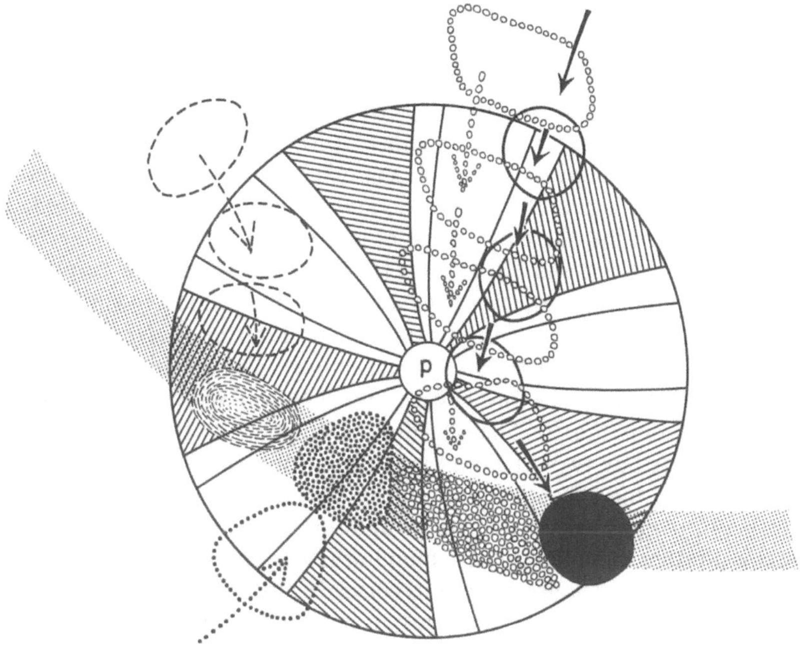

Fig. 8. Analysis of covering reaction in sea urchin *Lytechinus variegatus*. The diagram shows the placing of stones over localized, brightly illuminated areas of surface (stippled). Each stone, shaded by a distinctive convention (line, dots, circles or slid black) was moved approximately over the routes shown by successive outlines and arrows. The ambulacra are distinguished by cross-line shading. P : periproct. (From MILLOTT, 1956)

prevent the action of taking up gravel from the substratum. Coordination may be achieved through sequential reflexes triggered in combination with the dermal light sense and the mechanoreceptors of the tube feet and spines, though MILLOTT has carefully left the possibility that the nerve ring be involved in such highly coordinated responses shown by intact animals.

Last in this section, the role of symbiotic algae in locomotory behavior may be considered. Irrespective of the presence or absence of specialized photoreceptive systems, instances where algae-harboring animals show phototactic behavior are frequently encountered in various phyla, but the question regarding the role of algae in tactic behavior has not been answered. As mentioned above, some species of sea anemones become photosensitive when harboring symbiotic algae. According to PEARSE (1974a), *Anthopleura elegantissima* gathers in the light or shaded areas, depending on the light intensity. Differing from the reflex reactions, anemones that are genetically clonemates of phototactic individuals become indifferent to light and shade when they lose the symbionts. We also found in an acoela, *Convoluta* sp., that newly hatched worms were indifferent to light until they harbored green algae, at which time they became photopositive (M. GOTOW and YOSHIDA, unpublished). The mechanism with which the animals become phototactic by harboring symbiotic algae remains to be elucidated.

II. Action Spectra

Chemical analysis of photopigment(s) involved in the dermal light sense is practically impossible and one has to speculate about their properties through spectral sensitivity measurements. In most cases, wavelengths of the maximum sensitivity are found singly between 460 and 500 nm. GOLDSMITH (1972) has listed reliable data in this series of handbooks, and no more comments appear to be needed. I shall describe, therefore, some of the unusual cases.

According to CLARK and KIMELDORF (1970, 1971), the tentacle responses of a sea anemone, *Anthopleura xanthogrammica*, occur as a flexion toward the rim of the oral disk and a retraction from extended position upon exposure to wavelengths of 250–400 nm and 340–600 nm, respectively. The sensitivity maxima have been located at 280 nm in the former case, and at 500 nm with a secondary peak at 360 nm in the latter (Fig. 9). The question of whether the two opposite types of responses are based on two different kinds of photopigments or on different states of a pigment like the effects of photoconversion between rhodopsin and metarhodopsin (for review, see MENZEL, 1975; and this vol., Ch. 4), is not clear as yet.

Next is a rather complicated case found in *Hydra carnea* where culturing light conditions affect the response pattern to colored lights (ELLIS, 1970). The reaction times of contraction response are shortest to light of 650 nm in all groups cultured under white, blue, and green lights as well as in darkness, but not in the red-cultured group. This suggests that the photopigment involved absorbs the long wavelength, peaking at 650 nm, though the peaks of the action spectra for photo-induced contractions in hydras have mostly been located in the blue region; e.g., at 470 nm in *H. attenuata* (FREI, 1973). Complications arise when the sensitivities are compared among groups cultured under different conditions (Table 1). As expected, the sensitivities decrease significantly to the light under which the animals have been cultured, and to which the animals have been exposed for a long period. ELLIS found, however, that sensitivities to wavelengths which had not been used for the culture became higher than the sensitivities to those same

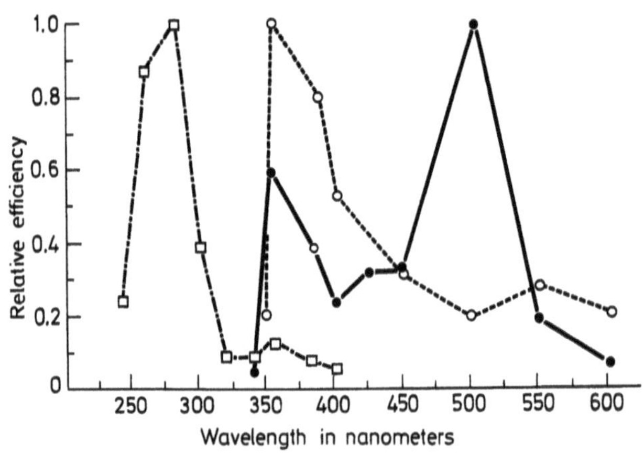

Fig. 9. Action spectra for tentacle flexion ———●, tentacle retraction —·—·—□, and oral disc flexion ----○ in the sea anemone *Anthopleura xanthogrammica*. (From CLARK and KIMELDORF, 1970)

Table 1. Comparison of sensitivity to colored lights between animals cultured under one of three cultured lights, and those of dark control. Figures in table are probabilities derived from students' *t* test. + and − indicate, respectively, shorter and longer reaction times than control series. (Compiled from ELLIS' data, 1970)

Cultured light	Wavelength of test light (nm)		
	450	550	650
Blue	− < 0.001	+ < 0.02	+ < 0.01
Green	+ < 0.001	− < 0.001	+ < 0.01
Red	+ < 0.02	+ < 0.02	− < 0.00

wavelengths for animals cultured in the dark. She interpreted this as follows: "Animals cultured in the dark show a generally decreased sensitivity to all light because of a decrease in the functioning of the photoreceptive system because of a lack of stimulation by light". If the photosystem concerned was but one, a decrease in sensitivity to one wavelength should have been accompanied by decreases to other wavelengths. If a dual system was concerned, the experiments should be considered to be of a type of selective adaptation, but we still could not expect any increase in sensitivity. Alternatively, if a pigment was photoconverted to another type or state, it appears to be more common to expect some kind of an opposite type of reaction to be induced by the photoconverted pigment. Further investigation should reveal the intimate mechanism.

III. Analysis of the Mechanisms

Some of the experimental analyses done more or less neurophysiologically on the mechanisms involved in eliciting responses that are mediated through the dermal light sense will be considered below.

The spines of the sea urchin, *Diadema antillarum*, respond by jerking to an increase (MILLOTT and YOSHIDA, 1959) or a decrease (MILLOTT, 1954) in light intensity and the responses are so consistent (Fig. 10; MILLOTT and YOSHIDA, 1960a), that they afford detailed experimental analyses. Scanning over the body surface by obliterating a spot of light revealed the maximal sensitivity at ambulacral margins, next at the ambulacral center and the least at the interambulacrum (MILLOTT, 1954; MILLOTT and YOSHIDA, 1960a). In echinoids, the superficial nervous system is formed by nerves that emerge through one of the pore pair at the ambulacral margin. Thus the gradations in sensitivity appear to correspond with the distribution of the superficial nervous system. It may be suggested that the dermal light sensitivity is a result of the superficial nerve fibers being photosensitive (MILLOTT, 1975). Though this has not yet been shown, the

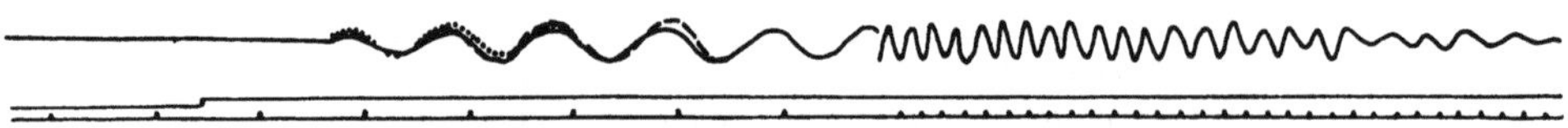

Fig. 10. Recordings of spine response of sea urchin, *Diadema antillarum*. Records of three separate reactions, recorded at approx. hourly intervals, superimposed in the left half. The moment of light-off shown by upward deflection in middle trace. Time in s. (From MILLOTT and YOSHIDA, 1960a)

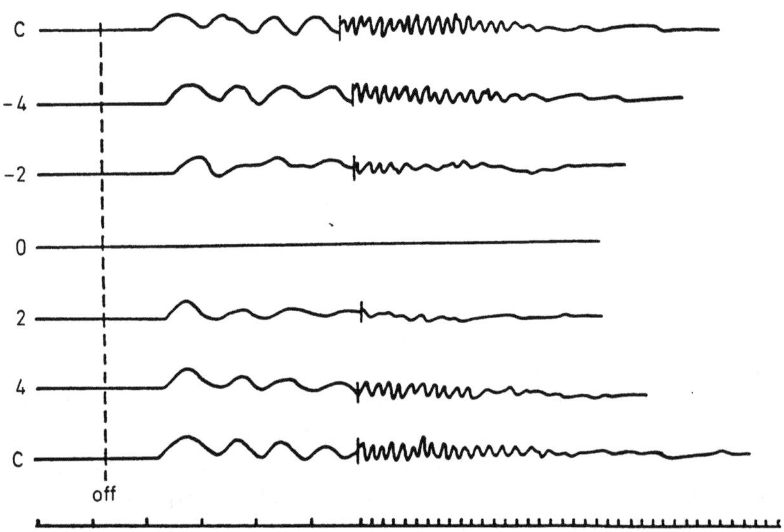

Fig. 11. Inhibitory effect of light on spine response of a sea urchin, *Diadema antillarum*. The number alongside each record shows spatial separation in mm of the readmitted light (L_2) from the conditioning one, extinction of which is shown by vertical broken line. C: control reactions obtained without readmission of L_2. (From MILLOTT and YOSHIDA, 1960b)

spine responses (YOSHIDA and MILLOTT, 1959) as well as electric activities (TAKAHASHI, 1964) can be elicited by direct stimulation of the radial nerve cord.

Studies on quantitative aspects of response parameters, such as reaction times, amplitudes of the spine jerks, the number of beats, etc., in relation to varying lighting conditions led MILLOTT and YOSHIDA (1960a) to a notion that the shadow response might be a rebound from inhibition due to light. Indeed, experiments as shown in Fig. 11 clearly indicated the inhibitory action of light (MILLOTT and YOSHIDA, 1960b). Here, the radial nerve was used as a receptive surface. A light (L_1) shone on a fixed point on the radial nerve. Upon shading this light, a second light (L_2) was admitted instantaneously onto a position shifted on either side of L_1. When L_1 and L_2 were made to coincide, the reaction was inhibited to a degree that was proportional to the intensity of L_2. Spatial separation of the two spots made the inhibitory effect weaker but the effects still appeared over distances as great as 6 mm.

The time course of development of excitation was studied by measuring the inhibitory threshold of L_2 which was admitted with different time intervals after L_1 was obliterated (Fig. 12; YOSHIDA, 1962). Each curve consisted of two portions, a flat part and a quickly rising phase. The former may be an expression of the

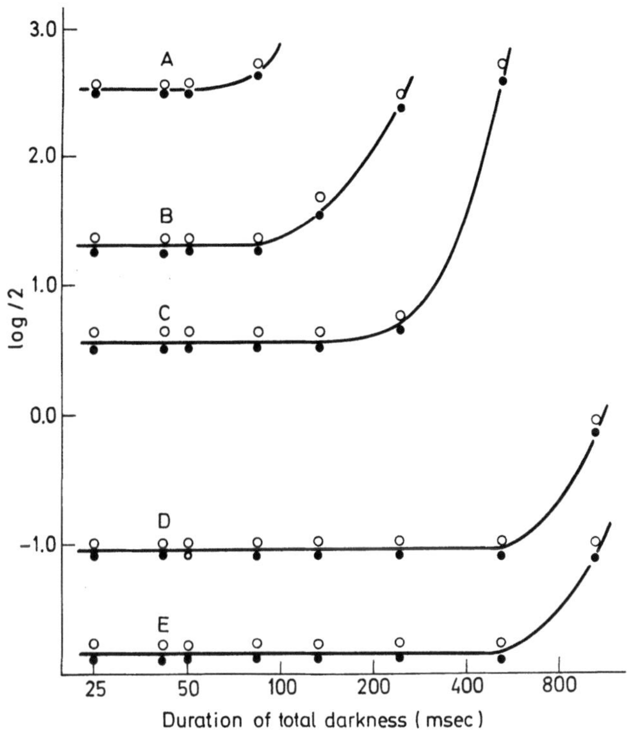

Fig. 12. Development of excitatory mechanism in darkness to elicit spine response of sea urchin, *Diadema setosum*, expressed in terms of changes in inhibitory threshold of light (abscissa), admitted with different time intervals (ordinate) after conditioning light is extinguished. Five examples at different intensities of conditioning light are shown. Open circles: minimum intensity to inhibit response. Solid circles: maximum intensity at which reaction was induced. (From YOSHIDA, 1962)

latent period of the primary receptive system. The mechanism leading to the overt response must consist of a chain of relay stations, and the quickly rising phase may reflect a stage whereby the excitation due to the primary interaction reached the second relay station, which is possibly associated with the so-called peripheral factor (MILLOTT and TAKAHASHI, 1963).

Treatments using dark and light adaptation affected the sensitivity level greatly, the degree of the effect depending on the size, intensity and duration of the wider field illumination (YOSHIDA, 1966). The interrelationship between the various lighting conditions and the proposed mechanism for eliciting the overt response may be schematized as shown in Fig. 13 (YOSHIDA, 1966).

A similar mechanism has been proposed to explain the myoneural photosensitivity in echinoderms by BOLTT and EWER (1963). The pharyngeal retractor muscle of *Cucumaria sykion* (POPLE and EWER, 1954, 1958) and the lantern retractor muscle of *Parechinus angulosus* (BOLTT and EWER, 1963) respond to light with a variety of contraction patterns, which consist of two basic types, quick and delayed. The contraction patterns may be a smooth transition from the quick to the delayed type, a clear differentiation between the two response types, a fall in tension below the resting level at *off*, a development of rhythmic activities under

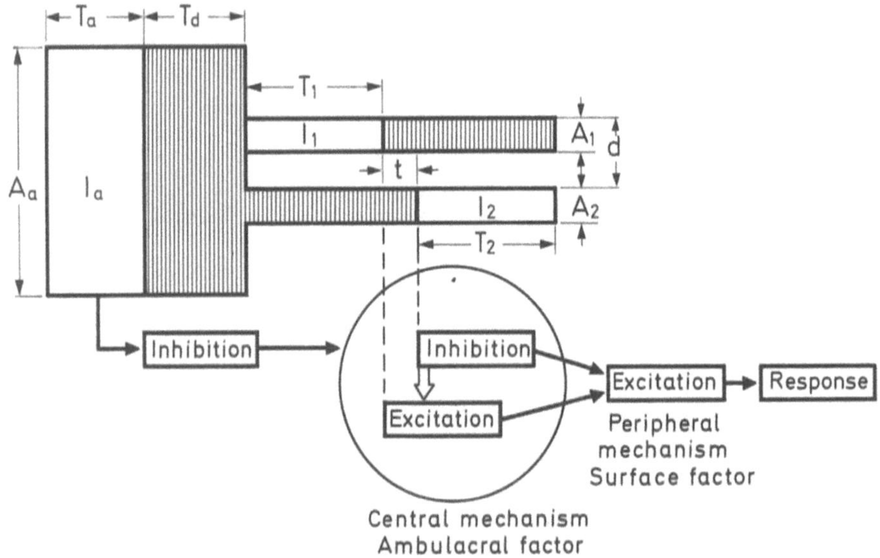

Fig. 13. Schematic representation of relation between photo-environmental factors and mechanism of shadow reaction of *Diadema setosum*. A: size of area illuminated. I: intensity of light. T: duration of illumination or darkness. Subscript a stands for adapting, 1 for L_1 and 2 for L_2, d for darkness. t: interval between extinction of L_1 and reillumination by L_2. d: spatial separation of the two lights. (From YOSHIDA, 1966)

prolonged illumination, etc. Evidence for the inhibitory action of light was obtained by making use of a mercury vapor lamp having a spectral composition different from that of the tungsten filament lamps. When two white lights were applied in succession, the muscle contracted in response to the second light as well as to the first. If, however, the second light was that of a mercury lamp, the muscle relaxed rather than contracted. They postulated a group of units in a motor complex: an inhibitory cyton (I) and two excitatory cytons, quick (Q) and delayed (D) (Fig. 14). Normally, the excitatory Q and D cytons are dominant. Under experimental conditions, the I cyton may come into play, either through its own photosensitivity or by a feedback mechanism fed in from the D cyton through its collaterals, resulting in inhibition of the activities of the two excitatory cytons to various degrees. The variety of response patterns to repetitive, paired, or prolonged stimuli has been considered to be a manifestation of the combined effects of the three interacting mechanisms.

Electrophysiologic analysis of events associated with purely dermal photosensitive systems has been done on *Hydra littoralis* (PASSANO and MCCULLOUGH, 1964, 1965). From time to time, the longitudinally arranged muscle sheet of the column showed a series of coordinated contractions which they called "contraction bursts" and recorded electric correlates of them. These bursts could be abruptly halted upon illumination. They have postulated the existence of a pacemaker which integrates various sensory inputs on the one hand and leads to coordinated responses in musculature on the other. Such pacemakers appear to be localized in the hypostome, hence being epidermal, and are directly inhibited by

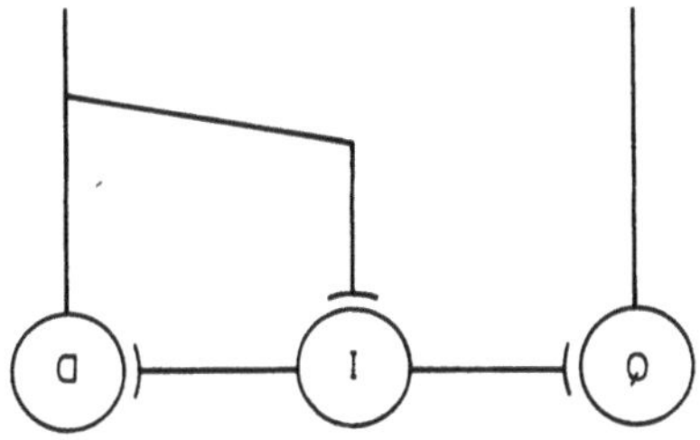

Fig. 14. Hypothetical motor unit to account for photoinduced contraction patterns in lantern retractor muscle of an echinoid, *Parechinus angulosus*, by interactions among quick (Q) and delayed (D) excitatory cytons, and inhibitory (I) cyton. (From BOLTT and EWER, 1963)

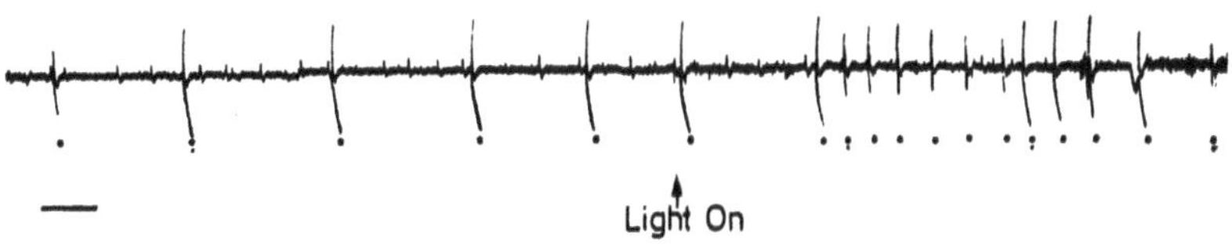

Fig. 15. Rhythmic potential (RP) responses to white light in dark-adapted *Hydra littoralis*. Note shifts in pacemaker *loci* during response, as shown by changes in recorded shape of RP impulse. Horizontal bar: 10 s. (From PASSANO and MCCULLOUGH, 1965)

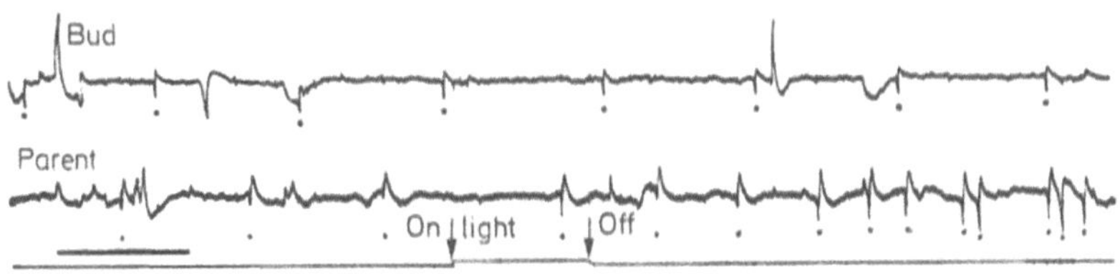

Fig. 16. Comparison of rhythmic potential (RP) systems between bud and parent of *Hydra pirardi*. Bud (*upper record*) shows rhythm independent of parent rhythm (*lower record*). Note rhythm of bud not altered by admission of light. (From PASSANO and MCCULLOUGH, 1965)

light. In the case of the sea anemone, *Calamactis praelongus*, however, the pacemakers which are located similarly in the margin of the oral disc appear to be all photo-excitable for inducing various types of co-ordinated bursts in the nerve net (MARKS, 1976).

PASSANO and MCCULLOUGH (1965) found another pacemaker activity which is gastrodermal in nature. This pacemaker elicited what they called rhythmic potentials (RPs). The RPs occurred regularly without apparent relation to movements of the animal, at frequencies varying from 0.5–12 cycles/min, and originated from pacemakers which were usually most active near the pedal base. Activities of the leading pacemaker was inhibited by sudden illumination. Since there were several pacemaker loci, other pacemakers would take over the inhibited leading pacemaker. The subsequent RP rhythm remained sometimes accelerated for some time (Fig. 15). Morphologic correlates of photoreceptive and pacemaker site(s) are not yet clear. It is interesting, however, that the photosensitivity gradually develops in the young after budding off from the parent, so that there is an intensity of light to which RP pacemakers in the parent are affected but those in buds are not (Fig. 16). In conclusion, the effect of light on the RP pacemakers of the adult *Hydra* is twofold when the animal is moderately illuminated; first it inhibits the initiation of RP impulses temporarily, and second it causes a subsequent enhancement of the RP rhythm when the firing is started.

D. Photosensitivities in the Nervous System

The foregoing accounts on the dermal photosensitivity point to the participation of nervous elements. BRAUN and JOB (1965) suggested such a mechanism for the shadow-induced reaction of siphons in *Mytilus edulis*, *Dreissensia polymorpha* and *Unio pictorum*. In the case of urchins, the degree of sensitivity gradient corresponded fairly well with the anatomic distribution of the superficial nervous system (MILLOTT and YOSHIDA, 1960b). Again, in *Hydra*, PASSANO and MCCULLOUGH (1964, 1965) used the term "neurons" and "protoneurons," respectively, for the possible sites of light-dependent pacemakers of contraction bursts and of rhythmic potentials. If functional evolution in sensory perception proceeds to acquire light as a modality of adequate stimuli from diffuse and nonspecialized organizations to localized and specialized ones, the photoreceptor cells scattered over the body surface as will be described later (Sec. E.II.1) may stand for a step toward the organized ocular organs. There appears to be another line, along which the dermal light sense that is associated with the nerves becomes specialized. This is neural or neuronal photosensitivity, which will be described below.

I. Neural Photosensitivity

This type of sensitivity has been noted not only in different forms of animals, but also from different parts of the nervous system in the same species. In crayfish, the caudal photosensitivity (see Sec. D.II.1) has long been known (WELSH, 1934), but the sensitivity appears not to be limited to the caudal region (CAMOUGIS, 1967). There are fibers in the abdominal nerve cord that respond either directly or indirectly to light stimulation by producing paired spikes or depolarizing by 3–4 mV (Fig. 17; WILKENS and LARIMER, 1972), but characterization of such units awaits future work. According to WELLS (1959), *Cambarus ayersii*, which live in subterranean cave streams, have extremely degenerate eyes, but the animals respond to photostimulation by leg movements, as do the surface species. Interestingly, the head region is the most sensitive to light. WELLS suggested that the caudal photoreceptor may be a precedent for photosensitivity of the CNS in crayfish. Again, the presence of nonocular and noncaudal photoreception has been

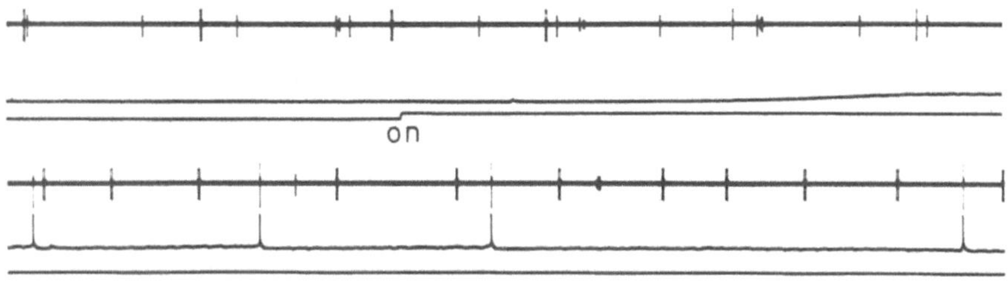

Fig. 17. Electric recordings from caudal photoreceptor and cells with related activity in crayfish, *Procambarus clarkii*. Medium-sized spikes in upper trace show existence of a second fiber in abdominal nerve cord. Characteristic firing pattern in doublet fashion increases under illumination. Middle traces are intracellular recordings of caudal photoreceptor. (From WILKENS and LARIMER, 1972)

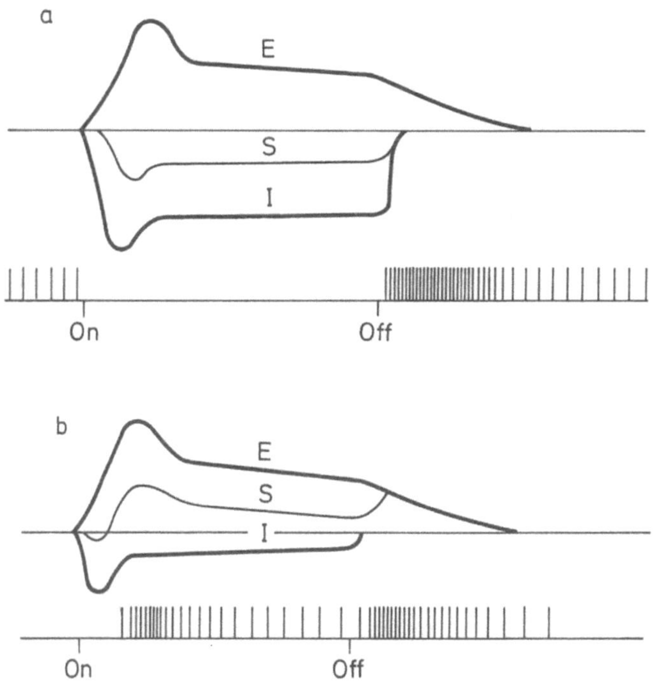

Fig. 18. Kennedy's hypothesis to explain neural photosensitivity of *Spisula solidissima*. A: stimulation with white light. B: stimulation with red light after prolonged illumination with blue light. E: excitatory process. I: inhibitory process. S: summed process, height above base line is related to frequency of impulses shown below each curve. (From KENNEDY, 1963b)

demonstrated to play a part in entrainment of the circadian rhythm of locomotor activities of *Procambarus clarkii* by PAGE and LARIMER (1972), who suggested that either the medulla interna or the medulla terminalis, or both, were involved.

Similarly, the photosensitivity in the nervous system in *Aplysia* is not limited to the visceral ganglion (see Sec. D.II.3). Thus, after the cessation of illumination of a whole unit of the head ganglia, activities can be recorded from all major branches of the posterior pedal nerves. By transecting connectives in various combinations,

BLOCK and SMITH (1973) have concluded that photoreceptors responsible for the *off*-response in pedal nerves are located in the cerebral ganglion and further suggested that the cerebral photoreceptors may participate in entrainment of the circadian rhythm of locomotion. The functional implications on the extraocularly entrainable circadian rhythm (BLOCK and LICKEY, 1973; BLOCK et al., 1974) will be dealt with in the following chapter of this volume by BENNETT.

In molluscs, the work of KENNEDY (1960) on the neural photosensitivity of a bivalve, *Spisula solidissima*, is worthy of being recalled to attention. He noted that spontaneous discharges of about 5/s at the proximal end of the pallial nerve near the visceral ganglion were inhibited by illumination on the nerve slightly distal to the visceral ganglion. Cessation of the illumination caused an increase in the discharge pattern and withdrawal responses of the siphon. KENNEDY proposed a scheme to explain the phenomena by assuming a difference in time courses between excitatory (slow) and inhibitory (fast) processes (Fig. 18).

The crucial point that made KENNEDY propose the scheme was based on differential effects of colored lights. He used a method of selective adaptation. After cessation of illumination with a blue light of 475 nm lasting for 5–30 s, the spontaneous discharges in darkness did not recover until more than 1 min later. If a red light of 630 nm was admitted 20–45 s after the cessation of the blue light, anomalous discharges were evoked upon admission of the red light. To account for these phenomena, KENNEDY proposed a dual mechanism differing in time constants, in function, and in responsible photopigments; one was fast, inhibitory and blue sensitive and the other, slow, excitatory and red sensitive. Thus, illumination with the blue light raises the threshold of the inhibitory mechanism so that the excitatory mechanism becomes easier to manifest, resulting in the appearance of the anomalous *on*-discharges with the red light.

Based on further microspectrophotometric measurements, KENNEDY suggested that two different kinds of pigments, a cytochrome and a carotenoid, were involved in eliciting the two opposing mechanisms. The suggestion of a heme-pigment participating in photic reactions is not new; it was shown, e.g., in photoexcitation of the *Aplysia* neurons by ARVANITAKI and CHALAZONITIS (1949a). On the other hand, it has been proposed recently that two different states of a pigment (its native and metastates) control two physiologically opposing mechanisms, such as in UV-sensitive photoreceptors of the median ocellus of *Limulus* (NOLTE and BROWN, 1972). In other words, any differences induced by selective adaptation with colored lights may not necessarily be taken to indicate the presence of multiple photopigments. Though a part of KENNEDY's results can be explained by the photoconversion theory, the choice of either one or two pigmentary system(s) is difficult to decide at the moment, since both schemes cannot account for the complicated pattern of responses in latency and frequency versus the intensity and the duration of stimulation.

Uncertainties in various aspects are encountered in so many instances of the neural photosensitivities that we must admit that it is often still to be determined whether the photoinduced activities picked up from the nervous system are in fact primarily triggered by the nerve cells themselves, or whether some kinds of structure that may be closely associated with the nervous elements are the primary site of phototransduction. Impulse activities recorded at light-*on* and -*off* from the

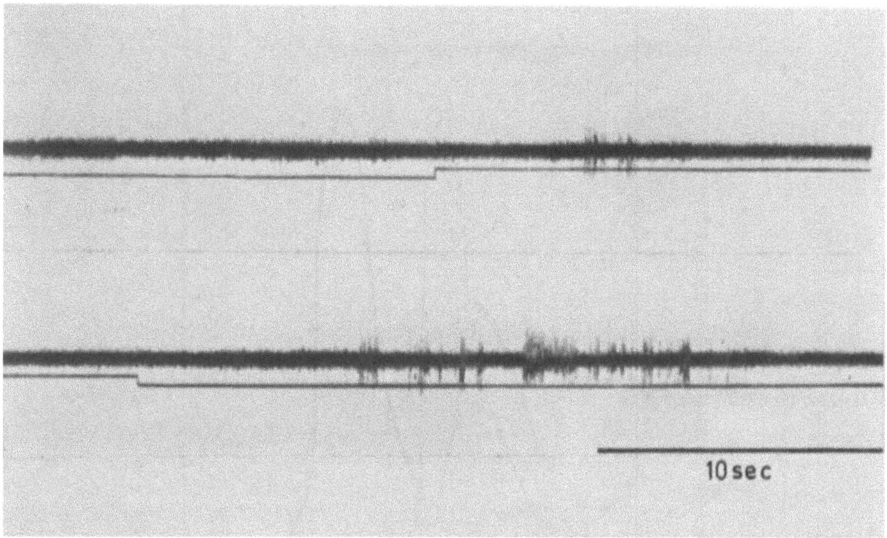

Fig. 19. Electric responses to on and off of light from isolated radial nerve cord of *Diadema setosum*. (From TAKAHASHI, 1964)

radial nerve cord of the sea urchin, *Diadema setosum*, by TAKAHASHI (1964) are one of such cases, since the initiation of the *off*-responses occurs sometimes as long as several seconds after the light-*off* (Fig. 19).

The situation of the neural photosensitivity in scorpions is far more controversial. Upon illumination of one of the five metasomatic segments in the Indian species *Heterometrus fulvipes* and *H. gravimanus*, spikes with frequencies of up to 24/s were evoked in the connectives in the mesosoma (GEETHABALI and RAO, 1973), not only on the ipsi- but also on the contra-lateral sides (GEETHABALI, 1976). Though the degree of sensitivity was variable within one species as well as between different species, a common feature in Indian species was that fifth and sixth ganglia located in the second and third metasomatic segments as well as the ganglion-free telsonic nerves were most sensitive. On the other hand, ZWICKY (1968) claimed earlier that the telsonic nerve was the structure solely responsible for the light sensitivity of an Australian species, *Urodacus novae-hollandiae*. The discrepancies were further extended by the peak position and the shape of action spectra (Fig. 20), which peaked at 480 nm with a narrow curve in the Australian species (ZWICKY, 1970), and at 568 nm with a secondary peak at 440 nm in the Indian species.

Last in this section, the work of WIEDERHOLD et al. (1973) on *Mercenaria mercenaria* may be noted. They recorded spikes from single axons in the siphonal nerve and, in contrast to KENNEDY (1960), found responses to cessation of illumination. They succeeded in locating the photosensitive area to within a circle of about 85 μm in diameter on the inner siphonal wall where the distal ends of the axons were distributed. Finding no photoreceptor cells distal to the axons from which they recorded, they postulated that the axonal branches were the site of phototransduction. Moreover, axonal processes contained well-organized penta-

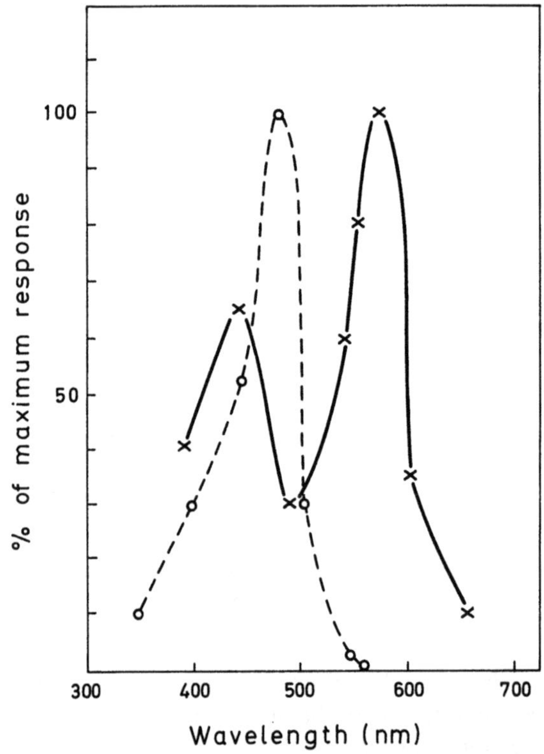

Fig. 20. Comparison of action spectra of scorpion tail responses between *Heterometrus fulvipes* (*solid line*) and *Urodacus novae-hollandiae* (*broken line*). (From Geethabali and Rao, 1973)

laminar whorls that they claimed were the site of photopigment concentration, and that reminds us of the structure in the photosensitive neuron soma of *Aplysia* (see Sec. D.II.3). This report is probably the only case in which the structural correlates for neural photosensitivities have been postulated. Most of the uncertainties mentioned above appear to be due partly to the lack of structural and biochemical facts and partly to labile properties of thin nerve fibers.

II. Neuronal Photosensitivity

We will consider below three kinds of identified photoexcitable neurons. Since it is here possible to study activities of individual neurons, the works can be more precise than those on any other types of extraocular systems.

1. Caudal Photoreceptors in Crayfish

Welsh (1934) demonstrated that illumination of the caudal ganglion affected the frequency of the leg movements in crayfish, *Procambarus clarkii* and *Orconectes virilis*, and in lobsters, *Panulirus argus* and *P. gattatus*, and that crayfish with eyes removed avoided the light, as did normal animals. A recent survey in marine decapods by WILKENS and LARIMER (1976) have revealed a wide, yet uneven occurrence of abdominal photoreceptor neurons, suggesting that their possible involvement in behavioral reflexes could be no less important than the major photoreceptive system.

PROSSER (1934a), working on the same species as WELSH (1934) did and on *Cambarus limosus* in addition, showed that electric responses to illumination of the caudal ganglion could be obtained in the abdominal and thoracic regions of the ventral nerve cord, in the pharyngeal commissures and in either the right or the left half of the cord alone. LARIMER (1966) showed further that caudal systems of cave-living species, *Cambarus setosus* and those of the cave population of *Procambarus simulans* responded well to photic stimulation. It thus appears that the system not only serves to warn the eye-equipped crayfish of exposure of the tail to light, but also to perceive photic information in cephalically blind cavernicolous crayfish.

The caudal receptor system consists of only two neurons, the axon of each travelling rostrally with considerable diameter changes in their course through the cord (KENNEDY, 1963a). The cell somata are not involved in transduction of photic signals, nor is it possible to modify ongoing activities by changing the membrane potential of the soma. Instead, as demonstrated by intracellular injection of Procion Yellow M-4RAM in combination with light probe experiments, the ganglionic area of light sensitivity corresponds closely to the dendritic tree originating from the contralateral cell body (Fig. 21; WILKENS and LARIMER, 1972).

Ultrastructural correlates for the photoreceptive site within the cell have not been clarified as yet. HAMA (1961) and UCHIZONO (1962) claimed that lamellated bodies and some other associated structures in the innermost sheath cell of the giant fiber of *Orconectes virilis* were responsible for light sensitivity. HERMANN and STARK (1963) pointed out that such structures were not restricted to the caudal ganglion but were also associated with the giant fiber system and yet not with B-fiber system, which is the conduction pathway for nerve impulses generated in the caudal ganglion. According to HESS (1940), however, *Crangon armillatus* and freshly moulted *Panulirus argus* are sensitive to light in many regions of the body. He also found that, whereas the sixth abdominal segment was the most sensitive region in *P. argus*, all the abdominal segments were equally sensitive in *C. armillatus* and *Homarus americanus*. In this connection, we must also be reminded of the fact that neural photosensitivities have been noted in various parts of the crayfish nervous system (see Sec. D.I). It appears to be too early to discard the possibility that the structures found independently by HAMA and by UCHIZONO are in fact photosensitive.

The caudal photoreceptive system has a slow adapting property (PROSSER, 1934a), and an excitatory state that is evoked by brief subthreshold flashes can exist and sum with the second flash given as long as 10 s later (KENNEDY, 1958). Synaptic potentials from different inputs can also sum to initiate spike discharge (PRESTON and KENNEDY, 1960). KENNEDY and PRESTON (1960) have differentiated intracellularly four classes of interneurons in the neuropil of the caudal ganglion, one of which responding both to tactile stimuli and to activation of the ganglionic photoreceptor. It may be assumed that, being a primary photoreceptor as well as

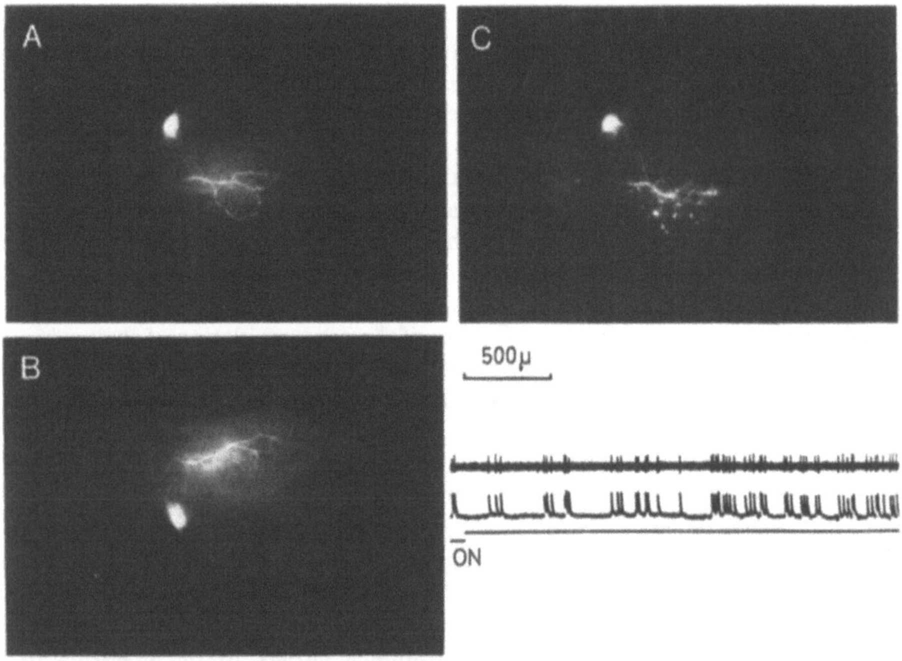

Fig. 21. Procion-Yellow-injected caudal photoreceptor cells in sixth abdominal ganglion of crayfish, *Procambarus clarkii*. A and B, dorsal and lateral view, respectively, of one preparation showing dendritic branching morphology and contralateral soma. C: Cell morphology of a second, dye-injected photoreceptor cell, demonstrates similar morphology, despite some leakage of dye at branch terminals. By recording from caudal photoreceptor axon in 5–6 abdominal connective with extracellular electrode (*lower inset*) and simultaneously with dye-filled microelectrode (*upper inset*), identity of cell shown in A and B was physiologically assured. (From WILKENS and LARIMER, 1972)

an interneuron, the cell serves to facilitate under illumination the transmission of other sorts of, possibly tactile, sensory information through subliminal effects on postsynaptic excitability.

2. Ganglionic Complex of *Onchidium*

A marine pulmonate, *Onchidium verruculatum*, responds to sudden increase or decrease in illumination by cessation of locomotion, withdrawal of the stalked eyes, or contraction of the mantle. Encouraged by finding large-sized nerve cells (100–300 µm in diameter) in the ganglion complex by HAGIWARA and SAITO (1959), HISANO et al. (1972a) have succeeded in identifying eight photoexcitable neurons. Of these, two are primarily (designated as Ep2 and Ep3) and one, secondarily (Es1) photoexcitable, and they are located in abdominal (Es1) and right (Ep2) and left (Ep3) pleuroparietal ganglia. Later, they found a fourth photoexcitable neuron in the pedal ganglion (HISANO et al., 1972b). The rest of the five neurons are all primarily photoinhibitory (Ip1–Ip5); they are located in abdominal (Ip1, Ip2) and right pleuroparietal (Ip3, Ip4, Ip5) ganglia (Fig. 22). Perhaps because they could find no secondary neurons associated with the

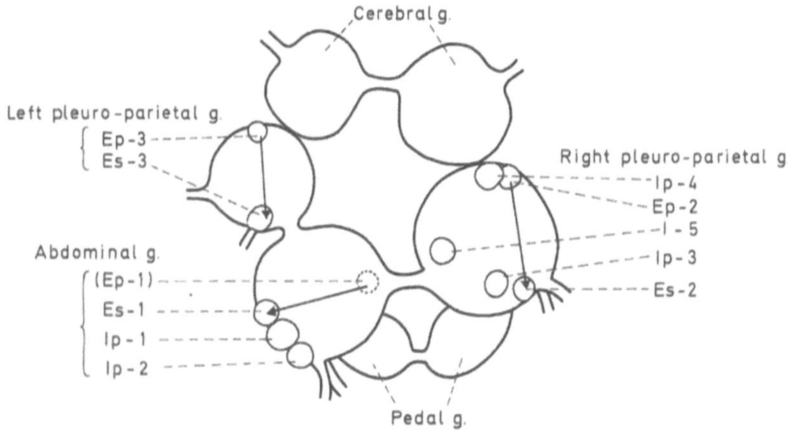

Fig. 22. Location of photoexcitive neurons in ganglion complex of *Onchidium verruculatum*. E: photo-excitable neuron. I: photoinhibitory neuron. p: primary neuron. s: secondary neuron. Arrows show synaptic connections between primary and secondary neurons. (From HISANO et al., 1972a)

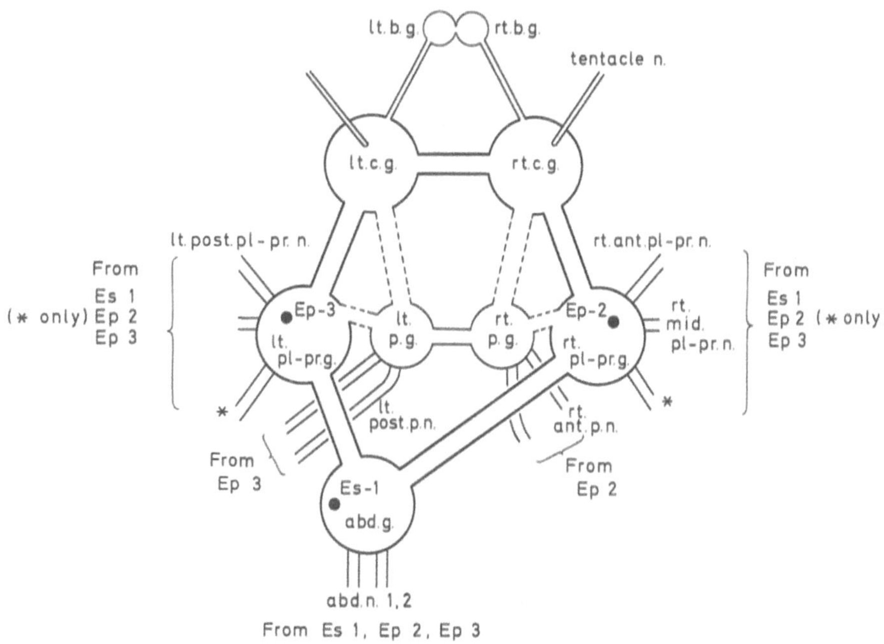

Fig. 23. Axonal pathways from photoexcitive neurons in ganglion complex of *Onchidium verruculatum*. (Modified from HISANO et al., 1972b)

inhibitory type, their succeeding work has mainly concentrated on the excitable ones, particularly on Es1, Ep2 and Ep3.

The axonal pathways from the photoexcitable cells have been studied 1) by recording antidromic spikes at the cell soma in response to electric stimulation to each nerve fiber running out of the ganglion complex (HISANO et al., 1972b), 2) by

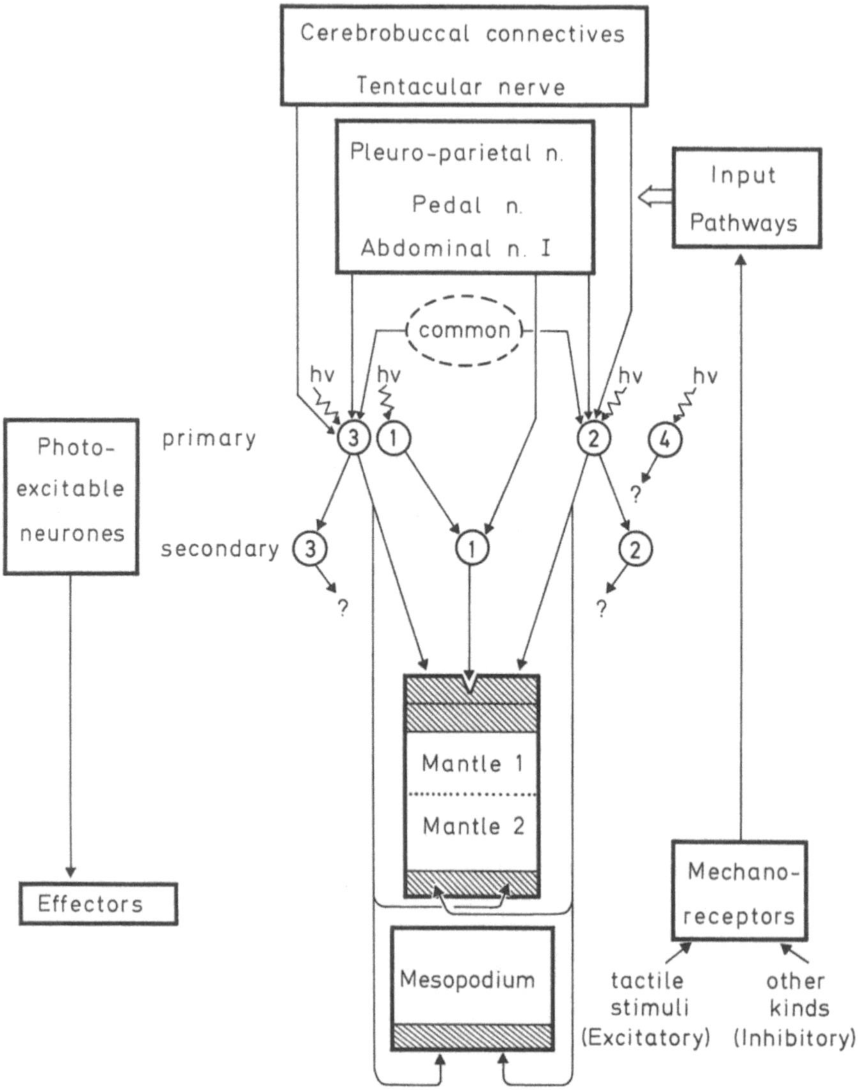

Fig. 24. Input and output organization of photoexcitable neurons of *Onchidium verruculatum*. The scheme constructed by YOSHIDA was slightly modified by Dr. T. S. GOTOW. (Personal communication)

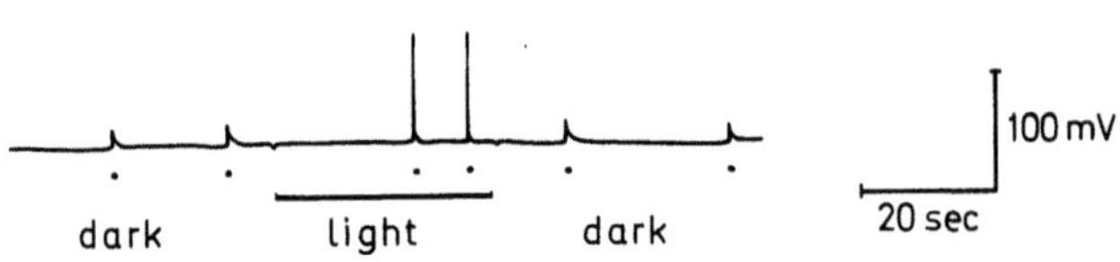

Fig. 25. Interaction between light and electric stimulation to right posterior pleuroparietal nerve in isolated preparation of *Onchidium verruculatum*. Intensity of the electric stimuli was subthreshold for spike initiation under dark conditions. Intensity of photic stimulation was also subthreshold. Full spikes were elicited under illumination by same electric stimuli, as result of summed depolarization. Underlying dot of each response indicates the electric stimulus. Black bar under record indicates subthreshold photic stimulation. (From GOTOW et al., 1973)

observing the levitation movements in response to direct electric stimulation (5/s for 10–20 s) of each photoexcitable neuron (Ts. GOTOW et al., 1973), 3) by injecting cobalt ions into each cell (Ts. GOTOW, 1975), and 4) by simultaneous recordings from the cell soma and from each nerve leaving the ganglion complex, upon depolarization of the soma by intracellularly applied current pulse (Ts. GOTOW, 1975). The results may be summarized as shown in Fig. 23.

These photoexcitable neurons are also sensory interneurons. The situation is similar to that in the crayfish caudal photoreceptors, but here the input organization has been studied more extensively by recording electric activities of each neuron in response 1) to tactile stimulation of the mantle surface by a brush, 2) to pressure on the mantle by a lucite bar (Ts. GOTOW et al., 1973), and 3) to orthodromic stimulation of each nerve leaving the ganglion complex (Ts. GOTOW, 1975). Figure 24 summarizes the whole results elucidated so far by the Kyushu group on the input and output organization in the ganglion complex. Ts. GOTOW et al. (1973) have clearly shown further that summation of two subthreshold stimuli, one from the input pathways and the other by light, does occur, resulting in firing only during the period of illumination (Fig. 25). Such a photofacilitatory mechanism may lead one to expect some advantages for the behavior of animals under natural conditions. It remains to be determined, however, what relationship this mechanism has with sensory inputs other than the mechanical stimuli, such as those through dorsal ocelli (YANASE and SAKAMOTO, 1965; FUJIMOTO et al., 1967), and what effects the photoinhibitory neurons exert on this mechanism.

3. Visceral Ganglion of *Aplysia*

Since the work of ARVANITAKI and CHALAZONITIS (1949a, b), giant neurons in the visceral (abdominal) ganglion of *Aplysia* have been studied widely. The terminology used by different authors has caused some confusion, however. Thus, photosensitive A, Br, and Gen cells called by ARVANITAKI and CHALAZONITIS (1960, 1961) correspond, respectively, to R2, R15 and L11 cells by FRAZIER et al. (1967) and B and H cells by the former, possibly to some of the rostral L series cells by the latter (Fig. 26, KANDEL et al., 1967).

Among those cells just mentioned, R2 cells have been most extensively studied. ARVANITAKI and CHALAZONITIS (1960) noted that if the cell showed spontaneous discharges in darkness, the spike frequencies increased at *on*, or if it was silent in darkness, it was depolarized and led to spiking at *on*. The cell also showed an increase in membrane potential and was led to spiking at *off*. The *on*-spiking phenomenon was considered to be an excitatory event, and the *off*-spiking a primary inhibition, each being associated, respectively, with a heme protein (λ_{max}: 579, 452 and 418 nm) and a carotenoid pigment (λ_{max}: 490 and 463 nm). It is also known that the R2 cell receives inputs from two excitatory interneurons, one of which responds briskly to light-*off* and that most of the identifiable neurons send one or more axons into peripheral nerves with ascending pathway from the R2 cell via the commissure to the higher ganglia (FRAZIER et al., 1967).

These findings on the European species, *A. fasciata* and *A. depilans*, by ARVANITAKI and CHALAZONITIS have partially been substantiated by BROWN and

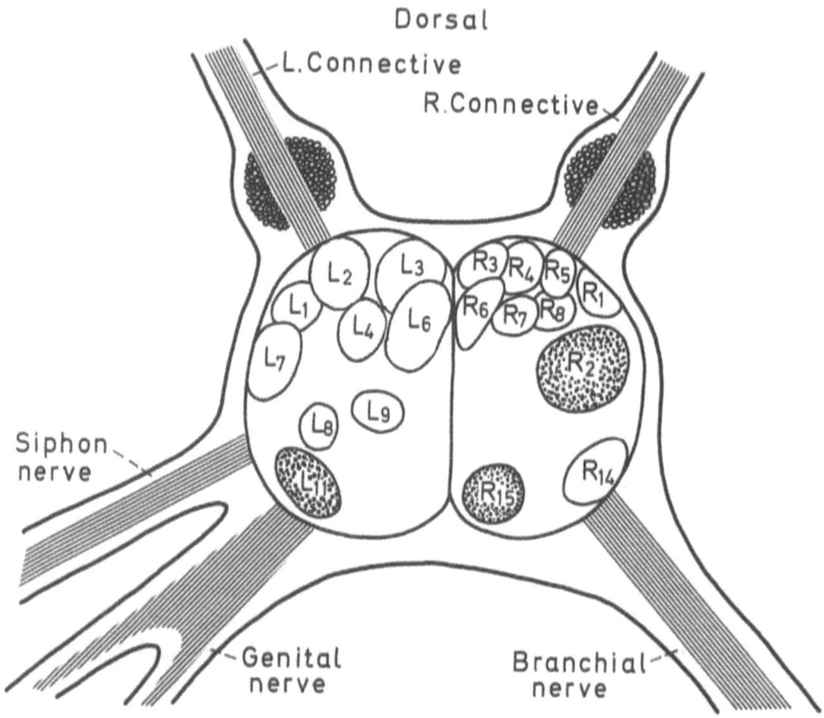

Fig. 26. Dorsal surface of visceral ganglion of *Aplysia californica*, indicating most common position of identified cells (slightly modified from FRAZIER et al., 1967). Photosensitive cells are stippled

BROWN (1973) on an American species, *A. californica*. The latter authors found that the cell became less negative upon illumination by 2–5 mV after a delay of 400 ms with about 15% reduction in membrane resistance. This cell can also be hyperpolarized by acetylcholine (FRAZIER et al., 1967), like H-cells (L11 and L6) in the visceral ganglion (BLANKENSHIP et al., 1971). Since the receptor potential has been reviewed in this volume by JÄRVILEHTO, it may be sufficient to state that, unlike the responses of H-cells which are dependent on Cl^- ions, this light-evoked hyperpolarization is brought about by an increase in K^+ conductance, a phenomenon which resembles the hyperpolarization of some of the atypical invertebrate photosensory cells, such as the distal retinal cells of a scallop, *Pecten irradians* (MCREYNOLDS and GORMAN, 1974) and microvilli-bearing cells of a pelagic tunicate, *Salpa democratica* (MCREYNOLDS and GORMAN, 1975). Mimicking the effects of light on the R2 cell by pressure injection of Ca^{2+} ions, BROWN and BROWN (1973) proposed that Ca^{2+} played a role in the phototransduction mechanism. Such a notion has recently been supported by findings of photoinduced ultrastructural changes of cytoplasmic inclusions from globules to membranelike lamellae (HENKART, 1975) and also by electron microprobe analysis which indicated a marked decrease in Ca content of pigmented granules by light (BROWN et al., 1975).

There are a few more photosensitive neurons in the visceral ganglion. Thus the R15 cell, which belongs to the D-cell according to the category of BLANKENSHIP et al. (1971), is hyperpolarized by light and decreases its burst rhythms which normally occur as a group of 15–20 spikes in every 5–20 s with silent periods between them (FRAZIER et al., 1967). Again, according to ARVANITAKI and CHALAZONITIS (1960), Gen (=L11=H-cell), B and H (=rostral L series) cells are hyperpolarized by light. It may be mentioned that the R15 neurons show a circadian rhythm in the frequency of spikes and the rhythm is shown to be entrained by some extraocular photoreceptors (LICKEY and ZACK, 1973). However, the functional significance of the pronounced photosensitivity of the abovementioned cell types is still obscure.

E. Unorganized Photoreceptive Cells

Since EAKIN (1963) proposed two lines for evolution of photosensory cells, a number of works have elucidated fascinating varieties in the peripheral structure where photon capture is assumed to occur (EAKIN, 1972). Considering that some of the rhabdomeric cells may develop from the ciliary cell as in *Helix aspersa* (EAKIN and BRANDENBURGER, 1967) and that the larval photoreceptors in tornaria (BRANDENBURGER et al., 1973) and the sensory cells in the pygidial ocelli in a polychaete, *Chone ecaudata* (ERMAK and EAKIN, 1976) bear ciliary and rhabdomeric characteristics in a single cell, the ciliary cell may be assumed to stand for

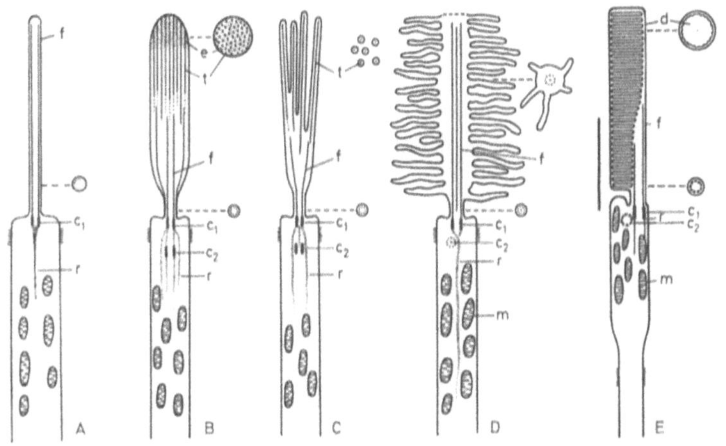

Fig. 27. Diagrams of distal nerve processes of epithelial ciliary receptor cells for various modalities of adequate stimuli. A: chemo- and/or mechanoreceptor of the auriculum of a planarian. B: mechanoreceptor of insect hair or campaniform sensillum. C: chemoreceptor of insect hair sensillum. D: photoreceptor of a hydromedusan. E: rod-photoreceptor of vertebrates. Hatched zones outside cells symbolize junctional complexes connecting to neighboring cells. $c_{1,2}$: centrioles. e: intensely stained material associated with microtubules. f: ciliary filaments. m: mitochondria. r: root filaments. t: microtubules. (From THURM, 1974)

the first step toward acquiring light as a modality of adequate stimulus. The structural resemblance associated with the ciliary structure in mechano-, chemo-, and photosensory cells has been neatly shown by THURM (1974) (Fig. 27). We will consider below the functional implications of such primitive photoreceptive cells.

I. Identification of Photosensitive Cells

A demonstration of a responsible photopigment located somewhere in the presumptive cell is an important step for determining that the cell in question is photosensitive. In organized receptors, there are a number of methods to perform such a task; these are extraction of pigment(s), microspectrophotometric analysis, osmic black method (EAKIN and BRANDENBURGER, 1970; GRIBAKIN, 1972), and utilization of either tritiated retinol (EAKIN and BRANDENBURGER, 1968; BRANDENBURGER and EAKIN, 1970), procion yellow M4RS (LATIES et al., 1976), or a dye called N,N′-didansyl cystine, which, combining specifically with rhodopsin, increases its fluorescent efficiency about 20-fold (YOSHIKAMI et al., 1974). For primitive photosensory cells, none of these methods are practically applicable. The methods commonly used, which are by no means highly effective are 1) the correspondence between the sensitivity gradient and the distribution of the

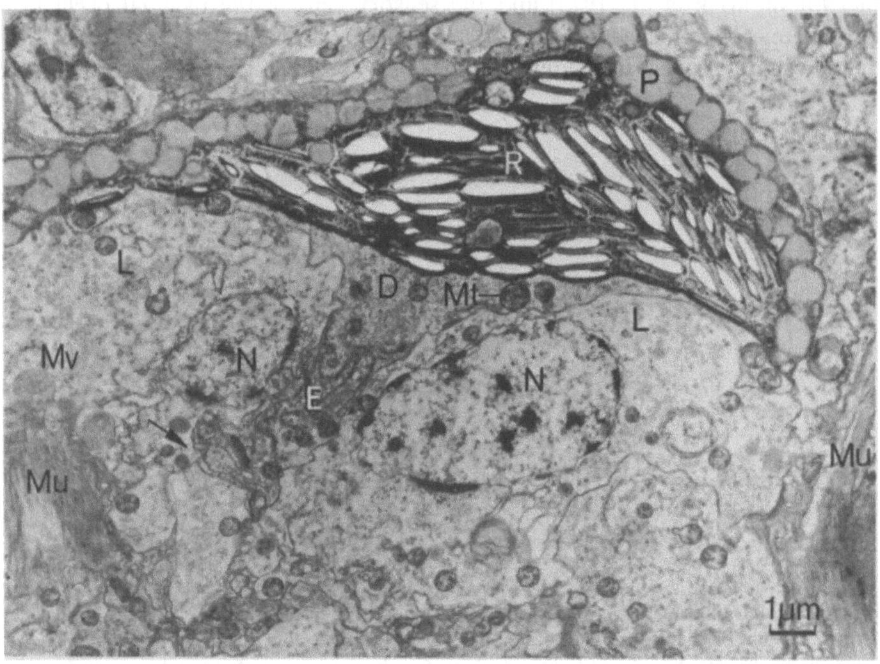

Fig. 28. Fine structure of ocellus of an acoela, *Convoluta* sp. D: dark cell (sensory). E: endoplasmic reticulum. N: nucleus. L: light cell (sensory). Mt: mitochondrion. Mu: muscle. Mv: multivesicular body. P: pigment layer. R: reflecting layer. Arrow: protrusion from light cell. (By courtesy of Dr. TERUFUMI YAMASU of Ochanomizu University, Tokyo)

candidate cells and 2) the morphologic resemblance in the peripheral parts and/or in the cytoplasmic inclusions with any one of the known photosensory cells. In some cases, the first method cannot be used and one is compelled to discuss the matter only by analogy in accordance with the second method.

There are still two more points to be mentioned. 1) Most of the morphologic work on the primitive photosensory cells has been concerned with the peripheral structure and only little has been mentioned as regards the relations with the secondary neurons, which are no less important for photic behavior of the animal in question. 2) The idea that modification must have occurred in the peripheral structure as an essential step towards evolution of a photosensory cell appears to have often occupied the workers' minds. As pointed out by FUORTES (1971), it is true that an increase in membrane surface fits well with the purpose of carrying a large number of photosensitive molecules. If we are reminded of the cases of the direct responses of chromatophores and the neuronal photosensitivity, however, it will be understood that the peripheral modification is not always required for a cell to become photoreceptive. Indeed, recent studies on the morphology and function of the eyespot of an acoela flatworm, *Convoluta* sp., (YAMASU, unpublished) have revealed that the supposed photosensory cells which are surrounded by a pigment cell and a reflecting layer have no peripheral modification and are synaptically connected to the brain (Fig. 28). I cannot help admitting, therefore, that the following descriptions are largely handicapped by the technical difficulties mentioned above for identification of photosensory cells.

II. Scattered Photoreceptive Cells

1. Externally Scattered Cells

The most well-known and classic example is the cells called "Lichtzellen" by HESSE (1896) in the earthworm. He noted a dark-staining terminal structure, which has long been considered to be an intracellular lens (often called phaosome), serving to focus light onto unknown photosensitive structures inside the cell. Such a notion was rightly doubted by STEVEN (1963). The recent work by RÖHLICH et al. (1970) on *Lumbricus terrestris* has clarified the situation (Fig. 29). The phaosome is in fact an assemblage of microvillar membranes and in this respect resembles that of the *Hirudo* photoreceptor cell. In the case of *Hirudo* receptors, it has been convincingly demonstrated that the phaosome is in fact not an intracellular structure but is connected to the extracellular space through clefts of about 160 Å wide and 2.7 µm long (WHITE and WALTHER, 1969; LASANSKY and FUORTES, 1969). Electrophysiologic studies revealed further that the resistance to the outside from the cytoplasm was $20 \times 10^6\ \Omega$, whereas that from the vacuole was $8 \times 10^6\ \Omega$ (LASANSKY and FUORTES, 1969), and that in darkness there was no potential difference between the vacuole and the outside (FIORAVANTI and FUORTES, 1972). Though RÖHLICH et al. (1970) could not find a cleft in these cells in the earthworm, they assumed by analogy the possibility that continuity between the microvillar membrane and the outer plasma membrane could exist. If we admit that the so-

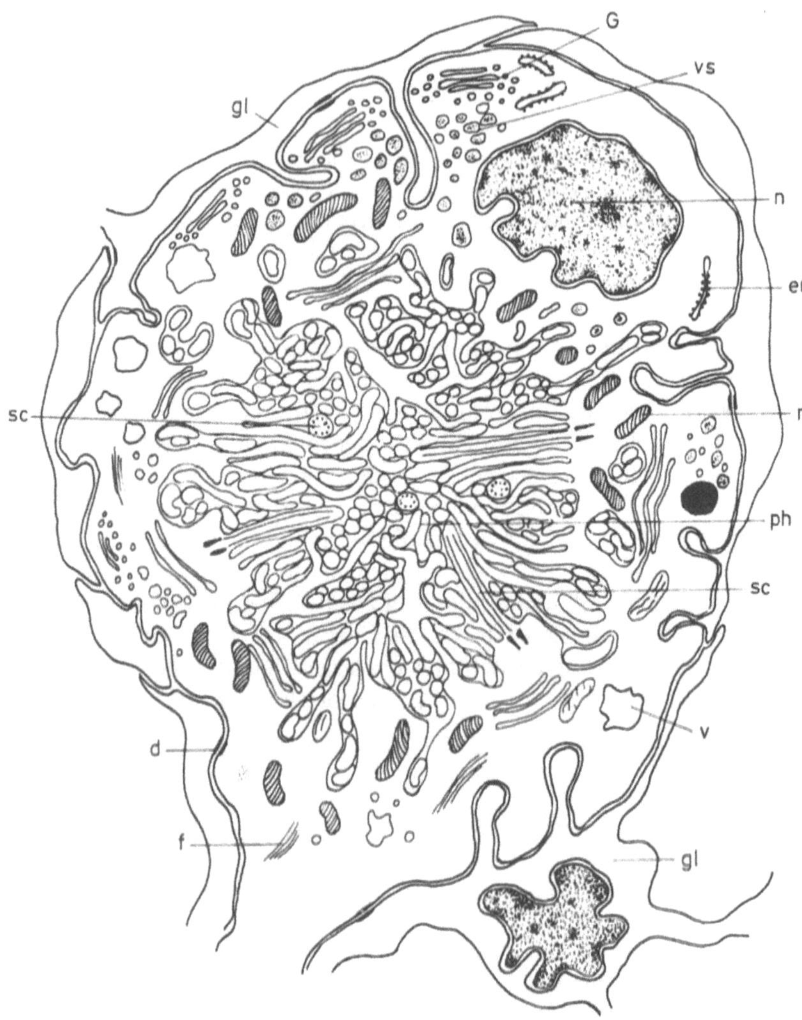

Fig. 29. Photoreceptor cell of *Lumbricus terrestris*. ph: phaosome (microvilli). sc: sensory cilia. m: mitochondria. N: nucleus. G: Golgi region. er: ergastoplasmic sacs. vs: vesicles having dense content. v: clear vacuoles. f: bundle of filaments. gl: glial cell and processes. d: desmosome. (From RÖHLICH et al., 1970)

called inside cavity is in fact an extracellular space, we have no reason to feel it peculiar that a number of cilia of 9+2 type project into the phaosome.

DHAINAUT-COURTOIS (1965) found ciliated photosensory cells in the brain of a polychaete, *Nereis pelagica*. Similarly, cells called "sensory sacs" or "photoreceptor-like cells", which are provided with cilia have been found embedded in the brain of some other polychaetes, *Armandia brevis* (HERMANS and CLONEY, 1966), *Nephtys caecoides*, *N. caeca* and *N. hombergi* (CLARK, 1956; ZAHID and GOLDING, 1974). We also have found them scattered near the cuticular surface of the prostomium of a scale worm, *Lepidonotus helotypus* (Fig. 30; TA. GOTOW, 1976). The cells that look intracellularly ciliated appear to be rather common in

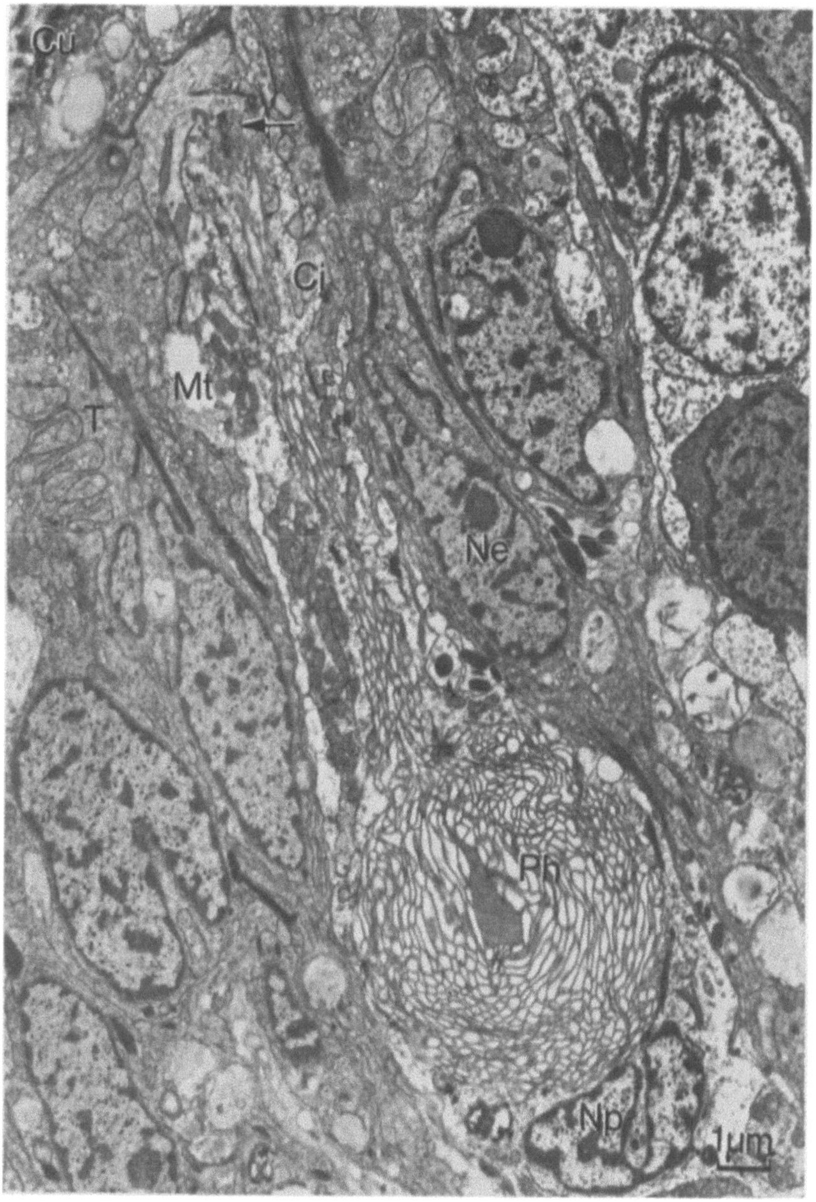

Fig. 30. Fine structure of the photoreceptor-like cell of scale worm, *Lepidonotus helotypus*. Ci: cilium. Cu: cuticle. Mt: mitochondria of neighboring cell. Ne: nucleus of epithelial cell. Np: nucleus of photoreceptor-like cell. Ph: phaosome. T: tonofilament. Arrow: basal structure of cilium. (T. A. GOTOW, 1976)

annelids, though the axonal structure and the functional significance are yet to be determined.

Photosensitivity in the earthworm has been studied by HESS (1924, 1925) on *L. terrestris* and by NOMURA (1926, 1927), SEGALL (1933) and PROSSER (1934b) on *Eisenia foetida*. Using negative phototaxis and reflex responses to light and shade as criteria, it has been found that there are marked differences in the degree of sensitivity, the prostomium and the anterior three segments being the most sensitive, extreme posterior segments the next, and the middle region of the body the least. The distribution of the above-mentioned photosensory cells corresponds well with the sensitivity gradients; 57 cells in the area of 200 × 300 μm in the prostomial region, 14 cells in the last segment, and one cell in the 40th segment.

As regards the information-processing mechanisms, HESS, NOMURA and PROSSER all agree that negative reactions that the earthworms normally show are under the control of the cerebral region, without which the ventral nerve cord dominates, resulting in positive responses. How such opposite activities can be exercised at different levels in the CNS in response to the information fed in from the same kind of photosensory cells is not yet known.

A marine neogastropod, *Nassarius reticulatus*, also possesses presumably photosensory cells that are scattered over the epithelial surface (CRISP, 1971, 1972). Intact and eyeless animals both respond well to shadows by a sudden contraction of the columellar muscle pulling the shell down to the substratum and of the longitudinal muscle of the siphon (CRISP, 1972). Earlier, CRISP (1971) distinguished two cell types distributed over the body surface; one (type-1), possibly chemo- and/or mechanoreceptors, having cilia that extend beyond the epithelial surface, and the other (type-2), possibly photoreceptors, whose cilia do not reach the general epithelial surface (Fig. 31). The former cells are distributed all over the surface, but the latter are restricted to the external siphonal epithelium. Removal of the outer siphonal surface reduces significantly the shadow reaction. The siphon appears not to be the sole receptive site, however, since the columellar response is not abolished by removal of the siphon (CRISP, 1972). She suggested, therefore, that photoreceptors were also distributed along the mantle edge, though it was not shown whether they were the type-2 cells or of a new type.

The distribution of the photosensitivity over the mantle surface has been noted in another gastropod; in *Helix pomatia*, FÖH (1932) found that shading the mantle edge elicited a sharp contraction, the most sensitive region being located at the forward edge of the shell. SCHULZ (1938), finding various types of sensory cells by the Golgi method, counted the number of the possible photoreceptor cells. He found that the density pattern of the presumed photoreceptor cells (9.4/mm^2 at the anterior head region, and 6.8 at the side of small tentacles) coincided well with the degree of responsiveness to light. Ultrastructural studies have not been made on these epidermal photoreceptors.

The siphonal photoreceptors are important in bivalves, particularly when they lack differentiated mantle ocelli (see Sec. F.I). In *Mya arenaria*, LIGHT (1930) reported the presence of pear-shaped cells, 9–15 μm long and 5–10 μm wide, immediately beneath the inner epithelial layer of the siphon. KOLLER and STUDNITZ (1934) proposed different receptor-effector systems for light and shadow reflexes as regards the photosensitive pigments. Though STEVEN (1963) cast some doubt on the photosensitivity of the pear-shaped cells, some kind of photosensitivity seems to exist in the siphonal region of the bivalves. KENNEDY (1960) recorded bursts of impulses not only from the entire siphonal nerve of *Mya arenaria* and *Venus mercenaria* but also from its more peripheral branches, both at the onset and the cessation of illumination. The ultrastructure of what LIGHT

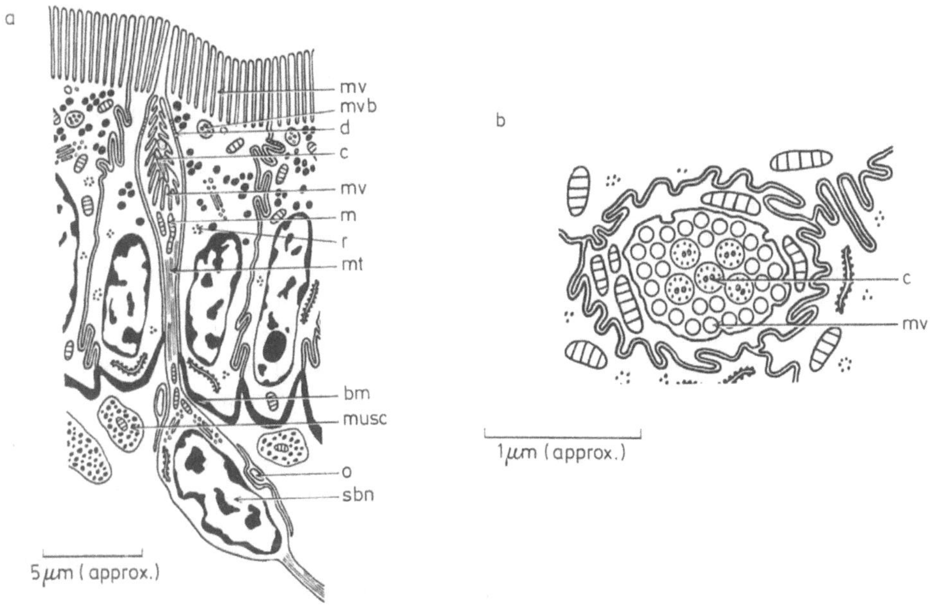

Fig. 31. Schematic diagram of the type-2 epithelial ending of *Nassarius reticulatus*. A: Longitudinal section. B: transverse section in distal one-third of epithelium. bm: basement membrane. c: cilium of sensory cell. d: septate desmosome. m: mitochondrion. mt: microtubules. musc: muscle fiber. mv: microvilli. mvb: multivesicular body. o: ovoid opaque body. r: free ribosomes. sbn: nucleus of sensory cell. (From CRISP, 1971)

(1930) called photoreceptors was studied later by BELL and ANDERSON (1970). They distinguished three separate structural entities among the so-called photoreceptors; one containing membrane-bound crystalline inclusions, about 0.5 µm in diameter and irregular in shape, and the other two being conceivably neural elements. They further suggested that the photoreceptor organs may receive efferent neurons.

There is no evidence against the notion that the scattered sensory cells discussed so far are not involved in photoreception, but lacking detailed knowledge of the nervous architecture which constitutes the afferent information pathways, we have to admit that little is known about the intimate mechanism underlying between the peripheral structure and the overt responses.

2. Internally Scattered Cells

Photosensory cells in the earthworm are not restricted to the epidermal surface but they are also found in the nerve enlargements within the prostomial and the caudal segments (HESS, 1925). Presumed photoreceptors associated similarly with the CNS have been postulated in rotifers, trematoda, and sipunculoids (see BULLOCK and HORRIDGE, 1965). The ultrastructure of some of them has been studied (a sipunculoid, *Phascolosoma agassizii* by HERMANS and EAKIN, 1969), but the most well-studied physiologically are the ventral photoreceptors in *Limulus polyphemus. Limulus* (see review: FAHRENBACH, 1975) has two external ocular systems called lateral and median, which, since the classic work of HARTLINE and GRAHAM (1932), have been studied extensively from a general physiologic viewpoint, and both of which play an important role for positive phototactic behavior under natural conditions (LALL and CHAPMAN, 1973). On the other hand, the ventral photoreceptor cells occur in pair in young forms and are located over the ventral surface of the protocerebrum. As *Limulus* grows, photoreceptor cells increase in number and some of them become separated from the brain. It appears that the physiology of these cells, which were long described as olfactory nerves, etc., was pioneered by HARTLINE'S unpublished experiments. Thus, MILLECCHIA et al. (1966) mentioned that HARTLINE first noticed the presence of light-induced signals from the "olfactory nerves", and also according to SNODDERLY (1971), HARTLINE recorded *off*-responses which were evoked by illumination of a brain without any afferent inputs.

The cells have highly modified infoldings of the plasmalemma (CLARK et al., 1969), contain a photolabile pigment peaking at 530 nm (MURRAY, 1966), and produce photoelectric potentials with λ_{max} also at 530 nm (BROWN et al., 1967). Since the cell is large, 200×60 µm, photoexcitable, and with no complicating factors due to lateral interactions, many works have been directed to the analysis of processes during adaptation (FEIN and DEVOE, 1973; SREBRO and BEHBEHANI, 1974), the membrane properties during excitation (LISMAN and BROWN, 1971; BROWN and LISMAN, 1972; SPIEGLER and YEANDLE, 1974; BROWN and MOTE, 1974; FEIN and CHARLTON, 1975), and the role of calcium ions in the photoexcitatory mechanism (BROWN and BLINKS, 1974; LISMAN and BROWN, 1972, 1975).

The cells are so sensitive that incidence of 10^3 quanta per second at the surface of a photoreceptor elicits slow potential fluctuations (MILLECCHIA and MAURO, 1969), yet no spike activities have been detected along the long ventral nerve cord (CLARK et al., 1969). This reminds us of the barnacle photoreceptive system which sends information electrotonically with a length constant of 4.9 mm (SHAW, 1972) to the supraesophageal ganglion (GWILLIAM, 1963, 1965). Here, the functional significance of this receptive system for inducing the well-known shadow reflex (BUDDENBROCK, 1930) appears to have been fully substantiated, since OZAWA et al. (1976) found, in the supraesophageal ganglion, cells that responded at "on" and "off" of light and since GWILLIAM (1976) recorded in the ventral ganglion two classes of burst-producing units; at light off one, located laterally, being excited and the other, in the mid-line, inhibited. The fact that propagated activity could not be recorded with light stimulation may be interpreted in one of the three ways: 1) The signal is conducted to the higher center by the electrotonic spread as barnacle photoreceptors do, though in the case of *Limulus* axons whose space constants range from 2–4 mm, the nearest synapse is located as far as 6 cm away (MILLECCHIA et al., 1966), 2) the failure is due simply to a technical difficulty in making responsive preparations, and 3) no information is in fact sent to the higher order neurons.

The possibility that the photoreceptor cell axons are extremely labile (CLARK et al., 1969) is plausible. Indeed, according to KAPLAN and BARLOW (1975), optic nerve activities from *Limulus* lateral eyes when recorded in situ cover over 10 log

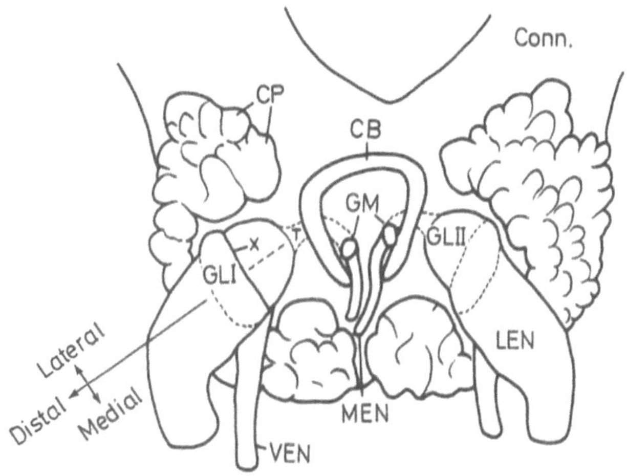

Fig. 32. *Limulus* brain. CB: central body. CONN: circumesophageal connective. CP: corpora pedunculata. GLI: first optic ganglion of lateral eye. GLII: second optic ganglion. GM: ganglion of median eye. LEN: lateral eye nerve. MEN: median eye nerve. T: optic tract connecting second optic ganglion, central body, and ganglion of ipsilateral median eye. VEN: ventral eye nerve. X: chiasma. (From SNODDERLY, 1971)

units of stimulus intensity, while those of the excised preparations, only 5 log units in a higher range of light intensity. Also, finding that illumination of the ventral region was effective in inducing downward or upward movements of the tail at light-*on* or -*off*, respectively, WASSERMAN (1973) suggested a possibility of production of propagated action potentials in situ in response to light. Upon illumination of the ventral eyes, SNODDERLY (1971) succeeded in recording impulses from the optic tract (T) between the second optic ganglion (GLII) and the central body (CB) after transection of the lateral eye nerves (LEN), which passed through the first and then the second optic ganglion (Fig. 32). It is possible, therefore, that the activities of the ventral photosensitive cells can affect neurons that are to be activated by the lateral eye. If so, ventral receptors "transduce light with great elegance" *not* "to absolutely no purpose" (CLARK et al., 1969), but with ample biologic significance.

III. Solitary Photoreceptive Cells

1. Adult Forms

Some of the free-living nematodes, *Deontostoma californicum* for example, are provided with paired spots in the pharyngeal muscle, each spot consisting of a pigment cell and a sensory cell which sends axons ventrolaterally into the lateral cephalic nerve (SIDDIQUI and VIGLIERCHIO, 1970). Within the cup-shaped pigment cell is found a part of the presumed photoreceptive site which is a stack of convoluted lamellae located in the anterior part of the sensory cell. This

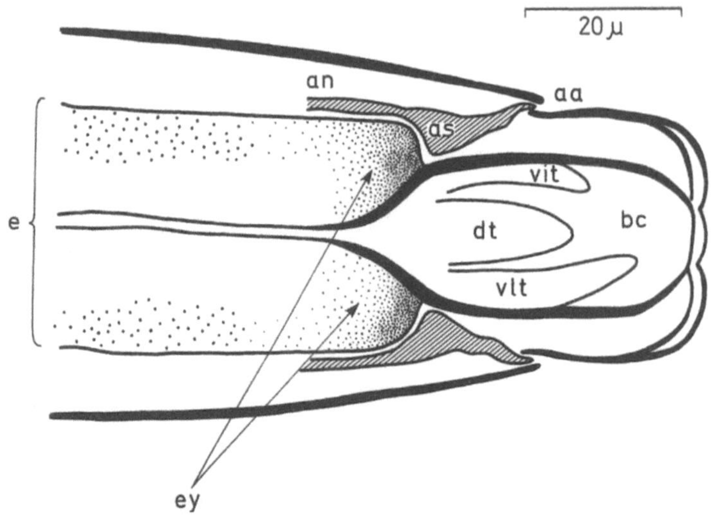

Fig. 33. Ventral view of head region of nematode, *Oncholaimus vesicarius*. aa: amphidial aperture. an: amphidial nerve. as: amphidial swelling. bc: buccal cavity. dt: dorsal tooth. e: esophagus. ey: pigmented eyespot. vlt: ventral tooth. (From BURR and WEBSTER, 1971)

organization represents the simplest type of inverted ocelli, which usually contain more than one sensory cells in more advanced forms, such as in Platyhelminthes (RÖHLICH and TÖRÖK, 1961). The sensory cells of this as well as two other nematode species, *Araeolaimus elegans* and *Chromadorina* sp. (CROLL et al., 1975) have been considered to belong to the so-called rhabdomeric type, since they lack ciliary structures in their periphery and the membranous lamellae resemble the phaosome of nauplius eyes (FAHRENBACH, 1964).

Not all the nematodes are provided with esophageal ocelli. Thus, the pigmented spot of *Enoplus communis* contains no rhabdomeric structures (CROLL et al., 1975). In *Oncholaimus vesicarius*, on the other hand, one of the cells in the amphidial organ appears to be a photoreceptor (BURR and WEBSTER, 1971). Amphids, which are the sense organs specific to the nematodes, are situated in pairs anterior to the pigmented spot (Fig. 33). According to BURR and BURR (1975), each organ is composed of six cells, of which four are sensory cells provided with cilia at their apices. Cilia from three of the amphidial dendrites are exposed to the external environment through a cuticular aperture, suggesting an olfactory function. Those from the remaining one dendrite, which they called "aberrant", project medially into a pouch that is located anterior to the pigmented area, the eye spot. The cilia, which usually number about 10, are packed into a platelike structure at their distal ends. Proximally, they contain 0–3 central singlets and 7–9 peripheral doublets but lack dynein arms, basal bodies, and rootlets, suggesting that they cannot be kinocilia of a usual sort. Their location is optimal for detecting shadows cast by the pigmented spot. Also, the cilia are oriented perpendicular to the body axis, hence perpendicular to the unshaded illumination during topophototaxis, since nematodes bend their frontal plane during locomotion. From these facts, BURR and BURR (1975) concluded that the aberrant cilia were the site of photoreception. Considering the structural similarities between the aberrant dendrite and the olfactory ones, they went on to suggest "the possibility that ciliary photoreceptors have evolved from chemoreceptors". At any rate, it appears that in some of the nematodes, single cells, whether they be esophageal sensory cells or aberrant dendrites, can send information sufficient for the animals to move properly with respect to the light source.

In a polychaete, *Armandia brevis*, ocelli are reported at the inner edge of the epidermis in front of the parapodia from the seventh to seventeenth segments (HERMANS and CLONEY, 1966; HERMANS, 1969). Each ocellus contains a single, supposedly photosensory, cell which is surrounded by about 30 pigment cells, but synaptic connections with the nervous system have not been found. Though the worm responds negatively to light during the benthic phase and the swimming epitokes are photopositive, the function of these ocelli in relation to such photic behavior has not been elucidated.

2. Larval Forms

There are a number of studies on the geo- and phototactic behavior of larvae that report changes in the sign of taxes with age from positive to negative or rarely in the reverse direction, depending on the mode of their life (THORSON, 1964). Mostly, such photosensitivities are mediated through eyes, the organization of which, though simple, is sometimes comparable to that of the adult forms as in three segmented young of polychaetes, *Neanthus succinea* (EAKIN and WESTFALL, 1964) and *Platynereis dumerilii* (FISCHER and BRÖKELMANN, 1966), but the response can be a simple shadow reaction resulting in avoidance behavior even in larval crustaceans which possess well organized eye(s) (FORWARD, 1976). In some forms, however, the structure of the eye is much simpler, the number of sensory cells being only one as in trematode larvae, cercaria of *Cryptocotyle lingua* (REES, 1975), trochophores of *Nereis succinea* (EAKIN and WESTFALL, 1964), bryozoan larvae of *Bugula neritina* (WOOLLACOTT and ZIMMER, 1972) and entoproct larvae (WOOLLACOTT and EAKIN, 1973), two to three cells in 5-day old tornaria larvae of *Ptychodera flava*, a hemichord (BRANDENBURGER et al., 1973) or up to 15–20 cells in ascidian tadpoles, *Amaroucium constellatum* (BARNES, 1971), *Distaplia occidentalis* and *Ciona intestinalis* (EAKIN and KUDA, 1971).

The important point which concerns us here is the question of whether or not the organs described above are in fact involved in photic reactions. There is a good series of progressive modification in the ciliary structure; from the typical kinocilia with 9+2 pattern of microtubules and dynein arms on a-tubules as in *Bugula neritina*, through the atypical ones which lack the arms as in entoproct larvae, to differentiation of a primitive grade as in tornaria, in which the ciliary sheath swells like a balloon and is borne on top of the presumed sensory cell that is provided proximally with additional rhabdomeric microvilli. EAKIN's group was careful enough, however, in calling those organs the "potential" photoreceptors.

It is suggested in cercaria of *Cryptocotyle lingua* that the paired ocelli can detect the direction of the light source, and the unpaired median ocellus which lies slightly ventral to the middle region of the body may help the cercaria to steer a straight course, once it has aligned itself with respect to the incident light (REES,

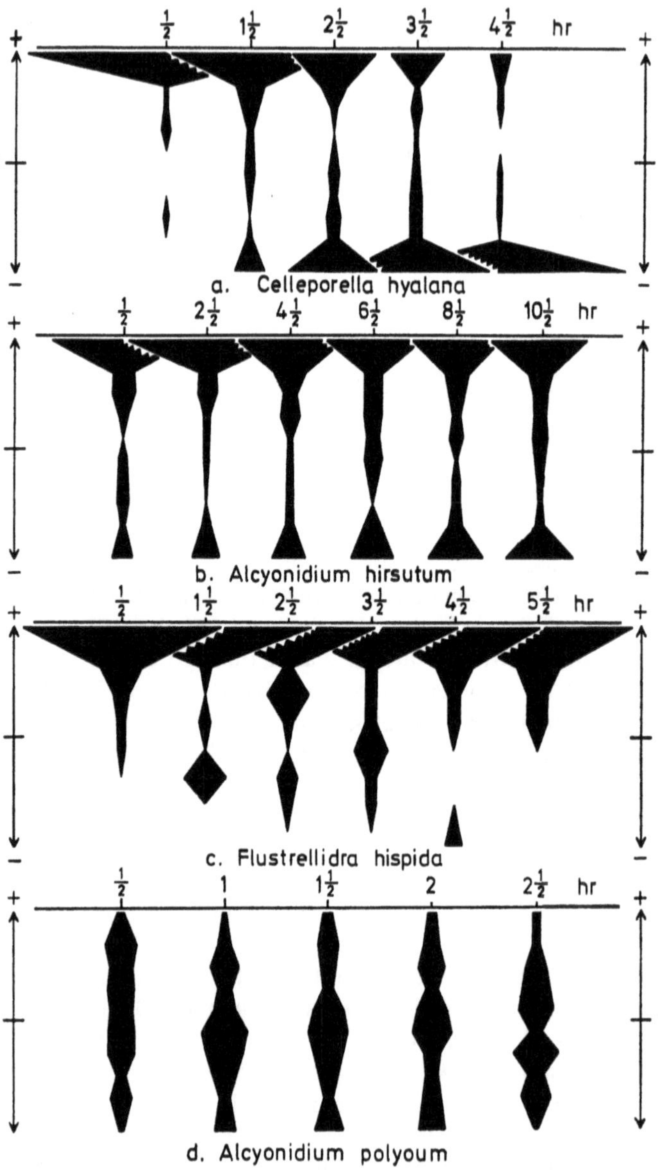

Fig. 34. Responses of polyzoan larvae to directional illumination. Kite diagrams show behavior of larvae of four species in troughs illuminated from one end. Troughs were marked into divisions and the distribution of larvae plotted at regular intervals. Four types of behavior were recorded; one example of each is figured. a: Type I. b: Type II. c: Type III. d: Type IV. (From RYLAND, 1960)

1975). In the case of *Bugula neritina*, the presumed photosensory cell, which WOOLLACOTT and ZIMMER (1972) called the basal sensory cell, is situated at the bottom of a cylindrical depression measuring 1–2 µm in diameter and 4–6 µm in depth, which is surrounded by a pigmented field about 40 µm wide and 80 µm high. Further, the receptoral basal cell is connected with locomotory coronal cells

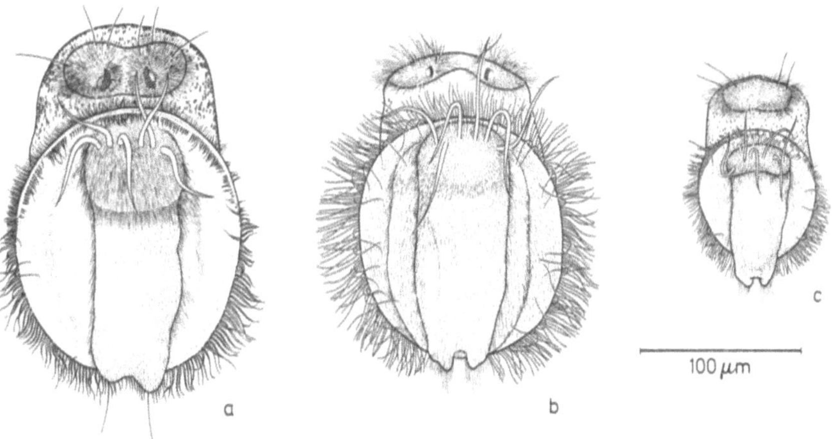

Fig. 35. Comparison of external appearance of pigmented spots in three species of *Loxosomella*, the entoproct. A: *L. harmeri*. B: *L. scaura*. C: *L. varians*. (From NIELSEN, 1971)

(effectors) by way of the equatorial nerve ring which underlies both types of the cells. According to RYLAND (1960), the majority of polyzoan larvae are photopositive when liberated, but become photonegative as the time of metamorphosis approaches. The presence of connections between receptors and effectors, the paired occurrence of the presumed organs, and the position of the basal cells, which is fitted for directional light perception, all suggest an aligning mechanism for the directional photopositive swimming of just released larvae. The arrangement resembles that described above for esophageal ocelli of nematodes which also consist of single sensory cells.

Difficulties remain, however, in assuming that the "potential" photoreceptors are in fact the sole mediators of photic responses. As both WOOLLACOTT and ZIMMER (1972) and WOOLLACOTT and EAKIN (1973) admitted, the occurrence of pigmented spots and the possession of phototactic behavior cannot always be correlated. In polyzoan larvae, RYLAND (1960) distinguished four types in the klinotactic behavior (Fig. 34): type I, released larvae are positive and then become negative after a few hours; type II, they are all positive at first and the majority of them remain so; type III, the positivity continues for several hours; and type IV, larvae are indifferent to light. *Bugula neritina* described above belongs to type II, but *Alcyonidium hirsutum* which also belongs to type II (Fig. 34b) lacks pigmented spots. In the case of entoprocts, among the 14 species belonging to the same genus, *Loxosomella*, five lack the pigmented eyes, and those provided with the pigmented spots appear to differ in the degree of differentiation (Fig. 35; NIELSEN, 1971).

The possibility left is that of general photosensitivity. Pluteus larvae of echinoids gather at a particular spot in a dish and it has been shown in *Hemicentrotus pulcherrimus* that the gathering is induced by combined effects of light on both the random movements at the water surface and the speed of vertical movements (NEYA, 1965). Since the plutei lack any, even "potential", photoreceptive cells, the light must have affected directly the ciliary activities of ectodermal cells. It may be that the larval photic responses, which are usually of kinetic nature and depend on dermal light sense in "blind" forms, would have become more refined by introduction of "potential" photoreceptorial organs so as to be able to steer or align the body axis with respect to the direction of incident light. The tactic behavior of adult forms which possess ocular as well as dermal receptive systems has been explained in a similar way.

F. Organized Photoreceptive Cells

When photosensitive cells are clustered into a discrete organ, it is called an eye, though various terms have been used to refer to various types of "eyes," such as eyespot, ocellus, stemmata, etc. (KOEHLER, 1957; DUKE-ELDER, 1958; MILNE and MILNE, 1959). They are usually few in number and located over the body surface near the CNS. An exceptional case may be noted here. Cephalopod molluscs have organs which are found deep in the body and yet have close structural resemblances with functional eyes. They are called epistellar bodies in *Octopus bimaculatus* (NISHIOKA et al., 1962), *O. vulgaris*, *Eledone moschata*, and *E. aldrovandii* and parolfactory vesicles in *Loligo vulgaris* and *Sepia officinalis* (NISHIOKA et al., 1966). Rhodopsins are detected with λ_{max} resembling that of the eye; 470 nm compared to 475 nm in *E. moschata* and 495 nm compared to 493 nm in *L. vulgaris* (NISHIOKA et al., 1966). Also intracellular recordings of the epistellar body of *E. moschata* reveal components which are of the type known in invertebrate receptors (MAURO and BAUMAN, 1968), but their physiologic meaning is still unknown.

I. Scattered Ocelli

There are instances where fairly well-differentiated ocelli are scattered over the body surface and function in a similar way as scattered primitive sensory cells or elements, and mediate photic information to the CNS.

Widely scattered eyespots are noted in turbellaria, nemertini, polyplacophora, and marine pulmonates. Among those, dorsal eyes of chitons are unique in that they are embedded in the outer shell layer. In *Onithochiton neglectus*, the distribution of the eyes on each valve shows a definite pattern (Fig. 36; BOYLE, 1969a) and one quarter of the total number is concentrated on the head valve (BOYLE, 1969b). Each eye having at its distal end an oval lens which is often covered with diatoms (BOYLE, 1976), sends an axon through the optic canal via the underlying dorsal epithelium to the pallial nerve cord. Although the eyes are sometimes called aesthetes (BULLOCK and HORRIDGE, 1965), the two are not entirely synonymous, since BOYLE (1969b) has observed that megalaesthetes branch from the peduncle of the eye and microaesthetes, from the distal end of the eye itself, and

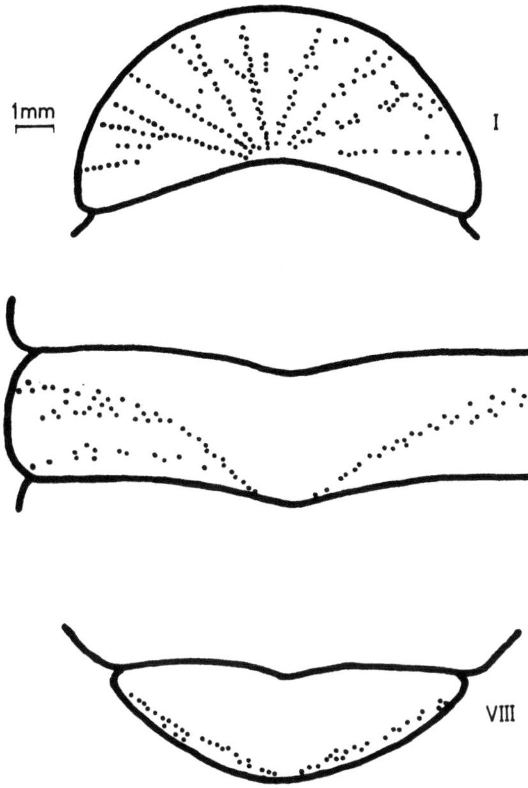

Fig. 36. Distribution of ocelli on shell valves, I, IV, and VIII of *Onithochiton neglectus*. (From BOYLE, 1969a)

since in aesthetes of *Lepidochitona cinereus* sensory elements such as microvillar and ciliary organizations are very poor, suggesting rather a secretory function (BOYLE, 1974). A similar situation, i.e., the co-existence of secretory activities, has been proposed for the early stage in evolution of photoreceptoral cells (ERMAK and EAKIN, 1976).

The eye is composed of two types of cells; one gives rise to a rhabdome with a closely packed arrangement of microvilli into the center of the eyecup, and the other is provided with lamellate bodies derived from ciliary membranes at the periphery of the eye.

Though the physiology of each type of cells is not evident at present, these shell eyes have been considered to be functional in inducing tactic and shadow responses, regardless of whether the eye belongs to the intrapigmental type as in *Chiton tuberculatus* (CROZIER and AREY, 1918; AREY and CROZIER, 1919) or to the extrapigmental type as in *Onithochiton neglectus* (BOYLE, 1969b). The animals are so sensitive that a dark spot of 2 mm at a distance of 2 cm was reported to be sufficient to elicit pronounced shadow reactions. Tactic movements also depend on the eyes but there occurs a change in sign with age, being negative in the young and positive in the old. AREY and CROZIER (1919) attributed this to the erosion of the tegumenta of the valves. It may be noted in passing that the shell eyes are not the sole site of photoreception but the ventral edge and the dorsal surface of the girdle as well as the surface of the proboscis are also sensitive, particularly to illumination. BOYLE (1972) confirmed the photoreceptive role of the aesthetes in *Ischnochiton maorianus*, *Onithochiton neglectus*, *Chiton rissori* and *C. siculus* with a possible participation of extra-tegumental surface for photo-orientation. Though the structure and the responses are as interesting as above, electrophysiologic work has not been successful (BOYLE, 1969b), probably because the various receptive elements are small (the eyecup, 40–60 × 25–40 μm; diameter of axon, 0.1–0.4 μm), and because the eyes are embedded in the hard shell.

Dorsal eyes of a pulmonate gastropod, *Onchidium*, are located at the apices of dorsal papillae, each bearing 1–5 ocelli. The ultrastructure was studied in *O. verruculatum* by YANASE and SAKAMOTO (1965) who observed lamellate bodies in the outer segment of the presumed visual cells, and the electroretinogram has been recorded by FUJIMOTO et al. (1967). According to OKUNO et al. (1976), receptor potentials recorded intracellularly hyperpolarize at light on and, in fresh preparations and on rare occasions, depolarize at light off, both being associated with an increase in membrane conductance. The results appear to indicate in all respects that the dorsal eyes are sensitive to light, but the functional significance is still obscure (see Sec. F.II.2).

Similarly scattered but more or less regionally restricted ocelli are found in some forms. The land leech, *Haemadipsa zeylanica agilis*, possesses eyes associated with the segmental papillae over the body surface (BHATIA, 1956). In scallops, about 60 eyes are arranged in a row along the mantle edge and they are shown to function as a single system with a high degree of coordination for eliciting orientation reactions in *Pecten jacobaeus* and *P. opercularia* (VON BUDDENBROCK and MOLLER-RACKE, 1953), and electrophysiologic evidence to support the ability of pattern recognition in such eyes has been given by LAND (1966).

Ocelli associated with tentacular structure are known in coelenterates, sabellid polychaetes, and synaptid holoturians. In medusae, the ocelli are situated either along the bell margin as in hydromedusae and scyphomedusae, or on the bell wall as in cubomedusae, and the number of such ocelli is sometimes several hundreds, as in *Spirocodon saltatrix*. The structure of these ocelli varies from the simple flat pigmented layer of the large ocellus of *Aurelia aurita* (YAMASU and YOSHIDA, 1973) to the highly organized lens-bearing eye of *Tamoya bursaria* (YAMASU and YOSHIDA, 1976). Various types of structures in between these two extremes have been reviewed by SINGLA (1974) and YOSHIDA (1974). In *A. aurita* (HORSTMANN, 1934), the pulsation rhythm is affected by light but it is not clear whether such a long-term effect on the rhythm is mediated through ocelli. Though electric activities of a type of electroretinogram have not been obtained from any one of the medusan ocelli, they appear to be functional photoreceptors. In *Spirocodon saltatrix*, removal of the ocelli abolishes shadow-induced pulsations (HISADA, 1956) and impulse activities to be evoked in the nerve ring by light-*off* require an intact connection with the ocellar structure (YOSHIDA and OHTSU, 1973). Such shadow reactions have been noted even in a rarely occurring pelagic species, *Pelagohydra mirabilis* (PILGRIM, 1967). *Polyorchis karafutoensis*, which is a statocyst-lacking species, bears 96 ocelli and it has been demonstrated that this

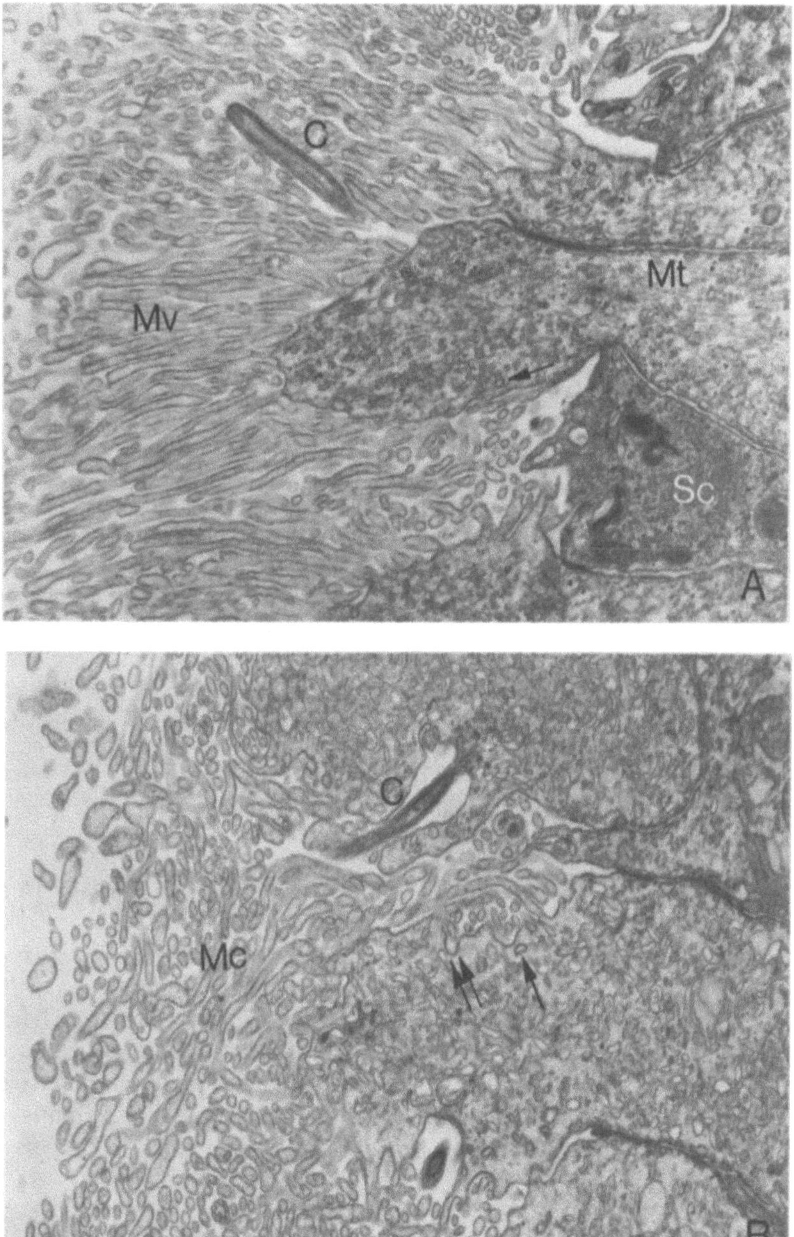

Fig. 37. Fine structure of the peripheral part of the sensory cells of a synaptid, *Opheodesoma spectabilis* (YAMAMOTO and YOSHIDA, unpublished); ×14000. A: dark adapted for 40 hrs. B: light adapted for 40 hrs with two fluorescent tubes (GE F40D) placed 50 cm above. Fixation: 3 hrs in 2.5% glutaraldehyde, 0.45 M sucrose and 0.1 M cacodylate (pH 7.4) at final concentration. Post-fixation: 1 hr in 1% osmium tetroxide in the same buffer. Note that microvillar (Mv) arrangement is more regular and microtubules (Mt) are more evident in the dark adapted preparation (A) and also that invaginating cell surface (double arrows) and fragmented membranous structure are seen often in the light adapted preparation (B). C: cilium. Sc: supportive cell. Single arrow: coated vesicle

medusa can keep the upright stance through ocellar photosensitivity (TAMASIGE and YAMAGUCHI, 1967). It may be said that such functional significance owes much to the circular distribution of photoreceptive organs.

A similar but more complicated situation has been noted in a Hawaiian synaptid, *Opheodesoma spectabilis* (BERRILL, 1966). This animal is photonegative and the sensitivity is highest in the region of the crown of tentacles, at the base of each of which is located a pair of ocelli. The ocelli are composed of two types of cells, sensory and supportive (YAMAMOTO and YOSHIDA, unpublished). Each sensory cell is provided at its distal end with a cilium and microvilli arising from the cell surface and in the apical half of the cell with numerous coated vesicles. The length and width of the microvilli and the number of the vesicles change by light adaptation (compare Fig. 37A and B). The ocelli thus appear to be in fact photoreceptive, but removal of them does not completely abolish the response, which suggests the presence of dermal light sense. The interesting feature is that the activity of the locomotory center is depressed by the input from statocysts. Through this depression of activity, the animal can stay in a favourable shaded area, while the statocyst-less animals run out of the place they have found through the ocellar function.

Sabellid polychaetes such as *Branchiomma vesiculosum*, *Eudistylia vancouveri*, *Pseudopotamilla ocellata*, *Schizobranchia insignis*, *Sabella crassicornia*, *Dasychone lucullana* and *D. bombyx* possess well-differentiated and unique ocelli at the tentacular tip of the branchial crown (KRASNE and LAWRENCE, 1966; KERNEIS, 1975). Their functional significance in relation to the dermal system will be discussed in the next section.

II. Coexistence of Multi-photoreceptive Systems

There are instances where two different types of photoreceptors, ciliary and rhabdomeric, exist in the same ocular organs, such as the double retina of *Pecten maximus* (BARBER et al., 1967) and the mixed retina of *Aplysia punctata* (HUGHES, 1970). In such cases, it is generally assumed that the ciliary receptors are concerned with *off*-responses (LAND, 1968). The complexity in the photoreceptive systems provides us with still other instances where the ocular and the dermal systems coexist. The ensuing responses may be divided into three types.

1. Dominant Role of the Dermal System

Curiously, removal of well-differentiated ocelli has no effect at all on the performance of some photic behavior. This may imply that the dermal photosystem is dominant over the ocular system. Elicitation of mantle contraction in *Aplysia limacina* is one such case (DIJKGRAAF, 1935). Similarly, though the stalked eyes are important for orientation of *Helix pomatia*, contractions of tentacles and the mantle are most effectively induced by shading the mantle surface along the anterior edge of the shell (FÖH, 1932).

In *Onchidium*, behavioral responses such as orientation to light, cessation of locomotion upon shading, etc., are mediated neither by the dorsal nor by the tentacular eyes (KATAGIRI, personal communication). It is pertinent to recall here the work of AREY and CROZIER (1921) on *O. floridanum* which has no differentiated dorsal eyes. In this species, the photic reactions as above are induced by shading any point over the whole dorsal surface of the mantle, and are not affected by amputation of tentacles. It thus seems likely that this animal relies entirely on the pure dermal receptive system and the role of the photoresponsive dorsal ocelli becomes problematic.

2. Subsidiary Role of the Dermal System

Removal of the most sensitive site sometimes makes remnant sensitivities manifest themselves, like the above-mentioned cases of a siphonal receptive site in *Nassarius reticulatus* (CRISP, 1972), paired ocelli in the tentacular crown of *Opheodesoma spectabilis* (BERRILL, 1966), and mantle ocelli of *Lima scabra* (MPITSOS, 1973). Manifestations of the dermal system occur even in those animals which are provided with well-differentiated eyes. Thus, in *Lymnaea stagnalis*, COOK (1975) observed that the responsiveness to shadows remained, though to a reduced extent, after denervation of the eyes, tentacles and most of the mantle.

Asteroids possess a compound ocellus at the tip of each arm. The ultrastructure and the role in phototactic movements have been well established, respectively, in *Asterias rubens* (VAUPEL-VON HARNACK, 1963) and in *A. gibbosa* (KALMUS, 1929) and *A. amurensis* (YOSHIDA and OHTSUKI, 1968). Coexistence with the dermal light sense was noted as early as 1911 by COWLES on the phototactic behavior of *Echinaster crassispina* and *Asterias forbesi*. ROCKSTEIN (1956) confirmed this notion in the latter species. *A. amurensis* shows, in addition, reflex movements of the arm tip, lifting it up at light-*off* and shifting it down at *on* (YOSHIDA and OHTSUKI, 1966). Removal of the ocellus does not abolish the reaction but raises the threshold about 10 times. On measuring spectral sensitivities for the downward movements, we obtained two separate action spectra, peaking at 485 nm with intact and at 504 nm with eyeless preparations (Fig. 38). The shift does not necessarily imply the existence of two different photopigments, however. It is possible that a pigment which absorbs longer wavelengths in ocelli could make the blue shift in the action spectrum of the intact preparation. A shift of the spectral sensitivity maxima in the combined receptive system due to participation of the dermal sensitivity has been suggested by BROWN et al. (1968). This would account for the discrepancies of the maxima measured behaviorally (at 475 nm in *Planaria lugubris* [VIAUD, 1949] and at 530 nm in *Dendrocoelum lacteum* [MARRIOTT, 1958]), when compared with their direct measurement of the ocellar potential which peaked at 508 nm.

It is plausible that removal of the ocular organs may affect the central processing mechanism. Disappearance of photosensitivity upon such surgical operations does not necessarily indicate the absence of a minor additional system. It is possible, therefore, that there are a number of hidden cases where the dermal system plays a subsidiary role in photic responses in situ.

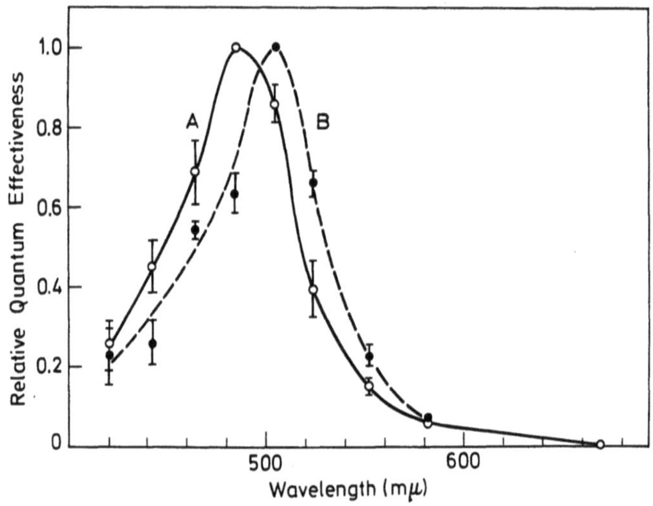

Fig. 38. Relative quantum efficiencies of photoreceptive systems of starfish, *Asterias amurensis*: arms having (A) and lacking (B) intact eyespots are compared. Vertical bars, ranges of SE. (From YOSHIDA and OHTSUKI, 1966)

3. Functional Differentiation

The common feature encountered as a functional differentiation between the ocular and the dermal systems is that the former concerns the orientation reaction and the latter, regardless of whether it is purely dermal or diffuse, is responsible for either kinetic or reflex activities (mainly to shadows), such as in *Helix pomatia* (FÖH, 1932). Similarly, five pairs of the cephalic photoreceptors in the medical leech, *Hirudo medicinalis*, have significantly sharper directional sensitivity than seven pairs of segmental ones (KRETZ et al., 1976). The differences in involvement of ocular and dermal systems in tactic and kinetic responses, respectively, are noted in vertical migration of *Daphnia magna* (HARRIS and MASON, 1956), though an endogeneous rhythm could be an additional factor in determining the diurnal migration cycle (HARRIS, 1963). It is interesting to note in passing that *Daphnia* from which compound eyes are surgically removed are much more sensitive to light than normal ones.

The eyes appear not to be the sole receptors in determining the locomotory reactions in *Planaria gonocephala* (PARKER and BURNETT, 1900), *P. maculata* (TALIAFERRO, 1920), *P. lugubris* (LEMKE, 1935) and *Dendrocoelum lacteum* (LEMKE, 1935; ULLYOTT, 1936a, b). Confirming these notions, VIAUD (1949) and VIAUD and MÉDIONI (1949) showed by tracing the responses during the course of regeneration in *P. lugubris* that positive phototropism and the photokinesis are due to dermal light sense, and negative phototropism and axial orientation to the ocular sense.

Differentiation also occurs in the type of reflex responses to be elicited. As described above, sabellid polychaetes possess well differentiated ocelli at the tip of the branchial crown. NICOL (1950) noticed in *Branchiomma vesiculosum* that sudden changes in intensity and moving objects induced different adaptation

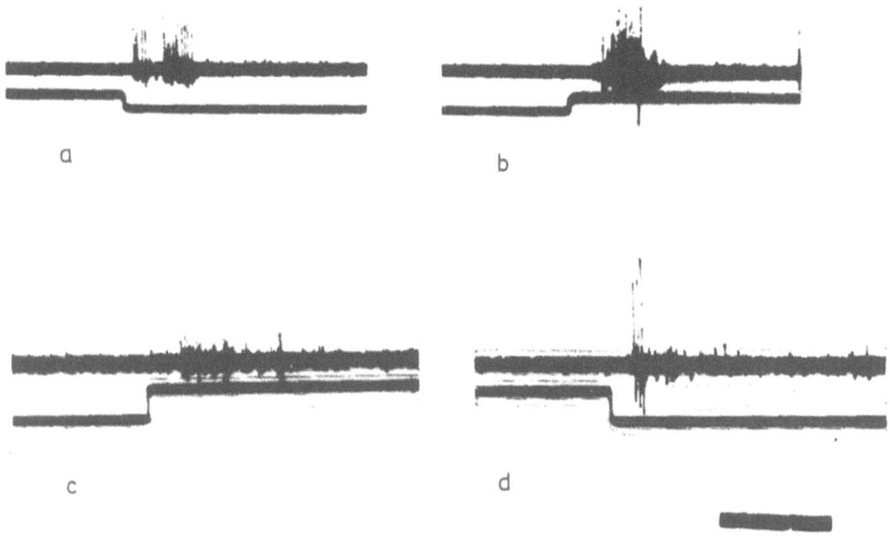

Fig. 39. Recordings taken with steel electrodes from ventral nerve cord of *Nereis diversicolor*. A and B: subesophageal ganglion. C and D: responses from cord posterior to subesophageal ganglion. Time calibration: 500 ms. (From GWILLIAM, 1969)

curves. He interpreted this phenomenon to indicate the presence of two distinct sensory processes. Later, KRASNE and LAWRENCE (1966) confirmed that after removal of eyes the worms reacted equally well to shadows but became less sensitive to moving objects.

Electrophysiologic demonstration of differential functioning of the ocular and the dermal systems has been made by GWILLIAM (1969) on polychaete worms, mainly on *Nereis diversicolor*. These animals have two pairs of prostomial ocelli, anterior and posterior. Though the two function differently as regards tactic movements (HERTER, 1926), both produce negative-going extracellular potentials upon illumination. Consistent recordings of bursts can be made neither from the supraesophageal ganglion nor from the circum-esophageal connectives but only from the subesophageal ganglion. The bursts consist of only fine-fiber activities (Fig. 39A, B) and the giant-fiber responses are recorded from the ventral nerve cord at light-*off*. This response is not abolished by ablation of the head region (removal of eyes) (Fig. 39C, D). The cut in the ventral nerve cord does not interfere with the fine-fiber responses at light-*on* and -*off* but it does so with the giant-fiber responses in the region posterior to the cut. These results suggest that the prostomial ocelli are not the sole receptors for mediating the giant-fiber responses. Since headless segments numbering more than seven induce the giant-fiber responses at light-*off*, provided that the parapodia are left intact, and since the same responses can be elicited from preparations with parapodia removed if the head is left intact, it is suggested that the dermal receptors are distributed on the pro- and peristomium as well as on the parapodia.

The *off*-responses of the giant-fiber system which are triggered by stimulation of the anterior part of the body, and the *on*- and *off*-responses of the fine-fiber system, which can be triggered by stimulation of only the posterior part of the body imply an important biologic significance. Thus, if a shadow, which may to the animal be a signal of approaching predators, is cast on the anterior part of the body, fast head withdrawal assisted by forward stepping of parapodia is elicited through the giant-fiber system. If the tail, which should have been retracted in the burrow, is accidentally exposed to light, the worm withdraws it with the assistance of backward stepping of the parapodia; this response is mediated through the fine-fiber system without the help of the ocular system. EVANS (1969a, b) noted a similar difference in response pattern to light-on and -off and showed further that responses to shadow, particularly if they were presented in pairs, habituated more slowly. The habituation process is independent of the presence of the brain, but the actual site of the process is still obscure.

G. Summary

We have considered structural and functional aspects of various types of extraocular photosensitivities. For the sake of convenience, I divided them categorically into dermal, diffuse, neural, and neuronal sensitivities. The gradations in the degree of structural differentiation may imply that evolution of photoreceptoral systems has proceeded from the direct responsiveness of cells via the dermal system 1) through the diffuse system to a localized organ and 2) through the neural to the neuronal system. However, the situation with which we are confronted is not so straightforward. The bewildering variabilities and complexities encountered in the primitive photoreceptive systems did not afford any formal and systematized approach, and it was unavoidable to choose each type of response separately, as if dealing with some unrelated phenomena. As a conclusion, I shall summarize below some of the principal problems of extraocular photoreception.

The structural basis for dermal and neural receptor systems remains to be elucidated. However, thanks largely to the contributions of EAKIN and associates, it is possible to observe a series of gradations in the peripheral structure of the diffuse systems, whether they be ciliary, rhabdomeric, or even phaosomal. In most of the diffuse systems, however, we have no means of either proving or disproving convincingly that any proposed receptive site is in fact involved in photoreception, because of the lack of proper techniques to demonstrate photopigment(s). On the other hand, it has been said that intracellular pigment granules are concerned with photoexcitation in some of the neurons. This leads us to a notion that no peripheral differentiation is fundamentally needed for a cell to be photosensitive. In this respect, there are additional small photosynthetic organisms, symbiotic algae, the presence of which lends animals photosensitivity. The quest for the minimal prerequisite for an animal to become photoresponsive may thus be a sort of Sphinx's riddle.

If, on the other hand, one tries to analyze electrophysiologic or reflex events in terms of neurophysiology, one will often find, even in the dermal or the neural light senses, instances of interactions between excitation and inhibition. Such a mechanism parallels in some respects that involved in information processing in differentiated retinae, but here again studies on structural and biochemical background are lacking.

Difficulties are even greater when we think of the chain of events leading to coordinated responses. Needless to say, the photosensitivity of a cell does not necessarily mean that the cell is involved in sensory perception. Seen from the other side, ill-defined receptive systems do not always form a subsidiary part of, but are sometimes more effective than coexisting well-differentiated receptoral organs in inducing some photic behavior. In some cases, moreover, sensory systems for modalities of adequate stimuli other than light may come into play. All those phenomena must involve an integrating mechanism. Some of the beautifully coordinated reactions such as tactic movements of *Hydra* and covering reaction of sea urchins, can be produced without centralization of the nervous system, but the integration usually occurs in the CNS. The situation becomes complicated by the fluidity in the activity of the central mechanism which may oscillate diurnally. We are thus far from having a complete understanding of the intimate mechanism involved in the extraocular photoreception.

H. Acknowledgement

I acknowledge with thanks Drs. W. WEBER, G. J. MPITSOS, V. B. PEARSE, L. J. MILNE, N. MILLOTT, D. F. KIMELDORF, L. M. PASSANO, L. A. WILKENS, D. KENNEDY, K. P. RAO, W. T. FRAZIER, U. THURM, P. RÖHLICH, M. CRISP, M. SNODDERLY, JR., A. H. BURR, J. S. RYLAND, C. NIELSEN, P. R. BOYLE and G. F. GWILLIAM, and also Professor ALLANSON on behalf of Dr. R. E. BOLTT, for permission to use their published figures. Thanks are particularly due to Drs. T. YAMASU, L. A. WILKENS and K. TAKAHASHI, for the generous gift of their original pictures. I also thank Drs. T. S. GOTOW, Y. KATAGIRI and K. NODA for their invaluable personal communications.

References

Arey, L. B., Crozier, W. J.: The sensory responses of *Chiton*. J. exp. Zool. **29**, 157—260 (1919).

Arey, L. B., Crozier, W. J.: On the natural history of *Onchidium*. J. exp. Zool. **32**, 443—502 (1921).

Arvanitaki, R., Chalazonitis, N.: Réactions bioélectriques neuroniques à la photoactivation spécifique d'une hème-proteine et d'une carotène-proteine. Arch. Sci. Physiol. **3**, 27—44 (1949a).

Arvanitaki, A., Chalazonitis, N.: Inhibition ou excitation des potentiéls neuroniques à la photo-activation distincte de deux chromoprotéides (caroténoïde et chlorophyllien). Arch. Sci. Physiol. **3**, 45—60 (1949b).
Arvanitaki, A., Chalazonitis, N.: Photopotentiéls d'excitation et d'inhibition de différents somata identifiables (*Aplysia*). Activations monochromatiques. Bull. Inst. Océanogr. (Monaco) **1164**, 1—83 (1960).
Arvanitaki, A., Chalazonitis, N.: Excitatory and inhibitory processes initiated by light and infra-red radiations in single identifiable nerve cells (giant ganglion cells of *Aplysia*). In: Nervous Inhibition, ed. E. Florey, pp. 194—231. Oxford: Pergamon Press 1961.
Barber, V.C., Evans, E.M., Land, M.E.: The fine structure of the eye of the mollusc *Pecten maximus*. Z. Zellforsch. **76**, 295—312 (1967).
Barnes, S.N.: Fine structure of the photoreceptor and cerebral ganglion of the tadpole larva of *Amaroucium constellatum* (Verrill) (Subphylum, Urochordata; Class, Asciaacea). Z. Zellforsch. **117**, 1—16 (1971).
Bell, A.L., Anderson, K.L.: The fine structure of "photoreceptors" in the siphon of the clam (*Mya arenaria*). Anat. Rec. **166**, 403 (1970).
Berrill, M.: The ethology of the synaptid holothurian, *Opheodesoma spectabilis*. Canad. J. Zool. **44**, 457—482 (1966).
Bhatia, M.L.: Extraocular photoreceptors in the land leech, *Haemadipsa zeylanica agilis* (Moore) from Nainital, Almora (India). Nature (Lond.) **178**, 420—421 (1956).
Blankenship, J.E., Wachtel, H., Kandel, E.F.: Ionic mechanisms of excitatory, inhibitory and dual synaptic actions mediated by an identified interneuron in the abdominal ganglion of *Aplysia*. J. Neurophysiol. **34**, 76—92 (1971).
Block, G.D., Hudson, D.J., Lickey, M.E.: Extraocular photoreceptors can entrain the circadian oscillator in the eye of *Aplysia*. J. comp. Physiol. **89**, 237—249 (1974).
Block, G.D., Lickey, M.E.: Extraocular photoreceptors and oscillators can control the circadian rhythm of behavioral activity in *Aplysia*. J. comp. Physiol. **84**, 367—374 (1973).
Block, G.D., Smith, J.T.: Cerebral photoreceptors in *Aplysia*. Comp. Biochem. Physiol. [A] **46**, 115—121 (1973).
Boltt, R.E., Ewer, D.W.: Studies on the myoneural physiology of Echinodermata. IV. The lantern retractor muscle of *Parechinus*: Responses to stimulation by light. J. exp. Biol. **40**, 713—726 (1963).
Boyle, P.R.: Rhabdomeric ocellus in a chiton. Nature **222**, 895—896 (1969a).
Boyle, P.R.: Fine structure of the eyes of *Onithochiton neglectus* (Mollusca: Polyplacophora). Z. Zellforsch. **102**, 313—332 (1969b).
Boyle, P.R.: The aesthetes of chitons. I. Role in the light response of whole animals. Mar. Behav. Physiol. **1**, 171—184 (1972).
Boyle, P.R.: The aesthetes of chitons. II. Fine structure in *Lepidochitona cinereus* (L). Cell Tissue Res. **153**, 383—398 (1974).
Boyle, P.R.: The aesthetes of chitons. III. Shell surface observations. Cell Tissue Res. **172**, 379—388 (1976).
Brandenburger, J.L.: Two new kinds of retinal cells in the eye of a snail, *Helix aspersa*. J. Ultrastruct. Res. **50**, 216—230 (1975).
Brandenburger, J.L., Eakin, R.M.: Pathway of incorporation of vitamin A^3H_2 into photoreceptors of a snail, *Helix aspersa*. Vision Res. **10**, 639—653 (1970).
Brandenburger, J.L., Woollacott, R.M., Eakin, R.M.: Fine structure of eye spots in Tornarian larvae (Phylum Hemichordata). Z. Zellforsch. **142**, 89—102 (1973).
Braun, R., Faust, I.: Weiteres zum Lichtsinn augenloser Muscheln. Experientia (Basel) **10**, 504—505 (1954).
Braun, R., Job, W.: Neues zum Lichtsinn augenloser Muscheln. Naturwissenschaften **16**, 482 (1965).
Brown, A.M., Baur, P.S.Jr., Tuley, F.H.Jr.: Phototransduction in *Aplysia* neurons: Calcium release from pigmented granules is essential. Science **188**, 157—160 (1975).
Brown, A.M., Brown, H.M.: Light response of a giant *Aplysia* neuron. J. Gen. Physiol. **62**, 239—254 (1973).
Brown, H.M., Ito, I., Ogden, T.E.: Spectral sensitivity of the planarian ocellus. J. Gen. Physiol. **51**, 255—260 (1968).
Brown, J.E., Blinks, J.R.: Changes in intracellular free calcium concentration during illumination of invertebrate photoreceptors. Detection with aequorin. J. Gen. Physiol. **64**, 643—665 (1974).

Brown, J. E., Lisman, J. E.: An electrogenic sodium pump in *Limulus* ventral photoreceptor cells. J. Gen. Physiol. **59**, 720—733 (1972).

Brown, J. E., Mote, M. I.: Ionic dependence of reversal voltage of the light response in *Limulus* ventral photoreceptors. J. Gen. Physiol. **63**, 337—350 (1974).

Brown, J. E., Murray, J. R., Smith, T. G.: Photoelectric potential from photoreceptor cells in ventral eye of *Limulus*. Science **158**, 665—666 (1967).

Von Buddenbrock, W.: Untersuchungen über den Schattenreflex. Z. Vergl. Physiol. **13**, 164—213 (1930).

Von Buddenbrock, W., Moller-Racke, I.: Über den Lichtsinn von *Pecten*. Pubbl. Staz. Zool. Napoli **24**, 217—245 (1953).

Bullock, T. H., Horridge, G. A.: Structure and function in the nervous system of invertebrates, I. San Francisco: Freeman 1965.

Burr, A. H., Burr, C.: The amphid of the nematode *Oncholaimus vesicarius*: Ultrastructural evidence for a dual function as chemoreceptor and photoreceptor. J. Ultrastruct. Res. **51**, 1—15 (1975).

Burr, A. H., Webster, J. M.: Morphology of the eyespot and description of two pigment granules in the esophageal muscle of a marine nematode, *Oncholaimus vesicarius*. J. Ultrastruct. Res. **36**, 621—632 (1971).

Camougis, G.: Unmasking of a photoinhibitory response in the crayfish caudal ganglion following partial anesthesia. Comp. Biochem. Physiol. **21**, 231—236 (1967).

Clark, A. W., Millecchia, R. M., Mauro, A.: Ventral photoreceptor cells of *Limulus*. I. The micro-anatomy. J. Gen. Physiol. **54**, 289—309 (1969).

Clark, E. D., Kimeldorf, D. J.: Tentacle responses of the sea anemone *Anthopleura xanthogrammica* to ultraviolet and visible radiations. Nature (Lond.) **227**, 856—857 (1970).

Clark, E. D., Kimeldorf, D. J.: Behavioral reactions of the sea anemone, *Anthopleura xanthogrammica*, to ultraviolet and visible radiations. Radiation Res. **45**, 166—175 (1971).

Clark, R. B.: The eyes and the photonegative behavior of *Nephtys* (Annelids, Polychaeta). J. exp. Biol. **33**, 461—477 (1956).

Cook, A.: The withdrawal response of a freshwater snail (*Lymnaea stagnalis* L.). J. exp. Biol. **62**, 783—796 (1975).

Cowles, R. P.: Reaction to light and other points in the behavior of the starfish. Carnegie Inst. Wash. **3**, 95—110 (1911).

Crisp, M.: Structure and abundance of receptors of the unspecialized external epithelium of *Nassarius reticulatus* (Gastropods, Prosobranchia). J. Mar. biol. Ass. U.K. **51**, 865—890 (1971).

Crisp, M.: Photoreceptive function of an epithelial receptor in *Nassarius reticulatus* (Gastropods, Prosobranchia). J. Mar. biol. Ass. U.K. **52**, 437—442 (1972).

Croll, N. A., Evans, A. A. F., Smith, J. M.: Comparative nematode photoreceptors. Comp. Biochem. Physiol. **51** A, 139—143 (1975).

Crozier, W. J.: The orientation of a holothurian by light. Amer. J. Physiol. **36**, 8—20 (1914).

Crozier, W. J.: The rhythmic pulsation of the cloaca of holothurians. J. exp. Zool. **20**, 297—356 (1916).

Crozier, W. J.: The behavior of holothurians in balanced illumination. Amer. J. Physiol. **43**, 510—513 (1917).

Crozier, W. J., Arey, L. B.: On the significance of the reaction to shading in *Chiton*. Amer. J. Physiol. **46**, 487—492 (1918).

Dambach, M., Jochum, F.: Zum Verlauf der Pigmentausbreitung beim Farbwechsel des Seeigels *Centrostephanus longispinus* Peters. Z. vergl. Physiol. **59**, 403—412 (1968).

Dambach, M., Weber, W.: Inhibition of pigment movement by cytochalasin B in the chromatophores of the sea urchin *Centrostephanus longispinus*. Comp. Biochem. Physiol. **50** C, 49—52 (1975).

Dhainaut-Courtois, N.: Sur la présence d'un organe photorécepteur dans le cerveau de *Nereis pelagica* L. (Annélide polychète). C. R. Acad. Sci. (Paris) **261**, 1085—1088 (1965).

Dijkgraaf, S.: Über Hautlichtempfindlichkeit bei *Aplysia limacina*. Zool. Anz. **111**, 254—256 (1935).

Dimelow, E. J.: Pigments present in arms and pinnules of the crinoid, *Antedon bifida* (Pennant). Nature (Lond.) **182**, 812 (1958).

Duke-Elder, S.: System of ophthalmology, I. London: Henry Kimpton 1958.

Eakin, R. M.: Lines of evolution of photoreceptors. In: General Physiology of Cell Specialization, eds. D. Mazia, A. Tyler. New York: McGraw-Hill 1963.

Eakin, R. M.: Structure of invertebrate photoreceptors. In: Handbook of Sensory Physiology: Photochemistry of Vision, ed. H. J. A. Dartnall, Vol. VII/1, pp. 625—684. Berlin-Heidelberg-New York: Springer 1972.

Eakin, R. M., Brandenburger, J. L.: Differentiation in the eye of a pulmonate snail *Helix aspersa*. J. Ultrastruct. Res. **18**, 391—421 (1967).

Eakin, R. M., Brandenburger, J. L.: Localization of vitamin A in the eye of a pulmonate snail. Proc. nat. Acad. Sci. (Wash.) **60**, 140—145 (1968).

Eakin, R. M., Brandenburger, J. L.: Osmic staining of amphibian and gastropod photoreceptors. J. Ultrastruct. Res. **30**, 619—641 (1970).

Eakin, R. M., Kuda, A.: Ultrastructure of sensory receptors in ascidian tadpoles. Z. Zellforsch. **112**, 287—312 (1971).

Eakin, R. M., Westfall, J. A.: Further observations on the fine structure of some invertebrate eyes. Z. Zellforsch. **62**, 310—332 (1964).

Ellis, V. L.: The spectral sensitivity of *Hydra carnea* L. Aggasiz (1850). Canad. J. Zool. **48**, 63—68 (1970).

Ermak, T. H., Eakin, R. M.: Fine structure of the cerebral and pygidial ocelli in *Chone ecaudata* (Polychaeta: Sabellidae). J. Ultrastruct. Res. **54**, 243—260 (1976).

Evans, S. M.: Habituation of the withdrawal response in nereid polychaetes. I. The habituation process in *Nereis diversicolor*. Biol. Bull. **137**, 95—104 (1969a).

Evans, S. M.: Habituation of the withdrawal response in nereid polychaetes. II. Rates of habituation in intact and decerebrate worms. Biol. Bull. **137**, 105—117 (1969b).

Fahrenbach, W. H.: The fine structure of a nauplius eye. Z. Zellforsch. **62**, 182—197 (1964).

Fein, A., Charlton, J. S.: Local membrane current in *Limulus* photoreceptors. Nature (Lond.) **258**, 250—252 (1975).

Fein, A., DeVoe, R. D.: Adaptation in the ventral eye of *Limulus* is functionally independent of the photochemical cycle, membrane potential, and membrane resistance. J. Gen. Physiol. **61**, 273—289 (1973).

Feldman, M., Lenhoff, H. M.: Phototaxis in *Hydra littoralis*: rate studies and localization of the "photoreceptor". Anat. Rec. **137**, 354—355 (1960).

Fioravanti, R., Fuortes, M. G. F.: Analysis of responses in visual cells of the leech. J. Physiol. (Lond.) **227**, 173—194 (1972).

Fischer, A., Brökelmann, J.: Das Auge von *Platynereis dumerilii* (Polychaeta); sein Feinbau im ontogenetischen und adaptiven Wandel. Z. Zellforsch. **71**, 217—244 (1966).

Föh, H.: Der Schattenreflex in *Helix pomatia*. Nebst Bemerkungen über den Schattenreflex in *Mytilus edulis*, *Limnaea stagnalis* and *Testudo ibera*. Zool. J. Abt. allg. Zool. Physiol. Tiere **52**, 1—78 (1932).

Forward, Jr., R. B.: A shadow response in a larval crustacean. Biol. Bull. **151**, 126—140 (1976).

Frazier, W. T., Kandel, E. R., Kupfermann, I., Waziri, R., Coggeshall, R. E.: Morphological and functional properties of identified neurons in the abdominal ganglion of *Aplysia californica*. J. Neurophysiol. **30**, 1288—1351 (1967).

Frei, E.: Untersuchungen über die allgemeine und die spektrale Photosensibilität von *Hydra attenuata* Pall. Inaugural-Dissertation, Univ. Zürich. 5—94 (1974).

Fujimoto, K., Yanase, T., Okuno, Y., Iwata, K.: Electrical response in the *Onchidium* eyes. Mem. Osaka Gakugei Univ. B **15**, 98—108 (1967).

Fuortes, M. G. F.: Generation of responses in receptor. In: Handbook of Sensory Physiology: Principles of Receptor Physiology, ed. W. R. Loewenstein, Vol. VII/1, pp. 243—268. Berlin-Heidelberg-New York: Springer 1971.

Geethabali: Central course of photic input in the ventral nerve cord of scorpion (*Heterometrus fulvipes*). Experientia (Basel) **15**, 345—347 (1976).

Geethabali, Rao, K. P.: A metasomatic neural photoreceptor in the scorpion. J. exp. Biol. **58**, 189—196 (1973).

Gladfelter, W. B.: Sea anemone with zooxanthellae: simultaneous contraction and expansion in response to changing light intensity. Science **189**, 570—571 (1975).

Goldsmith, T. H.: The natural history of invertebrate visual pigments. In: Handbook of Sensory Physiology: Photochemistry of Vision, ed. H. J. A. Dartnall, Vol. VII/1, pp. 685—719. Berlin-Heidelberg-New York: Springer 1972.

Gotow, Ta.: Photoreceptor-like cells in the prostomium of a scaleworm, *Lepidonotus helotypus*. Zool. Mag. Tokyo **85**, 265—269 (1976).

Gotow, Ts.: Morphology and function of the photo-excitable neurons in the central ganglia of *Onchidium verruculatum*. J. comp. Physiol. **99**, 139—152 (1975).
Gotow, Ts., Tateda, H., Kuwabara, M.: The function of photoexcitable neurons in the central ganglia for behavioral activity of the marine mollusc, *Onchidium verruculatum*. J. comp. Physiol. **83**, 361—376 (1973).
Gribakin, F. G.: The distribution of the long wave photoreceptors in the compound eye of the honey bee as revealed by selective osmic staining. Vision Res. **12**, 1225—1230 (1972).
Gwilliam, G. F.: The mechanism of the shadow reflex in cirripedia. I. Electrical activity in the supraesophageal ganglion and ocellar nerve. Biol. Bull. **125**, 470—485 (1963).
Gwilliam, G. F.: The mechanism of the shadow reflex in cirripedia. II. Photoreceptor cell response, second-order responses, and motor cell output. Biol. Bull. **129**, 244—256 (1965).
Gwilliam, G. F.: Electrical responses to photic stimulation in the eyes and nervous system of nereid polychaetes. Biol. Bull. **136**, 385—397 (1969).
Gwilliam, G. F.: The mechanism of the shadow reflex in cirripedia. III. Rhythmical patterned activity in central neurons and its modulation by shadows. Biol. Bull. **151**, 141—160 (1976).
Hagiwara, S., Saito, N.: Voltage-current relations in nerve cell of *Onchidium verruculatum*. J. Physiol. (Lond.) **148**, 161—179 (1959).
Hama, K.: A photoreceptor-like structure in the ventral nerve cord of the crayfish, *Cambarus virilis*. Anat. Rec. **140**, 329—336 (1961).
Harris, J. E.: The role of endogenous rhythms in vertical migration. J. Mar. biol. Ass. U.K. **43**, 153—166 (1963).
Harris, J. E., Mason, P.: Vertical migration in eyeless *Daphnia*. Proc. roy. Soc. B **145**, 280—290 (1956).
Hartline, H. K., Graham, C. H.: Nerve impulses from single receptors in the eye. J. cell. comp. Physiol. **1**, 277—295 (1932).
Haug, G.: Die Lichtreaktionen der Hydren (*Chlorohydra viridissima* und *Pelmatohydra oligactis* (P.) *typica*). Z. vergl. Physiol. **19**, 246—303 (1933).
Hecht, S.: The nature of photoreceptor process. In: A Handbook of General Experimental Psychology, ed. C. Murchison. Worcester (Mass.): Clark Univ. Press 1934.
Henkart, M.: Light-induced changes in the structure of pigmented granules in *Aplysia* neurons. Science **188**, 155—157 (1975).
Hermann, H., Stark, L.: Prerequisites for a photoreceptor structure in the crayfish tail ganglion. Anat. Rec. **147**, 209—217 (1963).
Hermans, C. O.: Fine structure of the segmental ocelli of *Armandia brevis* (Polychaeta: Opheliidae). Z. Zellforsch. **96**, 361—371 (1969).
Hermans, C. O., Cloney, R. A.: Fine structure of the prostomial eyes of *Armandia brevis* (Polychaeta: Opheliidae). Z. Zellforsch. **72**, 583—596 (1966).
Hermans, C. O., Eakin, R. M.: Fine structure of the cerebral ocelli of a sipunculid, *Phascolosoma agassizii*. Z. Zellforsch. **100**, 325—339 (1969).
Herter, K.: Versuche über die Phototaxis von *Nereis diversicolor* O. F. Müller. Z. vergl. Physiol. **4**, 103—141 (1926).
Hess, W. N.: Reactions to light in the earthworm, *Lumbricus terrestris* L. J. Morph. Physiol. **39**, 515—542 (1924).
Hess, W. N.: Photoreceptors of *Lumbricus terrestris*, with special reference to their distribution, structure, and function. J. Morph. Physiol. **41**, 63—93 (1925).
Hess, W. N.: Regional photosensitivity and photoreceptors of *Crangon armillatus* and the spiny lobster, *Panulirus argus*. Carnegie Inst. Wash. **517**, 153—161 (1940).
Hesse, R.: Untersuchungen über die Organe der Lichtempfindung bei niederen Thieren I. Die Organe der Lichtempfindung bei den Lumbriciden. Z. wiss. Zool. **61**, 393—419 (1896).
Hisada, M.: A study on the photoreceptor of a medusa, *Spirocodon saltatrix*. J. Fac. Sci. Hokkaido Univ. VI **12**, 529—533 (1956).
Hisano, N., Tateda, H., Kuwabara, M.: Photosensitive neurons in the marine pulmonate mollusc *Onchidium verruculatum*. J. exp. Biol. **57**, 651—660 (1972a).
Hisano, N., Tateda, H., Kuwabara, M.: An electrophysiological study of the photoexcitive neurons of *Onchidium verruculatum* in situ. J. exp. Biol. **57**, 661—671 (1972b).
Horstmann, E.: Nerven- und muskel-physiologische Studien zur Schwimmbewegung der Scyphomedusa. Pflügers Arch. **234**, 421—431 (1934).

Hughes, H.P.I.: A light and electron microscope study of some opisthobranch eyes. Z. Zellforsch. **106**, 79—98 (1970).
Kaissling, K.-E.: Die phototropische Reaktion der Zoide von *Bugula avicularia* L. Z. vergl. Physiol. **46**, 541—594 (1963).
Kalmus, H.: Versuche über die Bewegungen der Seesterne, besonders von *Asterias gibbosa*. Z. vergl. Physiol. **9**, 703—733 (1929).
Kandel, E.R., Frazier, W.T., Waziri, F., Coggeshall, R.E.: Direct and common connections among identified neurons in *Aplysia*. J. Neurophysiol. **30**, 1352—1376 (1967).
Kaplan, E., Barlow, Jr. R.B.: Properties of visual cells in the lateral eye of *Limulus* in situ. Extracellular recordings. J. Gen. Physiol. **66**, 303—326 (1975).
Kennedy, D.: Electrical activity of a "primitive" photoreceptor. Ann. N.Y. Acad. Sci. **74**, 329—336 (1958).
Kennedy, D.: Neural photoreception in a lamellibranch mollusc. J. Gen. Physiol. **44**, 277—299 (1960).
Kennedy, D.: Physiology of photoreceptor neurons in the abdominal nerve cord of the crayfish. J. Gen. Physiol. **46**, 551—572 (1963a).
Kennedy, D.: Inhibition in visual systems. Sci. Amer. **209**, 122—130 (1963b).
Kennedy, D., Preston, J.B.: Activity patterns of interneurones in the caudal ganglion of the crayfish. J. Gen. Physiol. **43**, 655—670 (1960).
Kerneis, A.: Etude comparée d'organes photorécepteurs de Sabellidae (Annélides Polychètes). J. Ultrastruct. Res. **53**, 164—179 (1975).
Kleinholz, L.H.: Color changes in echinoderms. Pubbl. Staz. Zool. Napoli **17**, 53—57 (1938).
Koehler, O.: Lichtempfindlichkeit und Lichtsinn niederer Tiere. Stud. gen. **10**, 191—203 (1957).
Koller, G., Von Studnitz, G.: Über den Licht- und Schattenreflex von *Mya arenaria*. Z. vergl. Physiol. **20**, 388—404 (1934).
Krasne, F.B., Lawrence, P.A.: Structure of the photoreceptors in the compound eyespots of *Branchiomma vesiculosum*. J. Cell Sci. **1**, 239—248 (1966).
Kretz, J.R., Stent, G.S., Kristan, Jr., W.B.: Photosensory input pathways in the medical leech. J. comp. Physiol. **106**, 1—37 (1976).
Lall, A.B., Chapman, R.M.: Phototaxis in *Limulus* under natural conditions: evidence for reception of near-ultraviolet light in the median dorsal ocellus. J. exp. Biol. **58**, 213—224 (1973).
Land, M.F.: Activity in the optic nerves of *Pecten maximus* in response to changes in light intensity, and to pattern and movement in the optical environment. J. exp. Biol. **45**, 83—99 (1966).
Land, M.F.: Functional aspects of the optical and retinal organization of the mollusc eye. Symp. Zool. Soc. Lond. **23**, 75—96 (1968).
Larimer, J.L.: A functional photoreceptor in blind cavernicolous crayfish. Nature (Lond.) **210**, 204—205 (1966).
Lasansky, A., Fuortes, M.G.F.: The site of origin of electrical responses in visual cells of the leech, *Hirudo medicinalis*. J. Cell Biol. **42**, 241—252 (1969).
Laties, A.M., Bok, D., Liebman, P.: Procion yellow: A marker dye for outer segment disc patency and for rod renewal. Exp. Eye Res. **23**, 139—148 (1976).
Lemke, G.: Beiträge zur Lichtorientierung und zur Frage des Farbensehens der Planarien. Z. vergl. Physiol. **22**, 298—345 (1935).
Lickey, M.E., Zack, S.: Extraocular photoreceptors can entrain the circadian rhythm in the abdominal ganglion of *Aplysia*. J. comp. Physiol. **84**, 361—366 (1973).
Light, V.E.: Photoreceptors in *Mya arenaria* with special reference to their distribution, structure and function. J. Morph. **49**, 1—42 (1930).
Lisman, J.E., Brown, J.E.: Two light-induced processes in the photoreceptor cells of *Limulus* ventral eye. J. Gen. Physiol. **58**, 544—561 (1971).
Lisman, J.E., Brown, J.E.: The effects of intracellular iontophoretic injection of calcium and sodium ions on the light response of *Limulus* ventral photoreceptors. J. Gen. Physiol. **59**, 701—719 (1972).
Lisman, J.E., Brown, J.E.: Effects of intracellular injection of calcium buffers on light adaptation in *Limulus* ventral photoreceptors. J. Gen. Physiol. **66**, 489—506 (1975).
Loewenstein, W.R.: Mechano-electric transduction in the Pacinian corpuscle. Initiation of sensory impulses in mechanoreceptors. In: Handbook of Sensory Physiology, ed. W.R. Loewenstein. Vol. I, Berlin-Heidelberg-New York: Springer 1971.
Marks, P.S.: Nervous control of light responses in the sea anemone, *Calamactis praelongus*. J. exp. Biol. **65**, 85—96 (1976).

Marriott,F.H.C.: The absolute light-sensitivity and spectral threshold curve of the aquatic flatworm *Dendrocoelum lacteum*. J. Physiol. (Lond.) **143**, 369—379 (1958).

Mauro,A., Baumann,F.: Electrical evidence of photoreceptors in the epistellar body of *Eledone moschata*. Nature (Lond.) **220**, 1332—1334 (1968).

McReynolds,J.S., Gorman,A.L.F.: Ionic basis of hyperpolarizing receptor potential in scallop eye: increase in permeability to potassium ions. Science **183**, 658—659 (1974).

McReynolds,J.S., Gorman,A.L.F.: Hyperpolarizing photoreceptors in the eye of a primitive chordate, *Salpa democratica*. Vision Res. **15**, 1181—1186 (1975).

Menzel,R.: Colour receptors in insects. In: The Compound Eye and Vision of Insects, ed. G.A.Horridge, Oxford: Clarendon Press 1975.

Millecchia,R., Bradbury,J., Mauro,A.: Simple photoreceptors in *Limulus polyphemus*. Science **154**, 1199—1201 (1966).

Millecchia,R., Mauro,A.: The ventral photoreceptor cells of *Limulus*. II. The basic photoresponse. J. Gen. Physiol. **54**, 310—330 (1969).

Millott,N.: Colour change in the echinoid, *Diadema antillarum* Philippi. Nature (Lond.) **170**, 325 (1952).

Millott,N.: Sensitivity to light and the reactions to changes in light intensity of the echinoid *Diadema antillarum* Philippi. Phil. Trans. **238**, 187—220 (1954).

Millott,N.: The covering reaction in a tropical sea urchin. Nature (Lond.) **175**, 561 (1955).

Millott,N.: The covering reaction of sea-urchins. I. A preliminary account of covering in the tropical echinoid *Lytechinus variegatus* (Lamarck), and its reaction to light. J. exp. Biol. **33**, 508—523 (1956).

Millott,N.: Pigmentary system of *Diadema antillarum* Philippi. Nature (Lond.) **203**, 206—207 (1964).

Millott,N.: The dermal light sense. Symp. Zool. Soc. Lond. **23**, 1—36 (1968).

Millott,N.: The photosensitivity of echinoids. Advanc. mar. Biol. **13**, 1—52 (1975).

Millott,N., Takahashi,K.: The shadow reaction of *Diadema antillarum* Philippi. IV. Spine movements and their implications. Phil. Trans. **246**, 437—470 (1963).

Millott,N., Yoshida,M.: The photosensitivity of the sea urchin *Diadema antillarum* Philippi. Responses to increase in light intensity. Proc. Zool. Soc. Lond. **133**, 67—71 (1959).

Millott,N., Yoshida,M.: The shadow reaction of *Diadema antillarum* Philippi. I. The spine response and its relation to the stimulus. J. exp. Biol. **37**, 363—375 (1960a).

Millott,N., Yoshida,M.: The shadow reaction of *Diadema antillarum* Philippi. II. Inhibition by light. J. exp. Biol. **37**, 376—389 (1960b).

Milne,L.J., Milne,M.J.: Invertebrate photoreceptors. In: Radiation Biology, ed. A.Hollaender, Vol. III. New York: McGraw-Hill 1956.

Milne,L.J., Milne,M.J.: Photosensitivity in invertebrates. In: Handbook of Physiology, ed. J.Field, Vol. I, Part 1. Neurophysiology. Baltimore: Williams and Wilkins 1959.

Morton,J.E.: The responses and orientation of the bivalve *Lasaea rubra* Montagu. J. Mar. Biol. Ass. U. K. **39**, 5—26 (1960).

Mpitsos,G.J.: Physiology of vision in the mollusk *Lima scabra*. J. Neurophysiol. **36**, 371—383 (1973).

Murray,G.C.: Intracellular absorption difference spectrum of *Limulus* extraocular photolabile pigment. Science **154**, 1182—1183 (1966).

Neya,T.: Photic behavior of sea urchin larvae, *Hemicentrotus pulcherrimus*. Zool. Mag. (Tokyo) **74**, 11—16 (1965).

Nicol,J.A.C.: Responses of *Branchiomma vesiculosum* (Montagu) to photic stimulation. J. Mar. Biol. Ass. U. K. **29**, 303—320 (1950).

Nielsen,C.: Entoproct life-cycles and the entoproct/ectoproct relationship. Ophelia **9**, 209—341 (1971).

Nishioka,R.S., Hagadorn,I.R., Bern,H.A.: Ultrastructure of the epistellar body of the octopus. Z. Zellforsch. **57**, 406—421 (1962).

Nishioka,R.S., Yasumasu,I., Bern,H.A.: Photoreceptive features of vesicles associated with the nervous system of cephalopods. Nature (Lond.) **212**, 1181 (1966a).

Nishioka,R.S., Yasumasu,I., Packard,A., Bern,H.A., Young,J.Z.: Nature of vesicles associated with the nervous system of cephalopods. Z. Zellforsch. **75**, 301—316 (1966b).

Nolte,J., Brown,J.E.: Ultraviolet-induced sensitivity to visible light in ultraviolet receptors of *Limulus*. J. Gen. Physiol. **59**, 186—200 (1972).

Nomura,E.: Effect of light on the movements of the earthworm, *Allolobophora foetida* (Sav.). Tohoku Imper. Univ. Sci. Rep. IV **1**, 294—409 (1926).

Nomura, E.: Effect of chemicals on phototaxis in *Allolobophora foetida* Sav., and its analysis based upon antagonistic reaction between the functionings of the brain and of the ventral nerve cord. Tohoku Imper. Univ. Sci. Rep. IV **2**, 1—53 (1927).

North, W.J.: Sensitivity to light in the sea anemone *Metridium senile* (L.). II. Studies on reaction time variability and effects of change in light intensity and temperature. J. Gen. Physiol. **40**, 715—733 (1957).

North, W.J., Pantin, C.F.A.: Sensitivity to light in the sea anemone *Metridium senile* (L.): adaptation and action spectra. Proc. roy. Soc. B**148**, 385—396 (1958).

Okuno, Y., Katagiri, Y., Fujimoto, K.: On-hyperpolarizing off-depolarizing receptor potential in the *Onchidium* dorsal eye. Zool. Mag. (Tokyo) **85**, 88—90 (1976).

Ozawa, S., Hagiwara, S., Nicolaysen, K., Stuart, A.E.: Signal transmission from photoreceptors to ganglion cells in the visual system of the giant barnacle. Cold Spr. Harb. Symp. quant. Biol. **40**, 563—570 (1976).

Page, T.L., Larimer, J.L.: Entrainment of the circadian locomotor activity rhythm in crayfish. The role of the eyes and caudal photoreceptor. J. comp. Physiol. **78**, 107—120 (1972).

Parker, G.H., Burnett, F.L.: The reactions of planarians, with and without eyes, to light. Amer. J. Physiol. **4**, 373—385 (1900).

Passano, L.M., McCullough, C.B.: The light response and the rhythmic potentials of Hydra. Proc. nat. Acad. Sci. (Wash.) **48**, 1376—1382 (1962).

Passano, L.M., McCullough, C.B.: Co-ordinating systems and behaviour in *Hydra*. I. Pacemaker system of the periodic contractions. J. exp. Biol. **41**, 643—664 (1964).

Passano, L.M., McCullough, C.B.: Co-ordinating systems and behaviour in *Hydra*. II. The rhythmic potential system. J. exp. Biol. **42**, 205—231 (1965).

Pearse, V.B.: Modification of sea anemone behavior by symbiotic zooxanthellae: Phototaxis. Biol. Bull. **147**, 630—640 (1974a).

Pearse, V.B.: Modification of sea anemone behavior by symbiotic zooxanthellae: Expansion and contraction. Biol. Bull. **147**, 641—651 (1974b).

Pilgrim, R.L.C.: Some responses to light in a specimen of *Pelagohydra mirabilis*. Dendy, 1902 (Coelenterata: Hydrozoa). J. exp. Biol. **46**, 491—497 (1967).

Pople, W., Ewer, D.W.: Studies on the myoneural physiology of echinodermata. I. The pharyngeal retractor muscle of *Cucumaria*. J. exp. Biol. **31**, 114—126 (1954).

Pople, W., Ewer, D.W.: Studies on the myoneural physiology of echinodermata. III. Spontaneous activity of the pharyngeal muscle of *Cucumaria*. J. exp. Biol. **35**, 712—730 (1958).

Preston, J.B., Kennedy, D.: Integrative synaptic mechanisms in the caudal ganglion of the crayfish. J. Gen. Physiol. **43**, 671—681 (1960).

Prosser, C.L.: Action potentials in the nervous system of the crayfish. II. Responses to illumination of the eye and caudal ganglion. J. cell. comp. Physiol. **4**, 363—377 (1934a).

Prosser, C.L.: Effect of the central nervous system on responses to light in *Eisenia foetida* Sav. J. comp. Neurol. **59**, 61—91 (1934b).

Rees, F.G.: Studies on the pigmented and unpigmented photoreceptors of the cercaria of *Cryptocotyle lingua* (Creplin) from *Littorina littorina* (L.). Proc. roy. Soc. B**188**, 121—138 (1975).

Rockstein, M.: Role of the terminal pigment spots of the starfish, *Asterias forbesi*, in light orientation. Nature (Lond.) **177**, 341—342 (1956).

Röhlich, P., Avros, B., Virágh, Sz.: Fine structure of photoreceptor cells in the earthworm *Lumbricus terrestris*. Z. Zellforsch. **104**, 345—357 (1970).

Röhlich, P., Török, L.J.: Elektronenmikroscopische Untersuchungen des Auges von Planarien. Z. Zellforsch. **54**, 362—381 (1961).

Rushforth, N.B., Burnett, A.L., Maynard, R.: Behavior in Hydra: Contraction responses of *Hydra pirardi* to mechanical and light stimuli. Science **139**, 760—761 (1963).

Rushforth, N.B., Krohn, I.T., Brown, L.K.: Behavior in Hydra: inhibition of the contraction responses of *Hydra pirardi*. Science **145**, 602—604 (1964).

Ryland, J.S.: Experiments on the influence of light on the behaviour of polyzoan larvae. J. exp. Biol. **37**, 783—800 (1960).

Schulz, F.: Bau und Funktion der Sinneszellen in der Körperoberfläche von *Helix pomatia*. Z. Morphol. Ökol. Tiere **33**, 555—581 (1938).

Segall, J.: Versuche über Lichtreaktionen und Lichtempfindlichkeit beim Regenwurm. Z. vergl. Physiol. **19**, 94—109 (1933).

Shaw, S. R.: Decremental conduction of the visual signal in barnacle lateral eye. J. Physiol. (Lond.) **220**, 145—175 (1972).

Siddiqui, I. A., Viglierchio, D. R.: Ultrastructure of photoreceptors in the marine nematode *Deontostoma californicum*. J. Ultrastruct. Res. **32**, 558—571 (1970).

Singla, C. L.: Ocelli of hydromedusae. Cell Tissue Res. **149**, 413—429 (1974).

Smith, J. E.: Some observations on the nervous mechanisms underlying the behaviour of starfishes. Symp. Soc. exp. Biol. **4**, 196—220 (1950).

Snodderly, Jr., D. M.: Processing of visual inputs by brain of *Limulus*. J. Neurophysiol. **34**, 588—611 (1971).

Spiegler, J. B., Yeandle, S.: Independence of location of light absorption and discrete wave latency distribution in *Limulus* ventral nerve photoreceptors. J. Gen. Physiol. **64**, 494—502 (1974).

Srebro, R., Behbehani, M.: Light adaptation in the ventral photoreceptor of *Limulus*. J. Gen. Physiol. **64**, 166—185 (1974).

Steven, D. M.: The dermal light sense. Biol. Rev. **38**, 204—240 (1963).

Takahashi, K.: Electrical responses to light stimuli in the isolated radial nerve of the sea urchin, *Diadema setosum* (Leske). Nature (Lond.) **201**, 1343—1344 (1964).

Taliaferro, W. H.: Reactions to light in *Planaria maculata*, with special reference to the function and structure of the eyes. J. exp. Zool. **31**, 59—116 (1920).

Tamasige, M., Yamaguchi, T.: Equilibrium orientation controlled by ocelli in an anthomedusa, *Polyorchis karafutoensis*. Zool. Mag. (Tokyo) **76**, 35—36 (1967).

Thorson, G.: Light as an ecological factor in the dispersal and settlement of larvae of marine bottom invertebrates. Ophelia **1**, 167—208 (1964).

Thurm, U.: Mechanisms of electrical membrane responses in sensory receptors, illustrated by mechanoreceptors. In: Biochemistry of Sensory Functions, ed. L. Jaenicke, Berlin-Heidelberg-New York: Springer 1974.

Uchizono, K.: The structure of possible photoreceptive elements in the sixth abdominal ganglion of the crayfish. J. Cell Biol. **15**, 151—154 (1962).

Von Uexküll, J.: Die Wirkung von Licht und Schatten auf die Seeigel. Z. Biol. **40**, 447—476 (1900).

Ullyott, P.: The behaviour of *Dendrocoelum lacteum*. I. Responses at a light and dark boundary. J. exp. Biol. **13**, 253—264 (1936a).

Ullyott, P.: The behaviour of *Dendrocoelum lacteum*. II. Responses in non-directional gradients. J. exp. Biol. **13**, 265—278 (1936b).

Vaupel-Von Harnack, M.: Über den Feinbau vom Nervensystem des Seesterns (*Asterias rubens* L.). III. Die Struktur der Augenpolster. Z. Zellforsch. **60**, 432—451 (1963).

Viaud, G.: Phototropisme et galvanotropisme de *Planaria lugubris* O. Sch. (individus entiers et segments). C. R. Soc. Biol. (Paris) **143**, 534 (1949).

Viaud, G., Médioni, J.: Phototropisme et régéneration chez *Planaria lugubris* O. Sch. C. R. Soc. Biol. (Paris) **143**, 1221—1223 (1949).

Wasserman, G. S.: Unconditioned response to light in *Limulus*: mediation by lateral, median and ventral eye loci. Vision Res. **13**, 95—105 (1973).

Weber, W., Dambach, M.: Amöboid bewegliche Pigmentzellen im Epithel des Seeigels *Centrostephanus longispinus*. Ein neuartiger Farbwechselmechanismus. Z. Zellforsch. **133**, 87—102 (1972).

Weber, W., Dambach, M.: Light-sensitivity of isolated pigment cells of the sea urchin *Centrostephanus longispinus*. Cell Tissue Res. **148**, 437—440 (1974).

Wells, P. H.: Response to light by the eyeless white crayfish, *Cambarus ayersii*. Anat. Rec. **113**, 613 (1959).

Welsh, J. H.: The caudal photoreceptor and responses of the crayfish to light. J. cell. comp. Physiol. **4**, 379—388 (1934).

White, R. H., Walther, J. B.: The leech photoreceptor cell: ultrastructure of clefts connecting the phaosome with extracellular space demonstrated by lanthanum deposition. Z. Zellforsch. **95**, 102—108 (1969).

Wiederhold, M. L., MacNichol, Jr., E. F., Bell, A. L.: Photoreceptor spike responses in the hardshell clam, *Mercenaria mercenaria*. J. Gen. Physiol. **61**, 24—55 (1973).

Wilkens, L. A., Larimer, J. L.: The CNS photoreceptor of crayfish: Morphology and synaptic activity. J. comp. Physiol. **80**, 389—407 (1972).

Wilkens, L. A., Larimer, J. L.: Photosensitivity in the sixth abdominal ganglion of decapod crustaceans: A comparative study. J. comp. Physiol. **106**, 69—75 (1976).

Woollacott, R. M., Eakin, R. M.: Ultrastructure of a potential photoreceptoral organ in the larva of an entoproct. J. Ultrastruct. Res. **43**, 412—425 (1973).
Woollacott, R. M., Zimmer, R. L.: Fine structure of a potential photoreceptor organ in the larva of *Bugula neritina* (Bryozoa). Z. Zellforsch. **123**, 458—469 (1972).
Yamasu, T., Yoshida, M.: Electron microscopy on the photoreceptors of an anthomedusa and a scyphomedusa. Publ. Seto Mar. Biol. Lab. **20**, 757—778 (1973).
Yamasu, T., Yoshida, M.: Fine structure of complex ocelli of a cubomedusan, *Tamoya bursaria* Haeckel. Cell Tiss. Res. **170**, 325—339 (1976).
Yanase, K., Sakamoto, S.: Fine structure of the visual cells of the dorsal eye in molluscan, *Onchidium verruculatum*. Zool. Mag. (Tokyo) **74**, 238—242 (1965).
Yoshida, M.: On the light response of the chromatophore of the sea urchin, *Diadema setosum* (Leske). J. exp. Biol. **33**, 119—123 (1956).
Yoshida, M.: Spectral sensitivity of chromatophores in *Diadema setosum* (Leske). J. exp. Biol. **34**, 222—225 (1957).
Yoshida, M.: Positive phototaxis in *Psammechinus microtuberculosis* (Blainville) Pubbl. Staz. Zool. Napoli **30**, 260—262 (1957).
Yoshida, M.: Further studies on the chromatophore response in *Diadema setosum* (Leske). Biol. J. Okayama Univ. **6**, 169—173 (1960).
Yoshida, M.: The effect of light on the shadow reaction of the sea urchin *Diadema setosum* (Leske). J. exp. Biol. **39**, 589—602 (1962).
Yoshida, M.: Photosensitivity. In: Physiology of Echinodermata, ed. R. A. Boolootian, New York: Interscience 1966.
Yoshida, M.: Photoreception in medusae. Mar. Sci. (Tokyo) **5**, 732—737 (1973).
Yoshida, M., Millott, N.: Light sensitive nerve in an echinoid. Experientia (Basel) **15**, 13—14 (1959).
Yoshida, M., Millott, N.: The shadow reaction of *Diadema antillarum* Philippi. III. Re-examination of the spectral sensitivity. J. exp. Biol. **37**, 390—397 (1960).
Yoshida, M., Ohtsu, K.: A preliminary note on the electrical response to shadows of the anthomedusa, *Spirocodon saltatrix*. Publ. Seto Mar. Biol. Lab. **20**, 647—651 (1973).
Yoshida, M., Ohtsuki, H.: Compound ocellus of a starfish: its function. Science **153**, 197—198 (1966).
Yoshida, M., Ohtsuki, H.: The phototactic behavior of the starfish, *Asterias amurensis* Lütken. Biol. Bull. **134**, 516—532 (1968).
Yoshikami, S., Robinson, W. E., Hagins, W. A.: Topology of the outer segment membranes of retinal rods and cones revealed by a fluorescent probe. Science **185**, 1176—1179 (1974).
Zahid, Z. R., Golding, D. W.: Structure and ultrastructure of the central nervous system of the polychaete *Nephtys*, with special reference to photoreceptor elements. Cell Tissue Res. **149**, 567—576 (1974).
Zwicky, K. T.: A light response in the tail of *Urodacus*, a scorpion. Life Sci. **7**, 257—262 (1968).
Zwicky, K. T.: The spectral sensitivity of the tail of *Urodacus*, a scorpion. Experientia (Basel) **26**, 317 (1970).

Chapter 11

Extraocular Light Receptors and Circadian Rhythms[1]

MIRIAM F. BENNETT, Waterville, Maine (USA)

With 9 Figures

Contents

A. Introduction

For organisms to which functioning in time with the solar day is vital, transitions from light to dark or from dark to light are some of the most reliable *Zeitgebers* or indicators of real time available to them in their physical environments. In its natural habitat, an animal or plant can use such changes to orient itself precisely to the day–night phases of a world which is characterized by many cyclic geophysical variations which influence and affect organisms and their behavior.

For researchers attempting to understand the circadian organization of living systems, changes in light intensity of the laboratory environments are among the finest analytical probes now available. In these artificial surroundings, changes from dark to light, from light to dark, or from one light intensity to another can be used as perturbations which may entrain the organismic cycles, shift phases of the

[1] The preparation of this chapter was aided in part by the Natural Sciences Grant Program, Colby College. That aid and the help of the Library Staff of the Marine Biological Laboratory, Woods Hole, are greatly appreciated.

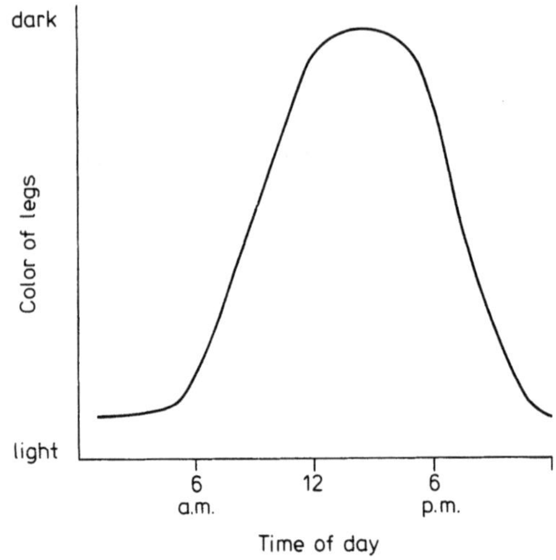

Fig. 1. Differences in the coloration of the legs of fiddler crabs during 1 solar day in the course of which the animals were exposed to natural day–night variations in light intensity. (BENNETT, 1974. Courtesy of Charles C. Thomas, Springfield, Illinois)

cycles, or affect the period of the cycles. The significance of light to the circadian rhythms of many organisms has been well reviewed during the last 15 years (ASCHOFF, 1960; BÜNNING, 1977; HOFFMANN, 1971; WINFREE, 1971).

Among earlier reports of persistent cycles in invertebrates are descriptions of the effects of variations in light–dark regimes on characteristics of the rhythms. MEGUSǍR (1912) reported that when fiddler crabs, *Uca pugnax*, are exposed to natural day–night illumination in the laboratory, the pigments of the chromatophores of their legs are dispersed, and consequently, the legs are very dark during real daytime. The pigments are concentrated and the legs are light during real night (Fig. 1). When the crabs are exposed to darkness during the day and bright light during the night, the phases of the color change cycles are reversed (Fig. 2). In a more detailed analysis of this rhythm, WEBB (1950) found that the real times during which light–dark changes are presented to *Uca* affect the amount and direction of shifts of the phases of their color change cycle. BÜNNING (1935) and BRETT (1955) both described the manner in which light regulates the phases of the circadian rhythm of eclosion of the fruit fly, *Drosophila melanogaster*. Studies of the eclosion cycle have contributed importantly to our understanding of the principles of organismic periodicity.

As the circadian rhythms of a broad range of animals were described, researchers realized that in some cases, as in cyclically functioning protists and plants in which there are no true eyes, the characteristic effects of light-dark changes and of light itself on the cycles were apparent even after the animals had been blinded or somehow deprived of the use of their eyes (BÜNNING, 1977; BENNETT, 1974). In such experimental animals, it is assumed that some type of extraocular light receptors, possibly including a light-sensitive cell membrane (NJUS et al., 1974), is the receiver of the photic stimulus which influences their circadian cycles. Nonocular photoreceptors and their functions were reviewed in general by STEVEN

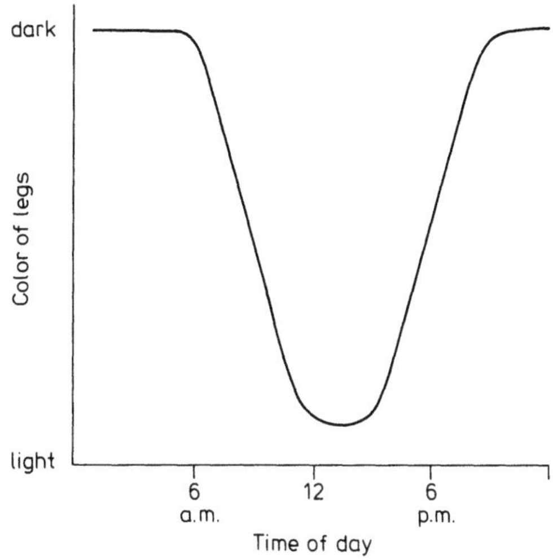

Fig. 2. Differences in the coloration of the legs of fiddler crabs during one 24-hr period in which the animals were exposed to light during the night and to darkness during the daytime. (BENNETT, 1974. Courtesy of Charles C. Thomas, Springfield, Illinois)

(1963), and the extraocular light receptors of invertebrates are discussed in detail in the present volume (YOSHIDA, this vol. Ch. 10).

Most analyses of extraocular light reception and circadian rhythms are based on the results of research with vertebrates, in which parts of the anterior central nervous system have been shown to function significantly in light reception (ADLER, 1970; MENAKER, 1972; AXELROD, 1974; BENNETT, 1974; WURTMANN, 1975). Other important findings, fewer in number, linking extraocular photoreception and rhythmicity in invertebrates have also been published. In these animals, as in vertebrates, parts of the central nervous system are suggested as the extraocular photoreceptors, stimulation of which affects circadian periodicity.

B. Cnidarians

Reports of persistent rhythms of approximate solar-day frequency in cnidarians are rare. BATHAM and PANTIN (1950) reviewed the few studies of cyclic activities in sea anemones that had been published previous to their own investigations of *Metridium senile* (L.). The cycles of these anemones, animals which lack discrete photoreceptors, are indeed affected by light, since their frequencies and phasing were regular under LD 12:12 and became much less precise when the cnidarians were kept in complete darkness.

Light–dark conditions also influence the rhythmic movements of sea pens, *Cavernularia obesa* Valenciennes. During the 1940s and 1950s, MORI and his associates published a series of papers describing their investigations of cyclic

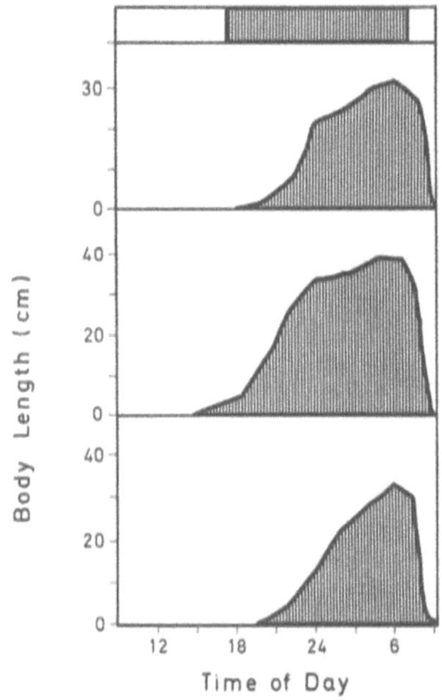

Fig. 3. Body lengths of three different specimens of *Cavernularia* during one 24-hr period. The *white and black bands* at the *top* of the figure indicate light and dark periods, respectively. (MORI, 1960. Copyright 1961 by Cold Spring Harbor Laboratory)

activity in this cnidarian under both natural and laboratory conditions. Those were summarized and discussed by MORI in 1960.

Cavernularia, which lacks specific photoreceptors, lives in the shallow waters off the coast of southern Japan. It is a colonial form, consisting of thousands of zooids, but the group usually behaves as an integrated unit. Generally, the rachis of the colony expands up and over the sandy bottom of the sea during nighttime and contracts down and into the sand during daytime. In the laboratory, colonies were maintained individually in glass cylinders through which sea water was circulated continuously. At 3-hr intervals, the lengths of the portions of the animals extending above the sand in the bottoms of the cylinders were measured. These lengths were used as indexes of activity.

Under natural lighting in the laboratory the activity cycle of the sea pens was much the same as it is in their natural environments (Fig. 3). That cycle also continued for as long as 100 days in DD or in LL.

Relationships between light–dark changes and the cnidarians' rhythms were determined by exposing colonies to light during real night and dark during real day, i.e., reversed-light periods, or to alternating light and dark periods of lengths varying from 5 min light: 5 min dark to 24 hrs light: 24 hrs dark, i.e., from total periods as short as 10 min to total periods as long as 48 hrs.

After 2–6 days under reversed lighting, most animals reversed their phases of activity, and extended their bodies during the daytime and contracted them at night. Under light–dark regimes of circadian (18–30 hrs) periodicity, e.g., 9 hrs: 9 hrs, 15 hrs: 15 hrs, 9 hrs: 15 hrs, 15 hrs: 9 hrs, the animals synchronized their phases of expansion with the dark periods. Under regimes departing from a circadian frequency, e.g., 5 min: 5 min, 6 hrs: 6 hrs, 24 hrs: 24 hrs, synchronization did not occur. Some animals became arrhythmic; others continued to function at circadian frequency. MORI and ONDO (1957) presented the detailed results of the studies of the effects of light on the *Cavernularia* cycles. On the basis of these findings, MORI (1960, p. 334) concluded "...that light is one of the most powerful environmental factors in controlling the rhythmic activity." The location and functioning of the light receptor mechanism, the existence of which is proved by MORI's work with the sea pens, have not been described. Knowledge of the integration of the photic stimuli and the mediation of the changes evoked in the circadian cycles of *Cavernularia* by them is completely lacking.

Very recent records made by SWEENEY (1976) during the "Alpha Helix" expedition to the Pacific provide examples of other cnidarians' circadian activities that are affected by light. Most of the polyps of corals, especially *Fungia* spp., were found to be contracted during the day when the animals were observed out on their reefs. While exposed to natural lighting in the laboratory, many of the polyps continued to contract during light periods and to expand during darkness. That general pattern of behavior was also apparent in DD as well as in LL of low intensity. However, in DD, the animals spent more time expanded than they did in LD. In addition, flashes from a strobe caused contraction of polyps that were in the expansion phases of their cycles. The light receptors and the mechanisms involved in all these reactions remain to be elucidated.

C. Echinoderms

Knowledge of the circadian organization of echinoderms is also fragmentary, but, once again, the importance of light and unidentified photoreceptors to cyclic behavior is suggested by the information which is available. MORI and MATUTANI (1952) described rhythmic activities in starfish, *Astropecten polyacanthus* Müller et Troschel. Under natural day–night changes in illumination, the animals were most active at dawn and moderately active at dusk. Around midday and midnight, they remained buried in the sand of their aquaria. That cycle continued through at least 3 days of DD after which it ceased.

A second circadian rhythm, one of varying sensitivity to light, was also suggested by MORI and MATUTANI (1952). When *Astropecten* live in constant darkness, they usually respond to sudden illumination by coming out of the sand. The tendency to do so increases as the real time of dawn approaches; it decreases between dawn and late afternoon and then increases again as dusk approaches.

These Japanese researchers attempted to find whether the eyespots or optic cushions are necessary for the regulation of the rhythmic activities of the starfish. Animals in which all of the eyespots had been severed functioned as did normal individuals, exhibiting the usual cyclic behavior. Consequently, other extraocular photoreceptors are assumed to influence circadian activities in this starfish.

D. Platyhelminthes and Annelids

Rhythms of circadian, circatidal, and circannual frequencies in platyhelminthes and annelids have been reviewed recently (BENNETT, 1974 and 1976). For the few examples of approximate solar-day cycles in freshwater planaria published to date, only descriptive information has been reported and the effects of light (and therefore the roles of photoreceptors) on the rhythms are unknown. BEST (1960) did suggest that a circadian cycle of "cannibalism"—*Dugesia tigrina* attacked *Cura foremani* (Turbellaria) more frequently at night than during the day—does not depend upon contemporary changes in illumination, but he did not describe the constant conditions or the changes in lighting used in his work. BECKER-CARUS (1970), who found circadian cycles of sensitivity to electric stimulation and of orientation in a multiple Y-labyrinth in *Dugesia dorotocephala*, has not reported whether or not changes in light–dark regimens affect those rhythms.

Variations in light–dark treatments are known to influence the circadian rhythm involving a parasitic flatworm, *Schistosoma mansoni* (VALLE et al., 1971), but whether photoreceptors of the host, the snail *Bromphalaria glabrata*, or of the cercariae of the worm or of both are operative is not clear. Under the natural conditions of light and temperature in the laboratory in Belo Horizonte, Brazil, 90% of the *Schistosoma* cercariae emerge from their hosts between 12:00 and 18:00 hrs.

Although changes in the ambient temperature of the snails have effects on the maintenance of this cycle, photoperiods also influence its expression. When the temperature was held constant and the snails were kept in LL or in DD, the parasites emerged arrhythmically. However, when the temperature was constant and light varied naturally, most of the cercariae left the host by day; when the snails were maintained at a constant temperature and were exposed to reversed photoperiods, artificial light from 19:00 to 7:00 hrs and darkness from 7:00 to 19:00 hrs, maximal emergence occurred during the light periods. Further analyses are needed to ascertain the contributions of the two species to the expression of this cycle before the roles of photoreceptors can be understood.

The circadian rhythm of color change in the polychaete, *Platynereis dumerilii*, reported by FISCHER (1965) and investigated further by RÖSELER (1970), is another cycle which is probably regulated in part by photic input received by several different structures. In these worms, the movements of three types of chromatophores, small red ones, large red ones and white ones, indicate phases of solar days. During light periods, the pigments are dispersed; in the dark, they are concentrated. These changes persist in LL and in DD in normal animals as well as in individuals from which the prostomium with the eyes has been removed (FISCHER, 1965). Certainly, structures other than the prostomial eyes of *Platynereis* are receptors of light changes that entrain this color change rhythm, since, as in intact worms, the cyclical phases of pigment movements in worms without prostomia were inverted by an inverted light cycle (RÖSELER, 1970). Synchronization of the phases of the worms' rhythms with those of the inverted light regime occurred in the majority of animals within 64 hrs after the beginning of the reversed lighting (Fig. 4). Proof that a factor, probably a neurosecretory one,

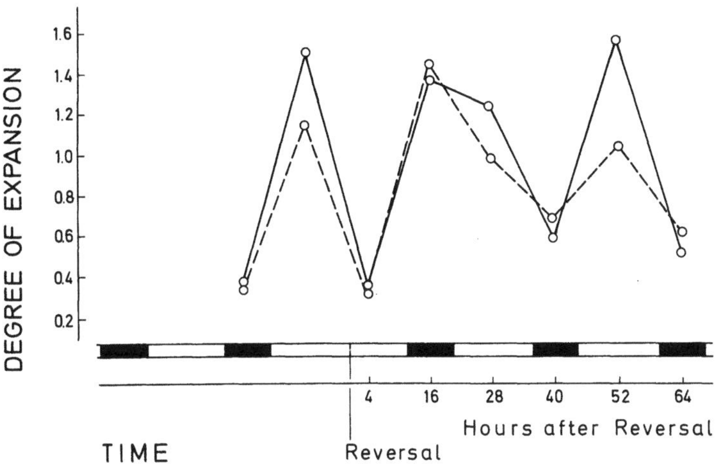

Fig. 4. Degrees of expansion of the chromatophores of *Platynereis dumerilii* before and after reversal of the light periods. The *white and black bands* show the periods of light and darkness. The *solid lines* trace out changes in normal animals; the *broken lines* indicate the changes seen in worms from which the prostomium had been removed. (RÖSELER, 1970)

of the posterior body of *Platynereis* is necessary for regulation of the color change cycle is also provided by RÖSELER's analyses. Chromatophores of anterior fragments of these polychaetes can be entrained to the inverted regime only in those cases in which neural connections with posterior segments are retained. RÖSELER also has evidence that the supraesophageal ganglia are somehow involved in the maintenance of the cycle, but no suggestions have been made regarding the nature of the receptors and the afferent pathways of the light changes, which affect the color change rhythm.

In addition to the definite color changes observed in *Platynereis* at circadian frequencies, there is the lunar monthly periodicity, i.e., about 28 days in length and obvious in worms living under LD 16:8. Modification of it, e.g., effecting dispersion of pigments at night by exposing the animals to bright light or to artificial moonlight during that time, occurs in normal worms as well as in worms without prostomia (RÖSELER, 1970). Once again, circumstantial evidence is provided for extraocular light reception.

This evidence is made even stronger by results of an earlier study by HAUENSCHILD (1961) of the lunar cycle of reproductive swarming in worms of the same species, *Platynereis dumerilii*. The setting of the most obvious phase of that cycle, the time of maximal swarming, with DD and LD manipulations occurs in the same ˙ ıanner in both normal and blinded worms.

Extraocular photoreceptors, the epidermal ones, are necessary for the expression of one indicator of circadian organization of the Oligochaete, *Lumbricus terrestris* (L). BENNETT and REINSCHMIDT (1965) proved that earthworms kept in DD under constant temperature withdraw their anterior ends from bright light directed at their heads at a faster rate during evening hours than around midday. Later studies (BENNETT, 1974) showed that the worms react fastest about 6:00 hrs, slowest about 17:00 hrs, and moderately fast during the hours of real night

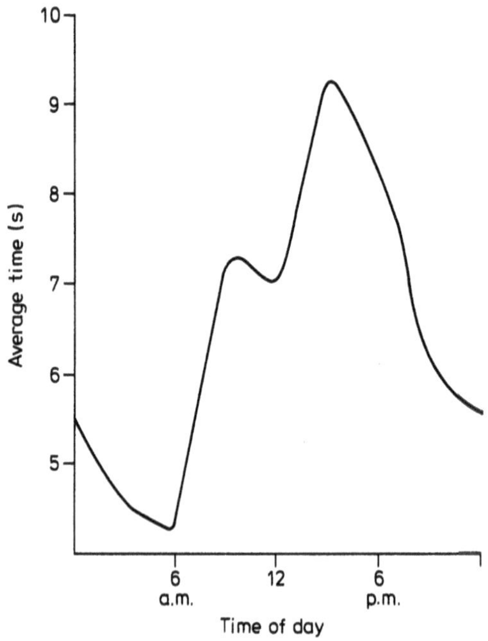

Fig. 5. Average light-withdrawal times for a group of ten earthworms during one solar period. (BENNETT, 1974. Courtesy of Charles C. Thomas, Springfield, Illinois)

(Fig. 5). We have also found that yellow or blue light is just as effective as is white in evoking light withdrawal, and further that epidermal photoreceptors of the posterior end function in the reflex removal of that portion of the worm from bright light. The rate at which that reaction occurs also varies during the circadian period.

Analyses of the light withdrawal cycle as well as of rhythms of the rate of crawling and of oxygen consumption of *Lumbricus* proved that the central nervous system of the anterior end of the worm is necessary for the mediation of the temporal differences, since those variations are lost when the brain is removed or after the circumpharyngeal connectives are cut (BENNETT, 1973). To date no information is available about the entrainment of these cycles by LD changes and, consequently, no evidence for or against the functioning of the epidermal photoreceptors in the phasing of earthworms' circadian rhythms of light withdrawal, crawling, or oxygen consumption exists.

However, the results of an investigation of the behavior of *Lumbricus terrestris* (L.) (GARDNER and RATNER, 1970) indicate that the epidermal photoreceptors do not function in phasing circadian differences in burrowing. When the depth of burrowing at various times of the solar day was compared in worms exposed to natural day–night changes in illumination and in *Lumbricus* living in complete darkness, no difference was found between the two groups. They all burrowed more deeply during the morning (8:00 to 12:00 hrs) than during the remainder of the day.

E. Molluscs

Among the molluscs, many examples of circadian cycles are known (BENNETT, 1974). However, in only a few cases is there information about the effects of light, and thereby the roles of photoreceptors, on these rhythms.

SALÁNKI and his associates have described periodic rest and activity, or the opening and closing of the valves, of the freshwater mussel, *Anodonta cygnea* (L.), maintained in the laboratory under natural light–dark conditions. After a long series of observations of this cycle, the researchers conclude that the alternation of rest and activity does not depend upon variations in any external factors (SALÁNKI, 1973). Thus, there is evidence to suggest that photoreceptors are not involved in the maintenance of this periodism. It has not been reported whether or not the phases of the cycle can be entrained by systematic changes in light and dark periods involving some photoreceptive mechanism.

There is experimental evidence that changes in light intensity affect cycles of activity of the quahog, *Mercenaria* (*Venus*) *mercenaria* (BENNETT, 1954). Under constant laboratory conditions, these bivalves open and close their shells at circadian, tidal, and lunar frequencies. After five consecutive 24-h periods of reversed photoperiod, i.e., light from 20:00 to 8:00 hrs and darkness from 8:00 to 20:00 hrs, the phases of each of these cycles were shifted. It is assumed that receptors of the siphon and mantle edges were operative in evoking those changes.

Comparable photoreceptors may function in regulating the patterns of shell banding in bivalves, aspects of which have been reported to be correlated with photoperiod. In Bay Scallops, *Argopecten irradians*, a daily pattern is seen in the horizontal extension of the shell and the formation of ridges, each of which is at right angles to the extension of the shell and forms a daily growth ring (WRENN, 1972). Scallops grown under natural photoperiod all formed their growth rings between 13:00 and 20:00 hrs. When the animals were maintained in LL, the formation of growth rings was aphasic (WRENN, 1972).

In a continuation of that investigation, WHEELER et al. (1975) found that shell calcification, in terms of the rate of incorporation of ^{14}C-carbonate, was three times greater from 10:00 to 14:00 hrs than it was from 19:00 to 20:30 hrs in *Argopecten* living under natural photoperiod. The difference continued even when the scallops were kept in darkness from 10:00 to 14:00 hrs. That finding correlates with a report from DODD (1969), who claimed that photoperiod had no effect on the banding pattern of the same scallop and some species of mussels. A similar conclusion based on her work with *Mercenaria mercenaria* was reached by THOMPSON (1970). However, in the case of *Argopecten*, WRENN'S (1972) evidence suggests that the circadian cycle of growth of the shell is affected by light–dark changes and therefore is dependent upon the stimulation of photoreceptors, even though the cycle of mineralization continues whether the photoperiod is the normal or an experimental one.

A more complete understanding of extraocular photoreception and circadian rhythms in molluscs comes from analyses of several persistent cycles in sea hares, *Aplysia californica*. Some of the earlier studies were reviewed by BENNETT (1974). These animals display cycles of spontaneous electric activity of neurons of several

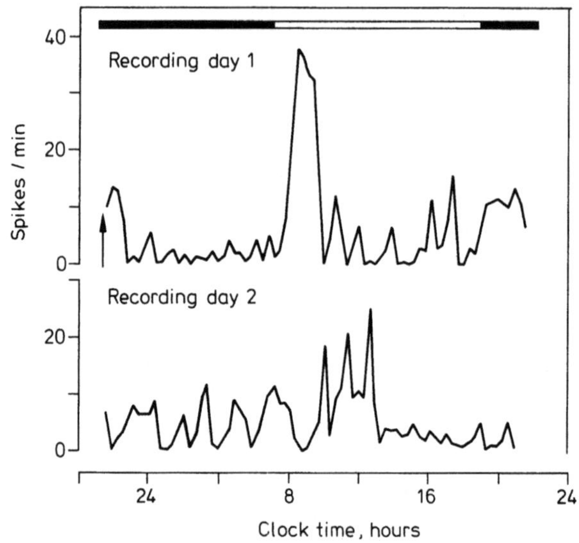

Fig. 6. Electric activity of a neuron of *Aplysia* in terms of real time. The *black and white bands* above the record of the first day indicate dark and light periods to which the donor animal had been exposed for 9 days before recording. The *arrow* indicates when electrodes were placed in the cell. (STRUMWASSER, 1965. Courtesy of North-Holland Publ. Co.)

different ganglia (STRUMWASSER, 1965; LICKEY, 1969) (Fig. 6). The ganglia can be removed from intact animals, which have been exposed previously to light–dark treatments of the researcher's choosing, and then maintained in seawater under experimental conditions. Recordings from the large neurons can be made for as long as 48 hrs after their removal from the sea hares.

BEISWANGER and JACKLET (1975) caution against considering these rhythms to be true circadian ones. Results of experiments in which they attempted to entrain cycles of spiking of a specific neuron, R15, to the 24-hr period in vitro were negative. Twenty-two of these cells were tested, and only one case showed evidence that the variations in electric activity, which were actually driven by light–dark changes, persisted. In all other cases, no rhythm was apparent after the first day of recording under constant conditions.

A circadian cycle of locomotion has been described for *Aplysia* (KUPFERMANN, 1968; JACKLET, 1972). When the animals were maintained in the laboratory under LD 12:12 they were much more active during the light than during the dark period. That phasing was retained during 48 hrs of complete darkness (KUPFERMANN, 1968). Another circadian rhythm of *Aplysia*, which is expressed in DD, LL, or under LD 12:12, is that of electric changes recorded from the stumps of the optic nerves of isolated eyes (JACKLET, 1969). Such preparations are also especially valuable to researchers, since they remain in healthy states for several days after their removal from donor animals. Entrainment of the cycles of optic activity can be accomplished by exposing the donors to the appropriate light–dark regimes before the eyes are removed and prepared for recording.

For all these *Aplysia* cycles there is evidence of phasing by light–dark changes, and therefore proof that photoreception is involved in the regulation of the rhythms (STRUMWASSER, 1965; KUPFERMANN, 1968; JACKLET, 1969; LICKEY, 1969; ESKIN, 1971). In addition, extraocular photoreception has been implicated in the entrainment of one of these cycles and is still believed to operate in two others. LICKEY and ZACK (1973) reported that the rhythm of spiking activity of neuron R15 could be entrained by exposing normal and eyeless donors to LD 12:12 before isolating their parietovisceral ganglia. They suggested that the extraocular photoreceptors, which appeared to function in phasing the blind animals' rhythms, are particular neurons in the central nervous system. However, a more recent report (AUDESIRK and STRUMWASSER, 1975) includes evidence that runs counter to that of LICKEY and ZACK (1973) and concludes that nonphotic time signals operated as *Zeitgebers* in the earlier work. AUDESIRK and STRUMWASSER were unable to photoentrain the R15 rhythms of blinded *Aplysia* as they did the cycles of normal animals. The average peak of electric activity of the neurons from the eyeless animals occurred about 5 hrs later than that of R15s from normal sea hares, all of which had been maintained under the same experimental light schedule and isolated from one another during entrainment. Those studied by LICKEY and ZACK (1973) were not. Intact and eyeless animals lived in the same aquarium where stimuli other than light changes, e.g., pheromones of the normal animals, locomotion of the normal animals, may have entrained the spiking activity of R15s of the eyeless animals (AUDESIRK and STRUMWASSER, 1975).

Light–dark control of the circadian rhythm of *Aplysia* locomotion itself may not require eyes as photoreceptors (BLOCK and LICKEY, 1973; BLOCK et al., 1974). Nevertheless, some of the characteristics of that rhythm do depend upon the functioning of the animals' true eyes, since in blinded *Aplysia*, the levels of activity and their exact times of onset differ from those of normal *Aplysia*.

BLOCK et al. (1974) believe that extraocular photoreceptors can also regulate photoentrainment of the cycle of electric changes recorded from the eyes of *Aplysia*. Their conclusion is based on the finding that denervation of the eyes does not prevent entrainment of the circadian rhythms of those eyes by white light cycles (Fig. 7). Incidentally, the extraoptic photoreceptors held to function in that entrainment are not sensitive to red light. Cycles of denervated eyes free-run, i.e., are not entrained, when red light is the stimulus.

What these extraoptic photoreceptors are and where they are located is unknown. BLOCK et al. (1974) did not succeed in locating any that entrained the cycle of electric changes in the eyes of *Aplysia* in the cerebral or buccal ganglia. STRUMWASSER, as cited by BLOCK and SMITH (1973), does not believe that neurons of the abdominal or parietovisceral ganglia are involved in the phasing of the cycles of activity by light–dark changes. However, cells which are sensitive to light have been located in the abdominal ganglia (ARVANITAKI and CHALAZONITIS, 1961), in the cerebral ganglia (ESKIN, 1971; BLOCK and SMITH, 1973) as well as in the mantle (LUKOWIAK and JACKLET, 1972) of sea hares. Very exacting analyses of the details of photoreception and transduction are being made using neurons of the pleural and the abdominal ganglia of *Aplysia* (BROWN et al., 1974; HENKART, 1974). Information derived from these and extended studies will certainly aid in working out the true relationships between extraocular photoreception and the many circadian cycles of *Aplysia*.

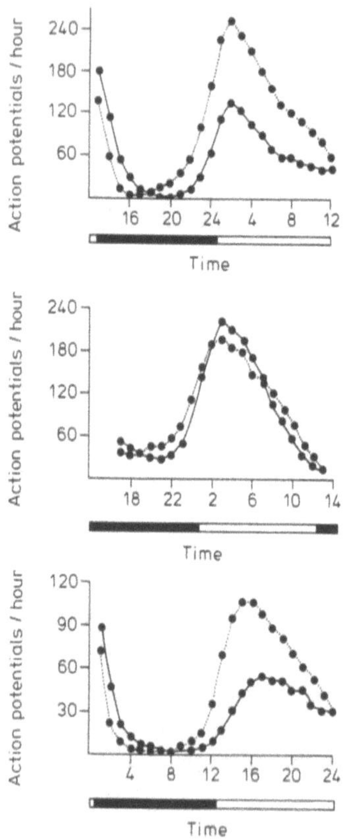

Fig. 7. Numbers of action potentials recorded per hour from eyes of *Aplysia* during 24-hr periods. The eyes were exposed to dark–light periods indicated by *black and white horizontal bars* after their removal from the animals. The *solid lines* show the changes in intact eyes. The *dotted lines* indicate the activity of denervated eyes. (BLOCK et al., 1974)

F. Arthropods

I. Crustaceans

The literature of rhythmicity in crustaceans contains numerous reports of effects of light and of light–dark treatments on the cycles of these animals. (For reviews, see BENNETT, 1974; PALMER, 1974.) Unfortunately, very few of the reports include results leading directly to the idea that extraoptic photoreceptors function in the regulation of these rhythms. A probable reason for that situation is this: to remove the compound eyes of crustaceans and thus to attempt to find evidence for or against extraoptic photoreception, requires the manipulation of their eyestalks. The eyestalks contain neural structures, which may affect rhythms, in addition to neurosecretory cells and sinuses (BROWN, 1973), the products of which are necessary in the mediation of many crustacean cycles. The color change rhythm of fiddler crabs is regulated by chromatotrophins, some of which are produced and stored in the eyestalks (FINGERMAN, 1970). Locomotory rhythms in *Uca* (BROWN et al., 1954) and in *Carcinus* (NAYLOR and WILLIAMS, 1968) depend on hormonal factors of the eyestalks for their normal expressions. After even the most skillful blinding of animals by removal of the ommatidia, researchers have found that parts of the eyestalks proximal to the ommatidia have been damaged (STEPHENS et al., 1964; PAGE and LARIMER, 1972). Thus, if after surgical blinding aberrations in cyclic activities result, how can we decide whether photoreception or the mediation of the rhythms has been set awry?

In general, blinding or eyestalk removal does effect some changes in crustacean circadian cycles. BROWN et al. (1954) reported that the amplitudes of the cycles of oxygen consumption of stalkless *Uca* were greater than those of normal crabs; STEPHENS et al. (1964) observed that the free-running period of the color change cycle of blinded *Uca pugnax* in low light intensity was different from that of the control animals. A more drastic effect, i.e., the apparent loss of persistent cycles of locomotor activity after destalking, has been found in crayfish (KALMUS, 1938a) and green crabs, *Carcinus maenas* (POWELL, 1965; NAYLOR and WILLIAMS, 1968).

However, the disappearance of rhythmicity is not always the result of the loss of the eyestalks. *Uca pugnax* and *Uca pugilator* retained their circadian and tidal rhythms of oxygen consumption after destalking (BROWN et al., 1954). WEBB et al. (1954) and FINGERMAN and YAMAMOTO (1964) proved that fiddler crabs' color change cycles continue in stalkless animals. BLISS (1962) saw cycles of locomotion in some of the *Gecarcinus* she had destalked. As was pointed out, the results of these various studies are neither easy to reconcile nor simple to explain, especially in terms of photoreception and its role in circadian rhythmicity.

Consequently, the results and analyses of well-conceived and skillfully executed investigations by PAGE and LARIMER (1972, 1975a, b) are welcome. In 1972 they described a rhythm of locomotion in the crayfish, *Procambarus clarkii*, which in LD 12:12 was bimodal with peaks at the times of "lights on" and shortly after "lights off". In constant darkness, the free-running circadian cycle has only one maximum, that associated with "lights off".

After the caudal photoreceptor, described by PROSSER (1934), was removed or isolated from the central nervous system, no effects on the crayfish cycles of activity were apparent. After the ommatidia of both eyes were removed or after the optic lobes were severed in both eyes between the lamina ganglionaris and medulla externa, the entrainment of the "lights off" response was normal. However, the "lights on" peak disappeared. Thus, even though the compound eyes are necessary for that latter response, they are not required for fundamental entrainment of the circadian rhythm of locomotion. Further, and most exciting to the present discussion, PAGE and LARIMER (1972) found that after removal of both the caudal photoreceptive ganglion and the ommatidia, the cycle could still be photo-entrained. The "lights off" evoked a typical response (Fig. 8). Which extraocular, extracaudal photoreceptor is operative? That question remains unanswered.

Later research on the circadian rhythm of locomotion in *Procambarus* (PAGE and LARIMER, 1975a, b) also proved that the eyestalks are not necessary for the

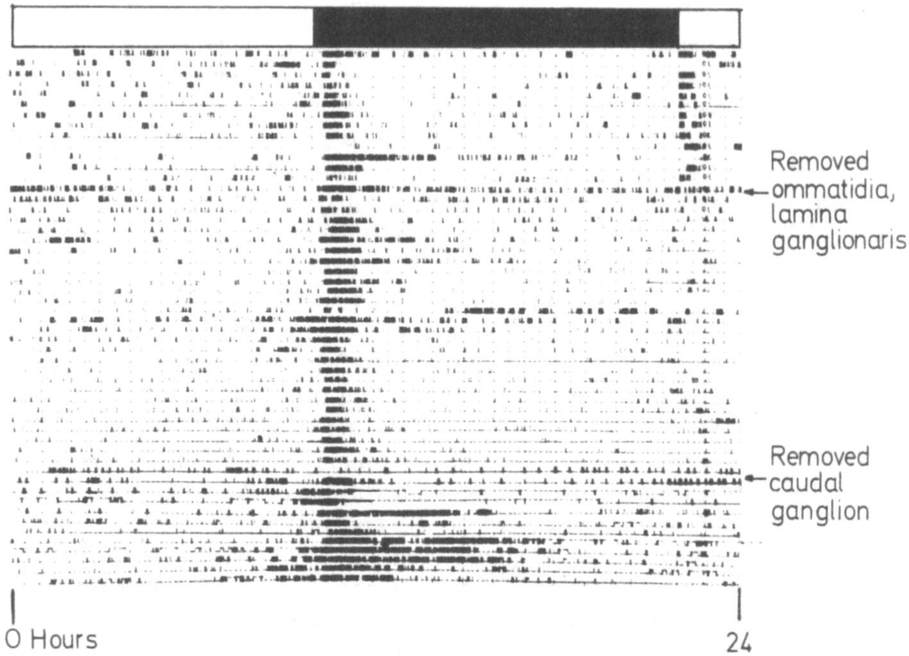

Fig. 8. Activity record of a crayfish that continued to show a "lights off" peak even after removal of its ommatidia and lamina ganglionaris. The "lights on" peak disappeared after the operation. Note that the entrained rhythm continued after the removal of the caudal ganglion. (The regular markings on the *right side* of the record are numbers printed on the chart paper and do not represent activity.) (PAGE and LARIMER, 1972)

maintenance of the cycle. Nevertheless, neural structures of the stalks, unidentified as yet, are somehow involved in *normal* entrainment of that rhythm.

Structures of the eyestalks are also necessary for the expression of a circadian cycle of glucose variations in the blood of crayfish (HAMANN, 1974). This rhythm with peak concentrations at night and minimal ones during the day demands the presence of the sinus gland and its secretions. If those glands are removed, the circadian variations disappear. However, the cycle apparently does not depend on stimulation of the compound eyes, since after the optic nerves are severed, it persists. It is not yet known whether extraocular photoreception is necessary (HAMANN, 1974).

II. Arachnids

In two different arachnids, *Urodacus*, an Australian scorpion, and *Leiobunum longipes*, an opilionid, circadian activities can be regulated by changes that stimulate extraocular photoreceptors. ZWICKEY (1970) described a pattern of motor activity for *Urodacus*, characterized by virtual inactivity during the day and by locomotion beginning near the time of sunset and continuing during the early hours of the night. This cycle persisted for 3 days, during which the scorpions were kept at a constant temperature and in darkness. After that period, the rhythm disappeared unless cued by light–dark changes.

Animals that were blinded by having their lateral and median eyes painted over with opaque paint behaved normally, i.e., their locomotor activity started at sunset and went on into the night when they were exposed to normal day–night changes. Normal animals living in continual darkness did not show the cycles. Photosensitive areas of the scorpions' tails, described by ZWICKEY in 1968, may play roles in entraining the cyclic behavior of the blinded animals.

Photoentrainment of a 24-hr rhythm of production of 5-hydroxytryptamine (serotonin) by cultured brain and gut tissues of *Leiobunum* has been discussed by FOWLER and GOODNIGHT (1975). This cycle is similar to those manifest in the same organs in intact animals (FOWLER and GOODNIGHT, 1966). The cultured tissues were maintained under various laboratory conditions while their serotonin contents were determined periodically. The phasing of the circadian cycles differed among cultures maintained in constant darkness, in constant light, or in LD 14:10, thus indicating that the cultured cells responded directly to changes in illumination. The photoreceptive mechanism is not known; its action spectrum is currently being determined (FOWLER, personal communication).

III. Insects

Circadian rhythms of insects were some of the first persistent cycles of animals to be studied in detail (BÜNNING, 1935) and some of the first for which a definite relationship to extraocular photoreception was proved (KALMUS, 1938b). BRADY reviewed the vast literature of persistent cycles in insects in 1974, and SAUNDERS (1976) discussed the many aspects of insect rhythmicity and photoperiodism in his recent book. BÜNNING (1935) described the now extremely well-known circadian cycle of eclosion in fruit flies, *Drosophila*. Under normal day–night conditions, the peak of emergence of the imagoes from the puparia occurs between 6:00 and 9:00 hrs. In cultures which have been exposed to light–dark changes, the cycle persists in constant darkness or in constant low-intensity light. In cultures reared in complete darkness, emergence is arrhythmic, but can be set to repeat at circadian frequency by a single exposure to light during late larval or pupal stages (BRETT, 1955).

Which photoreceptors are necessary for the cuing of emergence? They have not been located specifically. KALMUS (1938b) and ZIMMERMAN and IVES (1971) concluded that these photoreceptors are located in the anterior ends of pupal bodies, for only if those parts are exposed to light–dark regimens will emergence be entrained to the circadian period. Further, even after the compound eyes or ocelli of *Drosophila melanogaster* have differentiated, neither type of eye is the receptor necessary for photoentrainment. ENGELMANN and HONEGGER (1966) reported that the mutant *sine oculis* which has no eyes or ocelli can be entrained to LD 12:12 just as in the wild type fly. ZIMMERMAN and IVES (1971) also worked with that mutant and found its response to entraining treatments with dim monochromatic light to be the

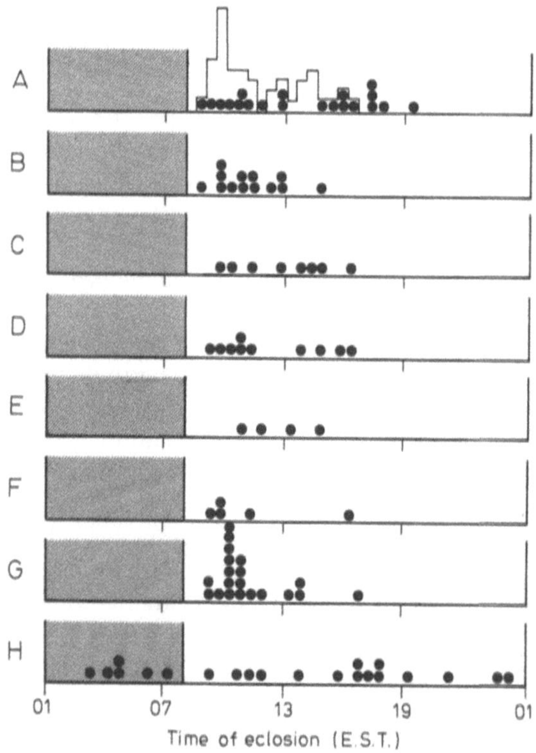

Fig. 9. Effects of different operations on emergence of moths. *A.* Transection of optic nerves; *B.* Transection of circumesophageal connectives; *C.* Removal of frontal ganglion; *D.* Removal of corpora cardiaca and corpora allata; *E.* Removal of imaginal discs of eyes; *F.* Severance of ventral nerve cord; *G.* Removal of subesophageal ganglion; *H.* Removal of brain. Each dot represents emergence of one moth. Histogram in *A.* shows the emergence times of moths not manipulated surgically. Cross-hatched areas indicate periods of 7 hrs of darkness. (From: TRUMAN, 1972)

same as that of wild type *Drosophila melanogaster.* In wild type and white-eyed mutant *Drosophila pseudoobscura,* the compound eyes of which are more sensitive to light than are those of normal flies, the cultures respond to light–dark regimes in the same manner. The difference in photosensitivity has no effect on phasing (ZIMMERMAN and IVES, 1971). These researchers believe, as do others, that the extraocular photoreceptors necessary for entrainment of the eclosion cycle are located in the brains of the flies.

Work with both *Drosophila melanogaster* and *Drosophila pseudoobscura* suggests that the photoreceptive pigment involved in the regulation of the circadian cycle of emergence is not a carotenoid derivative (FRANK and ZIMMERMAN, 1969; ZIMMERMAN and GOLDSMITH, 1971). Further and more detailed investigation of the action spectra of entrainment, as well as of the intensity characteristics of light which regulates the *Drosophila* eclosion rhythm, some of which is discussed by WINFREE (1974), should aid in characterizing more precisely the extraocular photoreceptors that can function in the phasing of emergence of fruit flies.

The brains of silk moths, *Antheraea pernyi*, *Antheraea polyphemus*, and *Hyalophora cecropis*, are sensitive to light–dark changes which entrain the animals' cycles of eclosion (TRUMAN and RIDDIFORD, 1970; TRUMAN, 1972). Using an elegant series of extirpation and implantation experiments, these researchers have located the photoreceptor mechanism involved in the phasing of the silk moth rhythms of emergence specifically in the cerebral lobe of the brain (Fig. 9).

Since circadian and photoperiodic phenomena are not only closely related in general (BÜNNING, 1960, 1971) but are both also necessary for normal development of many insects (TRUMAN, 1971; BECK, 1974a, b, 1975, 1976a), the knowledge of relationships between extraocular photoreception and insect photoperiodism is valuable to analyses of that type of light reception and the basic circadian organization of these animals. The photosensitivity of insect brains has long been emphasized by students of insect photoperiodism. In fact, in 1960 LEES emphasized that there was no evidence that the pathways of light changes, which mediated photoperiodic effects in arthropods, included their compound eyes or ocelli. He reported further that in silkworms, *Antheraea*, Colorado beetles, *Leptinotarsa*, and aphids, *Megoura*, photoperiodic responses continued in typical fashion after such photoreceptors had been cauterized or covered with black paint. Later work with *Megoura* led to the identification of the central region of the dorsum of the head as the area of photoreception (LEES, 1964) and showed that the light is absorbed by and acts upon the brain (LEES, 1966). WILLIAMS and ADKINSON (1964) reported that it is the brain of pupal *Antheraea pernyi* which absorbs the light necessary for normal development of that silkworm. Comparable situations exist in the butterfly, *Pieris brassicae* (CLARET, 1966), and the coddling moth, *Laspeyresia pomonella* (HAYES, 1971).

As was suggested above for the analysis of extraoptic photoreception and the circadian cycle of eclosion in *Drosophila*, studies of the action spectra of photoperiodic phenomena in insects may help characterize and locate even more precisely the extraocular photoreceptors involved. LEES (1971) and HAYES (1971) discussed the principal aspects and aims of their studies of the action spectra of insects' responses and the absorption spectra of their brains, the putative receptor tissues. The peaks of the two types of curves, lying between 400 and 500 nm, are at least compatible (HAYES, 1971).

In addition to their functioning in various periodic phenomena of insect development, extraocular photoreceptors are pathways of some light stimuli that regulate circadian cycles of motor activities of adult insects. A peak at 17:00 hrs in the 24-hr rhythm of oviposition of the grasshopper, *Chorthippus curtipennis*, was entrained by LD 12:12 in normal females, in those of which ocelli and/or compound eyes were rendered functionless, and in those the entire head capsules of which were covered with beeswax and resin (LOHER and CHANDRASHEKARAN, 1970). Not only extraocular but also extracephalic photoreception is indicated in this case. The specific location of the light sensitive elements has not been determined in *Chorthippus*. Other examples of noncephalic photoreception have been reported for the cockroach, *Periplaneta americana* (BALL, 1965), and for the locusts, *Schistocerca gregaria* and *Locusta migratoria* (NEVILLE, 1967). In these

Table 1. Effects of various experimental treatments on the flight behavior of male silk moths, *Hyalophora cecropia*, under LD regimens (TRUMAN, 1974)

Treatment	Number of animals	% showing activity:		
		Entrained to LD	Free-running	Arrhythmic
None	34	100	0	0
Brain removed	5	0	0	100
One cerebral lobe and both optic lobes removed	3	100	0	0
Compound eyes removed	13	100	0	0
Ocelli removed	4	100	0	0
Compound eyes and ocelli removed	8	100	0	0
Whole head covered with clear wax	7	100	0	0
Whole head covered with black wax	5	0	80	20
Whole head, except compound eyes, covered with black wax	23	30[a]	22	48

[a] Their heads probably had small areas on the dorsal surfaces which were not covered by black wax. Consequently, light could reach the brain via such openings. TRUMAN found actual slits in the wax in some of his animals.

latter insects, stimulation of the unidentified receptors by constant light uncouples lamellogenesis of chitin from the animals' clock mechanism (NEVILLE, 1967). Consequently, the circadian rhythm of exoskeleton deposition, described earlier by NEVILLE (1965), disappears.

Extraocular light receptors were also implicated in the entrainment of the stridulatory cycles of some other orthopterans, *Ephippiger ephippiger* (Fiebieg), *Ephippiger biterrensis* (Finot), and *Ephippiger provincialis* (Yerson) (DUMORTIER, 1972). When exposed to LD 13:15 : 10:45, these animals stridulate most actively during the first half of the light period, a pattern expressed also by *Ephippiger* from which the compound eyes and/or ocelli had been removed, by those that had sustained lesions in the optic lobes of the brain and by those in which both optic tracts had been cut. DUMORTIER (1972) suggested that the extraocular photoreceptors are located in the insects' heads.

Photoentrainment of circadian cycles of movements of many insects does require the animals' compound eyes and/or ocelli. That is indeed true for the cockroaches, *Periplaneta americana* (CLOUDSLEY-THOMPSON, 1953) and *Leucophaea madeirae* (NISHIITSUTSUJI-UWO and PITTENDRIGH, 1968), and for the crickets, *Gryllus domesticus* L. (NOWOSIELSKI and PATTON, 1963) and *Teleogryllus commodus* Walker (LOHER, 1972). On the other hand, those types of light receptors are not absolutely necessary, since extraocular photoreception may suffice in entrainment of locomotory cycles in the walking stick, *Carausius morosus* (EIDMANN, 1956), and the cockroach, *Blaberus craniifer* (BALL and CHAUDHURY, 1973). BALL and CHAUDHURY (1973), using implantation techniques, proved that *Blaberus* brains are photosensitive and can serve as direct pathways for light–dark changes that phase the circadian behavior of these roaches.

This system certainly exists in adult giant silk moths, *Antheraea pernyi*, *Hyalophora cecropia*, and *Samia cynthia*, which during their short lives exhibit obvious rhythms of flight activity characterized by greater activity in the dark than in the light of LD 12 : 12 (TRUMAN, 1974). In fact, the results of TRUMAN'S splendid study prove not only that extraocular receptors located in the head are stimulated by entraining light–dark regimens, but also that these cephalic receptors are pathways required for conduction of the photic information. If the moths' head capsules are made opaque while their compound eyes are exposed, phasing photoperiods do not in most cases entrain the activity cycles (Table 1). There is every reason to believe that the necessary extraocular photoreceptors are located in the cerebral lobe areas of the brain, since if only one of these is retained in a moth, entrainment of its rhythm by light–dark changes is still possible (Table 1).

G. Summary

Functions of extraocular photoreceptors in the regulation of circadian rhythms have been worked out in depth for only a few invertebrates, namely, for some molluscs and arthropods. In addition, important relationships between such receptors and cyclic behavior are suggested by the results of studies of cnidarians, starfish, flatworms, and annelids. In no case have the receptors been precisely located, but in many cases the evidence supports the conclusion that they are to be found in obvious parts of the central nervous system, e.g., the brain or another large ganglion. In vertebrates, areas of the brain and its dorsal derivatives are known to be stimulated by light changes that affect and regulate circadian rhythms of a variety of activities. Thus, similarities between the invertebrate and vertebrate mechanisms are apparent.

References

Adler, K.: The role of extraoptic photoreceptors in amphibian rhythms and orientation: A review. J. Herpetol. **4**, 99—122 (1970).

Arvanitaki, A., Chalazonitis, N.: Excitatory and inhibitory processes initiated by light and infrared radiations in single identifiable nerve cells. (Giant ganglion cells of *Aplysia*). In: Nervous Inhibition. ed. E. Florey, pp. 194—231. New York: Pergamon Press 1961.

Aschoff, J.: Exogenous and endogenous components in circadian rhythms. Cold Spr. Harb. Symp. quant. Biol. **25**, 11—28 (1960).

Audesirk, G., Strumwasser, F.: Circadian rhythm of neuron R15 of *Aplysia californica*: *In vivo* photoentrainment. Proc. nat. Acad. Sci. (Wash.) **72**, 2408—2412 (1975).

Axelrod, J.: The pineal gland: A neurochemical transducer. Science **184**, 1341—1348 (1974).

Ball, H. J.: Photosensitivity in the terminal abdominal ganglion of *Periplaneta americana* (L.). J. Insect Physiol. **11**, 1311—1315 (1965).

Ball, H. J., Chaudhury, M. F. B.: Photic entrainment of circadian rhythms by illumination of implanted brain tissues in the cockroach *Blaberus craniifer*. J. Insect Physiol. **19**, 823—830 (1973).

Batham, E. J., Pantin, C. F. A.: Phases of activity in the sea-anemone, *Metridium senile* (L.), and their relation to external stimuli. J. exp. Biol. **27**, 377—399 (1950).

Beck, S. D.: Photoperiodic determination of insect development and diapause. I. Oscillators, hour-glasses and a determination model. J. comp. Physiol. **90**, 275—295 (1974a).

Beck, S. D.: Photoperiodic determination of insect development and diapause. II. The determination gate in a theoretical model. J. comp. Physiol. **90**, 297—310 (1974b).

Beck, S. D.: Photoperiodic determination of insect development and diapause. III. Effects of nondiel photoperiods. J. comp. Physiol. **103**, 227—245 (1975).

Beck, S. D.: Photoperiodic determination of insect development and diapause. IV. Effects of skeleton photoperiods. J. comp. Physiol. **105**, 267—277 (1976a).

Beck, S. D.: Photoperiodic determination of insect development and diapause. V. Diapause, circadian rhythms, and phase response curves, according to the dual system theory. J. comp. Physiol. **107**, 97—111 (1976b).

Becker-Carus, C.: Die Bedeutung der Tageszeit für Sensibilität, Reizsättigung und Entscheidungsaktivität bei Planarien (*Dugesia dorotocephala*). Z. Tierpsychol. **27**, 761—770 (1970).

Beiswanger, C. M., Jacklet, J. W.: *In vitro* tests for a circadian rhythm in the electrical activity of a single neuron in *Aplysia californica*. J. comp. Physiol. **103**, 19—37 (1975).

Bennett, M. F.: The rhythmic activity of the quahog, *Venus mercenaria*, and its modification by light. Biol. Bull. **107**, 174—191 (1954).

Bennett, M. F.: The central nervous system and circadian differences in the earthworm. In: Neurobiology of Invertebrates. Mechanisms of Rhythm Regulation. ed. J. Salánki, pp. 353—361. Budapest: Akadémiai Kiadó 1973.

Bennett, M. F.: Living Clocks in the Animal World. Springfield: Charles C. Thomas 1974.

Bennett, M. F.: Rhythms in worms and worm-like organisms. In: Biological Rhythms in the Marine Environment. ed. P. J. DeCoursey, pp. 137—144. Columbia: Univ. South Carolina Press 1976.

Bennett, M. F., Reinschmidt, D. C.: The diurnal cycle and a difference in reaction times in earthworms. Z. vergl. Physiol. **49**, 407—411 (1965).

Best, J. B.: Diurnal cycles and cannibalism in planaria. Science **131**, 1884—1885 (1960).

Bliss, D.: Neuroendocrine control of locomotor activity in the land crab, *Gecarcinus lateralis*. In: Memoirs of the Society for Endocrinology, ed. H. Heller, R. B. Clark, pp. 391—410. London-New York: Academic Press 1962.

Block, G. D., Hudson, D. J., Lickey, M. E.: Extraocular photoreceptors can entrain the circadian oscillator in the eye of *Aplysia*. J. comp. Physiol. **89**, 237—249 (1974).

Block, G. D., Lickey, M. E.: Extraocular photoreceptors and oscillators can control the circadian rhythm of behavioral activity in *Aplysia*. J. comp. Physiol. **84**, 367—374 (1973).

Block, G. D., Smith, J. T.: Cerebral photoreceptors in *Aplysia*. Comp. Biochem. Physiol. **46**A, 115—121 (1973).

Brady, J.: The physiology of insect circadian rhythms. Advanc. Insect Physiol. **10**, 1—115 (1974).

Brett, W. J.: Persistent diurnal rhythmicity in *Drosophila* emergence. Ann. Entomol. Soc. Amer. **48**, 119—131 (1955).

Brown, A. M., Baur, P. S., Jr., Tuley, F. H., Jr.: Phototransduction in *Aplysia* neurons: Calcium release from pigmented granules is essential. Science **188**, 157—160 (1974).

Brown, F. A., Jr.: Chromatophores and color change. In: Comparative Animal Physiology, ed. C. L. Prosser, 3rd Ed., pp. 915—950. Philadelphia-London-Toronto: W. B. Saunders Co. 1973.

Brown, F. A., Jr., Bennett, M. F., Webb, H. M.: Persistent daily and tidal rhythms of O_2-consumption in fiddler crabs. J. cell. comp. Physiol. **44**, 477—505 (1954).

Bünning, E.: Zur Kenntnis der endogenen Tagesrhythmik bei Insekten und bei Pflanzen. Ber. dtsch. Botan. Ges. **53**, 594—623 (1935).

Bünning, E.: Circadian rhythms and the time measurement in photoperiodism. Cold Spr. Harb. Symp. quant. Biol. **25**, 249—256 (1960).
Bünning, E.: The adaptive value of circadian leaf movements. In: Biochronometry, ed. M. Menaker, p. 203—211. Washington: National Academy Sci. 1971.
Bünning, E.: Die physiologische Uhr. Dritte, gründlich überarbeitete Auflage. Berlin-Heidelberg-New York: Springer 1977.
Claret, J.: Mise en évidence du rôle photorécepteur du cerveau dans l'induction de la diapause chez *Pieris brassicae* (Lepidoptera). Ann. Endocr. (Paris) **27**, 311—320 (1966).
Cloudsley-Thompson, J. L.: LXIX.—Studies in diurnal rhythms.—III. Photoperiodism in the cockroach *Periplaneta americana* (L.). Ann. Mag. Nat. Hist. **6**, 705—712 (1953).
Dodd, J. R.: Effect of light on rate of growth of bivalves. Nature (Lond.) **224**, 617—618 (1969).
Dumortier, B.: Photoreception in the circadian rhythm of stridulatory activity in *Ephippiger* (Ins., Orthoptera). Likely existence of two photoreceptive systems. J. comp. Physiol. **77**, 80—112 (1972).
Eidmann, H.: Über rhythmische Erscheinungen bei der Stabheuschrecke *Carausius morosus* Br. Z. vergl. Physiol. **38**, 370—390 (1956).
Engelmann, W., Honegger, H. W.: Tagesperiodische Schlüpfrhythmik einer augenlosen *Drosophila melanogaster*-Mutante. Naturwissenschaften **53**, 588 (1966).
Eskin, A.: Properties of the *Aplysia* visual system: *in vitro* entrainment of the circadian rhythm and centrifugal regulation of the eye. Z. vergl. Physiol. **74**, 353—371 (1971).
Fingerman, M.: Comparative physiology: Chromatophores. Ann. Rev. Physiol. **32**, 345—372 (1970).
Fingerman, M., Yamamoto, Y.: Daily rhythm of color change in the eyestalkless fiddler crabs, *Uca pugilator*. Amer. Zool. **4**, 334 (1964).
Fischer, A.: Über die Chromatophoren und den Farbwechsel bei dem Polychäten *Platynereis dumerilii*. Z. Zellforsch. **65**, 290—312 (1965).
Fowler, D. J., Goodnight, C. J.: The cyclic production of 5-hydroxytryptamine in the opilionid. Amer. Zool. **6**, 187—193 (1966).
Fowler, D. J., Goodnight, C. J.: In vitro regulation by light of 24-hour rhythmic serotonin production. J. interdiscipl. Cycle Res. **6**, 121—128 (1975).
Frank, K. D., Zimmermann, W. F.: Action spectra for phase shifts of a circadian rhythm in *Drosophila*. Science **163**, 688—689 (1969).
Gardner, L. E., Ratner, S. C.: In-burrow behavior of earthworms. Psychol. Rec. **20**, 387—394 (1970).
Hamann, A.: Die neuroendokrine Steuerung tagesrhythmischer Blutzuckerschwankungen durch die Sinusdrüse beim Flußkrebs. J. comp. Physiol. **89**, 197—214 (1974).
Hauenschild, C.: Die Schwärmperiodizität von *Platynereis dumerilii* im DD-LD-Belichtungszyklus und nach Augenausschaltung. Z. Naturforsch. **16b**, 753—756 (1961).
Hayes, D. K.: Action spectra for breaking diapause and absorption spectra of insect brain tissue. In: Biochronometry, ed. M. Menaker, pp. 392—402. Washington: National Academy Sci. 1971.
Henkart, M.: Light-induced changes in the structure of pigmented granules in *Aplysia* neurons. Science **188**, 155—157 (1974).
Hoffmann, K.: Splitting of the circadian rhythm as a function of light intensity. In: Biochronometry, ed. M. Menaker, p. 134—151. Washington: National Academy Sci. 1971.
Jacklet, J. W.: Circadian rhythm of optic nerve impulses recorded in darkness from isolated eye of *Aplysia*. Science **164**, 562—563 (1969).
Jacklet, J. W.: Circadian locomotor activity in *Aplysia*. J. comp. Physiol. **79**, 325—341 (1972).
Kalmus, H.: Das Actogramm des Flußkrebses und seine Beeinflussung durch Organextrakte. Z. vergl. Physiol. **25**, 789—802 (1938a).
Kalmus, H.: Die Lage des Aufnahmeorganes für die Schlüpfperiodik von *Drosophila*. Z. vergl. Physiol. **26**, 362—365 (1938b).
Kupfermann, I.: A circadian locomotor rhythm in *Aplysia californica*. Physiol. Behav. **3**, 179—181 (1968).
Lees, A. D.: Some aspects of animal photoperiodism. Cold Spr. Harb. Symp. quant. Biol. **25**, 261—268 (1960).
Lees, A. D.: The location of the photoperiodic receptors in the aphid *Megoura viciae* Buckton. J. exp. Biol. **41**, 119—133 (1964).
Lees, A. D.: Photoperiodic timing mechanisms in insects. Nature (Lond.) **210**, 986—989 (1966).
Lees, A. D.: The relevance of action spectra in the study of insect photoperiodism. In: Biochronometry, ed. M. Menaker, pp. 372—380. Washington: National Academy Sci. 1971.

Lickey, M. E.: Seasonal modulation and non-24-hour entrainment of a circadian rhythm in a single neuron. J. comp. Physiol. Psychol. **68**, 9—17 (1969).

Lickey, M. E., Zack, S.: Extraocular photoreceptors can entrain the circadian rhythm in the abdominal ganglion of *Aplysia*. J. comp. Physiol. **84**, 361—366 (1973).

Loher, W.: Circadian control of stridulation in the cricket *Teleogryllus commodus* Walker. J. comp. Physiol. **79**, 173—190 (1972).

Loher, W., Chandrashekaran, M. K.: Circadian rhythmicity in the oviposition of the grasshopper *Chorthippus curtipennis*. J. Insect Physiol. **16**, 1677—1688 (1970).

Lukowiak, K., Jacklet, J. W.: Habituation: A peripheral and central nervous system process in *Aplysia*. Fed. Proc. **31**, 405 (1972).

Megusăr, F.: Experimente über den Farbwechsel der Crustaceen. Arch. Entwickl.-Mech. Org. **33**, 462—665 (1912).

Menaker, M.: Nonvisual light reception. Sci. Amer. **226(3)**, 22—29 (1972).

Mori, S.: Influence of environmental and physiological factors on the daily rhythmic activity of a sea-pen. Cold Spr. Harb. Symp. quant. Biol. **25**, 333—344 (1960).

Mori, S., Matutani, K.: Studies on the daily rhythmic activity of the starfish, *Astropecten polyacanthus* Müller et Troschel, and the accompanied physiological rhythms. Publ. Seto Mar. Biol. Lab. **II(2)**, 213—225 (1952).

Mori, S., Ondo, Y.: Daily rhythmic activity of the sea-pen, *Cavernularia obesa* Valenciennes. XV. Controlling of the activity by light. (3). Publ. Seto Mar. Biol. Lab. **6**, 79—98 (1957).

Naylor, E., Williams, B. G.: Effects of eyestalk removal on rhythmic locomotor activity in *Carcinus*. J. exp. Biol. **49**, 107—116 (1968).

Neville, A. C.: Circadian organization of chitin in some insect skeletons. Quart. J. Micr. Sci. **106**, 315—325 (1965).

Neville, A. C.: A dermal light sense influencing skeletal structure in locusts. J. Insect Physiol. **13**, 933—939 (1967).

Nishiitsutsuji-Uwo, J., Pittendrigh, C. S.: Central nervous system control of circadian rhythmicity in the cockroach. II. The pathway of light signals that entrain the rhythm. Z. vergl. Physiol. **58**, 1—13 (1968).

Njus, D., Sulzman, F. M., Hastings, J. W.: Membrane model for the circadian clock. Nature (Lond.) **248**, 116—120 (1974).

Nowosielski, J. W., Patton, R. L.: Studies on circadian rhythm of the house cricket, *Gryllus domesticus* L. J. Insect Physiol. **9**, 401—410 (1963).

Page, T. L., Larimer, J. L.: Entrainment of the circadian locomotor activity rhythm in crayfish. The role of the eyes and caudal photoreceptor. J. comp. Physiol. **78**, 107—120 (1972).

Page, T. L., Larimer, J. L.: Neural control of circadian rhythmicity in the crayfish. I. The locomotor activity rhythm. J. comp. Physiol. **97**, 59—80 (1975a).

Page, T. L., Larimer, J. L.: Neural control of circadian rhythmicity in the crayfish. II. The ERG amplitude rhythm. J. comp. Physiol. **97**, 81—96 (1975b).

Palmer, J. D.: Biological Clocks in Marine Organisms: The Control of Physiological and Behavioral Tidal Rhythms. New York-London-Sydney-Toronto: John Wiley and Sons 1974.

Powell, B. L.: The hormonal control of the tidal rhythm of locomotor activity in *Carcinus maenas*. Gen. Comp. Endocr. **5**, 705 (1965).

Prosser, C. L.: Action potentials in the nervous system of the crayfish. II. Responses to illumination of the eye and caudal ganglion. J. cell. comp. Physiol. **4**, 363—377 (1934).

Röseler, I.: Die Rhythmik der Chromatophoren des Polychaeten *Platynereis dumerilii*. Z. vergl. Physiol. **70**, 144—174 (1970).

Salánki, J.: Neural mechanisms in rhythm regulation of invertebrates. In: Neurobiology of Invertebrates. Mechanisms of Rhythm Regulation, ed. J. Salánki, pp. 17—31. Budapest: Akadémiai Kiadó 1973.

Saunders, D. S.: Insect Clocks. Oxford: Pergamon Press 1976.

Stephens, G. J., Halberg, F., Stephens, G. C.: The blinded fiddler crab: An invertebrate model of circadian desynchronization. Ann. N. Y. Acad. Sci. **117**, 386—406 (1964).

Steven, D. M.: The dermal light sense. Biol. Rev. **38**, 204—240 (1963).

Strumwasser, F.: The demonstration and manipulation of a circadian rhythm in a single neuron. In: Circadian Clocks, ed. J. Aschoff, pp. 442—462. Amsterdam: North-Holland Publ. Co. 1965.

Sweeney, B. M.: Circadian rhythms in corals, particularly Fungiidae. Biol. Bull. **151**, 236—246 (1976).

Thompson,I.L.: Biological clock control and shell growth in the bivalve *Mercenaria mercenaria*. Abstracts with Programs. Geol. Soc. Amer. **2**, 704 (1970).
Truman,J.W.: The role of the brain in the ecdysis rhythm of silkmoths: Comparisons with the photoperiodic termination of diapause. In: Biochronometry, ed. M.Menaker, pp. 483—504. Washington: National Academy of Sciences 1971.
Truman,J.W.: Physiology of insect rhythms. II. The silkmoth brain as the location of the biological clock controlling eclosion. J. comp. Physiol. **81**, 99—114 (1972).
Truman,J.W.: Physiology of insect rhythms. IV. Role of the brain in the regulation of the flight rhythm of the giant silkmoths. J. comp. Physiol. **95**, 281—296 (1974).
Truman,J.W., Riddiford,L.M.: Neuroendocrine control of ecdysis in silkmoths. Science **167**, 1624—1626 (1970).
Valle,C., Pellegrino,J., Alvarenga,N.: Ritmo circadiano de emergência de cercárias (*Schistosoma mansoni–Biomphalaria glabrata*). Rev. Brasil Biol. **31**, 53—63 (1971).
Webb,H.M.: Diurnal variations of response to light in the fiddler crab, *Uca*. Physiol. Zool. **23**, 316—337 (1950).
Webb,H.M., Bennett,M.F., Brown,F.A.,Jr.: A persistent diurnal rhythm of chromatophoric response in eyestalkless *Uca pugilator*. Biol. Bull. **106**, 371—377 (1954).
Wheeler,A.P., Blackwelder,P.L., Wilbur,K.M.: Shell growth in the scallop *Argopecten irradians*. I. Isotope incorporation with reference to diurnal growth. Biol. Bull. **148**, 472—482 (1975).
Williams,C.M., Adkinson,P.L.: Physiology of insect diapause. XIV. An endocrine mechanism for the photoperiodic control of pupal diapause in the oak silkworm, *Antheraea pernyi*. Biol. Bull. **127**, 511—525 (1964).
Winfree,A.T.: Corkscrews and singularities in fruitflies: Resetting behavior of the circadian eclosion rhythm. In: Biochronometry, ed. M.Menaker, pp. 81—109. Washington: National Acad. Sci. 1971.
Winfree,A.T.: Suppressing *Drosophila* circadian rhythm with dim light. Science **183**, 970—972 (1974).
Wrenn,S.A.: Daily increment formation and synchronization in the shell of the bay scallop. Amer. Zool. **12**, XXXVII (Abstract 417) (1972).
Wurtmann,R.J.: The effects of light on man and other mammals. Ann. Rev. Physiol. **37**, 467—483 (1975).
Yoshida,M.: Extraocular photoreception. In: Invertebrate-Photoreceptor Organs. Handbook of Sensory Physiology, Vol. VII/6A. Berlin-Heidelberg-New York: Springer 1979.
Zimmerman,W.F., Goldsmith,T.H.: Photosensitivity of the circadian rhythm and of visual receptors in carotenoid-depleted *Drosophila*. Science **171**, 1167—1169 (1971).
Zimmerman,W.F., Ives,D.: Some photophysiological aspects of circadian rhythmicity in *Drosophila*. In: Biochronometry, ed. M.Menaker, pp. 381—391. Washington: National Acad. Sci. 1971.
Zwickey,K.T.: A light response in the tail of *Urodacus*, a scorpion. Life Sci. [II] **7**, 257—262 (1968).
Zwickey,K.T.: Behavioural aspects of the extraocular light sense of *Urodacus*, a scorpion. Experientia (Basel) **26**, 747—748 (1970).

Chapter 12

Genetic Approach to a Visual System

MARTIN HEISENBERG, Würzburg, Germany

With 1 Figure

Contents

A. Introduction

Congenital diseases of vision have been investigated for a long time by ophthalmologists, and our knowledge about vision in man has been strongly influenced by these studies. But the usefulness of mutants in animal studies on vision has only recently been discovered. In the last decade fruitfly, mouse, cat, halobacterium, and even a white tiger entered the genetic literature on vision. Is the genetics of vision a new branch of science, is it a new (or newly sharpened) tool in neurobiology, or is it merely a new label for scattered work which happens to deal with genetic aspects of vision in animals?

Up to now, only the work carried out on *Drosophila* has been sufficiently systematic to be called a "genetic approach". *Drosophila* may be uniquely suited for this purpose. As the traditional animal in genetics, it offers a variety of techniques which simplify the search for mutations and their handling, the characterization of its genetic constitution, and the construction of mosaic animals. Its behavior seems to a large extent to be phylogenetically determined; even immediately after eclosion *Drosophila* possesses a fixed repertoire of behavioral responses irrespective of rearing conditions. Behavioral plasticity, although present, does not prevail. Hence, genetic defects may disintegrate behavior into meaningful subunits. The fly's various visual activities suggest parallels to general tasks of visual behavior. Thus their analysis can be thought of as a description of natural model solutions. The physiologic analysis of the nervous system with its some 10^5 neurons and its decentralized, well-structured building scheme is not completely out of reach.

Since the first discovery of visual mutants in *Drosophila* (BENZER, 1967), the number of genes known to be involved in vision has increased steadily. Most of the mutants are waiting to be used. Some mutants which have been studied provide puzzling phenomena; but so far the relation of these phenomena to any current problems can not be conceived. However, several mutants do contribute directly to various aspects of the analysis of visual systems. Those will be briefly reviewed here. I do not intend to give a complete and critical account of all current work in this field[1]. Several recent reviews on this topic are available or in press (PAK, 1975; HEISENBERG and GÖTZ, 1975; PAK and PINTO, 1976). Thorough accounts on the physiology of photoreceptors (EBREY and HONIG, 1975), invertebrate color vision (MENZEL, this volume), neuroanatomy of the visual system (STRAUSFELD, volume VII/3A) the dipteran retina (KIRSCHFELD, 1973), the optomotor behavior of *Drosophila* (GÖTZ and WENKING, 1973), and visual orientation of the housefly *Musca* (REICHARDT and POGGIO, 1976; POGGIO and REICHARDT, 1976) may provide sufficient background information.

I will consider three subjects. The first one (B) concerns mutants with defects in the photoreceptor cells (retinula cells), and their role in the analysis of the transducer process. In these mutants the genes act at the cellular level where microbial genetics has provided some experience as to what kinds of defects genetic changes might cause. The second chapter (C.I) will be concerned with spectral sensitivity of the retinula cells and with mutations specifically eliminating certain receptor types. The absence of rhabdomeres in these mutants facilitates experimentation since those structures can easily be made visible in the intact animals. The last chapter (C.II) describes dissection of behavior. This is still the most speculative and descriptive topic, since in the mutants neither the physical defects nor their location in the nervous system nor the mode of gene action are known. However for the analysis of brain function in particular, new methods are needed.

---

[1] The bibliography is also not intended to be complete. In many cases the most general references were chosen. The reviews cited below are recommended in this respect.

B. Mutations in the Transducer Process

I. Phototransduction

The chain of events in photoreceptors leading from the primary photoprocesses to the excitation of the cell membrane has been traced from the two ends by spectrophotometric and electrophysiologic techniques. In *Drosophila*, as in other invertebrates, the photopigment exists in two thermally stable isomers, the physiologically active rhodopsin (R) and the physiologically inactive metarhodopsin (M) (PAK and LIDDINGTON, 1974). Since the two forms have different absorption spectra, their ratio is determinded by the spectral composition of the incident light, provided the intensity is high enough that equilibrium is reached (HAMDORF et al., 1973). In *Drosophila* a slow regeneration of rhodopsin can also be observed in the dark (PAK and LIDDINGTON, 1974). Membrane excitation apparently depends upon the R to M conversion rate.

The retinula cells are assumed to build up a negative resting potential in the dark by transporting sodium ions against the concentration gradient out of the cell. Photoconversion of rhodopsin causes the membrane to become permeable for sodium and the resting potential to decrease (receptor potential) (see, e.g., EBREY and HONIG, 1975). Under continuous illumination equilibrium must be reached between the sodium transport system and the light-induced sodium influx. In vertebrate photoreceptors the time course and intensity dependence of the receptor potential has been used to derive theoretic models for its generation (BAYLOR et al., 1974; CERVETTO et al., 1976). The results suggest that many different functional components contribute to the receptor potential. This provides an indication of the complexity of the transducer process.

Two phenomena which illustrate this complexity have been discovered in invertebrate photoreceptors: One is the observation of so called "quantum bumps", which are small discrete receptor depolarizations of uniform size and time course apparently elicited by single quanta (YEANDLE, 1957; WU and PAK, 1975). It has been shown that the receptor potential results from the superposition of many such quantum bumps (WU and PAK, 1975). The other phenomenon is the so called prolonged depolarizing afterpotential (PDA) (NOLTE et al., 1968; HILLMAN et al., 1972; COSENS and BRISCOE, 1972). If the photoreceptor is stimulated by light that substantially reduces the R/M ratio, the depolarization can persist as long as many minutes after cessation of the stimulus. It is, however, quickly reverted to the resting potential by light that causes the inverse pigment change.

II. Phototransduction Mutants

Of all the known visual mutants of *Drosophila*, the majority show abnormalities in the electroretinogram (ERG). The ERG reflects the extracellularly recorded light-induced depolarization of the retinula cells and, superimposed, an "evoked potential" from the first optic ganglion, the lamina ganglionaris (evidence summarized by PAK, 1975). The ERG mutants can be classified according to these two components. Mutants without an ERG or an altered receptor component presumably have a defect in the retinula cells. Mutants with a normal receptor component of the ERG but a suppressed lamina potential may be altered in any of the steps leading to stimulation of the lamina cells. It appears from mosaic studies that in many mutants of the latter class (*non A*, *nbA*)[2] absence of the lamina potential is due to a defect originating from the retina anlage rather than from that of the lamina (HOTTA and BENZER, 1970). Thus, they too must be considered as retinula cell mutants.

Mutants with gross defects in the receptor potential are an obvious choice for the analysis of the transducer process, although it may be that, of the other type of mutants, those for which excitation cannot stimulate the subsequent neurons are equally relevant. Three groups of mutants of the first type have been studied so far.

In most of the mutants belonging to the gene *norp* A (*no* receptor potential A) the ERG is virtually absent, or at least very small. It has been shown that this is due to a block in phototransduction. Intracellular recordings have demonstrated that the retinula cells have a normal resting potential but their membrane resistance is not diminished upon illumination (ALAWI et al., 1972). Of the many *norp* A alleles isolated, two have been found to be temperature sensitive (ts) (HOTTA and BENZER, 1970; DELAND and PAK, 1973). One of these has served to demonstrate that inexcitability is neither a fault of the membrane nor of the photopigment (review by PAK and PINTO, 1976). Normal primary photoprocesses were microspectrophotometrically observed in vivo even at the nonpermissive temperature where excitability was blocked. The normal excitability of the membrane was deduced from experiments at an intermediate temperature where the receptor potential in the ts mutant is "sluggish": rising time and decay are much slower than at low temperature. This abnormal receptor potential can be shown to consist of quantum bumps of normal form and size but highly variable latencies. Thus an intermediate step of phototransduction, prior to the triggering of quantum bumps and posterior to the photoisomerization appears to deteriorate at elevated temperature. Finally, properties of the same ts allele and results from mosaic studies (HOTTA and BENZER, 1970) have led to the conclusion that the *norp* A gene product is a constituent of the receptor cells in the adult. A biochemical search has been started for the gene product (OSTROY and PAK, 1973), and a functional assay for it is presently being developed (PAK, personal communication).

The second mutant gene of interest in this context has been named *rdg* B (receptor degeneration B), since in mutants with defects in this gene rhabdomeres are already degenerated at eclosion or degenerate early in the life of the imago. (A second gene with similar phenotype has been found which is called *rdg* A.) One particular allele, *rdg* B^{KS222} has been studied since degeneration in this mutant is to some extent specific for one type of retinula cells (cells with large rhabdomeres R1–6, see Sec. C.I). At eclosion, the ERG and the structure of the eye appear

[2] A table of the mutants mentioned in the text summarizing some of their characteristics can be found at the end of this article. In order to have the table open while reading the article, this table is attached to this volume as individual sheet.

Table of mutants
(Numbers after "*location*" indicate mutant chromosome and—if known—position on genetic map)

norp A:	***no receptor potential A***; *location:* I–6.5; *reference:* PAK (1975); *phenotype:* ERG can be nearly absent; it varies in different alleles; in one allele in which ERG is slow at elevated temperature, latency of quantum bumps is longer and more variable; retina is autonomous for the defect.
rdg B^{KS222}:	***receptor degeneration B*** (allele: KS222); *location:* I–42.7; *references:* HARRIS and STARK (1977) and HARRIS et al. (1976); *phenotype:* large retinula cells R1–6 degenerate under illumination or at temperatures above 20° C; optomotor response varies from normal to nearly zero.
trp:	***transient receptor potential***; *location:* III not mapped; *synonym:* Cosens-Manning mutant; *references:* COSENS and MANNING (1969) and MINKE et al. (1975b); *phenotype:* receptor potential decays to nearly resting potential in the dark due to decrease in frequency of quantum bumps.
sev:	***sevenless***; *location:* I–33.2; *synonym:* LY3; *reference:* HARRIS et al. (1976); *phenotype:* rhabdomeres of retinula cells R7 are missing in all ommatidia; in differential phototaxis UV sensitivity is suppressed.
ora:	***outer rhabdomeres absent***; *location:* III–65.3; *synonym:* JK84; *references:* KOENIG and MERRIAM (1975) and HARRIS et al. (1976); *phenotype:* only distal tips of rhabdomeres develop in large retinula cells R1–6; optomotor response is strongly suppressed.
Hk_1:	***Hyperkinetic*** (allele No. 1); *location:* I–30.9; *reference:* KAPLAN and TROUT (1969); *phenotype:* legs shake in etherized flies; visual stimuli can cause an abnormal jump reflex; head is autonomous for visual defect.
nbA^{H18}:	***night blind A*** (allele H18); *location:* I–36.6; *synonyms:* non B, ldg A, opm 18; *reference:* HEISENBERG and GÖTZ (1975); *phenotype:* low intensity threshold of optomotor response and phototaxis is raised; lamina potential of ERG is suppressed; retina is autonomous for ERG defect.
elf:	***extra lamina fiber***: *location:* I not mapped; *synonym:* H37, opm 37; *reference:* HEISENBERG and GÖTZ (1975); *phenotype:* off-transient of lamina potential is reduced; low intensity threshold of optomotor response is raised; most lamina cartridges contain an additional large fiber.
non A^{H2}:	***no on-transient A*** (allele H2); *location:* I–52; *synonym:* opm 2; *references:* HOTTA and BENZER (1970), PAK (1975), and HEISENBERG (1972); *phenotype:* lamina potential of ERG is very small; optomotor response shows low visual acuity; in differential phototaxis UV sensitivity is suppressed.
P 37:	(laboratory name); *location:* I not mapped; *references:* PAK (1975) and HEISENBERG and GÖTZ (1975); *phenotype:* lamina potential of ERG is reduced; optomotor response shows low visual acuity.
apo:	***altered pattern orientation***; *location:* I not mapped; *reference:* HEISENBERG (unpublished); *phenotype:* pattern induced orientation in flying and walking animals is disturbed.

This table is attached to this volume as individual sheet.

normal if flies are grown at low temperature (below 20° C) and in darkness (HARRIS et al., 1976). Obvious visual defects in such flies are, however, revealed by behavioral experiments (KAPLAN, 1976). As in wild type, a bright flash of blue light elicits a receptor potential and a PDA, but in *rdg* B^{KS222} its decay is much faster, and red light does not restore sensitivity nor does prolonged dark adaptation (HARRIS and STARK, 1976). Although degeneration of the large retinula cells does not become obvious until many hours later, presumably it is triggered by this light stimulus. At first one might assume that reisomerization from M to R is blocked. This, however, is ruled out by spectrophotometric measurements in vitro and in vivo. The primary photoprocesses seem to be normal. The rhabdomeres of *rdg* B^{KS222} can be protected against degeneration by several means, for instance by introducing the *norp* A mutation in addition (HARRIS and STARK, 1976). Even an allele of *norp* A that seems to have a normal receptor potential suppresses degeneration. This shows that the cause for degeneration is somewhat independent of the receptor potential and is subsequent to the step blocked in *norp* A. These experiments of combining different transducer mutants document a special potential of the genetic approach: they may be the first step towards a genetic description of the sequential nature of the transducer pathway.

The mutant *rdg* B^{KS222} calls for recording intracellularly from the retinula cells and for studying degeneration quantitatively. Questions such as the following could also be answered. What are the immediate effects of the light stimulus on the membrane? Why should only the PDA and not any kind of depolarization lead to inexcitability and to degeneration? This would determine more closely the stages of the transducer pathway at which the *rdg* B gene defects might act.

Whereas the above mutants are located on the X chromosome, the third type of mutant belongs to a gene on the 3. chromosome. It is characterized by a transient receptor potential (trp) (COSENS and MANNING, 1969; name used here from MINKE et al., 1975). The receptor potential is normal at low stimulus light levels. With a long stimulus at higher intensity only the initial phase of the receptor potential is normal. Thereafter it decays to a low steady-state level, which above a certain intensity is independent of it. Steady-state level and decay time vary in different alleles and individual flies of one allele. In many experiments the receptor potential decays to the level of the resting potential in the dark. Recovery is at least ten times as slow as the decay.

Again, emphasis has been on demonstrating that the defects in the *trp* mutants interfere with the transducer process (MINKE et al., 1975). The primary photoprocesses can be checked by spectrophotometry, and seem to be normal. Intracellular recordings of mutant receptor cells provide strong arguments that the ability of the cell membrane to generate receptor potentials is unimpaired: the small steady-state response of the retinula cells exhibits the occurrence of normal quantum bumps. Only their frequency is reduced. Thus it seems that in *trp* the efficiency of the primary photoprocess in producing quantum bumps diminishes during the decay of the receptor potential.

In *trp*, as in wild type, a PDA can be elicited. It has an amplitude as small as the steady state of the receptor potential, and reveals that it is also composed of quantum bumps (MINKE et al., 1975). It had not been possible to observe this in wild type, since high-intensity light is needed to trigger a PDA. Thus a PDA can be described as the continued stimulation of quantum bumps even after cessation of the primary photoprocesses.

These results go beyond merely explaining mutant problems. It is quite likely that the opsin gene will be discovered soon. Mutants that are good candidates are being studied (PAK, personal communication). Relevant membrane mutants are also likely to be found in this system. For the intermediate steps, the most interesting mutants will be among those that have apparently minor or conditional defects in the receptor potential. But with a large store of potentially interesting mutants now at hand new techniques for the subcellular analysis of transduction are becoming most important.

C. Mutations for Network Analysis

In this chapter we turn from cell physiology to organ physiology and its genetics. Here we are in unknown territory as to gene action. Many of the useful mutations still act at the cellular level in the final state of differentiation. Obvious examples are the late somatic-cell-lethal genes (i.e., *rdg* A, *rdg* B). Secondly, one expects mutants to block or change certain signals during development, as in some cerebellar mutants of the mouse (SIDMAN, 1968). And, of course, there could be genes regulating general conditions such as extracellular ion concentrations, size, surface, or the nutritional state of an organ. The list reflects our ignorance of organ genetics. Visual mutants have not yet contributed significantly to this field. But even without this knowledge, mutants can be used to provide, in our context, flies lacking certain features of the visual system. This will be discussed in Section C.I.

I. Dissection of the Retina[3]

1. Two Retinal Subsystems

The *Drosophila* eye, like that of most dipterans, is composed of two types of retinula cells. Each ommatidium contains six cells with large rhabdomeres (R1–6) sending their axons to the first optic ganglion (lamina ganglionaris), and two cells with small rhabdomeres (R7/R8) extending the so called long visual fibers directly to the second synaptic region, the medulla (CAJAL and SANCHEZ, 1915) (see Fig. 1). Since this discovery, many attempts have been made to attribute functional specializations to these retinula cell types. However, a consistent interpretation did not evolve before the "neural superposition" principle of the dipteran eye was fully understood (KIRSCHFELD, 1967; BRAITENBERG, 1967; TRUJILLO-CENÓZ and MELAMED, 1966; VIGIER, 1909). This principle states that six of the large rhabdomeres in six different ommatidia receive light from the same part of the visual surround, and that the axons of the corresponding cells "superimpose" their

[3] Many of the concepts referred to in the following chapters have been developed in experiments with the housefly *Musca*. So far, it has been very useful to regard the visual systems of *Drosophila* and *Musca* as sufficiently alike to use basic results obtained for one system qualitatively in the other. Obviously, the large fly is more suitable for electrophysiologic experiments, while the small one is more suitable for genetics.

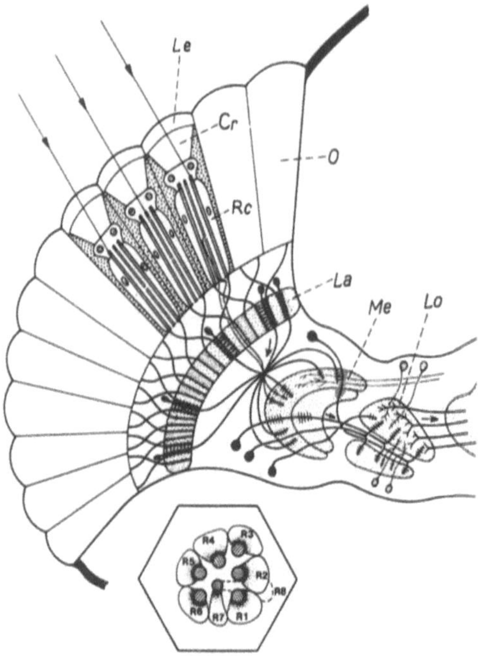

Fig. 1. Schematic drawing of the dipteran eye and optic ganglia; *Cr*, Crystalline cone; *La*, Lamina; *Le*, Lenslet; *Lo*, Lobula; *Me*, Medulla; *O*, Ommatidium; *Rc*, Receptor cell. Inset figure: cross-section of retinula cells in one ommatidium. (From HEISENBERG and BUCHNER, 1977)

excitations onto the same set of interneurons in one "cartridge" of the lamina. Retinula cells "looking" in the same direction, collectively will be called "visual element" (BUCHNER, 1976).

The sixfold visual input and the large size of the rhabdomeres immediately suggest that the subsystem of retinula cells R1–6 might have a higher absolute sensitivity than the subsystem of the retinula cells R7 and R8. KIRSCHFELD (1972a) estimated for *Musca* a sensitivity ratio of 24 : 48 for the two systems, provided that the quantum capture cross section and the packing of photopigment molecules in the two types of rhabdomeres is similar. Another functional distinction between the two systems can be deduced from anatomy and optical studies. The rhabdomeres function as optic fibers receiving light through their distal endings, which lie in the focal planes of the cornea lenses. Thus, if the resolution of the ommatidial optics is sufficiently high the spatial sensitivity distribution (horizontal halfwidth: $\Delta\varrho$) of a retinula cell should depend upon the size of the distal rhabdomere ending (KIRSCHFELD, 1972a). Sufficiently high resolution has been shown for *Musca* and *Drosophila* (KIRSCHFELD and FRANCESCHINI, 1968; FRANCESCHINI and KIRSCHFELD, 1971). Thus in *Drosophila* (as in *Musca*) the $\Delta\varrho$ of the cells R1–6 should be roughly twice as big as that of the rhabdomeres R7 and R8. This implies that the latter have an advantage in mediating the fine detail of visual patterns.

Aside from differences in sensitivity and acuity, two other visual properties have been proposed that could possibly be mediated by the two retinal subsystems. One is the sensitivity to polarized light. Most rhabdomeres show dichroic absorption. The directions of maximal absorption in cells R7 and R8 apparently are perpendicular to each other. In the subsystem of cell R1–6, however, the directions of maximal absorption differ in such a way that their summed excitations may be invariant to changes in E-vector orientation. Thus in each visual element polarized light can, in principle, be distinguished from unpolarized light (summary by KIRSCHFELD, 1972b, 1973). Finally, the receptors R1–6, R7, and R8 might be the basis for color vision if they have different spectral sensitivities.

2. Spectral Sensitivity of Receptors

In the past 15 years, action spectra with similar features were recorded in *Calliphora*, *Musca*, and *Drosophila* using a variety of techniques. They all show a peak close to or below 360 nm, a minimum around 400 nm, and a second somewhat broader peak, about equally high, near 480 nm (BURKHARDT, 1962; ECKERT, 1971; MCCANN and ARNETT, 1972; KIRSCHFELD, unpublished; HENGSTENBERG, unpublished; STARK and WASSERMAN, 1974; HEISENBERG and BUCHNER, 1977). This probably reflects the spectral sensitivity of cells R1–6, since in most cases low intensity threshold criteria or low intensity ERG recording were used. In one case (KIRSCHFELD, unpublished) stimulation and recording of R1–6 type cells was visually controlled.

Measuring spectral sensitivities of cells R7 and R8 was not successful until recently. The main difficulty was that of eliminating the responses of cells R1–6. This was made possible by the discovery of the PDA in white-eyed *Drosophila*. After stimulation of a PDA an ERG can be recorded that reflects mainly the activity of cells R7 and R8 (MINKE et al., 1975; STARK, 1975). At the same time, discussion about the two retinal subsystems had drawn attention to three mutants lacking certain receptor types (HARRIS et al., 1976).

One is the mutant *sev* (sevenless) in which the rhabdomeres of cells R7 are poorly developed. Since none of the rhabdomeres of cells R7 terminate in the focal plane of the respective cornea lens, visual input to the central cells R7/R8 is impaired. The second one, *rdg* B^{KS222}, was mentioned in Section B.2. In this mutant, cells R1–6 degenerate leaving R7 and R8 largely intact. The main difficulty for many experiments with *rdg* B^{KS222} is that degeneration is rarely complete and only statistically predictable. In the third mutant, called *ora* (outer rhabdomeres absent), rhabdomere formation in cells R1–6 is suppressed, although otherwise the cells seem to be normal. This mutant has the drawback that remnants of rhabdomeres (about 1–5% of the normal length) are still present at the distal side of the retinula cells.

With the mutants *ora* and *rdg* B^{KS222}, the contributions of cells R1–6 to the ERG can be suppressed. The result is essentially the same as that obtained inducing a PDA. The action spectrum has a shoulder in the near UV and a smaller peak in the blue. In the double mutants *rdg B sev* and *ora sev*, the shoulder in the UV is suppressed, leaving an action spectrum with a broad peak in the blue. Also, in wild type, this spectral sensitivity can be obtained by inducing a PDA with UV light (STARK, 1975). It was concluded that cells R7 contain a rhodopsin most sensitive in the UV and that cells R8 are blue receptors. Spectrophotometric studies of photopigment extracts of the mutants and double mutants have confirmed this for cells R7. The mutants *ora* and *rdg* B^{KS222} contain a photopigment that, like the photopigment in cells R1–6, has two stable forms. The action spectra for photoisomerization are compatible with a UV-sensitive rhodopsin (HARRIS et al., 1976).

In the work just summarized many problems about the spectral sensitivity of the three receptor types remain open. But this is the first convincing example of genetics being used for visual system analysis beyond the cellular level. The morphologic reduction of rhabdomeres and the concomitant change in spectral sensitivity are tentatively interpreted as a causal relation. Hopefully, results of this sort can also be obtained for genetic defects of the brain (see p. 677).

II. Dissection of Behavior

1. Receptor Mutants

As mentioned above, the structure of the retina and the morphology of the receptors suggest that cells R1–6 form a visual input system specialized for low light intensity, and that the input systems of cells R7 and R8 show higher acuity, due to the smaller diameter of the rhabdomeres. Since the proposal of this concept, it was assumed that these receptor properties were apparent in optomotor experiments. At low light levels the optomotor response was thought to be mediated predominantly by cells R1–6 and, using patterns of sufficiently narrow vertical stripes as movement stimuli, mainly by cells R7 and/or R8. Various experiments seemed to support this notion (ECKERT, 1971, 1973; KIRSCHFELD, 1975; HEISENBERG and GÖTZ, 1975; KIRSCHFELD and REICHARDT, 1970).

The receptor-deficient *Drosophila* mutants, however, lead to different conclusions. The mutant *sev* has an optomotor response about half as strong as wild type. But despite the strong reduction of rhabdomeres in cells R7, visual acuity of the optomotor response is entirely normal. The cells R8 that lack the normal rhabdomeres of cells R7 for a proper visual input most likely do not contribute to the optomotor response to narrow stripe patterns. In wild type the superior acuity of cells R7 and R8 does not show up in the optomotor response. The mutants *rdg* B^{KS222} and *ora* show "normal" ERG responses of cells R7 and R8. However, the optomotor response is suppressed to the degree to which R1–6 rhabdomeres are missing. In *ora* a very small optomotor response (about 1–5% of the normal response) can be detected; in *rdg* B^{KS222}, the response varies from normal to nearly absent, and is clearly correlated with degeneration. We have observed the state of cells R7 and R8 in flies that were virtually optomotor blind. In all cases, those cells were present and seemed morphologically intact. We have also used the PDA to render cells R1–6 insensitive to visual stimuli. It was possible to suppress the optomotor response after a bleaching stimulus of blue light, whereas an even brighter stimulus of red light immediately restored the responsiveness. As mentioned above, cells R7 and R8 are electrophysiologically functional after the blue bleaching stimulus. Collectively, these results indicate that cells R1–6 mediate the optomotor response at least to 95% in *Drosophila*. Whether the optomotor response amplitude can be modulated by the activity of cells R7 and R8 is not yet known. In most visual tests, such as visually controlled landing, pattern-induced orientation, and the jump reflex of the mutant Hk_1 (*H*yper*k*inetic 1) (KAPLAN, 1976), one finds that the mutant *sev* behaves similarly to wild type; *ora* and *rdg* B^{KS222} (in the fully degenerated state) turn out to be strongly affected (HEISENBERG and BUCHNER, 1977).

So far, only one visual test has been discovered in which *sev* differs qualitatively from wild type; this is the so-called "Schümperli effect" (SCHÜMPERLI, 1973): at medium and high light intensities *Drosophila* has an unproportional preference for UV light as compared to the low-intensity range. In contrast to wild type, the mutant *sev* avoids UV light relative to blue (HARRIS et al., 1976). This effect is observed in differential phototaxis in which flies have to compare two light sources. We have tried unsuccessfully to find a similar effect for the optomotor response, looking for differences between wild type and *sev*. The Schümperli effect seems to be unrelated to the optomotor response. This is not restricted to *Drosophila*. KAISER (1974) has found for the fleshfly *Phormia* and for the bee that periodic patterns of colored stripes did not elicit optomotor responses due to color contrast.

To summarize, it seems that the visual subsystem of cells R7 and R8 is used primarily for some kind of color vision (Schümperli effect), and that this system, surprisingly, is separate from most other visual functions, now experimentally accessible.

2. Acuity and Sensitivity-Deficient Mutants

Many of those visual mutants that have no obvious defects in the retina except abnormal optomotor properties can be classified into two groups with complementary properties.

In one group (nbA^{H18}, *elf*, etc.) the mutants are night blind in the optomotor response using broad stripe patterns for stimulation. In the other group ($nonA^{H2}$; P37; etc.) mutants are totally movement blind for periodic patterns of sufficiently narrow stripes. In the first group we assume a hypothetic mechanism has been disturbed, which in wild type can improve sensitivity at low light levels on account of acuity. Optomotor experiments on wild type in which visual acuity at very low light levels is determined seem to support this idea (HEISENBERG and BUCHNER, 1977). In the second group we assume the mechanism has been fixed at the low intensity state (mutant properties summarized by HEISENBERG and GÖTZ, 1975). These two complementary properties must be attributed entirely to the visual subsystem of retinula cells R1–6 since they are part of optomotor behavior.

The two mutant classes differ not only in the optomotor response, but also in pattern-induced flight-orientation behavior and in the Schümperli effect. Those mutants, which are night blind for the optomotor response to broad stripe patterns, perform normally in the pattern-induced orientation behavior and show an increased UV sensitivity in the Schümperli effect. Mutants with reduced acuity show very poor orientation behavior and a suppressed UV sensitivity (HEISENBERG and BUCHNER, 1977)[4].

It is not yet clear whether the two kinds of partial optomotor blindness are really two states of an adaptational mechanism or whether they reflect two favored breaking points of the visual system. These alternatives would have somewhat different implications for further work. Nevertheless, this result shows how the mutant approach can reveal properties of the visual system not previously recognized.

It is important to know whether these two types of defects are properties of the eye or of the optic ganglia. By the mosaic technique (HOTTA and BENZER, 1970), the ERG and phototaxis defects of some of these mutants were shown to originate from the retina anlage. But in another mutant (*elf*), which is of the night-blind type, we were able to separate the origin of the ERG defect from the retina anlage (HEISENBERG and HENGSTENBERG, unpublished). The optomotor response itself has not yet been sufficiently tested with the mosaic technique.

3. Dissection of Turning Behavior

Three different behavioral methods for visual mutant selection have been employed so far: fast phototaxis, optomotor response, and pattern-induced orientation behavior (HEISENBERG, unpublished). As expected, some mutant genes can be isolated by any of these methods; others have been found only by one of them. Mutants are available that are impaired in only one of the three behavioral responses. This is of particular interest since all three behavioral responses imply a turning reaction of the fly in response to a visual stimulus characterized by temporal and spatial intensity changes.

Recently a formalism has been developed that classifies various visual stimuli according to the neural interactions in the visual system that are necessary to generate the appropriate turning reactions (POGGIO and REICHARDT, 1976). A turning reaction in response to a stationary pattern that does not move relative to the fly requires only linear data processing steps and no lateral interactions between visual channels. If a flickering pattern is necessary to elicit a turning reaction a nonlinear data processing step must be involved, there are still no interactions necessary between different visual channels. A direction-sensitive response to a moving pattern can be obtained only by nonlinear asymmetric interactions between at least two visual channels. It was shown for *Musca* (VARJÚ and REICHARDT, 1967) and *Drosophila* (ZIMMERMANN, 1973) that this simplest possible model accounts quite well for the classic optomotor response (POGGIO and REICHARDT, 1973). The pattern-induced orientation behavior of *Musca* seems also to require symmetric nonlinear interactions of the second, fourth, and

[4] On the basis of their properties in the optomotor response the two mutant classes originally were thought to have defects in separate visual pathways belonging to the input subsystems of retinula cells R1–6 and R7/R8, respectively. The properties affected by the mutations were called High Sensitivity System (HSS) and High Acuity System (HAS) (HEISENBERG and GÖTZ, 1975). This nomenclature may have to be revised.

possibly sixth order (REICHARDT and POGGIO, 1976; POGGIO and REICHARDT, 1976).

Classification of behavior along these lines may provide evidence for the biologic relevance of this formalism, and may help to identify the underlying neural network. At present it appears that the turning response to intensity changes (flicker, symmetrical second-order interactions) is selectively suppressed in several mutants. For instance, the mutant *apo* (altered pattern orientation), which was recently selected for defects in pattern-induced orientation behavior, shows a normal optomotor response, but seems to be strongly affected in the turning reaction to flickering light (HEISENBERG, in preparation). I have previously described a mutant (*non* A^{H2}) in which the optomotor response to front-to-back movement is affected much more than the response to the opposite direction (HEISENBERG, 1972). In the above formalism, this defect would specifically concern one of the asymmetric second-order interactions.

It would be of particular interest to study the relation of the Schümperli effect to these various types of reactions. (Would it not be satisfying to have colored and non-colored interactions?) In such a study mutants would play an important part.

With these examples of *Drosophila* mutants influencing various aspects of visual system research I hope to have answered the questions posed at the beginning of this article. Comparative physiologic investigation at the cellular and network level using point mutants as the closest possible relatives to wild type is an innovation in this field. Since any study using mutants needs a rich field of problems, facts, and methods, the visual system of the fly is a very suitable object. The advantage at the network level is obvious: one can eliminate topologically unrelated but functionally related parts of the system. Equivalent methods, like pharmacology, have not yet been applied systematically. The science of how genes influence a visual system, however, has made little progress; presumably it can not be approached reasonably without the aid of embryology. Visual system genetics may eventually become a branch of science in itself. This would be the most rigid and most complete kind of questioning in visual physiology, unifying its developmental and functional aspects.

Our thinking about brains has been dominated in the last twenty years by analogy with the computer. Genetics emphasizes a complementary aspect.

Note added in proof

In the concluding remarque to chapter C I/2 the importance of finding behavioral defects with anatomical correlates in the brain has been emphasized. Now, two years later, several such mutants are available. In one of them a set of giant neurons in the lobula plate is missing or reduced and concomitantly optomotor turning responses are diminished while other movement dependent visual responses are intact (M. HEISENBERG, R. WONNEBERGER and R. WOLF: J. comp. Physiol. **124**, 287—296 (1978)). In another mutant the distal layer of the medulla is severely degenerated; all flies from this stock are movement blind but perform in visual orientation tasks. In still another group of mutants the number of neurons in the optic neuropile is reduced while the number of ommatidia in the eye and number of columns in the optic ganglia is normal. These mutants show altered visual orientation behavior and subtle differences in other visual responses (HEISENBERG, unpublished). These few examples may illustrate the special usefulness of structural brain mutants for network analysis.

References

Alawi, A. A., Jennings, V., Grossfield, J., Pak, W. L.: Phototransduction mutants of *Drosophila melanogaster*. In: Advances in experimental medicine and biology (ed. G. B. Arden), Vol. XXIV. New York: Plenum 1972.

Baylor, D. A., Hodgkin, A. L., Lamb, T. D.: Reconstruction of electrical responses of turtle cones to flashes and steps of light. J. Physiol. (Lond.) **242**, 759—791 (1974).

Benzer, S.: Behavioral mutants of *Drosophila* isolated by countercurrent distribution. Proc. nat. Acad. Sci. (Wash.) **58**, 1112—1119 (1967).

Braitenberg, V.: Patterns of projection in the visual system of the fly. I. Retina-lamina projections. Exp. Brain Res. **3**, 271—298 (1967).

Buchner, E.: Elementary movement detectors in an insect visual system. Biol. Cybernetics **24**, 85—101 (1976).

Burkhardt, D.: Spectral sensitivity and other response characteristics of single visual cells in the arthropod eye. Symp. Soc. Exp. Biol. **16**, 86—109 (1962).

Cajal, S. R., Sanchez, D.: Contribución al conocimiento de los centros nerviosos de los insectos. Trab. Lab. Invest. Biol. (Madrid) **13**, 1—164 (1915).

Cervetto, L., Pasino, E., Torre, V.: Electrical responses of rods in the retina of *Bufo marinus*. J. Physiol. (Lond.) **267**, 17—51 (1977).

Cosens, D. J., Briscoe, D.: A switch phenomenon in the compound eye of the white-eyed mutant of *Drosophila melanogaster*. J. Insect Physiol. **18**, 627—632 (1972).

Cosens, D. J., Manning, A.: Abnormal electroretinogram from *Drosophila* mutant. Nature (Lond.) **224**, 285—287 (1969).

Deland, M. C., Pak, W. L.: Reversibly temperature sensitive phototransduction mutant of *Drosophila melanogaster*. Nature (Lond.) **244**, 184—186 (1973).

Ebrey, T. G., Honig, B.: Molecular aspects of photoreceptor function. Quart. Rev. Biophys. **8**, 129—184 (1975).

Eckert, H.: Die spektrale Empfindlichkeit des Komplexauges von *Musca*. Kybernetik **9**, 145—156 (1971).

Eckert, H.: Optomotorische Untersuchungen am visuellen System der Stubenfliege *Musca domestica* L. Kybernetik **14**, 1—23 (1973).

Franceschini, N., Kirschfeld, K.: Etude optique in vivo des éléments photorécepteurs dans l'oeil composé de *Drosophila*. Kybernetik **8**, 1—13 (1971).

Götz, K. G.: Flight control in *Drosophila* by visual perception of motion. Kybernetik **4**, 199—208 (1968).

Götz, K. G., Wenking, H.: Visual control of locomotion in the walking fruitfly *Drosophila*. J. comp. Physiol. **85**, 235—266 (1973).

Hamdorf, K., Paulsen, R., Schwemer, J.: Photoregeneration and sensitivity control of photoreceptors in invertebrates. In: Biochemistry and physiology of visual pigments (ed. H. Langer). Berlin-Heidelberg-New York: Springer 1973.

Harris, W. A., Stark, W. S.: Hereditary retinal degeneration in *Drosophila melanogaster*: a mutant defect in the phototransduction process. J. gen. Physiol. **69**, 261—291 (1977).

Harris, W. A., Stark, W. S., Walker, J. A.: Genetic dissection of the photoreceptor system in the compound eye of *Drosophila melanogaster*. J. Physiol. (Lond.) **256**, 415—439 (1976).

Heisenberg, M.: Comparative behavioral studies on two visual mutants of *Drosophila*. J. comp. Physiol. **80**, 119—136 (1972).

Heisenberg, M., Buchner, E.: The role of retinula cell types in visual behavior of *Drosophila melanogaster*. J. comp. Physiol. **117**, 127—162 (1977).

Heisenberg, M., Götz, K. G.: The use of mutations for the partial degradation of vision in *Drosophila melanogaster*. J. comp. Physiol. **98**, 217—241 (1975).

Hillmann, P., Hochstein, S., Minke, B.: A visual pigment with two physiologically active stable states. Science **175**, 1486—1488 (1972).

Hotta, Y., Benzer, S.: Genetic dissection of the *Drosophila* nervous system by means of mosaics. Proc. nat. Acad. Sci. (Wash.) **67**, 1156—1163 (1970).

Kaiser, W.: The relationship between visual movement detection and colour vision in insects. In: The compound eye and vision of insects (ed. G. A. Horridge). Oxford: Clarendon Press 1974.

Kaplan, W. D.: The use of double mutants in the investigation of visual mutants of *Drosophila* (Abstract). Genetics **83**, 38—39 (1976).

Kaplan, W. D., Trout, W. E.: The behaviour of four neurological mutants of *Drosophila*. Genetics **61**, 399—409 (1969).

Kirschfeld, K.: Die Projektion der optischen Umwelt auf das Raster der Rhabdomere im Komplexauge von *Musca*. Exp. Brain Res. **3**, 248—270 (1967).

Kirschfeld, K.: The visual system of *Musca*: studies on optics, structure, and function. In: Information processing in the visual system of arthropods (ed. R. Wehner). Berlin-Heidelberg-New York: Springer 1972a.

Kirschfeld, K.: Vision of polarized light. Symposia Proceedings of the IV. International Biophysics Congress, Moscow, 1972b.

Kirschfeld, K.: Das neurale Superpositionsauge. Fortschr. Zool. **21**, 229—257 (1973).

Kirschfeld, K., Franceschini, N.: Optische Eigenschaften der Ommatidien im Komplexauge von *Musca*. Kybernetik **5**, 47—52 (1968).

Kirschfeld, K., Reichardt, W.: Optomotorische Versuche an *Musca* mit linear polarisiertem Licht. Z. Naturforsch. **25b**, 228 (1970).

Kirschfeld, K., Snyder, A. W.: Waveguide mode effects, birefringence, and dichroism in fly photoreceptors. In: Photoreceptor optics (eds. A. W. Snyder, R. Menzel). Berlin-Heidelberg-New York: Springer 1975.

Koenig, J., Merriam, J. R.: Autosomal ERG mutants. Drosophila Information Service **52**, 50—51 (1975).

McCann, G. D., Arnett, D. W.: Spectral and polarization sensitivity of the dipteran visual system. J. gen. Physiol. **59**, 534—558 (1972).

Minke, B., Wu, C.-F., Pak, W. L.: Isolation of light-induced response of the central retinula cells from the electroretinogram of *Drosophila*. J. comp. Physiol. **98**, 345—355 (1975).

Nolte, J., Brown, J. E., Smith, T. G.: A hyperpolarizing component of the receptor potential in the median ocellus of *Limulus*. Science **162**, 677—679 (1968).

Ostroy, S. E., Pak, W. L.: Protein differences associated with a phototransduction mutant of *Drosophila*. Nature (Lond.) **243**, 120—121 (1973).

Pak, W. L.: Mutations affecting the vision of *Drosophila melanogaster*. Handbook Gen. **3**, 703—733 (1975).

Pak, W. L., Liddington, K. J.: Fast electrical potential from a long-lived, long-wavelength photoproduct of fly visual pigment. J. gen. Physiol. **63**, 740—756 (1974).

Pak, W. L., Pinto, L. H.: Genetic approach to the study of the nervous system. Annu. Rev. Biophys. Bioeng. **5**, 397—448 (1976).

Poggio, T., Reichardt, W.: Considerations on models of movement detection. Kybernetik **13**, 223—227 (1973).

Poggio, T., Reichardt, W.: Visual control of orientation behaviour of the fly. II. Towards the underlying neural interactions. Quart. Rev. Biophys. **9**, 377—438 (1976).

Reichardt, W., Poggio, T.: Visual control of orientation behaviour of the fly. I. A quantitative analysis. Quart. Rev. Biophys. **9**, 311—375 (1976).

Schümperli, R.: Evidence for colour vision in *Drosophila melanogaster* through spontaneous phototactic choice behaviour. J. comp. Physiol. **86**, 77—94 (1973).

Sidman, R. L.: Development of interneuronal connections in brains of mutant mice. In: Physiological and biochemical aspects of nervous integration (ed. F. D. Carlson). New Jersey: Prentice-Hall 1968.

Stark, W. S.: Spectral selectivity of visual response alterations mediated by interconversions of native and intermediate photopigments in *Drosophila*. J. comp. Physiol. **96**, 343—356 (1975).

Stark, W. S., Wasserman, G. S.: Wavelength-specific ERG characteristics of pigmented- and white-eyed strains of *Drosophila*. J. comp. Physiol. **91**, 427—441 (1974).

Trujillo-Cenóz, O., Melamed, J.: Compound eye of dipterans: anatomical basis for integration—an electron microscope study. J. Ultrastruct. Res. **16**, 395—398 (1966).

Varjú, D., Reichardt, W.: Übertragungseigenschaften im Auswertesystem für das Bewegungssehen. II. Z. Naturforsch. **22b**, 1343—1351 (1967).

Vigier, M. P.: Mécanisme de la synthèse des impressions lumineuses recueillies par les yeux composés des Diptères. C. R. Soc. Biol. **64**, 1221—1223 (1908).

Wu, C.-F., Pak, W.: Quantal basis of photoreceptor spectral sensitivity of *Drosophila melanogaster*. J. gen. Physiol. **66**, 149—168 (1975).

Yeandle, S.: Studies on the slow potential and the effects of cation on the electrical responses of the *Limulus* ommatidium. Ph. D. Thesis: Johns Hopkins University, Baltimore, 1957.

Zimmermann, G.: Der Einfluß stehender und bewegter Musteranteile auf die optomotorische Reaktion der Fliege *Drosophila*. Doctoral Thesis: Eberhard-Karls-Universität, Tübingen, 1973.

Author Index

Page numbers in *italics* refer to the bibliography

Subject Index

Table of mutants

(Numbers after "*location*" indicate mutant chromosome and – if known – position on genetic map)

***norp A*:**	***no receptor potential A***; *location:* I–6.5; *reference:* Pak (1975); *phenotype:* ERG can be nearly absent; it varies in different alleles; in one allele in which ERG is slow at elevated temperature, latency of quantum bumps is longer and more variable; retina is autonomous for the defect.
rdg B^{KS222}:	***receptor degeneration B*** (allele: KS222); *location:* I–42.7; *references:* Harris and Stark (1977) and Harris et al. (1976); *phenotype:* large retinula cells R1–6 degenerate under illumination or at temperatures above 20 °C; optomotor response varies from normal to nearly zero.
***trp*:**	***transient receptor potential***; *location:* III not mapped; *synonym:* Cosens-Manning mutant; *references:* Cosens and Manning (1969) and Minke et al. (1975b); *phenotype:* receptor potential decays to nearly resting potential in the dark due to decrease in frequency of quantum bumps.
***sev*:**	***sevenless***; *location:* I–33.2; *synonym:* LY3; *reference:* Harris et al. (1976); *phenotype:* rhabdomeres of retinula cells R7 are missing in all ommatidia; in differential phototaxis UV sensitivity is suppressed.
***ora*:**	***outer rhabdomeres absent***; *location:* III–65.3; *synonym:* JK84; *references:* Koenig and Merriam (1975) and Harris et al. (1976); *phenotype:* only distal tips of rhabdomeres develop in large retinula cells R1–6; optomotor response is strongly suppressed.
Hk_1:	***Hyperkinetic*** (allele No. 1); *location:* I–30.9; *reference:* Kaplan and Trout (1969); *phenotype:* legs shake in etherized flies; visual stimuli can cause an abnormal jump reflex; head is autonomous for visual defect.
nbA^{H18}:	***night blind A*** (allele H18); *location:* I–36.6; *synonyms:* non B, ldg A, opm 18; *reference:* Heisenberg and Götz (1975); *phenotype:* low intensity threshold of optomotor response and phototaxis is raised; lamina potential of ERG is suppressed; retina is autonomous for ERG defect.
***elf*:**	***extra lamina fiber***: *location:* I not mapped; *synonym:* H37, opm 37; *reference:* Heisenberg and Götz (1975); *phenotype:* off-transient of lamina potential is reduced; low intensity threshold of optomotor response is raised; most lamina cartridges contain an additional large fiber.
non A^{H2}:	***no on-transient A*** (allele H2); *location:* I–52; *synonym:* opm 2; *references:* Hotta and Benzer (1970), Pak (1975), and Heisenberg (1972); *phenotype:* lamina potential of ERG is very small; optomotor response shows low visual acuity; in differential phototaxis UV sensitivity is suppressed.
***P 37*:**	(laboratory name); *location:* I not mapped; *references:* Pak (1975) and Heisenberg and Götz (1975); *phenotype:* lamina potential of ERG is reduced; optomotor response shows low visual acuity.
***apo*:**	***altered pattern orientation***; *location:* I not mapped; *reference:* Heisenberg (unpublished); *phenotype:* pattern induced orientation in flying and walking animals is disturbed.

This table is attached to this volume as individual sheet.
Handbook of Sensory Physiology, Volume VII/6A, Chapter 12, Heisenberg (p. 669).

Table of mutants

(Numbers after "*location*" indicate mutant chromosome and – if known – position on genetic map)

norp A*:** ***no receptor potential A; *location:* I–6.5; *reference:* PAK (1975); *phenotype:* ERG can be nearly absent; it varies in different alleles; in one allele in which ERG is slow at elevated temperature, latency of quantum bumps is longer and more variable; retina is autonomous for the defect.

rdg* B^{KS222}:** ***receptor degeneration B (allele: KS222); *location:* I–42.7; *references:* HARRIS and STARK (1977) and HARRIS et al. (1976); *phenotype:* large retinula cells R1–6 degenerate under illumination or at temperatures above 20 °C; optomotor response varies from normal to nearly zero.

trp*:** ***transient receptor potential; *location:* III not mapped; *synonym:* Cosens-Manning mutant; *references:* COSENS and MANNING (1969) and MINKE et al. (1975b); *phenotype:* receptor potential decays to nearly resting potential in the dark due to decrease in frequency of quantum bumps.

sev*:** ***sevenless; *location:* I–33.2; *synonym:* LY3; *reference:* HARRIS et al. (1976); *phenotype:* rhabdomeres of retinula cells R7 are missing in all ommatidia; in differential phototaxis UV sensitivity is suppressed.

ora*:** ***outer rhabdomeres absent; *location:* III–65.3; *synonym:* JK84; *references:* KOENIG and Merriam (1975) and HARRIS et al. (1976); *phenotype:* only distal tips of rhabdomeres develop in large retinula cells R1–6; optomotor response is strongly suppressed.

Hk_1: ***Hyperkinetic*** (allele No. 1); *location:* I–30.9; *reference:* KAPLAN and TROUT (1969); *phenotype:* legs shake in etherized flies; visual stimuli can cause an abnormal jump reflex; head is autonomous for visual defect.

nbA^{H18}: ***night blind A*** (allele H18); *location:* I–36.6; *synonyms:* non B, ldg A, opm 18; *reference:* HEISENBERG and GÖTZ (1975); *phenotype:* low intensity threshold of optomotor response and phototaxis is raised; lamina potential of ERG is suppressed; retina is autonomous for ERG defect.

elf*:** ***extra lamina fiber: *location:* I not mapped; *synonym:* H37, opm 37; *reference:* HEISENBERG and GÖTZ (1975); *phenotype:* off-transient of lamina potential is reduced; low intensity threshold of optomotor response is raised; most lamina cartridges contain an additional large fiber.

non* A^{H2}:** ***no on-transient A (allele H2); *location:* I–52; *synonym:* opm 2; *references:* HOTTA and BENZER (1970), PAK (1975), and HEISENBERG (1972); *phenotype:* lamina potential of ERG is very small; optomotor response shows low visual acuity; in differential phototaxis UV sensitivity is suppressed.

***P 37*:** (laboratory name); *location:* I not mapped; *references:* PAK (1975) and HEISENBERG and GÖTZ (1975); *phenotype:* lamina potential of ERG is reduced; optomotor response shows low visual acuity.

apo*:** ***altered pattern orientation; *location:* I not mapped; *reference:* HEISENBERG (unpublished); *phenotype:* pattern induced orientation in flying and walking animals is disturbed.

This table is attached to this volume as individual sheet.
Handbook of Sensory Physiology, Volume VII/6A, Chapter 12, HEISENBERG (p. 669).

MIX
Papier aus verantwortungsvollen Quellen
Paper from responsible sources
FSC® C105338

If you have any concerns about our products,
you can contact us on
ProductSafety@springernature.com

In case Publisher is established outside the EU,
the EU authorized representative is:
Springer Nature Customer Service Center GmbH
Europaplatz 3, 69115 Heidelberg, Germany

Printed by Libri Plureos GmbH
in Hamburg, Germany